Lecture Notes in Control and Information Sciences 433

Editors

Series Advisory Board

For further volumes:
http://www.springer.com/series/642

Carla Seatzu, Manuel Silva,
and Jan H. van Schuppen (Eds.)

Control of Discrete-Event Systems

Automata and Petri Net Perspectives

Springer

Editors
Carla Seatzu
Department of Electrical
and Electronic Engineering
University of Cagliari
Cagliari
Italy

Jan H. van Schuppen
Centrum Wiskunde & Informatica
Amsterdam
The Netherlands

Manuel Silva
Department of Computer Science
and Systems Engineering
University of Zaragoza
Zaragoza
Spain

ISSN 0170-8643
e-ISSN 1610-7411
ISBN 978-1-4471-4275-1
e-ISBN 978-1-4471-4276-8
DOI 10.1007/978-1-4471-4276-8
Springer London Heidelberg New York Dordrecht

Library of Congress Control Number: 2012941741

Printed on acid-free paper

Springer is part of Springer Science+Business Media (www.springer.com)

Preface

Control of discrete-event dynamic systems is the topic of this book. The aim is to provide an introduction to the field, starting at an elementary level and going to close to the current research front. The reader will find concepts, theorems, algorithms, and examples. Particularly addressed to Ph.D. students and junior researchers working on control of discrete-event dynamic systems (DEDS) and, more generally, on control theory, this monograph only presumes a little background of elementary topics of control theory. The chapters are almost all based on lectures of a summer school for Ph.D. students held in June 2011 in Cagliari, Sardinia, Italy.

Three related modeling formalisms of DEDS are covered: *automata*, *Petri nets*, and *systems in dioids*. The first focus of the book is on control of *decentralized* and of *distributed* DEDS, informally speaking composed by the interconnection of two or more subsystems. Most engineering systems are currently of this type. The second focus of the book deals with heavily loaded or populated DEDS, eventually distributed, for which the so called *state explosion* problem becomes particularly acute. Therefore it becomes important to consider 'coarse views' obtained through fluidization of the discrete event model. Those fluid or continuous over-approximated views of DEDS lead to special classes of hybrid systems.

Control theory for DEDS is motivated by the ordering of events or actions. Problems of control of DEDS arise in control engineering, computer engineering, and sciences. Areas in which DEDS control problems arise include manufacturing systems, automated guided vehicles, logistics, aerial vehicles, underwater vehicles, communication networks, mobile phones or chemical engineering systems, but also software systems on computers, laptops, and readers. The research into control of DEDS, heavily loaded or not, is thus well motivated by engineering but also theoretically quite deep.

A brief description of the book by parts and by chapters follows. The first part (nine chapters) concerns control of *automata*. After a chapter on modeling of engineering systems by automata (Chapter 1) there follows one with the concepts of automata, decidability, and complexity (Chapter 2). Supervisory control of automata is summarized first with respect to complete observations and subsequently with respect to partial observations (Chapters 3 and 4, respectively). Observers and diagnosers for automata, in particular distributed observers, are covered in Chapter 5. Supervisory control of distributed DEDS is introduced by three special cases in Chapter 6. That chapter is followed by one on distributed control with communication (Chapter 7) and one on coordination control (Chapter 8). Finally, Chapter 9 deals with timed automata.

The second part of the book addresses the control of *Petri nets*. It includes eleven chapters and can be seen as structured into two main parts: the first one, including eight chapters, dealing with discrete Petri nets; the second part, including three chapters, dealing with fluid relaxations. The former eight chapters can be divided into two blocks: the first six are devoted to *untimed* models, while the remaining two are related to *timed* models. In particular, Chapters 10 and 11 introduce basic concepts and structural analysis techniques. Chapters 12 and 13 are related to control. After considering supervisory control with languages specifications (Chapter 12), structural methods tailored for resource allocation problems are studied in the following one. Chapters 14 and 15 deals with diagnosis problems, the second one using net unfolding. Petri nets enriched with different temporal metrics and semantics are introduced in Chapter 16 and used in Chapter 17 for fault diagnosis, now on-line over timed models.

Chapters 18, 19 and 20 are based on fluid relaxation of DEDS. In particular, the fluid or continuous views of Petri nets are introduced in Chapter 18 on both untimed and timed models, dealing even with improvements of the relaxation with respect to the underlying discrete case. Finally, Chapters 19 and 20 are devoted to observability and diagnosis, and controllability and control, respectively.

The third and last part deals with the modeling and control of DEDS in dioids. A *dioid* is a mathematical structure with two operations, usually referred to as addition and multiplication, where the former, unlike in standard algebra, is idempotent. The most well known example for a dioid is the so-called max-plus algebra. Restricted classes of timed DEDS, in particular timed event graphs, become linear in a suitable dioid framework. For such systems, control does not involve logical decisions, only choices regarding the timing of events. They frequently appear in the context of manufacturing systems, but also arise in other engineering areas. Chapter 21 shows how to model timed event graphs in dioid frameworks and provides a detailed example from the area of high-throughput screening. Chapter 22 summarizes and illustrates control synthesis for systems in dioids.

Cagliari, Italy — *Carla Seatzu*
Zaragoza, Spain — *Manuel Silva*
Amsterdam, Netherlands — *Jan H. van Schuppen*
April 2012 — Editors

Acknowledgements

The editors thank the authors of the chapters of this book for their dedicated efforts to produce well readable chapters for the intended audience. They also thank the reviewers of the book for their efforts to provide comments to the authors on the chapters. A list of the reviewers is provided elsewhere in the book. They thank the Guest Editor Christoforos N. Hadjicostis for handling the review process of several chapters co-authored by one of the editors.

The European Commission is thanked for its financial support in part by the European Community's Seventh Framework Programme for the DISC Project under Grant Agreement number INFSO-ICT-224498. They also thank Alessandro Giua as the coordinator of the DISC project for his stimulation of the editors for this book.

The authors of Chapters 2, 3, 4, 6, and 8 acknowledge the financial support by the GAČR grants P103/11/0517 and P202/11/P028, and by RVO: 67985840.

The authors of Chapters 13, 18, 19 and 20 acknowledge the financial support by the Spanish CICYT under grant: DPI2010-20413.

Contents

List of Contributors

Béatrice Bérard
LIP6, Université Pierre & Marie Curie and CNRS UMR 7606
BC 169, 4 place Jussieu, 75005 Paris, France
e-mail: Beatrice.Berard@lip6.fr

René K. Boel
EESA SYSTeMS Research Group, Ghent University
Technologiepark 914, Zwijnaarde 9052, Belgium
e-mail: rene.boel@ugent.be

Olivier Boutin
Calle Santiago 2 – 4°C, 11005 Cadiz, Spain
e-mail: olivier.research@gmail.com

Thomas Brunsch
Fachgebiet Regelungssysteme, TU Berlin, Sekr. EN11, Einsteinufer 17, 10587 Berlin, Germany.
Also: Laboratoire d'Ingénierie des Systèmes Automatisés, Université d'Angers, 62 Avenue Notre-Dame du Lac, 49000 Angers, France
e-mail: brunsch@control.tu-berlin.de

Maria Paola Cabasino
Department of Electrical and Electronic Engineering, University of Cagliari
Piazza D'Armi, 09123 Cagliari, Italy
e-mail: cabasino@diee.unica.it

José-Manuel Colom
Instituto de Investigación en Ingeniería de Aragón (I3A), University of Zaragoza
María de Luna 1, E-50018, Zaragoza, Spain
e-mail: jm@unizar.es

Bertrand Cottenceau
LUNAM, University of Angers, LISA, ISTIA
62 Av. Notre-Dame du Lac, 49000 Angers, France
e-mail: bertrand.cottenceau@univ-angers.fr

Angela Di Febbraro
Department of Mechanical Engineering, Energetics, Production, Transportation and Mathematical Models, University of Genova
Via Montallegro 1, 16145 Genova, Italy
e-mail: angela.difebbraro@unige.it

Eric Fabre
INRIA Rennes Bretagne Atlantique
Campus de Beaulieu, 35042 Rennes Cedex, France
e-mail: eric.fabre@inria.fr

Alessandro Giua
Department of Electrical and Electronic Engineering, University of Cagliari
Piazza D'Armi, 09123 Cagliari, Italy
e-mail: giua@diee.unica.it

Stefan Haar
INRIA/LSV, CNRS & ENS de Cachan
61, avenue du Président Wilson, 94235 CACHAN Cedex, France
e-mail: Stefan.Haar@inria.fr

Laurent Hardouin
Laboratoire d'Ingénierie des Systèmes Automatisés, Université d'Angers
62 Avenue Notre-Dame du Lac, 49000 Angers, France
e-mail: laurent.hardouin@istia.univ-angers.fr

George Jiroveanu
Transelectrica SA - Romanian National Power Grid Company
Brestei 5, 200581, Craiova, Romania
e-mail: george.jiroveanu@transelectrica.ro

Jorge Júlvez
Instituto de Investigación en Ingeniería de Aragón (I3A), University of Zaragoza
María de Luna 1, E-50018, Zaragoza, Spain
e-mail: julvez@unizar.es

Juan-Pablo López-Grao
Department of Computer Science and Systems Engineering, University of Zaragoza
María de Luna 1, E-50018, Zaragoza, Spain
e-mail: jpablo@unizar.es

Jan Komenda
Institute of Mathematics, Academy of Sciences of the Czech Republic
Žižkova 22, 616 62 Brno, Czech Republic
e-mail: komenda@math.cas.cz

Cristian Mahulea
Instituto de Investigación en Ingeniería de Aragón (I3A), University of Zaragoza
María de Luna 1, E-50018, Zaragoza, Spain
e-mail: `cmahulea@unizar.es`

Tomáš Masopust
Institute of Mathematics, Academy of Sciences of the Czech Republic
Žižkova 22, 616 62 Brno, Czech Republic
e-mail: `masopust@math.cas.cz`

Jörg Raisch
Fachgebiet Regelungssysteme, TU Berlin, Sekr. EN11, Einsteinufer 17, 10587 Berlin, Germany.
Also: Fachgruppe System-und Regelungstheorie, Max-Planck-Institut für Dynamik komplexer technischer Systeme, Magdeburg
e-mail: `raisch@control.tu-berlin.de`

S. Laurie Ricker
Department of Mathematics & Computer Science, Mount Allison University
67 York St. Sackville, NB Canada E4L 1E6
e-mail: `lricker@mta.ca`

Carla Seatzu
Department of Electrical and Electronic Engineering, University of Cagliari
Piazza D'Armi, 09123 Cagliari, Italy
e-mail: `seatzu@diee.unica.it`

Manuel Silva
Instituto de Investigación en Ingeniería de Aragón (I3A), University of Zaragoza
María de Luna 1, E-50018, Zaragoza, Spain
e-mail: `silva@unizar.es`

Jan H. van Schuppen
CWI, P.O. Box 94079, 1090 GB Amsterdam, Netherlands
e-mail: `J.H.van.Schuppen@cwi.nl`

C. Renato Vázquez
Instituto de Investigación en Ingeniería de Aragón (I3A), University of Zaragoza
María de Luna 1, E-50018, Zaragoza, Spain
e-mail: `cvazquez@unizar.es`

Reviewers

René K. Boel	Ghent University, Belgium
José Manuel Colom	University of Zaragoza, Spain
Isabel Demngodin	University Aix-Marseille, France
Mariagrazia Dotoli	Polytechnique of Bari, Italy
Maria Pia Fanti	Polytechnique of Bari, Italy
Anne-Kathrin Hess	Technical University of Berlin, Germany
Richard Hill	University of Detroit Mercy, Michigan, USA
Jorge Júlvez	University of Zaragoza, Spain
Pia L. Kempker	VU University Amsterdam, Amsterdam, Netherlands
Jan Komenda	Academy of Sciences, Brno, Czech Republic
Xenofon Koutsoukos	Vanderbilt University, Nashville, Tennessee, USA
Dimitri Lefebvre	University Le Havre, Strasbourg, France
Feng Lin	Wayne State University, Detroit, Michigan, USA
Jan Lunze	University of Bochum, Germany
Kristian Lyngbaek	Palo Alto Research Center, Palo Alto, California, USA
Cristian Mahulea	University of Zaragoza, Spain
Tomáš Masopust	Academy of Sciences, Brno, Czech Republic
José Merseguer	University of Zaragoza, Spain
Behrang Monajemi Nejad	Technical University of Berlin, Germany
Geert Jan Olsder	Delft University of Technology, Delft, Netherlands
Laurie Ricker	Mount Allison University, Sackville, NB Canada
Karen Rudie	Queen's University, Kingston, Canada
Carla Seatzu	University of Cagliari, Italy
Manuel Silva	University of Zaragoza, Spain
Rong Su	Nanyang Technological University, Singapore
John Thistle	University of Waterloo, Ontario, Canada
Ton van den Boom	Delft University of Technology, Delft, Netherlands
Jan H. van Schuppen	CWI, Amsterdam, Netherlands
Renato Vázquez	University of Zaragoza, Spain
Tiziano Villa	University of Verona, Italy
Tae-Sic Yoo	Idaho National Laboratory, Idaho, USA
Liewei Wang	University of Zaragoza, Spain
Xu Wang	University of Zaragoza, Spain
Yorai Wardi	Georgia Institute of Technology, Atlanta, Georgia, USA
Thomas Wittmann	University of Erlangen, Germany

Acronyms

AGV	Automated Guided Vehicle
BIC	Bounded Input Controllable
BRG	Basis Reachability Graph
CDFG	Controlled Deterministic Finite Generator
CPN	Continuous Petri Net
CS	Controllability Space
CSS	Cooperating Sequential System
CTMC	Continuous Time Markov Chain
DcontPN	Distributed Continuous Petri Net
DEDS	Discrete Event Dynamic Systems
DES	Discrete Event System
DFA	Deterministic Finite Automaton
DFG	Deterministic Finite Generator
DFMeA	Deterministic Finite Mealy Automaton
DFMoA	Deterministic Finite Moore Automaton
DTPN	Deterministic Timed Petri Net
DuSCOP	Dual Supervisory Control and Observation
ELCP	Extended Linear Complementarity Problem
EQ	Equal Conflict
FMS	Flexible Manufacturing System
FSM	Finite State Machine
FSS	Finite Server Semantics
GTS	Graph Transformation System
HC	Hamiltonian Circuit
HTS	High-Throughput Screening
ILP	Integer Linear Programming
ILPP	Integer Linear Programming Problem
ISS	Infinite Server Semantics
JF	Join Free
LPP	Linear Programming Problem
L-S^3PR	Linear System of Simple Sequential Processes with Resources

MPC	Model Predictive Control
MPN	Markovian Petri Net
MRI	Magnetic Resonance Imaging
MTS	Mono-T-Semiflow
MVN	Modified Verifier Net
NDFG	Nondeterministic Finite Generator
NDFMeA	Nondeterministic Finite Mealy Automaton
NDFMoA	Nondeterministic Finite Moore Automaton
NFA	Nondeterministic Finite Automaton
NS-RAS	Non-Sequential Resource Allocation System
OCC	Output Control Consistent
PC^2R	Processes Competing for Conservative Resources
PDA	PushDown Automaton
PLC	Programmable Logic Controller
PN	Petri Net
QPP	Quadratic Programming Problem
RAP	Resource Allocation Problem
RAS	Resource Allocation System
SAT	Satisfiability
SB	Structurally Bounded
SCTdMG	Strongly Connected Timed Marked Graph
SL	Structurally Live
SPQR	System of Processes Quarrelling over Resources
S^3PR	System of Simple Sequential Processes with Resources
S-RAS	Sequential Resource Allocation System
STdPN	Stochastic Timed Petri Net
TCPN	Timed Continuous Petri Net
TdPN	Timed Petri Net
TEG	Timed Event Graph
TM	Turing Machines
TMG	Timed Marked Graph
TPN	Time Petri Net
VN	Verifier Net
WP	Weighted Path

Part I
Automata

Chapter 1
Modelling of Engineering Phenomena by Finite Automata

Jörg Raisch

1.1 Introduction

In "conventional" systems and control theory, signals "live" in $\mathbb{R}^n$ (or some other, possibly infinite-dimensional, vector space). Then, a signal is a map $T \to \mathbb{R}^n$, where T represents continuous or discrete time. There are, however, numerous application domains where signals only take values in a discrete set, which is often finite and not endowed with mathematical structure. Examples are pedestrian lights (possible signal values are "red" and "green") or the qualitative state of a machine ("busy", "idle", "down"). Often, such signals can be thought of as naturally discrete-valued; sometimes, they represent convenient abstractions of continuous-valued signals and result from a quantisation process.

Example 1.1. Consider a water reservoir, where $z : \mathbb{R}^+ \to \mathbb{R}^+$ is the (continuous-valued) signal representing the water level in the reservoir. The quantised signal

$$\tilde{y} : \quad \mathbb{R}^+ \to \{\texttt{Hi}, \texttt{Med}, \texttt{Lo}\} \ ,$$

where

$$\tilde{y}(t) = \begin{cases} \texttt{Hi} \text{ if } & z(t) > 2 \\ \texttt{Med} \text{ if } & 1 < z(t) \leq 2 \\ \texttt{Lo} \text{ if } & z(t) \leq 1 \end{cases}$$

represents coarser, but often adequate, information on the temporal evolution of the water level within the reservoir. This is indicated in Fig. 1.1, which also shows that the discrete-valued signal $\tilde{y}$ can be represented by a sequence or string of timed discrete events, e.g.,

$$(t_0, \texttt{Lo}), (t_1, \texttt{Med}), (t_2, \texttt{Hi}), \ldots ,$$

Jörg Raisch
Fachgebiet Regelungssysteme, TU Berlin, Sekr. EN11, Einsteinufer 17,
10587 Berlin, Germany. Also: Fachgruppe System- und Regelungstheorie,
Max-Planck-Institut für Dynamik komplexer technischer Systeme, Magdeburg
e-mail: `raisch@control.tu-berlin.de`

C. Seatzu et al. (Eds.): Control of Discrete-Event Systems, LNCIS 433, pp. 3–22.
springerlink.com

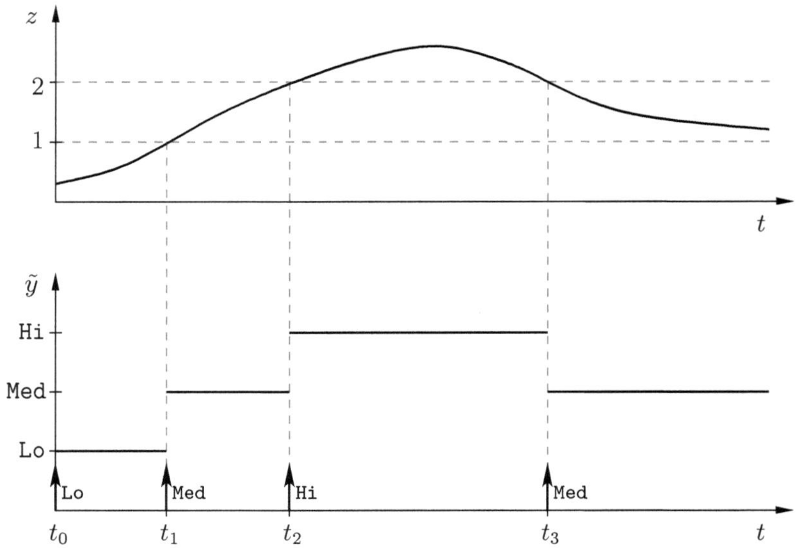

Fig. 1.1 Quantisation of a continuous signal

where $t_i \in \mathbb{R}^+$ are event times and $(t_i, \texttt{Med})$ means that at time t_i, the quantised signal $\tilde{y}$ changes its value to Med. ■

Note that an (infinite) sequence of timed discrete events can be interpreted as a map $\mathbb{N}_0 \to \mathbb{R}^+ \times Y$, where Y is an event set. Similarly, a finite string of timed discrete events can be seen as a map defined on an appropriate finite subset $I_j = \{0, \ldots, j\}$ of $\mathbb{N}_0$.

Often, even less information may be required. For example, only the temporal ordering of events, but not the precise time of the occurrence of events may be relevant. In this case, we can simply project out the time information and obtain a sequence (or string) of logical events, e.g.,

$$\texttt{Lo}, \texttt{Med}, \texttt{Hi}, \ldots,$$

which can be interpreted as a map $y : \mathbb{N}_0$ (resp. I_j) $\to Y$, where Y is the event set. It is obvious (but important) to note, that the domain $\mathbb{N}_0$ (respectively I_j) does in general *not* represent uniformly sampled time; i.e., the time difference $t_{k+1} - t_k$, with $k, k+1 \in \mathbb{N}_0$ (respectively I_j), between the occurrence of subsequent events $y(k+1)$ and $y(k)$ is usually not a constant.

Clearly, going from the continuous-valued signal z to the discrete-valued signal $\tilde{y}$ (or the corresponding sequence of timed discrete events), and from the latter to a sequence y of logical events, involves a loss of information. This is often referred to as signal aggregation.

We will use the following terminology: given a set U, the symbol $U^{\mathbb{N}_0}$ denotes the set of all functions mapping $\mathbb{N}_0$ into U, i.e., the set of all (infinite) sequences of elements from U. The symbol U^* denotes the set of all finite strings from elements of U. More specifically, $U^{I_j} \subset U^*$, $j = 0, 1, \ldots$, is the set of strings from U with length $j+1$, and ε is the empty string. The length of a string u is denoted by $|u|$, i.e., $|u| = j+1$ if $u \in U^{I_j}$, and $|\varepsilon| = 0$.

We will be concerned with models that explain sequences or strings of logical discrete events. In all but trivial cases, these models will be dynamic, i.e., to explain the k-th event, we will need to consider previous events. As in "conventional" systems and control theory, it is convenient to concentrate on state models.

1.2 State Models with Inputs and Outputs

The traditional control engineering point of view is based on the following assumptions (compare Fig. 1.2):

- the system to be controlled (plant) exhibits input and output signals;
- the control input is free, i.e., it can be chosen by the controller;
- the disturbance is an input that cannot be influenced; in general, it can also not be measured directly;
- the output can only be affected indirectly;
- the plant model relates control input and output signals;
- the disturbance input may or may not be modelled explicitly; if it is not, the existence of disturbances simply "increases the amount of nondeterminism" in the (control) input/output model;
- the design specifications depend on the output; they may additionally depend on the control input.

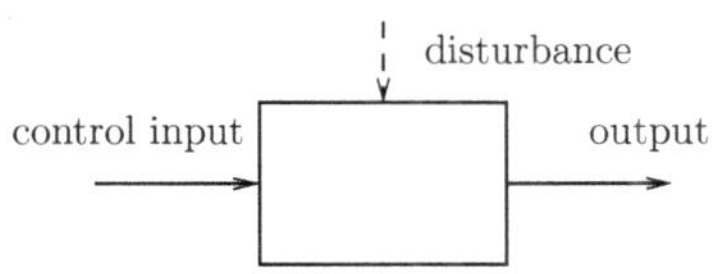

Fig. 1.2 System with inputs and outputs

1.2.1 Mealy Automata

The relation between discrete-valued input and output signals can often be modelled by finite Mealy automata:

Definition 1.1. *A deterministic finite Mealy automaton (DFMeA) is a sixtuple* (X, U, Y, f, g, x_0), *where*

- $X = \{\xi_1, \dots \xi_n\}$ *is a finite set of states,*
- $U = \{\mu_1, \dots \mu_q\}$ *is a finite set of input symbols,*
- $Y = \{\nu_1, \dots \nu_p\}$ *is a finite set of output symbols,*
- $f : X \times U \to X$ *represents the transition function,*
- $g : X \times U \to Y$ *represents the output function,*
- $x_0 \in X$ *is the initial state.*

Note that the transition function is a total function, i.e., it is defined for all pairs in $X \times U$. In other words, we can apply any input symbol in any state, and the input signal u is therefore indeed free. Furthermore, as x_o is a singleton and the transition function maps into X, any sequence (string) of input symbols will provide a unique sequence (string) of output symbols. Finally, from the state and output equations

$$x(k+1) = f(x(k), u(k)), \tag{1.1}$$

$$y(k) = g(x(k), u(k)), \tag{1.2}$$

it is obvious that the output symbols $y(0) \dots y(k-1)$ will not depend on the input symbols $u(k), u(k+1), \dots$. One says that the output *does not anticipate* the input, or, equivalently, that the input output relation is non-anticipating [19].

Remark 1.2. Sometimes, it proves convenient to extend the definition of DFMeA by adding a set $X_m \subseteq X$, the set of marked states. The role of X_m is to model acceptable outcomes when applying a finite string of input symbols. Such an outcome is deemed to be achieved when a string of input symbols "drives" the system state from x_0 into X_m. One could argue, of course, whether the definition of an acceptable outcome should be contained in the plant model or rather be part of the specification, which is to be modelled separately. ■

The following example illustrates how DFMeA can model discrete event systems with inputs and outputs.

Example 1.3. Let us consider a candy machine with (very) restricted functionality. It sells two kinds of candy, chocolate bars (product 1) and chewing gum (product 2). A chocolate bar costs 1 €, chewing gum costs 2 €. The machine only accepts 1 € and 2 € coins and will not give change. Finally, we assume that the machine is sufficiently often refilled and therefore never runs out of chocolate or chewing gum. In this simplified scenario, the customer can choose between the input symbols:

1 €	"insert a 1 € coin"
2 €	"insert a 2 € coin"
P1	"request product 1 (chocolate bar)"
P2	"request product 2 (chewing gum)"

The machine can respond with the following output symbols:

G1	"deliver product 1"
G2	"deliver product 2"
M1	"display message `insufficient credit`"
M2	"display message `choose a product`"
R1 €	"return 1 € coin"
R2 €	"return 2 € coin"

Clearly, to implement the desired output response, the machine needs to keep track of the customer's current credit. To keep the system as simple as possible, we restrict the latter to 2 €. Hence, we will have the following states:

ξ_1	"customer's credit is 0 €",
ξ_2	"customer's credit is 1 €",
ξ_3	"customer's credit is 2 €",

where ξ_1 is the initial state and the only marked state, i.e., $x_0 = \xi_1$ and $X_m = \{\xi_1\}$. The Mealy automaton shown in Fig. 1.3 models the functionality of our simple candy machine. In this figure, circles represent states, and arrows represent transitions between states. Transitions are labelled by a pair μ_i/ν_j of input/output symbols. For example, the arrow labelled 2 €/M2 beginning in ξ_1 and ending in ξ_3 is to be interpreted as $f(\xi_1, 2\,€) = \xi_3$ and $g(\xi_1, 2\,€) = \text{M2}$. Hence, if the customer's credit is 0 € and (s)he inserts a 2 € coin, (s)he will subsequently have 2 € credit, and the machine will display the message `choose a product`. The initial state is indicated by a small arrow pointing towards it "from the outside", and a marked state is shown by a small arrow pointing from the state "towards the outside". ■

The following notions characterise how a DFMeA responds to sequences (strings) of input symbols:

The set of all pairs of input/output signals (or, equivalently, the set of all pairs of sequences of input and output symbols) which are compatible with the automaton

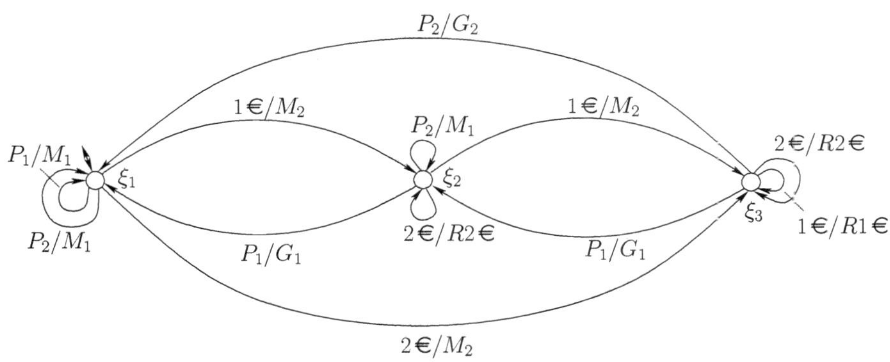

Fig. 1.3 Mealy automaton modelling a candy machine

dynamics is called the *behaviour generated by a DFMeA* and denoted by $\mathscr{B}$. Formally,

$$\mathscr{B} = \Big\{(u,y) \mid \exists x \in X^{\mathbb{N}_0} \text{ s.t. (1.1) and (1.2) hold } \forall k \in \mathbb{N}_0,\, x(0) = x_0\Big\}.$$

Hence, a pair (u,y) of input and output sequences is in the behaviour $\mathscr{B}$ if and only if there exists a state sequence x that begins in the initial state and, together with (u,y), satisfies the next state and output equations defined by f and g.

The *language generated by a DFMeA*, denoted by L, is the set of all pairs (u,y) of input and output strings with equal length such that applying the input string $u \in U^*$ in the initial automaton state x_0 makes the DFMeA respond with the output string $y \in Y^*$. Formally,

$$L = \Big\{(u,y) \mid |u| = |y|,\, \exists x \in X^{I_{|u|}} \text{ s.t. (1.1) and (1.2) hold for } \\ k = 0,\ldots,|u|-1,\, x(0) = x_0\big\}.$$

If a set of marked states is given, we can also define the *language marked by a DFMeA*, L_m. It represents the elements of the language L that make the state end in X_m. Formally,

$$L_m = \{(u,y) \mid (u,y) \in L,\, x(|u|) \in X_m\}.$$

Clearly, in our candy machine example, $((\text{P1},\text{P1},\ldots),(\text{M1},\text{M1},\ldots)) \in \mathscr{B}$, $((1\,€,\text{P2}),(\text{M2},\text{M1})) \in L$, and $((1\,€,\text{P1}),(\text{M2},\text{G1})) \in L_m$.

Modelling is in practice often intentionally coarse, i.e., one tries to model only phenomena that are important for a particular purpose (e.g., the synthesis of feedback control). In the context of input/output models, this implies that the effect of inputs on outputs is, to a certain extent, uncertain. This is captured by the notion of nondeterministic finite Mealy automata (NDFMeA).

Definition 1.2. *A nondeterministic finite Mealy automaton (NDFMeA) is a quintuple (X,U,Y,h,X_0), where*

- *X is a finite set of states,*
- *U is a finite set of input symbols,*
- *Y is a finite set of output symbols,*
- *$h : X \times U \to 2^{X\times Y} \setminus \emptyset$ represents a set-valued transition-output function,*
- *$X_0 \subseteq X$, $X_0 \neq \emptyset$ is a set of possible initial states.*

Hence, there are two "sources" of uncertainty in NDFMeA: (i) the initial state may not be uniquely determined; (ii) if the automaton is in a certain state and an input symbol is applied, the next state and/or the generated output symbol may be nondeterministic. The evolution is now governed by

$$(x(k+1),y(k)) \in h(x(k),u(k)). \tag{1.3}$$

As in the deterministic case, the output symbols $y(0)\ldots y(k-1)$ will not depend on the input symbols $u(k),u(k+1),\ldots$, i.e., the output does not anticipate the input.

Example 1.4. Consider the slightly modified model of our candy machine in Fig. 1.4, which includes a limited "change" policy. The only difference compared to Example 1.3 is when the machine is in state ξ_2 (i.e., the customer's credit is 1 €), and a 2 € coin is inserted. Depending on whether 1 € coins are available (which is not modelled), the machine may return a 1 € coin and go to state ξ_3 (i.e., the customer's credit is now 2 €), or it may return the inserted 2 € coin and remain in state ξ_2. ■

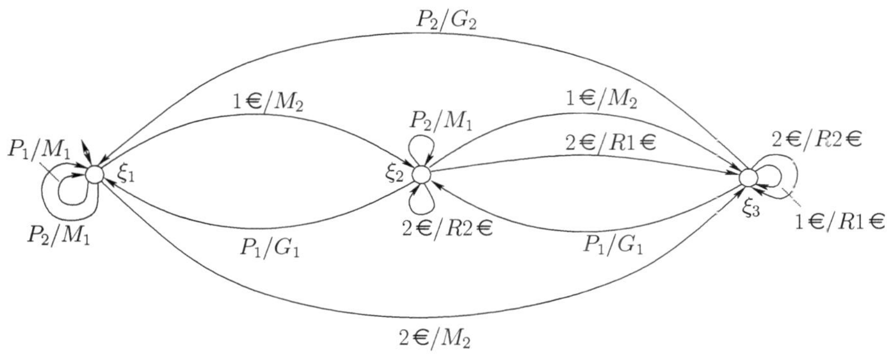

Fig. 1.4 Nondeterministic Mealy automaton modelling a candy machine

Remark 1.5. Note that a combined transition-output function $h : X \times U \to 2^{X\times Y} \setminus \emptyset$ is more expressive in the set-valued case than separate transition and output functions $f : X \times U \to 2^X \setminus \emptyset$ and $g : X \times U \to 2^Y \setminus \emptyset$. This is illustrated by Example 1.4 above. There, $h(\xi_2, 2\,€) = \{(\xi_3, \mathrm{R1}\,€), (\xi_2, \mathrm{R2}\,€)\}$. That is, in state ξ_2, after choosing input 2 €, there are two possibilities for the next state and two possibilities for the resulting output symbol; however, arbitrary combinations of these are not allowed. This could not be modelled by separate transition and output functions f, g. ■

Remark 1.6. The use of a combined transition-output function also leads to a more compact formulation of the behaviour and the language generated by an NDFMeA:

$$\mathscr{B} = \left\{(u,y) \;\middle|\; u \in U^{\mathbb{N}_0}, y \in Y^{\mathbb{N}_0},\ \exists x \in X^{\mathbb{N}_0} \text{ s.t. (1.3) holds } \forall k \in \mathbb{N}_0,\ x(0) \in X_0 \right\},$$

$$L = \Big\{(u,y) \mid u \in U^*, y \in Y^*, |u| = |y|,\ \exists x \in X^{I_{|u|}} \text{ s.t. (1.3) holds for } k = 0, \ldots, |u|-1,\ x(0) \in X_0\Big\}.$$

■

Remark 1.7. As in the deterministic case, we can extend the definition by adding a set of marked states, $X_m \subseteq X$. The language marked by an NDFMeA is then the subset of L that admits a corresponding string of states to end in X_m, i.e.,

$$L_m = \Big\{(u,y) \mid u \in U^*, y \in Y^*, |u| = |y|,\ \exists x \in X^{I_{|u|}} \text{ s.t. (1.3) holds for } k = 0, \ldots, |u|-1,\ x(0) \in X_0,\ x(|u|) \in X_m\Big\}.$$

Note that at least one string of states compatible with a $(u,y) \in L_m$ must end in X_m, but not all of them have to. ■

1.2.2 Moore Automata

Often, one encounters situations where the kth output event does not depend on the kth (and subsequent) input event(s). In this case, it is said that the output *strictly* does not anticipate the input. Clearly, this is achieved, if the output function of a DFMeA is restricted to $g : X \to Y$. The resulting state machine is called a Moore automaton:

Definition 1.3. *A deterministic finite Moore automaton (DFMoA) is a sixtuple* (X, U, Y, f, g, x_0), *where*

- X *is a finite set of states,*
- U *is a finite set of input symbols,*
- Y *is a finite set of output symbols,*
- $f : X \times U \to X$ *represents the transition function,*
- $g : X \to Y$ *represents the output function,*
- $x_0 \in X$ *is the initial state.*

The evolution of state and output is determined by

$$\begin{aligned} x(k+1) &= f(x(k), u(k)), \\ y(k) &= g(x(k)). \end{aligned}$$

Remark 1.8. As for Mealy automata, the above definition can be extended by adding a set of marked states, $X_m \subseteq X$, to model acceptable outcomes when applying a finite string of input symbols. ■

Remark 1.9. A nondeterministic version (NDFMoA) is obtained by the following straightforward changes in the above definition: the initial state is from a set $X_o \subseteq X$, $X_o \neq \emptyset$; the transition function and the output function map into the respective power sets, i.e., $f : X \times U \to 2^X \setminus \emptyset$ and $g : X \to 2^Y \setminus \emptyset$. ■

1.3 Automata with Controllable and Uncontrollable Events

Although distinguishing between control inputs and outputs is a natural way of modelling the interaction between plant and controller, in most of the work in the area of discrete event control systems a slightly different point of view has been adopted. There, plants are usually modelled as deterministic finite automata with an event set that is partitioned into sets of controllable (i.e., preventable by a controller) and uncontrollable (i.e., non-preventable) events. For example, starting your car is a controllable event (easily prevented by not turning the ignition key), whereas breakdown of a car is, unfortunately, uncontrollable. In this scenario, a string or sequence of events is in general not an input and therefore not free. Hence, the transition function is usually defined to be a partial function. The resulting machine is often referred to as a finite generator.

Definition 1.4. *A deterministic finite generator (DFG) is a sixtuple* (X, S, S_c, f, x_0, X_m), *where*

- $X = \{\xi_1, \dots \xi_n\}$ *is a finite set of states,*
- $S = \{\sigma_1 \dots \sigma_q\}$ *is a finite set of events (the "alphabet"),*
- $S_c \subseteq S$ *is the set of controllable events; consequently* $S_{uc} = S \setminus S_c$ *represents the set of uncontrollabe events,*
- $f : X \times S \to X$ *is a partial transition function,*
- $x_0 \in X$ *is the initial state,*
- $X_m \subseteq X$ *is a set of marked states.*

As for input/output automata, we distinguish the language generated and the language marked by DFG:

$$L = \Big\{ s \in S^* \mid \exists x \in X^{I_{|s|}} \text{ s.t. } x(k+1) = f(x(k), s(k)),\ k = 0, \dots, |s| - 1,$$
$$x(0) = x_0 \big\},$$
$$L_m = \Big\{ s \in S^* \mid \exists x \in X^{I_{|s|}} \text{ s.t. } x(k+1) = f(x(k), s(k)),\ k = 0, \dots, |s| - 1,$$
$$x(0) = x_0,\ x(|s|) \in X_m \big\}.$$

The set of events that can occur in state ξ is called the active event set, denoted by $\Gamma(\xi)$, i.e.,

$$\Gamma(\xi) = \{\sigma \mid f \text{ is defined on } (\xi, \sigma)\}.$$

Example 1.10. Consider the following simple model of a simple machine (taken from [21]): the model has three states, `idle`, `working`, and `broken`. The event set S consists of four elements:

a	`take a workpiece and start processing it,`
b	`finish processing of workpiece,`
c	`machine breaks down,`
d	`machine gets repaired.`

The events a and d are controllable in the sense that they can be prevented, i.e., $S_c = \{a, d\}$. It is assumed that neither the breakdown of the machine nor the finishing of a workpiece can be prevented by control, i.e., $S_{uc} = \{b, c\}$. The state `idle` is both the initial state and the only marked state. The transition structure is shown in Fig. 1.5, where, e.g., the arc from `idle` to `working` labelled by a means that the transition function is defined for the pair $(\texttt{idle}, a)$ and $f(\texttt{idle}, a) = \texttt{working}$. To distinguish controllable and uncontrollable events, arcs labelled with controllable events are equipped with a small bar. Clearly, in this example, the strings ab, aba,

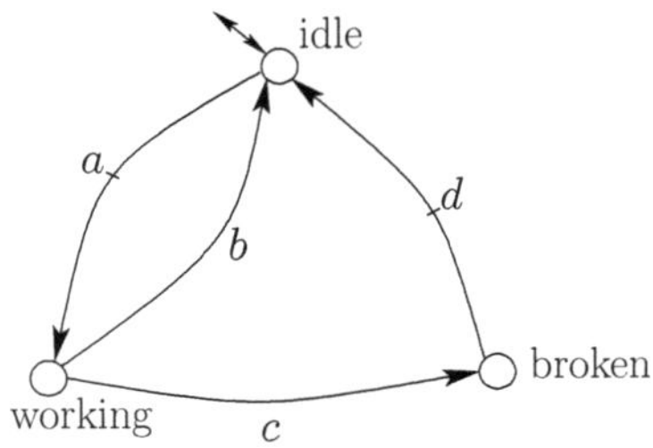

Fig. 1.5 Simple machine model

and *acd* are in the language generated by the DFG in Fig. 1.5, and *ab*, *acd* (but not *aba*) are also in the marked language. ∎

As indicated in the previous section, nondeterminism is in practice a feature that is often intentionally included to keep models simple.

Definition 1.5. *A nondeterministic finite generator (NDFG) is a sixtuple* (X, S, S_c, f, X_0, X_m)*, where*

- X, S, S_c *and* X_m *are as in Definition 1.4,*
- $f : X \times S \to 2^X$ *is the transition function,*
- $X_0 \subseteq X$ *is the set of possible initial states.*

Note that the transition function, as it maps into the power set 2^X (which includes the empty set), can be taken as a total function. That is, it is defined for all state-event pairs, and $\sigma \in \Gamma(\xi)$ if and only if $f(\xi, \sigma) \neq \emptyset$. The definitions of languages generated and marked by NDFGs carry over from the deterministic case with the obvious change that $x(k+1) = f(x(k), s(k))$ needs to be replaced by $x(k+1) \in f(x(k), s(k))$.

1.4 Finite Automata as Approximations of Systems with Infinite State Sets

We now discuss the question whether — for the purpose of control synthesis — finite automata can serve as approximate models for systems with an infinite state set. The motivation for investigating this problem stems from the area of hybrid dynamical systems. There, at least one discrete event system, e.g., in the form of a finite Moore or Mealy automaton, interacts with at least one system with an infinite state space, typically $\mathbb{R}^n$. A composition of infinite and finite state components would result in a hybrid system with state set $\mathbb{R}^n \times X$. This is neither finite nor a vector space, as the finite automaton state set X is typically without mathematical structure. Hence, neither established control synthesis methods from continuous systems theory nor from the area of supervisory control of *discrete event systems* (DES) can be applied. In this situation, it is then natural to ask whether the infinite state component can be approximated by a suitable finite state automaton, and if control for the resulting overall DES can be meaningful for the underlying hybrid system.

Assume that the component to be approximated can be represented by a state model P defined on a discrete, but not necessarily equidistant, time axis. Assume furthermore that the input is free and that both the input and output spaces are finite. This is natural in the described context, where the input and output signals connect the component's infinite state space to (the) finite DES component(s) in the overall hybrid system (Fig. 1.6).

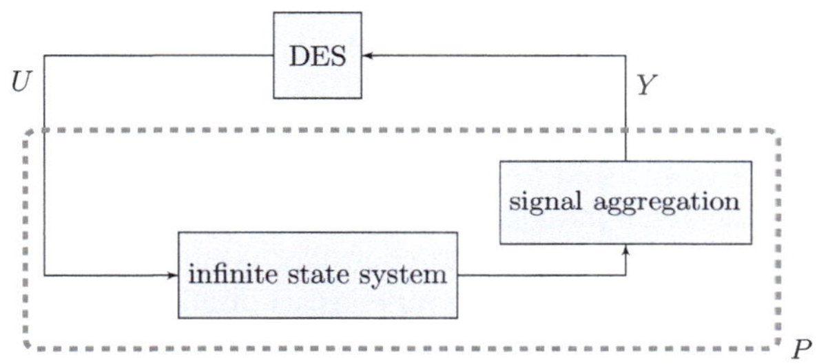

Fig. 1.6 Hybrid system

Furthermore, we do not assume any knowledge on the initial state. The resulting model is then the following infinite state machine:

$$P = (X, U, Y, f, g, X_0),$$

where $X = \mathbb{R}^n$ is the state space, U and Y are finite input and output sets, $f : X \times U \to X$ is the transition function, $g : X \times U \to Y$ is the output function, and $X_0 = \mathbb{R}^n$ is the set of possible initial states. In total analogy to Section 1.2, the input output relation of P is non-anticipating and its behaviour is

$$\mathscr{B} = \left\{ (u,y) \mid \exists x \in \mathbb{R}^{\mathbb{N}_0} \text{ s.t. } x(k+1) = f(x(k), u(k)), y(k) = g(x(k), u(k)) \ \forall k \in \mathbb{N}_0 \right\}.$$

Note that the above system is time invariant in the sense of [19], as shifting a pair $(u,y) \in \mathscr{B}$ on the time axis will never eliminate it from $\mathscr{B}$. Formally, $\sigma\mathscr{B} \subseteq \mathscr{B}$, where σ represents the backward or left shift operator, i.e., $(\sigma(u,y))(k) = (u(k+1), y(k+1))$.

To answer the question whether control synthesis for a finite state approximation can be meaningful for P, we obviously need to take into account specifications. For DES and hybrid systems, we often have inclusion-type specifications, i.e., the control task is to guarantee that the closed-loop behaviour is a (nonempty!)

subset of a given specification behaviour, thus ruling out signals that are deemed to be unacceptable. In this case, an obvious requirement for the approximation is that its behaviour contains the behaviour of the system to be approximated. The argument for this is straightforward: as linking feedback control to a plant provides an additional relation between the plant input and output sequences, it restricts the plant behaviour by eliminating certain pairs of input and output sequences. Hence, if there exists a controller that achieves the specifications for the approximation, it will — subject to some technical constraints regarding implementability issues — also do this for the underlying system (for a more formal argument see, e.g., [10, 11]). There is another issue to be taken into consideration, though. This is the question of whether the approximation is sufficiently accurate. In Willems' behavioural framework, e.g. [19, 20], a partial order on the set of all models which relate sequences of symbols from U and Y and which are unfalsified (i.e., not contradicted by available input/output data) is readily established via the $\subseteq$ relation on the set of the corresponding behaviours. In particular, if $\mathscr{B}_A \subseteq \mathscr{B}_B$ holds for two models A and B, the former is at least as accurate as the latter. B is then said to be an abstraction of A, and, conversely, A is said to be a refinement of B. It is well possible that an approximation of the given infinite state model with the required abstraction property is "too coarse" to allow successful controller synthesis. An obvious example is the trivial abstraction, whose behaviour consists of all pairs $(u,y) \in U^{\mathbb{N}_0} \times Y^{\mathbb{N}_0}$. It will allow all conceivable output sequences for any input sequence, and we will therefore not be able to design a controller enforcing any nontrivial specification. Hence, if the chosen approximation of P is too coarse in the sense that no suitable DES controller exists, we need to refine that approximation. Refinability is therefore another important feature we require on top of the abstraction property.

1.4.1 *l*-Complete Approximations

An obvious candidate for a family of approximations satisfying both the abstraction and refinability property are systems with behaviours

$$\mathscr{B}_l = \left\{ (u,y) \mid (\sigma^t(u,y))\big|_{[0,l]} \in \mathscr{B}\big|_{[0,l]} \ \forall t \in \mathbb{N}_0 \right\}, \ l = 1,2,\ldots \tag{1.4}$$

where σ^t is the backward t-shift, i.e., $(\sigma^t u)(k) = u(k+t)$, and $(\cdot)\big|_{[0,l]}$ is the restriction operator. The latter maps sequences, e.g., $u \in \mathbb{N}_0$, to finite strings, e.g., $u(0),\ldots,u(l) \in U^{l_l}$. Both the shift and the restriction operator can be trivially extended to pairs and sets of sequences (behaviours). The interpretation of (1.4) is that the behaviour $\mathscr{B}_l$ consists of all pairs of sequences (u,y) that, on any interval of length $l+1$, coincide with a pair of input/output sequences in the underlying system's behaviour $\mathscr{B}$. Clearly,

$$\mathscr{B}_1 \supseteq \mathscr{B}_2 \supseteq \mathscr{B}_3 \supseteq \ldots \supseteq \mathscr{B},$$

i.e., for any $l \in \mathbb{N}$, we have the required abstraction property, and refinement can be implemented by increasing l.

A system with behaviour (1.4) is called a *strongest l-complete approximation* ([10, 11]) of P : it is an l-complete system (e.g., [20]) as checking whether a signal pair (u,y) is in the system behaviour can be done by investigating the pair on intervals $[t,t+l]$, $t \in \mathbb{N}_0$. Formally, a time-invariant system defined on $\mathbb{N}_0$ with behaviour $\tilde{\mathscr{B}}$ is said to be l-complete, if $(u,y) \in \tilde{\mathscr{B}} \Leftrightarrow \sigma^t(u,y))\big|_{[0,l]} \in \tilde{\mathscr{B}}\big|_{[0,l]}$ $\forall t \in \mathbb{N}_0$. Furthermore, for any l-complete system with behaviour $\tilde{\mathscr{B}} \supseteq \mathscr{B}$, it holds that $\tilde{\mathscr{B}} \supseteq \mathscr{B}_l$. In this sense, $\mathscr{B}_l$ represents the most accurate l-complete approximation of P exhibiting the abstraction property.

We now need to decide whether an arbitrary pair of input and output strings of length $l+1$, denoted by $(\bar{u},\bar{y})\big|_{[0,l]}$, is an element in $\mathscr{B}\big|_{[0,l]}$, i.e., if the state model P can respond to the input string $\bar{u}\big|_{[0,l]}$ with the output string $\bar{y}\big|_{[0,l]}$. This is the case if and only if $\mathscr{X}((\bar{u},\bar{y})\big|_{[0,l]})$, the set of states of P that are reachable at time l when applying the input string $\bar{u}\big|_{[0,l]}$ and observing the output string $\bar{y}\big|_{[0,l]}$, is nonempty ([11]). $\mathscr{X}((\bar{u},\bar{y})\big|_{[0,l]})$ can be computed iteratively by

$$\mathscr{X}((\bar{u},\bar{y})\big|_{[0,0]}) = g_{\bar{u}_0}^{-1}(\bar{y}_0)),$$
$$\mathscr{X}((\bar{u},\bar{y})\big|_{[0,r+1]}) = f(\mathscr{X}((\bar{u},\bar{y})\big|_{[0,r]}),\bar{u}_r) \cap g_{\bar{u}_{r+1}}^{-1}(\bar{y}_{r+1}), \quad r = 0,\dots l-1,$$

where $g_{\bar{u}_r}^{-1}(\bar{y}_r)) := \{\xi \mid g(\xi,\bar{u}_r) = \bar{y}_r\}$ and $f(A,\bar{u}_r) := \{\xi \mid \xi = f(\xi',\bar{u}_r), \xi' \in A\}$.

As U and Y are finite sets, $\mathscr{B}\big|_{[0,l]} = \mathscr{B}_l\big|_{[0,l]}$ is also finite, and a nondeterministic finite Mealy automaton (NDFMeA)

$$P_l = (Z,U,Y,h,Z_0)$$

generating the behaviour $\mathscr{B}_l$ can be set up using the following procedure. It is based on the simple idea that the state of the NDFMeA memorises the past input and output data up to length l , i.e.,

$$z(k) := \begin{cases} \omega & \text{for } k=0, \\ (u(0)\dots u(k-1), y(0)\dots y(k-1)) & \text{for } 1 \le k \le l, \\ (u(k-l)\dots u(k-1), y(k-l)\dots y(k-1)) & \text{for } k > l, \end{cases}$$

where ω is a "dummy" symbol meaning "no input/output data recorded so far". Then

$$Z = \{\omega\} \bigcup_{1 \le r \le l} (U \times Y)^r$$
$$Z_0 = \{\omega\}$$

and

$$\underbrace{((\bar{u}_0,\bar{y}_0)}_{\tilde{z}_1},\ \bar{y}_0) \in h(\underbrace{\omega}_{\tilde{z}_0},\ \bar{u}_0) \ \text{ iff } (\bar{u}_0,\bar{y}_0) \in \mathscr{B}\big|_{[0,0]}$$

$$\underbrace{((\bar{u}_0\ldots\bar{u}_r,\bar{y}_0\ldots\bar{y}_r)}_{\tilde{z}_{r+1}},\ \bar{y}_r) \in h(\underbrace{(\bar{u}_0\ldots\bar{u}_{r-1},\bar{y}_0\ldots\bar{y}_{r-1})}_{\tilde{z}_r},\ \bar{u}_r)$$

$$\text{iff } (\bar{u}_0\ldots\bar{u}_r,\bar{y}_0\ldots\bar{y}_r) \in \mathscr{B}\big|_{[0,r]},\ \ 0<r<l$$

$$\underbrace{((\bar{u}_1\ldots\bar{u}_l,\bar{y}_1\ldots\bar{y}_l)}_{\tilde{z}'_l},\ \bar{y}_l) \in h(\underbrace{(\bar{u}_0\ldots\bar{u}_{l-1},\bar{y}_0\ldots\bar{y}_{l-1})}_{\tilde{z}_l},\ \bar{u}_l)$$

$$\text{iff } (\bar{u}_0\ldots\bar{u}_l,\bar{y}_0\ldots\bar{y}_l) \in \mathscr{B}\big|_{[0,l]}.$$

1.4.2 A Special Case: Strictly Non-anticipating Systems

It is instructive to briefly investigate the special case where the system to be approximated, P, is strictly non-anticipating, i.e., its output function is $g : \mathbb{R}^n \to Y$. This implies that the input $u(k)$ does not affect the output symbols $y(0),\ldots,y(k)$. Hence, as the input is free,

$$(\bar{u},\bar{y})\big|_{[0,l]} \in \mathscr{B}\big|_{[0,l]} \text{ iff } \mathscr{X}(\bar{u}\big|_{[0,l-1]},\bar{y}\big|_{[0,l]}) \neq \emptyset,$$

where $\mathscr{X}((\bar{u}\big|_{[0,l-1]},\bar{y}\big|_{[0,l]}))$ represents the set of states of P that are reachable at time l when applying the input string $\bar{u}\big|_{[0,l-1]}$ while observing the output string $\bar{y}\big|_{[0,l]}$. As before, we can readily come up with a recursive formula to provide $\mathscr{X}(\bar{u}\big|_{[0,l-1]},\bar{y}\big|_{[0,l]})$:

$$\mathscr{X}(\bar{y}\big|_{[0,0]}) = g^{-1}(\bar{y}_0),$$
$$\mathscr{X}((\bar{u}\big|_{[0,0]},\bar{y}\big|_{[0,1]})) = f(\mathscr{X}(\bar{y}\big|_{[0,0]}),\bar{u}_0) \cap g^{-1}(\bar{y}_1),$$
$$\mathscr{X}((\bar{u}\big|_{[0,r]},\bar{y}\big|_{[0,r+1]})) = f(\mathscr{X}(\bar{u}\big|_{[0,r-1]},\bar{y}\big|_{[0,r]}),\bar{u}_r) \cap g^{-1}(\bar{y}_{r+1}),\ \ 0<r<l.$$

We can now set up a nondeterministic finite Moore automaton (NDFMoA)

$$\tilde{P}_l = (\tilde{Z},U,Y,\tilde{f},\tilde{g},\tilde{Z}_0)$$

generating the behaviour $\mathscr{B}_l$. This procedure is based on the idea that the automaton state memorises the past input and output data up to length $l-1$ plus the present output symbol, i.e.,

$$\tilde{z}(k) := \begin{cases} y(0) & \text{for } k=0, \\ (u(0)\ldots u(k-1), y(0)\ldots y(k)) & \text{for } 1 \le k < l, \\ (u(k-l+1)\ldots u(k-1), y(k-l+1)\ldots y(k)) & \text{for } k \ge l. \end{cases}$$

Hence, for $l > 1$, the state set of the NDFMoA is

$$\tilde{Z} = Y \cup U \times Y^2 \cup \ldots U^{l-1} \times Y^l,$$

and $\tilde{Z}_0 = Y$. The transition function $\tilde{f}$ is defined by

$$\begin{aligned}
\underbrace{(\bar{u}_0, \bar{y}_0\bar{y}_1)}_{\tilde{z}_1} &\in \tilde{f}(\underbrace{\bar{y}_0}_{\tilde{z}_0}, \bar{u}_0) \ \text{ iff } (\bar{u}_0\%, \bar{y}_0\bar{y}_1) \in \mathscr{B}|_{[0,1]} \\
\underbrace{(\bar{u}_0 \ldots \bar{u}_r, \bar{y}_0 \ldots \bar{y}_{r+1})}_{\tilde{z}_{r+1}} &\in \tilde{f}(\underbrace{(\bar{u}_0 \ldots \bar{u}_{r-1}, \bar{y}_0 \ldots \bar{y}_r)}_{\tilde{z}_r}, \bar{u}_r) \\
&\quad \text{iff } (\bar{u}_0 \ldots \bar{u}_r\%, \bar{y}_0 \ldots \bar{y}_{r+1}) \in \mathscr{B}|_{[0,r+1]}, \quad 1 < r < l-1 \\
\underbrace{(\bar{u}_1 \ldots \bar{u}_{l-1}, \bar{y}_1 \ldots \bar{y}_l)}_{\tilde{z}'_{l-1}} &\in \tilde{f}(\underbrace{(\bar{u}_0 \ldots \bar{u}_{l-2}, \bar{y}_0 \ldots \bar{y}_{l-1})}_{\tilde{z}_{l-1}}, \bar{u}_{l-1}) \\
&\quad \text{iff } (\bar{u}_0 \ldots \bar{u}_{l-1}\%, \bar{y}_0 \ldots \bar{y}_l) \in \mathscr{B}|_{[0,l]},
\end{aligned}$$

where the "don't care" symbol % may represent any element of the input set U. For this realisation, the output function is deterministic, i.e., $\tilde{g} : \tilde{Z} \to Y$ and characterised by

$$\begin{aligned}
\tilde{g}(\bar{y}_0) &= \bar{y}_0 \\
\tilde{g}((\bar{u}_0 \ldots \bar{u}_{r-1}, \bar{y}_0 \ldots \bar{y}_r)) &= \bar{y}_r, \quad r = 1, \ldots l-1.
\end{aligned}$$

For $l = 1$, the state only memorises the current output symbol, i.e., $\tilde{Z} = \tilde{Z}_0 = Y$, the transition function $\tilde{f}$ is characterised by

$$\bar{y}_1 \in \tilde{f}(\bar{y}_0, \bar{u}_0) \text{ iff } (\bar{u}_0\%, \bar{y}_0\bar{y}_1) \in \mathscr{B}|_{[0,1]},$$

and the output function $\tilde{g}$ is the identity.

Example 1.11. We now introduce an example, whose *only* purpose is to illustrate the above procedure. Hence we choose it to be as simple as possible, although most problems will become trivial for this example. It represents a slightly modified version of an example that was first suggested in [15]. Consider the water tank shown in Fig. 1.7. Its cross sectional area is 100cm^2, its height $\hat{x} = 30$cm. The attached pump can be switched between two modes: it either feeds water into the tank at a constant rate of 1litre/min, or it removes water from the tank at the same flow rate. The pump is in feed mode if the control input is $u(k) =$ "+", and in removal mode if $u(k) =$ "−". We work with a fixed sampling rate, 1/min, and the control input remains constant between sampling instants. The output signal can take two values: $y(k) = E(mpty)$ if the water level $x(k)$ is less or equal to 15cm, and $y(k) = F(ull)$ if the water level is above 15cm. Hence $U = \{+, -\}$, $Y = \{E, F\}$, and $X = [0, 30\,\text{cm}]$. The behaviour $\mathscr{B}$ can be represented by an (infinite) state model $P = (X, U, Y, f, g, X_0)$, where $X_0 = X$, and f and g are defined by

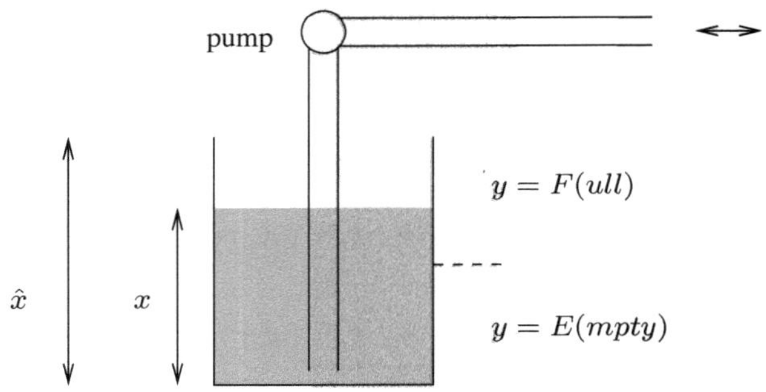

Fig. 1.7 Simple tank example

$$
\begin{aligned}
x(k+1) &= f(x(k),u(k)) \\
&= \begin{cases} x(k)+10\text{cm} & \text{if } u(k) = \text{“}+\text{” and } 0 \le x(k) \le 20\text{cm}, \\ 30\text{cm} & \text{if } u(k) = \text{“}+\text{” and } 20\text{cm} < x(k) \le 30\text{cm}, \\ x(k)-10\text{cm} & \text{if } u(k) = \text{“}-\text{” and } 10\text{cm} < x(k) \le 30\text{cm}, \\ 0\text{cm} & \text{if } u(k) = \text{“}-\text{” and } 0\text{cm} \le x(k) \le 10\text{cm}, \end{cases} \\
y(k) &= g(x(k)) \\
&= \begin{cases} F & \text{if } 15\text{cm} < x(k) \le 30\text{cm}, \\ E & \text{if } 0\text{cm} \le x(k) \le 15\text{cm}. \end{cases}
\end{aligned}
$$

As the system is strictly non-anticipating, we can use the procedure outlined in this section to construct NDFMoA $\tilde{P}_l$ that realise the strongest l-complete approximations. Let us first consider the case $l = 1$. We have to check whether strings $(\bar{u},\bar{y})\big|_{[0,1]} \in \mathscr{B}\big|_{[0,1]}$. This is the case if and only if

$$\mathscr{X}(\bar{u}_0, \bar{y}_0\bar{y}_1) = f(g^{-1}(\bar{y}_0), \bar{u}_0) \cap g^{-1}(\bar{y}_1) \neq \emptyset.$$

For our example, we obtain

$$
\begin{aligned}
\mathscr{X}(+,EE) = [10,25] \cap [0,15] &= [10,15] && \text{i.e.} \quad (+\%,EE) \in \mathscr{B}\big|_{[0,1]} \\
\mathscr{X}(+,EF) = [10,25] \cap (15,30] &= (15,25] && \text{i.e.} \quad (+\%,EF) \in \mathscr{B}\big|_{[0,1]} \\
\mathscr{X}(+,FF) = (25,30] \cap (15,30] &= (25,30] && \text{i.e.} \quad (+\%,FF) \in \mathscr{B}\big|_{[0,1]} \\
\mathscr{X}(+,FE) = (25,30] \cap [0,15] &= \emptyset && \text{i.e.} \quad (+\%,FE) \notin \mathscr{B}\big|_{[0,1]} \\
\mathscr{X}(-,EE) = [0,5] \cap [0,15] &= [0,5] && \text{i.e.} \quad (-\%,EE) \in \mathscr{B}\big|_{[0,1]} \\
\mathscr{X}(-,EF) = [0,5] \cap (15,30] &= \emptyset && \text{i.e.} \quad (-\%,EF) \notin \mathscr{B}\big|_{[0,1]} \\
\mathscr{X}(-,FF) = (5,20] \cap (15,30] &= (15,20] && \text{i.e.} \quad (-\%,FF) \in \mathscr{B}\big|_{[0,1]} \\
\mathscr{X}(-,FE) = (5,20] \cap [0,15] &= (5,15] && \text{i.e.} \quad (-\%,FE) \in \mathscr{B}\big|_{[0,1]}.
\end{aligned}
$$

The described realisation procedure then provides the NDFMoA $\tilde{P}_1$ shown in Fig. 1.8, where the output symbols associated to the states are indicated by dashed arrows. The (initial) state set is $\tilde{Z} = \tilde{Z}_0 = Y$, and there are six transitions. Clearly, $\mathscr{B}$ is a strict subset of the behaviour generated by $\tilde{P}_1$, i.e., $\mathscr{B} \subset \mathscr{B}_1$. For example, $\tilde{P}_1$ allows the sequence $EEE\ldots$ as a possible response to the input sequence $+++\ldots$, i.e., we can add water to the tank for an arbitrary period of time without ever observing the output symbol F. This would clearly not be possible for the system to be approximated.

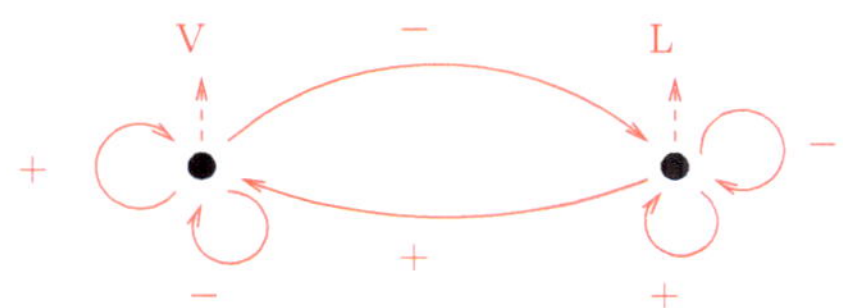

Fig. 1.8 Realisation $\tilde{P}_1$ of strongest 1-complete approximation

We now proceed to the case $l = 2$. To construct the strongest 2-complete approximation, we need to establish whether strings $(\bar{u},\bar{y})\big|_{[0,2]} \in \mathscr{B}\big|_{[0,2]}$. As stated above, this is true if and only if

$$\mathscr{X}(\bar{u}_0\bar{u}_1,\bar{y}_0\bar{y}_1\bar{y}_2) = f\left(f(g^{-1}(\bar{y}_0),\bar{u}_0) \cap g^{-1}(\bar{y}_1),\bar{u}_1\right) \cap g^{-1}(\bar{y}_2) \neq \emptyset.$$

For example, we obtain

$$\begin{aligned}\mathscr{X}(++,EEF) &= f\left(\underbrace{f(g^{-1}(E),+) \cap g^{-1}(E)}_{\mathscr{X}(+,EE)},+\right) \cap g^{-1}(F)\\ &= [20,25] \cap (15,30] \neq \emptyset \quad \text{i.e.} \quad (++\%,EEF) \in \mathscr{B}\big|_{[0,2]}.\end{aligned}$$

Repeating this exercise for other strings and using the realisation procedure described above, we obtain the NDFMoA $\tilde{P}_2$ shown in Fig. 1.9. Its state set is $\tilde{Z} = Y \cup U \times Y^2$, its initial state set $\tilde{Z}_0 = Y$. Note that only 8 out of the 10 elements of $\tilde{Z}$ are reachable. These are the initial states and states $(\bar{u}_C,\bar{y}_0\bar{y}_1)$ such that $(\bar{u}_0\%,\bar{y}_0\bar{y}_1) \in \mathscr{B}\big|_{[0,1]}$. To avoid unnecessary cluttering of the figure, the output symbols are not indicated for each state, but summarily for all states generating the same output.

We can readily check that for this simple example $\mathscr{B}_2 = \mathscr{B}$, i.e., the NDFMoA $\tilde{P}_2$ exhibits exactly the same behaviour as the underlying infinite state model P. In general, however, no matter how large l is chosen, we cannot expect that the behaviour generated by P is l-complete and, in consequence, we will *not* be able to recover it exactly by an l-complete approximation. ■

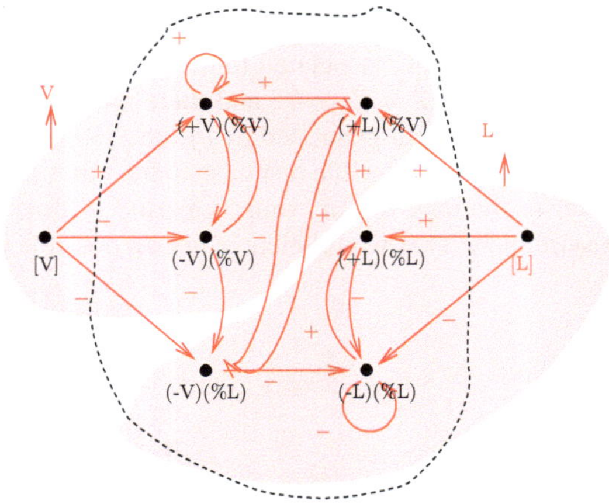

Fig. 1.9 Realisation $\tilde{P}_2$ of strongest 2-complete approximation

Remark 1.12. Suppose we design a feedback controller, or supervisor, for the approximating automaton P_l or $\tilde{P}_l$. Clearly, this controller must satisfy the following requirements: (i) it respects the input/output structure of P_l (respectively $\tilde{P}_l$), i.e., it cannot directly affect the output y; (ii) it enforces the (inclusion-type) specifications when connected to the approximation, i.e., $\mathscr{B}_l \cap \mathfrak{B}_{\text{sup}} \subseteq \mathfrak{B}_{\text{spec}}$, where $\mathfrak{B}_{\text{sup}}$ and $\mathfrak{B}_{\text{spec}}$ are the supervisor and specification behaviours, respectively, and where we assume that $\mathfrak{B}_{\text{spec}}$ can be realised by a finite automaton; (iii) the approximation and supervisor behaviours are nonconflicting, i.e., $\mathscr{B}_l\big|_{[0,k]} \cap \mathfrak{B}_{\text{sup}}\big|_{[0,k]} = (\mathscr{B}_l \cap \mathfrak{B}_{\text{sup}})\big|_{[0,k]}$ for all $k \in \mathbb{N}_0$, i.e., at any instant of time, the approximation and the controller can agree on a common future evolution. In this chapter, we will not discuss the solution of this control synthesis problem. [11] decribes how the problem can be rewritten to fit the standard supervisory control framework (e.g., [16, 17]), and Chapter 3 of this book discusses how to find the minimally restrictive solution to the resulting standard problem. If we find a nontrivial (i.e., $\mathfrak{B}_{\text{sup}} \neq \emptyset$) controller for the approximation, we would of course like to guarantee that it is also a valid controller for the underlying infinite state system P, i.e., items (i), (ii), and (iii) hold for P and its behaviour $\mathscr{B}$. As P and P_l (respectively $\tilde{P}_l$) exhibit the same input/output structure, (i) is straightforward and (ii) follows immediately from the abstraction property of P_l (respectively $\tilde{P}_l$). There is another (not very restrictive) technical requirement that ensures that (iii) also holds for $\mathscr{B}$, see [10, 11] for details. ■

Remark 1.13. From the construction of P_l (respectively $\tilde{P}_l$), it is immediately clear that the order of the cardinality of the approximation state set is exponential in the parameter l. In practice, one would therefore begin with the "least accurate" approximation P_1 (respectively $\tilde{P}_1$) and check whether a nontrivial control solution can be found. If this is not the case, one turns to the refined approximation P_2 (respectively $\tilde{P}_2$). In this way, refinement and control synthesis alternate until either a nontrivial control solution is found or computational resources are exhausted. Hence, failure of the control synthesis process to return a nontrivial solution triggers a "global" refinement step, although "local" refinements could well suffice. A more "intelligent" procedure suggested in [12] therefore analyses failure of the synthesis process and focuses its efforts on those aspects of the approximation that have "caused" the failure in synthesis. This procedure "learns from failure" in the synthesis step and thus implements a refinement that is tailored to the particular combination of plant and specification. ■

1.5 Further Reading

Finite automata, including Mealy and Moore automata are discussed in a vast number of standard textbooks. Examples are [3, 7]. The latter contains a number of instructive examples of how Moore and Mealy automata model various systems of practical interests. [4] is *the* standard textbook on discrete events systems, and also includes a number of examples that illustrate how finite generators can be used to model engineering problems.

The problem of approximating systems with infinite state space by finite state machines has attracted a lot of attention since hybrid systems theory became "en vogue". There are various approaches, and the reader is advised to consult special journal issues on hybid systems, e.g., [2, 5, 13], proceedings volumes such as [1, 9], or the recently published survey book [8] for an overview. In this chapter, we have concentrated on finding approximations with the so-called abstraction property. The closely related concepts of simulations and approximate bisimulations are discussed, e.g., in [6, 18].

References

1. Antsaklis, P.J., Kohn, W., Lemmon, M.D., Nerode, A., Sastry, S.S. (eds.): HS 1997. LNCS, vol. 1567. Springer, Heidelberg (1999)
2. Antsaklis, P.J., Nerode, A.: IEEE Transactions on Automatic Control–Special Issue on Hybrid Control Systems 43 (1998)
3. Booth, T.L.: Sequential Machines and Automata Theory. John Wiley, New York (1967)
4. Cassandras, C.G., Lafortune, S.: Introduction to Discrete Event Systems. Kluwer Academic Publishers, Boston (1999)
5. Evans, R., Savkin, A.V.: Systems and Control Letters–Special Issue on Hybrid Control Systems 38 (1999)
6. Girard, A., Pola, G., Tabuada, P.: Approximately bisimilar symbolic models for incrementally stable switched systems. IEEE Transactions on Automatic Control 55(1), 116–126 (2010)
7. Hopcroft, J.E., Ullman, J.D.: Introduction to Automata Theory, Languages and Computation. Addison-Wesley, Reading (1979)

8. Lunze, J., Lamnabhi-Lagarrigue, F. (eds.): The HYCON Handbook of Hybrid Systems Control: Theory, Tools, Applications. Cambridge University Press (2009)
9. Lynch, N.A., Krogh, B.H. (eds.): HSCC 2000. LNCS, vol. 1790. Springer, Heidelberg (2000)
10. Moor, T., Raisch, J.: Supervisory control of hybrid systems within a behavioural framework. Systems and Control Letters–Special Issue on Hybrid Control Systems 38, 157–166 (1999)
11. Moor, T., Raisch, J., O'Young, S.: Discrete Supervisory control of hybrid systems by l-Complete approximations. Journal of Discrete Event Dynamic Systems 12, 83–107 (2002)
12. Moor, T., Davoren, J.M., Raisch, J.: Learning by doing – Systematic abstraction refinement for hybrid control synthesis. IEE Proc. Control Theory & Applications–Special Issue on Hybrid Systems 153, 591–599 (2006)
13. Morse, A.S., Pantelides, C.C., Sastry, S.S., Schumacher, J.M.: Automatica–Special issue on Hybrid Control Systems 35 (1999)
14. Raisch, J., O'Young, S.D.: Discrete approximation and supervisory control of continuous systems. IEEE Transactions on Automatic Control-Special Issue on Hybrid Systems 43(4), 569–573 (1998)
15. Raisch, J.: Discrete abstractions of continuous systems–an Input/Output point of view. Mathematical and Computer Modelling of Dynamical Systems–Special Issue on Discrete Event Models of Continuous Systems 6, 6–29 (2000)
16. Ramadge, P.J., Wonham, W.M.: Supervisory control of a class of discrete event systems. SIAM Journal of Control and Optimization 25, 206–230 (1987)
17. Ramadge, P.J., Wonham, W.M.: The control of discrete event systems. Proceedings of the IEEE 77, 81–98 (1989)
18. Tabuada, P.: Verification and Control of Hybrid Systems: A Symbolic Approach. Springer-Verlag, New York Inc. (2009)
19. Willems, J.C.: Models for Dynamics. Dynamics Reported 2, 172–269 (1989)
20. Willems, J.C.: Paradigms and puzzles in the theory of dynamic systems. IEEE Transactions on Automatic Control 36(3), 258–294 (1991)
21. Wonham, W.M.: Notes on control of discrete event systems. Department of Electrical & Computer Engineering. University of Toronto, Canada (2001)

Chapter 2
Languages, Decidability, and Complexity

Stefan Haar and Tomáš Masopust

2.1 Introduction

Control problems for discrete-event systems or hybrid systems typically involve manipulation of languages that describe system behaviors. This chapter introduces basic automata and grammar models for generating and analyzing languages of the Chomsky hierarchy, as well as their associated decision problems, which are necessary for the understanding of other parts of this book. Notions of decidability of a problem (that is, is there an algorithm solving the given problem?) and of computational complexity (that is, how many computation steps are necessary to solve the given problem?) are introduced. The basic complexity classes are recalled. This chapter is not intended to replace a course on these topics but merely to provide basic notions that are used further in this book, and to provide references to the literature.

In the following, we introduce the basic terminology, notation, definitions, and results concerning the Chomsky hierarchy of formal languages to present the necessary prerequisites for understanding the topic of this chapter, that is, the devices recognizing and generating languages. As this is only an introductory material, not all details and proofs are presented here. Usually, only the basic ideas or sketches of proofs are presented. Further details can be found in the literature, see, e.g., [4, 5, 10, 12, 13, 15].

Stefan Haar
INRIA/LSV, CNRS & ENS de Cachan, 61, avenue du Président Wilson,
94235 CACHAN Cedex, France
e-mail: Stefan.Haar@inria.fr

Tomáš Masopust
Institute of Mathematics, Academy of Sciences of the Czech Republic, Žižkova 22,
616 62 Brno, Czech Republic
e-mail: masopust@math.cas.cz

C. Seatzu et al. (Eds.): Control of Discrete-Event Systems, LNCIS 433, pp. 23–43.
springerlink.com

2.2 Regular Languages and Automata

This section introduces the simplest type of languages and automata. First, however, let us define the fundamental concepts. The cardinality of a set A is denoted by $|A|$. For two sets A and B, we write $A \subseteq B$ to denote that A is a subset of B. If $A \subseteq B$ and $A \neq B$, we write $A \subsetneq B$. The notation 2^A denotes the set of all subsets of A.

2.2.1 Words and Languages

An *alphabet* (also called an *event set*) is a finite, nonempty set Σ of abstract elements, which are called *symbols* or *letters*. Let $\Sigma = \{a_1, a_2, \ldots, a_n\}$ be an alphabet. A *word* or *string* over Σ is a (finite or infinite) concatenation $w = a_1a_2a_3\ldots$ of letters $a_i \in \Sigma$. For instance, a and $ccbc$ are words over $\{a,b,c\}$. The *empty word* is a word consisting of zero letters, denoted by ε. It holds that $\varepsilon \cdot w = w \cdot \varepsilon = w$, for any word w. The set of all finite words over Σ is denoted by $\Sigma^* \triangleq \{a_1a_2a_3\ldots a_n \mid n \in \mathbb{N},\ a_i \in \Sigma\}$. The set of all nonempty words over Σ is denoted by $\Sigma^+ \triangleq \Sigma^* \setminus \{\varepsilon\}$. A set L is a *language* over Σ if $L \subseteq \Sigma^*$. The *length* of a word w is denoted $|w|$, that is, $|a_1a_2\ldots a_n| = n$. Let $|w|_a$ denote the number of a's in w. For instance, $|ccbc| = 4$ and $|ccbc|_b = 1$. Write w^R for the *mirror image* (or *reversal*) of w defined so that for $w = a_1a_2a_3\ldots a_n$, $w^R = a_na_{n-1}a_{n-2}\ldots a_1$; $ccbc^R = cbcc$. Word $u \in \Sigma^*$ is a *prefix* of $v \in \Sigma^*$ if there exists $u' \in \Sigma^*$ such that $v = uu'$. Dually, u is a *suffix* of v if there exists $u' \in \Sigma^*$ such that $v = u'u$. Furthermore, u is an *infix* or *factor* of v if there exist $u', u'' \in \Sigma^*$ such that $v = u'uu''$, and a *sub-word* of v if there exist $u_i, v_i \in \Sigma^*$ such that $v = v_0u_1v_1u_2\ldots u_nv_n$ with $u = u_1u_2\ldots u_n$.

For two languages $K, L \subseteq \Sigma^*$, we have the set theoretic operations $K \cup L$, $K \cap L$, $K \setminus L$, $K^c = \Sigma^* \setminus K$, etc. Define the *concatenation* of K and L as

$$K \cdot L \triangleq \{u \cdot v \mid u \in K,\ v \in L\}.$$

The *powers* of a language L are defined as follows: $L^0 \triangleq \{\varepsilon\}$, $L^{n+1} \triangleq L^n \cdot L = L \cdot L^n$,

$$L^* \triangleq \bigcup_{n \geqslant 0} L^n \quad \text{and} \quad L^+ \triangleq \bigcup_{n > 0} L^n.$$

Finally, we have the *quotient languages*

$$K^{-1} \cdot L \triangleq \{v \in \Sigma^* \mid \exists u \in K:\ u \cdot v \in L\} \text{ and } L \cdot K^{-1} \triangleq \{u \in \Sigma^* \mid \exists v \in K:\ u \cdot v \in L\}.$$

A *substitution* is a mapping $\sigma : \Sigma^* \to 2^{\Gamma^*}$ such that $\sigma(\varepsilon) = \{\varepsilon\}$ and $\sigma(xy) = \sigma(x)\sigma(y)$, where $x, y \in \Sigma^*$. A *(homo)morphism* is a substitution σ such that $\sigma(a)$ consists of exactly one string, for all $a \in \Sigma$. We write $\sigma(a) = w$ instead of $\sigma(a) = \{w\}$, i.e., $\sigma : \Sigma^* \to \Gamma^*$. A *projection* is a homomorphism $\sigma : \Sigma^* \to \Gamma^*$ with $\Gamma \subseteq \Sigma$ such that for all $a \in \Sigma$, $\sigma(a) \in \{a, \varepsilon\}$.

2.2.2 Regular Languages

The above language operations can be abstracted into a class $RE(\Sigma)$ of *regular expressions* over an alphabet Σ as follows.

1. $\underline{\emptyset}$ and $\underline{a}$, for $a \in \Sigma$, are in $RE(\Sigma)$;
2. if $E, F \in RE(\Sigma)$, then $(E+F)$, $(E \cdot F)$, $(E^*) \in RE(\Sigma)$;
3. nothing else is in $RE(\Sigma)$.

Regular expressions correspond to languages. To show this relation, we define the operation $L : RE(\Sigma) \to 2^{\Sigma^*}$ by $L(\underline{\emptyset}) \triangleq \emptyset$, $L(\underline{a}) \triangleq \{a\}$, for all $a \in \Sigma$, and

$$L(E+F) \triangleq L(E) \cup L(F), \quad L(E \cdot F) \triangleq L(E) \cdot L(F), \quad L(E^*) \triangleq (L(E))^* .$$

The class $\mathsf{Reg}(\Sigma^*)$ of *regular (also called rational) languages* over Σ is the smallest class of languages over Σ that *(i)* contains the empty language $\emptyset$ and all singletons $\{a\}$, for $a \in \Sigma$, and *(ii)* is closed under the operations $\cup$, $\cdot$, and $()^*$.

Example 2.1. The regular expression $a + ba^*b$ represents the language $L(a + ba^*b) = \{a\} \cup L(ba^*b) = \{a\} \cup \{ba^ib \mid i \geqslant 0\}$. ■

2.2.3 Automata

We now define the first model of dynamical systems to *generate* languages through their behaviors.

Deterministic Automata

Deterministic automata are finite-state machines, where the input changes the current state of the system. These machines have only finite memory.

Definition 2.1. *A* deterministic finite automaton *(DFA, for short) over alphabet Σ is a quintuple $A = (Q, \Sigma, f, q_0, Q_m)$, where Q is a* finite *set of states, $f : Q \times \Sigma \to Q$ is a* transition function *(total or partial), $q_0 \in Q$ is an* initial state, *and $Q_m \subseteq Q$ is a set of* final or marked states.

DFAs are associated to languages via the following central notions.

Definition 2.2. *For $q_1, q_2 \in Q$, and $a \in \Sigma$, write $q_1 \xrightarrow{a} q_2$ iff $f(q_1, a) = q_2$. A* accepts *word $w = a_1 a_2 \dots a_n \in \Sigma^*$ if there exist $q_i \in Q$ such that*

$$q_0 \xrightarrow{a_1} q_1 \xrightarrow{a_2} \dots \xrightarrow{a_n} q_n \in Q_m;$$

write $q_0 \xrightarrow{w} q_n$. *Using this definition, we can extend* f *to be a function from* $Q \times \Sigma^*$ *to* Q *so that* $f(q_0, w) = q_n$ *if* $q_0 \xrightarrow{w} q_n$, *where* $q \xrightarrow{\varepsilon} q$ *for all* $q \in Q$. *The* language accepted (or recognized) by A *is defined as the set*

$$L(A) \triangleq \{w \in \Sigma^* \mid \exists q \in Q_m : q_0 \xrightarrow{w} q\}.$$

The class $\mathsf{Rec}(\Sigma^*)$ *of* recognizable *languages in* Σ *is formed by those languages* $L \subseteq \Sigma^*$ *for which there exists a DFA* A *over* Σ *such that* $L = L(A)$.

Fig. 2.1 presents a simple example of a DFA which accepts or recognizes a language described by the regular expression $ba^*b + a$.

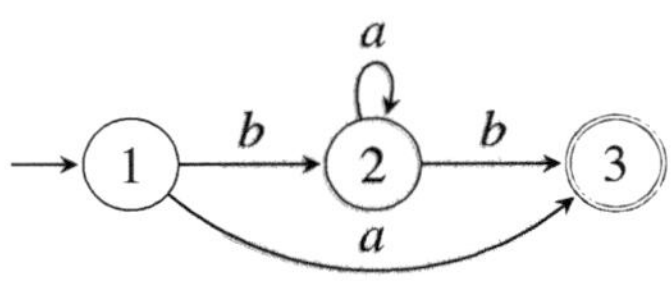

Fig. 2.1 A DFA $A = (\{1,2,3\}, \{a,b\}, f, 1, \{3\})$ which accepts the language $ba^*b + a$, where f is defined by the arrows

Nondeterministic Automata

It is often convenient or natural to work with nondeterministic models in the sense of the following definition, cf. state 2 in Fig. 2.2.

Definition 2.3. *A* nondeterministic finite automaton *(NFA, for short) over* Σ *is a quintuple* $A = (Q, \Sigma, T, \mathbf{I}, Q_m)$, *where* Q *is a* finite *set of states,* $T \subseteq Q \times \Sigma \times Q$ *is a set of* transitions, $\mathbf{I} \subseteq Q$ *is a set of* initial states, *and* $Q_m \subseteq Q$ *is a set of* final or marked states. *For* $q_1, q_2 \in Q$, *and* $a \in \Sigma$, *write* $q_1 \xrightarrow{a} q_2$ *if* $(q_1, a, q_2) \in T$. A accepts *word* $w = a_1 a_2 \ldots a_n \in \Sigma^*$ *if there exist* $q_i \in Q$ *such that* $q_0 \in \mathbf{I}$ *and* $q_0 \xrightarrow{a_1} q_1 \xrightarrow{a_2} \ldots \xrightarrow{a_n} q_n \in Q_m$; *write* $q_0 \xrightarrow{w} q_n$. *The* language accepted (or recognized) by A *is defined as the set* $L(A) \triangleq \{w \in \Sigma^* \mid \exists q_0 \in \mathbf{I},\ q \in Q_m : q_0 \xrightarrow{w} q\}$.

We shall see below that the classes of DFAs and NFAs are equivalent in terms of the recognized languages. First, however, we extend the notion of NFAs so that ε-transitions are allowed. This corresponds to a situation where the automaton changes the state, but reads no input letter.

ε-Automata

Definition 2.4. *An NFA* $A = (Q, \Sigma, T, \mathbf{I}, Q_m)$ *such that* $T \subseteq Q \times (\Sigma \cup \{\varepsilon\}) \times Q$ *is called an* ε*-automaton, or an NFA with* ε*-transitions.*

Similarly as for the deterministic automata, we can extend T so that $T \subseteq Q \times \Sigma^* \times Q$. Fig. 2.2 presents a simple example of an NFA with ε-transitions which accepts the language $a^*b + a$. The following theorem shows that we can always remove ε-transitions from the automaton.

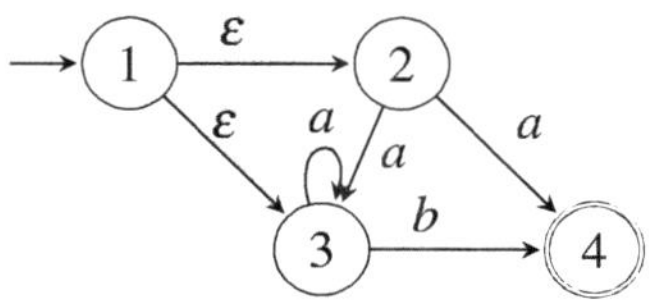

Fig. 2.2 An NFA with ε-transitions that accepts the language a^*b+a

Theorem 2.1. *For every ε-automaton A there exists an* ε-free *automaton A' such that $L(A) = L(A')$.*

Proof. The proof can be sketched as follows, see Fig. 2.3:

1. For every $q \xrightarrow{\varepsilon} q' \xrightarrow{a} q''$, where $a \in \Sigma$, add $q \xrightarrow{a} q''$;
2. For every $q \xrightarrow{\varepsilon} q'$ with $q' \in Q_m$, add q to Q_m;
3. Remove all ε-transitions. □

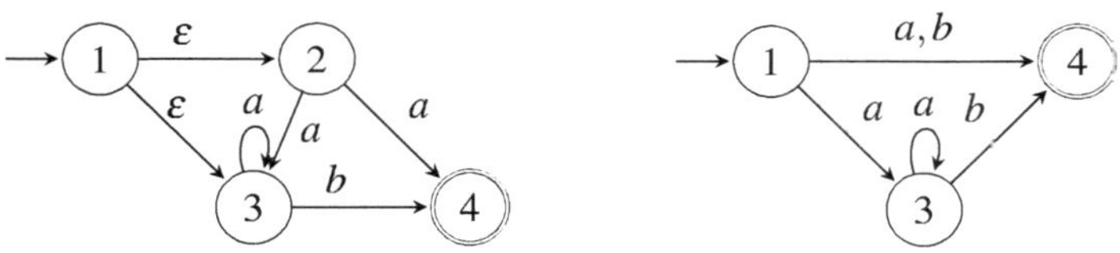

Fig. 2.3 Removing ε-transitions: The NFA of Fig. 2.2 and the ε-free NFA that accepts a^*b+a

Determinizing NFAs

The disadvantage of the nondeterminism is that we can have two or more choices how to continue from a state. In the right automaton of Fig. 2.3, this situation happens in state 1 for letter a. The following theorem shows that we can always eliminate the nondeterminism by transforming an NFA to a DFA.

Theorem 2.2. *For every NFA (with ε-transitions) $A = (Q, \Sigma, T, \mathbf{I}, Q_m)$ there exists a DFA $A' = (Q', \Sigma, f, q_0, Q_m{}')$ such that $L(A) = L(A')$.*

Proof. The key idea to obtaining the determinized version of A is the following powerset construction.

- $Q' \subseteq 2^Q$;
- initial state q_0 is given by the set of initial states $\mathbf{I}$ *and* those states that are ε-reachable from some initial state: $q_0 \triangleq \{s \in Q \mid \exists s' \in \mathbf{I} : s' \xrightarrow{\varepsilon} s\}$;
- $\forall U \subseteq Q$ and $a \neq \varepsilon : f(U,a) \triangleq \{q \in Q \mid \exists q_u \in U : q_u \xrightarrow{a\varepsilon} q\}$;
- $Q_m' = \{U \subseteq Q \mid U \cap Q_m \neq \emptyset\}$. □

Fig. 2.4 demonstrates the previous theorem. It should be noted that the automaton A' has, in general, a state space whose size is of the order of $2^{|Q|}$, that is, exponential with respect to the state space of A. There exist examples of NFAs for which it can be shown that any language-equivalent DFAs *must* be at least of this size, see [6] and the references therein. Determinization should therefore be used moderately.

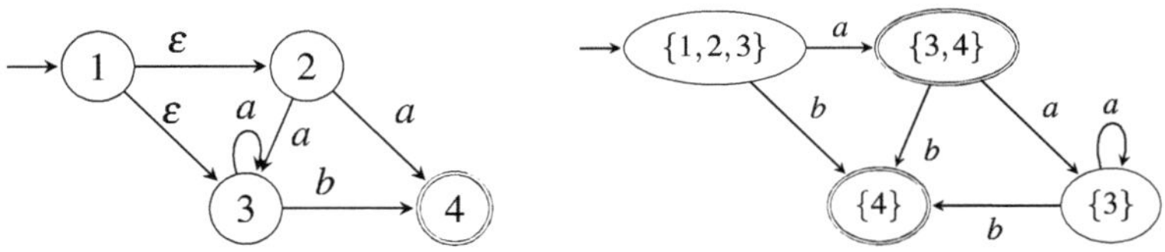

Fig. 2.4 Determinization: The NFA with ε-transitions of Fig. 2.2 and its DFA which accepts the language a^*b+a

2.2.4 Closure Properties of Regular Languages

The algebraic properties considered here for regular languages – and below for other language classes – are important results for testing whether or not a given language belongs to a given class.

Theorem 2.3. $\mathsf{Rec}(\Sigma^*)$ *is closed under the set operations union, intersection, and complement.*

Proof. Let $K, L \in \mathsf{Rec}(\Sigma^*)$ with DFAs A_K, A_L such that $L(A_K) = K$ and $L(A_L) = L$. An NFA to accept $K \cup L$ is obtained by combining A_K and A_L via fusion of initial states. Language $K \cap L$ is accepted by a DFA obtained as a synchronized product of A_K and A_L. In this product automaton, the state set is $Q_K \times Q_L$, and the transition function $f_{prod} : (Q_K \times Q_L) \times \Sigma \to (Q_K \times Q_L)$ is constructed componentwise so that

$$f_{prod}((q_K, q_L), a) = (q'_K, q'_L) \text{ if } f_K(q_K, a) = q'_K \text{ and } f_L(q_L, a) = q'_L.$$

Finally, to obtain a DFA for K^c, it suffices to replace Q_m by $Q \setminus Q_m$ in A_K. □

Theorem 2.4. $\mathsf{Rec}(\Sigma^*)$ *is closed under concatenation, iteration, substitutions, and projection to subalphabets.*

Proof. Let $K, L \in \mathsf{Rec}(\Sigma^*)$ with DFAs A_K, A_L such that $L(A_K) = K$ and $L(A_L) = L$. An automaton $A_{K \cdot L}$ such that $L(A_{K \cdot L}) = K \cdot L$ is obtained by combining A_K and A_L via extra ε-transitions from Q_{mK} to q_{0L}. For A_{L^*}, add to A_L a new state for recognizing ε, plus ε-transitions from Q_{mL} to q_{0L}. If $\sigma : \Sigma \to 2^{\Gamma^*}$ is a substitution, replacing all a-transitions in A_L by a copy of automaton $A_{\sigma(a)}$ yields $A_{\sigma(L)}$. Finally, for any $\Gamma \subseteq \Sigma$ and $a \in \Sigma \setminus \Gamma$, replace $q \xrightarrow{a} q'$ by $q \xrightarrow{\varepsilon} q'$. □

2.2.5 *Regularity and Recognizability*

The following theorem (Kleene 1936) summarizes the relation between regular expressions and recognizable languages.

Theorem 2.5. $\mathsf{Reg}(\Sigma^*) = \mathsf{Rec}(\Sigma^*)$.

Proof. To prove $\mathsf{Reg}(\Sigma^*) \subseteq \mathsf{Rec}(\Sigma^*)$, we have $\emptyset \in \mathsf{Rec}(\Sigma^*)$ and $\{a\} \in \mathsf{Rec}(\Sigma^*)$, for $a \in \Sigma$ (construct the DFAs). The closure of $\mathsf{Rec}(\Sigma^*)$ under $\cup$, $\cdot$, and $()^*$ then implies $\mathsf{Reg}(\Sigma^*) \subseteq \mathsf{Rec}(\Sigma^*)$. To prove the converse inclusion, $\mathsf{Reg}(\Sigma^*) \supseteq \mathsf{Rec}(\Sigma^*)$, we need to convert DFAs to regular expressions. The idea, depicted in Fig. 2.5, is to construct a regular expression by a step-by-step fusion of transitions. □

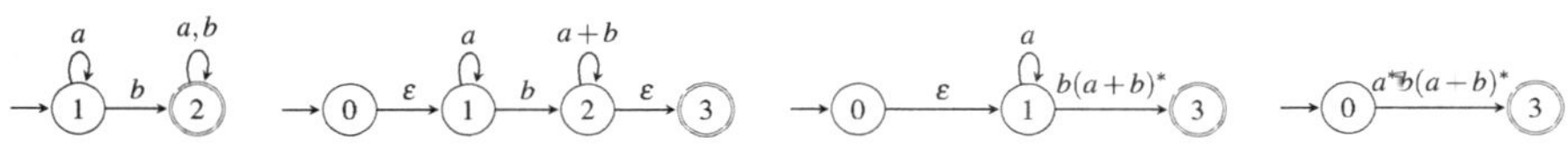

Fig. 2.5 Converting a DFA to a regular expression: The DFA on the left is transformed to the regular expression $a^*b(a+b)^*$

2.2.6 *Criteria for Regularity*

We now ask how to identify the limits of regularity. Is it true that all languages are regular? The answer is negative; for instance, the following languages are *not* regular:

- $L_1 = \{a^n b^n \mid n \geqslant 0\}$,
- $L_2 = \{u \in \Sigma^* \mid |u|_a = |u|_b\}$,
- $L_3 = \{w \in \Sigma^* \mid w = w^R\}$.

To prove the nonregularity of these and other languages, a very important tool is the use of results of the following type, which are called *pumping lemmas* or *star/iteration lemmas* in the literature. Of course, one may have a choice to apply one or the other result in cases where several such lemmas can be applied.

Lemma 2.1 (Pumping Lemma). *For every $L \in \mathsf{Reg}(\Sigma^*)$ there exists $N \geqslant 0$ such that for all $x \in L$*

- *If $|x| \geqslant N$, then $x = u_1u_2u_3$ with $u_i \in \Sigma^*$, $u_2 \neq \varepsilon$, $|u_1u_2| \leqslant N$, and $u_1u_2^*u_3 \in L$.*
- *If $x = w_1w_2w_3$ with $|w_2| \geqslant N$, then $x = w_1u_1u_2u_3w_3$ with $u_i \in \Sigma^*$, $u_2 \neq \varepsilon$, and $w_1u_1u_2^*u_3w_3 \in L$.*

Proof. The idea for the proofs of this and similar results is as follows. Take a DFA A_L that accepts L. Since the number of states of A_L is finite, every path of length greater than some N that depends on A_L has a loop. Iterating this loop allows to construct the sublanguages whose existence is claimed in Lemma 2.1. □

Example 2.2. To use Lemma 2.1 to show that $L_1 = \{a^nb^n \mid n \geqslant 0\}$ is not regular, assume for contradiction that L_1 is regular. Then there exists N as claimed in Lemma 2.1. Consider the word $w = a^Nb^N$. Then, by Lemma 2.1, $w = u_1u_2u_3$ with $u_1u_2^*u_3 \in L_1$. There are three possibilities: $u_2 \in a^*$, $u_2 \in b^*$, or $u_2 \in aa^*bb^*$. In all cases, however, $u_1u_2u_2u_3 \notin L_1$, which is a contradiction. ■

For proving that a given language L is *non-recognizable*, one often uses a pumping lemma *and* the established closure properties of $\mathsf{Reg}(\Sigma^*)$. For instance, for $L_2 = \{u \in \Sigma^* \mid |u|_a = |u|_b\}$, we have $L_2 \cap a^*b^* = L_1$, hence $L_2 \notin \mathsf{Reg}(\Sigma^*)$.

2.2.7 *Minimality*

In general, there are infinitely many DFAs and NFAs that accept a given language (for instance, aa^*). We shall see that one can find a *canonical* DFA which is minimal (up to isomorphism) in the number of states. This is not true for NFAs (construct two non-isomorphic two-state automata for the language aa^*).

Definition 2.5. *For $u \in \Sigma^*$ and $L \subseteq \Sigma^*$, the* residual language *of L with respect to u is the quotient $u^{-1}L = \{v \in \Sigma^* \mid uv \in L\}$. The* residual automaton *for L is $\mathscr{R}(L) = (Q_L, \Sigma_L, f_L, q_{0L}, Q_{mL})$ such that $Q_L = \{u^{-1}L \mid u \in \Sigma^*\}$, $f_L(u^{-1}L, a) = a^{-1}(u^{-1}L) = (ua)^{-1}L$, $q_{0L} = L = \varepsilon^{-1}L$, $Q_{mL} = \{u^{-1}L \mid \varepsilon \in u^{-1}L\} = \{u^{-1}L \mid u \in L\}$.*

Theorem 2.6. *Language $L \subseteq \Sigma^*$ is recognizable if and only if it has* finitely *many residuals. Moreover, any DFA A with $L(A) = L$ allows to inject $\mathscr{R}(L)$ into A by a morphism (*a fortiori, *$\mathscr{R}(L)$ has minimal size among those automata that accept L).*

However, for the construction, a more convenient algorithm is depicted in the following example.

Example 2.3. Consider a slightly modified automaton of Fig. 2.4. In the first step, we add the dead state, d, to make the automaton complete, i.e., to complete the transition function of the automaton. This is done by replacing all the transitions of the form $f(q,a) = \emptyset$ with $f(q,a) = d$. Then, in Step 1, we distinguish only final and non-final states. In Step 2, we distinguish any two states of the same class for which there exists a letter leading them to states from different classes. Step 2 is repeated until no new class is constructed, see Fig. 2.6. ■

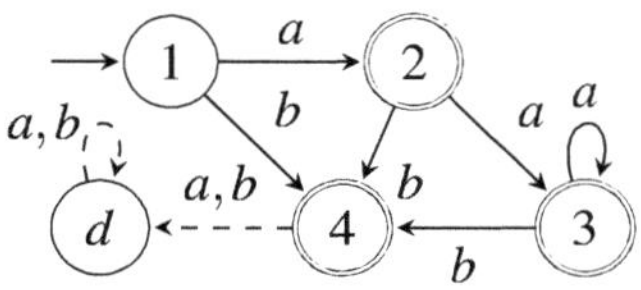

Step 1	Step 2
1	1
d	d
2 3 4	2 3
	4

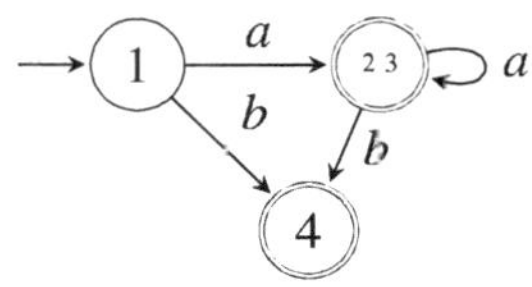

Fig. 2.6 Minimization: A modified DFA of Fig. 2.4 with the dead state d to be minimized, computation of the states, and the minimal automaton without the dead state d

2.3 Chomsky Grammars

Formal languages do not stop at the boundary of Reg; the concept of *grammars* for generating languages allows to classify a larger variety of language families in which Reg will be embedded.

2.3.1 Type 0 Grammars and Languages

We begin with the most general type of grammars.

Definition 2.6. *A* type 0 grammar *is a tuple* $G = (\Sigma, \mathcal{V}, \mathsf{S}, \mathsf{P})$, *where* Σ *is the* terminal *alphabet,* $\mathcal{V}$ *is the* nonterminal *alphabet (set of* variables*),* $\mathsf{S} \in \mathcal{V}$ *is the* axiom *(initial variable),* $\mathsf{P} \subseteq (\Sigma \cup \mathcal{V})^* \mathcal{V} (\Sigma \cup \mathcal{V})^* \times (\Sigma \cup \mathcal{V})^*$ *is a finite set of* rules *(productions) where at least one nonterminal appears on the left-hand side. A sentential form* $\beta \in (\Sigma \cup \mathcal{V})^*$ *is* derived *from a sentential form* $\alpha \in (\Sigma \cup \mathcal{V})^*$, *written* $\alpha \Rightarrow \beta$, *if there exists* $(\alpha_2, \beta_2) \in \mathsf{P}$ *such that* $\alpha = \alpha_1 \alpha_2 \alpha_3$ *and* $\beta = \alpha_1 \beta_2 \alpha_3$.

Rules $(\alpha, \beta) \in P$ are also written as $\alpha \to \beta$ (read α *is rewritten by* β).

Definition 2.7. *For a type 0 grammar* $G = (\Sigma, \mathcal{V}, \mathsf{S}, \mathsf{P})$ *and a sentential form* $\alpha \in (\Sigma \cup \mathcal{V})^*$, *let* $\stackrel{*}{\Rightarrow}$ *denote the reflexive and transitive closure of* $\Rightarrow$. *The* language generated by α *is the set* $L_G(\alpha) = \{u \in \Sigma^* \mid \alpha \stackrel{*}{\Rightarrow} u\}$. *The language generated by* G, *denoted* $L(G)$, *is defined as the set* $L_G(\mathsf{S})$. *A language is type 0 if it is generated by a type 0 grammar.*

Example 2.4. The following type 0 grammar $G = (\{a\}, \{S, D, X, F, Y, Z\}, S, P)$ with P defined as follows generates the language $\{a^{2^n} \mid n > 0\}$.

$$
\begin{array}{llll}
\mathsf{S} \to DXaF & Xa \to aaX & XF \to YF & XF \to Z \\
aY \to Ya & DY \to DX & aZ \to Za & DZ \to \varepsilon
\end{array}
$$

Note that the alphabet and the set of variables can be extracted from the productions, so the set of productions characterizes the whole grammar. One possible derivation of this grammar is $\mathsf{S} \Rightarrow DXaF \Rightarrow DaaXF \Rightarrow DaaZ \Rightarrow DaZa \Rightarrow DZaa \Rightarrow aa$. ■

2.3.2 Type 1: Context-Sensitive

We move "up" in the hierarchy by restricting the productions to be monotonic in the length of the sub-words, that is, the generated word can never be shortened.

Definition 2.8. *A grammar* $G = (\Sigma, \mathcal{V}, \mathsf{S}, \mathsf{P})$ *is* context-sensitive *(or* type 1*) if for all* $(\alpha, \beta) \in \mathsf{P}$, $|\alpha| \leqslant |\beta|$. *A language is context-sensitive (type 1 or in* Cse*) if it is generated by a type 1 grammar.*

Example 2.5. The following is a type 1 grammar that generates the language $\{xx \mid x \in \{a,b\}^*\}$:

$$\begin{array}{llll} \mathsf{S} \to aAS \mid bBS \mid T & Aa \to aA & Ba \to aB & Ab \to bA \\ Bb \to bB & BT \to Tb & AT \to Ta & T \to \varepsilon \end{array}$$

A derivation chain for *abab* is

$$\mathsf{S} \Rightarrow aAS \Rightarrow aAbBS \Rightarrow aAbBT \Rightarrow abABT \Rightarrow abATb \Rightarrow abTab \Rightarrow abab. \quad \blacksquare$$

Definition 2.9. *A context-sensitive grammar* $G = (\Sigma, \mathcal{V}, \mathsf{S}, \mathsf{P})$ *is in* normal form (NF) *if every rule is of the form* $\alpha_1 X \alpha_2 \to \alpha_1 \beta \alpha_2$ *with* $X \in \mathcal{V}$ *and* $\beta \neq \varepsilon$.

Theorem 2.7. *Every type 1 language can be generated by a context-sensitive grammar in NF.*

Example 2.6. A type 1 NF grammar for $\{a^{2^n} \mid n > 0\}$:

$$\begin{array}{llllll} \mathsf{S} \to aTa & T \to XA & XY \to Xa & Xa \to AAa & ZY \to ZX \\ \mathsf{S} \to aa & T \to AA & XY \to ZY & ZA \to AAA & aA \to aa \end{array} \quad \blacksquare$$

2.3.3 Type 2: Context-Free Languages and Pushdown Automata

Note that a context-sensitive grammar in NF rewrites a variable X by a nonempty word β according to contexts α_1 and α_2. The next level is reached by loosing the information about these contexts.

Definition 2.10. *A grammar* $G = (\Sigma, \mathcal{V}, \mathsf{S}, \mathsf{P})$ *is* context-free *or type 2 if* $\mathsf{P} \subseteq \mathcal{V} \times (\Sigma \cup \mathcal{V})^*$. *The class of languages generated by context-free grammars is denoted* Cfr. *A language is* linear *if it is generated by a context-free grammar with* $\mathsf{P} \subseteq \mathcal{V} \times (\Sigma^* \cup \Sigma^* \mathcal{V} \Sigma^*)$.

Lemma 2.2 (Fundamental Lemma). *If* $G = (\Sigma, \mathcal{V}, \mathsf{S}, \mathsf{P})$ *is context-free, then for all* $\alpha_1, \alpha_2, \beta \in (\Sigma \cup \mathcal{V})^*$, *and* $n \geqslant 0$

$$\alpha_1 \alpha_2 \stackrel{n}{\Rightarrow} \beta \iff \begin{cases} \alpha_1 \stackrel{n_1}{\Rightarrow} \beta_1 \\ \alpha_2 \stackrel{n_2}{\Rightarrow} \beta_2 \end{cases},$$

with $\beta = \beta_1 \beta_2$ *and* $n = n_1 + n_2$.

Example 2.7. The languages $L = \{a^n b^n \mid n \geqslant 0\}$ and the *nth Dyck language* of well-formed expressions with n brackets are in Cfr. ■

Definition 2.11. *For a context-free grammar* $G = (\Sigma, \mathscr{V}, \mathsf{S}, \mathsf{P})$, *a* derivation tree *is a tree labeled by* $\mathscr{V} \cup \Sigma$ *such that every interior node is* $\mathscr{V}$*-labeled, and if the sons of a node labeled* x *are labeled* $\alpha_1, \ldots, \alpha_k$, *then* $(x, \alpha_1 \ldots \alpha_k)$ *is a rule in* P.

Example 2.8. Consider the language $\{a^n b^n \mid n \geqslant 0\}$ generated by a context-free grammar with rules $\mathsf{S} \to A\mathsf{S}b$, $A \to a$, and $\mathsf{S} \to \varepsilon$. A derivation tree of the derivation $\mathsf{S} \Rightarrow A\mathsf{S}b \Rightarrow a\mathsf{S}b \Rightarrow ab$ is shown in Fig. 2.7. ■

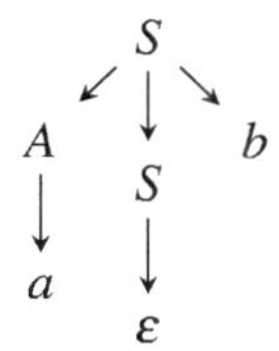

Fig. 2.7 Derivation tree of the word *ab*

Lemma 2.3. *If* $G = (\Sigma, \mathscr{V}, \mathsf{S}, \mathsf{P})$ *is a context-free grammar, then* (i) *for all* $x \overset{*}{\Rightarrow} \alpha$ *there exists a derivation tree with root* x *and yield* α; (ii) *if* t *is a derivation tree of* G, *then there exists a derivation* $root(t) \overset{*}{\Rightarrow} yield(t)$.

The class Cfr of context-free languages satisfies the following property of the same type as Lemma 2.1 for regular languages.

Lemma 2.4 (Pumping Lemma). *For every* $L \in$ Cfr *there exists* $N \geqslant 0$ *such that for all* $w \in L$ *with* $|w| \geqslant N$, $w = \alpha u \beta v \gamma$, *with* $|uv| > 0$ *and* $|u\beta v| \leqslant N$, *and* $\forall n \in \mathbb{N}$: $\alpha u^n \beta v^n \gamma \in L$.

As in the regular case, the pumping lemma allows to show *non*-inclusion of a language in Cfr; for instance, one can immediately see that $L_1 = \{a^n b^n c^n \mid n \geqslant 0\}$ is *not* context-free.

Closure Properties of Cfr**:** From the fact that L_1 is not context-free, we obtain that neither Cfr nor linear languages are closed under $\cap$; in fact, consider languages $\{a^n b^n c^p \mid n, p \geqslant 0\}$ and $\{a^p b^n c^n \mid n, p \geqslant 0\}$ whose intersection is the language L_1. On the other hand, Cfr *is* closed under $\cdot$ and $()^*$, as we will see below.

Pushdown Automata

As for regular languages, there exists a system model to recognize context-free languages.

Definition 2.12. *A* pushdown automaton *(PDA) is a tuple $A = (Q, \Sigma, Z, T, q_0, Q_m)$, where Q is a finite state set, Σ is an input alphabet, Z is a stack alphabet, $T \subseteq ZQ \times (\Sigma \cup \{\varepsilon\}) \times Z^*Q$ is a finite set of transitions, $q_0 \in ZQ$ is the initial configuration, and $Q_m \subseteq Q$ is the set of final states. The pushdown automaton A is called* real-time *if it is ε-free. The configurations of A are the words $hqw \in Z^*Q\Sigma^*$, where h is the content of the stack, q is the current state, and w is the unread part of the input.*

Example 2.9. Consider the configuration $\beta pw = A\gamma pax$ from the first part of Fig. 2.8. After the execution of transition $(Ap, a, \alpha q)$, the new configuration is $\alpha\gamma qx$. We denote this transition by $A\gamma pax \xrightarrow{a} \alpha\gamma qx$. ■

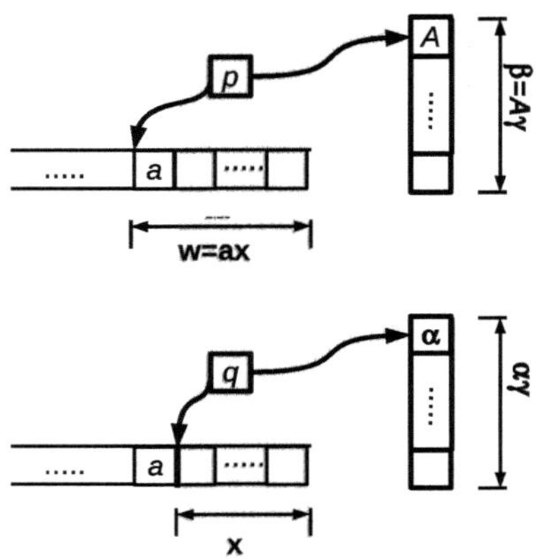

Fig. 2.8 Pushdown automaton executing $(Ap, a, \alpha q)$

We have several notions of K-acceptance, with different values for K. Let $L(A) = \{w \in \Sigma^* \mid \exists q_0 \xrightarrow{w} h,\ h \in K\}$, where

- $K = Z^*Q_m$: *final state* acceptance.
- $K = Q$: *empty stack* acceptance.
- $K = Q_m$: *empty stack and final state* acceptance.

Proposition 2.1. *Let $L \subseteq \Sigma^*$.*

1. *If L is K-accepted by a PDA A, there exists another PDA A' that accepts L by final state.*
2. *For every PDA A, the empty stack language L_A can be generated by a context-free grammar G that can be effectively constructed from A.*
3. *Conversely, for any context-free grammar G, a PDA A can be constructed that accepts $L(G)$ with empty stack.*

The closure for $\cdot$ and $()^*$ can be shown by combining the accepting automata.

2.3.4 Type 3 Languages

Finally, on the last level, we find regular languages as a subclass of Cfr by further restrictions of context-free rules so they cannot generate more than one nonterminal at a step which is always the last (resp. the first) symbol of any sentential form.

Definition 2.13. *A grammar* $G = (\Sigma, \mathscr{V}, \mathsf{S}, \mathsf{P})$ *is* left-linear *if* $\mathsf{P} \subseteq \mathscr{V} \times (\Sigma^* \cup \mathscr{V}\Sigma^*)$, *and* right-linear *if* $\mathsf{P} \subseteq \mathscr{V} \times (\Sigma^* \cup \Sigma^*\mathscr{V})$.

Proposition 2.2. *A language is regular if and only if it is generated by a left-linear (or right-linear) grammar.*

Example 2.10. There are linear languages that are not regular; in fact, languages $\{a^n b^n \mid n \geqslant 0\}$ and $\{a^n b^n c^p \mid n, p \geqslant 0\}$ are linear. ∎

2.3.5 The Chomsky Hierarchy

The following chain of inclusions summarizes the hierarchical relations between the language classes introduced here:

$$\mathsf{Reg} \subsetneq \mathsf{Cfr} \subsetneq \mathsf{Cse} \subsetneq \mathsf{REN}. \tag{2.1}$$

Here, we use REN to denote the class of type 0 languages, for reasons given below.

2.4 Turing Machines and Decidability

We now leave the realm of grammar generated languages and turn to the most fundamental model of computation, that of *Turing machines or TMs*. A TM consists of $k \geqslant 1$ two-way infinite tapes, each of whose cells are $\mathbb{Z}$-indexed. Every cell contains a symbol which is either blank (\$) or a letter from an auxiliary alphabet Σ. The k read/write heads (one for each tape) controlled by its internal state and the tape content move along the tapes. Formally, we have:

Definition 2.14. *A* k-tape Turing machine *(TM) is a tuple* $\mathscr{M} = (Q, \Sigma, \$, T, q_0, Q_m)$, *where* (i) Q *is a finite state set,* (ii) Σ *is an alphabet of* tape symbols, (iii) $\$ \in \Sigma$ *is the* blank *symbol; set* $\tilde{\Sigma} \triangleq \Sigma \setminus \{\$\}$; (iv) $q_0 \in Q$ *is the initial state,* (v) $Q_m \subseteq Q$ *is the set of final states, and* (vi) $T \subseteq Q \times \Sigma^k \times \tilde{\Sigma}^k \times \{L, R, S\}^k \times Q$ *is a set of* instructions (q, s, s', m, q') *for reading from/writing to the tapes, moving the k heads (left, right, stay) and state change. The TM* $\mathscr{M}$ *is* deterministic *(DTM) if* $(q_1, s_1, s_1', m_1, q_1'), (q_2, s_2, s_2', m_2, q_2') \in T$ *with* $(s_1', m_1, q_1') \neq (s_2', m_2, q_2')$ *implies* $(q_1, s_1) \neq (q_2, s_2)$.

There exist several structurally different definitions for TMs in the literature. In particular, there may be $k > 1$ tapes, one-sided tapes (i.e., the indices are restricted to $\mathbb{N}$), etc. All these models can be shown to be *equivalent* in the sense that the computability/decidability notions they define coincide. Of course, the concrete constructions proving each of these equivalences involve non-trivial modifications of alphabets, state spaces, etc., and produce very different machines.

The dynamics of TMs are illustrated in Fig. 2.9.

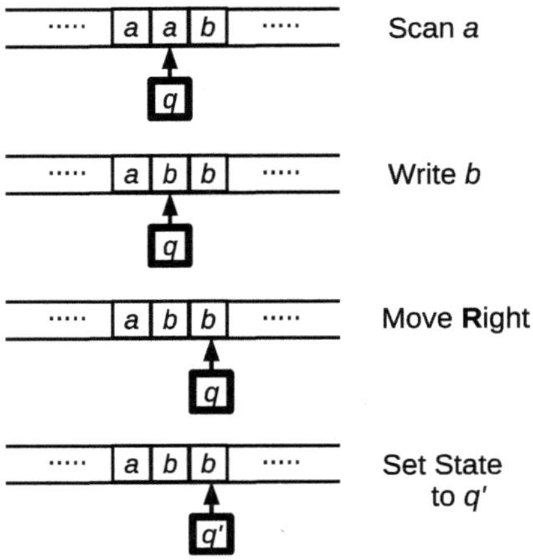

Fig. 2.9 One-Tape Turing Machine executing (q,a,b,R,q')

Definition 2.15 (TM Languages). *Let* $\mathcal{M} = (Q,\Sigma,\$,T,q_0,Q_m)$ *be a Turing machine. A global state/configuration* (q,w,z) *of* $\mathcal{M}$ *consists of the current state* q, *the content* w *of the tape cells that are* not \$, *and the index* $z \in \mathbb{Z}^k$ *of the current positions of heads. The* input *is the non-\$ tape content in the initial configuration. Computation* $\mathcal{M}(x)$ *of* $\mathcal{M}$ *on input* x halts *if it is finite and ends in a final state. Input* x *is* accepted *if* $\mathcal{M}(x)$ *halts in an* accepting *state, and* rejected *if* $\mathcal{M}(x)$ *halts in a* non-accepting (rejecting) *state. The set* $T(\mathcal{M}) \subseteq \tilde{\Sigma}^*$ *of inputs accepted by* $\mathcal{M}$ *is the* language accepted by $\mathcal{M}$. *Languages accepted by TMs are called* recursively enumerable; *denote the class of recursively enumerable languages by* REN. *A language* L *such that there exists a TM* $\mathcal{M}_L$ *which always halts when given a finite sequence of symbols from the alphabet of the language as input, and such that* $T(\mathcal{M}_L) = L$, *is called* recursive *or* decidable; *the class of these languages is denoted* REC.

The following is a small selection of results from a vast collection.

1. $L \subseteq \Sigma^*$ is type 0 if and only if it is recursively enumerable (and therefore REN appears rightfully in (2.1) above).
2. $\mathsf{Cse} \subsetneq \mathsf{REC} \subsetneq \mathsf{REN} \neq 2^{\Sigma^*}$.
3. For any multi-tape machine $\mathcal{M}$ there exists a single-tape machine $\mathcal{M}'$ that simulates $\mathcal{M}$, i.e., $L(\mathcal{M}) = L(\mathcal{M}')$, using only quadratically more computation time.

4. Many more equivalences hold; e.g., every DTM can be simulated by a DTM having only two states.

The following definition establishes TMs as a model for computation.

Definition 2.16 (Computability). *For an input x such that $\mathcal{M}(x)$ halts, denoted by $f_{\mathcal{M}}(x)$ the tape content on halting. Any function $f : \tilde{\Sigma}^* \to \tilde{\Sigma}^*$ for which there exists a TM $\mathcal{M}$ with $f(w) = f_{\mathcal{M}}(w)$, for all $w \in \tilde{\Sigma}^*$, is called* computable.

Theorem 2.8 (Decidability). *A language $L \subseteq \Sigma^*$ is* decidable *if and only if its characteristic function $\chi_L : \Sigma^* \to \{0,1\}$ is computable.*

We can now state the celebrated *Church-Turing-Rosser Thesis.*

EVERYTHING COMPUTABLE CAN BE COMPUTED BY A TM.

This thesis is not a claim about a mathematical object, and hence cannot have a proof. Instead, it postulates that the computability notion defined via the TM model is the *"right"* model. The reader willing to follow the thesis is therefore invited to continue with the remainder of the chapter.

2.4.1 Universal Turing Machine

One can see the TM-based computational model as a "programmable machine", in the sense that one strives to find a *universal* TM $\mathcal{U}$ whose inputs consist of a description $\langle \mathcal{M}, x \rangle$ of a TM $\mathcal{M}$ and an input x such that $\mathcal{M}(x)$ is simulated on $\mathcal{U}$.

Such a construction is actually possible: $\tilde{\Sigma}_{\mathcal{U}}$ encodes the i th symbol of $\tilde{\Sigma}_{\mathcal{M}}$ by the binary description of i; similarly for $Q_{\mathcal{M}}$ and x; L, R, S are $00, 01, 10$; parts of the encoding are separated by special symbols, e.g. ";". Three tapes are needed: one for the description of $\mathcal{M}$ and x, one for simulating $\mathcal{M}$'s tape, and one for the description of $\mathcal{M}$'s current state and the symbol read by $\mathcal{M}$'s tape head. If $\mathcal{M}$ halts, then $\mathcal{U}$ halts with the final state of $\mathcal{M}$. If the input is *not* a valid description of a TM and an input, $\mathcal{U}$ never halts.

Again, we give a selection of results on the *halting problem.*

Theorem 2.9. *The universal language $L_{\mathcal{U}} = \{\langle \mathcal{M}, w \rangle \mid w \in L(\mathcal{M})\}$ is recursively enumerable.*

Proof. Proof by construction of a universal TM. □

Let $\langle \mathcal{M} \rangle$ denote a string encoding of $\mathcal{M}$. The following theorem says that there exists a TM which can construct a description of $\mathcal{M}$ from its encoding.

Theorem 2.10. *The function $\langle \mathcal{M} \rangle \mapsto \langle \mathcal{M}, \langle \mathcal{M} \rangle \rangle$ is computable.*

Theorem 2.11. *If L and L^c are recursively enumerable, then L and L^c are recursive.*

Proof. A new TM to decide L runs TMs for both L and L^c in parallel. As each word w is either in L or in L^c, the machine always halts. □

Theorem 2.12. *$L^c_{\mathscr{U}}$ is not recursively enumerable, so $L_{\mathscr{U}}$ is not recursive.*

Proof. The proof is by contradiction. Let $\mathscr{M}_N$ be a TM with $L(\mathscr{M}_N) = L^c_{\mathscr{U}}$. Construct a TM D such that $L(D) = \{\langle \mathscr{M} \rangle \mid \langle \mathscr{M} \rangle \notin L(\mathscr{M})\}$. Apply $\mathscr{M}_N$ to $\langle D, \langle D \rangle \rangle$. Then we have the absurdity $\langle D \rangle \in L(D) \iff \langle D \rangle \notin L(D)$. □

Theorem 2.13. *The* halting problem *is undecidable, that is, the language $L_{halt} = \{\langle \mathscr{M}, x \rangle \mid \mathscr{M}(x) \text{ halts}\}$ is recursively enumerable but not recursive.*

Proof. The proof is by contradiction as in the above. □

2.4.2 Decidable and Undecidable Problems

Reduction is a strategy for showing that a problem $P \triangleq (D \in S)$ is undecidable. It consists of two steps:

1. Take an undecidable problem $P' \triangleq (D' \in S')$;
2. Reduce P' to P by giving a recursive function $f : S' \to S$ such that for all $D' \in S'$, $(D' \in S') \iff (f(D) \in S)$.

Examples of Decidable Problems

- The word problem for a DFA A: $w \in L(A)$?
- Emptiness problem for a DFA A: $L(A) = \emptyset$?
- Equivalence of regular expressions.
- Emptiness problem, word problem, finiteness problems for context-free languages.
- The word problem for context-sensitive languages.

Examples of Undecidable Problems

Rice's theorem: For $Q \subseteq \mathsf{REN}$ such that $\emptyset \neq Q \neq \mathsf{REN}$, the problem $P \triangleq (L \in Q)$ is undecidable.
Post's Correspondence Problem (PCP): Given two sequences of words $(u_1, \ldots, u_n)$, $(v_1, \ldots, v_n) \in (\Sigma^*)^n$. The question is whether there exist $k > 0$ and indices $i_1, \ldots, i_k$ such that $u_{i_1} \ldots u_{i_k} = v_{i_1} \ldots v_{i_k}$.

The PCP is a very popular undecidable problem in the literature for its reducibility to other problems since it often allows for shorter or more elegant proofs than reducing directly the halting problem. There exist several proofs of undecidability of the PCP; a reduction of the halting problem goes as follows (see, e.g., [15]): Given a TM, construct pairs (u_i, v_i) as blocks, each of which has two fields, with name u_i in the top field and v_i in the bottom field so that aligning these blocks, the top and bottom rows form words that code the moves of the TM, which halts exactly if and only if the PCP can be solved.

Turning to other formal languages, the *emptiness problem* for type 1 grammars is undecidable. Furthermore, for G_1, G_2 context-free, and $R \in \mathsf{Reg}$, the following problems are known to be undecidable:

- $L(G_1) = \Sigma^*$?
- $L(G_1) \cap L(G_2) = \emptyset$?
- $L(G_1) \subseteq L(G_2)$?
- $L(G_1) = L(G_2)$?
- $R \subseteq L(G_1)$?
- $L(G_1) \in \mathsf{Reg}$?
- Is $L(G_1) \cap L(G_2)$ context-free?
- Is $L(G_1)^c$ context-free?

2.5 Complexity Classes

While the previous section asked whether a given problem is possible to decide, we now ask, for problems known to be decidable, how difficult it is to solve. That is:

- What type of computation is needed (deterministic/nondeterministic TM)?
- What is the time and space required?

Let us formalize this.

Definition 2.17. *A non-decreasing function $f : \mathbb{N} \to \mathbb{N}$ is* constructible *if there exists a TM $\mathscr{M}$ such that for input x with $|x| = n$, $\mathscr{M}(x)$ produces output of length $f(x)$*

- *in time (number of transitions) $\mathscr{O}(n + f(n))$*
- *and space $\mathscr{O}(f(n))$ (number of cells visited).*[1]

For fixed f, language L is in

- $\mathsf{NTIME}(f)$ ($\mathsf{TIME}(f)$) *if there exists $n_0 \in \mathbb{N}$ and a TM (DTM) $\mathscr{M}$ that takes time $f(n)$ for every $x \in L$ such that $|x| \geqslant n_0$.*
- $\mathsf{NSPACE}(f)$ ($\mathsf{SPACE}(f)$) *if there exists $n_0 \in \mathbb{N}$ and a TM (DTM) $\mathscr{M}$ that takes space $f(n)$ for every $x \in L$ such that $|x| \geqslant n_0$.*

We have some obvious inclusions:

$$
\begin{aligned}
\mathsf{TIME}(f) &\subseteq \mathsf{NTIME}(f) \\
\mathsf{SPACE}(f) &\subseteq \mathsf{NSPACE}(f) \\
\forall\, k > 0:\ \mathsf{TIME}(kf) &\subseteq \mathsf{TIME}(f) \\
\mathsf{NTIME}(kf) &\subseteq \mathsf{NTIME}(f) \\
\forall\, k > 0:\ \mathsf{SPACE}(kf) &\subseteq \mathsf{SPACE}(f) \\
\mathsf{NSPACE}(kf) &\subseteq \mathsf{NSPACE}(f)
\end{aligned}
$$

[1] $\mathscr{O}(f(n))$ denotes the class of functions that do not grow faster than f, i.e., $\mathscr{O}(f(n)) = \{g : \mathbb{N} \to \mathbb{N} \mid \exists c \in \mathbb{N}, \forall n \in \mathbb{N} : |g(n)| \leqslant c \cdot |f(n)|\}$; see the literature for more details.

Other inclusions require more work, see [8, 15]. For every constructible f,

$$\mathsf{NTIME}(f) \subseteq \mathsf{SPACE}(f)$$
$$\mathsf{NSPACE}(f) \subseteq \bigcup_{i\in\mathbb{N}} \mathsf{TIME}\left(i^{f+\log(n)}\right)$$

The following list collects the classes most currently used, obtained by taking canonical functions for f, i.e. $f(n) \equiv \log_2(n)$ or $f(n) \equiv n^k$.

$$\mathsf{LOGSPACE} \triangleq \mathsf{SPACE}(\log_2(n))$$
$$\mathsf{NLOGSPACE} \triangleq \mathsf{NSPACE}(\log_2(n))$$
$$\mathsf{P} = \mathsf{PTIME} \triangleq \bigcup_{k\in\mathbb{N}} \mathsf{TIME}(n^k)$$
$$\mathsf{NP} = \mathsf{NPTIME} \triangleq \bigcup_{k\in\mathbb{N}} \mathsf{NTIME}(n^k)$$
$$\mathsf{PSPACE} \triangleq \bigcup_{k\in\mathbb{N}} \mathsf{SPACE}(n^k)$$
$$\mathsf{NPSPACE} \triangleq \bigcup_{k\in\mathbb{N}} \mathsf{NSPACE}(n^k)$$
$$\mathsf{EXP} = \mathsf{EXPTIME} \triangleq \bigcup_{k\in\mathbb{N}} \mathsf{TIME}(2^{n^k})$$
$$\mathsf{NEXP} = \mathsf{NEXPTIME} \triangleq \bigcup_{k\in\mathbb{N}} \mathsf{NTIME}(2^{n^k}).$$

We have the hierarchy of

$$\mathsf{LOGSPACE} \subseteq \mathsf{NLOGSPACE} \subseteq \mathsf{PTIME} \subseteq \mathsf{NPTIME} \subseteq \mathsf{PSPACE} \subseteq \mathsf{NPSPACE}.$$

Here, some inclusions may not be proper ones, see below.

2.5.1 Reduction

Reduction is used to prove that a particular problem belongs to a complexity class.

Definition 2.18. *A* reduction *from $f : A \to \{\mathbf{tt}, \mathbf{ff}\}$ to $f' : B \to \{\mathbf{tt}, \mathbf{ff}\}$ is a computable function $h : A \to B$ such that $\forall a \in A$, $f'(h(a)) = f(a)$.*

Complete and Hard Problems

Definition 2.19. *A problem P in a class C is* complete *for C if every P' in C reduces to P via a function h computable in C.*

One frequently encounters the notion of NP-hard problems. A problem P is NP-hard if there exists an NP-hard problem P' that is polynomial-time Turing reducible to P. Note that a problem can be NP-hard without being NP-complete; this is the case of the (undecidable!) *halting problem* discussed above. Thus, an NP-hard problem is NP-complete if it in addition belongs to NP.

Finally, the problem "Is $\mathsf{P} = \mathsf{NP}$?" is probably the most famous open question in computer science.

Examples from NP

For more details, the reader is referred to [3].

Satisfiability (SAT)

Let ϕ be a propositional formula, i.e., connecting atoms $p_i \in A_t$ by $\neg, \wedge, \vee$. The question is: Is ϕ *satisfiable*, i.e., is there $v : A_t \to \{\mathbf{tt}, \mathbf{ff}\}$ such that $v(\phi) = \mathbf{tt}$?

- SAT is in NP: a TM "guesses" v and computes $v(\phi)$ in polynomial time.
- SAT is NP-complete; to see this, let $L \in \mathsf{NP}$ and let $\mathscr{M}$ be a TM that solves L in (nondeterministic) polynomial time. One constructs a formula that holds if and only if $\mathscr{M}$ halts and accepts (Cook 1971).

Hamiltonian Circuit (HC)

Take a directed graph $G = (V, E)$ with $V = \{v_1, \dots, v_n\}$. Does there exist a permutation σ of $\{1, \dots, n\}$ such that $\forall\, 1 \leqslant i \leqslant n-1$, $(v_{\sigma(i)}, v_{\sigma(i+1)}) \in E$?

- HC is in NP since a TM can solve HC by trial-and-error in polynomial time.
- HC is NP-complete, which can be shown by reducing SAT to it.

The Knapsack Problem

Given $v_1, \dots, v_n, w \in \mathbb{N}$. Is there $I \subseteq \{1, \dots, n\}$ such that $\sum_{i \in I} v_i = w$?
NP completeness of Knapsack is shown by reduction of (a variant of) SAT to it.

Weighted Path (WP)

In a weighted graph $G = (V, E, p)$ with $p : E \to \mathbb{N}$ and $u, v \in V$, find a path from u to v such that the edge weights sum to a fixed value N.
NP-completeness follows here because Knapsack can be reduced to WP.

The Class PSPACE

First of all, we have the following result:

Theorem 2.14. [Savitch 1970] *The problem of accessibility in a directed graph with n vertices can be solved in space* $\mathcal{O}((\log n)^2)$.

As a corollary, we have that PSPACE = NPSPACE. The following problems are from PSPACE:

- Satisfiability of a Boolean formula with quantifiers ($\exists$, $\forall$).
- Universality of regular languages: $L = \Sigma^*$?

Note that $L = \emptyset$ is in PTIME but $L = \Sigma^*$ is not!

2.6 Further Reading

An introduction to automata and formal language theory can be found in [5, 10, 15, 16], and advanced material in [11, 14, 17]. More details on complexity, computability, and decidability can be found in [1, 2, 7, 8, 9], and a long list of NP-complete problems can be found in [3].

References

1. Börger, E., Grädel, E., Gurevich, Y.: The Classical Decision Problem, 2nd edn. Springer, Berlin (2001)
2. Bovet, D.P., Crescenzi, P.: Introduction to the Theory of Complexity. Prentice Hall, New York (1994)
3. Garey, M.R., Johnson, D.S.: Computers and Intractability: A Guide to the Theory of NP-Completeness. W.H. Freeman & Co., New York (1990)
4. Ginsburg, S.: Algebraic and Automata-Theoretic Properties of Formal Languages. North-Holland, Amsterdam (1975)
5. Hopcroft, J.E., Motwani, R., Ullman, J.D.: Introduction to Automata Theory, Languages and Computation, 3rd edn. Addison Wesley, Boston (2006)
6. Jirásková, G., Masopust, T.: Complexity in union-free regular languages. International Journal of Foundations of Computer Science 22(7), 1639–1653 (2011)
7. Jones, N.D.: Computability and Complexity from a Programming Perspective. MIT Press, Cambridge (1997)
8. Kozen, D.C.: Automata and Computability. Springer, Berlin (1997)
9. Papadimitriou, C.: Computational Complexity. Addison Wesley, Boston (1993)
10. Rozenberg, G., Salomaa, A.: Handbook of Formal Languages, pp. 1–3. Springer, Berlin (1997)
11. Sakarovitch, J.: Elements of Automata Theory. Cambridge University Press, New York (2009)
12. Salomaa, A.: Formal Languages. Academic Press, New York (1973)

13. Salomaa, A.: Computation and Automata. Cambridge University Press, Cambridge (1985)
14. Shallit, J.: A Second Course in Formal Languages and Automata Theory. Cambridge University Press, New York (2008)
15. Sipser, M.: Introduction to the Theory of Computation, Course Technology, 2nd edn., Boston, MA (2005)
16. Sudkamp, T.: Languages and Machines: An Introduction to the Theory of Computer Science, 3rd edn. Addison Wesley, Boston (2005)
17. Szepietowski, A.: Turing Machines with Sublogarithmic Space. Springer, Berlin (1994)

Chapter 3
Supervisory Control with Complete Observations

Tomáš Masopust and Jan H. van Schuppen

3.1 Introduction

The purpose of this chapter is to introduce to readers not yet familiar with the topic the problem of supervisory control, the required concepts, theorems, and algorithms.

Supervisory control is motivated by control engineering problems. An example is the control of a patient table of a magnetic resonance imaging (MRI) scanner in which a computer program controls all operations. The table is raised, shifted into the machine, and the reverse operations. A computer program based on supervisory control has been constructed to execute the requested operations in a safe order, see [19].

Control theory in general has been developed extensively since the early part of the twentieth century with a major contributions since the 1960's. Since about 1980, W.M. Wonham and many others have developed control theory of discrete-event systems motivated by the use of computers in engineering.

3.2 Motivation of Supervisory Control

Engineering design gives rise to many control problems and a subset of these may be formulated as control problems of discrete-event systems.

Tomáš Masopust
Institute of Mathematics, Academy of Sciences of the Czech Republic, Žižkova 22, 616 62 Brno, Czech Republic
e-mail: masopust@math.cas.cz

Jan H. van Schuppen
CWI, P.O. Box 94079, 1090 GB Amsterdam, Netherlands
e-mail: J.H.van.Schuppen@cwi.nl

C. Seatzu et al. (Eds.): Control of Discrete-Event Systems, LNCIS 433, pp. 45–64.
springerlink.com

Example 3.1. *Control of the patient table of an MRI scanner.* The example was mentioned in the introduction. The patient table is to execute operations in a safe manner. The table with a patient on it is to be raised to a prespecified level, shifted into the scanner, and stay put at the required location. In addition, the reverse operations are to be carried out.

A model of the operations of the table has been made in the form of a generator. The system consists of several components. The engineering control problem is transformed into a formal specification consisting of control objectives of safety and of required behavior. Safety describes that nothing bad should happen to the patient while she or he is on the table. Required behavior describes that the machine carries out the required movements to shift the patient in the scanner.

The control problem is to construct a supervisory control such that the closed-loop system meets the control objectives of safety and of required behavior. The closed-loop system consists of the interconnection of the generator and of the supervisory control.

In this chapter it is shown how to construct a supervisory control for such a system. The supervisory control can be implemented by a tuple consisting of a supervisor and a control law. The supervisor and the control law can easily be transformed into a computer code, say a C++ program, which then controls the machine. See [19] for more details on this example. ■

There follows a list of control engineering problems for which models in the form of discrete-event systems have been formulated and for which supervisory control has been investigated: (1) Control of a rapid thermal processor, [1]; (2) Databases, [4]; (3) Chemical pilot plant, [14, Ch. 7]; (4) Feature interaction in telephone networks, [22]; (5) Theme park vehicles, [8]; (5) A controller for a traffic light is discussed in the book by R.P. Kurshan, [11].

Research areas with examples of control laws for discrete-event systems include manufacturing systems, automated guided vehicles, aerial vehicles, underwater vehicles, communication networks, mobile phones, high-tech systems, but also software systems on computers, laptops, and readers.

Control objectives of control problems are properties which an engineer strives to attain for the closed-loop system. The main control objectives for control of discrete-event systems are: (1) safety; (2) required behavior; and (3) noblockingness. The *safety* control objective is motivated by the wish that the closed-loop system should not do actions which endanger humans or machines, like causing death or damage. That control objective may be formulated at the level of a discrete-event system as the language of *safe behavior* or *admissible behavior*. In terms of the system, it may be formulated by the partition of the state set into *safe states* and *forbidden states*.

The *required behavior* control objective is motivated by the wish that the system should perform certain actions. For example, the safety objective can be met by switching of the system or machine off completely. But then there is no production or required behavior. The required behavior specifies what the machine should produce. The *nonblockingness* control objective is motivated by the wish that the required behavior really has to be fully completed.

The control problem considered below in this chapter will take care of the above three control objectives.

3.3 Concepts of Automata and of Languages

The reader of this chapter is assumed to be familiar with the concepts of automata, languages, and complexity as described in Chapter 2. However, additional concepts are needed in this chapter.

The reader should distinguish between the *systems framework* and the *language* or *behavioral framework* for modeling of dynamic phenomena, see [16]. The frameworks are often described by the term *signals and systems.* For a dynamic phenomenon the systems framework uses the concept of a system, in discrete-event systems a system is called an automaton or generator. Correspondingly, the language framework uses the concept of a formal language. The reader best learns both frameworks and how to convert a representation from one framework to the other one.

Example 3.2. The *elementary deterministic finite generator.* Consider the following discrete-event system, see Fig. 3.1. The system is specified by $G = (\{1,2,3,4\}, \{a,b,c\}, f, 1, \{3\})$ where f is given by the arrows of Fig. 3.1. The language description is described by the language of the generator G, $L(G) = \{\varepsilon, a, ab, abc\}$, and its marked language $L_m(G) = \{ab\}$. Then $\text{prefix}(L_m(G)) = \{\varepsilon, a, ab\} \subsetneq L(G)$ hence G is blocking as defined below. ■

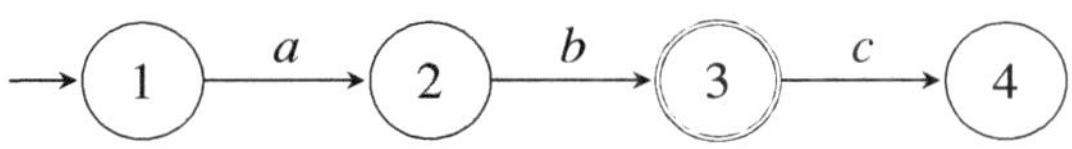

Fig. 3.1 The generator of the elementary deterministic finite generator

Definition 3.1. *A* deterministic finite generator *(DFG) is a tuple*[1]

$G = (Q, E, f, q_0, Q_m)$, *where*

Q *denotes the* finite state set *and* E *denotes the* finite event set,

$|E| \in \mathbb{Z}_+$ *denotes the number of elements of the set* E,

$f : Q \times E \to Q$ *denotes the* transition function, *a partial function,*

$q_0 \in Q$ *denotes the* initial state,

$Q_m \subseteq Q$ *denotes the* subset of marked states,

$f(q,e)!$ *denotes that the transition is defined at* $(q,e) \in Q \times E$,

[1] Compared to Definition 1.4 in Chapter 1, we do not consider controllable and uncontrollable events here. They are considered later in Definition 3.5.

$$\begin{aligned} E(q) &= \{e \in E \mid f(q,e)!\},\ \forall\, q \in Q,\ \textit{called the}\ \text{active event set}\ \textit{at}\ q \in Q, \\ E_{ac} &= \cup_{q \in Q} E(q),\ \textit{the}\ \text{active event set of the generator}, \\ L(G) &= \{s \in E^* \mid f(q_0,s) \in Q\},\ \textit{the}\ \text{language}\ \textit{of}\ G, \\ L_m(G) &= \{s \in E^* \mid f(q_0,s) \in Q_m\},\ \textit{the}\ \text{marked language}\ \textit{of}\ G. \end{aligned}$$

Recall from Chapter 2 the notation $E^* = \cup_{k=1}^{\infty} E^k$ and the extension of the transition function to $f : Q \times E^* \to Q$ by recursion.

It is of interest to describe the relation between a generator and the associated language. If G is a generator then $L(G) \subseteq E^*$ is its language. If for a language $K \subseteq E^*$ there exists a generator G_K such that $K = L(G_K)$ then call G_K *the recognizer* of K. In automata theory, it is customary to call G_K the recognizer of the language K if $K = L_m(G_K)$. It is a major result of automata theory that a language has a recognizer if and only if the language is regular, which concept will not be defined here. Define also the set of *recognizable languages* as

$$E^*_{DFG} = \{L(G) \subseteq E^* \mid G \text{ is a DFG over } E\}.$$

As usual, one constructs DFGs from smaller such generators. For this purpose, it is useful to know the concept of the product of two generators.

Definition 3.2. *Consider the generators* $G_1 = (Q_1, E_1, f_1, q_{1,0}, Q_{1,m})$ *and* $G_2 = (Q_2, E_2, f_2, q_{2,0}, Q_{2,m})$. *Define their* product generator *as the reachable part of the generator* $(Q_1 \times Q_2, E_1 \times E_2, f, (q_{1,0}, q_{2,0}), Q_{1,m} \times Q_{2,m})$, *where*

$$f((q_1,q_2),e) = \begin{cases} (f_1(q_1,e), f_2(q_2,e)), & \text{if } f_1(q_1,e)! \text{ and } f_2(q_2,e)! \\ (f_1(q_1,e), q_2), & \text{if } f_1(q_1,e)! \text{ and } e \in E_1 \backslash E_2, \\ (q_1, f_1(q_2,e)), & \text{if } f_2(q_2,e)! \text{ and } e \in E_2 \backslash E_1, \\ \text{undefined}, & \text{otherwise.} \end{cases}$$

In the remainder of the chapter several concepts are used for the issue of nonblockingness. Recall from Chapter 2 the concepts of a prefix of a string s and the set $\text{prefix}(s) \subset E^*$ of that string. The *prefix of a language* $K \subseteq E^*$ is defined as $\text{prefix}(K) = \cup_{s \in K} \text{prefix}(s)$. A language $K \subseteq E^*$ is called a *nonblocking language* (in the language framework) if $K = \text{prefix}(K)$. By definition of $L(G)$, $L(G) = \text{prefix}(L(G))$ hence $L(G)$ is prefix closed.

In the systems framework nonblockingness of a system is defined differently. Define for a generator G, initial state $q_0 \in Q$, the subset of marked states $Q_m \subseteq Q$, and event subset $E_1 \subseteq E$, the subsets

$$\begin{aligned} \text{reachset}(G, q_0, E_1) &= \{q \in Q \mid \exists s \in E_1^* \text{ such that } q = f(q_0, s)\}, \\ \text{coreachset}(G, Q_m, E_1) &= \{q \in Q \mid \exists s \in E_1^* \text{ such that } f(q,s) \in Q_m\}, \end{aligned}$$

called respectively the *reachable subset* and the *co-reachable subset* of the system. The DFG G is called *reachable* if $Q = \text{reachset}(G, q_0, E)$, *co-reachable* if $Q = \text{coreachset}(G, Q_m, E)$, and *trim* if it is both reachable and co-reachable. Note that $L_m(G) \neq \emptyset$ if $\text{reachset}(G, q_0, E) \cap \text{coreachset}(G, q_0, E) \neq \emptyset$.

A DFG G is called *nonblocking* if at any reachable state there exists a string which transfers the automaton to a marked state. Then nonblockingness of the DFG G holds if and only if $\text{reachset}(G, q_0, E) \subseteq \text{coreachset}(G, Q_m, E)$. This can be shown to be equivalent to $L(G) = \text{prefix}(L_m(G))$ which is nonblockingness in the corresponding language framework. A trim generator is thus nonblocking.

To establish nonblockingness the following results will be useful.

Definition 3.3. *Consider languages $K, L \subseteq E^*$. The language K is said to be* L-closed *if* $\text{prefix}(K) \cap L = K$. *The language K is said to be* L-marked *if* $\text{prefix}(K) \cap L \subseteq K$.

Proposition 3.1. *Consider languages $K, L \subseteq E^*$.*

(a) Assume that $K \subseteq L$. Then K is L-closed if and only if K is L-marked.
(b) If $K \subseteq L$ and if K is prefix closed then K is L-closed.
(c) If K is prefix closed then $K \cap L$ is L-closed.

Proof. (a) (Only if) This follows directly from the definition of K being L-closed.
(If) $K \subseteq \text{prefix}(K) \wedge K \subseteq L \Rightarrow K \subseteq \text{prefix}(K) \cap L$.
(b) $K = K \cap L = \text{prefix}(K) \cap L$.
(c) $\text{prefix}(K \cap L) \cap L \subseteq \text{prefix}(K) \cap L = K \cap L$, hence $K \cap L$ is L-marked, $K \cap L \subseteq L$, and from (a) follows that it is L-closed. □

Nonblockingness of a product of generators requires attention. There exists an example of two nonblocking generators whose product is blocking. Therefore it is of interest to know a characterization of nonblockingness of a product of generators.

Definition 3.4. *The tuple of languages $(L_1, L_2) \subseteq E^* \times E^*$ is called* nonconflicting *if* $\text{prefix}(L_1 \cap L_2) = \text{prefix}(L_1) \cap \text{prefix}(L_2)$. *A tuple of generators (G_1, G_2) is called* nonconflicting *if their marked languages, $L_m(G_1)$ and $L_m(G_2)$, are nonconflicting.*

Note that the inclusion $\text{prefix}(L_1 \cap L_2) \subseteq \text{prefix}(L_1) \cap \text{prefix}(L_2)$ always holds so the restriction of nonconflictingness imposes the converse inclusion.

3.4 Concepts of Control of Discrete-Event Systems

Definition 3.5. *A* controlled deterministic finite generator *(CDFG) is a tuple*

$$G_c = (Q, E, f, q_0, Q_m, \{E_c, E_{uc}\}, E_{cp}, f_c), \text{ where}$$
$$G = (Q, E, f, q_0, Q_m) \text{ is a deterministic finite generator,}$$

$\{E_c, E_{uc}\}$ *is a partition of E,*
$E_c \subseteq E$ *is called the subset of* controllable events,
$E_{uc} \subseteq E$ *is called the subset of* uncontrollable events,
$E_{cp} = \{E_{en} \subseteq E \mid E_{uc} \subseteq E_{en}\}$ *is called the set of* control patterns,
$E_{en} \in E_{cp}$ *is called the set of* enabled events,

$E \backslash E_{en}$ *is called the set of* disabled events,

$f_c : Q \times E \times E_{cp} \to Q$ *is called the* transition function *of* G_c,

$$f_c(q,e,E_{en}) = \begin{cases} f(q,e), & \text{if } f(q,e)! \wedge e \in E_{en}, \\ \text{undefined, otherwise.} \end{cases}$$

An example of a controllable event is the switching on of a machine. An example of an uncontrollable event is the switch of a machine from working to idle when a task is completed.

Example 3.3. The *elementary controlled deterministic finite generator.* Recall Example 3.2. In the following that example is modified so that the events are now controllable events. Consider the controlled discrete-event system described by Fig. 3.2 and the notations, $G_c = (Q, E, f, q_0, Q_m, \{E_c, E_{uc}\}, E_{cp}, f_c)$, $E = \{a,b,c\}$, $E_c = \{a,b,c\} = E$, $E_{uc} = \emptyset$, $E_{cp} = \text{Pwrset}(E)$. ■

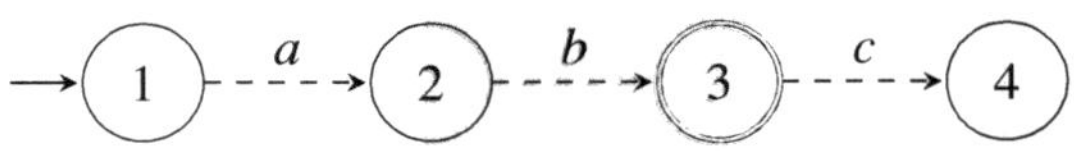

Fig. 3.2 The generator of the elementary controlled deterministic finite generator. Note that dashed arrows denote controllable events while solid arrows denote uncontrollable events

Control of discrete-event systems is based on the principle of feedback. According to the principle of feedback the observations of a system, in this case the string of all generated events, are used by a controller to generate the input of the system. Feedback is illustrated in Fig. 3.3.

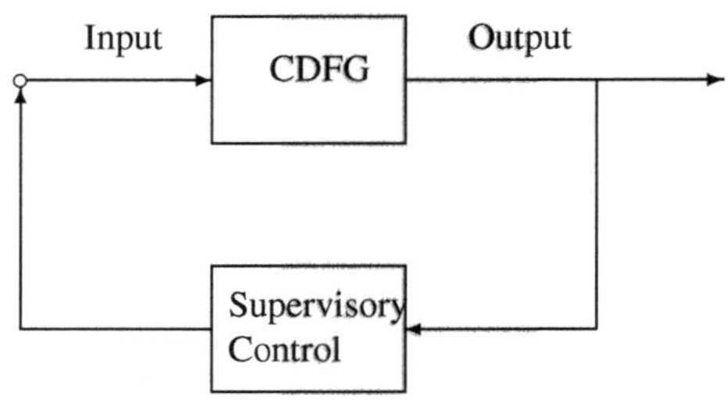

Fig. 3.3 A CDFG and a supervisory control in closed-loop

Below, a controller is defined by a supervisory control in the language framework.

Definition 3.6. *A* supervisory control *(with complete observations) for a CDFG* G_c *is a map* $v : L(G) \to E_{cp}$. *Thus for any string* $s \in E^*$, $v(s) \in E_{cp}$ *specifies the subset of events which are enabled.*

The supervisory control defined above is said to be with complete observations because at any stage of the generation of a string, the entire history of the system G_c in the form of the string of L(G) is available to the supervisory control.

Definition 3.7. *Define for a CDFG G_c and a supervisory control $v : L(G) \to E_{cp}$ their* closed-loop behavior *v/G_c as the smallest language $L(v/G_c)$ such that $L(v/G_c) \subseteq L(G)$, (1) $\varepsilon \in L(v/G_c)$, and (2) $se \in L(v/G_c)$ if $s \in L(v/G_c)$, $e \in v(s)$, and $se \in L(G)$. In addition, $L_m(v/G_c) = L(v/G_c) \cap L_m(G)$. A supervisory control v is said to be* nonblocking *for G_c if v/G_c is nonblocking, or equivalently, if $L(v/G_c) = \text{prefix}(L_m(v/G_c))$.*

From the definition of the closed-loop behavior v/G_c follows that $\{\varepsilon\} \subseteq L(v/G_c) \subseteq L(G)$, $L(v/G_c)$ is nonempty and prefix closed, and $L_m(v/G_c) \subseteq L_m(G)$.

Example 3.4. The *elementary controlled deterministic finite generator.* Recall Example 3.3. Consider the following supervisory control, see Fig. 3.2, and the mathematical specification, $v_1 : L(G) \to E_{cp}$, $v_1(\varepsilon) = \{a\}$, $v_1(a) = \{b\}$, $v_1(ab) = \emptyset$, $v_1(abc) = \emptyset$. Then, $L(v_1/G) = \{\varepsilon, a, ab\}$, $L_m(v_1/G) = L(v_1/G) \cap L_m(G) = \{ab\}$, $L(v_1/G) = \text{prefix}(L_m(v_1/G))$, hence v_1/G is nonblocking. ■

3.5 Problem of Existence of a Supervisory Control

Example 3.5. *The line specification.* Consider the generator $G_c = (Q, E, f, q_0, Q_m, \{E_c, E_{uc}\}, E_{cp}, f_c)$ with $E = \{a, b, c, d, e, f, g, h, i, \}$, $q_0 = 1$, $Q_m = \{1, 6, 7, 9\}$, and $E_c = \{a, d, f, g, h\}$. The transition function is displayed in Fig. 3.4. Note the marked states $1, 6, 7$, and 9. Note the difference between the uncontrollable events (solid arrow) and the controllable events (dashed arrow). Note that the specification is such that the states $7, 8, 9, 10$ are forbidden states. Denote the language of the specification by the sublanguage $K = \{abcde\} \subset L_m(G)$. The problem is thus to control the system in such a way that the forbidden states are never reached and such that the marked state 6 is reached. ■

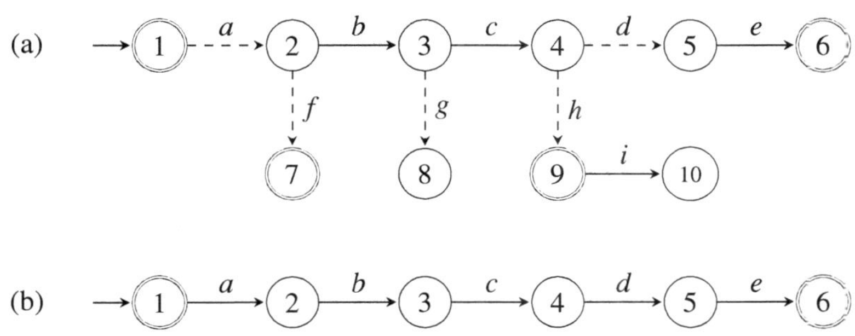

Fig. 3.4 (a) The diagram of the generator of the line plant and (b) the specification of the line plant

The modeling of a specification is an art which the reader has to learn from examples. In the language framework one works with the specification of the admissible language or, complementary, with the forbidden strings. In the system framework, one works with a forbidden subset of the state set whose elements will be called forbidden states.

Problem 3.1. *Supervisory control problem with complete observations - Existence of a supervisory control.* Consider a controlled deterministic finite generator G_c and a specification in terms of the marked languages: a *required language* $L_{rm} \subseteq E^*$ and an *admissible language* $L_{am} \subseteq E^*$ satisfying, $L_{rm} \subseteq L_{am} \subseteq L_m(G)$. Under which conditions does there exist a supervisory control $v : L(G) \to E_{cp}$ such that the closed-loop system v/G_c satisfies: (1) $L_{rm} \subseteq L_m(v/G_c) \subseteq L_{am}$; (2) v/G_c is nonblocking? ■

If all events of a CDFG are controllable then the solution of the problem is easy, one deduces directly the supervisory control from the specification. If all events are uncontrollable and if the specification is nontrivial then the specification can never be met. The interesting case of the problem is thus when there exist both controllable and uncontrollable events.

Supervisory control is inspired by the earlier research of W.M. Wonham on the geometric approach of control of time-invariant linear systems, see [28]. The control problem may be seen as a generalization of the verification problem of model checking. It is also related to control of a deterministic or of a stochastic system, for example, to optimal control on an infinite-horizon which results then in a time-invariant control law.

In the construction of a supervisory control one is often provided a specification in the form of a generator G_{spec}. Take then $L_{am} = L(G_{spec}) \cap L_m(G)$. Because $L(G_{spec})$ is prefix closed, it follows from Proposition 3.1(c) that L_{am} is $L_m(G)$-closed. By definition it holds that $L_{am} \subseteq L_m(G)$. If $L_{am} = L_m(G)$ then the uncontrolled plant meets the specification and no supervisory control is needed. If $L_{am} = L_m(v/G_c) = \emptyset$ then a nonblocking supervisory control implies that $L(v/G_c) = \text{prefix}(L_m(v/G_c)) = \emptyset$ which contradicts that $\varepsilon \in L(v/G_c)$.

Assumption 3.1. *Consider Problem 3.1. Assume that the admissible sublanguage $L_{am} \subseteq L(G_{spec})$ satisfies (1) $L_{am} \subseteq L_m(G)$, (2) L_{am} is $L_m(G)$-closed, and (3) $L_{am} \neq \emptyset$ and $L_{am} \neq L_m(G)$.*

3.6 Existence of a Supervisory Control

Consider the supervisory control problem with complete observations, Problem 3.1. Attention will first be restricted to the case where $L_{rm} = L_{am} \subseteq L_m(G)$. The general case will be discussed in Section 3.10. In this section, a necessary and sufficient condition will be formulated for the existence of a supervisory control. The condition is formulated below in terms of the control-theoretic concept of controllability.

Definition 3.8. *Consider a controlled deterministic finite generator G_c. The language $K \subseteq E^*$ is said to be* controllable *with respect to G_c if*

$$\text{prefix}(K)E_{uc} \cap L(G) \subseteq \text{prefix}(K); \tag{3.1}$$

equivalently, $\forall s \in \text{prefix}(K)$, $\forall e \in E_{uc}$, *$se \in L(G)$ implies that $se \in \text{prefix}(K)$.*

Controllability of a language differs from the system-theoretic concept of controllability of a linear system. Controllability of a language is equivalent with the language being invariant with respect to supervisory control. The terminology of a controllable language defined above is now standard in the literature.

Example 3.6. *The line specification (continued).* Consider Example 3.5 with the plant and the specification, see Fig. 3.4. Is the language of the specification controllable? By inspection of the plant it is clear that the specification language is controllable because the events which lead one out of the specification, $\{f,g,h\}$, are all controllable. In many other examples, the check of controllability is not that simple! ■

Proposition 3.2. *Consider a CDFG G_c. The following languages in E^* are controllable with respect to G_c: (a) $\emptyset \in E^*$; (b) $L(G) \in E^*$; (c) E^*.*

Proof. (a) $\text{prefix}(\emptyset)E_{uc} \cap L(G) = \emptyset E_{uc} \cap L(G) = \emptyset \cap L(G) = \emptyset = \text{prefix}(\emptyset)$. (b) $\text{prefix}(L(G))E_{uc} \cap L(G) = L(G)E_{uc} \cap L(G) \subseteq L(G) = \text{prefix}(L(G))$. □

Recall from Definition 3.1 the concept of an active event set at a state.

Algorithm 3.7. (Controllability of a regular language).

Consider a trim CDFG G and a trim DFG H such that $L_m(H) = L_{am}$ hence H is a recognizer of L_{am}.

1. *Construct* $J = (Q_J, E_J, f_J, q_{J,0}) = \text{trim}(H \times G)$.
2. *Check whether for all $(q_H, q_G) \in Q_J$ the following inclusion holds, $E(q_G) \cap E_{uc} \subseteq E(q_H, q_G)$.*
3. *If Condition (2) is satisfied then the language $L_{am} = L_m(H) \subseteq E^*$ is controllable with respect to G_c. If Condition (2) is not satisfied then the language $L_m(H)$ is not controllable.*

It can then be proven that the algorithm is correct. The time complexity of the algorithm is $O(|Q_G| \cdot |Q_H| \cdot |E|)$ where the notation is defined in Section 3.3.

Theorem 3.2. (Existence of a supervisory control with complete observations). [29, Th. 3.4.1] *Consider Problem 3.1 for a controlled deterministic finite generator G_c, which is such that the associated generator G is nonblocking. Consider the special case of the problem in which the required language and the admissible language are equal and denoted by $K = L_{rm} = L_{am}$, while the sublanguage K satisfies Assumption 3.1. There exists a supervisory control v for G_c such that the marked closed-loop language equals the specification, $L_m(v/G_c) = K$, and such that v/G_c is nonblocking if and only if (1) K is a controllable language with respect to G_c; and (2) K is $L_m(G)$-closed.*

Proof. (a) ($\Leftarrow$) (1) $K \subseteq L_m(G)$ implies that $\text{prefix}(K) \subseteq \text{prefix}(L_m(G)) = L(G)$, because G is nonblocking, and $K \neq \emptyset$ implies that $\varepsilon \in \text{prefix}(K)$.
(2) Define the supervisory control, $v : L(G) \to E_{cp}$, $v(s) = E_{uc} \cup \{e \in E_c \mid se \in \text{prefix}(K)\}$.
(3) It will be proven that, $L(v/G_c) \subseteq \text{prefix}(K)$, by induction on the length of the string. Note that $\varepsilon \in L(v/G_c) \cap \text{prefix}(K)$. Consider for $n \in \mathbb{Z}_+$,

$$\begin{aligned}
& se \in L(v/G_c), \text{ with } |s| = n, \\
\Rightarrow\ & s \in L(v/G_c), e \in v(s), \text{ and } se \in L(G), \text{ by definition of } L(v/G_c), \\
\Rightarrow\ & s \in \text{prefix}(K), e \in v(s), \text{ and } se \in L(G), \\
& \text{by the induction step } s \in L(v/G_c) \subseteq \text{prefix}(K), \\
\Rightarrow\ & \begin{cases} se \in \text{prefix}(K)E_{uc} \cap L(G) \subseteq \text{prefix}(K), & \text{if } e \in E_{uc} \text{ by controllability}, \\ se \in \text{prefix}(K), & \text{if } e \in E_c \text{ because } e \in v(s) \\ & \text{and because of the definition of } v(s). \end{cases}
\end{aligned}$$

(4) It will be shown that, $\text{prefix}(K) \subseteq L(v/G_c)$, by induction on the length of strings. Again $\varepsilon \in \text{prefix}(K) \cap L(v/G_c)$. Consider again for $n \in \mathbb{Z}_+$,

$$\begin{aligned}
& se \in \text{prefix}(K), \text{ with } |s| = n, \\
\Rightarrow\ & s \in \text{prefix}(K) \subseteq L(v/G_c) \wedge se \in \text{prefix}(K) \subseteq L(G), \\
& \text{by the induction step and by (1)}, \\
\Rightarrow\ & se \in L(v/G_c), \text{ if } e \in E_{uc} \text{ because then } e \in v(s), \\
& \qquad \text{and because of the definition of } L(v/G_c), \text{ or} \\
& se \in L(v/G_c), \text{ if } e \in E_c, \text{ because by definition of } v(s), \\
& \qquad se \in \text{prefix}(K) \Rightarrow e \in v(s), \\
& \qquad \text{and because of the definition of } L(v/G_c).
\end{aligned}$$

(5)

$$\begin{aligned}
L_m(v/G_c) &= L(v/G_c) \cap L_m(G), \text{ by def. of } L_m(v/G_c), \\
&= \text{prefix}(K) \cap L_m(G), \text{ by (3) and (4)}, \\
&= K, \text{ because K is } L_m(G)\text{-closed, see Definition 3.3}, \\
&\quad \text{which establishes the specification},
\end{aligned}$$

$$\text{prefix}(L_m(v/G_c)) = \text{prefix}(K) = L(v/G_c), \text{ by (3) and (4)},$$

hence v/G_c is nonblocking.
($\Rightarrow$) (6)

$$\begin{aligned}
\text{prefix}(K) &= \text{prefix}(L_m(v/G_c)), \text{ because } K = L_m(v/G_c), \\
&= L(v/G_c), \text{ because } v/G_c \text{ is nonblocking}, \\
K &= L_m(v/G_c) = L(v/G_c) \cap L_m(G) = \text{prefix}(K) \cap L_m(G),
\end{aligned}$$

by the above, hence K is $L_m(G)$-closed. Note that

$$\begin{aligned} & s \in \text{prefix}(K) \land e \in E_{uc} \land se \in L(G) \\ \Rightarrow\ & s \in \text{prefix}(K) = L(v/G_c) \land e \in v(s) \land se \in L(G) \\ \Rightarrow\ & se \in L(v/G_c) = \text{prefix}(K), \end{aligned}$$

by definition of $L(v/G_c)$ and by the above relation between $L(v/G_c)$ and K. Thus K is a controllable language with respect to G_c. □

Example 3.8. *The line specification (continued).* Recall the example of the line generator of Example 3.5. From Fig. 3.4 it follows by inspection that the language $K = L_m(G_K)$ is $L_m(G)$-closed. From Theorem 3.2 then follows that there exists a supervisory control which achieves the language K, for example

$$v(\varepsilon) = \{a\} \cup E_{uc},\ v(a) = E_{uc},\ v(ab) = E_{uc},\ v(abc) = E_{uc} \cup \{d\},\ v(abca) = E_{uc}.$$

■

3.7 Implementation of a Supervisory Control by a Supervisor

In this section, it is described how to implement a supervisory control by a supervisor in the form of a deterministic finite generator. Because such a supervisor is easily converted into a computer program, the practical usefulness to engineering design should be clear. This section is somewhat technical and outside the main line of the chapter. It could be skipped in a first reading.

Definition 3.9 (Supervisor). *Consider a controlled deterministic finite generator* $G_c = (Q, E, f, q_0, Q_m, \{E_c, E_{uc}\}, E_{cp}, f_c)$. *A* supervisor *for* G_c *is a tuple* (G_S, g), *where* $G_S = (Q_S, E, f_S, q_{S,0}, Q_{S,m})$ *is a DFG called the* supervisor generator *and* $g : Q_S \to E_{cp}$ *is the map* $g(q_S) = E_{uc} \cup E(q_S) = E_{uc} \cup \{e \in E_c \mid f_S(q_S, e)!\}$ *called the* control law *such that (1)* G_S *is trim; (2)* $L(G_S) \subseteq L(G)$*; and (3)* $L_m(G_S)$ *is a controllable language with respect to* G_c*. It is called a* proper supervisor *if in addition (4)* $(L_m(G_S), L_m(G))$ *is a nonconflicting pair of languages (see Definition 3.4).*

Example 3.9. Consider Example 3.8 of the line system. The supervisor associated with the supervisory control coincides with the automaton from Fig. 3.4.(b) and the control law $g : Q_S \to E_{cp}$ can be chosen as $g(1) = \{a\} \cup E_{uc}$, $g(2) = g(3) = g(5) = E_{uc}$, and $g(4) = \{d\} \cup E_{uc}$, where $E_{uc} = \{b, c, e, i\}$. ■

Definition 3.10. *Consider a CDFG* G_c *and a supervisor* (G_S, g) *with the same event set. Define the* closed-loop deterministic finite generator *(CLDFG) associated with* G_c *and* (G_S, g) *as,* $G_S/G_c = (Q_S \times Q, E, f_{cls}, (q_{S,0}, q_0), Q_{S,m} \times Q_m)$, *as in the product generator* $G_S \times G$ *but with*

$$f_{cls}((q_S, q), e) = \begin{cases} (f_S(q_s, e), f_c(q, e, g(q_S))), & \text{if } f_S(q_S, e)! \land f_c(q, e, g(q_S))!, \\ (f_S(q_S, e), f(q, e)), & \text{if } f_S(q_S, e)! \land f(q, e)! \land \{e \in g(q_S)\}, \end{cases}$$

where the equality follows from the definition of CDFG. The notation G_S/G_c signifies that the supervisor G_S supervises the generator G_c.

Proposition 3.3. [29, Prop. 3.6.2] *Consider a CDFG G_c and a supervisor (G_S, g) according to Definition 3.9.*

(a) Then $L(G_S/G_c) = L(G_S) \cap L(G)$ and $L_m(G_S/G_c) = L_m(G_S) \cap L_m(G)$.
(b) Assume that G is nonblocking. Then G_S/G_c is nonblocking if and only if $(L_m(G_S), L_m(G))$ is a nonconflicting pair of languages.

Next the relation between a supervisory control and a supervisor is discussed.

Definition 3.11. *Consider a CDFG G_c and a supervisory control v. A supervisor (G_S, g) is said to* implement *the supervisory control v if $L(v/G_c) = L(G_S/G_c)$ and $L_m(v/G_c) = L_m(G_S/G_c)$.*

It can then be proven that for any maximally permissive supervisory control there exists a supervisor which implements it and for any supervisor there exists a supervisory control in which tuple is related by implementation.

3.8 Computation of a Supervisor

How to compute a supervisor from the plant and the specification?

Procedure 3.3. Design of a supervisor in the case that the marked admissible language is controllable.

Data *Consider a controlled deterministic finite generator G_c, representing the plant, and a specification G_{spec}, $G_c = (Q, E, f, q_0, Q_m, \{E_c, E_{uc}\}, E_{cp}, f_c)$, $G_{spec} = (Q_{spec}, E, f_{spec}, q_{spec,0}, Q_{spec,m})$.*

1. *Compute $G_1 = \text{trim}(G)$ and $G_2 = (Q_2, E_2, f_2, q_{2,0}, Q_{2,m}) = \text{trim}(G_{spec})$.*
2. *Compute the DFG G_3 by $G_3 = (Q_2, E_2, f_2, q_{2,0}, Q_2) \times G_1$. Then, by definition of G_3, $L_m(G_3) = L(G_2) \cap L_m(G_1)$, and from Proposition 3.1(b) follows that $L(G_3)$ is $L_m(G_1)$-closed.*
3. *Compute $G_a = \text{trim}(G_3)$.*
4. *Check if $L_m(G_a) \neq \emptyset$. If so proceed with Step 5 else output a message that the marked admissible language is empty and stop.*
5. *Check if the language $L_m(G_a)$ is controllable with respect to G_c by Algorithm 3.7. If so proceed with Step 6 else output a message and stop.*
6. *Define $G_S = G_a$ and the map $g : Q_S \to E_{cp}$, $g(q_S) = E_{uc} \cup \{e \in E \mid f_S(q_S, e)!\}$. Then (G_S, g) is a nonblocking proper supervisor such that $L_m(G_S/G_c) = L(G_{spec}) \cap L_m(G)$.*

It can then be proven that the supervisor constructed by Procedure 3.3 is a solution of Problem 3.1 and that it implements the corresponding supervisory control.

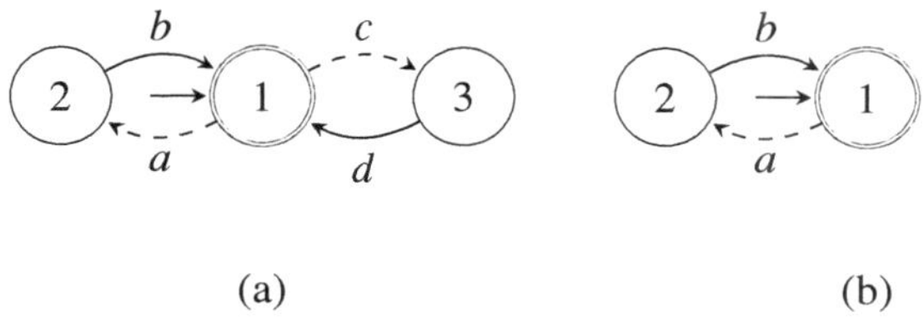

Fig. 3.5 (a) Diagram of the generator of the two-ring system and (b) the diagram of the specification of the two-ring system

Example 3.10. *The two-ring example.* Consider the CDFG G_c with the dynamics as specified in Fig. 3.5.(a). Note that $G_c = (Q, E, f, q_0, Q_m, \{E_c, E_{uc}\}, E_{cp}, f_c)$, $Q_m = \{1\}$, and $E_c = \{a, c\}$. Consider the specification in terms of the generator G_{spec} of which the discrete dynamics is specified in Fig. 3.5.(b). Let the corresponding generator be denoted by $G_{spec} = (Q_{spec}, E, f_{spec}, q_{spec,0}, Q_{m,spec})$, where the same event set is used as in G_c. ■

Consider the problem of synthesizing a supervisor G_S such that $L_m(G_S/G_c) = L_m(G_{spec})$. The construction of the supervisor is so simple that it can be illustrated. The steps of Procedure 3.3 will be followed.

(1) Let $G_1 = \text{trim}(G) = G$, $G_2 = \text{trim}(G_{spec}) = G_{spec}$.
(2) Note that $L_m(G) = ((ab) + (cd))^*$, $L_m(G_{spec}) = (ab)^* \subseteq L_m(G)$, $L(G_{spec}) = (ab)^*, (ab)^*a$, $L_m(G_{spec}) = L(G_{spec}) \cap L_m(G) = \{(ab)^*\}$, and $G_a = G_{spec}$.
(3) $L_m(G_a) = L_m(G_{spec}) \neq \emptyset$.
(4) $L_m(G_{spec})$ is controllable with respect to G_c because $\text{prefix}(L_m(G_{spec}))E_{uc} \cap L(G) \subseteq \text{prefix}(L_m(G_{spec}))$. This follows from,

$$\text{prefix}(L_m(G_{spec})) = \{\varepsilon, a, ab, \ldots\},\ \varepsilon a \notin \text{prefix}(L_m(G_{spec}))E_{uc},$$
$$ab \in \text{prefix}(L_m(G_{spec}))E_{uc} \cap L(G) \cap \text{prefix}(L_m(G_{spec})),\ \text{etc.}$$

(5) $G_S = G_a = G_{spec}$, $g : Q_S \to E_{cp}$, $g(1) = E_{uc} \cup \{e \in E \mid f(1_S, e)!\} = \{b, d, a\}$, $g(2) = \{b, d\}$, and $g(3) = \{b, d\}$. Then (G_S, g) is a nonblocking supervisor for G_c such that $L_m(G_S/G_c) = L(G_{spec}) \cap L_m(G) = L_m(G_{spec})$.

3.9 Partially-Ordered Sets

The problem of supremal supervision formulated in Section 3.11 requires several concepts from universal algebra. It is useful to review these concepts briefly in this chapter and to illustrate the concepts by the example of the set of subsets of a set.

Definition 3.12. *A relation $R \subseteq X \times X$ is said to be:*

- reflexive *if $(x, x) \in R$, $\forall x \in R$;*
- anti-symmetric *if $(x, y) \in R$ and $(y, x) \in R$ imply that $x = y$; and*
- transitive *if (x, y), $(y, z) \in R$ imply that $(x, z) \in R$.*

A partial order *R on a set X is a relation which is reflexive, anti-symmetric, and transitive. A* partially-ordered set *or a* poset *is a set with a partial-order relation. Notation* (X,R).

Example 3.11. *(Powerset, inclusion)* If X is a set then the set of all subsets of X with the inclusion relation, $(\text{Pwrset}(X), \subseteq)$, is a poset. ■

Definition 3.13. *Consider a poset* $(X, \leq)$ *and a subset* $Y \subseteq X$.

- *An* upper bound *of* $x,y \in Y$ *is an element* $u \in X$ *such that* $x \leq u$ *and* $y \leq u$.
- *A* lower bound *of* $x,y \in Y$ *is an element* $l \in X$ *such that* $l \leq x$ *and* $l \leq y$.

Example 3.12. *(Powerset, inclusion)* (continued). Consider for the set X, the poset $(\text{Pwrset}(X), \subseteq)$. $X_1 \cup Y_1$ is an upper bound of the sets X_1, Y_1, and $X_1 \cap Y_1$ is a lower bound of the same sets. ■

Definition 3.14. *Consider a poset* $(X, \leq)$ *and a subset* $X_1 \subseteq X$.

- *The* supremum *of* X_1 *(also called* join *or* lowest upper bound*) is an element* $\sup(X_1) \in X$*, denoted by* $\sup(X_1) = \sup_{x_1 \in X_1} x_1 = \vee_{x_1 \in X_1} x_1 \in X$*, such that (1)* $x_2 \leq \sup(X_1)$, $\forall x_2 \in X_1$*; and (2)* $\sup(X_1) \leq u$, $\forall\, u \in X$ *which are upper bounds of* X_1.
- *The* infimum *of* X_1 *(also called* meet *or* greatest lower bound*) is an element* $\inf(X_1) \in X_1$*, denoted by* $\inf(X_1) = \inf_{x_1 \in X_1} x_1 = \wedge_{x_1 \in X_1} x_1 \in X$*, such that (1)* $\inf(X_1) \leq x_2$, $\forall x_2 \in X_1$*; and (2)* $l \leq \inf(X_1)$, $\forall\, l \in X$ *which are lower bounds of* X_1.

Example 3.13. *(Powerset, inclusion)* (continued) Consider the set X, the associated poset $(\text{Pwrset}(X), \subseteq)$, and a subset $X_1 \subseteq \text{Pwrset}(X)$. Then X_1 has the supremum $\sup(X_1) = \cup_{x_1 \in X_1} x_1$ and the infimum $\inf(X_1) = \cap_{x_2 \in X_1} x_2$. ■

Definition 3.15. *Consider a partially-ordered set* $(X, \leq)$*. It is called:*

- *a* lattice *if for all* $x,y \in X$, $\sup(x,y) = x \vee y \in X$*, and* $\inf(x,y) = x \wedge y \in X$*;*
- *a* complete upper semi-lattice *if* $\sup(X_1) \in X_1$ *for all* $X_1 \subseteq X$.
- *a* complete lower semi-lattice *if* $\inf(X_1) \in X_1$ *for all* $X_1 \subseteq X$.
- *a* complete lattice *if it is both a complete upper semi-lattice and a complete lower semi-lattice.*

3.10 Supremal Controllable Sublanguages

Recall the notation of Problem 3.1.

Definition 3.16. *Consider a CDFG* $G_c = (Q, E, f, q_0, Q_m, \{E_c, E_{uc}\}, E_{cp}, f_c)$ *and assume that the sublanguage* $L_{am} \subseteq L_m(G)$ *satisfies Assumption 3.1. Define the following sets of controllable languages:*

(a) The set of controllable sublanguages *of* L_{am},

$$\mathrm{C}(L_{am}, G_c) = \{K \subseteq L_{am} \mid \mathrm{prefix}(K)E_{uc} \cap L(G) \subseteq \mathrm{prefix}(K)\}.$$

(b) The set of prefix-closed controllable sublanguages *of* L_{am},

$$\mathrm{C}_{\mathrm{pc}}(L_{am}, G_c) = \{K \subseteq L_{am} \mid K = \mathrm{prefix}(K), KE_{uc} \cap L(G) \subseteq K\}.$$

The tuple $(\mathrm{C}(L_{am}, G_c), \subseteq)$ of controllable sublanguages of a tuple (L_{am}, G_c) is a poset. Is it also a complete upper semi-lattice or even a lattice?

Proposition 3.4. [29, Prop. 3.5.1] *Consider a CDFG G_c and a language $L_{am} \subseteq L_m(G)$ satisfying Assumption 3.1.*

(a) $\mathrm{C}(L_{am}, G_c)$ is nonempty, in fact $\emptyset \in \mathrm{C}(L_{am}, G_c)$.
(b) $(\mathrm{C}(L_{am}, G_c), \subseteq)$ is a poset.
(c) $\mathrm{C}(L_{am}, G_c)$ is closed with respect to countable or arbitrary unions.
(d) There exists a unique supremal element $\sup \mathrm{C}(L_{am}, G_c) \in \mathrm{C}(L_{am}, G_c)$ *which will be called the* supremal controllable sublanguage *of* L_{am}.
(e) $\mathrm{C}(L_{am}, G_c)$ is a complete upper semi-lattice of the lattice $(Pwrset(L_{am}), \subseteq)$.

Proof. (a) $\emptyset \in \mathrm{C}(L_{am}, G_c)$ because of Proposition 3.2.
(b) $\mathrm{C}(L_{am}, G_c)$ is a collection of sets that is easily verified to be a poset with respect to set inclusion.
(c) Let $\{K_i \subseteq L_{am}, i \in I\} \subseteq \mathrm{C}(L_{am}, G_c)$ with $I \subseteq \mathbb{Z}$ and let $K = \cup_{i \in I} K_i \subseteq L_{am}$. Then

$$\begin{aligned}
\mathrm{prefix}(K) &= \mathrm{prefix}(\cup K_i) = \cup_{i\in I}\mathrm{prefix}(K_i), \text{ because of a result} \\
&\quad\ \text{on the commutativity of union and prefix closure,} \\
\mathrm{prefix}(K)E_{uc} &= \cup_{i\in I}\mathrm{prefix}(K_i)E_{uc}, \\
\mathrm{prefix}(K)E_{uc} \cap L(G) &= [\cup_{i\in I}\mathrm{prefix}(K_i)E_{uc}] \cap L(G) = \cup_{i\in I}[\mathrm{prefix}(K_i)E_{uc} \cap L(G)] \\
&\subseteq \cup\mathrm{prefix}(K_i), \text{ because } K_i \in \mathrm{C}(L_{am}, G_c),\ \forall i \in I, \\
&= \mathrm{prefix}(\cup K_i) = \mathrm{prefix}(K).
\end{aligned}$$

(d) This follows from (c), Example 3.13, and from Definition 3.14.
(e) This follows from the definition of complete upper semi-lattice and from (d). □

The set $\mathrm{C}(L_{am}, G_c)$ is not closed with respect to intersection as is shown below.

Example 3.14. *Set of controllable sublanguages not intersection closed.* [29, Ex. 3.5.1]. There exists an example of a CDFG G_c and languages $K_1, K_2 \subseteq E^*$ such that:

(a) K_1, K_2 are controllable sublanguages with respect to G_c.
(b) $K_1 \cap K_2$ is not a controllable sublanguage with respect to G_c.
(c) $\mathrm{prefix}(K_1) \cap \mathrm{prefix}(K_2)$ is a controllable sublanguage with respect to G_c.

Proof of the above three properties. Consider the CDFG of Fig. 3.6,

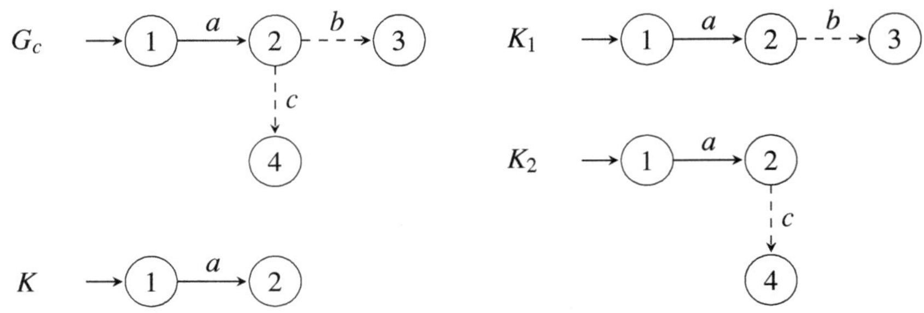

Fig. 3.6 Diagram of the generator and of the sublanguages K, K_1, and K_2

$$G_c = (Q,E,f,q_0,Q_m,\{E_c,E_{uc}\},E_{cp},f_c), \; E=\{a,b,c\}, \; E_c=\{b,c\}, \; E_{uc}=\{a\},$$
$$K_1 = \{\varepsilon,ab\}, \; \text{prefix}(K_1)=\{\varepsilon,a,ab\}, \; K_2=\{\varepsilon,ac\}, \text{prefix}(K_2)=\{\varepsilon,a,ac\}.$$

(a) K_1 is controllable with respect to G_c because, $a=\varepsilon a \in \text{prefix}(K_1)E_{uc} \cap L(G) \subseteq \text{prefix}(K_1)$. Similarly K_2 is controllable with respect to G_c.
(b) $K_1 \cap K_2 = \{\varepsilon\}$, $\text{prefix}(K_1 \cap K_2) = \{\varepsilon\}$, $a = \varepsilon a = \in \text{prefix}(K_1 \cap K_2)E_{uc} \cap L(G)$, and $a \notin \text{prefix}(K_1 \cap K_2)$, hence $K_1 \cap K_2$ is not controllable with respect to G_c.
(c) $K = \text{prefix}(K_1) \cap \text{prefix}(K_2) = \{\varepsilon,a\}$, $\text{prefix}(K) = K = \{\varepsilon,a\}$, $KE_{uc} \cap L(G) \subseteq K$, or K is controllable with respect to G_c. ■

Proposition 3.5. [29, Prop. 3.5.2] *Consider a CDFG G_c and a language $L_{am} \subseteq L_m(G)$ satisfying the conditions of Assumption 3.1. Then* $\mathrm{C}(L_{am},G_c)$ *is an upper-semilattice of the lattice $(Pwrset(L_{am}),\subseteq)$. Hence there exists a unique* supremal controllable and prefix-closed sublanguage $\sup\mathrm{C}(L_{am},G_c) \in \mathrm{C}(L_{am},G_c)$ *which also displays the notation of this supremum.*

A recursive algorithm to compute $\sup\mathrm{C}(L_{am},G_c) \in \mathrm{C}(L_{am},G_c)$ is presented in [17] but no explicit formula was given there. For the prefix-closed case, there exists an explicit formula for the supremal controllable sublanguage due to the collection of authors: R.D. Brandt, V. Garg, R. Kumar, F. Lin, S.I. Marcus, W.M. Wonham, see [2]. The time complexity of the algorithm is $O(|Q|^2|Q_a|)$. Another way to view the problem of determining the supremal controllable sublanguage is to consider it as a fixed point of an equation. Then Tarski's fixed point theorem for a poset can be used to prove the existence and to provide an algorithm.

Proposition 3.6. [29, Prop. 3.5.3], [4, pp. 153–155] *Consider the CDFG G_c and a language $L_{am} \subseteq L_m(G)$ as in Assumption 3.1.*

(a) Because L_{am} is $L_m(G)$-closed, so is $\sup\mathrm{C}(L_{am},G_c)$.
(b) $\text{prefix}(\sup\mathrm{C}(L_{am},G_c)) \subseteq \sup\mathrm{C}(\text{prefix}(L_{am}),G_c)$.

3.11 Supremal Supervision

Recall Assumption 3.1 and the notation defined there.

Problem 3.2. *The supremal supervisory control problem with complete observations.* Consider a CDFG G_c and a specification in terms of a sublanguage $L_{rm} = L_{am} \subseteq L_m(G)$ which is $L_m(G)$-closed. Does there exist and, if so, construct a supervisory control $v : L(G) \to E_{cp}$ and a language $K \subseteq E^*$ such that: (a) $K \subseteq L_{am}$; (b) $K = L_m(v/G_c)$; (c) K is supremal with respect to (a-b); (d) v/G_c is nonblocking. A *supremal supervisory control* with respect to G_c and L_{am} is defined to be a supervisory control which satisfies the conditions (a–c) above. A *nonblocking supremal supervisory control* is a supervisory control which satisfies the conditions (a–d) above. ■

The difference between Problem 3.1 and Problem 3.2 is that in the first problem one requires that $K = L_{am}$ while in the second that only $K \subseteq L_{am}$. It follows from Theorem 3.2 as a corollary that there exists a supervisory control v_1 such that $L_m(v_1/G_c) = L_{am}$ if and only if L_{am} is a controllable language with respect to G_c. If the language L_{am} is not controllable with respect to G_c then the supremal controllable language within L_{am} is the best attainable goal. Another way to think of this is to note that the supremal sublanguage $\sup \mathrm{C}(L_{am}, G_c)$ imposes the least-restrictive behavioral constraints on the closed-loop language $L(v/G_c)$.

Theorem 3.4. (Existence of the nonblocking supremal supervisory control with complete observations). [29, Th. 3.5.1] *Consider a CDFG G_c and a language $L_{am} \subseteq L_m(G)$ as in Problem 3.2. Consider* $\sup \mathrm{C}(L_{am}, G_c)$ *which exists because of Proposition 3.5. Assume that* $\sup \mathrm{C}(L_{am}, G_c) \neq \emptyset$. *Then there exists a nonblocking supremal supervisory control v for G_c and L_{am} such that* $L_m(v/G_c) = \sup \mathrm{C}(L_{am}, G_c)$.

Proof. From Proposition 3.4 follows that $\sup \mathrm{C}(L_{am}, G_c) \in \mathrm{C}(L_{am}, G_c)$ hence is controllable with respect to G_c. From Proposition 3.6(a) follows that $\sup \mathrm{C}(L_{am}, G_c)$ is $L_m(G)$-closed. From Theorem 3.2 then follows that there exists a supervisory control v for G_c such that $\sup \mathrm{C}(L_{am}, G_c) = L_m(v/G_c)$, and v/G_c is nonblocking. Moreover, $\sup \mathrm{C}(L_{am}, G_c) \subseteq L_{am}$. Because of the definition of $\sup \mathrm{C}(L_{am}, G_c)$ as the supremal element of $\mathrm{C}(L_{am}, G_c)$ it follows that v is a supremal nonblocking supervisory control. □

Recall that in Problem 3.1 the problem was to determine a supervisory control v for the controlled discrete-event system G_c such that $L_{rm} \subseteq L(v/G_c) \subseteq L_{am}$. That problem can now be solved by the methods of the previous sections.

Proposition 3.7. *Consider Problem 3.1 in which the required L_{rm} and the admissible language L_{am} may be different. There exists a solution of this problem if and only if $L_{rm} \subseteq$* $\sup \mathrm{C}(L_{am}, G_c)$.

In addition to the above problem, there has been formulated the infimal supervision problem. In the corresponding problem one wants to determine the infimal controllable sublanguage in the range $L_{rm} \subseteq K \subseteq L_{am}$ and the corresponding supervisory control. That problem has been solved, the theory corresponds to supremal supervision, and the reader will find it described in the book [4, Ch. 3].

3.12 Further Reading

The reader may want to continue with the following chapters of the collection in which this chapter appears: Chapter 4 *Supervisory Control with Partial Observations* and Chapter 6 *Supervisory Control of Distributed Discrete-Event Systems*. The latter chapter provides the theory for supervisory of complex or large-scale discrete-event systems.

Books and lecture notes on supervisory control with complete observations. The lecture notes of W.M. Wonham, [29], are available on the web and are regularly updated, `http://www.control.utoronto.ca/~wonham/`

The second edition of the book by C. Cassandras and S. Lafortune is [4]. See Chapter 3 for supervisory control with complete observations.

Early papers. An early paper on the topic by P.J. Ramadge and W.M. Wonham is [17]. An expository paper on the supervisory control with complete observations by these authors is [18]. An expository paper published later is by John Thistle, [21]. Early work by S. Lafortune is [5, 12].

Algorithms etc. The main publication on the computation of the supremal controllable sublanguage and the associated supervisory control is the six author paper [2]. Computation of supremal controllable sublanguages is covered by [9].

Supervisory control of nondeterministic systems is treated in [7, 15].

Control of infinite string automata. Büchi automata and other infinite-string automata are described in the paper [25]. A game theoretic approach to control of infinite-string automata is the paper [3]. Control of infinite-string automata is treated by John Thistle, see [20, 23, 24, 27].

Algebra of discrete-event systems. Supervisory control of discrete-event systems is based on concepts and theory of algebra. Books on algebra include [13, 26]. A book on lattice theory is [6] and [10, Section 8.1].

Software. The following software programs and tools are recommended for supervisory control with complete observations:

- DESUMA and UMDES.
  `http://www.mta.ca/giddes/desuma.html`
- IDES. Developed by K. Rudie and co-workers.
  `http://qshare.queensu.ca/Users01/rudie/www.`
- LibFAUDES and DESTool. Developed by T. Moor, K. Schmidt, and co-workers. This package is recommended by the authors of this chapter.
  `http://www.rt.eei.uni-erlangen.de/FGdes/faudes/`
- LTCT, XPTCT, etc. Developed by W.M. Wonham and his students.
  `http://www.control.utoronto.ca/~wonham/`
- STSLib. Developed by Chuan Ma for state tree structures.
  `http://www.control.utoronto.ca/~wonham/`

References

1. Balemi, S., Hoffmann, G.J., Gyugi, P., Wong-Toi, H., Franklin, G.F.: Supervisory control of a rapid thermal processor. IEEE Transactions on Automatic Control 38, 1040–1059 (1993)
2. Brandt, R.D., Garg, V., Kumar, R., Lin, F., Marcus, S.I., Wonham, W.M.: Formulas for calculating supremal controllable and normal sublanguages. Systems & Control Letters 15, 111–117 (1990)
3. Büchi, J.R., Landweber, L.H.: Solving sequential conditions by finite-state strategies. Transactions of the Amererican Mathematical Society 138, 295–311 (1969)
4. Cassandras, C.G., Lafortune, S.: Introduction to Discrete Event Systems, 2nd edn. Springer, New York (2008)
5. Chen, E., Lafortune, S.: Dealing with blocking in supervisory control of discrete event systems. In: Proc. of the 28th IEEE Conference on Decision and Control, New York, USA (1989)
6. Davey, B.A., Priestley, H.A.: Introduction to Lattices and Order. Cambridge University Press, Cambridge (1990)
7. Fabian, M., Lennartson, B.: A class of nondeterministic specification for supervisory control. European Journal of Control 3, 81–90 (1997)
8. Forschelen, S.T.J., van de Mortel-Fronczak, J.M., Su, R., Rooda, J.E.: Application of supervisory control theory to theme park vehicles. In: Proc. 10th Workshop on Discrete Event Systems, Berlin, Germany (2010)
9. Hashtrudi-Zad, S., Kwong, R.H., Wonham, W.M.: Supremum operators and computation of supremal elements in system theory. SIAM Journal of Control and Optimization 37, 695–709 (1999)
10. Jacobson, N.: Basic Algebra, 2nd edn., vol. 1. W.H. Freeman and Company, New York (1985)
11. Kurshan, R.P.: Computer-Aided Verification of Coordinating Processes. Princeton University Press, Princeton (1994)
12. Lafortune, S., Chen, E.: The infimal closed controllable superlanguage and its application in supervisory control. IEEE Transactions on Automatic Control 35, 398–405 (1990)
13. Lawvere, F.W., Roseburgh, R.: Sets for Mathematics. Cambridge University Press, Cambridge (2003)
14. Lee, S.H.: Structural Decentralised Control of Concurrent Discrete-Event Systems. PhD thesis. Australian National University, Canberra, Australia (1998)
15. Overkamp, A.: Supervisory control using failure semantics and partial specifications. IEEE Transactions on Automatic Control 42, 498–510 (1997)
16. Polderman, J.W., Willems, J.C.: Introduction to mathematical system theory–A behavioral approach. Texts in Applied Mathematics, vol. 26. Springer, New York (1997)
17. Ramadge, P.J., Wonham, W.M.: Supervisory control of a class of discrete event processes. SIAM Journal of Control and Optimization 25, 206–230 (1987)
18. Ramadge, P.J., Wonham, W.M.: The control of discrete event systems. Proceedings IEEE 77, 81–98 (1989)
19. Theunissen, R.J.M., Schiffelers, R.R.H., van Beek, D.A., Rooda, J.E.: Supervisory control synthesis for a patient support system. In: Proc. European Control Conference, Budapest, Hungary (2009)
20. Thistle, J.G.: On control of systems modelled as deterministic Rabin automata. Discrete Event Dynamics Systems 5, 357–381 (1995)
21. Thistle, J.G.: Supervisory control of discrete event systems. Mathematical and Computer Modelling 23, 25–53 (1996)

22. Thistle, J.G., Malhamé, R.P., Hoang, H.H., Lafortune, S.: Feature interaction modelling, detection and resolution: A supervisory control approach. In: Feature Interactions in Telecommunications and Distributed Systems IV. IOS Press, Amsterdam (1997)
23. Thistle, J.G., Wonham, W.M.: Control of infinite behavior of finite automata. SIAM Journal of Control and Optimization 32, 1075–1097 (1994)
24. Thistle, J.G., Wonham, W.M.: Supervision of infinite behavior of discrete-event systems. SIAM Journal of Control and Optimization 32, 1098–1113 (1994)
25. Thomas, W.: Automata on infinite objects. In: van Leeuwen (ed.) Handbook of Theoretical Computer Science, vol. B, pp. 133–191. Elsevier Science Publishers, Amsterdam (1990)
26. Thomas, W.: Universal Algebra for Computer Scientists. Springer, Berlin (1992)
27. Wong-Toi, H., Hoffmann, G.: The control of dense real-time discrete event systems. In: Proc. 30th IEEE Conference on Decision and Control, New York (1991)
28. Wonham, W.M.: Linear Multivariable Control: A Geometric Approach. Springer, Berlin (1979)
29. Wonham, W.M.: Supervisory Control of Discrete-Event Systems. University of Toronto (2008)

Chapter 4
Supervisory Control with Partial Observations

Jan Komenda

4.1 Introduction

This chapter presents supervisory control of partially observed discrete-event systems represented as finite automata. In engineering systems it is often not realistic to assume that all events are observable. For instance, some hidden (internal) actions or failures are typically not directly observable. Sometimes it is simply too costly to install the necessary equipment (sensors) to observe all events that are occurring in the system. The partial observations are not only due to lack of some of the sensors but due to economic reasons, it is too costly to install the sensors.

Therefore, supervisory control theory has been extended to deal with systems, where only a part of the events that occur in the system is observable. Two independent papers were the first to treat the important problem of supervisory synthesis with partial observations: [6] and [10].

A closed-loop system under partial observations is defined using a supervisory feedback map that specifies the enabled actions (events) after a string of (observable) events has been observed.

Due to partial observations strings that are not distinguishable by the observations require the same control action. There are two possible control laws that satisfy this requirement: one is called permissive and the other antipermissive.

Partial observation about the state of the system is encoded by the observer automaton defined over the observable alphabet. These are helpful for implementing supervisory control laws and for many algorithms solving supervisory control problems.

An additional property, called observability, is required of a specification language to be achievable as the language of the closed-loop system (no matter which

Jan Komenda
Institute of Mathematics, Academy of Sciences of the Czech Republic,
Žižkova 22, 616 62 Brno, Czech Republic
e-mail: komenda@math.cas.cz

C. Seatzu et al. (Eds.): Control of Discrete-Event Systems, LNCIS 433, pp. 65–84.
springerlink.com

of the two policies is chosen) in addition to controllability and relative closedness conditions (known as $L_m(G)$-closedness) needed in the case of complete observations. Basic facts about observability will be presented together with the main existential result, the existence theorem of a supervisor in terms of observability, and controllability conditions. Since observability is not in general preserved by language unions, a stronger notion called normality of a language will be described.

Finally, algorithms for checking observability and normality, and for computing supremal normal sublanguages and computing supremal normal and controllable sublanguages will be presented. The algorithm for checking observability will be of polynomial complexity, while the ones for constructing supremal sublanguages are of exponential complexity. This is because they require an observer automaton. Moreover, the representation of the specification language must have a special property called the state-partition automaton. Such a representation requires that states with indistinguishable past are partially disambiguated in the sense that different states of the observer do not overlap. For any finite automaton such a state-partition automaton always exists as a finite automaton.

Both the language-based (behavioral) [17] and the state-based based framework [14] have been developed for supervisory control with partial observations. Since both frameworks are equivalent, the state-based approach does not add much to the originally developed language framework and there exist bidirectional transformations among them, only the language framework is presented in this chapter.

The concepts and results presented in this chapter will be illustrated by simple examples.

4.2 Concepts of Supervisory Control with Partial Observations

The notation for (deterministic) generators $G = (Q,E,f,q_o,Q_m)$ and controlled generators (CDFG G_c) is the same as in the previous chapter on supervisory control with complete observations. In addition, it is assumed that the set of events E is partitioned into the disjoint union of observable events (denoted E_o) and unobservable events (denoted E_{uo}), i.e. $E = E_o \cup E_{uo}$. The concept of *(natural) projection* is associated with the observable event subset as the morphism of monoids $P : E^* \to E_o^*$ defined by

$$\begin{aligned} P(\varepsilon) &= \varepsilon \text{ and} \\ P(se) &= \begin{cases} P(s)e, & \text{if } e \in E_o, \\ P(s), & \text{if } e \in E_{uo}. \end{cases} \end{aligned}$$

The morphism property ensures that natural projection is catenative, i.e. $P(uv) = P(u)P(v)$ for any $u, v \in E^*$. The projection is extended to languages in a natural way, for $L \subseteq E^*$: $P(L) = \bigcup_{w \in L} P(w)$.

Given a set Q we use the notation $Pwr(Q)$ to denote the set of all subsets of Q including the empty set. Since the projection is not injective, the *inverse projection*

$P^{-1}: E_o^* \to Pwr(E^*)$ of P takes values on $Pwr(E^*)$ and is defined as $P^{-1}(w) = \{s \in E^* \mid P(s) = w\}$, where $w \in E_o^*$ is an observed sequence of events and $P^{-1}(w)$ is the set of all possible sequences of the plant events that yield the observation w. The following well-known properties of natural projections [3] are useful.

Lemma 4.1. *Natural projection has the following properties:*

1. *Both projections and inverse projections are morphisms for language unions:*

$$P(\cup_{i\in I} L_i) = \cup_{i\in I} P(L_i) \tag{4.1}$$
$$P^{-1}(\cup_{i\in I} L_i) = \cup_{i\in I} P^{-1}(L_i) \tag{4.2}$$

2. *For language intersections it holds that:*

$$P(\cap_{i\in I} L_i) \subseteq \cap_{i\in I} P(L_i) \tag{4.3}$$
$$P^{-1}(\cap_{i\in I} L_i) = \cap_{i\in I} P^{-1}(L_i) \tag{4.4}$$

Note that the converse inclusion in 4.3 does not hold in general.

It appears that partial observations cause a sort of nondeterminism, because after a given string of observed events $w \in E_o^*$ the generator can in general be in several possible states. An important construction is then the observer automaton that enables effective implementation using finite automata notions and concepts of supervisory control with partial observations. It is the reachable part of the so-called projected automaton defined below using a modified subset construction known from determinization of a nondeterministic automaton.

4.2.1 Observer Automaton

Let $G = (Q, E, f, q_o, Q_m)$ be a generator with partial observations $E = E_o \cup E_{uo}$. The notation $Pwr^+(Q)$ is reserved for the set of all nonempty subsets of Q, i.e. $B \in Pwr^+(Q)$ means that $B \subseteq Q :\ B \neq \emptyset$. Additional notation is needed for the unobservable reach set, which can be as an instance of reach set defined in the previous chapter, but with the event set restricted to the unobservable one. We define for $B \in Pwr^+(Q)$

$$UnobsReach(B) = \{q' \in Q \mid \exists q \in B \text{ and } \exists w \in E_{uo}^* \text{ with } f(q,w) = q'\}.$$

In words, for a considered nonempty subset B of states of G, the unobservable reach set $UnobsReach(B)$ is the set of states that are reachable from a state of B by a sequence of unobservable events.

Definition 4.1 (Projected generator). *The projected generator over the observable alphabet E_o is:*

$P(G) = (Pwr^+(Q), E_o, f_P, UnobsReach(q_o), Q_P^m)$, where
for $e \in E_o$ and $B \in Pwr^+(Q)$ we have

$f_P(B,e) = UnobsReach(q \in Q : (\exists x \in B)\ s.\ t.\ q = f(x,e))$ *and*
$Q_P^m = \{X \subseteq Q : X \cap Q_m \neq \emptyset\}$.

Note that $f_P(B,e) = \cup_{\{q\in B\}\times\{w\in E^*|P(w)=e\}} f(q,w)$.

Definition 4.2 (Observer). *The observer is a generator over E_o that is defined as the reachable part (subautomaton) of $P(G)$, i.e.*

$Obs(G) = (Q_{obs}, E_o, f_{Obs}, UnobsReach(q_o), Q_{Obs}^m)$ *with* $Q_{obs} \subseteq Pwr^+(Q)$ *defined inductively as follows:*
1) $UnobsReach(q_o) \in Q_{obs}$
2) If $Q \in Q_{obs}$ *then* $\forall e \in E_o$: $f_P(Q,e) \in Q_{obs}$.
The transition structure of the observer automaton is simply the structure of $P(G)$ restricted to Q_{obs}, i.e. $\forall e \in E_o$ and $\forall Q \in Q_{obs}$: $f_{obs}(Q,e) = f_P(Q,e)$ and $Q_{Obs}^m = \{X \in Q_{obs} : X \cap Q_m \neq \emptyset\}$

Note that if $f(x,w)$ is not defined for any $x \in B$ and $w \in E^* : P(w) = e$, then $f_{obs}(B,e)$ is not defined.

Finally, let us mention that an alternative, equivalent, and well-known construction of the observer consists in replacing all events from E_{uo} by ε and building the corresponding deterministic automaton (using ε- removal).

Let us denote by $\|Q\|$ the number of reachable states of an automaton G. A major problem with the observer automaton is that the number of reachable states, denoted $\|Q_{obs}\|$, can be exponential, up to $c.2^{\|Q\|}$ with a positive constant $c \in \mathbb{R}_+$. In special cases, e.g. when P has the so-called observer property (see the definition below) with respect to $L(G)$, it is known that the observer has at most as many states as the original automaton, i.e. $\|Q_{obs}\| \leq \|Q\|$.

Definition 4.3 (Observer property). *The projection $P : E^* \to E_o^*$ is an L-*observer *for a language $L \subseteq E^*$ if the following holds: for all strings $s \in$* prefix(L), *if $P(s)t \in P(L)$, then there exists $u \in E^*$ such that $su \in L$ and $P(u) = t$, see Fig. 4.1.*

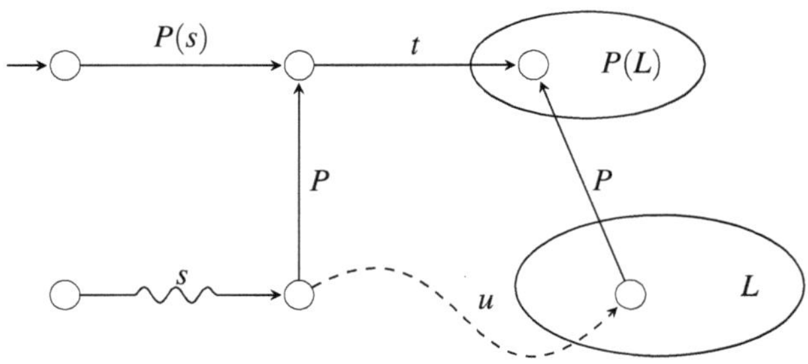

Fig. 4.1 Illustration of the observer property

It is well known that the projected generator and the observer automata recognize the same language, namely the projected plant language [3].

Theorem 4.1. *Both $P(G)$ and $Obs(G)$ recognize $P(L(G))$, i.e.*

$L_m(P(G)) = L_m(Obs(G)) = P(L_m(G))$, *and* $L(P(G)) = L(Obs(G)) = P(L(G))$.

The states of the observer automaton represent the states, in which the original automaton can be after a string with a given projection has occurred. It is seen from Example 4.1 below that there exist different states of the observer automaton $Obs(G)$ that have nonempty intersections (as subset of states of G). There is an auxiliary concept of observational indistinguishability relation on states of G induced by partial observations. It is defined as follows: two states $q, q' \in Q$ are said to be observationally indistinguishable (denoted $\langle q, q' \rangle \in Aux(G)$) if they can be reached by two indistinguishable strings, i.e. if there exist two strings $s, s' \in E^*$ with $P(s) = P(s')$ such that $f(q_0, s) = q$ and $f(q_0, s') = q'$. This relation is reflexive and symmetric, but it is not in general an equivalence relation, because it might be non transitive. Such a relation is called a tolerance relation. Automata representations, where observational indistinguishability is an equivalence relation are very important for most of the algorithms presented in this chapter. Essentially, they enable to disambiguate partly the ambiguities (nondeterminism) caused by partial observations, which helps in proving correctness of many algorithms in supervisory control of partial observations.

There is a condition called state-partition automaton or $M-$recognizability for G, see [4], to ensure that $Aux(G)$ is an equivalence relation. It has been shown in [9] that if G is a state-partition automaton then $Aux(G)$ is an equivalence relation.

Definition 4.4 (State-partition automaton). $G = (Q, E, f, q_o, Q_m)$ is called a *state-partition automaton* if different states of its observer $Obs(G)$ are non overlapping, i.e. $\forall B \neq B' \in Q_{obs} : B \cap B' = \emptyset$.

Below is an example of an automaton that does not satisfy the above condition.

Example 4.1. Let us consider the example from Fig. 4.2 with $E_o = \{a\}$ and $E_{uo} = \{\tau\}$. The problem is that there are two different states of the observer automaton, namely the initial state $\{0, 2\}$ and the state $\{0, 1, 2, 3\}$, that have nonempty intersection and still form two different states in the observer. ■

A natural question that arises for non-state-partition automata is how to compute an automaton that recognizes the same language and is a state-partition automaton.

The standard construction for the computation of a state-partition automaton is known as the *Schützenberger covering*. It has been introduced in the more general setting of automata with multiplicities in semirings, cf. [13]. For the special case of logical automata that we consider this construction amounts to computing the synchronous product of the original automaton with its observer.

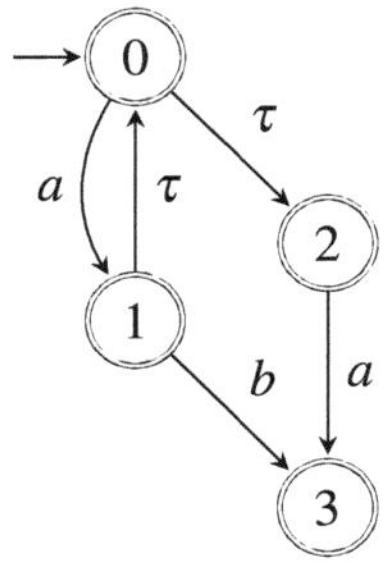

Fig. 4.2 A generator G

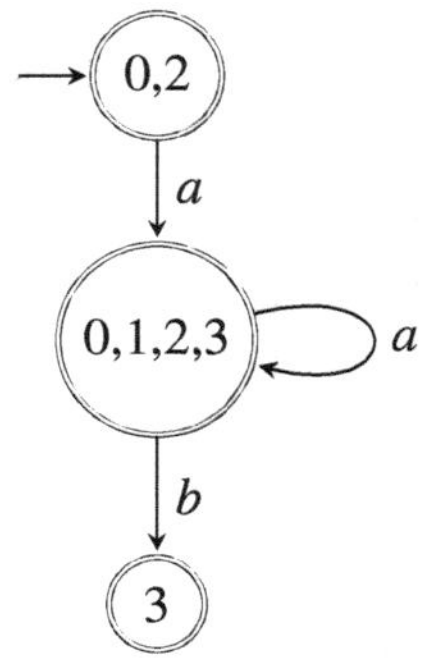

Fig. 4.3 Its observer $Obs(G)$

For generators $G_1 = (Q_1, E_1, f_1, q_{0,1}, Q_{m,1})$ and $G_2 = (Q_2, E_2, f_2, q_{0,2}, Q_{m,2})$, their synchronous product $G_1 \| G_2$ is defined as the accessible part of the generator $(Q_1 \times Q_2, E_1 \cup E_2, f, (q_{0,1}, q_{0,2}), Q_{m,1} \times Q_{m,2})$, where

$$f((x,y),e) = \begin{cases} (f_1(x,e), f_2(y,e)), & \text{if } f_1(x,e) \in Q_1 \text{ and } f_2(y,e) \in Q_2; \\ (f_1(x,e), y), & \text{if } f_1(x,e) \in Q_1 \text{ and } e \notin E_2; \\ (x, f_2(y,e)), & \text{if } e \notin E_1 \text{ and } f_2(y,e) \in Q_2; \\ \text{undefined}, & \text{otherwise.} \end{cases}$$

Theorem 4.2 (Schützenberger covering). *For any* $G = (Q, E, f, q_o, Q_m)$*:* $G \| Obs(G)$ *is always a state-partition automaton .*

A proof of this result for finite state automata can be found in [4].

The construction of a state-partition automaton is now illustrated for Example 4.1.

Example 4.2. The corresponding observer that satisfies the state-partition automaton condition is on the right of Fig. 4.5. ∎

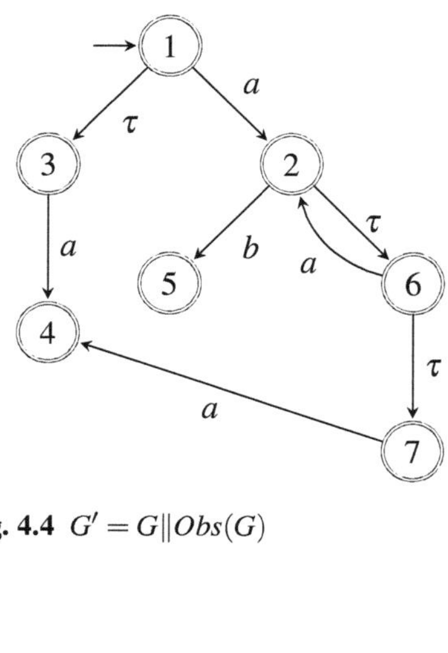

Fig. 4.4 $G' = G \| Obs(G)$

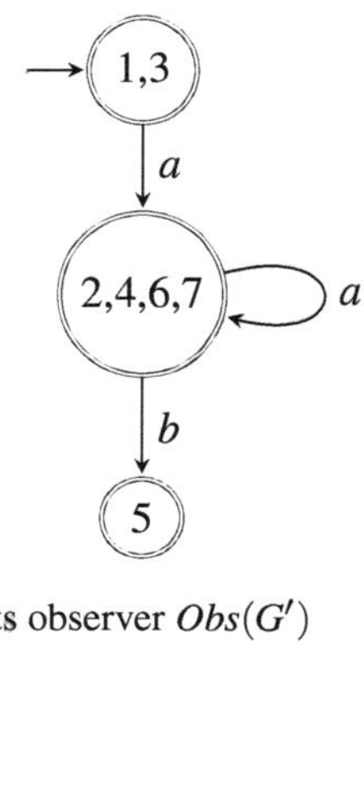

Fig. 4.5 Its observer $Obs(G')$

4.2.2 Supervisor with Partial Observations

A *controlled generator (with partial observation)* is a structure $G_c = (G, E_c, P, E_{cp})$, where G is a generator, $E_c \subseteq E$ is the set of *controllable events*, $E_u = E \setminus E_c$ is the set of *uncontrollable events*, $P : E^* \to E_o^*$ is the projection, and $E_{cp} = \{E_{en} \subseteq E \mid E_u \subseteq E_{en}\}$ is the *set of control patterns* that consists of all possible sets of enabled events E_{en}. Note that the set of enabled events must always include all uncontrollable events, because these events cannot be disabled by the supervisor. In concrete plants only feasible events that can actually occur in the plant are enabled. The definition of controlled generator is almost the same as in the case of complete observations (only the natural projection is added). In particular, any control pattern as a subset of enabled event should always contain all uncontrollable events which, by definition, cannot be prevented from happening.

The definition of a supervisory control law changes, because due to unobservable events the supervisor has only access to the projected behavior, whence the following definition. Note that a supervisor S is not formally defined, because it is an automaton representation of the supervisory control law v_P. Moreover, for supervisors that satisfy certain properties (essentially their language is controllable and observable) the closed-loop system $S/G = S \| G$. However, in this chapter we only provide a language definition of the closed-loop system.

Definition 4.5 (Supervisory control with partial observations). *A supervisory control with partial observations for a controlled generator G_c is a map $v_P : P(L(G)) \to E_{cp}$. In words, for any observed string $s \in P(L(G)) \subseteq E_o^*$, $v_P(s)$ specifies corresponding control patterns, i.e. the subset of the enabled events.*

The definition by induction of the closed-loop system language $L(S/G)$ under partial observations is then the same as in the previous chapter with v replaced by v_P. It is defined as the smallest language $L(S/G) \subseteq E^*$ which satisfies

1. $\varepsilon \in L(S/G)$,
2. if $s \in L(S/G)$, $se \in L(G)$, and $e \in v_P(P(s))$, then $se \in L(S/G)$.

The marked language is again defined as $L_m(S/G) = L(S/G) \cap L_m(G)$.

4.3 Existence of Supervisors

Given a plant automaton $G = (Q, E, f, q_o, Q_m)$ and a specification language $K \subseteq E^*$, the problem is to determine whether there exists a supervisor S such that $L(S/G) = \text{prefix}(K)$ and $L_m(S/G) = K$. Note that this means that the marked specification is achieved in a nonblocking way (closed-loop system is then nonblocking: $\text{prefix}(L_m(S/G)) = L(S/G)$.) It is not surprising that similarly as in the case of complete observations not every language can be achieved as the behavior of the closed-loop system. Indeed, it is sufficient to consider a specification which requires for two strings of the plant with the same projection that one is included in the specification and the other is not. From the definition of the supervisor with partial observations such a specification can never be matched as the behavior of the closed-loop system. Another condition, called observability, is needed in addition to controllability. It is called observability in the literature, but it has not much to do with observability for linear or nonlinear discrete-time and/or continuous-time systems. It is not dual to controllability, but it may be best viewed as a complementary condition needed for achievability of the specification in addition to controllability in the presence of partial observations. Note that it is not restrictive to assume that $\text{prefix}(K) \subseteq L$, because for a specification language that is not included in the plant language, it suffices to consider the intersection $K \cap L$ as a new specification that satisfies this assumption.

Definition 4.6. *Let $K \subseteq E^*$ be a specification language and $L = L(G)$ be a prefix-closed plant language over an event set E. Let $E_c \subseteq E$ be the subset of controllable events, and let $E_o \subseteq E$ be the set of* observable *events with P as the corresponding projection from E^* to E_o^*. The specification language K is said to be* observable *with respect to L, E_c, and P if for all $s \in \text{prefix}(K)$ and all $\sigma \in E_c$,*

$$(s\sigma \notin \text{prefix}(K)) \textit{ and } (s\sigma \in L) \Rightarrow P^{-1}[P(s)]\sigma \cap \text{prefix}(K) = \emptyset.$$

There is the following equivalent formulation. K is said to be *observable* with respect to L, E_c, and P if for all $s, s' \in \text{prefix}(K)$ and all $\sigma \in E_c$ such that $s\sigma \in L$, $s'\sigma \in \text{prefix}(K)$, and $P(s) = P(s')$ we have $s\sigma \in \text{prefix}(K)$.

Note that observability requires that if there is a specification string that can be extended by a controllable event within the plant, but the continuation by this event exits the specification then all observationally indistinguishable strings followed by this event must be outside the specification as well. Thus, strings $s' \in \text{prefix}(K) : P(s') = P(s)$ must satisfy $s'\sigma \notin \text{prefix}(K)$. We should point out here that the original definition of observability from [10] does not require the event σ to be controllable. However, from an application viewpoint both the controllability and the observability of a language are needed in a DES with partial observations, and by the controllability condition introduced in the previous chapter one cannot leave a specification language by an uncontrollable event while remaining within the plant language. Hence, it is reasonable to consider only extension by controllable events in the definition of language observability.

It is easily seen that observability of a language is a property of its prefix closure, meaning that K is observable with respect to L, E_c, and P if and only if $\text{prefix}(K)$ is observable with respect to L, E_c, and P.

Example 4.3. Let us consider the following specification and plant languages.

The only unobservable event is τ, i.e. $E_{uo} = \{\tau\}$, while all other events are observable. We assume that all events are controllable, i.e. $E_c = E$. We can see that specification language K is not observable with respect to the plant language L and projection P, because there are strings $s = a$ and $s' = \tau a$ that violate the definition of observability: $s\tau = a\tau \in L \setminus \text{prefix}(K)$, while $s'\tau = \tau a\tau \in \text{prefix}(K)$ and $P(s) = P(s')$. ■

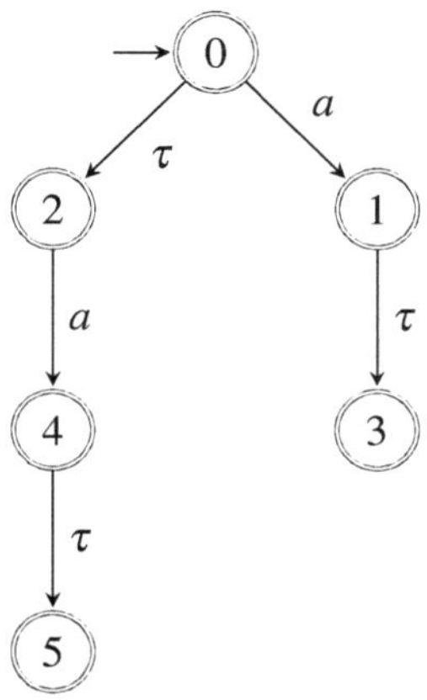

Fig. 4.6 Generator for L with $E_{uo} = \{\tau\}$

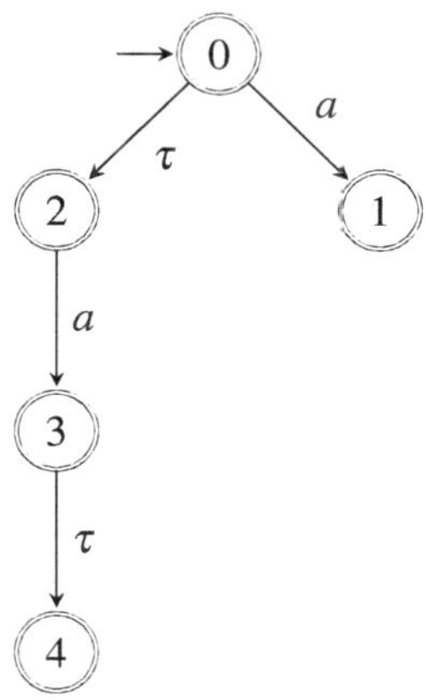

Fig. 4.7 Generator for specification K

The main existential theorem, known as the controllability and the observability theorem, follows.

Theorem 4.3. [F. Lin and W.M. Wonham, [10]] *There exists a nonblocking supervisory controller with partial observations S such that $L_m(S/G) = K$ and $L(S/G) =$* prefix(K) *if and only if*

(i) *K is controllable with respect to $L(G)$ and E_u,*
(ii) *K is observable with respect to $L(G)$ and P, and*
(iii) *$K =$ prefix$(K) \cap L_m(G)$ (K is $L_m(G)-$closed).*

Proof. ($\Leftarrow$)(Sufficiency) Let $K \subseteq L$ be observable with respect to L and P, controllable with respect to $L(G)$ and E_u, and $L_m(G)-$closed. Define the supervisory control for $s \in P(L(G))$,

$$\nu_P : P(L(G)) \to E_{cp}, \ with$$

$$\nu_P(s) = E_u \cup \{e \in E_c : \ \exists s' \in \text{prefix}(K) \text{ with } P(s') = P(s) \text{ and } s'e \in \text{prefix}(K)\}.$$

Then we show by induction that $L(S/G) = \text{prefix}(K)$. The basic case is for strings of length 0 : by respective definitions of prefix closure and of closed-loop systems we have $\varepsilon \in L(S/G)$ as well as $\varepsilon \in \text{prefix}(K)$. The induction hypothesis is that for all strings s such that $|s| \leq n$, $s \in L(S/G)$ iff $s \in \text{prefix}(K)$. Let $|s| = n$ and consider se for $e \in E$. Both inclusions are easy to show:

"$\subseteq$": If $se \in L(S/G)$ then $s \in L(S/G)$, $e \in \nu_P(P(s))$ and $se \in L(G)$. Hence $s \in \text{prefix}(K)$ using the induction hypothesis. For $e \in E_u$ controllability implies $se \in \text{prefix}(K)$.
For $e \in E_c$, by definition of ν_P there exists $s' \in \text{prefix}(K)$ with $P(s') = P(s)$ and $s'e \in \text{prefix}(K)$.
Since $s \in L(G)$, observability of K implies that $se \in \text{prefix}(K)$.

"$\supseteq$": If $se \in \text{prefix}(K)$ then $se \in L(G)$ since $\text{prefix}(K) \subseteq \text{prefix}(L_m(G) \subseteq L(G)$.
If $e \in E_u$, then $e \in \nu_P(P(s))$ by the definition of ν_P (admissibility).
If $e \in E_c$, then by definition of ν_P, $se \in \text{prefix}(K)$ implies $e \in \nu_P(P(s))$.
Overall, $s \in \text{prefix}(K)$, $e \in \nu_P(P(s))$, and $se \in L(G)$, by the induction hypothesis $s \in L(S/G)$, $e \in \nu_P(P(s))$, and $se \in L(G)$. Therefore, $se \in L(S/G)$.
The rest of the proof is similar to the proof of controllability theorem from complete observations case. Namely, $L_m(G)-$closure of $\text{prefix}(K)$ will imply $L_m(S/G) = K$, i.e. that S is nonblocking.

($\Rightarrow$)(Necessity) Let $L(S/G) = \text{prefix}(K)$. We prove that observability holds.
Let $s, s' \in \text{prefix}(K)$ and $e \in E_c$:
$se \in L$, $s'e \in \text{prefix}(K)$, and $P(s) = P(s')$.
We have $s, s', s'e \in L(S/G)$. Since $P(s) = P(s')$ we have $\nu_P(P(s)) = \nu_P(P(s'))$.
Hence, $se \in L(S/G) = \text{prefix}(K)$ as well, which proves the observability condition.
Controllability and $L_m(G)-$closedness can be proved in the same way as in the case of complete observations. □

In the proof of the controllability and observability theorem, the supervisor map defined in the sufficiency part is called the permissive control policy. It is defined for arbitrary specification languages, not necessarily observable and controllable ones. Note that $\text{prefix}(K) \subseteq L(S/G)$ always holds if S/G is computed using permissive policy. This is because events can be enabled that exit the prefix of the specification language whenever this event is uncontrollable or there exists at least one string with the same projection that can be extended within $\text{prefix}(K)$, while the extension of a given string itself might exit $\text{prefix}(K)$. A natural counterpart of the permissive control policy is the so-called antipermissive control policy. Here, it is required for an event to be enabled after a given observation that all distinguishable strings from the system (plant) yielding the given observation can be continued by this event within $\text{prefix}(K)$, cf. the formula below.

$$\nu_A(s) = E_u \cup \{e \in E_c : \forall s' \in E^* : P(s') = P(s) : s'e \in L \Rightarrow s'e \in \text{prefix}(K)\}.$$

It should be noted that in the proof of the controllability and the observability theorem, we could have replaced the permissive control policy by the antipermissive one without affecting the result, i.e. there is a unique notion of observability that is common for achievability of a given language as the closed-loop language under both antipermissive and permissive control policies. This is a significant property of the centralized control under partial observations as opposed to the decentralized control studied in the following chapters.

Returning to Example 4.3 we have:
$\nu_P(a) = E_{uc} \cup \{e \in E : \exists s' \in \text{prefix}(K) \text{ with } P(s') = P(s) \text{ and } s'e \in \text{prefix}(K)\} = \{\tau\}$, while
$\nu_A(a) = E_{uc} \cup \{e \in E : \forall s' \in E^* : P(s') = P(s) : s'e \in L \Rightarrow s'e \in \text{prefix}(K)\} = \emptyset$!

In accordance with the controllability and the observability theorem the closed-loop languages under both policies depicted in Figs. 4.8 and 4.9 are different from the specification which is not observable. Note that since $\tau \in \nu_P(a)$, the event τ is enabled in all states of the plant that correspond to observation a, namely in states 1

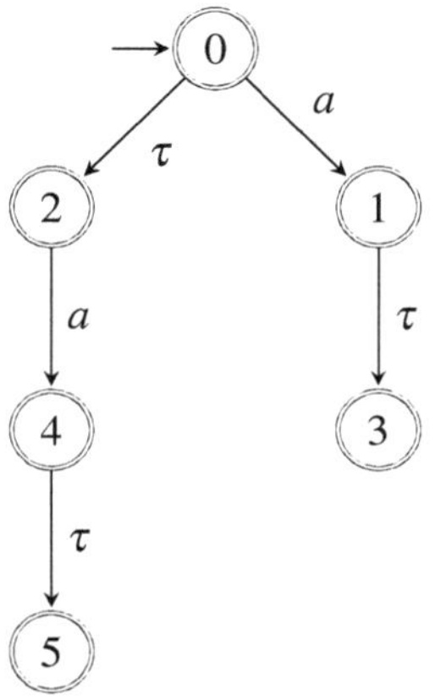

Fig. 4.8 Generator for $L(S/G)$: permissive policy

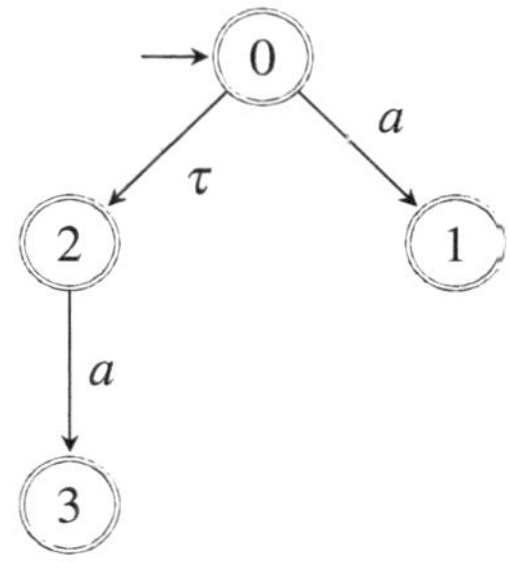

Fig. 4.9 Generator for $L(S/_A G)$: antipermissive policy

and 4. Similarly, as $\tau \notin v_A(a)$, the event τ is disabled in all states that correspond to observation a, namely in states 1 and 3 as is easily seen from Fig. 4.8.

4.4 Algorithm for Verification of Observability and Automata Implementation of Supervisors

In the last section existential results have been presented, namely necessary and sufficient conditions for the specification language to be achievable as the language of the closed-loop system. It remains to investigate whether observability can effectively be verified and what is the computational complexity of checking it.

The time complexity of checking observability of the specification language K with respect to the plant language L and projection P is $O(n_K^2 n_L)$ [3], where n_K is the number of states of recognizer of K and n_L is the number of states of recognizer of L. In fact, originally a different algorithm for checking observability has been known, based on the observer construction, but it suffers from exponential worst case complexity due to the construction of the observer automaton. It is known already from [15] that a test of observability can be done in polynomial time. There are two similar polynomial-time algorithms. One of them is based on nondeterministic automaton known as simplified M-machine [12], that has been introduced for checking co-observability in the decentralized supervisory control. The second one is a deterministic variant of this algorithm that uses a special alphabet composed of event triples. We present here the variant of the polynomial algorithm that has appeared in [3]. The construction of a deterministic automaton below will capture all violations of observability. It is a variant of the M-machine, where instead of having two copies of the specification and one copy of the plant with nondeterministic transition function, a special alphabet of event triples is used, and the test automaton remains deterministic.

The polynomial algorithm for checking observability is presented below.

Input: plant generator $G = (Q, E, f, q_o, Q_m)$: $L(G) = L$ and generator for the specification $H = (X, E, f_s, x_0, X_m)$: $L(H) = \text{prefix}(K)$.

Let $E_\varepsilon = E \cup \{\varepsilon\}$. The generator $OTest(H,G)$ for testing observability of $L(H)$ with respect to $L(G)$ and P is constructed below. The notation $Ac(.)$ refers to the accessible part of the associated automaton.

Algorithm 4.4

1. $OTest(H,G) := Ac((X \times X \times Q) \cup \{fail\}, E_\varepsilon \times E_\varepsilon \times E_\varepsilon, f_{test}, (x_0, x_0, q_0), \{fail\})$
2. $f_{test} : [X \times X \times Q \cup \{fail\}] \times E_\varepsilon \times E_\varepsilon \times E_\varepsilon \to [X \times X \times Q \cup \{fail\}]$
3. (a) For $a \in E_o$:

$$f_{test}((x_1,x_2,x_3),(a,a,a)) = (f_s(x_1,a), f_s(x_2,a), f(x_3,a))$$
$$f_{test}((x_1,x_2,x_3),(a,\varepsilon,a)) = fail \text{ if Condition } C \text{ holds.}$$

An auxiliary condition C is used. It is defined as follows.
C: $(a \in E_c)$ *and* $f_s(x_1,a)!$ *and (not* $f_s(x_2,a)!$*) and* $f(x_3,a)!$*. (b) For* $a \in E_{uo}$:

$$f_{test}((x_1,x_2,x_3),(a,\varepsilon,\varepsilon)) = (f_s(x_1,a), x_2, x_3)$$
$$f_{test}((x_1,x_2,x_3),(\varepsilon,a,a)) = (x_1, f_s(x_2,a), f(x_3,a))$$
$$f_{test}((x_1,x_2,x_3),(a,\varepsilon,a)) = fail \text{ if Condition } C \text{ holds.}$$

The idea of this algorithm is that the test automaton captures all violations of observability using marked state " fail": $s,s' \in \text{prefix}(K) : P(s) = P(s')$ and $sa \in \text{prefix}(K) \cap L$ while $s'a \notin \text{prefix}(K)$.

Note that $(x_1,x_2,x_3) \in X \times X \times Q$ is reachable by a string (s_1,s_2,s_3) iff $f_s(x_0,s_1) = x_1$, $f_s(x_0,s_2) = x_2$, $f(q_0,s_3) = x_3$, $s_i \in \text{prefix}(K)$, $i = 1,2,3$, $P(s_1) = P(s_2)$, and $s_2 = s_3$.

The theorem below states that the specification is not observable if there exists a string leading to the only marked state fail and otherwise it is observable.

Theorem 4.4. *(Polynomial test for observability)* $K = L(H)$ *is observable w.r.t.* $L = L(G)$ *and* P *if and only if* $L_m(OTest(H,G)) = \emptyset$.

Proof. Let (s_1,s_2,s_2) be any triple of strings such that $f_{test}((x_0,x_0,q_0),(s_1,s_2,s_2)) = (x_1,x_2,x_3)$.

If $f_{test}((x_1,x_2,x_3),(a,\varepsilon,a)) = fail$ then ($s_2 \in \text{prefix}(K)$, $a \in E_c$: $s_2a \in \text{prefix}(K)$, $s_2a \in L$, $P(s_1) = P(s_2)$) and ($s_1a \notin \text{prefix}(K)!$) witnesses violation of observability.

Conversely, any $s_1,s_2 \in \text{prefix}(K), a \in E_c : s_2a \in \text{prefix}(K)$, $s_2a \in L$, $P(s_1) = P(s_2)$ and ($s_1a \notin \text{prefix}(K)!$) correspond to having state

$(x_1,x_2,x_3) = f_{test}((x_0,x_0,q_0),(s_1,s_2,s_2))$ with $f_s(x_1,a)!$, not $f_s(x_2,a)!$,

and $f(x_3,a)!$. Hence, state fail is reachable. □

Yet another related problem is how to efficiently implement the supervisory control v_P with partial observations using automata. It would be interesting to read the control actions directly from automata representations as in the case of deterministic systems with complete observations, where the supervisor can be viewed as the generator of the specification or of its supremal controllable sublanguage and control actions $v(s)$ are given by the active event set of the supervisor automaton after the string s has been read.

Unlike the complete observations case it is not possible to read the control actions directly from automata representations (generators) of the plant and of the specification languages. In addition, the observer automaton of the specification is needed. Effective computation of a supervisor with partial observations requires the three generators for the plant, the specification and the observer of the specification. Formally, let $H = (X, E, f_s, x_0)$ be a generator of K and $Obs(H) = (X_{obs}, E_o, f_{obs}, UnobsReach(x_o))$ be its observer. Then for $w \in E_o^*$: $Obs(H)$ is in the state $f_{obs}(UnobsReach(x_o), w) \subseteq X$ meaning that after observing w the specification automaton H is in one of the states from this set.

Put $v_P(w) = \cup_{x \in f_{obs}(UnobsReach(x_o),w)} A_H(x)$, where $A_H(x)$ is the active event set at state x of H, i.e. $A_H(x) = \{a \in E : f(x,a)!\}$.

Note that only feasible uncontrollable events are enabled, namely those occurring in the plant generator G. We see that the situation is more complicated than in the case of complete observations, where the product automaton of the specification and the plant automata serve as a suitable supervisor.

Once the supervisor S under partial observation is given, the corresponding supervisory control $v_P : P(L(G)) \to E_{cp}$ is constructed as above using its observer and union of active event sets of all states S that form a given observer state.

4.5 General Case of Unobservable Specifications

In the case, where the specification language is either not controllable or not observable, a controllable and observable sublanguage of the specification has to be found. In the previous chapter it has been established that the supremal controllable sublanguage of a given language as the union of controllable sublanguages always exists.

Unfortunately, observability is not preserved by language unions, unlike controllability. In order to see this, it suffices to consider the plant language $L = \{\varepsilon, \tau, a, \tau a\}$ with $E_o = \{a\}$. Then both $K_1 = \{\varepsilon, \tau\}$ and $K_2 = \{\varepsilon, a\}$ are observable with respect to L and $P : E^* \to E_o^*$, but $K_1 \cup K_2$ is not observable with respect to L and P.

Consequently, the supremal observable sublanguage does not always exist.

If specification K fails to satisfy controllability, observability, and/or $L_m(G)$-closedness, a search for a sublanguage satisfying these conditions should be performed [16]. There are only maximal observable sublanguages, which are not unique in general, cf. [5] or [16]. Therefore, a stronger property, called normality, has been introduced.

Definition 4.7. *$K \subseteq E^*$ is said to be* normal *with respect to $L = \text{prefix}(L)$ and P if* $\text{prefix}(K) = P^{-1}[P(\text{prefix}(K))] \cap L$.

According to the definition, normality (as well as observability) of a language is a property of its prefix closure. However, the original definition of normality that can be found in the literature does not use the prefix-closure.

Normality is conceptually simpler than observability, because it requires that two indistinguishable strings are compatible with respect to the membership in the prefix-closed specification and plant languages. Namely, if $s, s' \in L$ are such that $P(s) = P(s')$ and $s \in \text{prefix}(K)$ then $s' \in \text{prefix}(K)$ as well. Since the inclusion $\text{prefix}(K) \subseteq P^{-1}[P(\text{prefix}(K))] \cap L$ is always satisfied for $\text{prefix}(K) \subseteq L$, the inclusion $P^{-1}[P(\text{prefix}(K))] \cap L \subseteq \text{prefix}(K)$, characterized by the string condition above, is in fact equivalent to normality.

It is easy to show that normality implies observability and moreover it coincides with observability in the case when all controllable events are observable, see [3] and [10].

Proposition 4.1. *Language normality implies observability. Conversely, if $E_c \subseteq E_o$, K is observable with respect to L and P, and controllable with respect to L and E_u, then K is normal with respect to L and P.*

Proof. Let K be normal with respect to L and P and let $s, s' \in \text{prefix}(K)$ be such that $P(s) = P(s')$, $e \in E_c$, $se \in L$, and $s'e \in \text{prefix}(K)$. Then $P(se) = P(s'e)$, hence $se \in P^{-1}[P(\text{prefix}(K))] \cap L = \text{prefix}(K)$ as well by normality. Now, let $E_c \subseteq E_o$ and K be observable with respect to L and P. We show by contradiction that K must be normal with respect to L and P. If there are $s, s' \in L$ with $P(s) = P(s')$ such that $s \in \text{prefix}(K)$ and $s' \notin \text{prefix}(K)$, then let t' be the longest prefix of s' that is still in $\text{prefix}(K)$. Clearly, such a t' exists, because $\varepsilon \in \text{prefix}(K)$. Hence, $s' = t'u'$ with $u' = u'_1 \dots u'_k$ and $t'u'_1 \notin \text{prefix}(K)$. Now there are two possibilities: either $u'_1 \in E_o$ or $u'_1 \in E_{uo}$. If $u'_1 \in E_{uo}$ then from $E_c \subseteq E_o$ we have $u'_1 \in E_u$, hence by controllability of K it cannot happen that $t'u'_1 \notin \text{prefix}(K)$. Now, let $u'_1 \in E_o$. Since $P(s) = P(s')$ and t' is the prefix of s', there must exist the longest prefix t of s such that $P(t) = P(t')$. We get $t, t' \in \text{prefix}(K)$, $P(t) = P(t')$, $u'_1 \in E_o$: $tu'_1 \in \text{prefix}(K)$ (since $s \in \text{prefix}(K)$) and $t'u'_1 \notin \text{prefix}(K)$ and $t'u'_1 \in L$ (because L is prefix-closed). Hence, by observability of K we conclude that $t'u'_1 \in \text{prefix}(K)$, which is a contradiction. □

Note that some authors define observability, where the condition $e \in E_c$ is not required, i.e. $e \in E$. In this case it is not necessary to assume that K is controllable with respect to L and E_u in order to get the second implication in Proposition 4.1.

Another important property is that language normality is preserved by language unions as is shown below.

Proposition 4.2. *Language normality is preserved under arbitrary unions.*

Proof. Let K_i, $i \in I$ be normal with respect to L and P. Then

$$P^{-1}[P(\cup_{i \in I}\text{prefix}(K_i))] \cap L = P^{-1}(\cup_{i \in I} P(\text{prefix}(K_i))) \cap L =$$
$$\cup_{i \in I}[P^{-1}P(\text{prefix}(K_i)) \cap L] = \cup_{i \in I}\text{prefix}(K_i).$$

We have used above that both the projection and the inverse projection distribute with language unions, cf. Lemma 4.1 and the last equation holds because K_i, $i \in I$ are all normal with respect to L and P. □

An important consequence of Proposition 4.2 is that the supremal normal sublanguage of a given specification language always exists. It is simply given by the union of all normal sublanguages. Unfortunately, supremal normal sublanguages are difficult to compute, especially in the distributed framework. Only exponential time algorithms are known for their computation. One such algorithm will be presented in Section 4.6.

Problem 4.1. Basic Supervisory Control and Observation (BSCO)
Given generator G with partial observations, corresponding natural projection $P : E^* \to E_o^*$, and safety specification language $K \subseteq L(G)$, find a P-supervisor S such that:

(1) $L(S/G) \subseteq \text{prefix}(K)$,
(2) $L(S/G)$ is the largest it can be, i.e. for any other P-supervisor S' such that $L(S'/G) \subseteq \text{prefix}(K)$ we have $L(S'/G) \subseteq L(S/G)$. ■

Unfortunately, there is no single supremal solution for this problem, because supervisors (and the associated closed-loop systems) are given by observable sublanguages of K and it is well known that there does not always exist the supremal observable sublanguage of K. This is because observability is not preserved by language unions (unlike controllability and normality). A natural way to solve (BSCO) is to compute a maximal observable and controllable sublanguage of K with respect to L, E_u, and P. A notion of maximal element of an ordered set is different from the supremal element. Maximality means that there does not exist an observable and controllable sublanguage that is strictly larger than M. There might be several such maximal sublanguages [16].

Similarly as in the case of complete observations there is a dual problem, where instead of safety (maximal behavior), a minimal behavior is required.

Problem 4.2. Dual Supervisory Control and Observation (DuSCOP)
Given G with partial observations, corresponding natural projection $P : E^* \to E_o^*$, and a minimal required specification $M \subseteq L(G)$, find a P-supervisor S such that:

(1) $L(S/G) \supseteq M$,
(2) $L(S/G)$ is the smallest it can be, i.e. for any other P-supervisor S' such that $L(S'/G) \supseteq M$ we have $L(S/G) \subseteq L(S'/G)$. ■

It has been shown in [10] that, similar to controllability, observability of prefix-closed languages is also closed with respect to intersections. Therefore, the infimal observable superlanguage ([11]) as the intersection of all observable (prefix-closed) superlanguages always exist!

The unique solution to DuSCOP is then to take S such that $L(S) = \inf\{N \supseteq M : N \text{ is observable \& controllable w.r.t. } L(G), P, \text{ and } E_u\}$.

Interestingly, the above language corresponds to the permissive control policy, which naturally yields this language as $L(S/G)$, cf. [9].

4.6 Algorithms

In this section a construction procedure for supremal normal and supremal normal and controllable sublanguages is presented. The notation $\sup \mathrm{N}(K,L,P)$ is adopted to denote the supremal normal sublanguage of a specification language K with respect to the plant language L and natural projection P. Similarly, $\sup \mathrm{CN}(K,L,P)$ denotes the supremal normal and controllable sublanguage and $\sup \mathrm{C}(K,L)$ denotes the supremal controllable sublanguage of K with respect to L and a given set of uncontrollable events that is for simplicity not specified. In the case of prefix-closed languages, closed formulas are known for computation of $\sup \mathrm{N}(K,L,P)$ and $\sup \mathrm{CN}(K,L,P)$. Namely, it has been shown in [2] that

$$\sup \mathrm{N}(K,L,P) = \mathrm{prefix}(K) - P^{-1}(P(L - \mathrm{prefix}(K)))E^* \text{ and}$$

$$\sup \mathrm{CN}(K,L,P) = L \cap P^{-1}(\sup \mathrm{C}(P(\sup \mathrm{N}(K,L,P)), P(L)))\,.$$

Since all known algorithms for computing $\sup \mathrm{CN}(K,L,P,E_u)$ are of exponential worst-case complexity, we present the original algorithm of [4] based on a particular representation of the specification language K by a state-partition automaton and a subautomaton of the plant automaton at the same time.

The algorithm below relies on the notion of subautomaton of an automaton (generator) that is recalled below.

Definition 4.8. *Let* $G = (Q, E, f, q_o, Q_m)$ *be a generator.* $G' = (Q', E, f', q'_o, Q'_m)$ *is called a subautomaton of* G *if the following hold true:*

(1) $q_0 = q'_0$,
(2) $Q' \subseteq Q$,
(3) $\forall q \in Q'$ *and* $e \in E$: *if* $f'(q,e)!$ *then* $f'(q,e) = f(q,e)$.

In words, subautomaton G' has the same initial state as G, its state set Q' is a subset (possibly a proper subset) of Q. Finally, the transition function f' coincides with f on Q', i.e. $f' = f_{|Q'}$. There exist other definitions of subautomata in the literature that are not all equivalent, in particular for nondeterministic automata, which are however not the subject of this chapter.

The very first algorithm for computation of $\sup \mathrm{N}(K,L,P)$ appeared in [4].

Algorithm 4.5. *Given languages* $\mathrm{prefix}(K) \subseteq L = L(G)$ *assume that* H *representing* K *is a state-partition automaton (cf. Schutzenberger covering) and subautomaton of* G*. Below are the steps of the algorithm.*

(1) Compute the observer automata $Obs(G)$ *and* $Obs(H)$
(2) If $Obs(H)$ *is a subautomaton of* $Obs(G)$ *then* K *is normal, i.e.* $\sup \mathrm{N}(K,L,P) = K$
(3) Otherwise construct H_s^{obs}*, the largest subautomaton of* $Obs(H)$ *that is also a subautomaton of* $Obs(G)$.
(4) Compute the product $Z = G \times P^{-1}(H_s^{obs})$
Then $L(Z) = \sup \mathrm{N}(K,L,P)$.

Remark 4.6. The algorithm deserves a little explanation. It relies on the notion of subautomaton and the construction of the largest subautomaton of an automaton, which can be done in a standard way by eliminating states and edges of $Obs(H)$ so that H_s^{obs} is the largest subautomaton of G that is also subautomaton of $Obs(H)$. The notion of subautomaton is subtle, especially in the particular context of observer automata with powerset state sets that it is used. Two states of observer automata are in fact equal if and only if they are composed of the same states of G. Otherwise they are different and edges and states must be removed such that the largest subautomaton in the step (3) is obtained. Finally, particular representation of the specification automaton (that must be a state-partition automaton) is needed in order to ensure the correctness of the algorithm. ■

Complexity of the computation of $\sup\mathrm{N}(K,L,P)$ and $\sup\mathrm{CN}(K,L,E_u,P)$ is exponential in the worst case (unlike the algorithms for checking observability and normality). This is because all known algorithms require to construct the observer of H representing specification K and observer of G representing the plant language and observers cannot be computed in polynomial time in general. These algorithms are polynomial in special cases : e.g. if P satisfies the observer property (Definition 4.3) with respect to K and L. In this case the observer automata $Obs(H)$ and $Obs(G)$ are even guaranteed to be smaller or equal in size of the state set than H and G.

There follows an example of computation of the supremal normal sublanguage.

Example 4.7. Let us consider the following plant and specification languages.

Since the specification automaton of Fig. 4.11 is not a state-partition automaton, another representation of the specification is needed and illustrated in Fig. 4.12. Figure 4.13 then illustrates the resulting supremal normal sublanguage. ■

Now an algorithm is recalled for computation of supremal normal and controllable sublanguages based on iterative computations of supremal normal and supremal controllable sublanguages.

Algorithmic computation of the supremal normal and controllable sublanguages has been presented in [4].

Algorithm 4.8. *Given $K \subseteq L = L(G)$ assume that H representing K is a state-partition automaton and subautomaton of G.*

(1) Compute $Obs(G)$ and $Obs(H)$ and set $i = 0$ and let $G_{cn}^i = H$,

(2) Compute G_c : $L(G_c) = \sup\mathrm{C}(L(H), L(G), E_u)$ using standard algorithm presented in the previous chapter,

(3) If G_c is not state-partition automaton then $G_c^i := G_c^i \| Obs(G_c^i)$ and compute the automaton for $\sup\mathrm{N}(L(G_c), L(G), P) = L(G_c^i)$ using Algorithm 4.5,

(4) If $G_{cn}^i = G_{cn}^{i-1}$ then $G_{cn}^i = \sup\mathrm{CN}(L(G_c), L(G), E_u, P)$
Else put $i := i+1$ and Go to step (2).

The above algorithm relies on alternating computations of supremal controllable sublanguage and supremal normal sublanguage. It has been shown that iterations are not needed and a single-step algorithm based on direct coinduction-like definition

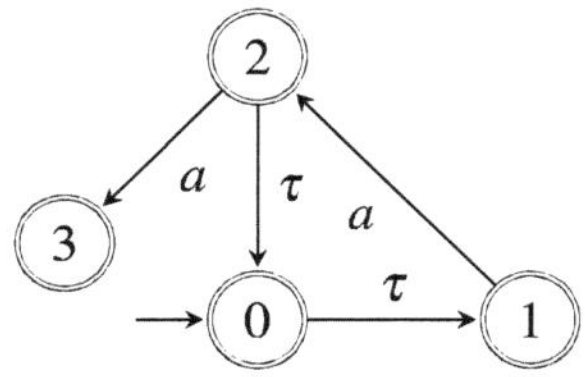

Fig. 4.10 $L(G)$ over $E = \{a,\tau\}$ with $E_{uo} = \{\tau\}$

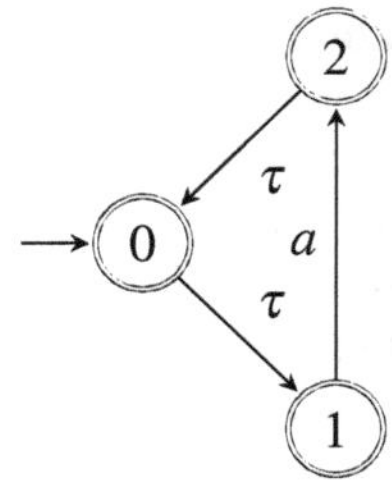

Fig. 4.11 K is not normal

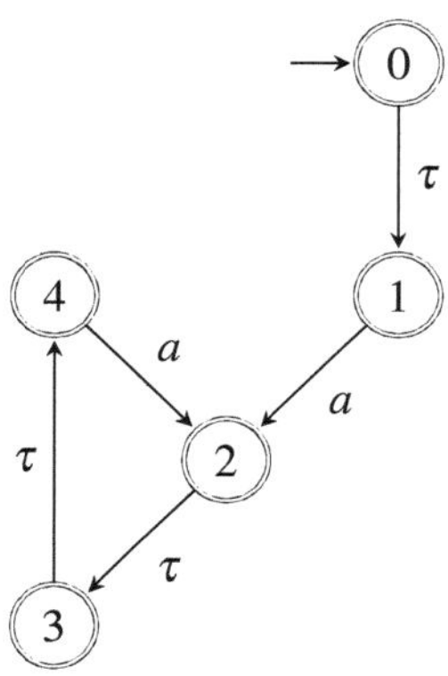

Fig. 4.12 Generator for the specification

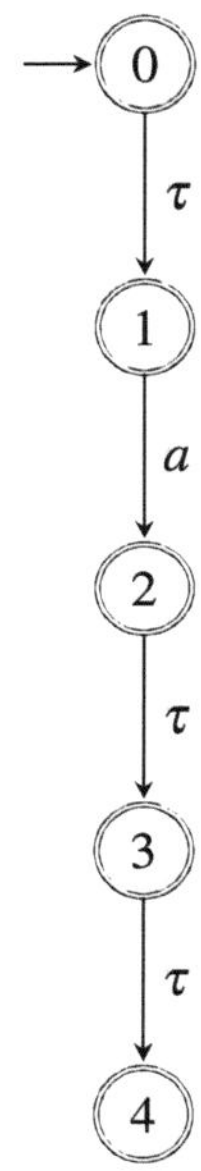

Fig. 4.13 $\sup \mathrm{N}(K,L,P)$

of $\sup \mathrm{CN}(K,L,E_u,P)$ has been proposed in [9]. The advantage of the coalgebraic framework is that there is a uniform treatment of automata and their languages in the sense that languages are considered as automata with special transition functions, which makes it easier to translate language properties into automata algorithms. Another algorithm that avoids costly iterations on the supremal controllable sublanguage and supremal normal sublanguage operators has been presented in [8]. Note however that even without iterations between computing $\sup \mathrm{C}$ and $\sup \mathrm{N}$ the time complexity remains exponential (in the worst case), because the algorithm subsumes computation of $\sup \mathrm{N}$ and no polynomial algorithm is known for that computation.

4.7 Extensions

There have been many extensions and/or alternative approaches to supervisory control. The state-based framework has been developed as opposed to the language (behavioral) framework. Since both frameworks are equivalent and there exist well-known bidirectional transformations between them, state-based framework is not presented in this chapter. We limit ourselves to explain that instead of language specifications predicates on states are used and supervisors are state-based, i.e. state-feedbacks. There exist notions of controllable, observable, and normal predicates similar to corresponding language properties.

State-based framework is however useful for nondeterministic system and/or specification, where language equality is too weak and therefore bisimilarity or weak bisimilarity are required of supervised closed-loop systems with specification automata instead of language equality.

State-based observability and controllability [14] are then sufficient conditions for existence of bisimilarity enforcing supervisors with partial observations. Essentially these conditions require that the property of controllability or observability must be verified in "every branch", i.e. in every state that is reached by a given string.

Other extensions of supervisory control theory have been developed to infinite state systems such as timed automata, where general control synthesis problem under partial observations is undecidable [1], or to push-down automata [7], where decidable cases have been identified.

4.8 Further Reading

Below are listed the main books on supervisory control with partial observations, where more information can be found.

- W.M. Wonham, Lecture notes on control of discrete-event systems, 2011 edition, U. Toronto, Toronto, Canada. See Chapter 6 on supervisory control with partial observations.
- C. Cassandras, S. Lafortune, Discrete-event systems, Springer, New York, 2008. See Chapters 2 and 3 on supervisory control.

References

1. Bouyer, P., D'Souza, D., Madhusudan, P., Petit, A.: Timed Control with Partial Observability. In: Hunt Jr., W.A., Somenzi, F. (eds.) CAV 2003. LNCS, vol. 2725, pp. 180–192. Springer, Heidelberg (2003)
2. Brandt, R.D., Garg, V., Kumar, R., Lin, F., Marcus, S.I., Wonham, W.M.: Formulas for calculating supremal controllable and normal sublanguages. Systems Control Letters 15(2), 111–117 (1990)

3. Cassandras, C.G., Lafortune, S.: Introduction to Discrete Event Systems, 2nd edn. Springer (2008)
4. Cho, H., Marcus, S.I.: On supremal languages of classes of sublanguages that arise in supervisor synthesis problems with partial observations. Mathematics of Control, Signal and Systems 2, 47–69 (1989)
5. Cho, H., Marcus, S.I.: Supremal and maximal sublanguages arising in supervisor synthesis problems with partial observations. Mathematical Systems Theory 22, 171–211 (1989)
6. Cieslak, R., Desclaux, C., Fawaz, A., Varaiya, P.: Supervisory control of a class of discrete event processes. IEEE Transactions on Automatic Control 33, 249–260 (1988)
7. Griffin, C.: Decidability and Optimality in Pushdown Control Systems: A New Approach to Discrete Event Control. PhD Thesis. Penn State University, USA (2007)
8. Hashtrudi Zad, S., Moosaeib, M., Wonham, W.M.: On computation of supremal controllable, normal sublanguages. Systems Control Letters 54(9), 871–876 (2005)
9. Komenda, J., van Schuppen, J.H.: Control of discrete-event systems with partial observations using coalgebra and coinduction. Discrete Event Dynamic Systems 15, 257–315 (2005)
10. Lin, F., Wonham, W.M.: On observability of discrete event systems. Information Sciences 44(3), 173–198 (1988)
11. Rudie, K., Wonham, W.M.: The infimal prefix-closed and observable superlanguage of a given language. Systems Control Letters 15, 361–371 (1990)
12. Rudie, K., Willems, J.C.: The computational complexity of decentralized discrete-event control problems. IEEE Transactions on Automatic Control 40(7), 1313–1319 (1995)
13. Schutzenberger, M.P.: On the definition of a family of automata. Information and Control 4, 145–270 (1961)
14. Takai, S., Kodama, S.: M -controllable subpredicates arising in state feed back control of discrete event systems. International Journal of Control 67(4), 553–566 (1997)
15. Tsitsiklis, J.N.: On the control of discrete-event dynamical systems. Mathematics of Control, Signal and Systems 2, 95–107 (1989)
16. Takai, S., Ushio, T.: Effective computation of an $L_m(G)$-closed, controllable, and observable sublanguage arising in supervisory control. In: Proc. of 6th Workshop on Discrete Event Systems, Zaragoza, Spain (2002)
17. Wonham, W.M.: Supervisory control of discrete-event systems. Lecture Notes University of Toronto (2011), `http://www.control.utoronto.ca/DES`

Chapter 5
Diagnosis and Automata

Eric Fabre

5.1 Diagnosis vs. State Estimation

Automaton. Our starting point is a nondeterministic automaton $\mathscr{A} = (S, \Sigma, I, \delta)$, with S the finite state set, $I \subseteq S$ possible initial states, action alphabet Σ and transition function $\delta : S \times \Sigma \to 2^S$. The latter extends naturally into $\delta : 2^S \times \Sigma^* \to 2^S$ by union on the first variable and by iteration on the second one. As usual, the action alphabet is partitioned into $\Sigma = \Sigma_o \uplus \Sigma_u$, observable and unobservable (or silent, or invisible) labels, respectively The transition set of $\mathscr{A}$ is denoted as $T = \{(s, \alpha, s') \in S \times \Sigma \times S : s' \in \delta(s, \alpha)\}$, and for a transition $t = (s, \alpha, s')$, we denote $s^-(t) = s, s^+(t) = s'$, and $\sigma(t) = \alpha$. If $|I| = 1$ and $\forall (s, \alpha) \in S \times \Sigma$, $|\delta(s, \alpha)| \leq 1$, automaton $\mathscr{A}$ is said to be deterministic.

A path or trajectory π of $\mathscr{A}$ is a sequence of transitions $\pi = t_1 \ldots t_n$ such that $s^-(t_1) \in I$ and $s^+(t_i) = s^-(t_{i+1})$, for $1 \leq i < n$. We adopt notations $s^-(\pi) = s^-(t_1)$, $s^+(\pi) = s^+(t_n)$, $|\pi| = n$, the length of π, $\sigma(\pi) = \sigma(t_1) \ldots \sigma(t_n)$ and $\sigma_o(\pi) = \Pi_{\Sigma_o}(\sigma(\pi))$ where Π_{Σ_o} is the natural projection of Σ^* on Σ_o^*. The language of $\mathscr{A}$ is $\mathscr{L}(\mathscr{A}) = \{\sigma(\pi) : \pi \text{ path of } \mathscr{A}\}$, and its observable language is $\mathscr{L}_o(\mathscr{A}) = \Pi_{\Sigma_o}(\mathscr{L}(\mathscr{A}))$. $\mathscr{A}$ is Σ_o-live iff from every state of $\mathscr{A}$, one can reach a transition labeled by Σ_o.

State estimation. Assume the system $\mathscr{A}$ performs some hidden run π, over which one only gets a partial knowledge by means of the observed sequence of labels $w = \sigma_o(\pi)$ produced by π. A natural question is *'What are the possible current states of $\mathscr{A}$ given that w was observed?'* So one wishes to build a function $f : \Sigma_o^* \to 2^S$ such that $f(w) = \{s^+(\pi) : \pi \text{ path of } \mathscr{A}, \sigma_o(\pi) = w\}$. There exists an obvious way of building f, recusively on the length of π: from $f^n : \Sigma_o^n \to 2^S$ solving the problem for observed sequences of length n, one can derive f^{n+1} by the so-called 'guided simulation' of $\mathscr{A}$.

Eric Fabre
INRIA Rennes Bretagne Atlantique, Campus de Beaulieu, 35042 Rennes cedex, France
e-mail: eric.fabre@inria.fr

C. Seatzu et al. (Eds.): Control of Discrete-Event Systems, LNCIS 433, pp. 85–106.
springerlink.com

Alternately, one can build an *observer* $Obs(\mathscr{A})$ of $\mathscr{A}$ as a pair $(\mathscr{O},\phi)$, where $\mathscr{O}=(Q,\Sigma_o,q_0,\delta')$ is a *deterministic* automaton over alphabet Σ_o, and $\phi: Q \to 2^S$ is a label function over its states. For any observed $w \in \Sigma_o^*$, let $q=\delta'(q_0,w)$ be the unique state reached by w in $\mathscr{O}$, then ϕ satisfies $\phi(q)=f(w)\subseteq S$. An observer can thus be seen as a precompiled and finite version of the recursive function f. The derivation of observers is detailed later in this chapter, which also examines some of their properties.

Diagnosis. The problem is usually stated as follows. One first associates types to the transitions of $\mathscr{A}$. This is done simply by setting $T=T_1\cup\ldots\cup T_K$, where each T_k gathers transitions of 'type k'. Sets T_k need not be disjoint, although the literature generally makes this assumption [3, 15]: transition types are usually interpreted as distinct failure modes, with one of them, say T_1, representing the 'safe' (*i.e.* non faulty) transitions. On wishes to build K diagnosis functions $f_k:\Sigma_o^*\to\{y,a,n\}$, $1\le k\le K$, such that

$$f_k(w)=\begin{cases} y & \text{if } \forall\pi\in\sigma_o^{-1}(w),\ \pi\notin(T\setminus T_k)^* \\ n & \text{if } \forall\pi\in\sigma_o^{-1}(w),\ \pi\in(T\setminus T_k)^* \\ a & \text{otherwise}\end{cases} \tag{5.1}$$

In other words, $f_k(w)$ answers 'yes' if all runs of $\mathscr{A}$ explaining w use a transition of T_k, it answers 'no' if none of these runs uses a transition of T_k, and answers 'ambiguous' otherwise. A *diagnoser* of $\mathscr{A}$ is now a pair $(\mathscr{D},\psi)$ where $\mathscr{D}=(Q,\Sigma_o,q_0,\delta')$ is again a deterministic automaton over alphabet Σ_o, and $\psi:\{1,...,K\}\times Q\to\{y,a,n\}$ is a (collection of) label function(s) over its states. For any observed $w\in\Sigma_o^*$, let $q=\delta'(q_0,w)$ be the unique state reached by w in $\mathscr{D}$, then $\psi(k,q)=f_k(w)$, $1\le k\le K$. A diagnoser is thus a finite and precompiled version of the K diagnosis functions.

Relation between the two problems. Despite an apparent difference, observers and diagnosers are similar objects. To build a diagnoser, the first step consists in augmenting the states of $\mathscr{A}$ with some memory $\mu\subseteq\{1,...,K\}$ to keep track of transition types that are fired by $\mathscr{A}$ along its trajectory. This yields $\bar{\mathscr{A}}=(\bar{S}=S\times 2^{\{1,...,K\}},\Sigma,I\times\{\emptyset\},\bar{\delta})$ where

$$(s',\mu')\in\bar{\delta}((s,\mu),\alpha)\Leftrightarrow\begin{cases} s'\in\delta(s,\alpha) \\ \mu'=\mu\cup\{k:(s,\alpha,s')\in T_k\}\end{cases} \tag{5.2}$$

In words, this 'state augmentation trick' does the following: as soon as $\mathscr{A}$ fires a transition of T_k, the memory set μ stores index k (forever). Equivalently, the above construction can be seen as computing the synchronous product (see Section 5.5 for a definition) of $\mathscr{A}$ with K elementary memory automata[1].

[1] The memory automaton for T_k only has two states 0 and 1, and $\{1,...,K\}$ as label set. It is deterministic and complete, and the only transition from 0 to 1 is labeled by k. Transitions of $\mathscr{A}$ must of course be relabeled by their type before the synchronous product can be computed, using types as labels. Details are left to the reader.

Observe now that deriving a diagnoser for $\mathscr{A}$ is equivalent to computing an observer for the augmented automaton $\bar{\mathscr{A}}$. Given $w \in \mathscr{L}_o(\mathscr{A}) \subseteq \Sigma_o^*$ and the unique state q reached by w in $Obs(\bar{\mathscr{A}})$, the diagnosis function ψ is then given by

$$\psi(k,q) = \begin{cases} y & \text{if } \forall(s,\mu) \in q,\ k \in \mu \\ n & \text{if } \forall(s,\mu) \in q,\ k \notin \mu \\ a & \text{otherwise} \end{cases} \tag{5.3}$$

Given this similarity, the chapter focuses on the derivation of observers and will only mention diagnosers to comment specific aspects.

5.2 Observer and Diagnoser

Given $\Sigma = \Sigma_o \uplus \Sigma_u$, an observer of $\mathscr{A}$ is obtained by first performing an ε-reduction, and then determinizing the result: $Obs(\mathscr{A}) = Det(Red(\mathscr{A}))$. The label function ϕ is trivial (the identity), and thus can be omitted.

ε-reduction. The ε-reduction $\mathscr{A}' = Red(\mathscr{A}) = (S, \Sigma_o, I', \delta')$ amounts to bypassing all transitions of $\mathscr{A}$ labeled by Σ_u (or equivalently the generic silent label ε). It can be performed either to the left of visible transitions, or to their right. Without loss of generality, in this chapter we focus on the reduction to the right. The ε-reduction to the right (see Fig. 5.1) is defined by $\delta'(s,\alpha) = \delta(s, \alpha\Sigma_u^*) = \cup_{w \in \alpha\Sigma_u^*} \delta(s,w)$. For the initial states, one has $I' = \delta(I, \Sigma_u^*) = \cup_{s \in I} \delta(s, \Sigma_u^*)$. Observe that the resulting automaton $\mathscr{A}'$ has the same states as $\mathscr{A}$, operates on the reduced alphabet Σ_o, but is generally nondeterministic. By construction, one has $\mathscr{L}(\mathscr{A}') = \mathscr{L}_o(\mathscr{A})$. By contrast, the ε-reduction to the left would take $I' = I$ and $\delta'(s,\alpha) = \delta(s, \Sigma_u^*\alpha) = \cup_{w \in \Sigma_u^*\alpha} \delta(s,w)$, still preserving $\mathscr{L}(\mathscr{A}') = \mathscr{L}_o(\mathscr{A})$.

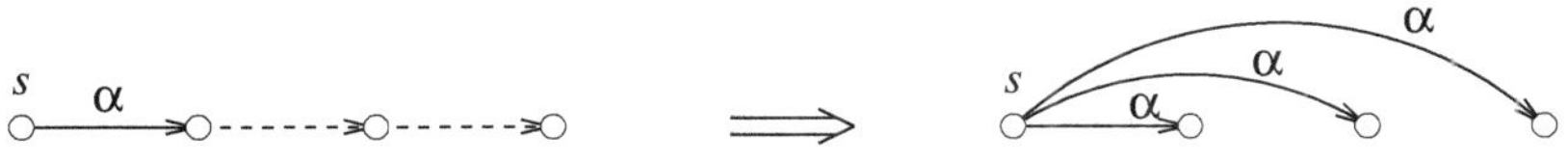

Fig. 5.1 Epsilon-reduction to the right. Dashed arrows represent silent (epsilon) transitions

Determinization. The determinization $\mathscr{A}'' = Det(\mathscr{A}') = (Q, \Sigma_o, q_0, \delta'')$ of $\mathscr{A}'$ is obtained by the standard subset construction. One has $Q = 2^S$, $q_0 = I'$, and for $q \in Q$ and $\alpha \in \Sigma_o$, the unique state $q' = \delta''(q,\alpha)$ in $\mathscr{A}''$ is defined as $q' = \delta'(q,\alpha) \triangleq \cup_{s \in q} \delta'(s,\alpha)$. Not all states in 2^S are reachable, so one often directly takes for Q the reachable part of 2^S, starting from $q_0 = I'$ and exploring recursively the $\delta'(q,\alpha)$ for all $\alpha \in \Sigma_o$ until no new q is found (Fig. 5.2). Determinization obviously has an exponential space complexity, in the worst case. Automaton $\mathscr{A}''$ directly yields a state estimator, or an observer of $\mathscr{A}$, by taking the identity for ϕ. By abuse of notation, one thus say that $\mathscr{A}''$ is an observer of $\mathscr{A}$, rather than $(\mathscr{A}'', \phi)$.

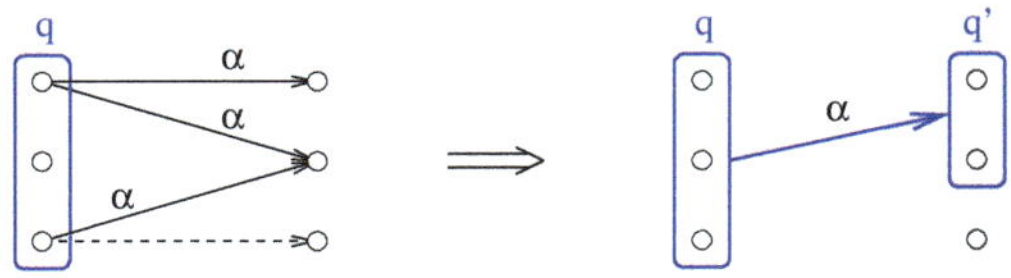

Fig. 5.2 Determinization. The dashed arrow represents a transition not labeled by α in $\mathscr{A}'$

Remarks

1. As mentioned in the previous section, building a diagnoser boils down to building an observer. Without loss of generality, for the diagnosis problem one can directly assume that states of $\mathscr{A}$ are partitioned into $S = S_1 \uplus \ldots \uplus S_L$ with $L = 2^K$, corresponding to the 2^K possible values of the memory in $\bar{\mathscr{A}}$. The diagnosis then reduces to checking whether all final states compatible with observation $w \in \mathscr{L}_o(\mathscr{A})$ lie into the union of some selected S_l.
2. If one is only interested in diagnosing independently the occurrence of each T_k, it is simpler to build K diagnosers, one for each T_k, by augmenting $\mathscr{A}$ with a simpler binary memory, or equivalently by assuming a partition $S = S_s \uplus S_f$ into safe and faulty states. In terms of complexity, this clearly saves an exponential in K. The diagnoser derived in the previous section is much more powerful, since it can also test for the simultaneous presence of several transition types in each trajectory explaining an observed word w.
3. Surprisingly, the ε-reduction to the right is frequent in the literature about state estimation (see the notion of 'unobservable reach'), but the ε-reduction to the left is preferred to derive a diagnoser. In other words, it is admitted that the supervised system changes its state silently, but not that it produce a fault silently. For diagnosis, one is generally interested in the occurrence or not of some transition type *before* the last observation of w (and not necessarily in the silent moves that follow w), which can be considered as an optimistic assumption. This choice is not bothering for Σ_o-live systems, since it only delays by one observation the detection of a fault occurrence, and in particular it does not change the notions of diagnosability. However, it makes a difference for non Σ_o-live systems, in the case where some states necessarily lead to a failure after which no more observation is collected (system crash).
4. Some contributions introduced so-called 'observation filters' [16, 17]: rather than a partition $\Sigma = \Sigma_o \uplus \Sigma_u$, one gives a filter $\lambda : S \times \Sigma \times S \to 2^{\Lambda \cup \{\varepsilon\}}$, and when $t = (s, \alpha, s')$ is fired, one label $\beta \in \lambda(t) \subseteq \Lambda \cup \{\varepsilon\}$ is observed (possibly none if $\beta = \varepsilon$). This does not change the expressive power of the model, that can be recoded in the classical setting by replacing (s, α, s') by (s, β, s') for every $\beta \in \lambda(t)$. The only difficulty introduced by such a recoding is that two versions of (s, β, s') may co-exist, one faulty and the other not. But this is captured by the possibility that a transition belong to several T_k.
5. An extended notion of diagnoser was proposed in [9]. It tests the occurrence (or not) of more complex properties on the partially observed trajectory of $\mathscr{A}$, such as the crossing of specific states interleaved with the crossing of specific

transitions. As long as these properties are regular and can thus be described by an automaton, the ideas of this chapter adapt naturally: one simply has to replace the simple state augmentation described previously by the product of $\mathscr{A}$ with the automaton describing the property to check.

5.3 Probabilistic Observer

Probabilistic automata [14] form a subclass of weighted automata. The idea is that the transitions rooted at any state are associated to firing probabilities. The state estimation problem now evolves into computing the probability to stop in state $s \in S$ given that $w \in \Sigma_o^*$ was observed (for diagnosis one wishes the conditional probability that a transition in T_k was fired). Again, one can design a recursive way to compute this conditional distribution over S; it takes the form of a filtering equation 'à la Kalman.' This section rather examines the precompiled version of this filter, that has great similarities with the non stochastic case.

Probabilistic automaton. We define it as $\mathscr{A} = (S, \Sigma, \mathbb{P}_0, \mathbb{P})$ where $\mathbb{P}_0 : S \to [0,1]$ is an initial probability on states, with initial states $I = supp(\mathbb{P}_0)$[2], and $\mathbb{P} : S \times \Sigma \times S \to [0,1]$ a transition probability, *i.e.* $\forall s \in S$, $\mathbb{P}(s,\cdot,\cdot)$ is a probability distribution over next labels and next states, given the current state s. Transitions are given by $T = supp(\mathbb{P})$, and the transition function by $\delta(s,\alpha) = supp(\mathbb{P}(s,\alpha,\cdot))$. This definition assumes that $\mathscr{A}$ is live, for simplicity[3]. $\mathscr{A}$ is said to be deterministic iff its support $supp(\mathscr{A}) = (S, \Sigma, I, \delta)$ is a deterministic (non stochastic) automaton. The probability of a path $\pi = t_1 ... t_n$ is equal to the product of the probabilities of its events $\mathbb{P}(\pi) = \mathbb{P}(t_1)...\mathbb{P}(t_n)$. And the language of $\mathscr{A}$ is defined as the formal power series $\mathscr{L}(\mathscr{A}) = \sum_{w \in \Sigma^*} \mathscr{L}(\mathscr{A}, w) \cdot w$ where coefficients are given by $\mathscr{L}(\mathscr{A}, w) = \sum_{\pi,\, \sigma(\pi)=w} \mathbb{P}_0(s^-(\pi))\mathbb{P}(\pi)$.

Observable (or stopped) language. To define the observable language of $\mathscr{A}$ as a weighted language, one must choose an appropriate notion of stopping time for $\mathscr{A}$, in order to define where runs should stop when they perform silent transitions. We adopt the following: $\mathscr{A}$ stops immediately before the production of the next observation, assuming $\mathscr{A}$ is Σ_o-live *i.e.* can reach an observable transition from any reachable state. Equivalently, $\mathscr{A}$ stops when it has been decided that the next step would produce an observation, but it is not yet decided which one[4]. This definition allows one to consume all silent steps after each observation. It contrasts with the usual choice of stopping immediately after an observable transition, which is slightly easier to handle and thus has often been chosen. It corresponds to the

[2] $supp$ = support of; this operation selects the elements with non zero probability

[3] One can easily extend this setting to include stopping probabilities at each state, just like standard weighted automata include stopping weights.

[4] For systems that have final states and stopping probabilities, one can choose to assimilate (or not) the choice to terminate in some state to the production of an observation, for the definition of the stopping time.

'optimistic' assumption that the system does not evolve silently by itself. Or at least that this evolution is ignored until there is evidence of it. Technically, the only impact is on the ε-reduction below, performed to the right (our case) instead of to the left (as in [16] for example).

We define the stopped language of $\mathscr{A}$ as follows: for a path π we take $\mathbb{P}^s(\pi) = \mathbb{P}_0(s^-(\pi))\mathbb{P}(\pi)\mathbb{P}(s^+(\pi),\Sigma_o,S)$, where $\mathbb{P}(s^+(\pi),\Sigma_o,S)$ is the probability of firing an observable transition from state $s^+(\pi)$. Observe that a path cannot stop at a state which has no observable outgoing transition, *i.e.* such a path has a vanishing probability. Therefore some states in $\mathscr{A}$ are 'unstable.' Then $\mathscr{L}^s(\mathscr{A},w) = \sum_{\pi,\,\sigma(\pi)=w} \mathbb{P}^s(\pi)$ and the stopped language of $\mathscr{A}$ is $\mathscr{L}^s(\mathscr{A}) = \sum_{w\in\Sigma^*} \mathscr{L}^s(\mathscr{A},w)\cdot w$. The observable language of $\mathscr{A}$ is given by $\mathscr{L}_o(\mathscr{A}) = \sum_{w\in\Sigma_o^*} \mathscr{L}_o(\mathscr{A},w)\cdot w$ where

$$\mathscr{L}_o(\mathscr{A},w) = \sum_{v\in\Sigma^*,\,\Pi_{\Sigma_o}(v)=w} \mathscr{L}^s(\mathscr{A},v). \tag{5.4}$$

Notice that the support of the stopped language may be strictly smaller than the support of the ordinary language, since some states of $\mathscr{A}$ may forbid stopping. However, the observed stopped and non-stopped languages of $\mathscr{A}$ have identical supports.

Probabilistic observer. Given partitions $\Sigma = \Sigma_o \uplus \Sigma_u$ and $S = S_1 \uplus \ldots \uplus S_L$, the objective is to derive a deterministic probabilistic automaton $\mathscr{O} = (Q,\Sigma_o,\mathbb{P}_0^{\mathscr{O}},\mathbb{P}^{\mathscr{O}})$, and a labeling $\phi : Q \to \mathscr{P}(L)$ of its states, where $\mathscr{P}(L)$ is the set of probability distributions over $\{1,\ldots,L\}$. Given $w \in \Sigma_o^*$ produced by $\mathscr{A}$, and $q \in Q$ the unique state reached by w in $\mathscr{O}$, we want $(\phi(q))(l) = \mathbb{P}(\mathscr{A}$ stops in $S_l \mid w$ was observed) for $l \in \{1,\ldots,L\}$. Every such probabilistic observer can trivially be derived from a universal one assuming the finest partition of S, *i.e.* we actually aim at building an observer that computes $\mathbb{P}(\mathscr{A}$ stops in $s \mid w$ was observed) for every $s \in S$.

In fact, the probabilistic observer derived below exhibits more properties than requested above: it guarantees that $\mathscr{L}(\mathscr{O}) = \mathscr{L}_o(\mathscr{A})$, *i.e.* for any observed word $w \in \Sigma_o^*$, it can also compute the probability of this observed word in $\mathscr{A}$. If one is simply interested in the conditional distribution over S given w, the labeling function ϕ is sufficient and $\mathscr{O}$ can be reduced to its support.

ε-reduction. We look for a probabilistic automaton $\mathscr{A}' = Red(\mathscr{A}) = (S,\Sigma_o,\mathbb{P}'_0,\mathbb{P}')$ such that $\mathscr{L}(\mathscr{A}') = \mathscr{L}_o(\mathscr{A}') = \mathscr{L}_o(\mathscr{A})$. Structurally, the automaton will be the same as in the non probabilistic case, and obtained by ε-reduction to the right. The difficulty lies in the computation of transition probabilities, since an unbounded number of silent steps may be performed until $\mathscr{A}$ decides to stop (and commits to fire a visible transition at the next step). This requires to integrate probabilities over a possibly infinite set of silent paths. The difficulty can addressed by different methods, of equivalent complexities. We give two of them below; see [8] for a graphical one.

The transition probability of automaton $\mathscr{A}'$ can be expressed as $\mathbb{P}'(s,\alpha,s') = \sum_{s''\in S} \mathbb{P}(s,\alpha,s'')\mathbb{P}^{\varepsilon}(s'',s')\mathbb{P}(s',\Sigma_o,S)$, where

$$\mathbb{P}^{\varepsilon}(s,s') = \mathbb{P}(s,\Sigma_u^*,s') = \sum_{\substack{\pi,\ \sigma(\pi)\in\Sigma_u^* \\ s^-(\pi)=s,\ s^+(\pi)=s'}} \mathbb{P}(\pi) \tag{5.5}$$

and similarly for the initial probability: $\mathbb{P}'_0(s') = \sum_{s''\in S}\mathbb{P}_0(s'')\mathbb{P}^{\varepsilon}(s'',s')\mathbb{P}(s',\Sigma_o,S)$. Notice that (5.5) does not represent the probability to reach s' from s after an arbitrary number of silent steps, because s' can be met several times along such paths. One must take into account a stopping or exit probability at s' to turn this quantity into a probability, as it is done in the definition of $\mathbb{P}'(s,\alpha,s')$: the term $\mathbb{P}^{\varepsilon}(s'',s')\mathbb{P}(s',\Sigma_o,S)$ does correspond to the probability of going from s'' to s' through an arbitrary number of silent steps *and* to exit at s' towards a visible label. Similarly, $\mathbb{P}^{\varepsilon}(s,s)\geq 1$ yields the inverse of the probability to leave state s for a visible label (after performing an arbitrary number of silent loops around s).

The *pseudo* transition matrix $\mathbb{P}^{\varepsilon}$ can be obtained through a Floyd-Warshall procedure. It is usually applied to compute minimum distances between all pairs of nodes in a graph. By replacing the $(min,+)$ setting by the $(+,*)$ setting one obtains a simple way to integrate probabilities over all paths relating two nodes [4, 13]. Specifically, denoting $S=\{s_1,\ldots,s_N\}$, one defines $\mathbb{P}^{\varepsilon}_n(s,s')$ as in (5.5), excepted that the sum is limited to paths that go through states in $\{s_1,\ldots,s_n\}$. The desired $\mathbb{P}^{\varepsilon}$ corresponds to $\mathbb{P}^{\varepsilon}_N$, and $\mathbb{P}^{\varepsilon}_0(s,s')=\mathbb{P}(s,\Sigma_u,s')$: probability of a direct silent step from s to s'. The $\mathbb{P}^{\varepsilon}_n$ satisfy the following recursion: for $s\neq s_{n+1}\neq s'$

$$\mathbb{P}^{\varepsilon}_{n+1}(s,s') = \mathbb{P}^{\varepsilon}_n(s,s_{n+1})\,\mathbb{P}^{\varepsilon}_n(s_{n+1},s_{n+1})^*\,\mathbb{P}^{\varepsilon}_n(s_{n+1},s') + \mathbb{P}^{\varepsilon}_n(s,s') \tag{5.6}$$

$$\mathbb{P}^{\varepsilon}_n(s_{n+1},s_{n+1})^* = \sum_{k\geq 0}\mathbb{P}^{\varepsilon}_n(s_{n+1},s_{n+1})^k = \frac{1}{1-\mathbb{P}^{\varepsilon}_n(s_{n+1},s_{n+1})} \tag{5.7}$$

where (5.7) integrates over paths that make an arbitrary number of silent loops around s_{n+1}. For completeness, one must add to (5.6) three following specific cases: $\mathbb{P}^{\varepsilon}_{n+1}(s_{n+1},s_{n+1}) = \mathbb{P}^{\varepsilon}_n(s_{n+1},s_{n+1})^*$, then $\mathbb{P}^{\varepsilon}_{n+1}(s,s_{n+1}) = \mathbb{P}^{\varepsilon}_n(s,s_{n+1})\mathbb{P}^{\varepsilon}_n(s_{n+1},s_{n+1})^*$, and finally $\mathbb{P}^{\varepsilon}_{n+1}(s_{n+1},s') = \mathbb{P}^{\varepsilon}_n(s_{n+1},s_{n-1})^*\,\mathbb{P}^{\varepsilon}_n(s_{n+1},s')$. The complexity of the ε-reduction by this method is $\mathcal{O}(|S|^3)$.

Instead of the Floyd-Warshall procedure, one may also consider a fix-point relation satisfied by matrix $\mathbb{P}^{\varepsilon}$. Let $\bar{\mathbb{P}}^{\varepsilon}=\mathbb{P}^{\varepsilon}-I$; this corresponds to Definition (5.5) where Σ_u^* is replaced by Σ_u^+, *i.e.* paths must cross at least one unobservable transition. One then has

$$\forall s,s'\in S,\quad \bar{\mathbb{P}}^{\varepsilon}(s,s') = \sum_{s''\in S}\bar{\mathbb{P}}^{\varepsilon}(s,s'')\mathbb{P}(s'',\Sigma_u,s') + \mathbb{P}(s,\Sigma_u,s') \tag{5.8}$$

Still denoting by $\mathbb{P}^{\varepsilon}_0$ the matrix with entries $\mathbb{P}(s,\Sigma_u,s')$, (5.8) means $\bar{\mathbb{P}}^{\varepsilon}=\bar{\mathbb{P}}^{\varepsilon}\cdot\mathbb{P}^{\varepsilon}_0+\mathbb{P}^{\varepsilon}_0$ (notice that $\bar{\mathbb{P}}^{\varepsilon}=\mathbb{P}^{\varepsilon}_0\cdot\bar{\mathbb{P}}^{\varepsilon}+\mathbb{P}^{\varepsilon}_0$ holds as well). This entails $\mathbb{P}^{\varepsilon}=I+\mathbb{P}^{\varepsilon}\cdot\mathbb{P}^{\varepsilon}_0=I+\mathbb{P}^{\varepsilon}_0\cdot\mathbb{P}^{\varepsilon}$ whence $\mathbb{P}^{\varepsilon}=(I-\mathbb{P}^{\varepsilon}_0)^{-1}$ (assuming invertibility holds). Deriving $\mathbb{P}^{\varepsilon}$ by this methods reveals again a generic complexity in $\mathcal{O}(|S|^3)$, due to the matrix inversion.

Determinization. To determinize a probabilistic automaton $\mathcal{A}'=(S,\Sigma_o,\mathbb{P}'_0,\mathbb{P}')$, one can rely on the standard determinization procedure of weighted automata, that adapts the recursive subset construction given in the previous section [2, 10, 11]. One has $\mathscr{A}'' = Det(\mathscr{A}') = (Q, \Sigma_o, \mathbb{P}_0'', \mathbb{P}'')$ where $Q \subset 2^{S\times[0,1]}$ and can be infinite. $\mathbb{P}_0''$ assigns probability 1 to the unique state $q_0 = \{(s, \mathbb{P}_0'(s)) : s \in supp(\mathbb{P}_0')\}$. Successive states are obtained recursively as follows. Let $q = \{(s_1, p_1), ..., (s_M, p_M)\} \in Q$ and $\alpha \in \Sigma_o$, one has $\delta''(q, \alpha) = q' = \{(s_1', p_1'), ..., (s_N', p_N')\}$ iff $\{s_1', ..., s_N'\} = \delta'(\{s_1, ..., s_M\}, \alpha) \neq \emptyset$, and for $1 \leq n \leq N$

$$p_n'' = \sum_{1\leq m\leq M} p_m \cdot \mathbb{P}'(s_m, \alpha, s_n') \tag{5.9}$$

$$p_n' = p_n''/C \quad \text{where} \quad C = \sum_{1\leq k\leq N} p_k'' = \mathbb{P}''(q, \alpha, q') \tag{5.10}$$

Proposition 5.1. *Let $w \in \Sigma_o^*$ and $\delta''(q_0, w) = q = \{(s_1, p_1), ..., (s_M, p_M)\}$ in $\mathscr{A}''$, then*

$$p_m = \mathbb{P}(\mathscr{A}' \text{ is in state } s_m | w \text{ was observed}) \tag{5.11}$$

Proof. This is obviously true at q_o for $w = \varepsilon$. Assume it is true at $q = \delta''(q_o, w)$ and let $q' = \delta''(q, \alpha)$. Eq. (5.9) is a standard filtering equation for $\mathscr{A}'$ (based on Bayes rule and the Markov property), so p_n'' is the probability that $\mathscr{A}'$ produces $\alpha \in \Sigma_o$ and reaches state $s_n' \in S$ given that w was observed. Consequently, C is the probability to fire α given w was observed, and the $(p_n')_{1\leq n\leq N}$ give the conditional probability of the current state of $\mathscr{A}'$ given the observed sequence $w\alpha$. □

Corollary 5.1. *The probabilistic automaton $Det(Red(\mathscr{A}))$ built above yields a universal probabilistic observer. For state $q = \{(s_1, p_1), ..., (s_M, p_M)\}$, the index function $\phi(q) \in \mathscr{P}(S)$ is defined by $[\phi(q)](s_m) = p_m$, for $1 \leq m \leq M$, and by $[\phi(q)](s) = 0$ otherwise.*

Proof. If $\mathscr{A}' = Red(\mathscr{A})$, p_m is the probability that $\mathscr{A}$ stops in s_m given that w was observed. To make $\mathscr{A}''$ an observer for partition $S = S_1 \uplus ... \uplus S_L$, one simply has to take the distribution defined by the $[\phi(q)](S_l)$, for $1 \leq l \leq L$. Notice also that $\mathscr{L}(\mathscr{A}'') = \mathscr{L}(\mathscr{A}') = \mathscr{L}_o(\mathscr{A})$. □

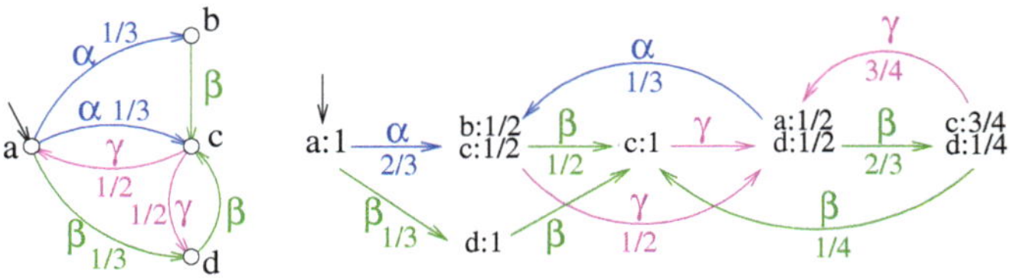

Fig. 5.3 A probabilistic automaton (left) and its determinized version (right). Transition probabilities are only mentioned when they differ from 1

Example 5.1. Figure 5.3 illustrates the determinization procedure. This simple example seems to suggest that the conditional probabilities appear as extra information attached to a standard (*i.e.* non-probabilistic) observer. This is not the case, and the determinization procedure may very well not terminate, as revealed by the counter-example in Fig. 5.4. While for weighted automata taking values in the $(\mathbb{R}^+, \min, +)$ semiring there exist sufficient conditions to guarantee termination (see the twin property in [12]), to our knowledge it is still not clear what these conditions could be for probabilistic automata. ■

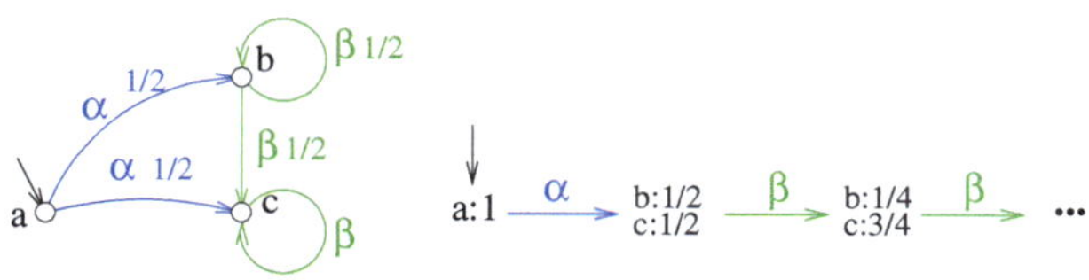

Fig. 5.4 Determinization may not terminate

Probabilistic diagnoser. Using the same technique as in the previous section, a probabilistic diagnoser for $\mathscr{A}$ is nothing else than a probabilistic observer for an augmented automaton $\bar{\mathscr{A}}$, that keeps track of which transition types have been crossed along the run of $\mathscr{A}$:

$$\bar{\mathbb{P}}((s,\mu),\alpha,(s',\mu')) = \mathbb{P}(s,\alpha,s') \cdot \mathbb{I}_{\mu'=\mu\cup\{k:(s,\alpha,s')\in T_k\}} \tag{5.12}$$

where $\mathbb{I}$ is the indicator function. From the conditional distribution on states of $\bar{\mathscr{A}}$ given some observation $w \in \Sigma_o^*$, one then easily derives the conditional distribution on memory values μ, and further on transition classes T_k that were crossed by $\mathscr{A}$. Again, if one is only interested in this probability distribution, the observer can be turned into an ordinary (non stochastic) deterministic automaton, by taking the support of $\mathscr{O}$.

Remark. The 'observation filters,' that randomly modify the labels of Σ produced by transitions of $\mathscr{A}$, can be processed in a similar manner as in Remark 4 of Section 5.2. The slight difference here is that a given observed label $\beta \in \Lambda \cup \{\varepsilon\}$ may correspond to several underlying transition types T_k, that have different probabilities. This case is captured simply as follows: one replaces the deterministic memory represented by the $\mathbb{I}$ term in (5.12) by a 'randomized' memory. Specifically, given $T = T_1 \cup \ldots \cup T_K$ and for $\mu' = \mu \uplus \mu$", (5.12) becomes $\bar{\mathbb{P}}((s,\mu),\beta,(s',\mu')) = \mathbb{P}(s,\beta,s') \cdot \mathbb{P}(\bigwedge_{k\in\mu''} T_k \wedge \bigwedge_{k\notin\mu'} \bar{T}_k \,|\, (s,\beta,s'))$. The first term is the probability to move from s to s' and produce label β, the second one is the (conditional) probability that this move crosses a transition lying in all T_k for $k \in \mu$", and in none of the T_k for $k \notin \mu'$.

5.4 Diagnosability

For simplicity, and without loss of generality, this section assumes an automaton $\mathscr{A}$ with a partition $S = S_s \uplus S_f$ of its states into safe and faulty ones (recall the state augmentation trick), and such that no safe state is reachable from a faulty one. It is also assumed that $\mathscr{A}$ is Σ_o-live. For $w \in \Sigma_o^*$, let us consider again the diagnosis function $f(w)$ for $\mathscr{A}$ defined as

$$f(w) = \begin{cases} y & \text{if } \forall \pi \in \sigma_o^{-1}(w),\ s^+(\pi) \in S_f \\ n & \text{if } \forall \pi \in \sigma_o^{-1}(w),\ s^+(\pi) \in S_s \\ a & \text{otherwise} \end{cases} \tag{5.13}$$

Definition 5.1. *$\mathscr{A}$ is diagnosable iff there exists some integer N such that*

$$\forall \pi : s^+(\pi) \in S_f,\ \forall \pi' : s^-(\pi') = s^+(\pi),\ [\, |\pi'|_{\Sigma_o} > N \Rightarrow f(\sigma_o(\pi\pi')) = y\,]$$

where $|\pi'|_{\Sigma_o}$ denotes the number of visible transitions in path π'.

In other words, as soon as a path hits a faulty state, at most N observations later the fault will be diagnosed. Ambiguity cannot last forever.

Remarks

1. Some definitions rather take $|\pi'|$ rather than $|\pi'|_{\Sigma_o}$, with the assumption that $\mathscr{A}$ has no silent circuit, which yields an equivalent definition for this smaller class of systems. This restriction is not really necessary, and $|\pi'|_{\Sigma_o}$ makes more sense since it gives an observable criterion to position the fault.
2. Some authors do not require that $\mathscr{A}$ is Σ_o-live, and then extend the condition to extensions π' that contain $M \leq N$ observations after which no more visible transition is reachable (deadlock or silent live-lock). This generalization introduces minor technical changes, that are left to the reader.

Proposition 5.2. *If $\mathscr{A}$ is not diagnosable, then a diagnoser $(\mathscr{D}, \psi)$ of $\mathscr{A}$ necessarily contains a circuit of ambiguous states.*

Proof. In Def. 5.1, $f(\sigma_o(\pi\pi'))$ can only take values y or a. If $\mathscr{A}$ is not diagnosable, let N be greater than the number of states in $\mathscr{D}$. There exists π, π' with $s^+(\pi) \in S_f$, $|\pi'|_{\Sigma_o} > N$ and $f(\sigma_o(\pi\pi')) = a$. Let q be the unique state reached by $\sigma_o(\pi)$ in $\mathscr{D}$, then $\psi(q) = a$ and any state q' crossed by $\sigma_o(\pi')$ after q in $\mathscr{D}$ is also ambiguous, $\psi(q') = a$, since by construction the values of ψ can only evolve from n to a and then y. And $\sigma_o(\pi')$ necessarily crosses twice some state of $\mathscr{D}$. □

This proposition gives a sufficient condition for diagnosability, which unfortunately is not necessary. Consider the counter-example in Fig. 5.5, where safe states are represented as a white circle, and faulty states as a colored one. The diagnoser contains an ambiguous circuit, because for any sequence $(\alpha\beta)^n$ or $(\alpha\beta)^n\alpha$ it is not certain that a fault occured. However, any path π leading to the faulty state s_3 will necessarily produce a γ as second next observation, which characterizes the occurrence of the fault. So $\mathscr{A}$ is 2-diagnosable ($N = 2$).

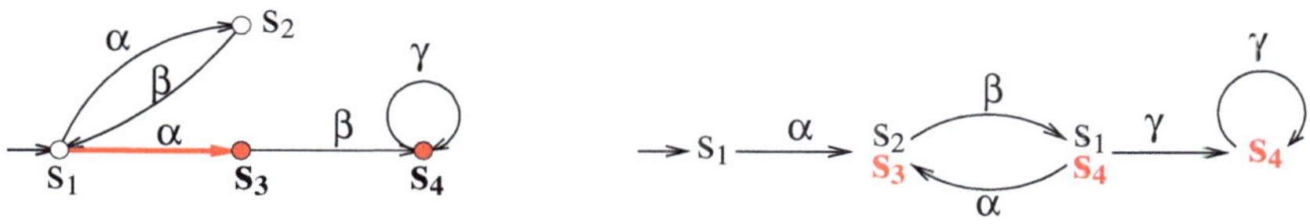

Fig. 5.5 A diagnosable system (left), and its observer/diagnoser (right)

Proposition 5.3. *$\mathscr{A}$ is not diagnosable iff, for any integer N,*

$$\exists \pi, \pi', \pi'' : \ s^+(\pi) = s^-(\pi') \in S_f, \ |\pi'|_{\Sigma_o} > N, \ s^+(\pi'') \in S_s, \ \sigma_o(\pi\pi') = \sigma_o(\pi'')$$

The proof is straightforward by logical inversion of Def. 5.1. One can easily reinforce the result into the existence of a faulty π (the same for all N) for which one can find arbitrarily long extensions π' such that there exists a non faulty π'' with the same visible signature: $\sigma_o(\pi\pi') = \sigma_o(\pi'')$ (hint: use the pumping lemma) Further, if the result holds for some large enough N, then it holds for any N (same trick).

This simple result has interesting practical consequences. First of all, it allows one to check if the ambiguous circuits of a diagnoser of $\mathscr{A}$ really entail non-diagnosability. Specifically, $\mathscr{A}$ is not diagnosable iff a diagnoser of $\mathscr{A}$ contains an *indeterminate cycle*, following the vocabulary in [15]. Such cycles are ambiguous circuits of $\mathscr{D}$ where both a safe run of $\mathscr{A}$ and a faulty one are nested. Specifically, the circuit from q to q in $\mathscr{D}$, following the visible sequence $w \in \Sigma_o^*$, is indeterminate iff there exist two circuits π_1, π_2 in $\mathscr{A}$ such that π_1 only crosses faulty states, π_2 only crosses safe states, and $\sigma_o(\pi_1) = \sigma_o(\pi_2) = w^n$ for some n. In such situations, using the pumping lemma, one can easily build the π, π', π'' of Prop. 5.3 that prove the non diagnosability. And conversely.

A second consequence of Prop. 5.3 is to provide a direct and more practical means of checking diagnosability (and actually polynomial rather than exponential). The idea is based on the *twin machine* construction. Consider $\mathscr{A}_s$, the restriction of $\mathscr{A}$ to the safe states S_s: all transitions involving states in S_f are discarded. The twin machine is obtained as the synchronous product of the ε-reductions of $\mathscr{A}$ and $\mathscr{A}_s$: $\mathscr{T} = Red(\mathscr{A}) \times Red(\mathscr{A}_s)$.

Proposition 5.4. *$\mathscr{A}$ is diagnosable iff no cycle of the twin machine $\mathscr{T} = Red(\mathscr{A}) \times Red(\mathscr{A}_s)$ contains a faulty state of $\mathscr{A}$.*

The proof is directly based on Prop. 5.3, using again the pumping lemma (details are left to the reader). Applied to the counter-example of Fig. 5.5, assuming all labels are observable, this yields the construction of Fig. 5.6, where the unique circuit crosses only safe states of $\mathscr{A}$, which makes $\mathscr{A}$ diagnosable.

Probabilistic diagnosability. For the sake of completeness, let us briefly examine how diagnosability extends to probabilistic automata. For simplicity, we assume that $\mathscr{A}$ is already ε-reduced, based on some stopping time definition.

As a (live) stochastic automaton, $\mathscr{A}$ defines a probability $\mathbb{P}_n$ on the set $\mathscr{F}_n$ of runs of length n. These $(\mathscr{F}_n)_n$ define a natural filtration over the set of infinite runs

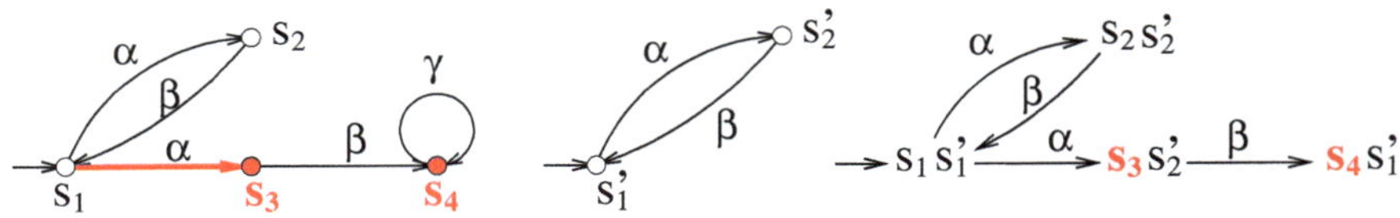

Fig. 5.6 System $\mathscr{A}$ (left), its safe restriction $\mathscr{A}_s$ (center), and the twin machine (right)

of $\mathscr{A}$, and since $\mathbb{P}_n$ is the restriction of $\mathbb{P}_{n+1}$ to $\mathscr{F}_n$, there exists a unique distribution $\mathbb{P}$ over the set $\mathscr{F}$ of infinite runs of $\mathscr{A}$, by the Kolmogorov extension theorem.

On this probability space, let us define two diagnosis indicators, as random variables. We denote by ω an infinite run (path) of $\mathscr{A}$, and by ω^n its restriction to the first n transitions. The first diagnosis indicator is $X_n(\omega) = \mathbb{P}(\{\omega' : \sigma_o(\omega'^n) = \sigma_o(\omega^n), s^+(\omega'^n) \in S_f\})$, which is an $\mathscr{F}_n$-measurable random variable. X_n is thus a failure likelyhood, and $X_n(\omega) = 0$ iff all runs of length n that produce the same observations as ω^n finish in S_{nf}. The second indicator is $D_n(\omega) \in \{0,1\}$, another $\mathscr{F}_n$-measurable random variable, defined by $D_n(\omega) = 0$ iff $\exists \omega', \sigma_o(\omega'^n) = \sigma_o(\omega^n) \wedge s^+(\omega'^n) \in S_s$. So $D_n(\omega)$ switches to one when *all* runs of length n that produce the same observation as ω^n contain a failure, which corresponds to the detection of that failure.

In [16], two notions of diagnosability were proposed. The *A*-diagnosability corresponds to the following: for all k, conditioned on $X_k > 0$, D_{k+n} converges to 1 in probability. The *AA*-diagnosability only requires that X_{k+n} converges to 1 in probability, again conditioned on $X_k > 0$. The first criterion expresses that the 'hard' detector D_n will ultimately switch to 1 (certain detection) after a failure has occurred, while the second criterion means that the detection probability X_n will converge to 1 (the more one waits, the more the detection is certain). These convergences are in probability: the more one waits, the more these events are likely. (There is still space for defining and characterizing a diagnosability based on an almost sure convergence.)

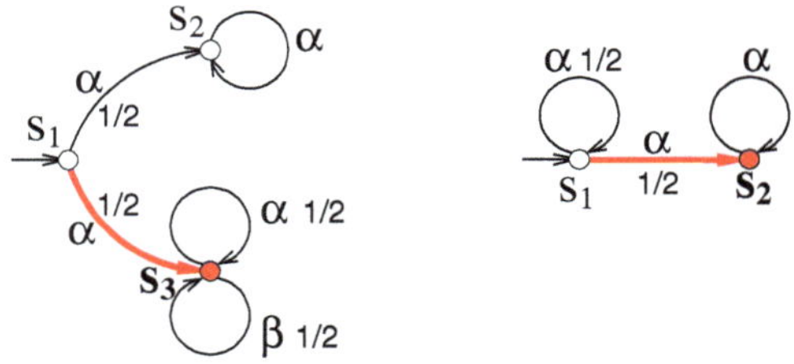

Fig. 5.7 Left: a non-diagnosable probabilistic automaton that is A-diagnosable. Right: an AA-diagnosable automaton that is not A-diagnosable

Fig. 5.7 (left) shows a probabilistic automaton that is not diagnosable, if probabilities are ignored: after the faulty state s_3 has been reached, one can observe an arbitrary long sequence of α^n that will not allow to discriminate between s_2 and s_3, *i.e.* safe and faulty. However, with probability 1 a β will ultimately be fired after the faulty state s_3, which will make the hard detector D_n jump to 1. The right hand side in the figure shows a probabilistic automaton that is not diagnosable nor A-diagnosable, since after the faulty transition, whatever the number of observations, one can not be certain to be in s_2. However, the longer one waits, the more likely the system jumped from s_1 to s_2. So with probability 1 the soft detector X_n will converge to 1.

One of the main results in [16] is to translate the A-diagnosability of $\mathscr{A}$ into a *structural* property of its *non probabilistic* diagnoser/observer. Equivalently, this amounts to first replacing $\mathscr{A}$ by its support $\bar{\mathscr{A}}$. The A-diagnosability criterion then observes the product $\mathscr{B} = Det(Red(\bar{\mathscr{A}})) \times Red(\bar{\mathscr{A}})$, where the ε-reduction corresponds to the chosen notion of stopping time in $\mathscr{A}$. This non-probabilistic automaton $\mathscr{B}$ has $Q \times S$ as state set. One then has to examine the recurrent states of $\mathscr{B}$, *i.e.* the states (q,s) that belong to a terminal connected component of $\mathscr{B}$, regarded as a directed graph. This is standard in convergence analysis of Markov chains, since with probability one the chain – here the probabilistic diagnoser of $\mathscr{A}$ – will terminate in one of its recurrent components, and A-diagnosability deals with the limit behaviors of the diagnoser of $\mathscr{A}$. Theorem 3 in [16] then expresses that $\mathscr{A}$ is A-diagnosable iff any recurrent state (q,s) in $\mathscr{B}$ with $s \in S_f$ satisfies $\psi(q) = y$ or equivalently $q \subseteq S_f$. In other words, after a faulty state s is crossed, the (probabilistic) diagnoser will terminate with probability one in states where the failure is unambiguous. As $\mathscr{B}$ requires a determinization, the complexity of the A-diagnosability test proposed in [16] is exponential. But as for standard diagnosability, one recovers a polynomial complexity by performing the same test on the recurrent states of the twin-machine derived for $\bar{\mathscr{A}}$.

5.5 Modular Observers

A compound system is obtained by assembling components by means of a composition operator. Here we chose the usual synchronous product of automata. The section proves that composition and derivation of an observer are two operations that commute, under some circumstances. This has important consequences to design efficient observers for some compound systems. Results are presented in the simple case of two components, but extend to larger compound systems through the notion of interaction graph between components (see for example [7] where this notion is used for distributed planning purposes).

Composition of Automata

Definition 5.2. *The* synchronous product *(see Fig. 5.8) of two automata $\mathscr{A}_1$ and $\mathscr{A}_2$ is the automaton $\mathscr{A}_1 \times \mathscr{A}_2 = (S,\Sigma,I,\delta)$ such that: $S = S_1 \times S_2, \Sigma = \Sigma_1 \cup \Sigma_2, I = I_1 \times I_2$, and the transition function δ is defined by $\forall (s_1,s_2) \in S, \forall \alpha \in \Sigma$, $\delta((s_1,s_2),\alpha) = \delta_1^+(s_1,\alpha) \times \delta_2^+(s_2,\alpha)$, where $\delta_i^+(s_i,\alpha)$ coincides with δ_i for $\alpha \in \Sigma_i$, and $\delta_i^+(s_i,\alpha) \triangleq \{s_i\}$ for $\alpha \notin \Sigma_i$.*

In other words, a transition carrying a shared label in one component must be fired jointly with a transition carrying the same label in the other component. By contrast, a transition carrying a private label only changes the state of one component, while the other remains idle. Observe that a firable sequence u of transitions in $\mathscr{A}_1 \times \mathscr{A}_2$ leads to a unique firable sequence u_1 in $\mathscr{A}_1$ (for example) by simply removing private moves of $\mathscr{A}_2$, *i.e.* transitions $((s_1,s_2),\alpha,(s_1,s_2'))$ with $\alpha \notin \Sigma_1$, and then erasing the states of component $\mathscr{A}_2$.

A plain synchronous product may yield an automaton that is not trimmed, *i.e.* that contains unaccessible states. So we define the composition as a synchronous product followed by a trimming operation, and still denote it by $\times$, with a light abuse of notation. This definition naturally extends to observers (or diagnosers) by simply gathering the label functions on states, for example $\phi(q_1,q_2) \triangleq \phi_1(q_1) \times \phi_2(q_2)$.

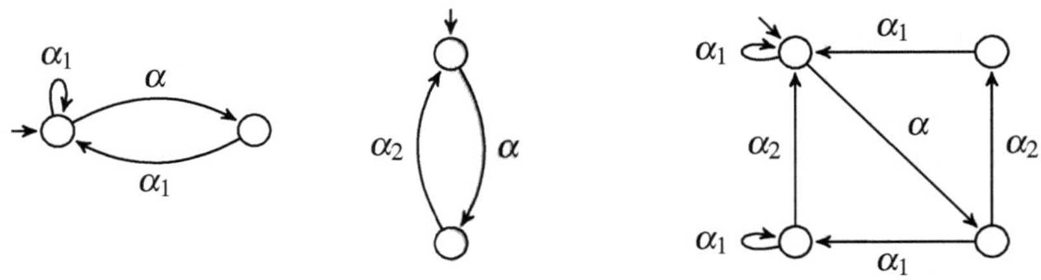

Fig. 5.8 Two automata (left) sharing only label α, and their synchronous product (right)

Observer of compound systems. We consider here the 'canonical' observers derived in the previous section by epsilon-reduction and determinization. Such an observer for $\mathscr{A}$ has $Q = 2^S$ as state set, if S represents states of $\mathscr{A}$.

Proposition 5.5. *Let $\mathscr{A}_1, \mathscr{A}_2$ be two automata, with $\mathscr{A}_i = (S_i, \Sigma_i, I_i, \delta_i)$ and $\Sigma_{o,i}$ as set of observable labels, $i \in \{1,2\}$. If all synchronizations are observable by each automaton,* i.e. $\Sigma_1 \cap \Sigma_2 = \Sigma_{o,1} \cap \Sigma_{o,2}$ *, then $Obs(\mathscr{A}_1 \times \mathscr{A}_2)$ and $Obs(\mathscr{A}_1) \times Obs(\mathscr{A}_2)$ are isomorphic.*

Proof. We consider the epsilon-reduction to the right. For the proof, we show by induction the bisimilarity of the two deterministic automata $Obs(\mathscr{A}_1 \times \mathscr{A}_2)$ and $Obs(\mathscr{A}_1) \times Obs(\mathscr{A}_2)$. Given the one to one correspondence of state sets, this will induce isomorphism.

Our recursion assumption is the following: let $w \in \Sigma_o^*$, and $w_i = \Pi_{\Sigma_{o,i}}(w)$, then $\bar{\delta}(I',w) = \bar{\delta}_1(I_1',w_1) \times \bar{\delta}_2(I_2',w_2)$. In other words, all reachable states q in $Obs(\mathscr{A}_1 \times \mathscr{A}_2)$ have a product form $q = q_1 \times q_2$.

This is obviously true for $w = \varepsilon$. If nothing has been observed, then only (unobservable) private events of the $\mathscr{A}_i$ can have been fired. Let u_i be any unobservable transition sequence in $\mathscr{A}_i$, starting from an initial state. So $s^-(u_i) = s_{i,0} \in I_i$ and $s^+(u_i) = s_i \in I_i' = \delta_i(I_i, \Sigma_{u,i}^*)$ where $\Sigma_{u,i} \triangleq \Sigma_i \setminus \Sigma_{o,i}$ denote unobservable labels of $\mathscr{A}_i$. Any interleaving u of sequences u_1 and u_2 is a firable sequence of transitions in

$\mathscr{A}_1 \times \mathscr{A}_2$, with $s^-(u) = (s_{1,0}, s_{2,0}) \in I$ and so $s^+(u) = (s_1, s_2) \in I' = \delta(I, \Sigma_u^*)$. Conversely, starting from an unobservable sequence u in $\mathscr{A}_1 \times \mathscr{A}_2$, one easily derives the associated u_i in $\mathscr{A}_i$. So this proves $I' = I'_1 \times I'_2$, *i.e.* that the initial state q_0 of $Obs(\mathscr{A}_1 \times \mathscr{A}_2)$ is the product of the initial states $q_{i,0}$ of the $Obs(\mathscr{A}_i)$.

For the recursion, let $q = \bar{\delta}(q_0, w) = (q_1, q_2)$ be an accessible state in $Obs(\mathscr{A}_1 \times \mathscr{A}_2)$, and let $\alpha \in \Sigma_o$. Three cases must be considered.

Case 1: $\alpha \in \Sigma_{o,1} \cap \Sigma_{o,2}$, which corresponds to a synchronization event. Let $s_i \in q_i$, $t_i = (s_i, \alpha, s'_i)$ be a transition of $\mathscr{A}_i$ and u_i be a sequence of silent transitions in $\mathscr{A}_i$, with $s^-(u_i) = s'_i$ and $s^+(u_i) = s''_i$. Then $((s_1, s_2), \alpha, (s'_1, s'_2))$ is a transition of $\mathscr{A}_1 \times \mathscr{A}_2$ and any interleaving u of sequences u_1, u_2 is firable in $\mathscr{A}_1 \times \mathscr{A}_2$ after (s'_1, s'_2) and leads to (s''_1, s''_2). This proves that $q' = \bar{\delta}(q, \alpha)$ in $Obs(\mathscr{A}_1 \times \mathscr{A}_2)$ contains $q'_1 \times q'_2$ where $q'_i = \bar{\delta}_i(q_i, \alpha)$ in $Obs(\mathscr{A}_i)$. For the converse inclusion, consider as above a visible transition $((s_1, s_2), \alpha, (s'_1, s'_2))$ in $\mathscr{A}_1 \times \mathscr{A}_2$ followed by some unobservable sequence u leading to (s''_1, s''_2), and split u into u_1 and u_2 as above. Then $(s_i, \alpha, s'_i)u_i$ is firable in $\mathscr{A}_i$ and leads from s_i to s''_i. This proves $q' \subseteq q'_1 \times q'_2$. So one concludes $\bar{\delta}(q, \alpha) = \bar{\delta}_1(q_1, \alpha) \times \bar{\delta}_2(q_2, \alpha)$.

Case 2: $\alpha \in \Sigma_{o,1} \setminus \Sigma_{o,2}$, which corresponds to a private observable event of $\mathscr{A}_1$. Let $s_1 \in q_1$, $t_1 = (s_1, \alpha, s'_1)$ be a private (observable) transition of $\mathscr{A}_1$ and u_1 be a sequence of silent transitions in $\mathscr{A}_1$, with $s^-(u_1) = s'_1$ and $s^+(u_1) = s''_1$. For any $s_2 \in q_2$, the private sequence $t_1 u_1$ of $\mathscr{A}_1$ is mapped into a sequence tu of $\mathscr{A}_1 \times \mathscr{A}_2$ leading from (s_1, s_2) to (s''_1, s_2), thus leaving $\mathscr{A}_2$ idle. So $q' = \bar{\delta}(q, \alpha) \supseteq q'_1 \times q_2$ where $q'_1 = \bar{\delta}_1(q_1, \alpha)$. And the converse inclusion is derived again as above, which proves $\bar{\delta}(q, \alpha) = \bar{\delta}_1(q_1, \alpha) \times q_2$.

Case 3: $\alpha \in \Sigma_{o,2} \setminus \Sigma_{o,1}$, which corresponds to a private observable event of $\mathscr{A}_2$. Similar to case 2.

The three cases above allow one to extend by one letter the recursion assumption, which induces the desired bisimilarity. □

Remarks

1. The above proposition assumed an epsilon-reduction to the right, but it remains valid with a reduction to the left.
2. Although canonical observers were assumed in the proof, it extends to general observers with minor modifications. One simply has that the product $Obs(\mathscr{A}_1) \times Obs(\mathscr{A}_2)$ yields *one* observer for $\mathscr{A}_1 \times \mathscr{A}_2$. In other words, the bisimilarity holds, but not necessarily the isomorphism.

Application. The application of this proposition to modular/distributed observation is direct: from a given observed sequence $w \in \Sigma_o^*$, derive projections $w_i = \Pi_{\Sigma_{o,i}}(w)$ and feed them to observers $Obs(\mathscr{A}_i)$ to get local state estimates q_i. Then assemble the latter by $q = q_1 \times q_2$ to get a state estimate of the global system. Reading this property in the reverse direction, this means that the interleaving of private events in w_1 and w_2 does not carry information to estimate the state of $\mathscr{A} = \mathscr{A}_1 \times \mathscr{A}_2$. In other words, it is equivalent to observe a total order of events, as sequence w or a *partial order* of events under the form of two partially synchronized sequences (w_1, w_2).

The assumption $\Sigma_1 \cap \Sigma_2 = \Sigma_{o,1} \cap \Sigma_{o,2}$ is crucial and one can check that the proof fails without this argument. Actually, without this assumption, it is possible to build examples where there exists no pair of *finite* deterministic machines that would play the role of local observers $Obs(\mathscr{A}_i)$, and such that their 'composition' in any way would have the same power as an observer of the global system. This is due to the possibly infinite number of assumptions that must be stored in each local observer about the way the two components synchronize in their unobserved sequences.

This is briefly illustrated by the example in Figure 5.9, where $\mathscr{A}_1$ and $\mathscr{A}_2$ are two automata such that $\Sigma_{o,1} = \{\alpha_1, \alpha_2\}$, $\Sigma_{o,2} = \{\beta_1, \beta_2\}$ and $\Sigma_1 \cap \Sigma_2 = \{\gamma_1, \gamma_2\} \subseteq \Sigma_{uo}$. Observe that, by construction, the production of α's and β's in $\mathscr{A}_1 \times \mathscr{A}_2$ alternate[5]. In other words, from a sequence w_1 of α's observed on $\mathscr{A}_1$ and a sequence w_2 of β's observed on $\mathscr{A}_2$, one can recover their interleaving w. So dealing with distributed observations here does not bother state estimation: no interleaving information is lost.

Then observe that as long as the indexes of the α's observed in w_1 match those observed on the β's in w_2, the component $\mathscr{A}_2$ will be in one of the states of its 'upper part,' $\{s'_o, s'_1, s'_2\}$. As soon as these two index sequences differ, $\mathscr{A}_2$ becomes trapped in the states of its lower part $\{s'_3, s'_4\}$. A global observer of $\mathscr{A}_1 \times \mathscr{A}_2$ fed with w is of course able to determine where $\mathscr{A}_2$ finished. But there is no pair of finite local observers for $\mathscr{A}_1$ and $\mathscr{A}_2$ that would have this power, since they would have to store the sequences of indexes of the α's and of the β's they have seen along w_1 and w_2 in order to compare them and decide.

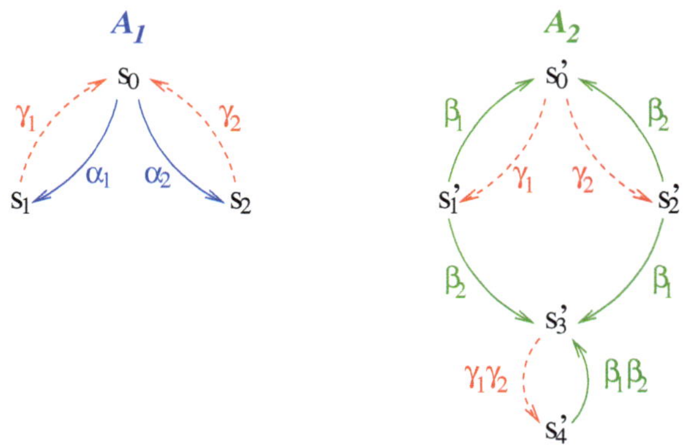

Fig. 5.9 For these two interacting components $\mathscr{A}_1$ and $\mathscr{A}_2$, with non observable interactions, there are no finite local observers that would jointly have the same power as an observer of their product $\mathscr{A}_1 \times \mathscr{A}_2$

[5] This is not strictly the case, since between two consecutive γ the labels α and β can appear in any order. This detail can be easily fixed, at the expense of a more complex example, and does not really bother the rest of the reasoning.

5.6 Distributed State Estimation

The lesson of the previous section is that distributed/modular state estimation (or diagnosis) is easy when synchronizations are observable. What about the general case? Assume a modular system $\mathscr{A} = \mathscr{A}_1 \times \mathscr{A}_2$, with $\Sigma_1 \cap \Sigma_2 \not\subseteq \Sigma_{o,1} \cap \Sigma_{o,2}$, performs a hidden run π that is observed by two sensors, one per component $\mathscr{A}_i$. This yields the pair of observed words (w_1, w_2), with $w_i = \Pi_{\Sigma_{o,i}}(\sigma(\pi))$, where shared events may be observed by only one or none of the two sensors. Notice that the exact interleaving of w_1 and w_2, that would be $w = \Pi_{\Sigma_{o,1} \cup \Sigma_{o,2}}(\sigma(\pi))$, is definitely lost, so all possible interleavings must be assumed to estimate the current state of $\mathscr{A}$. In other words, one truly observes a partial order of events, without the possibility to come back to a unique sequence as in Section 5.5. This constitutes a major change.

In this section we consider automata with marked states $\mathscr{A} = (S, \Sigma, I, \delta, F)$ where $F \subseteq S$. A path π of $\mathscr{A}$ is accepted by $\mathscr{A}$ iff $s^-(\pi) \in I$ and $s^+(\pi) \in F$. The language of $\mathscr{A}$ becomes $\mathscr{L}(\mathscr{A}) = \{\sigma(\pi) : \pi \text{ is accepted by } \mathscr{A}\}$. And for the product of automata, one takes $F = F_1 \times F_2$. To address the state estimation (or diagnosis) problem, we extend it into computing all paths π of $\mathscr{A}$ that can explain (w_1, w_2). Without loss of generality, we also assume that $\mathscr{A}_1$ and $\mathscr{A}_2$ are deterministic, so the problem amounts to finding words of $\mathscr{L}(\mathscr{A})$ that match (w_1, w_2).

Product of languages

Definition 5.3. *Let the $\mathscr{L}_i \subseteq \Sigma_i^*$ be two languages, $i = 1, 2$, their product is defined as $\mathscr{L}_1 \times_L \mathscr{L}_2 = \Pi_{\Sigma_1}^{-1}(\mathscr{L}_1) \cap \Pi_{\Sigma_2}^{-1}(\mathscr{L}_2)$, where the $\Pi_{\Sigma_i} : (\Sigma_1 \cup \Sigma_2)^* \to \Sigma_i^*$ are the natural projections.*

For example, with $\mathscr{L}_1 = \{\alpha\gamma\alpha\} \subseteq \{\alpha, \gamma\}^*$ and $\mathscr{L}_2 = \{\gamma\beta, \gamma\beta\gamma\} \subseteq \{\beta, \gamma\}^*$, one has $\mathscr{L} = \mathscr{L}_1 \times_L \mathscr{L}_2 = \{\alpha\gamma\alpha\beta, \alpha\gamma\beta\alpha\} \subseteq \{\alpha, \beta, \gamma\}^*$. The second word in $\mathscr{L}_2$ matches no word of $\mathscr{L}_1$, while the first one interleaves in two different ways with the only word of $\mathscr{L}_1$. Observe the following result:

Lemma 5.1. *Let $\mathscr{L} = \mathscr{L}_1 \times_L \mathscr{L}_2$, and $\mathscr{L}_i' = \Pi_{\Sigma_i}(\mathscr{L})$, then $\mathscr{L}_i' \subseteq \mathscr{L}_i$ and one has $\mathscr{L} = \mathscr{L}_1' \times_L \mathscr{L}_2'$. The $\mathscr{L}_i'$ are the minimal sublanguages of the $\mathscr{L}_i$ that allow one to recover $\mathscr{L}$.*

The proof is left to the reader as an exercise. As a direct application of $\times_L$, observe that $\{w_1\} \times_L \{w_2\}$, denoted $w_1 \times_L w_2$ for short, yields all interleavings w of w_1 and w_2 that must be considered.

The products of automata and of languages are related by the following property:

Proposition 5.6. *Let $\mathscr{A} = \mathscr{A}_1 \times \mathscr{A}_2$, then $\mathscr{L}(\mathscr{A}) = \mathscr{L}(\mathscr{A}_1) \times_L \mathscr{L}(\mathscr{A}_2)$.*

The proof is left to the reader as an exercise. Notice that $\Pi_{\Sigma_i}(\mathscr{L}(\mathscr{A})) \subseteq \mathscr{L}(\mathscr{A}_i)$ represents the behaviors of $\mathscr{A}_i$ that remain possible once the other component is connected.

Application to hidden run recovery. Assume centralized observation: one collects $w = \Pi_{\Sigma_o}(\sigma(\pi)) \in \Sigma_o^*$. The runs (or equivalently the words) of $\mathscr{A}$ that explain w are given by $\mathcal{E} = \mathcal{L}(\mathcal{A}) \times_L w$. Let $\mathcal{W}$ be a deterministic automaton such that $\mathcal{L}(\mathcal{W}) = \{w\}$, one also has $\mathcal{E} = \mathcal{L}(\mathcal{A} \times \mathcal{W})$. So one can take $\mathcal{A} \times \mathcal{W}$ as a compact and finite representation of this possibly infinite language (set of runs).

With distributed observations, one has (see Prop. 5.6)

$$\mathcal{E} = \mathcal{L}(\mathcal{A}_1 \times \mathcal{A}_2) \times_L (w_1 \times_L w_2) \tag{5.14}$$
$$= (\mathcal{L}(\mathcal{A}_1) \times_L w_1) \times_L (\mathcal{L}(\mathcal{A}_2) \times_L w_2) \tag{5.15}$$
$$= \mathcal{E}_1 \times_L \mathcal{E}_2 \tag{5.16}$$

where $\mathcal{E}_i = \mathcal{L}(\mathcal{A}_i) \times_L w_i$ represents local explanations to observation w_i in component $\mathcal{A}_i$. One is actually interested in building a *distributed explanation*, under the form $(\mathcal{E}'_1, \mathcal{E}'_2)$, where $\mathcal{E}'_i = \Pi_{\Sigma_i}(\mathcal{E})$ represents the local view in component $\mathcal{A}_i$ of runs of $\mathcal{A}$ that explain all observations w_1 and w_2. Again, one has $\mathcal{E} = \mathcal{E}'_1 \times_L \mathcal{E}'_2$ (see Lemma 5.1).

Distributed computation of a distributed explanation. The objective here is to determine directly the elements $\mathcal{E}'_i$ of a distributed explanation... without computing $\mathcal{E}$ itself! This can be done in a distributed manner, by message exchanges between the local 'supervisors' in charge of each component. The key idea is a notion of *conditional independence* on languages:

Proposition 5.7. *Let $\mathcal{L}_i \subseteq \Sigma_i^*$, $i = 1, 2$, be two languages, and let $\Sigma_1 \cap \Sigma_2 \subseteq \Sigma' \subseteq \Sigma_1 \cup \Sigma_2$, then $\Pi_{\Sigma'}(\mathcal{L}_1 \times_L \mathcal{L}_2) = \Pi_{\Sigma'}(\mathcal{L}_1) \times_L \Pi_{\Sigma'}(\mathcal{L}_2)$.*

In our setting, taking $\Sigma' = \Sigma_1$, this induces for example

$$\mathcal{E}'_1 = \Pi_{\Sigma_1}(\mathcal{E}_1 \times_L \mathcal{E}_2) = \mathcal{E}_1 \times_L \Pi_{\Sigma_1 \cap \Sigma_2}(\mathcal{E}_2) \tag{5.17}$$

and symmetrically for $\mathcal{E}'_2$. Equation (5.17) expresses that the local view $\mathcal{E}'_1$ of global explanations $\mathcal{E}$ are obtained by synchronizing the local explanations $\mathcal{E}_1$ on component $\mathcal{A}_1$ with the *message* $\Pi_{\Sigma_1 \cap \Sigma_2}(\mathcal{E}_2)$ from component $\mathcal{A}_2$. This message propagates the constraints that explanations $\mathcal{E}_2$ impose on synchronizations. Given the small alphabet $\Sigma_1 \cap \Sigma_2$ and the projection operation that removes private events of $\mathcal{A}_2$, (5.17) generally involves smaller objects than $\mathcal{E}$.

Example 5.2. The above computations involve possibly infinite languages. But again, they can be translated into automata computations thanks to Prop. 5.6 and to the fact that projection as well preserves the regularity (recall the construction of observers in Section 5.2, where $\mathcal{L}(\mathcal{A}'') = \mathcal{L}(\mathcal{A}') = \Pi_{\Sigma_o}(\mathcal{L}(\mathcal{A}))$).

Consider the example in Fig. 5.10, with two components and a distributed observations (b, d) represented as two single word automata. Fig. 5.11 computes the local explanations $\mathcal{E}_i$ by product $\mathcal{E}_i = \mathcal{L}(\mathcal{A}_i) \times_L w_i$, represented as $\mathcal{A}_i \times \mathcal{W}_i$.

Eq. (5.17) is illustrated in Fig. 5.12 (top), where the central automaton (obtained by projection) represents the message from $\mathcal{A}_2$ to $\mathcal{A}_1$. The bottom figure illustrates the message propagation and integration in the reverse direction (bottom), *i.e.* the symmetric version of (5.17).

The final distributed explanations are obtained by taking the languages of the automata at the top left and bottom right in Fig. 5.12. One has $\mathcal{E}'_1 = \{a\alpha b\alpha, \beta b\}$ and

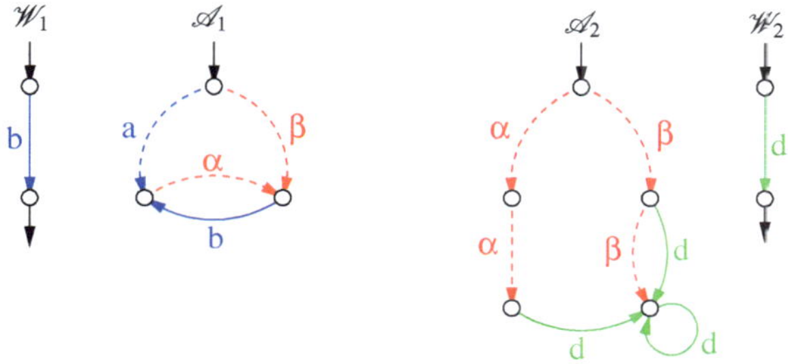

Fig. 5.10 Components $\mathscr{A}_1, \mathscr{A}_2$, on $\Sigma_1 = \{a, b, \alpha, \beta\}$ and $\Sigma_2 = \{\alpha, \beta, d\}$ resp. Only labels b and d are observable (dashed transitions are unobservable). All states are final in the $\mathscr{A}_i$. Automata $\mathscr{W}_1, \mathscr{W}_2$ encode the observed words w_1, w_2; only their bottom states are final

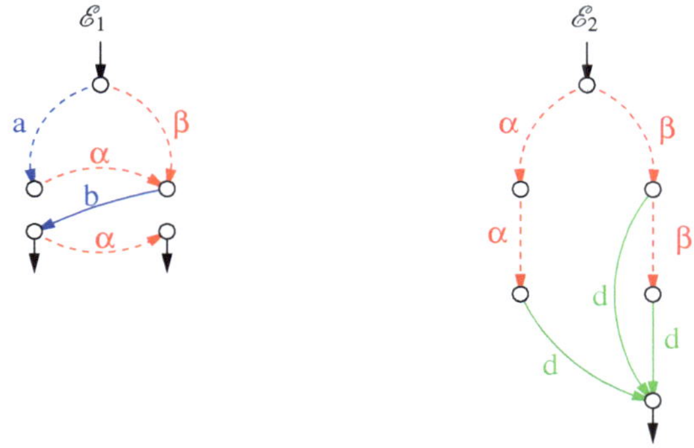

Fig. 5.11 Local explanations to the observed word of w_i in each component $\mathscr{A}_i$, represented as the language of $\mathscr{E}_i = \mathscr{A}_i \times \mathscr{W}_i$

$\mathscr{E}'_2 = \{\alpha\alpha d, \beta d\}$. Each word in $\mathscr{E}'_1$ matches at least one word of $\mathscr{E}'_2$, and *vice versa*. This yields two pairs of runs of $\mathscr{A}_1$ and $\mathscr{A}_2$ that explain the distributed observation (b, d): $(a\alpha b\alpha, \alpha\alpha d)$ and $(\beta b, \beta d)$. Observe that each pair can be interleaved in several manners to produce explaining runs of $\mathscr{A}_1 \times \mathscr{A}_2$. This shows the interest of distributed state/run computations: useless interleavings need not be explored, which can greatly reduce the search space. ■

Remarks

1. If component $\mathscr{A}_1$ is not deterministic, one can easily recover its explaining runs in the 'automaton version' of (5.17): $\mathscr{A}'_1 \triangleq \mathscr{A}_1 \times [\mathscr{W}_1 \times \Pi_{\Sigma_1 \cap \Sigma_2}(\mathscr{A}_2 \times \mathscr{W}_2)]$. The bracketted term simply constrains the runs of $\mathscr{A}_1$. Any run of $\mathscr{A}'_1$ (restricted to its component in $\mathscr{A}_1$) is a local view of a run π of $\mathscr{A}$ that explains the distributed observation (w_1, w_2).
2. From the runs of the $\mathscr{A}_i$ that match the distributed observation (w_1, w_2), one easily recovers the possible final states of $\mathscr{A}_i$, and consequently can establish a diagnosis, relying on the state augmentation trick.

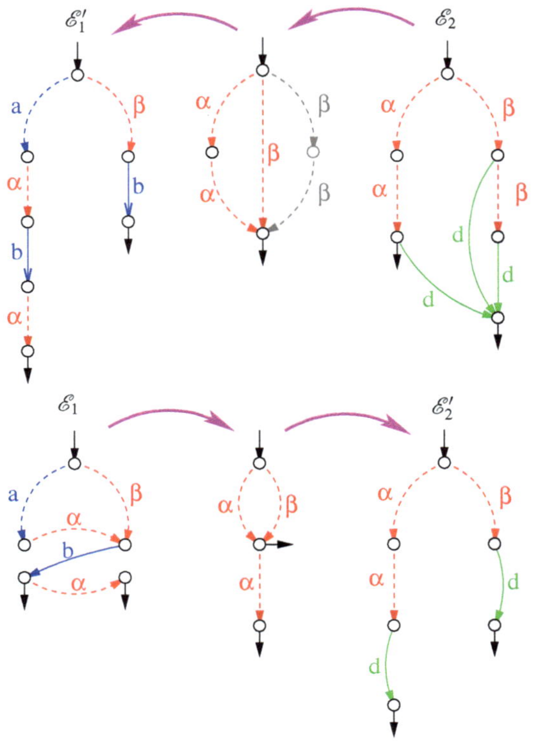

Fig. 5.12 Message propagation from $\mathscr{A}_2$ to $\mathscr{A}_1$ (top) and from $\mathscr{A}_1$ to $\mathscr{A}_2$ (bottom) to compute the local views $\mathscr{E}_1', \mathscr{E}_2'$ of global explanations

3. Observe that, by contrast with the counter-example of the previous section, we have a single observation and with limited length here. This does not prevent however having to check an infinite number of assumptions on the possible (unobserved) synchronizations that take place between the two components. What makes the approach work is that this possibly infinite set of explanations can be condensed into an automaton, thanks to its regularity.

5.7 Conclusion and Further Reading

What are the lessons of the above developments? First of all, diagnosis and state estimation are closely related problems, if not equivalent. Diagnosers and observers are obtained by simple and similar operations on automata. They extend without difficulty to weighted automata, and in particular to probabilistic automata, with the main difference that determinization may not yield a finite automaton. But if it is the case, they yield structures that output detection probabilities. Diagnosability, or observability of a state property, represents the ability to detect a fault/property not long after it occurs/becomes true. It can be checked on the observer/diagnoser, but is more efficiently tested by a direct method, using the so-called twin-machine. Observability/diagnosability extends to the probabilistic case, and takes the form of the convergence (with probability 1) of some numerical indicator after this property becomes true. In the simplest case, this indicator switches suddenly to 1 (A-diagnosability), while for AA-diagnosability it converges to 1. The first case is rather simple, and actually A-diagnosability can be translated into a structural property of the non-probabilistic observer, or of the twin-machine of $\mathscr{A}$.

Other important lessons relate to distributed or modular systems. Having distributed observations amounts to considering a global observation as a partial orders of events. And similarly, representing runs of a distributed system as a tuple of partially synchronized sequences amounts to considering them as partial orders of events. This somehow invisible change of semantics greatly saves in complexity when dealing with distributed systems. It still allows one to perform state estimation or diagnosis, possibly with distributed methods. Notice that for distributed diagnosis, the properties one wishes to characterize must also be expressible as a product of local properties (one per component). Or they should be separable, following the vocabulary of [20]. Distributed/modular diagnosability has been examined by different authors [18], assuming or not that synchronization events are observable, which greatly simpifies the problem, as mentioned above. Modular state estimation or diagnosis for probabilistic systems remains an open issue. A central difficulty in such settings is to define a probabilistic setting that is compatible with the concurrency of events: a not careful way of combining probabilistic automata generally produces weird phenomena, for example private events of some component may change the occurrence probability of private events in another component... A next step towards the management of distributed systems consists in adopting true concurrency semantics. See [6] for a discussion, and Chapter 15 or [1, 5] for a detailed treatment.

References

1. Benveniste, A., Fabre, E., Jard, C., Haar, S.: Diagnosis of asynchronous discrete event systems a net unfolding approach. IEEE Transactions on Automatic Control 48(5), 714–727 (2003)
2. Buchsbaum, A.L., Giancarlo, R., Westbrook, J.R.: On the determinization of weighted finite automata. SIAM Journal on Computing 30, 1502–1531 (1998)
3. Cassandras, C.G., Lafortune, S.: Introduction to Discrete Event Systems, 2nd edn. Springer (2008)
4. Cortes, C., Mohri, M., Rastogi, A., Riley, M.D.: Efficient Computation of the Relative Entropy of Probabilistic Automata. In: Correa, J.R., Hevia, A., Kiwi, M. (eds.) LATIN 2006. LNCS, vol. 3887, pp. 323–336. Springer, Heidelberg (2006)

5. Fabre, E., Benveniste, A., Haar, S., Jard, C.: Distributed monitoring of concurrent and asynchronous Systems. Journal of Discrete Event Systems 15(1), 33–84 (2005)
6. Fabre, E., Benveniste, A.: Partial order techniques for distributed discrete event systems: why you can't avoid using them. Journal of Discrete Events Dynamical Systems 17(3), 355–403 (2007)
7. Fabre, E., Jezequel, L.: Distributed optimal planning: an approach by weighted automata calculus. In: Proc. 48th Conference on Decision and Control, Shangai, China (2009)
8. Fabre, E., Jezequel, L.: On the construction of probabilistic diagnosers. In: Proc. 10th Workshop on Discrete Event Systems, Berlin, Germany (2010)
9. Jeron, T., Marchand, H., Pinchinat, S., Cordier, M.O.: Supervision patterns in discrete event systems diagnosis. In: Proc. 8th Workshop on Discrete Event Systems, Ann Arbor, Michigan (2006)
10. Kirsten, D., Murer, I.: On the determinization of weighted automata. Journal of Automata, Languages and Combinatorics 10(2/3), 287–312 (2005)
11. Mohri, M.: Weighted automata algorithms. In: Kuich, W., Vogler, H., Droste, M. (eds.) Handbook of Weighted Automata. Springer (2009)
12. Mohri, M.: Finite-state transducers in language and speech processing. Computational Linguistics 23, 269–311 (1997)
13. Mohri, M.: Generic epsilon-removal and input epsilon-renormalization algorithms for weighted transducers. International Journal on Foundations of Computer Sciences 13(1), 129–143 (2002)
14. Paz, A.: Introduction to Probabilistic Automata. Academic Press, New-York (1971)
15. Sampath, M., Sengupta, R., Lafortune, S., Sinnamohideen, K., Teneketzis, D.: Diagnosability of discrete-event systems. IEEE Transactions on Automatic Control 40(9), 1555–1575 (1995)
16. Thorsley, D., Teneketzis, D.: Diagnosability of stochastic discrete-event systems. IEEE Transactions on Automatic Control 50(4), 476–492 (2005)
17. Thorsley, D., Yoo, T.S., Garcia, H.E.: Diagnosability of stochastic discrete-event systems under unreliable observations. In: Proc. American Control Conference, Seattle, USA (2008)
18. Ye, L., Dague, P.: An optimized algorithm for diagnosability of component-based systems. In: Proc. 10th Workshop on Discrete Event Systems, Berlin, Germany (2010)
19. Yoo, T.S., Lafortune, S.: Polynomial-time verification of diagnosability of partially observed discrete-event systems. IEEE Transactions on Automatic Control 47(9), 1491–1495 (2002)
20. Zhou, C., Kumar, R., Sreenivas, R.S.: Decentralized modular diagnosis of concurrent discrete event systems. In: 9th Workshop on Discrete Event Systems, Goteborg, Sweden (2008)

Chapter 6
Supervisory Control of Distributed Discrete-Event Systems

Jan Komenda, Tomáš Masopust, and Jan H. van Schuppen

6.1 Introduction

The purpose of this chapter is to introduce to the reader the problem of supervisory control of distributed discrete-event systems. This problem area is highly relevant for current engineering and for companies developing high-tech systems. There are no fully satisfactory solutions yet. This chapter has therefore more the character of an introduction to the problem area and a research program, rather than the exposition of a fully investigated research topic. This chapter may also be regarded as an introduction to the Chapters 7 and 8.

Control engineering of high-tech systems nowadays is about systems consisting of very many components or subsystems. Examples of distributed discrete-event systems are large printers, an MRI scanner, a chemical plant, automobiles, aerial vehicles, etc. Each component was separately designed and has a controller developed for that component exclusively. But a large high-tech system has to meet a specification which requires the interaction of all the subsystems. Control engineering and control theory therefore have to address the cooperation or the coordination of these subsystems.

In this chapter distributed discrete-event systems and the related concept of a decentralized discrete-event system are defined. Four control architectures are then defined: (1) distributed control, (2) distributed control with communication, (3) coordination control, and (4) hierarchical control. The control problems and the research approaches are then discussed. Results are presented for several subtopics

Jan Komenda · Tomáš Masopust
Institute of Mathematics, Academy of Sciences of the Czech Republic, Žižkova 22, 616 62 Brno, Czech Republic
e-mail: komenda@ipm.cz, masopust@math.cas.cz

Jan H. van Schuppen
CWI, P.O. Box 94079, 1090 GB Amsterdam, Netherlands
e-mail: J.H.van.Schuppen@cwi.nl

C. Seatzu et al. (Eds.): Control of Discrete-Event Systems, LNCIS 433, pp. 107–126.
springerlink.com

of decentralized and distributed control. The chapter ends with research issues for control of distributed discrete-event systems and advice of further reading.

6.2 Motivation

Example 6.1. *The alternating bit protocol.* The protocol or variants of it are used in many communication networks. It is a canonical model of a distributed discrete-event system.

The engineering model of the communication system consists of a *sender*, a *receiver*, and a *communication channel.* The sender gets packets from the host at which it is located. The sender sends the package to the communication channel. The communication channel delivers a packet to the receiver either fully, or in a distorted way, or does not deliver the packet at all.

The alternating bit protocol operates as described below. The sender sends a packet and attaches a bit, either a zero or a one, to the header of the packet. The receiver receives information. If the receiver finds that the header of the packet is undamaged then it sends an acknowledgement to the sender including the protocol bit. The sender, if it does not receive an acknowledgement in a prespecified period, sends the packet again with the same bit. The sender, if it receives an acknowledgement of the receiver with the protocol bit, sends the next message with the next alternate protocol bit.

Note that the sender possesses information about the packets to be sent and that have been sent but not about the packets received by the receiver. Similarly, the receiver knows what it has received but not what was sent by the sender. The distributed character of this system is in the different information and the different control actions at the two locations, at the sender and at the receiver. There is also communication from the receiver to the sender. The alternating bit protocol is a form of distributed control with communication. An extension of the above protocol exists in which, instead of only one bit, two or more bits are used so four, eight, or more messages can be in process simultaneously, see [54]. ■

Example 6.2. *Underwater vehicles.* At the university of Porto in Portugal there is a laboratory in which control engineering of underwater vehicles is developed. This is a form of control of distributed systems. Part of the control deals with continuous space systems, but part of the control addresses the distributed discrete-event system.

Consider then a group of vehicles consisting of a surface vehicle, which acts as the local command center, and two or more underwater vehicles. The communication between the surface vehicle and the underwater vehicle is sonar communication, by sound waves. This form of communication requires relatively much energy from the underwater vehicle having limited battery power on board.

The operation of the vehicles is a characteristic example of a distributed system. Each vehicle is a subsystem of the group system. Each vehicle has a local observed event stream of its position and speed, and of its control actions. An underwater vehicle may communicate its position, speed, and action to the surface vehicle, either regularly, or when requested, or when required by a protocol.

A particular operation is formation flying. The surface vehicle follows a path specified by the command center. It then sends instructions to the underwater vehicles for a position and a speed to be reached at a particular time. The underwater vehicles then carry out the instructions and react only if they cannot meet the requirements, see [12].

Characteristic of this example is the local availability of the state information of the subsystem and the communication. ■

There follows a list of control engineering problems for which models in the form of decentralized/distributed discrete-event systems have been formulated and for which supervisory control has been investigated:

1. The alternating bit protocol described in Example 6.1. See [54], while discrete-event models are described in [40, 38].
2. Communication networks are a rich source of control of distributed and decentralized systems, see [8, 54].
3. Feature interaction in telephone networks [50].
4. Manufacturing cells.
5. Control of high-speed printers with many sensors and actuators [26].
6. A chemical pilot plant for which modular control is applied [21, Ch. 7].
7. Distributed algorithms are studied by computer scientists, see [25, 31, 48, 53]. Supervisory control is not studied in most of those references.

6.3 Systems

In this section, the reader is provided a classification of the system architectures for decentralized and for distributed discrete-event systems. In the literature there is no standardized nomenclature while one is needed. The concepts formulated below should be regarded as preliminary. The terms and notations used below have been introduced in the Chapters 2, 3, and 4.

Definition 6.1. (Overview of system architectures). *The acronym DES stands for a discrete-event system.*

1. A decentralized discrete-event system *is a global plant modeled as a DES with two or more observed event streams and two or more inputs of enabled events. Each controller receives an observed event stream defined by either a projection or a mask, and inputs a subset of enabled events.*
2. A distributed discrete-event system *is a DES consisting of the interconnection of two or more subsystems. Each subsystem has an observed event stream of local events and an input of enabled events to the local subsystem.* A modular

discrete-event system *is the same as a distributed DES, the term is still used in the literature. The authors prefer the more general term of a distributed DES.*
3. *An* asynchronous timed distributed discrete-event system. *This is defined as a distributed discrete-event system in which each subsystem has its own clock and the clocks may drift with respect to each other. An example is an audio set consisting of a tuner, a CD player, and an amplifier, see [3]. Asynchronous systems require a timed DES model.*

The main difference between a decentralized DES and a distributed DES is the use of the decomposition of the system into an interconnection of subsystems in a distributed DES while in a decentralized DES such a distinction is not made. For future research it is expected that the decomposition of the system into its subsystems can be usefully exploited for control synthesis.

Definition 6.2. *A* decentralized discrete-event systems, *see Fig. 6.1, is a tuple denoted by* $G_{dec} = (Q, E, f, q_0, Q_m, \{E_{i,c}, i \in I\}, \{E_{i,o}, i \in I\})$, *where* $I = \{1, 2, \ldots, n\}$ *denotes the* index set of the observed event streams, $\{E_{i,c},\ E_{i,uc} \subseteq E,\ i \in I\}$ *is a partition of* E, $E_{i,c}$ *and* $E_{i,uc}$ *denote, respectively, the subset of* controllable events *and of* uncontrollable events *of Controller i*, $\{E_{i,o},\ E_{i,uo} \subseteq E,\ i \in I\}$ *is a partition of* E, *and* $E_{i,o}$ *and* $E_{i,uo}$ *denote, respectively, the* subsets of observable events *and* of unobservable events *of Controller i.*

Definition 6.3. *A* distributed discrete-event system, *see Fig. 6.2, is a set of discrete-event systems denoted by* $G_{dis} = \{G_i,\ i \in I\}$, *with* $G_i = (Q_i, E_i, f_i, q_{i,0}, Q_{i,m})$ *and* $E_{i,c}$, $E_{i,o}$, *where* $I = \{1, 2, \ldots, n\}$ *denotes the* index set of the subsystems, G_i *for each* $i \in I$ *is an automaton,* E_i *denotes the* event set of Subsystem i, $E_{i,c} \subseteq E_i$ *denotes the* subset of controllable events of Subsystem i, *and* $E_{i,o} \subseteq E_i$ *denotes the* subset of observable events of Subsystem i.

Definition 6.4 (Control architectures)

1. Decentralized/distributed control, *see Fig. 6.3. The observed event stream of a controller consists of a projection or mask of the strings of the system which for a distributed system are restricted to the local subsystem. Each controller inputs a subset of enabled events to the subsystem. There is no direct communication whatsoever with other controllers though the controllers communicate indirectly with other controllers via the system.*
2. Distributed control with communication, *see Fig. 6.4. Controllers may send part of their observed event stream or of their states to other controllers. Each controller uses besides its observed event stream received directly from the plant also the other observed event stream received from other controllers. An example is the class of nearest neighbor controls which use for the supervisory control the state of the local subsystem and the states of the nearest neighbors of the local subsystem.*

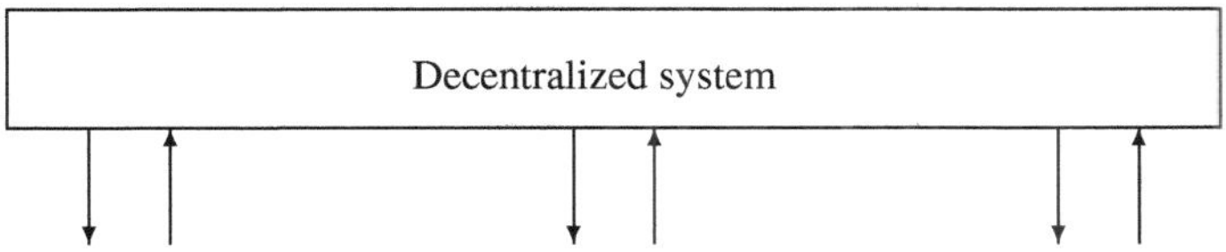

Fig. 6.1 Diagram of a decentralized system

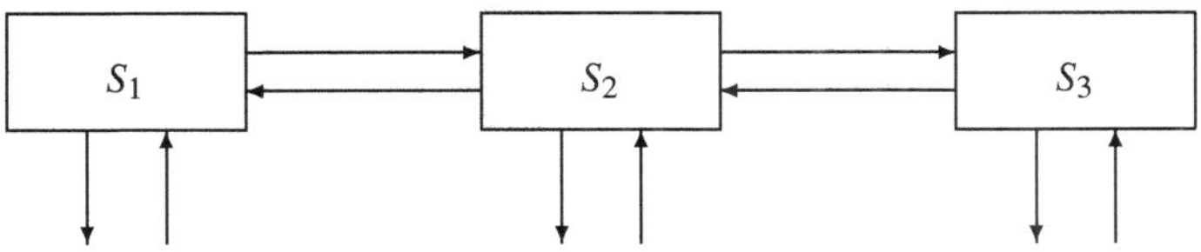

Fig. 6.2 Diagram of a distributed system

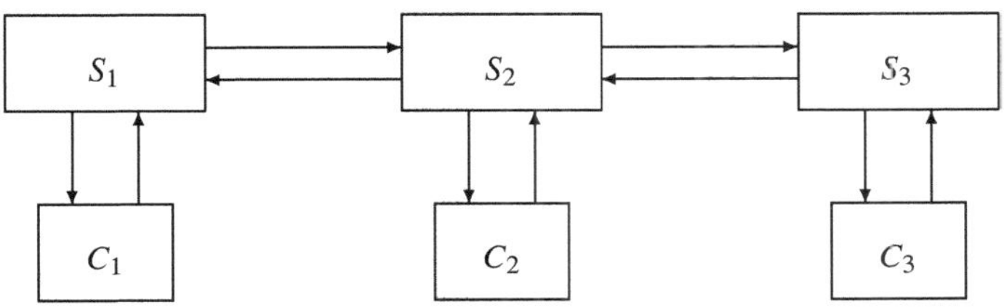

Fig. 6.3 Diagram of distributed control of a distributed discrete-event system

3. Coordination control *of a coordinated DES, see Fig. 6.5. A coordinated DES is defined as a distributed DES distinguished into a* coordinator *and the remaining* subsystems. *The coordinator has the control-theoretic task to coordinate the actions of the other subsystems. There is a controller for the coordinator and for each of the subsystems.*
4. Hierarchical control. *There is a controller for each subsystem at each level of the hierarchy.*

Guidelines for the appropriate choice of a control architecture are not much discussed in the literature and deserve more attention. The tradeoff between more central control or more distributed control has to be made on a case-by-case basis. The principle is often stated that it is best to have the subsystems operate as independently as possible. But in many examples of distributed control problems a degree of coordination or centralized control is necessary to achieve the control objectives.

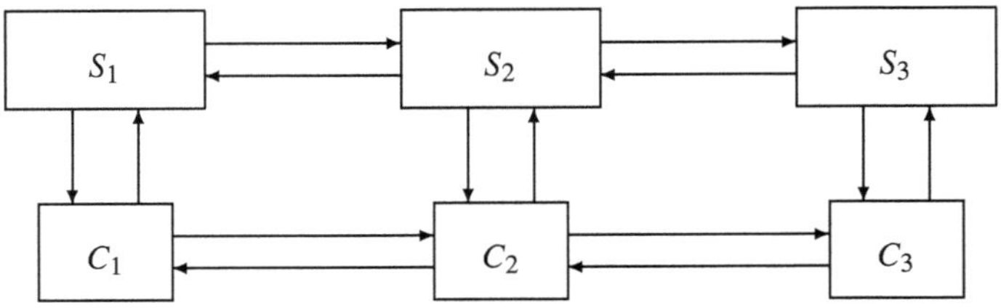

Fig. 6.4 Diagram of distributed control with communication. In general, there is also communication between the controllers C_1 and C_3 which however is not displayed in the figure.

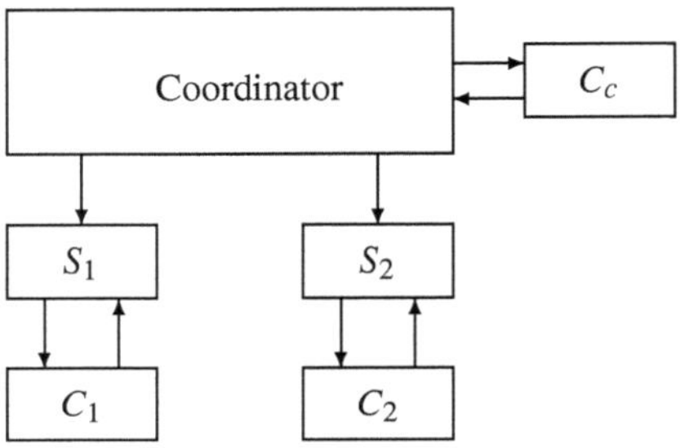

Fig. 6.5 Diagram of a controlled coordinated discrete-event system

6.4 Problem Formulation

There follows a verbal general problem statement which will be refined in the subsequent sections. The reader is assumed to be familiar with supervisory control of a discrete-event system both with complete observations and with partial observations as described the Chapters 3 and 4.

Consider a distributed or a decentralized DES with two or more observed event streams. Consider a specification language, either a global specification or a set of local specifications. Determine a set of supervisory controls, as many as there are inputs of enabled subsets of events, such that the closed-loop system meets the control objectives of safety, required behavior, nonblockingness, and of fairness.

Definition 6.5 (Control objectives of control of distributed discrete-event systems). *There are of course the control objectives of supervisory control of discrete-event systems such as safety, required behavior, and nonblockingness. In addition, there are control objectives particular for control of distributed DES:*

- Fairness. *Each subsystem is regularly provided access to all shared resources.*
- Non-starvation. *No subsystem is denied access forever to a shared resource.*

The main difficulties of distributed control are: (1) control synthesis and design: the construction of a tuple of supervisory controls; more specifically, the fact that the partial observations of the subsystems differ and are not nested; and (2) the decidability and the complexity issues of the problem which are enormous.

In the next sections follow special cases of the above problem.

6.5 Decentralized Control – Existence and Construction of a Tuple of Supervisors

In this section, the solution is presented to the supervisory control problem of a decentralized DES. It is remarkable that a necessary and sufficient condition for the existence of a tuple of supervisory controls can be presented as will be explained later. The main reference of this section is [39] and the results are extendable from two to three or more supervisors.

Definition 6.6. [39] *Consider a decentralized DES. For the supervisors* (S_1, g_1) *and* (S_2, g_2) *over the event sets* E_1 *and* E_2*, respectively, where* $P_1 : E_o^* \to E_{1,o}^*$ *and* $P_2 : E_o^* \to E_{2,o}^*$ *are natural projections, define a supervisor* $(S_1 \wedge S_2, g_1 * g_2)$ *over the global observable event set* $E_o = E_{1,o} \cup E_{2,o}$ *with* $(g_1 * g_2)(s) = g_1(P_1(s)) \cap g_2(P_2(s))$ *for* $s \in E_o^*$.

Read this as: enable the event if both S_1 and S_2 enable the event. Hence it is called the *conjunctive* ($\cap$) *and permissive control-implementation architecture*. The alternative is called the *disjunctive and antipermissive control-implementation architecture*. The supervisors working with partial observations are permissive while the global fusion rule is conjunctive.

Proposition 6.1. [39]. *Consider a distributed discrete-event system. The closed-loop system has the following properties:* $L(S_1 \wedge S_2/G) = L(S_1/G) \cap L(S_2/G)$ *and* $L_m(S_1 \wedge S_2/G) = L_m(S_1/G) \cap L_m(S_2/G)$.

Recall from Chapter 3 that $E_{1,cp}$ denotes the set of control patterns of Subsystem 1.

Definition 6.7 (Global and local supervisors). [39] *Let* (S_1, g_1), $g_1 : E_{1,o}^* \to E_{1,cp}$ *be a* local supervisor. *Its extension to the global event set* E*, called the* global supervisor, *is defined as* $(\tilde{S}_1, \tilde{g}_1)$*, where* $\tilde{g}_1 : E_o^* \to E_{cp}$, $\tilde{g}_1(s) = g_1(s)\ \forall\ s \in E_{1,o}^*$*, and it enables all events of* $E_o \backslash E_{1,o}$*. The corresponding extension is defined for* (S_2, g_2).

Problem 6.1. [39]. *Decentralized control for a global legal specification.* Consider a decentralized DES G with a regular specification language $\emptyset \neq K \subseteq L_m(G)$. Here $E_{uc} = E \backslash (E_{1,c} \cup E_{2,c})$. Construct supervisors (S_1, g_1) and (S_2, g_2) such that (1) $L_m(\tilde{S}_1 \wedge \tilde{S}_2/G) = K$ and (2) $(\tilde{S_1} \wedge \tilde{S_2}/G)$ is nonblocking.

The solution to the above formulated problem requires introduction of concepts.

Definition 6.8. [39] *Consider Problem 6.1. Define the relations* next action, *denoted* $\text{nextact}_K \subseteq E^* \times E \times E^*$*, so that* $(s_1, e, s_2) \in \text{nextact}_K$ *if* $s_1 e \in \text{prefix}(K)$,

$s_2 \in \text{prefix}(K)$, *and* $s_2 e \in L(G)$ *imply that* $s_2 e \in \text{prefix}(K)$; *and* mark action, *denoted* $\text{markact}_K \subseteq E^* \times E^*$, *so that* $(s_1, s_2) \in \text{markact}_K$ *if* $s_1 \in K$ *and* $s_2 \in \text{prefix}(K) \cap L_m(G)$ *imply that* $s_2 \in K$.

Definition 6.9 (Coobservability). [39] *Consider Problem 6.1. The sublanguage* $K \subseteq L_m(G)$ *is called* coobservable *with respect to* (G, P_1, P_2) *if*

$$\begin{aligned}
&\forall s, s_1, s_2 \in E^* \text{ with } P_1(s) = P_1(s_1),\ P_2(s) = P_2(s_2),\\
&\quad \Rightarrow \forall e \in E_{1,c} \cap E_{2,c},\ (s,e,s_1) \in \text{nextact}_K \vee (s,e,s_2) \in \text{nextact}_K; && (6.1)\\
&\quad \wedge\ \forall e \in E_{1,c} \backslash E_{2,c},\ (s,e,s_1) \in \text{nextact}_K; && (6.2)\\
&\quad \wedge\ \forall e \in E_{2,c} \backslash E_{1,c},\ (s,e,s_2) \in \text{nextact}_K; && (6.3)\\
&\quad \wedge\ (s,s_1) \in \text{markact}_K \vee (s,s_2) \in \text{markact}_K. && (6.4)
\end{aligned}$$

The concept of next action describes that the events are related if the corresponding observed strings are indistinguishable (Conjuncts 6.1-6.3). The marking relation describes that marking actions are related (Conjunct 6.4). Coobservability of a decentralized system corresponds to invariance of the closed-loop system with respect to control with distributed partial observations. There exists an algorithm to check whether a language is coobservable, see [37].

(a)
(b)

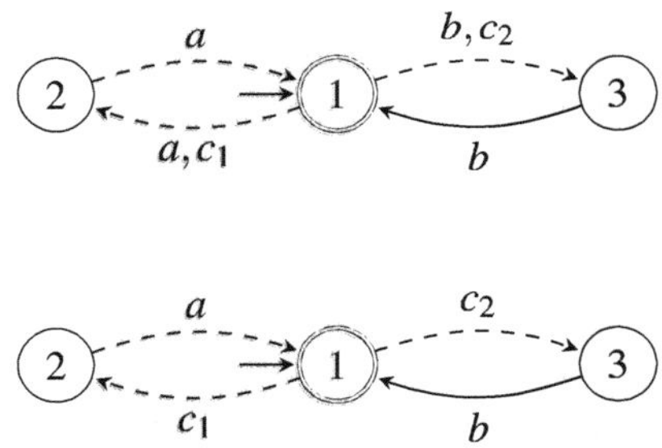

Fig. 6.6 (a) A two-ring discrete-event system and (b) its specification

Example 6.3. *Coobservability compared with observability.* [39]. Consider the plant G of which the diagram is displayed in Fig. 6.6(a) and its specification K of which the diagram is displayed in Fig. 6.6(b). Denote the event sets by $E_{1,c} = E_{1,o} = \{a, b, c_1\}$, $E_{2,c} = E_{2,o} = \{a, b, c_2\}$. Then K is coobservable; it is neither observable with respect to (G, P_1) nor with respect to (G, P_2). There exists a tuple of supervisory controls such that $L_m(\tilde{S}_1 \wedge \tilde{S}_2 / G) = K$, see [39, Ex. 4.2]. ■

Definition 6.10 (C & P Coobservability). *Consider Problem 6.1. The sublanguage* $K \subseteq L_m(G)$ *is called* C & P coobservable *with respect to* $(L(G), \{E_{o,i}, i \in I\})$ *if*

$$\begin{aligned}
&\forall\, s \in \text{prefix}(K),\ \forall\, e \in E_c \text{ such that } se \in L(G),\\
&\exists\, i \in I \text{ such that } e \in E_{i,c},\ s' \in \text{prefix}(K),\ P_i(s) = P_i(s'),\ s'e \in \text{prefix}(K)\\
&\Rightarrow se \in \text{prefix}(K).
\end{aligned}$$

The concept of C & P coobservability stated above is based on a corresponding concept defined in [60] and [15]. C & P coobservability can be proven to be equivalent to coobservability. Several researchers prefer its interpretation over that of Definition 6.9.

Theorem 6.1 (Existence supervisors). See [39, Th. 4.1]. *There exists a solution to Problem 6.1 of decentralized control for a global regular legal specification language $K \subseteq L_m(G)$ if and only if (1) the language K is controllable with respect to (G, E_{uc}) and (2) the language K is coobservable with respect to (G, P_1, P_2).*

An alternative proof is provided in [60]. There is a similar result in case of the D&A control implementation architecture, see [60].

Note that the language $K \subseteq L_m(G)$ is fixed by the problem statement and is not to be determined in the problem. This is the main difference with respect to decentralized control of nonlinear systems and is the main reason why this decentralized control problem admits an explicit solution.

Algorithm 6.4. (Construction of a tuple of supervisors). [39, p. 1701].
Data algorithm. *Consider Problem 6.1 with a decentralized DES and with the regular language $K \subseteq L_m(G)$ having a recognizer $G_K = (Q_K, E_K, f_K, q_{K\,0}, Q_{K,m})$. Compute*

$$
\begin{aligned}
& (S_i, g_i),\ S_i = (Q_i, E_{i,o}, f_i, q_{i,0}, Q_{i,m}),\ i = 1,2, \\
Q_i &= Pwrset(Q_K)\backslash\{\emptyset\},\ Q_{i,m} = \{q_i \in Q_i \mid q_i \cap Q_{K,m} \neq \emptyset\}, \\
q_{i,0} &= \{f_K(q_{K,0}, s) \in Q_K \mid s \in E^*,\ P_i(s) = \varepsilon\}, \\
& \text{the unobservable (wrt. } P_i\text{) reachset from the initial state,} \\
f_i(q_i, e) &= \begin{cases} \{f_K(q,e) \in Q_K \mid s \in E^*,\ P_i(s) = e,\ q \in q_i\}, & \text{if not empty,} \\ \text{undefined}, & \text{else.} \end{cases} \\
& g_i : Q_i \times E_{i,c} \to E_{i,cp}, \\
g_i(q_i, e) &= E_{i,uc} \cup \{e \in E_{i,c} \mid \exists\, q \in q_i \in Q_i \text{ such that } f_i(q,e) \text{ is defined}\}.
\end{aligned}
$$

Then $\{g_1, g_2\}$ is a solution of Problem 6.1.

Proposition 6.2. [39, Prop. 4.1]. *Consider Problem 6.1.*

(a) *If the sublanguage $\emptyset \neq K \subseteq L_m(G)$ is controllable, coobservable, and prefix-closed then the supervisors constructed in the above algorithm satisfy (1) $L_m(\tilde{S}_1 \wedge \tilde{S}_2/G) = K$ and (2) $(\tilde{S}_1 \wedge \tilde{S}_2/G)$ is nonblocking.*
(b) *The time complexity of Algorithm 6.4 is exponential in the size of the state set of G_K.*

The new condition for the existence of a tuple of decentralized supervisors is the concept of coobservability. A sufficient condition of coobservability is decomposability.

Definition 6.11. [39, p. 1696, p. 1702] *Consider Problem 6.1*

(a) *The language* $K \subseteq L_m(G)$ *is called* decomposable with respect to (G, P_1, P_2) *if* $K = P_1^{-1}(P_1(K)) \cap P_2^{-1}(P_2(K)) \cap L(G)$.

(b) *The language* K *is called* strongly decomposable with respect to (G, P_1, P_2) *if* $K = L(G) \cap \left[P_1^{-1}(P_1(K)) \cup P_2^{-1}(P_2(K))\right]$.

Proposition 6.3. [39, Prop. 4.2]. *Let* $K \subseteq L_m(G)$ *be* $L_m(G)$*-closed. If* prefix(K) *is strongly decomposable then* K *is coobservable.*

There exists an example showing that K being coobservable does not imply that prefix(K) is strongly decomposable, see [39, Ex. 4.1].

Proposition 6.4. [39, Prop. 4.3]. *Assume that* $K \subseteq L_m(G)$ *is regular,* $L_m(G)$*-closed, and controllable,* $E_{i,c} \subseteq E_{i,o}$*, for* $i = 1,\ 2$*,* $E_{1,o} \cap E_{2,c} \subseteq E_{1,c}$*, and* $E_{2,o} \cap E_{1,c} \subseteq E_{2,c}$*. Then* K *is coobservable if and only if* K *is decomposable.*

The following example shows that there exists a discrete-event system for which decentralized control can never meet the control objective. The control objective can only be met if the two controllers communicate with each other. The focus of research should therefore be extended or redirected to decentralized/distributed control with communication as described in Chapter 7.

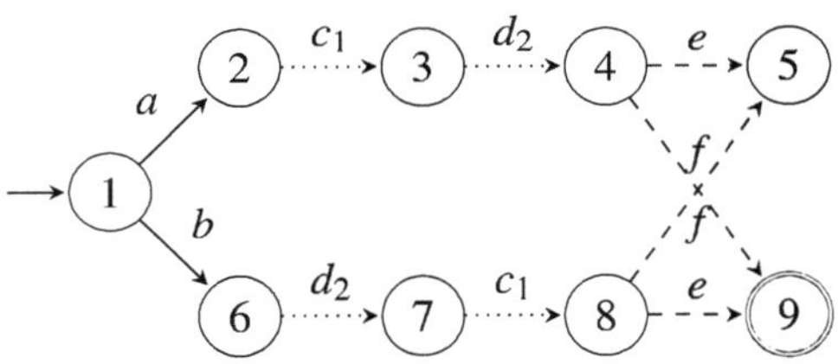

Fig. 6.7 The diagram of a distributed DES requiring communication between the supervisors. The dotted arrows denote observable events.

Example 6.5. *Decentralized control requiring communication.* The distributed DES is specified in Fig. 6.7. Note that $E_{1,c} = E_{2,c} = \{e, f\}$, $E_{1,o} = \{c_1, e, f\}$, $E_{2,o} = \{d_2, e, f\}$, q_5 is the forbidden state, q_9 is an accepting/marked state. From the diagram it is then clear that neither of the supervisors knows whether in state 4 and in state 8 to enable e, or f, or both because neither of them observes in which order the events c_1 and d_2 occur. They must communicate with each other to determine this order. ■

6.6 Decentralized Control – Undecidability

Next follows a result on undecidability of a decentralized control problem. In [39] it is proven that the decentralized control problems with the following relations are decidable: (1) $L_m(\tilde{S}_1 \wedge \tilde{S}_2/G) = K$, [39, Th. 4.1]; and (2) $L(\tilde{S}_1 \wedge \tilde{S}_2/G) \subseteq K$ [39, Th. 4.2]. Does the decidability still hold true in a further generalization of the problem?

Problem 6.2. [51, Def. 5.1]. Consider the setting of Problem 6.1 with a regular specification language $K \subseteq L_m(G)$. Does there exist a tuple of supervisors $((S_1, g_1), (S_2, g_2))$ such that $L_m(\tilde{S}_1 \wedge \tilde{S}_2/G) \subseteq K$ and such that the closed-loop system is nonblocking?

Note the difference between Problem 6.1 and Problem 6.2 in the relation of the marked language of the closed-loop system with the language of the specification.

Theorem 6.2. [51, Th. 5.1]. *Problem 6.2 is undecidable.*

The proof of the above theorem proceeds by reduction of an observability problem of [30] to Problem 6.2. A result corresponding to the above theorem for an ω-language setting was proven in [20]. See further [36, 49].

6.7 Decentralized Control – Maximal Languages

In supervisory control of a discrete-event system if the specification is not controllable one considers the supremal controllable sublanguage. In the setting of control of a distributed DES this cannot be done directly because there are always two or more supervisors. Therefore, other concepts are needed, a maximal solution and a Nash equilibrium as described below. The theory below was published before Theorem 6.2 listed above. In this section, a slightly different definition of a supervisor is used than in Section 6.5 borrowed from [27]. The reasons to do so are only notational, there is no inherent restriction.

Definition 6.12. *Consider a discrete-event system denoted by $G = (Q, E, f, q_0, Q_m)$. Denote a* supervisory control *based on partial observations of this system by $g : P(L(G)) \to Pwrset(E_c)$, where $P : E^* \to E_o^*$ is a natural projection. Unlike the previous sections, $g(P(s))$ denotes the subset of disabled controllable events,*

$$L(g/G) = \{s \in L(G) \mid \forall\, we \in \text{prefix}(s),\ e \notin g(P(w))\}.$$

Definition 6.13. *Consider two supervisory controls $g_a, g_b : P(L(G)) \to Pwrset(E_c)$. Define the* implementation relation, *denoted by $g_a \sqsubseteq g_b$, and say that g_a* implements *g_b, if $g_b(s) \subseteq g_a(s)$, $\forall\, s \in P(L(g_a/G))$. In words, g_a disables more than g_b.*

Definition 6.14. *Consider a discrete-event system G and two local supervisory controls, $g_1 : P_1(L(G)) \to Pwrset(E_{1,c})$ and $g_2 : P_2(L(G)) \to Pwrset(E_{2,c})$, and let $P : E^* \to (E_{1,o} \cup E_{2,o})^* = E_o^*$ be a natural projection. Define the* composition *of these two supervisory controls as the supervisory control, $g_1 \wedge g_2 : P(L(G)) \to Pwrset(E_{1,c} \cup E_{2,c})$, $(g_1 \wedge g_2)(s) = g_1(P_1(s)) \cup g_2(P_2(s))$, $\forall s \in P(L(G))$.*

Proposition 6.5. [27, Prop. 2.10]. *Consider the setting of Definition 6.14 with a discrete-event system and two local supervisory controls g_1 and g_2. Then $L(g_1 \wedge g_2/G) = L(g_1/G) \cap L(g_2/G)$.*

Problem 6.3. The *decentralized supervision safety problem.* Consider a discrete-event system G, a specification language $K \subseteq L_m(G)$, $E = E_1 \cup E_2$, and the centralized optimal supervisory control $g^{\uparrow}$ such that $L(g^{\uparrow}/G) = \sup C(K,G)$. Determine a tuple of local supervisory controls, (g_1, g_2), such that $g_1 \wedge g_2 \sqsubseteq g^{\uparrow}$.

Definition 6.15. *Consider Problem 6.3. The tuple of local supervisory controls $(g_1^{\uparrow}, g_2^{\uparrow})$ is called an* optimal decentralized solution *if (1) $g_1^{\uparrow} \wedge g_2^{\uparrow} \sqsubseteq g^{\uparrow}$; (2) $\forall (g_1, g_2)$: $g_1 \wedge g_2 \sqsubseteq g^{\uparrow}$ implies that $L(g_1 \wedge g_2/G) \subseteq L(g_1^{\uparrow} \wedge g_2^{\uparrow}/G)$. Thus, the closed-loop language of the optimal decentralized solution is least restrictive.*

An optimal decentralized solution may not exist. Therefore, the attention is restricted to a maximal solution defined next.

Definition 6.16. *Consider Problem 6.3. The tuple of local supervisory controls $(g_1^{\square}, g_2^{\square})$ is called a* maximal solution *if (1) $g_1^{\square} \wedge g_2^{\square} \sqsubseteq g^{\uparrow}$; (2) $\nexists$ (g_1, g_2) such that $g_1 \wedge g_2 \sqsubseteq g^{\uparrow}$ and $L(g_1^{\square} \wedge g_2^{\square}/G) \subsetneq L(g_1 \wedge g_2/G)$. Equivalently, there does not exist another tuple of supervisory controls with a strictly larger closed-loop language.*

How to determine a maximal solution? There do not exist general results on how to determine all maximal solutions. A way to proceed is to use the concept of a Nash equilibrium.

Definition 6.17. *The tuple of local supervisory controls (g_1^o, g_2^o) is called a* Nash equilibrium *if*

$$\begin{aligned}
&(1) \quad (g_1^o \wedge g_2^o) \sqsubseteq g^{\uparrow};\\
&(2.1) \quad g_1^o \wedge g_2 \sqsubseteq g^{\uparrow} \Rightarrow L(g_1^o \wedge g_2/G) \subseteq L(g_1^o \wedge g_2^o/G), \ \forall\, g_2\\
&(2.2) \quad g_1 \wedge g_2^o \sqsubseteq g^{\uparrow} \Rightarrow L(g_1 \wedge g_2^o/G) \subseteq L(g_1^o \wedge g_2^o/G), \ \forall\, g_1.
\end{aligned}$$

The concept of a Nash equilibrium is named after the mathematician/economist John Nash who introduced the concept into game theory. Decentralized control is a special case of a dynamic game problem, all players have the same cost function though different observations. There exists an example such that a Nash equilibrium is not a maximal solution. Therefore, the concept of Nash equilibrium has to be strengthened. After that the equivalence condition of a maximal solution can be stated.

Definition 6.18. *The tuple of local supervisory controls (g_1^o, g_2^o) is called a* strong Nash equilibrium *if (1) it is a Nash equilibrium and (2) $\forall (g_1, g_2)$: $L(g_1 \wedge g_2/G) = L(g_1^o \wedge g_2^o/G)$ implies that (g_1, g_2) is a Nash equilibrium.*

Theorem 6.3. [27, Th. 3.4]. *Consider Problem 6.3. The tuple of local supervisory controls (g_1^o, g_2^o) is a maximal element if and only if it is a strong Nash equilibrium.*

Procedure 6.4. Construction of a maximal tuple of supervisory controls.

1. *Determine a tuple of supervisory controls which is a strong Nash equilibrium, see below.*
2. *Then conclude with Theorem 6.3 that the considered tuple of supervisory controls is a maximal element.*

Example 6.6. *Joint action.* Consider the discrete-event system of which the diagram is displayed in Fig. 6.8. Note that $E_1 = \{a_1, b_1\}$, $E_2 = \{a_2, b_2\}$, $E_1 \cap E_2 = \emptyset$, $E_c = \{a_1, a_2\}$. The centralized supervisory control which produces the specification sublanguage is described by the language $K = \{(a_2 b_1 + a_1 b_2)^*\}$. Consider the following tuple of supervisory controls (g_1, g_2). Supervisory control g_1 always disables controllable event a_1 and the uncontrollable event b_1 is always enabled. Supervisory control g_2 initially enables a_2, and disables a_2 only after the first occurrence of b_2 and then a_2 remains always disabled. Uncontrollable event b_2 is always enabled.

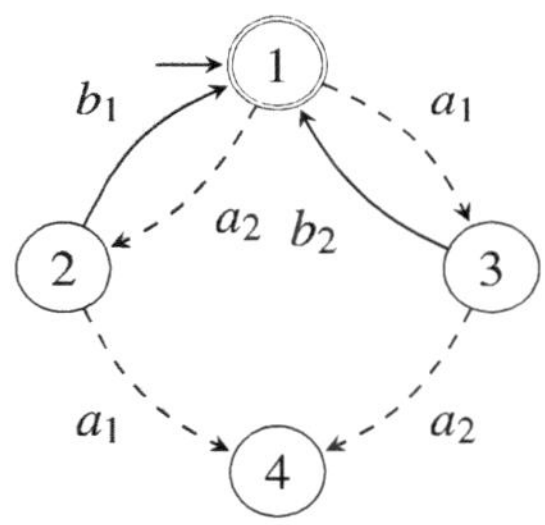

Fig. 6.8 The diagram of a distributed DES for which there exists a strong Nash equilibrium

This tuple of supervisory controls in closed-loop, with the plant achieves the specification, is a strong Nash equilibrium, and hence a maximal solution. See [27, Ex. 4.5.]. ■

There exists a procedure for computing Nash equilibria, see [27], but there is an example of a controlled DES for which this procedure does not converge in a finite number of steps while there exists a strong Nash equilibrium, see [27, Ex. 4.6].

6.8 Distributed Control of Distributed DESs

In this section distributed control is investigated. The older term used for this approach is *modular control* and both terms will be used in this section. By going from a decentralized system, as discussed in the three preceding sections, to a distributed system, the decomposition of the global system into two or in general many subsystems is to be noted.

The motivation for the investigation of distributed/modular discrete-event systems is the complexity of control design. The time complexity of the computation of the supervisor increases exponentially with respect to the number of subsystems. If the computation of the supervisor can be carried out in parallel for each subsystem then the overall time complexity will be much less. This is the main motivation for the investigation of the control synthesis of distributed/modular systems.

Modular control synthesis of DES was first investigated in [57]. But approaches of distributed/modular control have been investigated in control theory and other areas of engineering and the sciences for many centuries.

Specifications for the control of distributed systems can be distinguished into a global specification and a set of local specifications, where the latter contain one specification per subsystem.

Problem 6.4. *Distributed control of a distributed/modular discrete-event system.* Consider a distributed/modular discrete-event system and a specification. Restrict the attention to the distributed control architecture introduced in Definition 6.4. Determine a set of supervisors, one for each subsystem, such that the closed-loop system meets the specification.

Define the *distributed/modular control synthesis* as to synthesize a supervisor for each of the local subsystems separately. The supervisor of the distributed system is then the set of the local supervisors.

The *global control synthesis* is defined in two steps: (1) compose all subsystems into one system and then (2) compute the supervisory control for the composed system according to the method described in Chapter 3. The global control synthesis is used only for a theoretical comparison.

Problem 6.4 leads to the following research issues:

1. Is the closed-loop system nonblocking? There exist examples in which the system is blocking. The quest is therefore to find equivalent or sufficient conditions for nonblockingness of the closed-loop distributed system.
2. Can a distributed control synthesis as described below achieve the same closed-loop language as the global control synthesis?

The research issue of nonblockingness of a composition of two or more subsystems has been discussed in Chapter 3. See also [57]. If the languages of two subsystems satisfy the condition of nonconflicting languages, see Chapter 3, then the product of their languages is nonblocking. But the time complexity of checking nonconflictingness is almost as high as the time complexity of checking nonblockingness of the global systems. Alternative approaches are to use the observer property, see [28] or abstractions with particular properties [7]. Coordination control, see Chapter 8, is another approach to deal with nonblockingness of the interconnection of a distributed system.

The second research issue is whether the closed-loop systems obtained by the global control synthesis and by the modular control synthesis are equal. The answer to the question depends on the interaction of the subsystems. It will be stated below that the equality of the global control synthesis and of the modular control synthesis is equivalent to the concept of modular controllability.

Recall from Definition 6.3 the notation of a distributed discrete-event system, in this section also referred to as a modular DES,

$$G_{dis} = \{G_i,\ i \in I\},$$
$$G_i = (Q_i, E_i, f_i, q_{i,0}, Q_{i,m}),\ E_{i,c},\ E_{i,o} \subseteq E_i.$$

where the index set of the subsystems is denoted by $I = \{1,2,\ldots,n\}$, G_i is an automaton for each $i \in I$, $E_i \subseteq E$ denotes the subset of the events of Subsystem i, $E_{i,c} \subseteq E_i$ denotes the subset of controllable events of Subsystem i, and $E_{i,o} \subseteq E_i$ denotes the subset of observable events of Subsystem i. The interaction of the subsystems of the distributed system is via common events of the local subsystems. Denote further $L_i = L(G_i)$, $L = ||_{i=1}^n L(G_i)$, $E_c = \cup_{i=1}^n E_{i,c}$, and $E_{uc} = \cup_{i=1}^n E_{i,uc}$. $P_i : E^* \to E_i^*$ is a natural projection and P_i^{-1} : Pwrset$(E_i^*) \to$ Pwrset(E^*).

Definition 6.19. *Consider a modular DES. The subsystems are said to* agree on the controllability of their common events *if $E_{i,c} \cap E_j = E_{j,c} \cap E_i$ for all $i,j \in I$ with $i \neq j$. The subsystems are said to* agree on the observability of their common events *if $E_{i,o} \cap E_j = E_{j,o} \cap E_i$ again for all $i,j \in I$ with $i \neq j$.*

The definition that the subsystems agree on the controllability status of their events implies that $E_{i,c} = E_c \cap E_i$ and $E_{uc} = \cup_i (E_i \setminus E_{i,c})$.

Definition 6.20 (Local specification languages). *Define the* local specification languages *corresponding to a global specification language $K \subseteq E^*$ as $\{K_i,\ i \in I\}$, $K_i = K \cap P_i^{-1}(L_i) \subseteq E^*$, for all $i \in I$. Then, $K \cap L = \cap_{i=1}^n K_i$. One then says that K_i* locally overapproximates *K.*

Definition 6.21 (Modular control synthesis and global control synthesis). *Define* global control synthesis *by the associated closed-loop language of the plant and the supervisory control,* $\sup\text{C}(K \cap L, L, E_{uc})$. *Define a* modular control synthesis *by the associated closed-loop language of the plant and the associated supervisory controls, $\cap_{i=1}^n \sup\text{C}(K_i, L_i, E_{uc})$.*

Problem 6.5. *Does the modular control synthesis equal the global control synthesis?* Which conditions imply that the modular control synthesis equals the global control synthesis? Equivalently, when does the equality in the following expression $\cap_{i=1}^n \sup\text{C}(K_i, L_i, E_{uc}) = \sup\text{C}(K \cap L, L, E_{uc})$ hold true?

Definition 6.22. *The modular system is called* modularly controllable *if $LE_{uc} \cap P_i^{-1}(L_i) \subseteq L,\ \forall\, i \in I$.*

Theorem 6.5 (Modular equals global control synthesis in case of locally complete observations). [14, Th. 6.7] *Assume that the subsystems agree on the controllability of their common events.*

(a) If the modular system is modularly controllable then $\cap_{i=1}^n \sup\text{C}(K_i, L_i, E_{uc}) = \sup\text{C}(K \cap L, L, E_{uc})$.
(b) If the above equality holds for all $K \subseteq E^*$ then the modular system is modularly controllable.

In the literature there are other sufficient conditions that are much less complex to check, called mutual controllability and global mutual controllability, [14].

Proposition 6.6. [14, Prop. 8.3, Prop. 8.4] *The time complexity of the computation of the supremal controllable sublanguage by the modular control synthesis is $O(n_m^2(n^*)^2 n_K^2 m^2)$ where $n_m =$ the number of modules, $n^* = \max_{i \in I} n_i$, $m = \max_{i \in I} m_i$, $n_i = |Q_i|$, $m_i = |E_i|$, and $n_K = |Q_{spec}|$, the size of the state set of a minimal recognizer of the specification. The time complexity of the computation of the supremal controllable sublanguage by global control synthesis is $O((n^*)^{2n_m} n_K)$. The gain in time complexity is then clear.*

There exist concepts and theorems for modular control with only local partial observations. The main references for this section are [13, 14, 17, 16]; see also [55].

6.9 Research Issues

Engineering has a need for further research on control of distributed systems because the existing theory is far from being satisfactory.

- *Decentralized control.* The results presented earlier in this chapter require further analysis of control theoretic interpretations of the operation of the supervisors which may then be useful for the development of decentralized control also for other classes of systems. Based on control theory of stochastic systems one expects to note how the structure of the control law deals with the common and the private information of the subsystems.
- *Distributed control of a distributed DES.* An open problem is to find a condition for nonblockingness of the closed-loop system which is of low time complexity compared with the global control synthesis. What is decidable, computationally attractive, and achievable by distributed control of distributed DES?
- *Distributed control with communication.* For particular engineering control problems this form of control is attractive and this motivates investigations of the following research topics: (1) Experience with the forms of this type of control. Examples of control of communication networks will be useful. (2) Synthesis and design of communication laws and control laws. What, when, and to whom to communicate? (3) A theoretical framework for synthesis of distributed control laws with communication has to be developed using the concepts of common and of private information, observers, and control. (4) The tradeoffs between control and communication on the overall performance have to be investigated.
- *Coordination control.* Research issues needing attention include: Decomposition of large systems into coordinated systems may be useful depending on the application. The tradeoffs between control and communication also play a role in this setting. The reader is referred to Chapter 8 for further information.
- *Hierarchical control.* The class of hierarchical systems to be considered will be quite general. Research issues include decomposition, abstraction, system reduction, the algebra of hierarchical-distributed systems, the relations between controllers of adjacent hierarchical levels, and the communication and computational aspect of the different hierarchical layers.

6.10 Further Reading

The reader is advised to read the following chapters of this book on other aspects of control of distributed systems: Chapter 7 *An Overview of Synchronous Communication for Control of Decentralized Discrete-Event Systems*, and Chapter 8 *Coordination Control of Discrete-Event Systems*.

Books and lecture notes. The lecture notes of W.M. Wonham, see the Chapters 4, 5, and 6 of [58]. These are available on the web, see `http://www.control.utoronto.ca/~wonham/`. For a book on supervisory control of decentralized DES, see [4, Chapter 3].

Decentralized control. The problem of decentralized control of discrete-event systems was formulated in [5]. The concept of coobservability and the theorem that it is equivalent to existence of a decentralized supervisor is due to [39]. The algorithm to check coobservability is described in [37]. The space complexity of the algorithm and related matters were investigated in [33, 34, 35]. Early papers on decentralized control include [23, 22, 24, 18]. Other papers on decentralized control include [11, 19, 29, 42, 56, 59].

Decentralized control for non DES. To assist the reader, there follow several references on decentralized control of systems which are *not* discrete-event systems. Books on decentralized control include [6, 45, 46]. Survey papers are [10, 41]. Papers on complexity of decentralized control problems include [52, 2].

Hierarchical control of hierarchical discrete-event systems. The concept of a hierarchical system is rather old. In artificial intelligence the study of these systems was stimulated by Herbert Simon. An early framework for hierarchical DES was proposed in [9] and is called state charts. Other references on the hierarchical control synthesis of DES include [61, 47]. The approach of hierarchical and distributed DES is described in [44, 43].

Asynchronous timed DES. There are engineering distributed systems in which each subsystem has its own clock and the clocks may drift with respect to each other. An example is a set of audio equipment consisting of a tuner, a CD player, and an amplifier, see the paper [3]. An approach to diagnosis of distributed asynchronous systems is described in [1]; see also [32].

References

1. Benveniste, A., Fabre, E., Haar, S., Jard, C.: Diagnosis of asynchronous discrete-event systems: a net unfolding approach. IEEE Transactions on Automatic Control 48, 714–727 (2003)
2. Blondel, V.D., Tsitsiklis, J.N.: A survey of computational complexity results in systems and control. Automatica 36, 1249–1274 (2000)
3. Bosscher, D., Polak, I., Vaandrager, F.: Verification of an Audio Control Protocol. In: Langmaack, H., de Roever, W.-P., Vytopil, J. (eds.) FTRTFT 1994 and ProCoS 1994. LNCS, vol. 863, pp. 170–192. Springer, Heidelberg (1994)

4. Cassandras, C.G., Lafortune, S.: Introduction to Discrete Event Systems, 2nd edn. Springer, New York (2008)
5. Cieslak, R., Desclaux, C., Fawaz, A.S., Varaiya, P.: Supervisory control of discrete-event processes with partial observations. IEEE Trans. Automatic Control 33, 249–260 (1988)
6. Findeisen, W., Bailey, F.N., Brdys, M., Malinowski, K., Tatjewski, P., Wozniak, A.: Control and Coordination in Hierarchical Systems. John Wiley & Sons, Chichester (1980)
7. Flordal, H., Malik, R.: Compositional verification in supervisory control. SIAM J. Control & Optimization 48, 1914–1938 (2009)
8. Gallager, R.G.: Information Theory and Reliable Communication. John Wiley & Sons, New York (1968)
9. Harel, D.: Statecharts: A visual formalism for complex systems. Science of Computer Programming 8, 231–274 (1987)
10. Ho, Y.C.: Team decision theory and information structures. Proceedings of the IEEE 68, 644–654 (1980)
11. Inan, K.: An algebraic approach to supervisory control. Math. Control Signals Systems 5, 151–164 (1992)
12. Kempker, P.L., Ran, A.C.M., van Schuppen, J.H.: A formation flying algorithm for autonomous underwater vehicles. In: Proc. 50th IEEE Conference on Decision and Control, Orlando, USA (2011)
13. Komenda, J., van Schuppen, J.H.: Conditions structurelles dans le contrôle modulaire des systèmes à événements discrets concurrents. In: Proc. Modélisation des Systèmes Reactifs. Ecole Normale Superior de Lyon, Hermès, Lavoisier (2007)
14. Komenda, J., van Schuppen, J.H.: Control of discrete-event systems with modular or distributed structure. Theoretical Computer Science 388, 199–226 (2007)
15. Komenda, J., van Schuppen, J.H.: Decentralized supervisory control with coalgebra. In: Proc. European Control Conference 2003, Cambridge, United Kingdom, CD-ROM (2003)
16. Komenda, J., van Schuppen, J.H.: Modular control of discrete-event systems with coalgebra. IEEE Transactions on Automatic Control 53, 447–460 (2008)
17. Komenda, J., van Schuppen, J.H., Gaudin, B., Marchand, H.: Supervisory control of modular systems with global specification languages. Automatica 44, 1127–1134 (2008)
18. Kozák, P., Wonham, W.M.: Fully decentralized solutions of supervisory control problems. Report 9310, Systems Control Group Report. Department of Electrical Engineering. University of Toronto (1993)
19. Kumar, R., Shayman, M.A.: Centralized and decentralized supervisory control of nondeterministic systems under partial observation. SIAM J. Control & Optimization 35, 363–383 (1997)
20. Lamouche, H., Thistle, J.: Effective control synthesis for des under partial observations. In: Proc. 44th IEEE Conference on Decision and Control, Sydney, NSW, Australia (2000)
21. Lee, S.H.: Structural Decentralised Control of Concurrent Discrete-Event Systems. PhD Thesis. Australian National University, Canberra (1998)
22. Lin, F., Wonham, W.M.: Decentralized control and coordination of discrete-event systems. In: Proc. 27th Conference on Decision and Control, Austin, TX, USA (1988)
23. Lin, F., Wonham, W.M.: Decentralized supervisory control of discrete-event systems. Information Sciences 44, 199–224 (1988)
24. Lin, F., Wonham, W.M.: Decentralized control and cooperation of discrete event systems with partial observations. IEEE Transactions on Automatic Control 35, 1330–1337 (1990)
25. Lynch, N.A.: Distributed Algorithms. Kaufman, San Francisco (1996)

26. Markovski, J., Jacobs, K.G.M., van Beek, D.A., Somers, L.J.A.M., Rooda, J.E.: Coordination of resources using generalized state-based requirements. In: Proc. 10th Int. Workshop on Discrete Event Systems, Berlin (2010)
27. Overkamp, A., van Schuppen, J.H.: Maximal solutions in decentralized supervisory control. SIAM J. Control & Optimization 39, 492–511 (2000)
28. Pena, P.N., Cury, J., Lafortune, S.: Verification of nonconflict of supervisors using abstractions. IEEE Transactions on Automatic Control 54, 2803–2815 (2009)
29. Pnueli, A., Rosner, R.: Distributed reactive systems are hard to synthesize. In: Proc. 1990 IEEE Symposium on the Foundations of Computer Science, St. Louis, MO, USA (1990)
30. Puri, A., Tripakis, S., Varaiya, P.: Problems and examples of decentralized observation and control for discrete event system. In: Proc. Symposium on the Supervisory Control of Discrete-Event Systems, Paris, France (2001)
31. Raynal, M., Helary, J.M.: Synchronization and Control of Distributed Systems and Programs. Wiley, New York (1990)
32. Ricker, S.L., van Schuppen, J.H.: Decentralized failure diagnosis with asynchronous communication between supervisors. In: Proc. 2001 European Control Conference, Sydney, Australia (2001)
33. Rohloff, K., Lafortune, S.: On the computational complexity of the verification of modular discrete-event systems. In: Proc. 41th IEEE Conference on Decision and Control, Las Vegas, NV, USA (2002)
34. Rohloff, K., Lafortune, S.: Recent results on computational issues in supervisory control. In: ATPN-Workshop on Discrete Events Systems Control. Eindhoven University of Technology, The Netherlands (2003)
35. Rohloff, K., Lafortune, S.: The control and verification of similar agents operating in a broadcast network environment. In: Proc. 42th IEEE Conference on Decision and Control, Maui, HI, USA (2003)
36. Rudie, K.: A Summary of Some Discrete-Event System Control Problems. In: Domaratzki, M., Salomaa, K. (eds.) CIAA 2010. LNCS, vol. 6482, pp. 4–16. Springer, Heidelberg (2011)
37. Rudie, K., Willems, J.C.: The computational complexity of decentralized discrete-event control problems. IEEE Transactions on Automatic Control 40, 1313–1318 (1995)
38. Rudie, K., Wonham, W.M.: Protocol verification using discrete-event systems. In: Proc. 31st IEEE Conference on Decision and Control, Tucson, AZ, USA (1992)
39. Rudie, K., Wonham, W.M.: Think globally, act locally: Decentralized supervisory control. IEEE Transactions on Automatic Control 37, 1692–1708 (1992)
40. Rudie, K., Wonham, W.M.: Supervisory control of communicating processes. In: Logrippo, L., Probert, R.L., Ural, H. (eds.) Protocol Specification, Testing and Verification X, pp. 243–257. Elsevier Science Publishers B.V., North Holland (1990)
41. Sandell Jr., N.R., Varaiya, P., Athans, M., Safonov, M.G.: Survey of decentralized control methods for large scale systems. IEEE Transactions on Automatic Control 23, 108–128 (1978)
42. Schmidt, K., Breindl, C.: Maximally permissive control of decentralized discrete event systems. IEEE Transactions on Automatic Control 56, 723–737 (2011)
43. Schmidt, K., Gaudin, B., Marchand, H.: Modular and decentralized supervisory control of concurrent discrete event systems using reduced system models. In: Proc. Int. Workshop on Discrete-Event Systems, Ann-Arbor, MI, USA (2006)
44. Schmidt, K., Moor, T., Perk, S.: Nonblocking hierarchical control of decentralized discrete event systems. IEEE Transactions Automatic Control 53, 2252–2265 (2008)
45. Šiljak, D.D.: Decentralized Control of Complex Systems. Academic Press, New York (1990)

46. Šiljak, D.D.: Large-Scale Dynamic Systems–Stability and Structure. New York. Dover Publications, Inc., Mineola (2007)
47. Su, R., van Schuppen, J.H., Rooda, J.E.: Aggregative synthesis of distributed supervisors based on automata abstraction. IEEE Transactions on Automatic Control 55, 1627–1640 (2010)
48. Tel, G.: Introduction to Distributed Algorithms, 2nd edn. Cambridge University Press, Cambridge (2000)
49. Thistle, J.: Undecidability in decentralized supervision. Systems & Control Lett. 54, 503–509 (2005)
50. Thistle, J.G., Malhamé, R.P., Hoang, H.-H., Lafortune, S.: Feature interaction modelling, detection and resolution: A supervisory control approach. In: Dini, P., Boutaba, R., Logrippo, L. (eds.) Feature Interactions in Telecommunications and Distributed Systems IV, pp. 93–120. IOS Press, Amsterdam (1997)
51. Tripakis, S.: Undecidable problems of decentralized observation and control. In: Proc. 40th IEEE Conference on Decision and Control, Orlando, FL, USA (2001)
52. Tsitsiklis, J.N., Athans, M.: On the complexity of decentralized decision making and detection problems. IEEE Transactions on Automatic Control 30, 440–446 (1985)
53. Veríssimo, P., Raynal, M.: Time in Distributed System Models and Algorithms. In: Krakowiak, S., Shrivastava, S.K. (eds.) BROADCAST 1999. LNCS, vol. 1752, pp. 1–32. Springer, Heidelberg (2000)
54. Walrand, J., Varaiya, P.P.: High-Performance Communication Networks, 2nd edn. Morgan Kaufmann, San Francisco (2000)
55. Wong, K.C., Wonham, W.M.: Modular control and coordination of discrete-event systems. Discrete Event Dynamics Systems 8, 247–297 (1998)
56. Wong, K.C., Lee, S.: Structural decentralized control of concurrent discrete-event systems. European J. Control 8, 477–491 (2002)
57. Wonham, W.M., Ramadge, P.J.: Modular supervisory control of discrete-event systems. Math. Control Signals Systems 1, 13–30 (1988)
58. Wonham, W.M.: Supervisory Control of Discrete-Event Systems. University of Toronto, Canada (2008)
59. Yoo, T., Lafortune, S.: New results on decentralized supervisory control of discrete-event systems. In: Proc. 39th IEEE Conference on Decision and Control, Sydney, NSW, Australia (2000)
60. Yoo, T.-S., Lafortune, S.: A general architecture for decentralized supervisory control of discrete-event systems. Discrete Event Dynamics Systems 12, 335–377 (2002)
61. Zhong, H., Wonham, W.M.: On the consistency of hierarchical supervisors in discrete-event systems. IEEE Transactions on Automatic Control 35, 1125–1134 (1990)

Chapter 7
An Overview of Synchronous Communication for Control of Decentralized Discrete-Event Systems

Laurie Ricker

7.1 Introduction

Communication plays a crucial role in the operation of many large decentralized and distributed systems. In the classical formulation of the decentralized discrete-event control problem, there are no communication channels between controllers. Recall that decentralized controllers have only a partial observation of system behavior, and thus, for each controller, there is some uncertainty as to the precise state of the system. When a global disablement decision must be taken, a control strategy succeeds if there is at least one controller, with the ability to take the decision that can unambiguously determine that disablement is the correct decision. If no such controller exists, then there is no control solution.[1] To ameliorate this situation, communication between controllers can occur when the communicated information, in tandem with local observations, allows all the correct control decisions to be taken.

The role of communication has been investigated within the context of synthesizing control strategies for decentralized discrete-event systems. This chapter presents a brief overview of some of the ways in which communication protocols have been incorporated into the decentralized control domain.

Some of the fundamental issues for incorporating communication into decentralized control problems were identified in [12]. In particular, the answers to these questions affect the model design and the subsequent synthesis of control and communication solutions (adapted from [12]):

Laurie Ricker
Department of Mathematics & Computer Science, Mount Allison University, Sackville, NB Canada
e-mail: lricker@mta.ca

[1] Note that all discussions about decentralized control in this chapter are made with respect to the C&P decentralized architecture of Rudie and Wonham [11]. Analogous statements can be made for the D&A architecture of Yoo and Lafortune [17].

C. Seatzu et al. (Eds.): Control of Discrete-Event Systems, LNCIS 433, pp. 127–146.
springerlink.com

- Why should communication be introduced?
- Who should communicate with whom and when?
- What information should be communicated?
- Who should know what and when?

With inspiration from the work by Witsenhausen [15], the initial proposals for information structures with which the above questions can be addressed, focussed primarily on controllers either keeping track of estimates of sequences, based on observed strings [14, 16], or keeping track of state estimates of the system, based on observed strings [2, 10].

We will examine two approaches to the synthesis of communication protocols: state-based communication and event-occurrence communication. In each case, the communication is presumed to take place with zero delay, i.e., synchronous communication. The literature includes some preliminary examination of communication with non-zero delay [6, 13]; however, the communication protocol under consideration requires no synthesis as every observation of a controller is communicated to all other controllers. For the development of communication protocols beyond synchrony in distributed control architectures, some considerations and possible future directions are presented in [3].

7.2 Notation and Definitions

We assume that the uncontrolled system is described by a finite automaton $G_L = (Q, E, \delta, q_0)$ — with finite state set Q, finite alpabet E, partial transition function $\delta : Q \times E \to Q$, and initial state q_0 — which generates a regular language L. The corresponding specification automaton $G_K = (Q^K, E, \delta^K, q_0)$, where $Q^K \subseteq Q$ and $\delta^K \subseteq \delta$, generates a language $K \subseteq L$. Alternatively, we denote the transition set of L by T, and the transition set of K by $T^K \subseteq T$. The prefix closure of a language L is defined as follows: $\overline{L} := \{s \in E^* \mid \exists t \in E^* \text{ such that } st \in L\}$. We assume, for the rest of this chapter, that all languages are prefix-closed. To discuss decentralized control for a set of controllers $I = \{1, \ldots, n\}$, the event set E is partitioned into controllable events E_c and uncontrollable events E_{uc}. Similarly, E is partitioned into observable events E_o and unobservable events E_{uo}. To describe events that each decentralized controller $i \in I$ controls and observes, respectively, we use the notation $E_{c,i} \subseteq E_c$ and $E_{o,i} \subseteq E_o$. (The transition set T can similarly be partitioned into T_c and T_o, based on the controllable and observable properties of the transition labels.) We refer to the set of controllers that observe $e \in E_o$ by $I_o(e) := \{i \in I \mid e \in E_{o,i}\}$. Analogously, we refer to the set of controllers that control $e \in E_c$ by $I_c(e) := \{i \in I \mid e \in E_{c,i}\}$.The natural projection describing the partial view of each controller is denoted by $\pi_i : E^* \to E_{o,i}^*$, for $i \in I$.

A decentralized controller is an automaton $\mathscr{S}_i$, for $i \in I$, (see Fig. 7.1) that has only a partial view of the system behavior. Each controller issues its own local control decision based on its current view of the system and a final control decision is taken by fusing or combining all the local decisions with a particular fusion rule. The rule varies depending on the decentralized architecture in use. We can synthesize decentralized controllers that cooperate to ensure that the supervised system generates exactly the behavior in the specification K if K is controllable (Definition 7.1) and satisfies one of the versions of co-observability (Definition 7.2 or 7.3).

Definition 7.1. *A language $K \subseteq L$ is* controllable *wrt L and E_{uc} if $\overline{K}E_{uc} \cap L \subseteq \overline{K}$.*

Definition 7.2. *A language $K \subseteq L = \overline{L}$ is* unconditionally co-observable *[11] with respect to L, π_i, and E_c if*

$$\forall t \in \overline{K}, \forall \sigma \in E_c : \; t\sigma \in L \setminus \overline{K} \Rightarrow \exists i \in I_c(\sigma) : \; \pi_i^{-1}[\pi_i(t)]\sigma \cap \overline{K} = \emptyset.$$

In this scenario, decentralized controllers take local control decisions based on their partial observations and there must be at least one controller that has sufficient information from its own view of the system to take the correct control decision when the system leaves K (i.e., to disable) for each $\sigma \in E_c$. When $I = \{1\}$, this condition is called *observability*.

Definition 7.3. *A language $K \subseteq L = \overline{L}$ is* conditionally co-observable *[18] with respect to L, π_i, and E_c, if*

$$\forall t \in \overline{K}, \forall \sigma \in E_c : \; t\sigma \in L \setminus \overline{K} \Rightarrow \; \exists i \in I_c(\sigma) : \forall t'\sigma \in \pi_i^{-1}[\pi_i(t)]\sigma \cap \overline{K} \Rightarrow \\ \exists j \in I_c(\sigma) : \pi_j^{-1}[\pi_j(t')]\sigma \cap L \subseteq \overline{K}.$$

In this scenario, decentralized controllers that are incapable of taking the correct disable decision can *infer* that there is at least one controller that will correctly know when the system remains in K (i.e., to take an enable decision), leaving the uncertain controller with the opportunity to take a conditional control decision "disable unless another knows to enable". That is, the enable decision, if taken by one controller, overrides conditional decisions of any of the other controllers.

When K is neither unconditionally nor conditionally co-observable but *is* observable, it may be possible to construct a communication protocol between controllers such that when communication occurs all the correct control decisions are taken.

Definition 7.4. *A language $K \subseteq L = \overline{L}$ is* articulate *wrt L, π_i and E_c if*

$$(\exists t \in \overline{K})(\exists \sigma \in E_c) t\sigma \in L \setminus \overline{K} \Rightarrow \bigcap_{i \in I_c(\sigma)} \pi_i^{-1}[\pi_i(t)]\sigma \cap \overline{K} \neq \emptyset.$$

This property corresponds to a complete absence of information that is available to be inferred from other controllers in $I_c(\sigma)$, thereby leaving communication as the only means of acquiring information from which to take the correct control decisions.

To discuss the various approaches to synthesizing communication protocols, we will refer to the following common terminology. Communicating controllers can be any controller in I. For simplicity, unless otherwise stated, we assume point-to-point communication, i.e., controller i sends a message to controller j, for $i, j \in I$.

Although the content varies from one approach to another, we refer to the finite set of messages involved in a communication protocol by $\Delta := \cup_{i \in I} \Delta_i$, where Δ_i is the set of messages that controller i sends out to other controllers as directed by its communication protocol.

Definition 7.5. *A* communication protocol *for decentralized controller i is* $\mu^!_{ij} : L \to \Delta_i \cup \{\varepsilon\}$ *and represents the information that controller i sends to controller j following the occurrence of some sequence* $s \in L$. *The information that controller i receives from controller j, following the occurrence of some sequence* $s \in L$ *is denoted by* $\mu^?_{ij} : L \to \Delta_j \cup \{\varepsilon\}$.

When $\mu^?_{ij}(s) \neq \varepsilon$, we can construct an alphabet of received information $E^?_i \subseteq E_o \setminus E_{o,i}$. To incorporate the received information into the observed information of a controller, we extend the natural projection as follows: $\pi^?_i : E^* \to (E_{o,i} \cup E^?_i)^*$, for $i \in I$. As a consequence, communicating controllers take local control decisions based on $\pi^?_i(L)$.

The decentralized architecture that we will assume for this chapter is shown in Fig. 7.1. The decentralized control problem that we will consider is described below.

Problem 7.1. Given regular languages K, L defined over a common alphabet E, controllable events $E_c \subseteq E$, observable events $E_{o,1}, \ldots, E_{o,n} \subseteq E$, and a finite set of messages Δ. We assume that $K \subseteq L \subseteq E^*$ is controllable wrt L, E_{uc}, observable wrt L, π, E_c and articulate wrt L, π_i, E_c. Construct communication protocols $\mathsf{M}^!_i = \langle \mu^!_{i,1}, \ldots, \mu^!_{i,j}, \ldots, \mu^!_{i,n} \rangle$ (for $i, j \in I$) such that either

1. K is unconditionally co-observable wrt L, $\pi^?_i$ (for $i \in I$), and E_c; OR
2. K is conditionally co-observable wrt L, $\pi^?_i$ (for $i \in I$), and E_c. ■

We will examine two main approaches to this problem: when messages are state estimates (i.e., $\Delta_i \subseteq Pwr(Q)$) and when messages are constructed from event occurrences (i.e., $\Delta_i \subseteq E_{o,i}$ or $\Delta_i \subseteq T_{o,i}$). In particular, we consider only approaches which synthesize a communication protocol, as opposed to approaches which assume that part of the input is a set of communications that must be subsequently reduced to satisfy some notion of optimality.

The motivation for introducing communication is independent of the message content or the mode of communication: in the techniques examined here, communication is introduced to eliminate *illegal configurations* from the finite state structure used to determine whether K is co-observable wrt the natural projection that has been updated to include each controller's received messages.

Example 7.1. We will use the following example (from [9]) to illustrate different ways to synthesize a decentralized communication protocol. The joint automaton for G_L and G_K is shown in Fig. 7.2. Here, we assume that $I = \{1,2\}, E = \{\mathsf{a,b,c},\sigma\}$ such that $E_{o,1} = \{\mathsf{a,c},\sigma\}$ and $E_{o,2} = \{\mathsf{b},\sigma\}$. Further, $E_c = \{\sigma\}$, where $E_{c,1} = \{\sigma\}$ and $E_{c,2} = \emptyset$. Note that K is neither unconditionally nor conditionally co-observable. Since $I_c(\sigma) = \{1\}$, we just need to check the co-observability definitions wrt controller 1. For the former case, let $t = \mathsf{ac}$ then $\pi_1^{-1}[\pi_1(t)]\sigma = \{\mathsf{ac}\sigma, \mathsf{bac}\sigma\}$. To satisfy

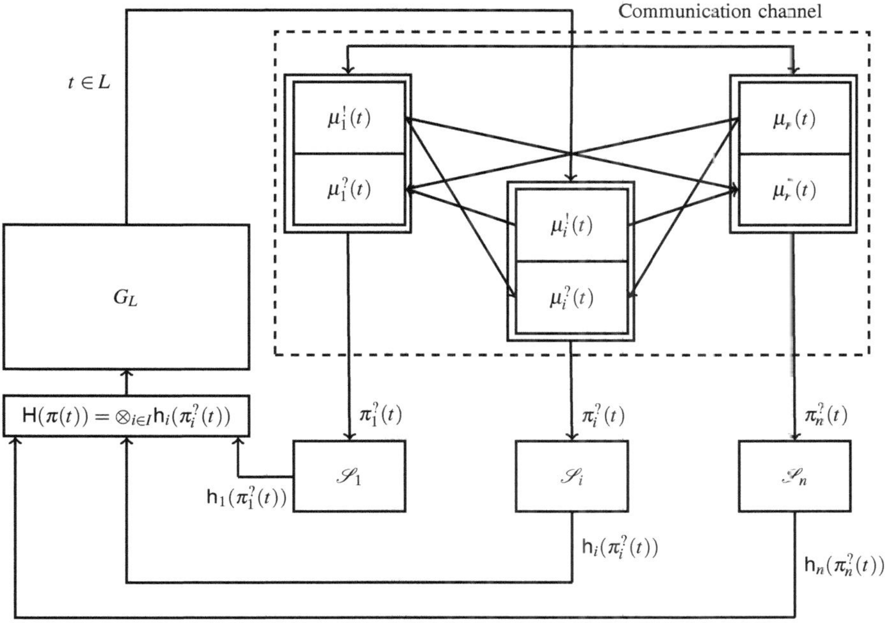

Fig. 7.1 Decentralized architecture for communication and control, where decentralized controllers $\mathscr{S}_i$ (for $i \in I$) make control decisions $\mathsf{h}_i(t)$ that are combined by a fusion rule (denoted here by $\otimes$) to produce a global control decision $\mathsf{H}(t)$ to either enable or disable events after observing sequence t generated by G_L and receiving communication from other controllers

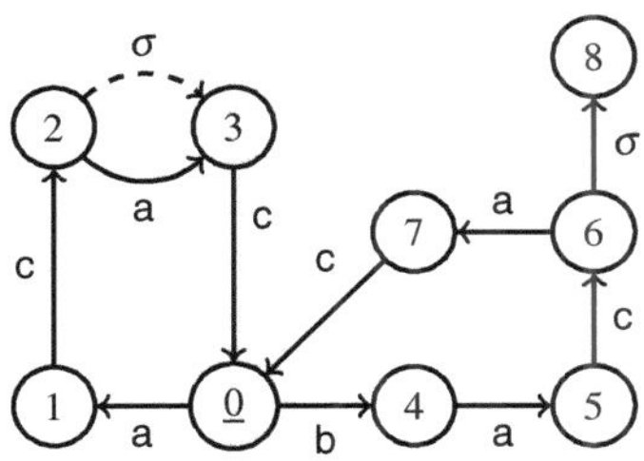

Fig. 7.2 Automata for joint G_L (all transitions) and G_K (only solid-line transitions). Initial state is underlined

unconditional co-observability, $\{\mathsf{ac}\sigma, \mathsf{bac}\sigma\} \cap \overline{K}$ must be empty; however, the intersection is $\{\mathsf{bac}\sigma\}$. It is trivial to show that K is not conditionally co-observable. It suffices to note that $I_c(\sigma) = \{1\}$ and thus there is no other $j \in I_c(\sigma)$ to take correctly the enable decisions regarding σ. It remains to show that K is articulate.

Again, this follows in a straightforward manner from K not being unconditionally co-observable. Let $t = \mathsf{ac}$. Then $\bigcap_{i\in I_c(\sigma)} \pi_i^{-1}[\pi_i(\mathsf{ac})]\sigma \cap \overline{K} = \{\mathsf{bac}\sigma\} \neq \emptyset$. ■

We synthesize communication protocols using this example as illustrated by two state-based approaches described in [2] and [9] and the event-occurrence approach introduced in [8].

7.3 State-Based Communication Protocols

When the communication protocol features messages that consist of local *information states* (i.e., local state estimates), there are two main synthesis approaches that have been proposed. The central idea is straightforward: build a finite structure that contains illegal configurations (i.e., states that correspond to violations of co-observability). Identify states in the structure where communication would eliminate the illegal configurations (i.e., refine the structure so that such states are no longer defined). The first approach requires an iterative update of the structure to reflect the effect of each identified communication [2, 10], ideally converging to a structure free of illegal configurations, whereas the second approach uses a much larger structure which, by construction, takes into account the effect of communication at all states where the reception of a message improves the local state estimates of the receiver and communications are chosen in such a way as to make illegal configurations unreachable [9].

We begin with the communication strategy of [2], where we have taken the liberty to adjust their notation for ease of comparison to the other models. To calculate the information state for controller i at state $q \in Q$ wrt $E_{o,i}$, we use the algorithm for subset construction [7], which is based on the notion of ε-closure.

Definition 7.6. *The* $\varepsilon\text{-closure}_i(X)$*, where* $X \subseteq Pwr(Q)$ *and* $i \in I$*, is the least set such that*

(i) $X \subseteq \varepsilon\text{-}closure_i(X)$;
(ii) $\forall x \in \varepsilon\text{-}closure_i(X), \forall \sigma \notin E_{o,i}, (\delta(x,\sigma) = x' \Rightarrow x' \in \varepsilon\text{-}closure_i(X))$.

When considering communication of information states, we build a structure $\mathcal{V}_0$ to monitor the progress of automaton G_L and each controller's state-based partial view of G_L. To describe the transition function for $\mathcal{V}_0$, we also need to calculate the set of states that can be reached in one step via a transition of an event σ from a given set of states $X \subseteq Pwr(Q)$:

$$step_\sigma(X) = \{x' \in Q \mid \exists x \in X \text{ such that } \delta(x,\sigma) = x'\}.$$

Thus, in $\mathcal{V}_0$, a transition from $(q, X_1, \ldots, X_n, \sigma')$ to $(q', X_1', \ldots, X_n', \sigma)$ via transition label $\sigma \in E$ is defined as follows: $\delta(q,\sigma) = q$; if $\sigma \in E_{o,i}$ then $X_i' = \text{step}_\sigma(\varepsilon\text{-closure}_i(X_i))$, otherwise $X_i' = X_i$.

There are several characteristics of $\mathcal{V}_0$ that set it apart from subsequent structures described in this chapter. First, because communication is synchronous, for a given state in the state set of $\mathcal{V}_0$, say $(q, X_1, \ldots, X_n, \sigma)$, the local state estimates for controller i, namely X_i, do not include the ε-closure$_i$ of the incoming observation σ. As a consequence, the initial state is always $(0, \{0\}, ..., \{0\}, \varepsilon)$. Second, the states include the incoming transition σ to make clear which event observation triggers a communication.

Formally, $\mathcal{V}_0 = (X, E, \delta_{\mathcal{V}}, x_0, B_{uncond}, B_{cond})$, where the finite state set $X \subseteq Q \times Pwr(Q)^n \times E$; the transition function is $\delta_{\mathcal{V}} : X \times E \to X$; the initial state is $x_0 = (q_0, \{q_0\}, \ldots, \{q_0\}, \varepsilon)$; $B_{uncond} \subseteq X$ is a set of *illegal configurations* that correspond to violations of unconditional co-observability, where $B_{uncond} = \{(q, X_1, \ldots, X_n, \gamma) \in X \mid \exists \sigma \in E_c$ such that $\delta(q, \sigma) \in Q \setminus Q_K$ and for all $i \in I_c(\sigma)$ $\exists q' \in X_i$ such that $\delta(q', \sigma) \in Q_K\}$; and $B_{cond} \subseteq X$ is an additional set of illegal configurations that correspond to violations of conditional co-observability, where $B_{cond} = \{(q, X_1, \ldots, X_n, \gamma) \in X \mid \exists \sigma \in E_c$ such that $\delta(q, \sigma) \in Q_K$ and for all $i \in I_c(\sigma)$ $\exists q' \in X_i$ such that $\delta(q', \sigma) \in Q \setminus Q_K\}$.

For all three communication protocol synthesis approaches discussed here, the common goal is to eliminate illegal configurations so that the resulting system satisfies one of the notions of co-observability. Proofs of these theorems (in various forms) can be found in the original papers [2, 8, 9].

Theorem 7.1. *$B_{uncond} = \emptyset \Leftrightarrow K$ is unconditionally co-observable wrt L, $\pi_i^?$, and E_c.*

This theorem can be extended to include conditional co-observability, even though only unconditional co-observability is considered in [2, 8, 9].

Theorem 7.2. *$B_{uncond} \neq \emptyset$ and $B_{cond} = \emptyset \Leftrightarrow K$ is conditionally co-observable wrt L, $\pi_i^?$, and E_c.*

The structure $\mathcal{V}_0$ for Example 7.1 is shown in Fig. 7.3. Although $B_{cond} = \emptyset$, there are two illegal configurations: $B_{uncond} = \{(2, \{2,6\}, \{0\}, \mathsf{c}), ((2, \{2,6\}, \{4\}, \mathsf{c})\}$. These states are identified by their double box outline in Fig. 7.3.

The process for transforming $\mathcal{V}_0$ into a structure that contains no illegal configurations begins by identifying *communication states* that will lead to the elimination of illegal configurations by refining a controller's set of local state estimates after taking into account communicated information. We first define the set of states $\Omega \subseteq Q$ from which states in $Q \setminus Q_K$ are reachable. Let $\Omega = \{q' \in Q \mid \exists s_1, s_2 \in E^*$ where $\delta(q_0, s_1) = q'$ and $\delta(q', s_2) \in Q \setminus Q_K\}$.

Definition 7.7. *A state $x = (q, X_1, \ldots, X_n, \sigma)$ is a* communication state *if*

$$\exists i \in I_o(\sigma) \text{ s.t. } (X_i \cap (\cap_{j \in I \setminus \{i\}} \varepsilon\text{-}closure_j(X_j))) \setminus \Omega = \emptyset.$$

In [2] communication occurs "as late as possible" and thus the search for a communication state begins at each $b \in B_{uncond}$. (Note that conditional control decisions came about after [2] appeared; however, it is straightforward to extend the model to detect violations of conditional co-observability.) If b is not suitable, then a backwards reachability is performed until a communication state is found. The proof of

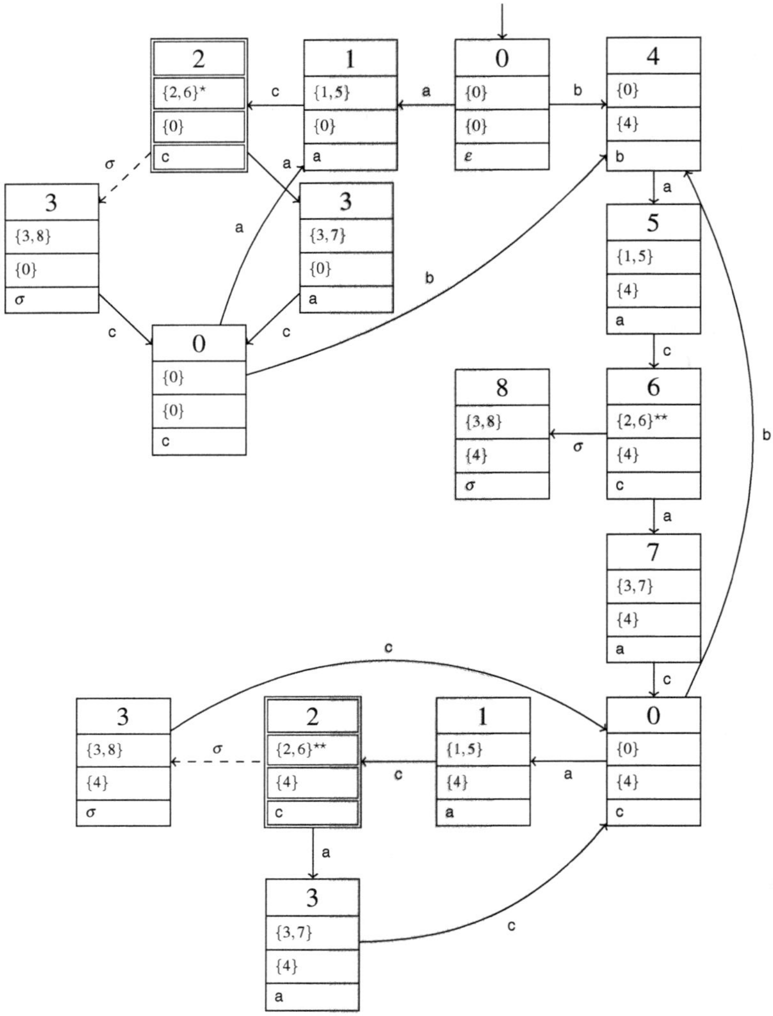

Fig. 7.3 The automaton $\mathcal{V}_0$ constructed from G_L and G_K in Fig. 7.2. Illegal configurations are indicated by states with a double box. State where communication is initiated to resolve illegal configuration $\langle 2, \{2,6\}, \{0\}, \mathsf{c} \rangle$ according to [2] is indicated by a $\star$. States where communication occurs to satisfy feasibility are indicated by a $\star\star$

guaranteed existence of a communication state for each $b \in B_{uncond}$ can be found in [2].

At a communication state, communication is initiated by a controller that has just observed the most recently occurred event γ, as indicated by the event component of the communication state. For simplicity, the initiator is fixed to be one of the controllers in $I_o(\gamma)$. The initiator then broadcasts its information state to the others, who respond by sending the initiator their information state. The communication protocol for the sending and receiving of messages at a communication state $x \in X$, for $s\gamma \in L$ such that $\delta_{\mathscr{V}}(x_0, s\gamma) = x$, is defined as follows:

$$\text{for initiator } i \in I_o(\gamma), \forall j \in I \setminus \{i\}, \quad \mu^!_{ij}(s\gamma) = \mu^?_{ji}(s\gamma) = X_i,$$
$$\mu^!_{ji}(s\gamma) = \mu^?_{ij}(s\gamma) = \varepsilon\text{-closure}_j(X_j).$$

The communication state in $\mathscr{V}_0$ wrt illegal configuration $(2,\{2,6\},\{0\},\mathsf{c})$ for initiator controller 1 is $(2,\{2,6\},\{0\},\mathsf{c})$ itself. We can now specify the communication protocol for $\mathscr{V}_0$: $\mu^!_{12}(\mathsf{ac(acac)^*}) = \{2,6\}$, $\mu^!_{21}(\mathsf{ac(acac)^*}) = \{0,1,2,3\}$.

It must be the case that communication occurs at all states that are indistinguishable to the initiator of the communication.

Definition 7.8. *Two states are* indistinguishable *to initiator i if the incoming event is identical and the local state estimate is the same:*

$$(q, X_1, \ldots, X_i, \ldots X_n, \gamma) \sim_i (q', X'_1, \ldots, X'_i, \ldots X'_n, \gamma') \Leftrightarrow X_i = X'_i \text{ and } \gamma = \gamma' \in E_{o,i}.$$

Thus, in addition to incorporating the effect of communication at a communication state, one must also add the effect of communication at states that the initiator finds indistinguishable from the communication state. This is called a *feasible* communication state.

There are two feasible communication states in $\mathscr{V}_0$ wrt communication state $(2,\{2,6\},\{0\},\mathsf{c})$, namely $(6,\{2,6\},\{4\},\mathsf{c})$ and $(2,\{2,6\},\{4\},\mathsf{c})$. Extending the communication protocol for $\mathscr{V}_0$, we have $\mu^!_{12}(\mathsf{(acac)^*bac((acbac)^*(acacbac)^*)^*}) = \{2,6\}$ and $\mu^!_{12}(\mathsf{ac(acac)^+}) = \{2,6\}$; $\mu^!_{21}(\mathsf{(acac)^*bac((acbac)^*(acacbac)^*)^*}) = \{0..8\}$ and $\mu^!_{21}(\mathsf{ac(acac)^+}) = \{0..8\}$.

When the controllers receive information after the occurrence of $s\gamma \in L$, they update their local state estimates according to

$$(\forall i \in I) \quad X_i = X_i \cap \mu^?_{i,1}(s\gamma) \cap \mu^?_{i,2}(s\gamma) \cap \ldots \; \mu^?_{i,i-1}(s\gamma) \cap \mu^?_{i,i+1}(s\gamma) \cap \ldots \mu^?_{i,n}(s\gamma).$$

This gives rise to the construction of a new version of $\mathscr{V}_0$, which we denote by $\mathscr{V}_1$, where the effect of the communication is calculated and then propagated through the calculation of a new state set and transition function, as well as an updated B_{uncond}.

Taking into account the communication performed at the communication states identified in $\mathscr{V}_0$, the next iteration $\mathscr{V}_1$ is shown in Fig. 7.4. In keeping with the notational conventions in [2], the communication state and the transformed state after

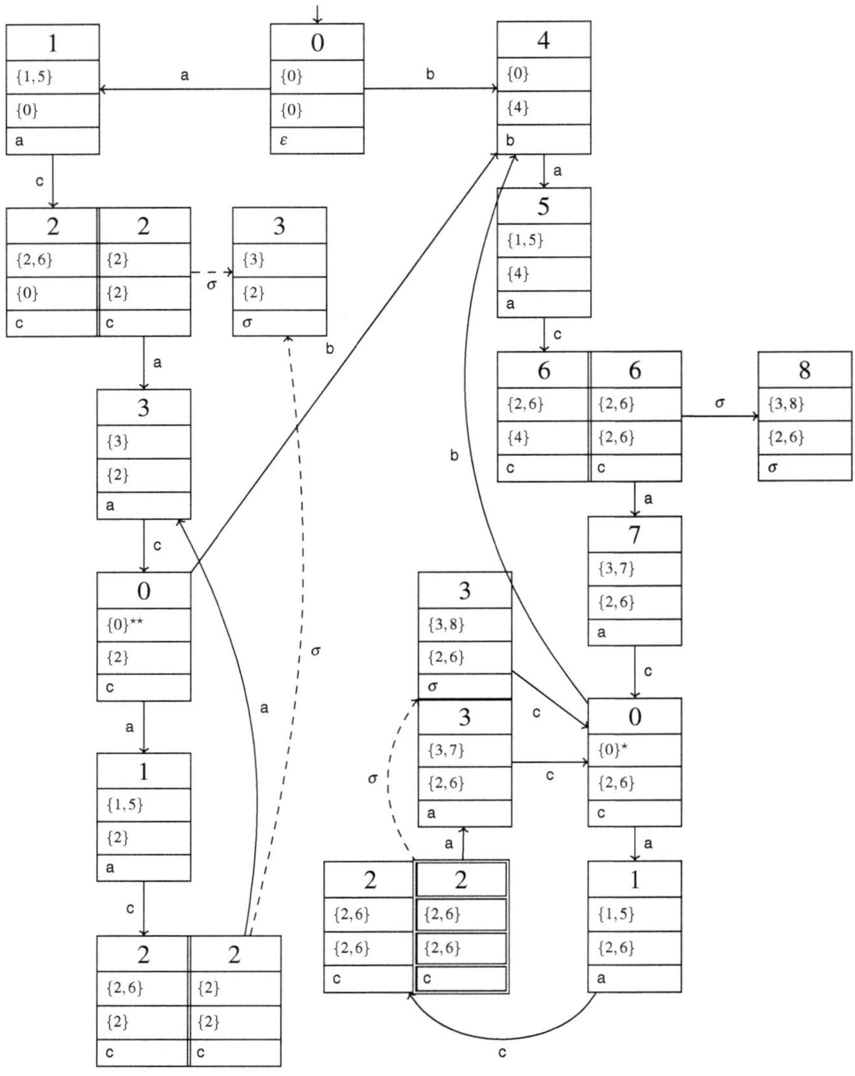

Fig. 7.4 The automaton $\mathscr{V}_1$ after the effects of the communication to eliminate illegal configuration $\langle 2, \{2,6\}, \{0\}, \mathsf{c}\rangle$ is taken into consideration. Additional communication states identified during this iteration are noted by a $\star$ whereas feasible communication states are indicated by $\star\star$

communication are shown as double states. Subsequent states in $\mathscr{V}_1$ are calculated based on the post-communication state (the rightmost box of a double state). Note that $\mathscr{V}_1$ still contains an illegal configuration. Thus, we must identify an additional communication state by examining, states that precede the illegal configuration. It can be verified that it is necessary to backtrack two states, to state $(0,\{0\},\{2,6\},\mathsf{c})$, to identify another communication state, where the initiator is controller 1. The corresponding feasible communication state is $(0,\{0\},\{2\},\mathsf{c})$.

The final iteration, $\mathscr{V}_2$, is shown in Fig. 7.5. Communication has resolved all occurrences of illegal configurations. The communication protocol for the initiator controller 1 wrt $\mathscr{V}_2$ is $\mu^!_{12}(\mathsf{ac}) = \{2,6\}$, $\mu^!_{12}(((\mathsf{bacac})^*(\mathsf{acac})^+)^*) = \{0\}$, $\mu^!_{12}((\mathsf{acac})^*\mathsf{bac}((\mathsf{acbac})^*(\mathsf{acacbac})^*)^*) = \{6\}$, $\mu^!_{12}(((\mathsf{acac})^*(\mathsf{bacac})^+)^*) = \{0\}$. The proof of convergence of this algorithm (i.e., that the iteration of $\mathscr{V}_0$ will terminate in a finite number of steps) is presented in [2].

Synthesizing communication protocols using the results of [2] assumes that communication to eliminate an illegal configuration occurs "as late as possible" and only along sequences that eventually leave K. To explore a wider range of communication opportunities, a different model was proposed in [9]. This model, denoted by $\mathscr{W}$, is more complex because, by construction, it explicitly contains communication and subsequent effect of communication on the receiver's information state, whenever communication leads to the introduction of new information for a controller. As a result, $\mathscr{W}$ is built only once and requires no further iterations; however, in the worst case, it is significantly larger than $\mathscr{V}_0$. Other differences between the two models include the definition of an information state (the trailing incoming event is no longer needed in states of $\mathscr{W}$) and communication occurs between a single sender and a single receiver as a point-to-point communication and not as a two-way broadcast between the initiator and the other controllers.

One of the most significant differences between the two models is the introduction of three different state types: ○ represents an *update* state, □ represents a *configuration* state, and ◇ represents a *communication* state. An update state reflects the changes to information states as a result of a communication from sender i to receiver j, thus avoiding the need for subsequent iterations of $\mathscr{W}$. A configuration state is equivalent to a state of $\mathscr{V}_0$ without the trailing incoming event. A communication state in the context of $\mathscr{W}$ encapsulates the information states just prior to a message being sent from sender i to receiver j.

The second significant difference is the introduction of three different kinds of transition labels: an *update mapping*, an event occurrence, and a *communication directive*. An update mapping Υ provides details of the mechanics of communication. In particular, given the event triggering the communication, the identity of the sender, and the sender's message (i.e., its local information state without ε-closure), the update mapping indicates the identity of the receiver. For example, a transition label of $\Upsilon(b,\mathsf{2},\{4\}) = \mathsf{1}$ means that at its information state $\{4\}$, the sender, controller 2, will send information regarding the occurrence of event b (as encoded by its information state $\{4\}$) to the receiver, controller 1. A communication directive

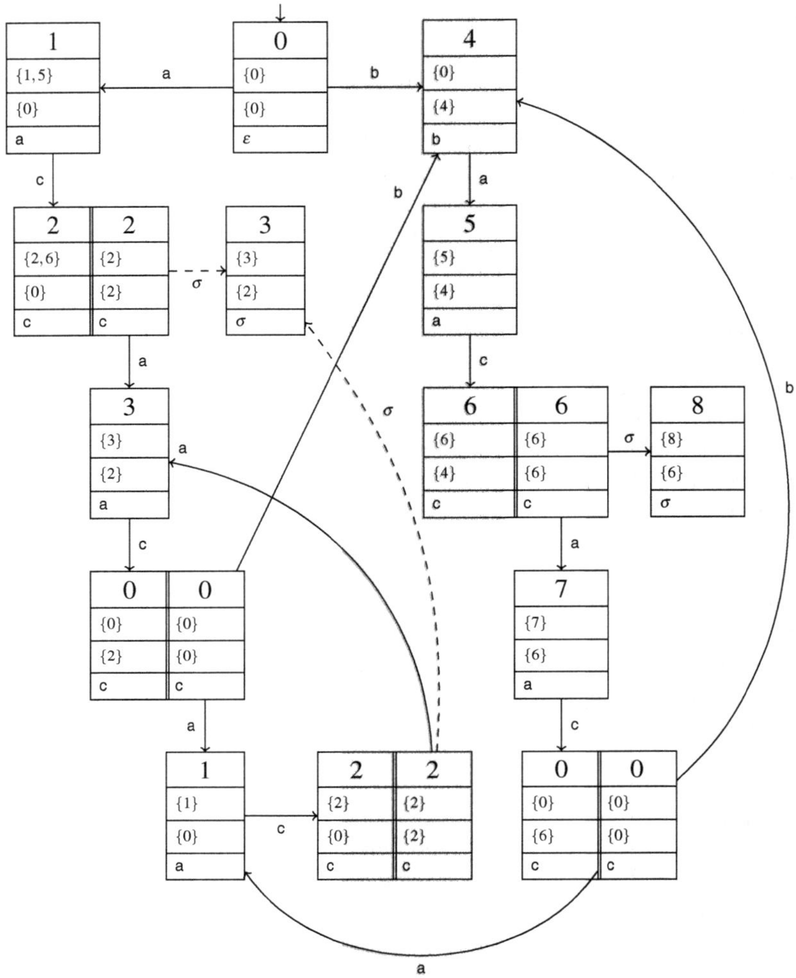

Fig. 7.5 The automaton $\mathcal{V}_2$ after the effects of the communication to eliminate illegal configuration $\langle 2, \{2,6\}, \{2,6\}, \mathtt{c} \rangle$ is taken into consideration

Φ simply indicates the identity of the sender and the receiver and is used to update the information states after a communication occurs.

Formally, $\mathscr{W} = (R, E, \delta_{\mathscr{W}}, r_0, \Phi, \Upsilon, B_{uncond}, B_{cond})$, where the finite state set $R \subseteq (\{\circ, \square\} \cup (\{\diamond\} \times E)) \times Q \times Pwr(Q)^n$; the transition set is $\delta_{\mathscr{W}} \subseteq R \times (E \cup \Phi \cup \Upsilon) \times R$; the initial state is $r_0 = \langle \circ, q_0, \{q_0\}, \ldots, \{q_0\} \rangle \in R$; the communication directive is $\Phi \subseteq (I \cup \emptyset) \times (I \cup \emptyset)$; the update mapping is $\Upsilon : E \times I \times Pwr(Q) \to I \cup \emptyset$; $B_{uncond} \subseteq X$ and $B_{cond} \subseteq X$ are sets of illegal configurations, as defined previously for structure $\mathscr{V}_0$.

Before continuing with a closer examination of the different transition and state types, we first update the definition of ε-closure in light of received information. We need to calculate the set of states that are reachable via unobservable events with the exception of those unobservable events that were just received in a communication.

Definition 7.9. *The* $\boldsymbol{\varepsilon}$***-closure***$_{i,u}(\boldsymbol{X})$, *where* $X \subseteq Pwr(Q)$, $i \in I$ *and* $u \in I \cup \emptyset$ *is the least set such that*

(i) $X \subseteq \varepsilon\text{-}closure_{i,u}(X)$;
(ii) $\forall x \in \varepsilon\text{-}closure_{i,u}(X), \forall \sigma \notin E_{o,i}, (\forall j \in I, \exists y \subseteq Pwr(Q), i \notin u = \Upsilon(\sigma, j, y))$ *and* $(\delta(x, \sigma) = x' \Rightarrow x' \in \varepsilon\text{-}closure_{i,u}(X))$.

The three types of transitions—communications, updates, and a move of the system—are now described in more detail.

1. An update transition goes from an update state to a configuration state:
$$\langle \circ, q, X_1, \ldots, X_n \rangle \xrightarrow{u} \langle \square, q, X_1', \ldots, X_n' \rangle,$$
where $u \in I \cup \emptyset, \forall i \in I\ X_i' = \varepsilon\text{-closure}_{i,u}(X_i)$.
2. A system transition goes from a configuration state to a communication state:
$$\langle \square, q, X_1, \ldots, X_n \rangle \xrightarrow{\sigma} \langle (\diamond, \sigma), q', X_1', \ldots, X_n' \rangle,$$
where $\delta(q, \sigma) = q'$, and $\forall i \in I$ $(\sigma \in E_{o,i} \Rightarrow X_i' = step_\sigma(X_i))$ and $(\sigma \notin E_{o,i} \Rightarrow X_i' = X_i)$.
3. A communication transition goes from a communication state to an update state:
$$\langle (\diamond, \sigma), q, X_1, \ldots, X_n \rangle \xrightarrow{\phi} \langle \circ, q, X_1', \ldots, X_n' \rangle,$$
where $\phi = (i, j) \in \Phi$ such that $\sigma \notin E_{o,j}$ $(X_j' = step_\sigma(X_j))$ and $(\forall i \in I \setminus \{j\}, X_i' = X_i)$.

Update transitions are unobservable to all $i \in I$. An update is merely an automatic consequence of a communication. A communication transition is observable to the sender and to the receiver j. We abuse notation and define the set of communications observable to controller i by $\Phi_i := \{\phi \in \Phi \mid (\phi = (i,j), \text{ for } i \neq j) \text{ and } (\phi = (k,i), \text{ for } i \neq k)\}$.

To build a communication protocol for $\mathscr{W}$ we choose an update mapping at an update state such that illegal configurations are unreachable. Having chosen a specific update mapping transition (i.e., either a specific sender or no communication at all), it is necessary to propagate this choice to the relevant communication state, making the feasible choice for the communication directive. In lieu of pruning the transitions not chosen at update states and at communication states, it may be simpler to think of the update mapping and communication directives as being controllable events only for the senders involved. Then by taking a choice of a particular update mapping transition, the sender enables the chosen transition and simply disables all others at that particular update state. The sender then follows a similar strategy at communication states. Subsequently, the communication protocol $\mathsf{M}_i^!$ for each $i \in I$ consists of all enabled communication directives.

Although, the construction of $\mathscr{W}$ assumes that there is a general pattern of update state, followed by configuration state, followed by communication state, for every transition in G_K, we can substantially reduce the size of the model as follows. For a given violation of conditional or unconditional co-observability, we will consider update transitions and communication states for only those transitions that correspond to events in $E_{o,j} \setminus E_{o,i}$, for $j \in I \setminus \{i\}$ and $i \in I_c(\sigma)$.

Figure 7.6 shows $\mathscr{W}$ for Example 7.1. Here, $B_{uncond} = \{(\square, 2, \{2,6\}, \{0,1,2,3\}), (\square, 2, \{2,6\}, \{0..8\})\}$, and, as before, are indicated by a double box. By enabling the update mapping transition $\Upsilon(\mathsf{b}, 2, \{4\}) = 1$ at update state $(\circ, 0, \{0\}, \{0\})$, the illegal configuration $(\square, 2, \{2,6\}, \{0,1,2,3\})$ is no longer reachable. As a consequence, at communication state $(\diamond, \mathbf{b}, 4, \{0\}, \{4\})$, controller 2 must choose to communicate to controller 1, as indicated by the mapping transition, thereby enabling communication directive $(2,1)$.

To satisfy feasibility, controller 2 must also choose to communicate to controller 1 at communication state $(\diamond, \mathbf{b}, 4, \{0,4\}, \{4\})$. To ensure that all other communication directives are consistent with these two communication directives, i.e., when controller 2 has a local state of $\{4\}$ after the occurrence of event b. Thus, controller 2 must enable any transition $\Upsilon(\mathsf{b}, 2, \{4\}) = 1$ at any other update state that has such an outgoing transition label. Hence, $\Upsilon(\mathsf{b}, 2, \{4\}) = 1$ is enabled at update states $(\circ, 4, \{4\}, \{4\})$, $(\circ, 0, \{0\}, \{0..8\})$, $(\circ, 4, \{0\}, \{4\})$ and $(\circ, 4, \{0,4\}, \{4\})$; however, by enabling transition $(2,1)$ at each of the communication states, the update states $(\circ, 4, \{0\}, \{4\})$ and $(\circ, 4, \{0,4\}, \{4\})$ become unreachable, even though their outgoing transition $\Upsilon(\mathsf{b}, 2, \{4\}) = 1$ is enabled.

Thus, for Example 7.1, $\mu_{12}^!(L) = \emptyset$ and $\mu_{21}^!(((\mathsf{acac})^*\mathsf{b})^+) = \{4\}$, and for all $s \in L \setminus ((\mathsf{acac})^*\mathsf{b})^+$, $\mu_{21}^!(s) = \emptyset$, where we interpret the message $\emptyset$ to correspond to silence. The behavior of $\mathscr{W}$ operating under communication protocol $\mathsf{M}^!$ is shown in Fig. 7.6 by the collection of transitions in bold.

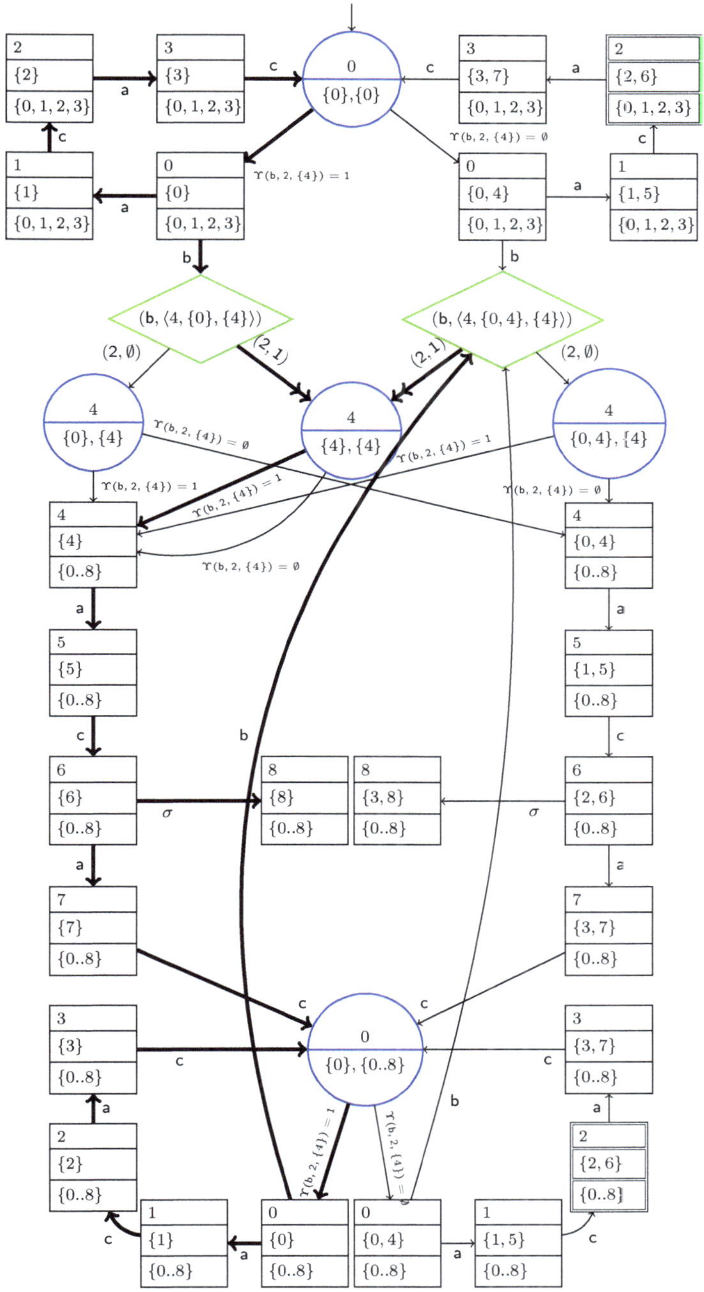

Fig. 7.6 The structure $\mathscr{W}$ (all transitions) for the automaton in Fig. 7.2. The collection of transitions in bold represent $\mathscr{W}$ operating under communication protocol $\mathsf{M}^!$. The communication protocol $\mathsf{M}^!$ is indicated by the collection of transitions with double arrowheads

7.4 Event-Based Communication Protocols

The strategy for synthesizing communication protocols based on the communication of event occurrences to distinguish sequences in $L \setminus K$ from those in K is significantly different from the information state strategy. Like $\mathscr{V}_0$ and $\mathscr{W}$, the structure $\mathscr{U}$ described below is finite-state. Like $\mathscr{V}_0$ there is only one type of state and one type of transition label; however, it is this set of transition labels that distinguish $\mathscr{U}$ from the other models.

The alphabet of $\mathscr{U}$ is based on vector labels of [1]. To begin the construction of the alphabet, we use an augmented set of controllers $I_0 = \{0\} \cup I$, where 0 represents the system. As well, we use an augmented alphabet $E^\varepsilon = \{\varepsilon\} \cup E$. A label $\ell : I_0 \longrightarrow E^\varepsilon$ is a mapping from each controller to either an event from E or ε. We will sometimes refer to the ith element of label $\ell = \langle \ell(0), \ell(1), \ldots, \ell(n) \rangle$ by $\ell(i)$, where $i \in I_0$. The empty label is $\langle \varepsilon, \ldots, \varepsilon \rangle$, i.e., for all $i \in I_0, \ell(i) = \varepsilon$.

Labels are generated from a given finite set of *atoms* denoted by A. The set of atoms is defined as the union of the following sets of labels, based on the observability of events in E:

- $\sigma \in E_{uo,i} \Rightarrow \ell(i) = \sigma$ and $\forall j \in I_0 \setminus \{i\}, \ell(j) = \varepsilon$; and
- $\forall i \in I_0$ s.t. $\sigma \in E_{o,i} \Rightarrow \ell(i) = \sigma$, otherwise $\ell(i) = \varepsilon$.

The set of atoms for Example 7.1 is $A = \{\langle \mathsf{a}, \mathsf{a}, \varepsilon \rangle, \langle \varepsilon, \varepsilon, \mathsf{a} \rangle, \langle \mathsf{b}, \varepsilon, \mathsf{b} \rangle, \langle \varepsilon, \mathsf{b}, \varepsilon \rangle, \langle \mathsf{c}, \mathsf{c}, \varepsilon \rangle, \langle \varepsilon, \varepsilon, \mathsf{c} \rangle\}$. We define $A^\varepsilon := A \cup \{\langle \varepsilon, \ldots, \varepsilon \rangle\}$.

We require the following three properties of labels.

Definition 7.10. *Two labels* ℓ_1, ℓ_2 *are* compatible, *denoted by* $\ell_1 \uparrow \ell_2$, *iff* $\forall i \in I_0$, $\ell_1(i) = \varepsilon$ *or* $\ell_2(i) = \varepsilon$ *or* $\ell_1(i) = \ell_2(i)$.

Definition 7.11. *The* least upper bound *of two compatible labels, denoted by* $\ell_1 \vee \ell_2$, *is computed as follows.*

$$\forall i \in I_0, (\ell_1 \vee \ell_2)(i) = \begin{cases} \ell_1(i), & \text{if } \ell_1(i) \neq \varepsilon; \\ \ell_2(i), & \text{if } \ell_2(i) \neq \varepsilon; \\ \varepsilon, & \text{otherwise.} \end{cases}$$

Definition 7.12. *Two labels,* ℓ_1 *and* ℓ_2, *are* independent, *denoted* $\ell_1 | \ell_2$, *iff* $\forall i \in I_0$ $\ell_1(i) = \varepsilon$ *or* $\ell_2(i) = \varepsilon$.

The alphabet for $\mathscr{U}$ is the least upper bound of compatible elements in A^ε:

$$\mathbb{A} := \{a \vee \ell \mid a \in A^\varepsilon,\ \ell \in \mathbb{A} \text{ and } a \uparrow \ell\}.$$

To construct $\mathscr{U}$, we build an augmented version of G_L and G_K as follows. Update their alphabets to be $E \cup \{\varepsilon\}$ and add a self-loop of ε at each state of Q_L and Q_K. We refer to the augmented automaton as G_L^ε and G_K^ε. We replace δ and δ_K with T and T^K, the transition sets for G_L and G_K, respectively. Finally, we add a set of special transitions that correspond to whether or not the transition is part of $L \setminus K$ or K: in G_K

we add $F_K := \{(q,e,q') \in T \mid \exists s \in \overline{K}$ such that $se \in \overline{K}$ and $(q_0,s,q)(q,e,q') \in T_K^*\}$ and in G_L we add $F_L := T \setminus F_K$. We continue by composing G_L^ε with n copies of G_K^ε, one for each decentralized controller:

$$\mathscr{U} = G_L^\varepsilon \times \prod_{i=1}^{n} G_K^\varepsilon = (X, \mathbb{A}, T^{\mathscr{U}}, x_0, B_{uncond}, B_{cond}),$$

where $(x(0), x(1), \ldots, x(n)) \in X \subseteq (Q)^{n+1}$; $\mathbb{A}$ is a finite alphabet of labels; the transition relation $T^{\mathscr{U}}$ is defined according to: $x \xrightarrow{\ell} x' \in T^{\mathscr{U}}$ iff $\ell \in \mathbb{A}$ and $\forall i \in I_0$, $x(i) \xrightarrow{\ell(i)} x'(i) \in T$; the initial state $x_0 = (q_0)^{n+1}$; and $B_{uncond}, B_{cond} \subseteq T^{\mathscr{U}}$, where the set of illegal configurations wrt transitions that correspond to violations of unconditional co-observability is $B_{uncond} := \{x \xrightarrow{\ell} x' \in T^{\mathscr{U}} \mid x(0) \xrightarrow{\ell(0)} x'(0) \in F_L$ and $(\forall i \in I_c(\ell(0)))x(i) \xrightarrow{\ell(i)} x'(i) \in F_K\}$ and the set of illegal configurations wrt transitions that correspond to violations of conditional co-observability is $B_{cond} := \{x \xrightarrow{\ell} x' \in T^{\mathscr{U}} \mid |I_c(\ell(0))| > 1$ and $x(0) \xrightarrow{\ell(0)} x'(0) \in F_K$ and $(\forall i \in I_c(\ell(0)))x(i) \xrightarrow{\ell(i)} x'(i) \in F_L\}$.

The resulting $\mathscr{U}$ structure for Example 7.1 has 1513 states, 54 labels, and 3929 transitions. As was the case for the corresponding $\mathscr{V}_0$ and $\mathscr{W}$, in $\mathscr{U}$, $B_{cond} = \emptyset$; however, $B_{uncond} = \{\langle(1,5,2), \langle\sigma,\sigma,\sigma\rangle, (2,6,2)\rangle, \langle(1,5,6), \langle\sigma,\sigma,\sigma\rangle, (2,2,6)\rangle\}$. The portion of $\mathscr{U}$ containing the transitions in B_{uncond} is shown in Fig. 7.7.

A communication protocol $M^!$ is synthesized using $\mathscr{U}$ by choosing transitions representing potential communications. We rely on an architectural property of $\mathscr{U}$ that provides a straightforward means of identifying communication transitions:

Definition 7.13. (Adapted from [4].) *The* diamond/step property *holds at $x_1 \in X$ if there exist labels $\ell_1, \ell_2 \in \mathbb{A}$ that satisfy the following axioms:*

(i) $x_1 \xrightarrow{\ell_1} x_2, x_1 \xrightarrow{\ell_2} x_3 \in T^{\mathscr{U}}$ and $\ell_1 | \ell_2 \Rightarrow x_1 \xrightarrow{\ell_1 \vee \ell_2} x_4 \in T^{\mathscr{U}}$ [Forward step];

(ii) $x_1 \xrightarrow{\ell_1 \vee \ell_2} x_4 \in T^{\mathscr{U}}$ and $\ell_1 | \ell_2 \Rightarrow x_1 \xrightarrow{\ell_1} x_2, x_2 \xrightarrow{\ell_2} x_4 \in T^{\mathscr{U}}$ [Step decomposition];

(iii) $x_1 \xrightarrow{\ell_1} x_2, x_2 \xrightarrow{\ell_2} x_4 \in T^{\mathscr{U}}$ and $\ell_1 | \ell_2 \Rightarrow x_1 \xrightarrow{\ell_1 \vee \ell_2} x_4 \in T^{\mathscr{U}}$ [Independent step].

Definition 7.14. *A* communication transition *for $(b, \ell, b') \in B_{uncond} \cup B_{cond}$ wrt $i \in I_c(\ell(0))$ is a transition $x_1 \xrightarrow{\ell_1 \vee \ell_2} x_4 \in T^{\mathscr{U}}$ such that x_1, ℓ_1, and ℓ_2 satisfy the forward step axiom (axiom (i) of Definition 7.13) where $\ell_1(0), \ell_2(i) \in E_o \setminus E_{o,i}$, and $\exists s \in \mathbb{A}^*$ such that $x_4 \xrightarrow{s} b$.*

At a communication transition for some illegal configuration $b \in B_{uncond} \cup B_{cond}$, we interpret label ℓ_1 as the occurrence and observation of event $\ell_1(0)$, an event that is not observed by controller i, and ℓ_2 controller i's "guess" that $\ell_1(0)$ has occurred. A label for an unobservable event for controller i acts merely as a placeholder. Then $\ell_1 \vee \ell_2$ represents the synchronous communication to controller i that $\ell_1(0)$ has just occurred. Thus, by choosing this communication transition, the two other transitions

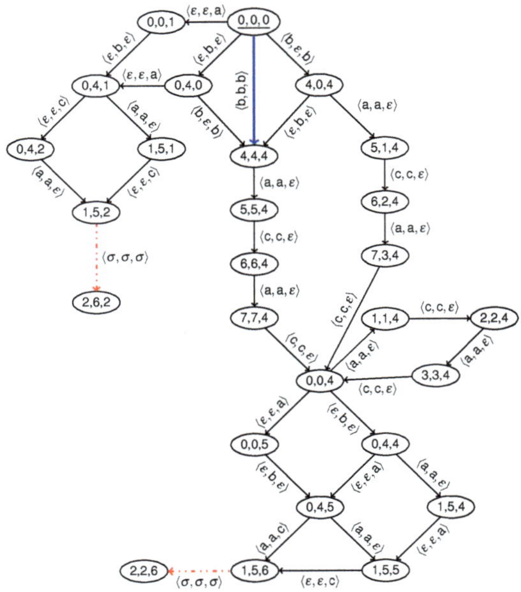

Fig. 7.7 A portion of $\mathcal{U}$ that contains all the illegal configurations for the example from Fig. 7.2. Initial state is underlined. Transitions in B_{uncond} are denoted by a dashed/dotted line (red). Potential communication transitions are indicated in bold (blue)

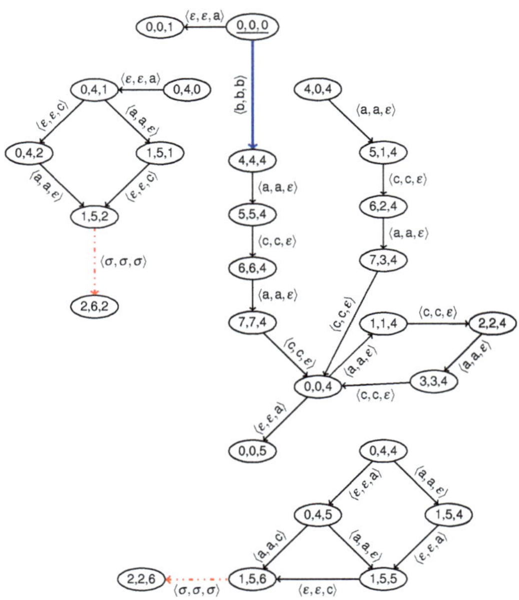

Fig. 7.8 Result of pruning the portion of $\mathcal{U}$ from Fig. 7.7. The transitions in B_{uncond} are no longer reachable

must be pruned from $\mathscr{U}$. That is, when we prune $x_1 \xrightarrow{\ell_1} x_2$ and $x_1 \xrightarrow{\ell_2} x_3$, we must prune *all* transitions $x' \xrightarrow{\ell_1} x''$ such that $x'(i) = x_1(i)$, $x''(i) = x_2(i)$ and all transitions $x' \xrightarrow{\ell_2} x'''$ such that $x'(i) = x_1(i)$, $x'''(i) = x_3(i)$.

There is only one communication transition in $\mathscr{U}$, namely $(0,0,0) \xrightarrow{\langle \mathsf{b},\mathsf{b},\mathsf{b} \rangle} (4,4,4)$. We prune $(0,0,0) \xrightarrow{\langle \varepsilon,\mathsf{b},\varepsilon \rangle} (0,4,0)$ and $(0,0,0) \xrightarrow{\langle \mathsf{b},\varepsilon,\mathsf{b} \rangle} (4,0,4)$. It is now the case that illegal configuration $(1,5,2) \xrightarrow{\langle \sigma,\sigma,\sigma \rangle} (2,6,2)$ is unreachable. Pruning of $(0,0,4) \xrightarrow{\langle \mathsf{b},\mathsf{b},\mathsf{b} \rangle} (0,4,4)$ and $(0,0,5) \xrightarrow{\langle \mathsf{b},\mathsf{b},\mathsf{b} \rangle} (0,4,5)$ makes the other illegal configuration $(1,5,6) \xrightarrow{\langle \sigma,\sigma,\sigma \rangle} (2,2,6)$ unreachable.

To ensure that communication in $\mathscr{U}$ is feasible, we must also choose communication transitions at states of $\mathscr{U}$ that are indistinguishable to the sender.

Definition 7.15. *Two states $x = (x(0),\ldots,x(n))$, $x' = (x'(0),\ldots,x'(n)) \in \mathscr{X}$ are* indistinguishable *to controller i, denoted $x \sim_i x'$, where $\sim_i$ is the least equivalence relation such that*
i. $x \xrightarrow{\langle \ell(0),\ldots,\ell(i)=\varepsilon,\ldots,\ell(n) \rangle} x' \Rightarrow x \sim_i x'$;
ii. $x \xrightarrow{\langle \varepsilon,\ldots,\varepsilon,\ell(i)\neq\varepsilon,\varepsilon,\ldots,\varepsilon \rangle} x' \Rightarrow x \sim_i x'$;
iii. if $x \sim_i x'$ *and* $(x,\ell,x''),(x',\ell,x''') \in T^{\mathscr{U}} \Rightarrow x'' \sim_i x'''$.

As there is only one potential communication transition in $\mathscr{U}$ in Fig. 7.7, there are no additional communications that must be identified to satisfy feasibility. The final communication protocol is $\mu^!_{12}(L) = \varepsilon$ and $\mu^!_{21}(((\mathsf{acac})^*\mathsf{b})^+) = \mathsf{b}$ whereas $\mu^!_{21}(L \setminus ((\mathsf{acac})^*\mathsf{b})^+) = \varepsilon$.

7.5 Further Reading

Although this chapter has focussed on communication protocol synthesis for control, there are additional strategies to calculate optimal communication sets from a given set of communications. This literature focuses on state disambiguation, where the analysis is performed on the original state space (in contrast to the synthesis techniques presented in this chapter). The problem of dynamic sensing is also closely related to the synthesis of decentralized communication protocols, where one can think of turning a sensor on and off as equivalent to communicating an event occurrence. Finally, communication has been examined in the context of decentralized diagnosis. Some representative papers are noted below.

- K. Rudie, S. Lafortune and F. Lin, "Minimal Communication in a Distributed Discrete-Event System", *IEEE Trans. Autom. Control*, vol. 48, no. 6, 957–975, 2003.
- W. Wang, S. Lafortune and F. Lin, "Minimization of Communication of Event Occurrences in Acyclic DES," *IEEE Trans. Autom. Control*, vol. 53, no. 9, pp. 2197–2202, 2008.
- F. Cassez and Tripakis, S., "Fault Diagnosis with Static and Dynamic Observers," *Fundamenta Informaticae*, vol. 88, no. 4, pp. 497–540, 2008.

- R. Debouk, S. Lafortune and D. Teneketzis, "Coordinated Decentralized Protocols for Failure Diagnosis of DES," *Discrete Event Dyn. S.*, vol. 10, no. 1/2, pp. 33–86, 2000.
- Qiu, W. and Kumar, R., "Distributed Diagnosis Under Bounded-Delay Communication of Immediately Forwarded Local Observations," *IEEE Trans. Sys. Man Cyber Part A*, vol. 38, no. 3, pp 628–643, 2008.

References

1. Arnold, A.: Finite Transition Systems. Prentice-Hall (1994)
2. Barrett, G., Lafortune, S.: Decentralized supervisory control with communicating controllers. IEEE Transactions on Automatic Control 45(9), 1620–1638 (2000)
3. Darondeau, P., Ricker, L.: Towards distributed control of discrete-event systems. In: Proc. Workshop Applications of Region Theory, Newcastle upon Tyne, UK (2011)
4. Diekert, V., Rozenberg, G. (eds.): The Book of Traces. World Scientific (1995)
5. Mannani, A., Gohari, P.: Decentralized supervisory control of DES over communication networks. IEEE Transactions on Automatic Control 53(2), 547–559 (2008)
6. Hiraishi, K.: On solvability of a decentralized supervisory control problem with communication. IEEE Transactions on Automatic Control 54(3), 468–480 (2009)
7. Rabin, M.O., Scott, D.: Finite automata and their decision problems. IBM Journal of Research and Development 3(2), 114–125 (1959)
8. Ricker, S.L.: Asymptotic minimal communication for decentralized discrete-event control. In: Proc. 9th Workshop on Discrete Event Systems, Göotegorg, Sweden (2008)
9. Ricker, S.L., Caillaud, B.: Revisiting state-based models for synthesizing optimal communicating decentralized discrete-event controllers. In: Proc. European Control Conference, Budapest, Hungary (2009)
10. Ricker, S.L., Rudie, K.: Incorporating communication and knowledge into decentralized discrete-event systems. In: Proc. 38th IEEE Conference on Decision and Control, Phoenix, Arizona, USA (1999)
11. Rudie, K., Wonham, W.M.: Think globally, act locally: decentralized supervisory control. IEEE Transactions on Automatic Control 37(11), 1692–1708 (1992)
12. Teneketzis, D.: On information structures and nonsequential stochastic control. CWI Quarterly 10(2), 179–199 (1997)
13. Tripakis, S.: Decentralized control of discrete event systems with bounded or unbounded delay communication. IEEE Transactions on Automatic Control 49(9), 1489–1501 (2004)
14. van Schuppen, J.H.: Decentralized supervisory control with information structures. In: Proc. 4th Workshop on Discrete Event Systems, Cagliari, Italy (1998)
15. Witsenhausen, H.: Separation of estimation and control for discrete time systems. Proceedings of the IEEE 59, 1557–1566 (1971)
16. Wong, K.C., van Schuppen, J.H.: Decentralized supervisory control of DES with communication. In: Proc. 3rd Workshop on Discrete Event Systems, Edinburgh, Scotland (1996)
17. Yoo, T.S., Lafortune, S.: A general architecture for decentralized supervisory control of discrete-event systems. Discrete Event Dynamic Systems 12(3), 335–377 (2002)
18. Yoo, T.S., Lafortune, S.: Decentralized supervisory control with conditional decisions: supervisor existence. IEEE Transactions on Automatic Control 49(11), 1886–1904 (2004)

Chapter 8
Coordination Control of Distributed Discrete-Event Systems

Jan Komenda, Tomáš Masopust, and Jan H. van Schuppen

8.1 Introduction

This chapter discusses supervisory control of distributed discrete-event systems with a coordinator. Discrete-event systems represented as finite automata have been studied by P.J. Ramadge and W.M. Wonham in [21]. Large discrete-event systems are typically formed as a synchronous composition of many small components (subsystems) that are modeled by finite automata and run in parallel. Such systems are called distributed. The aim of supervisory control is to ensure that the control objectives are satisfied by the closed-loop system. As only controllable specifications can be achieved, one of the key issues is to compute the supremal controllable sublanguage of a given specification from which the supervisor can be constructed.

Supervisory control of distributed discrete-event systems with a global specification and local supervisors is a difficult problem. There exist restrictive conditions under which distributed control is maximally permissive, hence we introduce the coordination control architecture to handle the general case. In a coordinated distributed discrete-event system, one distinguishes a coordinator and two or more subsystems. The coordinator directly influences the dynamics of the other subsystems but the subsystems do not directly influence each other. Coordination control of a distributed system is to synthesize supervisors for the coordinator and for each subsystem so that the closed-loop system meets the specification. A necessary and sufficient condition on a specification to be achieved in the coordination control architecture is presented. Moreover, the supremal conditionally-controllable

Jan Komenda · Tomáš Masopust
Institute of Mathematics, Academy of Sciences of the Czech Republic, Žižkova 22, 616 62 Brno, Czech Republic
e-mail: `komenda@ipm.cz`, `masopust@math.cas.cz`

Jan H. van Schuppen
CWI, P.O. Box 94079, 1090 GB Amsterdam, The Netherlands
e-mail: `J.H.van.Schuppen@cwi.nl`

C. Seatzu et al. (Eds.): Control of Discrete-Event Systems, LNCIS 433, pp. 147–167.
springerlink.com

sublanguage of a given specification always exists and is included in the supremal controllable sublanguage. A procedure for its computation is proposed.

8.2 Definitions

Discrete-event systems are modeled as deterministic generators that are finite automata with partial transition functions. In this chapter, the notion of a *language* stands for a regular language, see Chapters 2, 3, 4, 6 for more details.

A *generator* is a quintuple $G = (Q, E, f, q_0, Q_m)$, where Q is a finite set of states, E is a finite set of events, $f : Q \times E \to Q$ is a partial transition function, $q_0 \in Q$ is the initial state, and $Q_m \subseteq Q$ is a set of marked states. Usually, f is extended to $\hat{f} : Q \times E^* \to Q$ so that $\hat{f}(q, \varepsilon) = q$ and $\hat{f}(q, aw) = \hat{f}(f(q,a), w)$, for $a \in E$ and $w \in E^*$. The language generated by G is defined as the set $L(G) = \{s \in E^* \mid \hat{f}(q_0, s) \in Q\}$, and the marked language of G as the set $L_m(G) = \{s \in E^* \mid \hat{f}(q_0, s) \in Q_m\}$. The set of all *reachable events* of G, denoted by $E_r(G)$, is defined as the set $E_r(G) = \{e \in E \mid \exists s_1, s_2 \in E^*, s_1 e s_2 \in L(G)\}$.

For two sets $E_0 \subseteq E$, a *(natural) projection* $P : E^* \to E_0^*$ is a morphism defined by $P(a) = \varepsilon$, for $a \in E \setminus E_0$, and $P(a) = a$, for $a \in E_0$. The *inverse image* $P^{-1} : E_0^* \to 2^{E^*}$ of P is defined as $P^{-1}(a) = \{s \in E^* \mid P(s) = a\}$. These definitions can be extended to languages in a natural way. Given sets E_i, E_j, E_k, E_ℓ, we denote by $P^{i+j}_{k \cap \ell}$ the projection from $E_i \cup E_j$ to $E_k \cap E_\ell$. We use the notation $E_{i+j} = E_i \cup E_j$.

The *synchronous product* of two languages $L_1 \subseteq E_1^*$ and $L_2 \subseteq E_2^*$ is defined as $L_1 \| L_2 = P_1^{-1}(L_1) \cap P_2^{-1}(L_2) \subseteq (E_1 \cup E_2)^*$, where $P_i : (E_1 \cup E_2)^* \to E_i^*$ are projections, for $i = 1, 2$.

For generators $G_1 = (Q_1, E_1, \delta_1, q_{0,1}, Q_{m,1})$ and $G_2 = (Q_2, E_2, \delta_2, q_{0,2}, Q_{m,2})$, the generator $G_1 \| G_2$ is defined as the accessible part of the generator $(Q_1 \times Q_2, E_1 \cup E_2, \delta, (q_{0,1}, q_{0,2}), Q_{m,1} \times Q_{m,2})$, where

$$\delta((x,y),e) = \begin{cases} (\delta_1(x,e), \delta_2(y,e)), & \text{if } \delta_1(x,e) \in Q_1 \text{ and } \delta_2(y,e) \in Q_2; \\ (\delta_1(x,e), y), & \text{if } \delta_1(x,e) \in Q_1 \text{ and } e \notin E_2; \\ (x, \delta_2(y,e)), & \text{if } e \notin E_1 \text{ and } \delta_2(y,e) \in Q_2; \\ \text{undefined}, & \text{otherwise.} \end{cases}$$

It is known that $L(G_1 \| G_2) = L(G_1) \| L(G_2)$ and $L_m(G_1 \| G_2) = L_m(G_1) \| L_m(G_2)$.

A *distributed discrete-event system* is a concurrent system formed by the synchronous product of several local subsystems. For simplicity, we consider only a synchronous product of two subsystems in this chapter.

A *controlled generator* is a structure (G, E_c, Γ), where G is a generator, $E_c \subseteq E$ is the set of *controllable events*, $E_u = E \setminus E_c$ is the set of *uncontrollable events*, and $\Gamma = \{\gamma \subseteq E \mid E_u \subseteq \gamma\}$ is the set of *control patterns*. A *supervisory control* for the controlled generator (G, E_c, Γ) is a map $v : L(G) \to \Gamma$. A *closed-loop system* associated with the controlled generator (G, E_c, Γ) and the supervisory control v is defined as the minimal language $L(v/G) \subseteq E^*$ satisfying (1) $\varepsilon \in L(v/G)$ and (2) if $s \in L(v/G)$, $a \in v(s)$, and $sa \in L(G)$, then $sa \in L(v/G)$.

In the automata framework, where the supervisory control v is represented by a generator (*supervisor*), S, cf. Chapter 3, the closed-loop system can be recast as a synchronous product of the supervisor and the plant, i.e., $L(v/G) = L(S) \| L(G)$. In what follows, for a supervisor S corresponding to a supervisory control v, we use the notation $L(S/G) = L(S) \| L(G) = L(v/G)$.

The prefix closure, prefix(L), of a language L is the set of all prefixes of all its words; L is prefix-closed if $L = \text{prefix}(L)$. Let $L = \text{prefix}(L) \subseteq E^*$ be a language and $E_u \subseteq E$ be a set of uncontrollable events. A language $K \subseteq L$ is *controllable* with respect to L and E_u if $\text{prefix}(K)E_u \cap L \subseteq \text{prefix}(K)$.

Consider a language $K = \text{prefix}(K) \subseteq E^*$, the goal of supervisory control is to find a supervisory control v and its corresponding supervisor S such that $L(S/G) = K$. Such a supervisor exists if and only if K is controllable [21]. For uncontrollable languages, supremal controllable sublanguages are considered. The notation $\sup C(K, L, E_u)$ denotes the supremal controllable sublanguage of K with respect to L and E_u, which always exists and equals to the union of all controllable sublanguages of K [3]. In the following, we extend this notions to the case of distributed discrete-event systems.

Definition 8.1 (Coordinator). *A* coordinator, *denoted G_k, is a generator with the specific control-theoretic task to coordinate the actions of the other subsystems (for example, for safety, nonblockingness, etc.). In this paper, the focus is on coordinators for safety.*

Definition 8.2 (Conditional Independence). *Generators G_1 and G_2 are* conditionally independent *with respect to a coordinator G_k if there is no common transition of both G_1 and G_2 without the coordinator G_k being also involved in the global system. In other words, $E_r(G_1) \cap E_r(G_2) \subseteq E_r(G_k)$.*

Example 8.1. Database transactions are typical examples of discrete-event systems that should be controlled to avoid incorrect behaviors. Transactions are modeled by a sequence of request (r), access (a), and write (w) events. Often, several users access the database at the same time, which can lead to inconsistencies when operations of different users are executed concurrently. Consider two users and the corresponding events r_i, a_i, w_i, for $i = 1, 2$. Possible schedules are given by the language $\{r_1 a_1 w_1\} \| \{r_2 a_2 w_2\}$ of the generator $G = G_1 \| G_2$ over the event set $E = \{r_1, r_2, a_1, a_2, w_1, w_2\}$, where G_1 and G_2 are defined as in Fig. 8.1, and $E_c = \{a_1, a_2, w_1, w_2\}$ is the set of controllable events. Then $E_r(G_1) = \{r_1, a_1, w_1\}$ and $E_r(G_2) = \{r_2, a_2, w_2\}$. This means that for this example, G_1 and G_2 are conditionally independent for any G_k because $E_r(G_1) \cap E_r(G_2) = \emptyset \subseteq E_r(G_k)$. ■

Definition 8.3 (Conditional Decomposability). *A language $K \subseteq (E_1 \cup E_2 \cup E_k)^*$ is* conditionally decomposable *with respect to event sets E_1, E_2, E_k if it can be written as $K = P_{1+k}(K) \| P_{2+k}(K) \| P_k(K)$.*

Fig. 8.1 Generators G_1 and G_2

Note that there always exists an event set E_k with the corresponding projection $P_k : (E_1 \cup E_2 \cup E_k)^* \to E_k^*$ such that K is conditionally decomposable with respect to E_1, E_2, E_k. The choice $E_k = E_1 \cup E_2$ satisfies the property. However, preferably a smaller subset $E_k \subset E_1 \cup E_2$ is chosen. Polynomial-time algorithms to verify whether K is conditionally decomposable and to extend the event set E_k are discussed in [9].

Example 8.2. Consider Example 8.1. The specification language K, defined by the generator depicted in Fig. 8.2, describes the correct behavior consisting in finishing the transaction in the write stage before another transaction can proceed to the write phase. For $E_k = \{a_1, a_2\}$, the language K is conditionally decomposable. ■

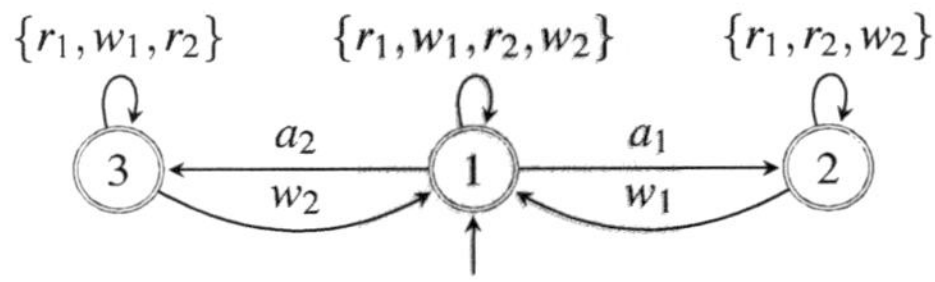

Fig. 8.2 Generator for the specification language K

Lemma 8.1. *A language $M \subseteq E^*$ is conditionally decomposable with respect to event sets E_1, E_2, E_k if and only if there exist languages $M_{1+k} \subseteq E_{1+k}^*$, $M_{2+k} \subseteq E_{2+k}^*$, $M_k \subseteq E_k^*$ such that $M = M_{1+k} \| M_{2+k} \| M_k$.*

Proof. If $M = P_{1+k}(M) \| P_{2+k}(M) \| P_k(M)$, define $M_{i+k} = P_{i+k}(M)$, $i = 1,2$, and $M_k = P_k(M)$. On the other hand, assume that there exist languages $M_{1+k} \subseteq E_{1+k}^*$, $M_{2+k} \subseteq E_{2+k}^*$, and $M_k \subseteq E_k^*$ such that $M = M_{1+k} \| M_{2+k} \| M_k$. Obviously, $P_{i+k}(M) \subseteq M_{i+k}$, $i = 1,2$, and $P_k(M) \subseteq M_k$, which implies that $P_k(M) \| P_{1+k}(M) \| P_{2+k}(M) \subseteq M$. As it also holds that $M \subseteq P_{i+k}^{-1}[P_{i+k}(M)]$, the definition of the synchronous product implies that $M \subseteq P_k(M) \| P_{1+k}(M) \| P_{2+k}(M)$. □

If $K = M_1 \| M_2 \| M_k$, then $P_{1+k}(K) \subseteq M_1$, $P_{2+k}(K) \subseteq M_2$, and $P_k(K) \subseteq M_k$. This means that even though several languages M_i may exist in general, the triple $P_{1+k}(K)$, $P_{2+k}(K)$, $P_k(K)$ is the smallest decomposition of K.

Definition 8.4 (Coordinated System). *Let G_1 and G_2 be two subsystems, and let G_k be a coordinator. A* coordinated *discrete-event system is a composition of both subsystems with the coordinator, i.e., the system $G_1 \| G_2 \| G_k$.*

The fundamental question is the construction of a coordinator. One possible way is presented in the following algorithm.

Algorithm 8.3. (Construction of a Coordinator). *Let G_1 and G_2 be two subsystems over the event sets E_1 and E_2, respectively. Construct the event set E_k and the coordinator G_k as follows:*

1. *Set $E_k = E_r(G_1) \cap E_r(G_2)$.*
2. *Extend E_k so that K is conditionally decomposable.*
3. *Define $G_k = P_k(G_1) \parallel P_k(G_2)$.*

The aim is to extend E_k so that the coordinator is minimal. However, the concept of minimality of the coordinator needs to be investigated. Furthermore, projections in Step 3 can be of exponential size. To overcome this, E_k can be further extended to satisfy an observer property that ensures that the projections are smaller (see Definition 8.6 and Theorem 8.2 below). An advantage of this choice is that the coordinator does not change the original plant, that is, $G_1 \| G_2 \| G_k = G_1 \| G_2$.

Example 8.4. Continue Example 8.2. For $E_k = \{a_1, a_2\}$, the condition of Step 2 is satisfied. Thus, the coordinator G_k can be chosen as $G_k = P_k(G_1) \| P_k(G_2)$, see Fig. 8.3. ■

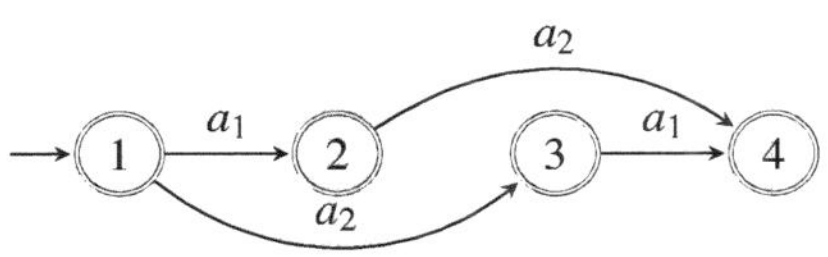

Fig. 8.3 Coordinator G_k

8.3 Problem Statement

The problem with the computation of the global system, $G = G_1 \| G_2 \| \ldots \| G_n$, composed of n subsystems is that the number of states of G is up to exponential with respect to the number of subsystems. Therefore, another (distributed) method must be used. A natural method is to find supervisors S_i so that S_i/G_i meets the corresponding part $P_i(K)$ of the specification K. Unfortunately, this approach does not work.

Example 8.5. Consider Example 8.2. Even if you could find supervisors S_1 and S_2 such that $L(S_1/G_1) = P_1(K)$ and $L(S_2/G_2) = P_2(K)$, these supervisors do not solve the problem because $L((S_1 \| S_2)/(G_1 \| G_2)) = P_1(K) \| P_2(K) \supsetneq K$, which shows that the solution is not safe because it permits behaviors that are not in the specification language K. ■

Coordination control is useful for systems such as autonomous underwater vehicles, uninhabited aerial vehicles, road networks, automated guided vehicles, complex machines consisting of many different sensors, actuators, and local control computers (such as high-speed printers), see [19], or a paint factory that produces cups of colored fluids, see [1].

Problem 8.1. Consider a distributed system $G_1 \| G_2$, where G_1 and G_2 are generators over E_1 and E_2, respectively. Let G_k be a coordinator over $E_k \supseteq E_1 \cap E_2$. Let $K \subseteq L(G_1 \| G_2 \| G_k)$ be a prefix-closed specification. Assume that the coordinator G_k makes G_1 and G_2 conditionally independent and that K is conditionally decomposable with respect to E_1, E_2, E_k. The problem is to determine supervisors S_1, S_2, S_k for the respective generators so that the closed-loop system with the coordinator satisfies

$$L(S_1/[G_1\|(S_k/G_k)]) \; \| \; L(S_2/[G_2\|(S_k/G_k)]) \; \| \; L(S_k/G_k) = K \, .$$

We consider only such supervisors for which it holds that $L(S_k/G_k) \subseteq P_k(K)$, $L(S_1/[G_1\|(S_k/G_k)]) \subseteq P_{1+k}(K)$, and $L(S_2/[G_2\|(S_k/G_k)]) \subseteq P_{2+k}(K)$. ■

8.4 Coordination Control with Complete Observations: Existence

Example 8.6. Consider event sets $E_k = \{a,b,e,\varphi\}$, $E_1 \cup E_2 = \{a,d,e,\varphi\} \cup \{b,f,\varphi\}$, where the set of controllable events is $E_c = \{e,b,\varphi\}$. Define generators G_1, G_2, and the coordinator G_k as in Fig. 8.4. Define the specification language K as the behavior of the generator D given in Fig. 8.5. It can be verified that G_k makes G_1 and G_2 conditionally independent and that K is conditionally decomposable with respect to event sets E_1, E_2, E_k. ■

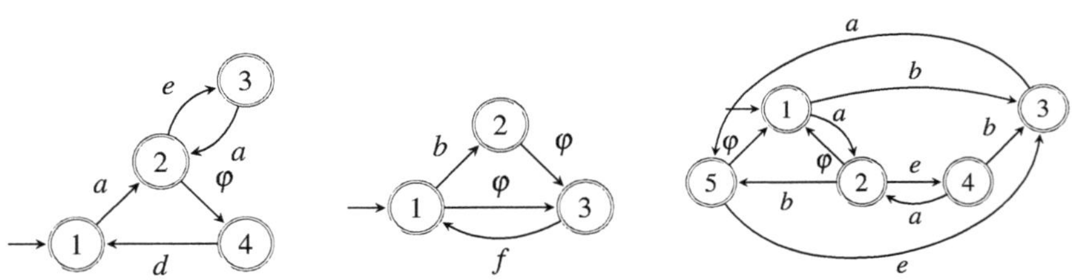

Fig. 8.4 Generators G_1, G_2, and G_k

This section presents a condition for the existence of a solution.

Lemma 8.2. Exercise 3.3.7 in [27] *Let $E_k \subseteq E_1 \cup E_2$ be such that $E_1 \cap E_2 \subseteq E_k$. Let $L_1 \subseteq E_1^*$ and $L_2 \subseteq E_2^*$ be languages. Let $P_k : (E_1 \cup E_2)^* \to E_k^*$ be a projection, then $P_k(L_1 \| L_2) = P_k(L_1) \| P_k(L_2)$.*

The following lemma shows that the synchronous product of a language with its projection does not change the language.

Lemma 8.3. *Let $L \subseteq E^*$ be a language, $E_k \subseteq E$, and $P_k : E^* \to E_k^*$ be a projection. Then $L \| P_k(L) = L$.*

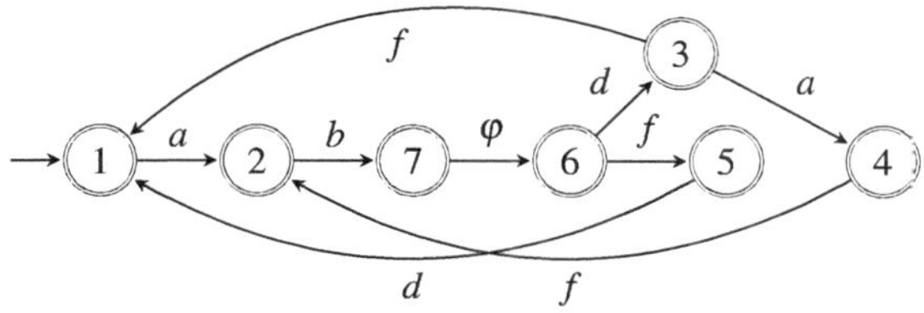

Fig. 8.5 Generator D

Proof. As $L \subseteq P_k^{-1}P_k(L)$, $L \| P_k(L) = L \cap P_k^{-1}P_k(L) = L$. □

The solution of Problem 8.1 uses the following concepts.

Definition 8.5 (Conditional Controllability). *A prefix-closed language K is said to be* conditionally controllable *for generators G_1, G_2, G_k and locally uncontrollable event sets $E_{1+k,u}$, $E_{2+k,u}$, $E_{k,u}$, where $E_{i,u} = E_u \cap E_i$, for $i = 1+k, 2+k, k$, if*

1. *$P_k(K)$ is controllable with respect to $L(G_k)$ and $E_{k,u}$; equivalently,*

$$P_k(K)E_{k,u} \cap L(G_k) \subseteq P_k(K) .$$

2. *$P_{1+k}(K)$ is controllable with respect to $L(G_1) \| P_k(K) \| P_k^{2+k}[L(G_2) \| P_k(K)]$ and $E_{1+k,u}$; equivalently,*

$$P_{1+k}(K)E_{1+k,u} \cap L(G_1) \| P_k(K) \| P_k^{2+k}[L(G_2) \| P_k(K)] \subseteq P_{1+k}(K) .$$

3. *$P_{2+k}(K)$ is controllable with respect to $L(G_2) \| P_k(K) \| P_k^{1+k}[L(G_1) \| P_k(K)]$ and $E_{2+k,u}$; equivalently,*

$$P_{2+k}(K)E_{2+k,u} \cap L(G_2) \| P_k(K) \| P_k^{1+k}[L(G_1) \| P_k(K)] \subseteq P_{2+k}(K) .$$

Definition 8.5 can be extended to more subsystems with one central coordinator, whose event set contains all shared events. Conditions 2 and 3 then result in: $P_{i+k}(K)$ is controllable with respect to $P_{i+k}(\|_{i=1}^{n} L(G_i) \| P_k(K)) = \|_{i=1}^{n} P_{i+k}[L(G_i) \| P_k(K)]$ and $E_{i+k,u}$.

If K is conditionally controllable, then there exists a supervisor S_k such that $P_k(K) = L(S_k/G_k)$ because $P_k(K)$ is controllable. The conditions of Definition 8.5 can be checked by classical algorithms with polynomial computational complexity with respect to the number of states discussed in Chapter 3. The complexity of checking conditional controllability is thus less than that of the global system $G_1 \| G_2 \| G_k$. This is because instead of checking controllability with the global specification and the global system, we check it only on the corresponding projections to E_{1+k} and E_{2+k}. The projections are smaller when they satisfy the observer property (see Definition 8.6 below).

The following theorem presents the necessary and sufficient condition on a specification language to be exactly achieved in the coordination control architecture.

Theorem 8.1. *Consider Problem 8.1. There exist supervisors S_1, S_2, and S_k such that*

$$L(S_1/[G_1\|(S_k/G_k)]) \parallel L(S_2/[G_2\|(S_k/G_k)]) \parallel L(S_k/G_k) \;=\; K \tag{8.1}$$

if and only if K is conditionally controllable for generators G_1, G_2, G_k and locally uncontrollable event sets $E_{1+k,u}$, $E_{2+k,u}$, $E_{k,u}$.

Proof. To simplify the notation, denote $L_i = L(G_i)$, $i = 1,2,k$, and $L = L_1\|L_2\|L_k$. By Lemma 8.1, $P(L) = P_{1+k}(L)\|P_{2+k}(L)\|P_k(L)$ because $L = (L_1\|L_k)\|(L_2\|L_k)\|L_k$, which follows from the fact that the synchronous product is idempotent.

To prove sufficiency, let $K \subseteq L$ be conditionally controllable. Then $P_k(K) \subseteq L_k$ is controllable with respect to L_k, and there exists a supervisor S_k such that $L(S_k/G_k) = P_k(K)$ [20]. Furthermore, $K \subseteq L$ implies $P_{1+k}(K) \subseteq L_1\|L_k\|P_k^2(L_2)$, by Lemma 8.2. This, together with $P_{1+k}(K) \subseteq (P_k^{1+k})^{-1}P_k(K)$, $P_k(K) \subseteq L_k$, and $P_k^2(L_2)\|P_k(K) = P_k^{2+k}(L_2\|L(S_k/G_k))$, implies the inclusion $P_{1+k}(K) \subseteq L_1\|P_k(K)\|P_k^{2+k}(L_2\|P_k(K))$. By conditional controllability of K, there exists a supervisor S_1 such that $P_{1+k}(K) = L(S_1/[G_1\|(S_k/G_k)\|P_k^{2+k}(G_2\|(S_k/G_k))])$, where for a generator H and a projection P, $P(H)$ denotes a generator such that $L(P(H)) = P(L(H))$. Similarly, there exists a supervisor S_2 such that $L(S_2/[G_2\|(S_k/G_k)\|P_k^{1+k}(G_1\|(S_k/G_k))]) = P_{2+k}(K)$. Since $L = L\|P_k(L)$, by Lemma 8.3, $L(G_i\|(S_k/G_k)\|P_k^{i+k}(G_i\|(S_k/G_k))) = L(G_i\|(S_k/G_k))$. Notice that $L(S_1/[G_1\|(S_k/G_k)]) \parallel L(S_2/[G_2\|(S_k/G_k)]) \parallel L(S_k/G_k) = P_{1+k}(K) \parallel P_{2+k}(K) \parallel P_k(K) = K$ because K is conditionally decomposable. This proves (8.1).

To prove necessity, projections P_k, P_{1+k}, and P_{2+k} will be applied to (8.1). First, note that $K = L(S_1\|S_2\|S_k)\|L$, which follows from (8.1) by replacing $/$ with $\|$, and by the commutativity of the operation $\|$. This yields $P_k(K) \subseteq L(S_k)\|L_k = L(S_k/G_k)$. On the other hand, recall that $L(S_k/G_k) \subseteq P_k(K)$, cf. Problem 8.1. Hence, $L(S_k/G_k) = P_k(K)$, which means that $P_k(K)$ is controllable with respect to $L(G_k)$, i.e., item (1) of Definition 8.5 is satisfied. Now, we prove item (2); item (3) is symmetric. As $E_{1+k} \cap E_{2+k} = E_k$, $L(S_2)\|L(G_2\|(S_k/G_k)) = L(S_2) \cap L(G_2\|(S_k/G_k))$, because the components are over the same event set E_{2+k}, and $P_{1+k}^{2+k} = P_k^{2+k}$, we get that $P_{1+k}(K) \subseteq L(S_1\|G_1\|(S_k/G_k)\|P_k^{2+k}(G_2\|(S_k/G_k))) \subseteq L(S_1)\|L(S_k)\|L_1\|L_k = L(S_1/[G_1\|(S_k/G_k)]) \subseteq P_{1+k}(K)$. Then, we can take the system $G_1\|(S_k/G_k)\|P_k^{2+k}(G_2\|(S_k/G_k))$ as a new plant, i.e., the language $P_{1+k}(K)$ is controllable with respect to $L(G_1\|(S_k/G_k)\|P_k^{2+k}(G_2\|(S_k/G_k)))$. Thus, (2) is satisfied. □

Example 8.7. Consider Example 8.6. Languages $P_{1+k}(K)$, $P_{2+k}(K)$, and $P_k(K)$ are controllable with respect to the plant languages $L(G_1)\|P_k(K)\|P_k^{2+k}[L(G_2)\|P_k(K)]$, $L(G_2)\|P_k(K)\|P_k^{1+k}[L(G_1)\|P_k(K)]$, and $L(G_k)$, respectively. Hence, there exist supervisory controls v_1, v_2, and v_k, and respective supervisors S_1, S_2, and S_k such that the closed-loop languages equal the languages $P_{1+k}(K)$, $P_{2+k}(K)$, and $P_k(K)$, respectively, see Fig. 8.6. Thus, $L(S_1/[G_1\|(S_k/G_k)]) \parallel L(S_2/[G_2\|(S_k/G_k)]) \parallel L(S_k/G_k)) = K$. ■

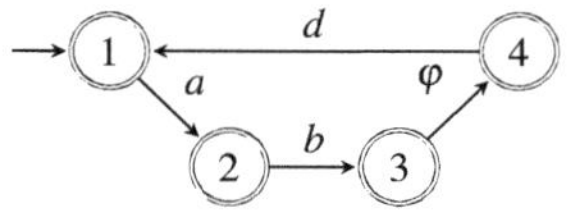

(a) $P_{1+k}(K)$.

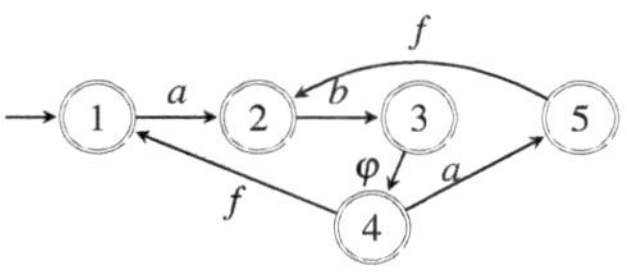

(b) $P_{2+k}(K)$.

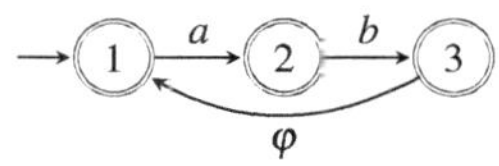

(c) $L(S_k/G_k) = P_k(K)$.

Fig. 8.6 Supervisors S_k, S_1, and S_2

8.5 Coordination Control with Complete Observations: Supremal Supervision

It may happen that a specification language K is not achievable using the coordination control scheme. By Theorem 8.1, this occurs whenever K is not conditionally controllable. In this case, the supremal conditionally-controllable sublanguage is considered.

This section presents a procedure for the computation of the supremal conditionally-controllable sublanguage of the specification language K.

Lemma 8.4. [4] *Consider event sets $E = E_1 \cup E_2$ and $E_u \subseteq E$. For $i = 1,2$, let $K_i \subseteq L_i$ be prefix-closed languages such that K_i is controllable with respect to L_i and $E_{i,u}$. Then $K_1 \| K_2$ is controllable with respect to $L_1 \| L_2$ and E_u.*

Proof. Let $s \in K_1 \| K_2$, $e \in E_u$, and $se \in L_1 \| L_2$, then $P_i(s) \in K_i$ and $P_i(se) \in L_i$. For $e \in E_i$, $P_i(se) = P_i(s)e \in L_i$. As K_i is controllable with respect to L_i and $E_{i,u}$, $P_i(s)e \in K_i$. For $e \notin E_i$, $P_i(se) = P_i(s) \in K_i$. Thus, for any $e \in E_u$, $se \in K_1 \| K_2$. □

Lemma 8.5 (Transitivity of Controllability). *Let $K \subseteq L \subseteq M$ be prefix-closed languages over E where K is controllable with respect to L and E_u, and L is controllable with respect to M and E_u. Then K is controllable with respect to M and E_u.*

Proof. By the assumptions, $KE_u \cap L \subseteq K$ and $LE_u \cap M \subseteq L$. To show that $KE_u \cap M \subseteq K$, assume that $s \in K$, $a \in E_u$, and $sa \in M$. Then, $K \subseteq L$ implies that $s \in L$. As $sa \in M$, it follows from the controllability of L with respect to M that $sa \in L$. However, $sa \in L$ implies that $sa \in K$, by controllability of K with respect to L. □

The following properties (Definitions 8.6 and 8.7), adopted from hierarchical supervisory control and introduced by K. C. Wong and W. M. Wonham [25], play a significant role in the rest of this chapter.

Definition 8.6 (Observer Property). *Let $E_k \subseteq E$ be event sets. The projection $P_k : E^* \to E_k^*$ is an* L-observer *for a language $L \subseteq E^*$ if the following holds: for all strings $t \in P(L)$ and $s \in \text{prefix}(L)$, if $P(s)$ is a prefix of t, then there exists $u \in E^*$ such that $su \in L$ and $P(su) = t$, see Fig. 8.7.*

If G is a generator with n states, then the time and space complexity of the verification whether P is an $L(G)$-observer are both $O(n^2)$, see [18]. Moreover, there is an

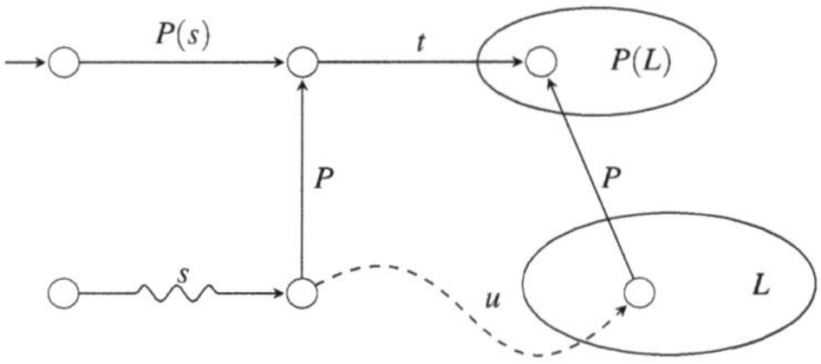

Fig. 8.7 Diagram of the concept of an L-observer (cf. [4])

algorithm with the time and space complexities $O(n^3)$ and $O(n)$, respectively, that enlarges the event set so that the projection becomes an $L(G)$-observer, see [4].

The most significant consequence of this definition is the following theorem.

Theorem 8.2. [18, 26] *If the projection P is an $L(G)$-observer, for a generator G, then the minimal generator for the language $P(L(G))$ has no more states than G.*

The other property is the output control consistency condition.

Definition 8.7 (Output Control Consistency). *The projection $P : E^* \to E_k^*$, for $E_k \subseteq E$, is* output control consistent *(OCC) for $L \subseteq E^*$ if for every $s \in \text{prefix}(L)$ of the form $s = \sigma_1\sigma_2 \dots \sigma_\ell$ or of the form $s = s'\sigma_0\sigma_1 \dots \sigma_\ell$, $\ell \geq 1$, where $\sigma_0, \sigma_\ell \in E_k$ and $\sigma_i \in E \setminus E_k$, for $i = 1, 2, \dots, \ell - 1$, if $\sigma_\ell \in E_u$, then $\sigma_i \in E_u$, for all $i = 1, 2, \dots, \ell - 1$.*

The time and space complexities of the verification whether P is OCC for L are $O(n^2)$ and $O(n)$, respectively, see [4].

Existence of Supremal Supervisors

Theorem 8.3. *The supremal conditionally-controllable sublanguage of a language K exists and is equal to the union of all conditionally-controllable sublanguages of K. The notation* $\sup\text{cC}(K, L, E_u)$ *denotes the supremal conditionally-controllable sublanguage of K with respect to $L = L(G_1 \| G_2 \| G_k)$ and uncontrollable event sets $E_{1+k,u}$, $E_{2+k,u}$, $E_{k,u}$.*

Proof. Let I be an index set, and let K_i, $i \in I$, be conditionally-controllable sublanguages of $K \subseteq L(G_1 \| G_2 \| G_k)$ with respect to generators G_1, G_2, G_k and uncontrollable event sets $E_{1+k,u}$, $E_{2+k,u}$, $E_{k,u}$. We prove that $\cup_{i \in I} K_i$ is conditionally controllable by showing that the items of Definition 8.5 hold. (1) Language $P_k(\cup_{i \in I} K_i)$ is controllable with respect to $L(G_k)$ and $E_{k,u}$ because $P_k(\cup_{i \in I} K_i) E_{k,u} \cap L(G_k) = \cup_{i \in I} (P_k(K_i) E_{k,u} \cap L(G_k)) \subseteq \cup_{i \in I} P_k(K_i) = P_k(\cup_{i \in I} K_i)$ where the inclusion is by controllability of $P_k(K_i)$ with respect to $L(G_k)$ and $E_{k,u}$, for $i \in I$. (2) Note that, $L(G_1) \| P_k(\cup_{i \in I} K_i) \| P_k^{2+k}[L(G_2) \| P_k(\cup_{i \in I} K_i)] = L(G_1) \| P_k(\cup_{i \in I} K_i) \| P_k^{2+k}(L(G_2))$ because, by Lemma 8.2, $P_k^{2+k}[L(G_2) \| P_k(\cup_{i \in I} K_i)] = P_k^{2+k}(L(G_2)) \| P_k(\cup_{i \in I} K_i)$, and the second element is already included in the equation. Thus, we need to show that $P_{1+k}(\cup_{i \in I} K_i) E_{1+k,u} \cap L(G_1) \| P_k(\cup_{i \in I} K_i) \| P_k^{2+k}(L(G_2)) \subseteq P_{1+k}(\cup_{i \in I} K_i)$. However,

$$
\begin{aligned}
&P_{1+k}(\cup_{i\in I}K_i)E_{1+k,u}\cap L(G_1)\|P_k(\cup_{i\in I}K_i)\|P_k^{2+k}(L(G_2))\\
&\quad=\cup_{i\in I}\,(P_{1+k}(K_i)E_{1+k,u})\cap\cup_{i\in I}[L(G_1)\|P_k(K_i)\|P_k^{2+k}(L(G_2))]\\
&\quad=\cup_{i\in I}\cup_{j\in I}(P_{1+k}(K_i)E_{1+k,u}\cap L(G_1)\|P_k(K_j)\|P_k^{2+k}(L(G_2)))\,.
\end{aligned}
$$

For the sake of contradiction, assume that there are two different indexes $i,j\in I$ such that $P_{1+k}(K_i)E_{1+k,u}\cap L(G_1)\|P_k(K_j)\|P_k^{2+k}(L(G_2))\not\subseteq P_{1+k}(\cup_{i\in I}K_i)$. Then, there exist $x\in P_{1+k}(K_i)$ and $u\in E_{1+k,u}$ such that $xu\in L(G_1)\|P_k(K_j)\|P_k^{2+k}(L(G_2))$, and

$$xu\notin P_{1+k}(\cup_{\ell\in I}K_\ell)\,. \tag{8.2}$$

It follows that (i) $P_k(x)\in P_kP_{1+k}(K_i)=P_k(K_i)$, (ii) $P_k(xu)\in P_k(K_j)$, and (iii) $P_k(xu)\notin P_k(K_i)$. Namely, (i) and (ii) are clear and (iii) can be shown as follows. If $P_k(xu)\in P_k(K_i)$, then $xu\in L(G_1)\|P_k(K_i)\|P_k(L(G_2))$, and by controllability of $P_{1+k}(K_i)$ with respect to $L(G_1)\|P_k(K_i)\|P_k(L(G_2))$ we get $xu\in P_{1+k}(K_i)\subseteq P_{1+k}(\cup_{i\in I}K_i)$, which does not hold by (8.2). Now, assume that $u\notin E_{k,u}$. Then, $P_k(xu)=P_k(x)\in P_k(K_i)$, which does not hold. Thus, $u\in E_{k,u}$. As $P_k(K_i)\cup P_k(K_j)\subseteq L(G_k)$, we get that $P_k(xu)=P_k(x)u\in L(G_k)$. However, controllability of $P_k(K_i)$ with respect to $L(G_k)$ and $E_{k,u}$ implies that $P_k(x)u=P_k(xu)$ is in $P_k(K_i)$. This is a contradiction. (3) The last item of the definition is proven in the same way. □

Computation of Supremal Supervisors

This subsection presents a procedure for the computation of supremal conditionally-controllable sublanguages.

Theorem 8.4. [10] *Let $K\subseteq L=L_1\|L_2\|L_k$ be two prefix-closed languages over an event set $E=E_1\cup E_2\cup E_k$, where $L_i\subseteq E_i^*$, $i=1,2,k$. Assume that K is conditionally decomposable, and define the languages*

$$
\begin{aligned}
\sup \mathrm{C}_k &= \sup \mathrm{C}(P_k(K),L_k,E_{k,u})\,,\\
\sup \mathrm{C}_{1+k} &= \sup \mathrm{C}(P_{1+k}(K),L_1\|\sup \mathrm{C}_k,E_{1+k,u})\,,\\
\sup \mathrm{C}_{2+k} &= \sup \mathrm{C}(P_{2+k}(K),L_2\|\sup \mathrm{C}_k,E_{2+k,u})\,.
\end{aligned}
$$

Let the projection P_k^{i+k} be an $(P_i^{i+k})^{-1}(L_i)$-observer and OCC for $(P_i^{i+k})^{-1}(L_i)$, for $i=1,2$. Then, $\sup \mathrm{C}_k\|\sup \mathrm{C}_{1+k}\|\sup \mathrm{C}_{2+k}=\sup \mathrm{cC}(K,L,E_u)$.

It follows from Theorem 8.4 and Lemma 8.1 that the supremal conditionally-controllable sublanguage is conditionally decomposable. The consequence of this is stated in the next result.

Theorem 8.5. *In the setting of Theorem 8.4, the supremal conditionally-controllable sublanguage* $\sup \mathrm{cC}(K,L,E_u)$ *of K is controllable with respect to L and E_u, hence* $\sup \mathrm{cC}(K,L,E_u)\subseteq \sup \mathrm{C}(K,L,E_u)$.

Proof. It is sufficient to show that $\sup \mathrm{cC}=\sup \mathrm{cC}(K,L,E_u)$ is controllable with respect to $L=L_1\|L_2\|L_k$ and E_u. There exist $\sup \mathrm{C}_k\subseteq E_k$, $\sup \mathrm{C}_{1+k}\subseteq E_{1+k}$, and

$\sup C_{2+k} \subseteq E_{2+k}$ of Theorem 8.4 so that $\sup cC = \sup C_k \| \sup C_{1+k} \| \sup C_{2+k}$. In addition, $\sup C_k$ is controllable with respect to L_k and $E_{k,u}$, $\sup C_{1+k}$ is controllable with respect to $L_1 \| \sup C_k$ and $E_{1+k,u}$, and $\sup C_{2+k}$ is controllable with respect to $L_2 \| \sup C_k$ and $E_{2+k,u}$. By Lemma 8.4, $\sup cC$ is controllable with respect to $L_k \| (L_1 \| \sup C_k) \| (L_2 \| \sup C_k) = L \| \sup C_k$ and E_u. Analogously, we can obtain that $L \| \sup C_k$ is controllable with respect to $L \| L_k = L$ and E_u. Finally, by the transitivity of controllability, Lemma 8.5, we obtain that $\sup cC$ is controllable with respect to L and E_u, which was to be shown. □

Note that if the observer and OCC assumptions of Theorem 8.4 are not satisfied, it still holds that the computed language is controllable.

Corollary 8.1. *The sublanguage* $\sup C_k \| \sup C_{1+k} \| \sup C_{2+k}$ *of* K *is controllable with respect to* L *and* E_u.

If additional conditions are satisfied, the supremal conditionally-controllable sublanguage is also optimal, i.e., it coincides with the supremal controllable sublanguage of K with respect to L and E_u.

Lemma 8.6. [4] *Let* $L_i \subseteq E_i^*$, $i = 1,2$, *be two (prefix-closed) languages, and let* $P_i : (E_1 \cup E_2)^* \to E_i^*$, *where* $i = 1,2,k$ *and* $E_k \subseteq E_1 \cup E_2$, *be projections. If* $E_1 \cap E_2 \subseteq E_k$ *and* $P^i_{k \cap i}$ *is an* L_i*-observer, for* $i = 1,2$, *then the projection* P_k *is an* $L_1 \| L_2$*-observer.*

In the following lemma, we prove that conditions of Theorem 8.4 imply that the projection P_k is OCC for L.

Lemma 8.7. *Let* $L_i \subseteq E_i^*$, $i = 1,2$, *be two (prefix-closed) languages, and let* $P_i : (E_1 \cup E_2)^* \to E_i^*$, *where* $i = 1,2,k$ *and* $E_k \subseteq E_1 \cup E_2$, *be projections. Denote by* $E_u \subseteq E_1 \cup E_2$ *the set of uncontrollable events. If* $E_1 \cap E_2 \subseteq E_k$ *and* P_k^{i+k} *is OCC for* $(P_i^{i+k})^{-1}(L_i)$, *for* $i = 1,2$, *then the projection* P_k *is OCC for* $L = L_1 \| L_2 \| L_k$.

Proof. Let $s \in L$ be of the form $s = s'\sigma_0\sigma_1 \dots \sigma_{k-1}\sigma_k$, for some $k \geq 1$, and assume that $\sigma_0, \sigma_k \in E_k$, $\sigma_i \in E \setminus E_k$, for $i = 1,2,\dots,k-1$, and $\sigma_k \in E_u$. We need to show that $\sigma_i \in E_u$, for all $i = 1,2,\dots,k-1$. However, $P_{i+k}(s) = P_{i+k}(s')\sigma_0 P_{i+k}(\sigma_1 \dots \sigma_{k-1})\sigma_k \in (P_i^{i+k})^{-1}(L_i)$ and the OCC property implies that $P_{i+k}(\sigma_1 \dots \sigma_{k-1}) \in E_u^*$, for $i = 1,2$. Consider $\sigma \in \{\sigma_1, \sigma_2, \dots, \sigma_{k-1}\}$. Then, $\sigma \in (E_1 \cup E_2) \setminus E_k$. Without loss of generality, assume that $\sigma \in E_1$. Then, $P_{1+k}(\sigma) = \sigma \in E_u$ and $P_{2+k}(\sigma) = \varepsilon \in E_u^*$. Thus, $\{\sigma_1, \sigma_2, \dots, \sigma_{k-1}\} \subseteq E_u$, which was to be shown. □

Theorem 8.6. *Consider the setting of Theorem 8.4. If, in addition,* $L_k \subseteq P_k(L)$ *and* P_{i+k} *is OCC for* $P_{i+k}^{-1}(L_i \| L_k)$, *for* $i = 1,2$, *then* $\sup cC(K, L, E_u) = \sup C(K, L, E_u)$.

Proof. The inclusion $\subseteq$ is proven in Theorem 8.5. Thus, we prove the other inclusion. From the assumptions, P_k^{i+k} is the $(P_i^{i+k})^{-1}(L_i)$-observer, for $i = 1,2$, and P_k^k is an L_k-observer because the observer property always holds for the identity projection. Now, Lemma 8.6 applied to projections P_k^{1+k} and P_k^{2+k} implies that P_k is an $(P_1^{1+k})^{-1}(L_1) \| (P_2^{2+k})^{-1}(L_2) = L_1 \| L_2$-observer. Another application of this lemma to projections P_k and P_k^k implies that P_k is an $(L_1 \| L_2) \| L_k = L$-observer. In addition, by Lemma 8.7, the projection P_k is also OCC for L. For short, denote $\sup C = \sup C(K, L, E_u)$. We now prove that $P_k(\sup C)$ is controllable with respect to L_k and $E_{k,u}$. To do this, assume that $t \in P_k(\sup C)$, $a \in E_{k,u}$, and $ta \in L_k \subseteq P_k(L)$. Then, there exists $s \in \sup C$ such that $P_k(s) = t$. As P_k is the L-observer, there exists $v \in E^*$ such that $sv \in L$ and $P_k(sv) = P_k(s)P_k(v) = ta$, i.e., $v = ua$, for some $u \in (E \setminus E_k)^*$. Furthermore, from the OCC property of P_k, $u \in E_u^*$. From controllability of $\sup C$ with respect to L and E_u, this implies that $sua \in \sup C$, which means that $P_k(sua) = ta \in P_k(\sup C)$. Hence, (1) of Definition 8.5 is satisfied.

Next, we have that P_{i+k}^{i+k} (the identity) is the $(P_i^{i+k})^{-1}(L_i)$-observers, for $i = 1, 2$, and that $P_{j+k}^{i+k} = P_k^{i+k}$ is the $(P_i^{i+k})^{-1}(L_i)$-observer, for $\{i, j\} = \{1, 2\}$, and $P_k^k = P_{i+k}^k$ is the L_k-observer, for $i = 1, 2$. Then, similarly as above, Lemma 8.6 applied to projections P_{i+k}^{i+k}, P_{i+k}^{j+k}, $j \neq i$, and P_{i+k}^k implies that the projection P_{i+k} is an L-observer, for $i = 1, 2$. Thus, to prove (2) and (3) of Definition 8.5, assume that, for some $1 \leq i \leq 2$, $t \in P_{i+k}(\sup C)$, $a \in E_{i+k,u}$, and $ta \in L_i \| P_k(\sup C) \| P_k^{j+k}(L_j \| P_k(\sup C))$, for $j \neq i$. Then, there exists $s \in \sup C$ such that $P_{i+k}(s) = t$. As P_{i+k} is the L-observer, and $L_i \| P_k(\sup C) \| P_k^{j+k}(L_j \| P_k(\sup C)) \subseteq P_{i+k}(L) = L_i \| L_k \| P_k^{j+k}(L_j \| L_k)$, $j \neq i$, because $P_k(\sup C) \subseteq P_k(K) \subseteq P_k(L) \subseteq L_k$, there exists $v \in E^*$ such that $sv \in L$ and $P_{i+k}(sv) = P_{i+k}(s)P_{i+k}(v) = ta$, i.e., $v = ua$, for some $u \in (E \setminus E_{i+k})^*$. Since P_{i+k} is OCC for $P_{i+k}^{-1}(L_i \| L_k)$ and $sua \in L \subseteq P_{i+k}^{-1}(L_i \| L_k)$, we obtain that $u \in E_u^*$. Finally, from the controllability of $\sup C$ with respect to L and E_u, we obtain that $sua \in \sup C$. This means that $P_{i+k}(sua) = ta \in P_{i+k}(\sup C)$, which was to be shown. □

The complexity of the computation of the supremal controllable sublanguage of a specification language K with respect to the plant language L with n and m states in their minimal generator representations, respectively, is shown (for prefix-closed languages) to be $O(mn)$ [2, 15]. We denote the number of states of the minimal generators for $L(G_1)$, $L(G_2)$, and $L(G_k)$ by m_1, m_2, and m_k, respectively. As the specification K is conditionally decomposable, $K = P_{1+k}(K) \| P_{2+k}(K) \| P_k(K)$, we denote the number of states of the minimal generators for $P_{1+k}(K)$, $P_{2+k}(K)$, and $P_k(K)$ by n_1, n_2, and n_k, respectively. Then, in the worst case, $m = O(m_1 m_2 m_k)$ and $n = O(n_1 n_2 n_k)$. The computational complexity of $\sup C_k$, $\sup C_{1+k}$, and $\sup C_{2+k}$ gives the formula $O(m_k n_k + m_1 n_1 m_k n_k + m_2 n_2 m_k n_k)$, which is better than $O(mn) = O(m_1 m_2 m_k n_1 n_2 n_k)$ of the monolithic case.

Example 8.8. Continue Example 8.4. The assumptions of Theorem 8.4 are satisfied for $E_k = \{a_1, a_2\}$ and for the coordinator $G_k = P_k(G_1) \| P_k(G_2)$. Thus, we can compute $\sup C_k$, $\sup C_{1+k}$, $\sup C_{2+k}$, see Fig. 8.8. The solution is optimal: the supremal conditionally-controllable sublanguage of K coincides with the supremal controllable sublanguage of K.

An extension to more subsystems means to add new events a_i and w_i into E_c, a_i into E_k, for $i \geq 3$, and to modify the specification in a natural way. The required space results in a square root of the number of states needed by the global plant. ■

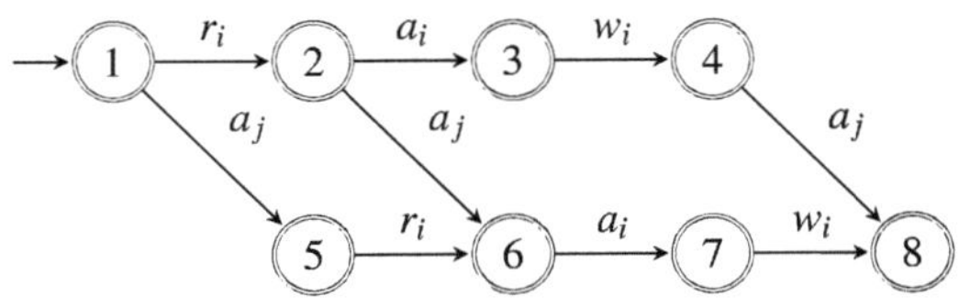

Fig. 8.8 Supervisors $supC_{i+k}$, $i = 1,2$, where $j = 1,2$, $j \neq i$; $supC_k = G_k$

8.6 Coordination Control with Partial Observations: Existence

In supervisory control with partial observations, the specification language must be observable and controllable for a supervisor to exist so that the closed-loop system meets the specification. See Chapter 4 for more details. An extension for distributed discrete-event systems with partial observations is described in this section. The results are based on the concepts of conditional controllability and conditional normality.

Let K and M be prefix-closed languages over an event set E. Let $E_c \subseteq E$ be the set of controllable events and $E_o \subseteq E$ be the set of *observable* events with P as the corresponding projection from E^* to E_o^*. The specification language K is *observable* with respect to M, E_c, and P if for all $s \in K$ and $\sigma \in E_c$,

$$(s\sigma \notin K) \text{ and } (s\sigma \in M) \Rightarrow P^{-1}[P(s)]\sigma \cap K = \emptyset .$$

The language K is *normal* with respect to M and P if $K = P^{-1}[P(K)] \cap M$. Note that, normality implies observability [3].

A *controlled generator (with partial observations)* is a structure (G, E_c, P, Γ), where G is a generator, $E_c \subseteq E$ is the set of *controllable events*, $E_u = E \setminus E_c$ is the set of *uncontrollable events*, $P : E^* \to E_o^*$ is the corresponding projection (partial observation), and $\Gamma = \{\gamma \subseteq E \mid E_u \subseteq \gamma\}$ is the *set of control patterns*. A *supervisory control* for the controlled generator (G, E_c, P, Γ) is a map $v : P(L(G)) \to \Gamma$. A *closed-loop system* associated with the controlled generator (G, E_c, P, Γ) and the supervisory control v is defined as the smallest language $L(v/G) \subseteq E^*$ which satisfies (1) $\varepsilon \in L(v/G)$ and (2) if $s \in L(v/G)$, $sa \in L(G)$, and $a \in v(P(s))$, then $sa \in L(v/G)$. Again, a supervisor S is a generator representation of the supervisory control v such that $L(v/G) = L(S) \| L(G)$. We write $L(S/G)$ to denote $L(S) \| L(G)$.

Let $\sup \mathrm{CN}(K, L, E_u, P)$ denote the supremal sublanguage of K which is both controllable with respect to L and E_u and normal with respect to L, E_u, and P.

Problem Statement

The supervisory control problem with partial observations is formulated exactly in the same way as Problem 8.1, therefore we do not recall it here. The only difference is that the meaning of used symbols is now as defined in this section.

Existence of Supervisors

Conditional observability, along with conditional controllability, are necessary and sufficient conditions for a specification language to be exactly achieved according to Problem 8.1.

Definition 8.8 (Conditional Observability). *Call the specification language* $K \subseteq E^*$ conditionally observable *for generators* G_1, G_2, G_k*, controllable subsets* $E_{1+k,c}, E_{2+k,c}, E_{k,c}$*, and projections* Q_{1+k}, Q_{2+k}, Q_k*, where* $Q_i : E_i^* \to E_{i,o}^*$*, for* $i = 1+k, 2+k, k$*, if*

1. $P_k(K)$ *is observable with respect to* $L(G_k)$*,* $E_{k,c}$*, and* Q_k*; equivalently, for all* $s \in P_k(K)$ *and for all* $\sigma \in E_{k,c}$*,*

$$(s\sigma \notin P_k(K)) \text{ and } (s\sigma \in L(G_k)) \Rightarrow Q_k^{-1}[Q_k(s)]\sigma \cap P_k(K) = \emptyset.$$

2. $P_{1+k}(K)$ *is observable with respect to* $L(G_1)\|P_k(K)\|P_k^{2+k}[L(G_2)\|P_k(K)]$*,* $E_{1+k,c}$ *(*$E_{1+k,c} = E_c \cap (E_1 \cup E_k)$*), and* Q_{1+k}*; equivalently, for all* $s \in P_{1+k}(K)$ *and for all* $\sigma \in E_{1+k,c}$*,*

$$(s\sigma \notin P_{1+k}(K)) \text{ and } (s\sigma \in L(G_1)\|P_k(K)\|P_k^{2+k}[L(G_2)\|P_k(K)])$$
$$\Rightarrow Q_{1+k}^{-1}[Q_{1+k}(s)]\sigma \cap P_{1+k}(K) = \emptyset.$$

3. $P_{2+k}(K)$ *is observable with respect to* $L(G_2)\|P_k(K)\|P_k^{1+k}[L(G_1)\|P_k(K)]$*,* $E_{2+k,c}$*, and* Q_{2+k}*; equivalently, for all* $s \in P_{2+k}(K)$ *and for all* $\sigma \in E_{2+k,c}$*,*

$$(s\sigma \notin P_{2+k}(K)) \text{ and } (s\sigma \in L(G_2)\|P_k(K)\|P_k^{1+k}[L(G_1)\|P_k(K)])$$
$$\Rightarrow Q_{2+k}^{-1}[Q_{2+k}(s)]\sigma \cap P_{2+k}(K) = \emptyset.$$

Theorem 8.7. [11] *Consider Problem 8.1. There exist supervisors* S_1*,* S_2*, and* S_k *such that* $L(S_1/[G_1\|(S_k/G_k)])\|L(S_2/[G_2\|(S_k/G_k)])\| L(S_k/G_k) = K$ *if and only if the specification language K is both (1) conditionally controllable with respect to generators* G_1, G_2, G_k *and locally uncontrollable event sets* $E_{1+k,u}, E_{2+k,u}, E_{k,u}$*, and (2) conditionally observable with respect to generators* G_1, G_2, G_k*, locally controllable event sets* $E_{1+k,c}, E_{2+k,c}, E_{k,c}$*, and projections* Q_{1+k}, Q_{2+k}, Q_k *from* E_i^* *to* $E_{i,o}^*$*, for* $i = 1+k, 2+k, k$*.*

Supremal observable sublanguages do not exist in general and it is also the case of conditionally-observable sublanguages. Therefore, this section introduces an analogous notion to normality, so-called conditional normality, and shows that conditional normality along with conditional controllability are sufficient conditions for the specification language to solve Problem 8.1.

Definition 8.9 (Conditional Normality). *Call the specification language* $K \subseteq E^*$ conditionally normal *for generators* G_1*,* G_2*,* G_k *and projections* Q_{1+k}, Q_{2+k}, Q_k*, where* $Q_i : E_i^* \to E_{i,o}^*$*, for* $i = 1+k, 2+k, k$*, if*

1. *The language* $P_k(K) \subseteq E_k^*$ *is normal with respect to* $L(G_k)$ *and* Q_k*; equivalently,*

$$Q_k^{-1}Q_k(P_k(K)) \cap L(G_k) = P_k(K).$$

2. *The language* $P_{1+k}(K) \subseteq (E_1 \cup E_k)^*$ *is normal with respect to the language* $L(G_1)\|P_k(K)\|P_k^{2+k}[L(G_2)\|P_k(K)]$ *and the projection* Q_{1+k}*; equivalently,*

$$Q_{1+k}^{-1}Q_{1+k}(P_{1+k}(K)) \cap L(G_1)\|P_k(K)\|P_k^{2+k}[L(G_2)\|P_k(K)] = P_{1+k}(K).$$

3. *The language* $P_{2+k}(K) \subseteq (E_2 \cup E_k)^*$ *is normal with respect to the language* $L(G_2)\|P_k(K)\|P_k^{1+k}[L(G_1)\|P_k(K)]$ *and the projection* Q_{2+k}*; equivalently,*

$$Q_{2+k}^{-1}Q_{2+k}(P_{2+k}(K)) \cap L(G_2)\|P_k(K)\|P_k^{1+k}[L(G_1)\|P_k(K)] = P_{2+k}(K).$$

Theorem 8.8. *Consider Problem 8.1. If the specification language* K *is conditionally controllable with respect to* G_1, G_2, G_k *and* $E_{1+k,u}$, $E_{2+k,u}$, $E_{k,u}$ *of locally uncontrollable events, and conditionally normal with respect to* G_1, G_2, G_k *and* Q_{1+k}, Q_{2+k}, Q_k *of projections from* E_i^* *to* $E_{i,o}^*$*, for* $i = 1+k, 2+k, k$*, then there exist supervisors* S_1, S_2, S_k *such that* $L(S_1/[G_1\|(S_k/G_k)])\|L(S_2/[G_2\|(S_k/G_k)])\|L(S_k/G_k) = K$.

Proof. As normality implies observability, the proof of this theorem follows immediately from Theorem 8.7. □

Example 8.9. Controllability is discussed in the previous examples, so only conditional normality is considered here. Let $G = G_1\|G_2$ be a plant over $E = E_1 \cup E_2 = \{a_1,c,t,t_1\} \cup \{a_2,c,t,t_2\} = \{a_1,a_2,c,t,t_1,t_2\}$, where G_1 and G_2 are given in Fig. 8.9, and the set of unobservable events is $E_{uo} = \{t,t_1,t_2\}$. The specification $K = \text{prefix}(\{t_2t_1, t_2a_1, a_1t_2, a_1a_2t, t_1t_2, t_1a_2, a_2a_1t, a_2t_1\})$. Let the coordinator G_k over $E_k = \{c,t,t_1\}$ be as in Fig. 8.9. Projections of K are $P_k(K) = \text{prefix}(\{t,t_1\})$, $P_{1+k}(K) = \text{prefix}(\{a_1t, t_1\})$, $P_{2+k}(K) = \text{prefix}(\{t_2t_1, a_2t, a_2t_1, t_1a_2, t_1t_2\})$. It can be verified that K is conditionally controllable and conditionally normal as required in Theorem 8.8. The supervisors S_1, S_2, S_k then correspond to generators for $P_{1+k}(K)$, $P_{2+k}(K)$, $P_k(K)$. ■

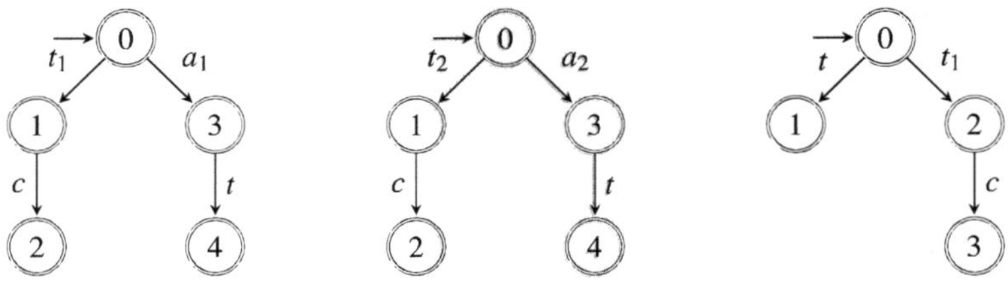

Fig. 8.9 Generators G_1, G_2, and the coordinator G_k

8.7 Coordination Control with Partial Observations: Supremal Supervision

If the specification language is not conditionally controllable or not conditionally normal, the supremal sublanguage is to be considered. We present a procedure for the computation of the supremal conditionally-controllable and conditionally-normal sublanguage for a prefix-closed specification. Because of space restrictions, the proofs are omitted and can be found in [11].

Theorem 8.9. [11] *The supremal conditionally-normal sublanguage of a given language K exists and is equal to the union of all conditionally-normal sublanguages of K.*

Consider generators G_1, G_2, G_k. Let $\sup\mathrm{cCN}(K,L,E_u,Q)$ denote the supremal conditionally-controllable and conditionally-normal sublanguage of K with respect to $L = L(G_1 \| G_2 \| G_k)$, the sets of uncontrollable events $E_{1+k,u}$, $E_{2+k,u}$, $E_{k,u}$, and the projections Q_{1+k}, Q_{2+k}, Q_k, where $Q_i : E_i^* \to E_{i,o}^*$, for $i = 1+k, 2+k, k$.

Theorem 8.10. [11] *Consider Problem 8.1. Define the local languages*

$$\begin{aligned}
\sup\mathrm{CN}_k &= \sup\mathrm{CN}(P_k(K), L(G_k), E_{k,u}, Q_k)\,,\\
\sup\mathrm{CN}_{1+k} &= \sup\mathrm{CN}(P_{1+k}(K), L(G_1) \| \sup\mathrm{CN}_k, E_{1+k,u}, Q_{1+k})\,,\\
\sup\mathrm{CN}_{2+k} &= \sup\mathrm{CN}(P_{2+k}(K), L(G_2) \| \sup\mathrm{CN}_k, E_{2+k,u}, Q_{2+k})\,.
\end{aligned}$$

Let P_k^{i+k} be a $(P_i^{i+k})^{-1}L(G_i)$-observer and OCC for $(P_i^{i+k})^{-1}L(G_i)$, $i = 1,2$, and the language $P_k^{1+k}(\sup\mathrm{CN}_{1+k}) \cap P_k^{2+k}(\sup\mathrm{CN}_{2+k})$ be normal with respect to $L(G_k)$ and Q_k. Then, $\sup\mathrm{CN}_k \| \sup\mathrm{CN}_{1+k} \| \sup\mathrm{CN}_{2+k} = \sup\mathrm{cCN}(K,L,E_u,Q)$.

The assumptions that P_k^{i+k} is a $(P_i^{i+k})^{-1}L(G_i)$-observer and OCC are needed only for controllability. Let $\sup\mathrm{N}(K,L,Q)$ denote the supremal normal sublanguage of K with respect to L and Q, and let $\sup\mathrm{cN}(K,L,Q)$ denote the supremal conditionally-normal sublanguage of K with respect to L and Q. Then we have:

Corollary 8.2. [11] *Consider Problem 8.1. Define the local languages*

$$\begin{aligned}
\sup\mathrm{N}_k &= \sup\mathrm{N}(P_k(K), L(G_k), Q_k)\,,\\
\sup\mathrm{N}_{1+k} &= \sup\mathrm{N}(P_{1+k}(K), L(G_1) \| \sup\mathrm{N}_k, Q_{1+k})\,,\\
\sup\mathrm{N}_{2+k} &= \sup\mathrm{N}(P_{2+k}(K), L(G_2) \| \sup\mathrm{N}_k, Q_{2+k})\,.
\end{aligned}$$

Assume that the language $P_k^{1+k}(\sup\mathrm{N}_{1+k}) \cap P_k^{2+k}(\sup\mathrm{N}_{2+k})$ is normal with respect to $L(G_k)$ and Q_k. Then, $\sup\mathrm{N}_k \| \sup\mathrm{N}_{1+k} \| \sup\mathrm{N}_{2+k} = \sup\mathrm{cN}(K,L,Q)$.

The computational complexity of the supremal controllable and normal sublanguage is $O(2^{mn})$ [2]. Denote the number of states of the minimal generators for $L(G_1)$, $L(G_2)$, and $L(G_k)$ by m_1, m_2, and m_k, respectively. As the specification language K is conditionally decomposable, denote the number of states of the minimal generators for $P_{1+k}(K)$, $P_{2+k}(K)$, and $P_k(K)$ by n_1, n_2, and n_k, respectively. Then, in the worst case, $m = O(m_1 m_2 m_k)$ and $n = O(n_1 n_2 n_k)$. The computational complexity of $\sup \mathrm{CN}_k$, $\sup \mathrm{CN}_{1+k}$, and $\sup \mathrm{CN}_{2+k}$ gives the formula $O(2^{m_k n_k} + 2^{m_1 n_1 2^{m_k n_k}} + 2^{m_2 n_2 2^{m_k n_k}})$, which is better than $O(2^{m_1 m_2 m_k n_1 n_2 n_k})$ of the monolithic case if $m_i n_i > \frac{2^{m_k n_k}}{m_k n_k}$, for $i = 1, 2$, i.e., if the coordinator is significantly smaller than the subsystems. As the coordinator (and its event set) can be chosen to be minimal, there is a possibility to choose the coordinator so that it, in addition, satisfies the condition that the number of states of the minimal generator of $\sup \mathrm{CN}_k$ is in $O(m_k n_k)$ or even in $O(\min\{m_k, n_k\})$. This requires further investigation.

Theorem 8.11. [11] *The language* $\sup \mathrm{CN}_k \| \sup \mathrm{CN}_{1+k} \| \sup \mathrm{CN}_{2+k}$ *is controllable with respect to* $L = L(G_1 \| G_2 \| G_k)$ *and* E_u, *and normal with respect to* L *and* $Q : (E_{1+k} \cup E_{2+k})^* \to E_o^*$.

Theorem 8.11 says that in the setting of Theorem 8.10, $\sup \mathrm{cCN}(K, L, E_u, Q)$ is controllable and normal with respect to L, E_u, and Q. If additional conditions are satisfied, the constructed supremal conditionally-controllable and conditionally-normal sublanguage is optimal.

Theorem 8.12. [11] *Consider the setting of Theorem 8.10. If, in addition, it holds that* $L_k \subseteq P_k(L)$ *and* P_{i+k} *is OCC for the language* $P_{i+k}^{-1}(L_i \| L_k)$, *for* $i = 1, 2$, *then* $\sup \mathrm{CN}(K, L, E_u, Q) = \sup \mathrm{cCN}(K, L, E_u, Q)$ *if and only if*

$$P_k[Q^{-1}Q(\sup \mathrm{CN}) \cap L] = P_k[Q^{-1}Q(\sup \mathrm{CN})] \cap L_k \tag{8.3}$$

and

$$\begin{aligned} &P_{i+k}[Q^{-1}Q(\sup \mathrm{CN}) \cap L_1 \| L_2 \| P_k(\sup \mathrm{CN})] \\ &= P_{i+k}[Q^{-1}Q(\sup \mathrm{CN})] \cap P_{i+k}(L_1 \| L_2 \| P_k(\sup \mathrm{CN})), \end{aligned} \tag{8.4}$$

for $i = 1, 2$, *where* $\sup \mathrm{CN} = \sup \mathrm{CN}(K, L, E_u, Q)$.

To verify this condition, we need to compute the plant language L. However, we do not want to compute this language because of complexity reasons. It is an open problem how to verify the conditions of Theorem 8.12 based only on the local languages L_1, L_2, and L_k.

Example 8.10. Consider Example 8.9 with a different specification language K defined by the generator shown in Fig. 8.10. We construct the coordinator G_k as described in Algorithm 8.3. To do this, E_k has to contain both shared events c and t. To ensure that K is conditionally decomposable, at least one of t_1 and t_2 has to be added to E_k. Assume t_1 is added; $E_k = \{c, t, t_1\}$. Set $G_k = P_k(G_1) \| P_k(G_2)$, see Fig. 8.9. The projections of K are $P_k(K) = \mathrm{prefix}(\{t, t_1 c\})$, $P_{1+k}(K) = \mathrm{prefix}(\{a_1 t, t_1 c\})$, $P_{2+k}(K) = \mathrm{prefix}(\{t_2 t_1, a_2 t, a_2 t_1, t_1 a_2, t_1 t_2 c\})$. We compute $\sup \mathrm{N}_k = \mathrm{prefix}(\{t, t_1 c\})$, $\sup \mathrm{N}_{1+k} = \mathrm{prefix}(\{t_1 c, a_1 t\})$, $\sup \mathrm{N}_{2+k} = \mathrm{prefix}(\{t_2 t_1, t_1 t_2, t_1 a_2, a_2 t_1, a_2 t\})$. Then, the supremal conditionally-normal sublanguage $\sup \mathrm{N}_k \| \sup \mathrm{N}_{1+k} \| \sup \mathrm{N}_{2+k}$ of K results in $\mathrm{prefix}(\{t_2 t_1, t_2 a_1, a_1 t_2, a_1 a_2 t, t_1 t_2, t_1 a_2, a_2 a_1 t, a_2 t_1\})$, which is also normal by Theorem 8.11. It can be verified that the resulting language coincides with the supremal normal sublanguage of K with respect to $L(G)$ and Q. ■

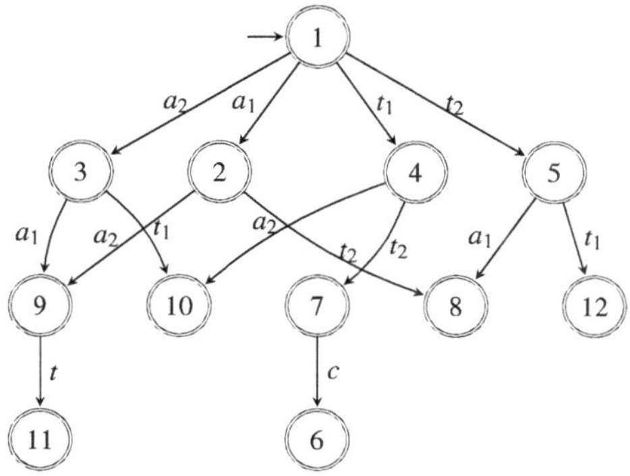

Fig. 8.10 Generator for the specification language K

8.8 Further Reading

The theory presented here is based on papers [11, 10]. For information about coordination control of distributed discrete-event systems with non-prefix-closed specifications, the reader is referred to [8]. This topic is, however, still under development. The procedures will be implemented in libFAUDES [17]. For other structural conditions on local plants under which it is possible to synthesize the supervisors locally, but which are quite restrictive, see [6, 14]. Among the most successful approaches to supervisory control of distributed discrete-event systems are those that combine distributed and hierarchical control [23, 24], or the approach based on interfaces [16]. Distributed computations of supremal normal sublanguages were further studied in [13] for local specification languages and in [12] for global specification languages. For further information on observers, the reader is referred to [5, 26]. For coordination control of linear and of stochastic systems, the reader is referred to [7, 22].

References

1. Boutin, O., van Schuppen, J.H.: On the control of the paint factory scale model. In: CWI MAC-1103, Amsterdam, The Netherlands (2011), `http://oai.cwi.nl/oai/asset/18598/18598D.pdf`
2. Brandt, R.D., Garg, V., Kumar, R., Lin, F., Marcus, S.I., Wonham, W.M.: Formulas for calculating supremal controllable and normal sublanguages. Systems Control Letters 15(2), 111–117 (1990)
3. Cassandras, C.G., Lafortune, S.: Introduction to Discrete Event Systems, 2nd edn. Springer (2008)

4. Feng, L.: Computationally Efficient Supervisor Design for Discrete-Event Systems. PhD Thesis. University of Toronto, Canada (2007), http://www.kth.se/polopoly_fs/1.24026thesis.zip
5. Feng, L., Wonham, W.M.: On the computation of natural observers in discrete-event systems. Discrete Event Dynamic Systems 20(1), 63–102 (2010)
6. Gaudin, B., Marchand, H.: Supervisory control of product and hierarchical discrete event systems. European Journal of Control 10(2), 131–145 (2004)
7. Kempker, P.L., Ran, A.C.M., van Schuppen, J.H.: Construction of a coordinator for coordinated linear systems. In: Proc. of Euroepan Control Conference, Budapest, Hungary (2009)
8. Komenda, J., Masopust, T., van Schuppen, J.H.: Coordinated control of discrete event systems with nonprefix-closed languages. In: Proc. of IFAC World Congress 2011, Milano, Italy (2011)
9. Komenda, J., Masopust, T., van Schuppen, J.H.: On conditional decomposability. CoRR, abs/1201.1733 (2012), http://arxiv.org/abs/1201.1733
10. Komenda, J., Masopust, T., van Schuppen, J.H.: Supervisory control synthesis of discrete-event systems using a Coordination Scheme. Automatica 48(2), 247–254 (2011)
11. Komenda, J., Masopust, T., van Schuppen, J.H.: Synthesis of controllable and normal sublanguages for discrete-event systems using a coordinator. Systems and Control Letters 60(7), 492–502 (2011)
12. Komenda, J., van Schuppen, J.H.: Control of discrete-event systems with modular or distributed structure. Theoretical Computer Sciences 388(3), 199–226 (2007)
13. Komenda, J., van Schuppen, J.H.: Modular control of discrete-event systems with coalgebra. IEEE Transactions on Automatic Control 53(2), 447–460 (2008)
14. Komenda, J., van Schuppen, J.H., Gaudin, B., Marchand, H.: Supervisory control of modular systems with global specification languages. Automatica 44(4), 1127–1134 (2008)
15. Kumar, R., Garg, V., Marcus, S.I.: On controllability and normality of discrete event dynamical systems. Systems and Control Letters 17(3), 157–168 (1991)
16. Leduc, R.J., Pengcheng, D., Raoguang, S.: Synthesis method for hierarchical interface-based supervisory control. IEEE Transactions on Automatic Control 54(7), 1548–1560 (2009)
17. Moor, T., et al.: Libfaudes–Discrete Event Systems Library (2006), http://www.rt.eei.uni-erlangen.de/FGdes/faudes/index.html
18. Pena, P.N., Cury, J.E.R., Lafortune, S.: Polynomial-time verification of the observer property in abstractions. In: Proc. American Control Conference, Seattle, USA (2008)
19. Polycarpou, M.M., Panayiotou, C., Lambrou, T., van Schuppen, J.H.: CON4COORD WP3 Aerial Vehicles–Final Report (2011), http://www.c4c-project.eu
20. Ramadge, P.J., Wonham, W.M.: Supervisory control of a class of discrete event processes. SIAM Journal of Control Optimization 25(1), 206–230 (1987)
21. Ramadge, P.J., Wonham, W.M.: The control of discrete event systems. Proceedings of the IEEE 77(1), 81–98 (1989)
22. Ran, A.C.M., van Schuppen, J.H.: Control for coordination of linear systems. In: Proc. of Mathematical Theory of Networks and Systems, Blacksburg, USA (2008)
23. Schmidt, K., Breindl, C.: Maximally permissive hierarchical control of decentralized discrete event systems. IEEE Transactions on Automatic Control 56(4), 723–734 (2011)

24. Schmidt, K., Moor, T., Perk, S.: Nonblocking hierarchical control of decentralized discrete event systems. IEEE Transactions on Automatic Control 53(10), 2252–2265 (2008)
25. Wong, K.C., Wonham, W.M.: Hierarchical control of discrete-event systems. Discrete Event Dynamic Systems 6(3), 241–273 (1996)
26. Wong, K.C.: On the complexity of projections of discrete-event systems. In: Proc. 4th Workshop on Discrete Event Systems, Cagliari, Italy (1998)
27. Wonham, W.M.: Supervisory control of discrete-event systems. Lecture Notes. University of Toronto (2011), `http://www.control.utoronto.ca/DES`

Chapter 9
An Introduction to Timed Automata

Béatrice Bérard

9.1 Motivation and Example

The main verification problem is expressed by the question: does a system satisfy a specification ? If $\mathscr{P}$ and $\mathscr{S}$ are two models, describing respectively a specification and a system implementation, this question can be answered by checking the inclusion:

$$\mathscr{L}(\mathscr{S}) \subseteq \mathscr{L}(\mathscr{P}) \text{ which is equivalent to } \mathscr{L}(\mathscr{S}) \cap \overline{\mathscr{L}(\mathscr{P})} = \emptyset.$$

Therefore, intersection, complementation and emptiness check are interesting operations, which are easy to perform when $\mathscr{P}$ and $\mathscr{S}$ are finite automata. In this chapter, we investigate issues related to this problem for the model of timed automata.

Timed models are needed to represent and analyze systems where explicit time constraints must be integrated. This includes for instance time-out mechanisms for a system:

> After 3 time units without any command, the system returns in an idle state.

or response times as a specification:

> When a request has been issued, it is granted in less than 5 time units.

The model of timed automata has been designed by Alur and Dill in the early 90s [3, 5] to deal with such requirements. The principle consists in associating with a finite automaton a finite set of real valued variables called *clocks*. These clocks evolve synchronously with time and can be reset or compared with constant values when transitions are fired.

Béatrice Bérard
LIP6, Université P. & M. Curie and CNRS, Paris, France
e-mail: `Beatrice.Berard@lip6.fr`

C. Seatzu et al. (Eds.): Control of Discrete-Event Systems, LNCIS 433, pp. 169–187.
springerlink.com

A very simple example describes cooking some roast lamb, as depicted in the timed automaton of Fig. 9.1.

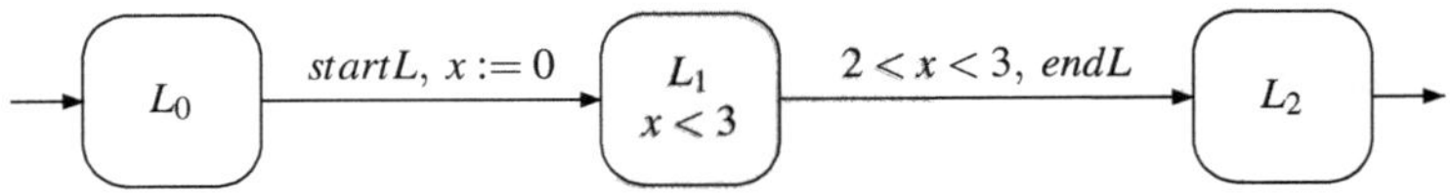

Fig. 9.1 A basic timed automaton for cooking a roast lamb

The start action consists in putting the dish in the oven and starting the clock x (by the reset $x := 0$). In state L_1, the clock value progresses and the roast is ready between 2 and 3 time units, hence action *endL* is guarded by the constraint $2 < x < 3$. State L_1 is equipped with the constraint $x < 3$ to express that the clock value must not reach 3 in this state. Of course, the time unit must be adequately defined. A configuration of such a model is a pair (q, v) where $q \in \{L_0, L_1, L_2\}$ and v is the value of clock x, with $(L_0, 0)$ the initial configuration. A run of this model could be:

$$(L_0,0) \xrightarrow{1.5} (L_0,1.5) \xrightarrow{startL} (L_1,0) \xrightarrow{2.7} (L_1,2.7) \xrightarrow{endL} (L_2,2.7)$$

Note that in this case, there are only finite runs ending in state L_2.

9.2 Definition and Timed Semantics

We now give formal definitions for the model and its semantics, described by timed transition systems.

We write respectively $\mathbb{N}$, $\mathbb{Z}$ and $\mathbb{R}_{\geq 0}$ for the sets of natural numbers, integers and non-negative real numbers. We denote by $[a, b[$ the interval of $\mathbb{R}$ which is open on the right and closed on the left, *i.e.* contains a but not b, and we use similar notations for the various types of intervals. Given an alphabet Σ, the set Σ^* (respectively Σ^ω) contains all finite (respectively infinite) words over Σ and a language is a subset of $\Sigma^* \cup \Sigma^\omega$.

9.2.1 Timed Transition Systems

We first define the notion of timed words, where a date is associated with each action.

Definition 9.1. *A* timed word *on alphabet* Σ *is a sequence* $w = (a_1, t_1)(a_2, t_2)\ldots$, *where* a_i *is in* Σ *for all* $i \geq 1$ *and* $(t_i)_{i \geq 1}$ *is a nondecreasing sequence of real numbers. A* timed language *is a set of timed words.*

Hence, a timed word is a word over the (infinite) alphabet $\Sigma \times \mathbb{R}_{>0}$. For a timed word w, the projection on Σ, removing the time component of actions, yields $Untime(w) = a_1 a_2 \ldots$, hence a (standard) word over Σ. Similarly, for a timed language L, we define $Untime(L) = \{Untime(w) \mid w \in L\}$.

Recall now that a transition system over a set Lab of labels is a triple $\mathcal{T} = (S, s_0, E)$, where:

- S is the set of configurations;
- s_0 is the initial configuration;
- E is the transition relation, given as a subset of $S \times Lab \times S$.

A transition (s, ℓ, s') in E is denoted by $s \xrightarrow{\ell} s'$.

Definition 9.2. *A* Timed Transition System *over an alphabet Σ is a transition system $\mathcal{T}$ over the set of labels $\Sigma \cup \{\varepsilon\} \cup \mathbb{R}_{>0}$, such that the transitions with label in $\mathbb{R}_{>0}$ have the following properties:*

- zero delay: $s \xrightarrow{0} s'$ *if and only if* $s' = s$;
- additivity: *if* $s \xrightarrow{d} s'$ *and* $s' \xrightarrow{d'} s''$, *then* $s \xrightarrow{d+d'} s''$.

Transitions $\xrightarrow{a}$ with a in $\Sigma \cup \{\varepsilon\}$ correspond to usual actions or events and are instantaneous. *Silent transitions*, labeled by the empty word $\varepsilon \in \Sigma^*$, are useful for modeling purposes: they represent internal actions of the system. A transition $\xrightarrow{d}$ with $d \in \mathbb{R}_{>0}$ represents a duration of d time units. The two additional conditions above, satisfied by delay transitions, express the consistency of the system evolution with respect to elapsing time.

Moreover, the system is sometimes required to be:

- *time deterministic:* if $s \xrightarrow{d} s_1$ and $s \xrightarrow{d} s_2$ then $s_1 = s_2$;
- *continuous:* if $s \xrightarrow{d} s'$ then for all d_1 and d_2 such that $d = d_1 + d_2$, there exists s_1 such that $s \xrightarrow{d_1} s_1$ and $s_1 \xrightarrow{d_2} s'$.

A run of $\mathcal{T}$ is a path $\rho = s_0 \xrightarrow{d_1} s'_0 \xrightarrow{a_1} s_1 \xrightarrow{d_2} s'_1 \xrightarrow{a_2} \cdots$ starting from the initial configuration and where durations and actions strictly alternate. This specific form can be ensured for systems which satisfy the zero delay and additivity properties. With a run ρ are associated:

- the sequence $(t_i)_{i \geq 0}$ of *absolute dates*, defined by $t_0 = 0$ and $t_i = \Sigma_{j=0}^{i} d_j$;
- the timed word $w = (a_{i_1}, t_{i_1})(a_{i_2}, t_{i_2}) \ldots$ corresponding to the actions labeling the runs with their dates, where all pairs such that $a_i = \varepsilon$ are removed.

For instance, the timed word associated with the execution of the introductory example is $w = (startL, 1.5)(endL, 4.2)$.

9.2.2 The Model of Timed Automata

For a set X of clocks, $\mathcal{C}(X)$ denotes the set of *clock constraints* which are conjunctions of atomic constraints of the form $x \bowtie c$ where x is a clock, c a constant (usually in $\mathbb{N}$) and $\bowtie$ an operator in $\{<, \leq, =, \geq, >\}$.

Definition 9.3. *A* timed automaton *over alphabet Σ (of* actions*) is a tuple $\mathcal{A} = (X, Q, q_0, \Delta, Inv)$, where:*

- *X is a finite set of clocks;*
- *Q is a finite set of states (or control nodes);*
- *$q_0 \in Q$ is the initial state;*
- *Δ is a subset of $Q \times \mathcal{C}(X) \times (\Sigma \cup \{\varepsilon\}) \times 2^X \times Q$;*
- *$Inv : Q \mapsto \mathcal{C}(X)$ associates with each state a constraint in $\mathcal{C}(X)$, called* invariant, *using only the operators $<$ and $\leq$.*

A transition in Δ, written as $q \xrightarrow{g,a,r} q'$, expresses a change from q to q' with action a, if *guard* g is satisfied. Clocks in $r \subseteq X$ are then reset, which is also written as $x := 0$ for all $x \in r$, like in Fig. 9.1 for the reset of clock x. In all figures depicting timed automata, *true* constraints are omitted, both for invariants and guards.

9.2.3 Semantics of Timed Automata

Given a set X of clocks, a valuation is a mapping $v : X \to \mathbb{R}_{>0}$, with $\mathbf{0}$ the null valuation assigning zero to all clocks in X. Note that, for a set X of cardinality n, a geometric view is obtained by considering a valuation v as the tuple $(v(x))_{x \in X}$, hence as a point in $\mathbb{R}^n_{>0}$. We define the following operations on valuations.

- For $d \in \mathbb{R}_{>0}$, time elapsing of d time units from valuation v results in valuation $v + d$ defined by: $(v + d)(x) = v(x) + d$ for each clock x: all clocks evolve at the rate of time. We write $v \leq v'$ if there exists d such that $v' = v + d$.
- For $r \subseteq X$, reset of the clocks in r from valuation v results in valuation $v[r \mapsto 0]$ defined by: $v[r \mapsto 0](x) = 0$ if $x \in r$ and $v(x)$ otherwise.

Clock constraints are interpreted on valuations: valuation v satisfies the atomic constraint $x \bowtie c$, denoted by $v \models x \bowtie c$, if $v(x) \bowtie c$. The notation is extended to general clock constraints by conjunction.

Definition 9.4. *The semantics of a timed automaton $\mathcal{A} = (\Sigma, X, Q, q_0, \Delta, Inv)$ is then given as the timed transition system $\mathcal{T}_{\mathcal{A}} = (S, s_0, E)$, over Σ with:*

- *$S = \{(q, v) \in Q \times \mathbb{R}^X_{>0} \mid v \models Inv(q)\}$. Hence, configurations are pairs (q, v) where $q \in Q$ and v is a clock valuation satisfying the state invariant;*
- *The initial configuration is $s_0 = (q_0, \mathbf{0})$;*
- *The transitions in E are:*

- *either* $(q,v) \xrightarrow{d} (q,v+d)$, *a delay of* $d \in \mathbb{R}_{>0}$, *possible if* $v+d \models Inv(q)$;
- *or* $(q,v) \xrightarrow{a} (q',v')$, *a discrete transition with label* $a \in \Sigma \cup \{\varepsilon\}$, *possible if there exists* $q \xrightarrow{g,a,r} q'$ *in* Δ *such that valuation* v *satisfies guard* g *and* $v' = v[r \mapsto 0]$.

A delay transition thus corresponds, as expected, to the time spent in a state of the automaton. For such a transition system $\mathscr{T}_{\mathscr{A}}$ with time domain $\mathbb{R}_{>0}$, the four properties mentioned above hold: zero delay, additivity, time determinism and continuous delays.

Consider for instance the timed automaton with two clocks on the left of Fig. 9.2. Starting from valuation $v_0 = \mathbf{0} = [0,0]$, the run

$$(q_0,[0,0]) \xrightarrow{1.2} (q_0,[1.2,1.2]) \xrightarrow{a} (q_0,[1.2,0]) \xrightarrow{1.8} (q_0,[3,1.8]) \xrightarrow{b} (q_0,[0,1.8])$$

can be described in $\mathbb{R}^2_{>0}$ by the trajectory on the right, where the dashed lines represent the reset operations that occurred when transitions a and b were fired.

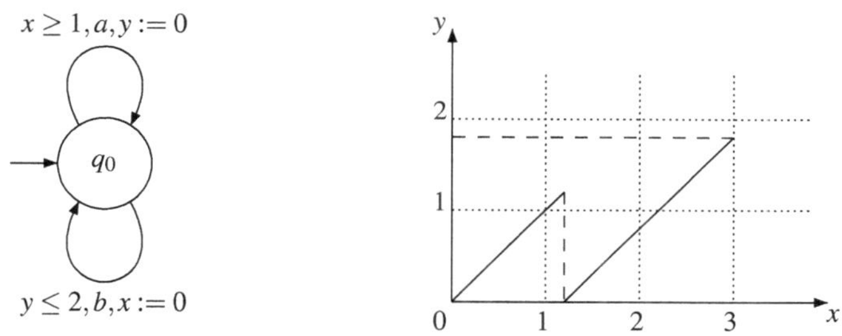

Fig. 9.2 A trajectory of a timed automaton

An important feature that should be noticed about the transition system of a timed automaton interpreted with $\mathbb{R}_{>0}$ as time domain is the infinite noncountable number of its configurations. This means that specific techniques must be developed for the analysis of such systems (see Sections 9.4 and 9.5).

9.2.4 Languages of Timed Automata

In order to associate a timed language with a timed automaton $\mathscr{A}$, several mechanisms are possible, based on the definition of acceptance conditions on runs. The timed word labeling an accepting run is then said to be accepted by $\mathscr{A}$.

For instance, if a subset $F \subseteq Q$ of *final* states is added to a timed automaton $\mathscr{A} = (\Sigma, X, Q, q_0, \Delta, Inv)$, an accepting run can be defined as a finite run

$$\rho = (q_0, v_0) \xrightarrow{d_1} (q_0, v_0 + d_1) \xrightarrow{a_1} (q_1, v_1) \cdots \xrightarrow{d_n} (q_{n-1},, v_{n-1} + d_n) \xrightarrow{a_n} (q_n, v_n)$$

ending in a state $q_n \in F$. The timed language $\mathscr{L}_{fin}(\mathscr{A})$ is then the corresponding set of accepted timed words. We write $\mathscr{A} = (X, Q, q_0, \Delta, Inv, F)$ in this case.

Definition 9.5. *A timed language L is* timed regular *if there is a timed automaton* $\mathscr{A} = (X, Q, q_0, \Delta, Inv, F)$ *such that* $L = \mathscr{L}_{fin}(\mathscr{A})$.

This is what was done in the example of Fig. 9.1 where L_2 was the final state. In this case, the timed language is:

$$\{(startL, t_1)(endL, t_2) \mid t_1 \in \mathbb{R}_{>0} \text{ and } t_1 + 2 < t_2 < t_1 + 3\}.$$

Regarding infinite runs, Büchi or Muller conditions can be defined in a similar way with conditions on the subset of Q containing the states appearing infinitely often in the runs, leading to *timed ω-regular languages.*

Next, we present syntactical extensions of timed automata, with a special focus on networks of timed automata.

9.3 Networks of Timed Automata

With the aim to make the modeling process easier without changing the expressive power of the model, several extensions have been proposed. Recovering the original basic model is usually done by unfolding the set of states, which can result in an exponential blow-up. We mention the extensions described below.

- Addition to $\mathscr{C}(X)$ of *diagonal* constraints of the form $x - y \bowtie c$ was proposed in early versions of the model. It is revisited in [14], where it is shown to lead to exponentially more concise models.
- In another direction, variables with finite range which do not evolve with time can also be added to the model. This extension is used in particular in the tool UPPAAL [10], where arrays are also permitted.
- As explained in more details below, a network of timed automata, along with a synchronization function, can be unfolded in a synchronized product which is still a timed automaton. This feature appears in all analysis tools for timed automata, making modular design possible.

To give an example of this product operation, we add a second timed automaton to the one of Fig. 9.1. Suppose now that Alice and Bob want to prepare two different dishes for a dinner, both being served at the same time. Alice will cook the roast lamb while Bob will prepare a vegetable curry. For this, he first fries the vegetables between 1 and 2 time units, then he adds some coconut milk several times between

0 and 1 time unit, and finally the preparation has to slowly boil more than 3 time units. This is modeled by the network of two automata depicted in Fig. 9.3.

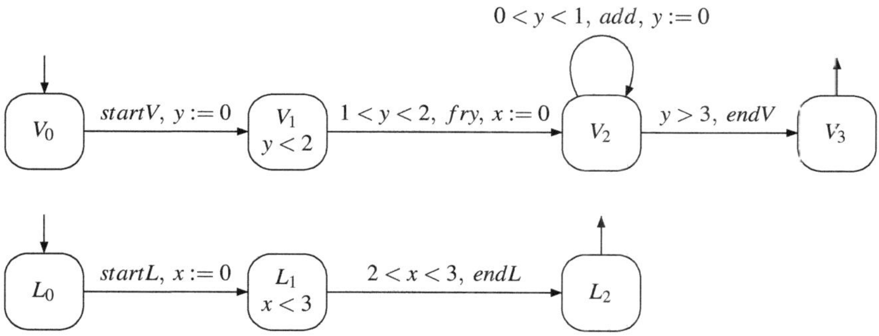

Fig. 9.3 A network of 2 timed automata for preparing dinner

In this simple case, the only synchronization concerns the two actions *endV* and *endL*, which occur as a single action *end* in the product, while all other actions are performed asynchronously. The resulting automaton is given in Fig. 9.4, where the following can be observed.

- In state (V_1, L_1) the invariant is the conjunction of those related respectively to V_1 and L_1.
- For the synchronized action *end*, the guard is the conjunction of those related respectively to actions *endV* and *endL*. No clock reset appears here, but in the general case, the union of clocks to be reset by both transitions would be associated with the synchronized transition.

We now give the formal definition for the product of two timed automata.

Definition 9.6. *Let* $\mathscr{A}_1 = (X_1, Q_1, i_1, \Delta_1, Inv_1)$ *and* $\mathscr{A}_2 = (X_2, Q_2, i_2, \Delta_2, Inv_2)$ *be two timed automata, on alphabets* Σ_1 *and* Σ_2 *respectively, such that* $X_1 \cap X_2 = \emptyset$. *Let* $-$ *be a new symbol and let* f *be a partial mapping from* $(\Sigma_1 \cup \{\varepsilon, -\}) \times (\Sigma_2 \cup \{\varepsilon, -\})$ *into* $\Sigma \cup \{\varepsilon\}$, *for some alphabet* Σ, *such that* $f(-,-)$ *is not defined. The synchronized product is defined by* $(\mathscr{A}_1 \otimes \mathscr{A}_2)_f = (X_1 \cup X_2, Q_1 \times Q_2, (i_1, i_2), \Delta, Inv)$, *with:*

- $Inv((q_1, q_2)) = Inv_1(q_1) \wedge Inv_2(q_2)$ *for all pairs* $(q_1, q_2) \in Q_1 \times Q_2$,
- *A transition* $(q_1, q_2) \xrightarrow{g,a,r} (q'_1, q'_2)$ *is in* Δ *if:*
  - *either there exist transitions* $q_1 \xrightarrow{g_1,a_1,r_1} q'_1$ *in* Δ_1 *and* $q_2 \xrightarrow{g_2,a_2,r_2} q'_2$ *in* Δ_2 *with* $g = g_1 \wedge g_2$, $a = f(a_1, a_2)$ *and* $r = r_1 \cup r_2$,
  - *or* $q_1 = q'_1$ *and* $q_2 \xrightarrow{g,a_2,r} q'_2$ *is in* Δ_2 *with* $f(-, a_2) = a$,
  - *or* $q_2 = q'_2$ *and* $q_1 \xrightarrow{g,a_1,r} q'_1$ *is in* Δ_1 *with* $f(a_1, -) = a$.

The operation can be extended to any finite number of timed automata. It should be used carefully, however, to avoid *timed deadlocks*. This is illustrated in the next

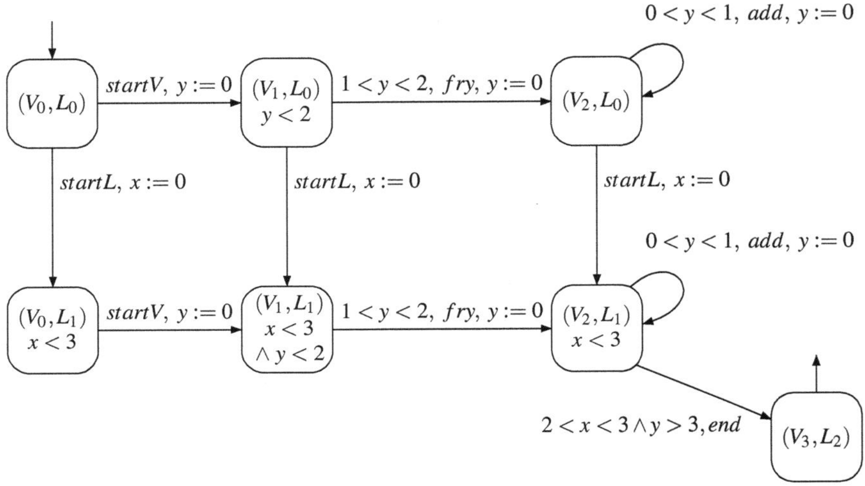

Fig. 9.4 The resulting product automaton

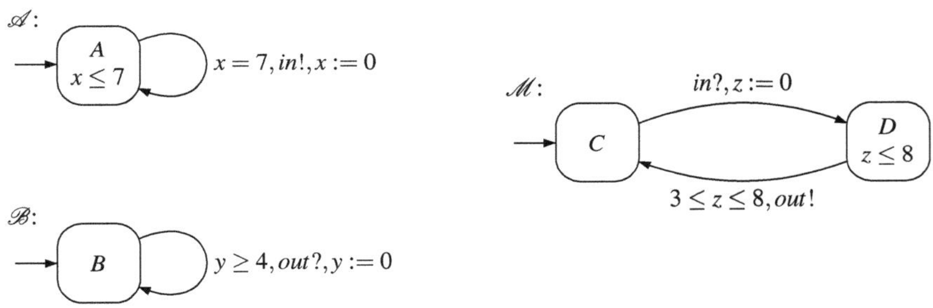

Fig. 9.5 Three communicating components

example from [21], where two components $\mathscr{A}$ and $\mathscr{B}$ communicate through a third one, the medium $\mathscr{M}$. Component $\mathscr{A}$ produces an *in*! message every 7 time units, $\mathscr{B}$ emits *out*? messages, with at least 4 time units between them, and $\mathscr{M}$ performs the communication by transmitting to $\mathscr{B}$ (*out*!) the messages received from $\mathscr{A}$ (*in*?), with a delay between 3 and 8 time units.

The synchronized product $(\mathscr{A} \otimes \mathscr{M} \otimes \mathscr{B})_f$ shown in Fig. 9.6 simulates the two channels between $\mathscr{A}$ and $\mathscr{M}$ and between $\mathscr{M}$ and $\mathscr{B}$ respectively, with function f defined by: $f(in!, in?, -) = in$ and $f(-, out!, out?) = out$.

Suppose now that the transmission interval $[3,8]$ is replaced by $]7,8]$. Then, a timed deadlock occurs in configuration (A,D,B) with clock values $x = 7$ and $z = 7$, since time progress is restricted by the invariant $x \leq 7$ so action *out* cannot be taken.

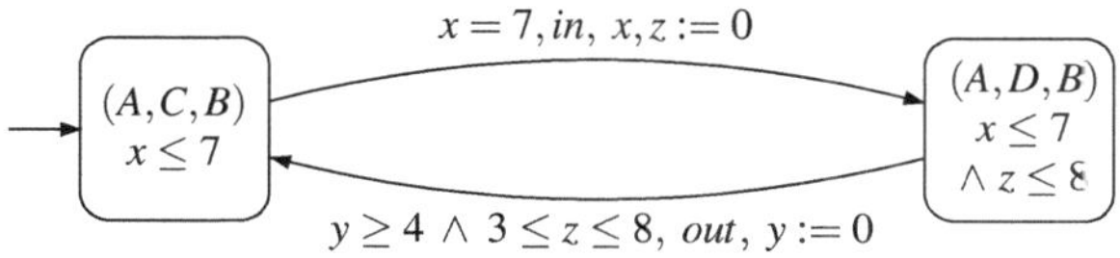

Fig. 9.6 Composition $(\mathscr{A} \otimes \mathscr{M} \otimes \mathscr{B})_f$

We now present analysis techniques for timed automata. The basic idea underlying these techniques is to *fold* the transition system by defining subsets of valuations, keeping together those for which the behavior of the automaton is the same (in a sense that must be precisely stated). The two main categories of such subsets are called respectively *zones* and *regions*.

9.4 Zone Graph of a Timed Automaton

Definition 9.7. *Let X be a finite set of clocks. A* zone *is a subset of $\mathbb{R}^X_{>0}$ defined by a conjunction of atomic clock constraints of the form $x \bowtie c$ or $x - y \bowtie c$ (diagonal constraints), with $x, y \in X$, $c \in \mathbb{Z}$ and $\bowtie \in \{\leq, <, =, >, \geq\}$. For such a constraint g, the corresponding zone is $[[g]] = \{v \in \mathbb{R}^X_{>0} \mid v \models g\}$. The set of all zones is denoted by $\mathscr{Z}(X)$.*

Operations on valuations are extended to zones as follows.

- The future of zone Z corresponds to the set of valuations reached from Z by letting time elapse. It is defined by: $\overrightarrow{Z} = \{v + d \mid v \in Z, d \in \mathbb{R}_{>0}\}$. Note that, $\overrightarrow{Z}$ contains Z itself since d can be equal to 0.
- The reset of zone Z with respect to a subset $r \subseteq X$ of clocks corresponds to a reset by r for all valuations in Z. It is defined by: $Z[r \mapsto 0] = \{v[r \mapsto 0] \mid v \in Z\}$.

Observe that, seen as a subset of $\mathbb{R}^n_{>0}$, a zone is a convex set. Therefore, the union of zones is not necessarily a zone. The following proposition describes the main properties of zones. The first two points are easy to obtain and the closure under future is essentially due to the addition of diagonal constraints.

Proposition 9.1. • *The future of a zone is a zone;*
- *The reset of a zone is a zone;*
- *The intersection of two (and a finite number of) zones is a zone.*

For a set $X = \{x, y\}$ with two clocks, a zone Z defined by:

$$2 < x < 4 \ \wedge\ 1 < y < 3 \ \wedge\ x - 2 < y$$

is depicted on the left of Fig. 9.7, with its future $\overrightarrow{Z}$ on the right, defined by the constraint: $2 < x \wedge 1 < y \wedge x - 2 < y < x + 1$. Zones obtained by reset are in darker

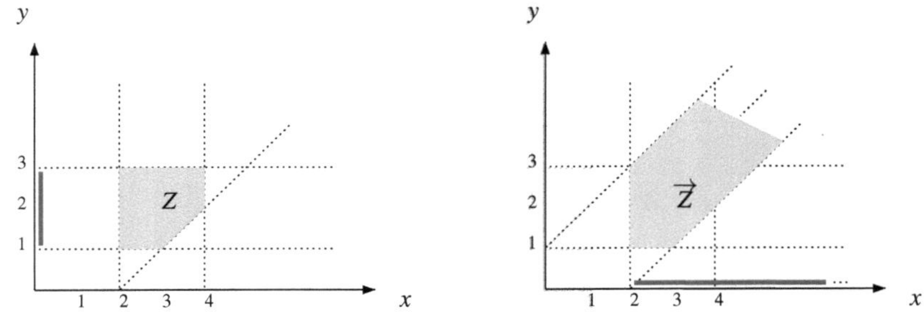

Fig. 9.7 A zone and its future, with some resets

gray: $Z[x \mapsto 0]$ is the open segment defined by $x = 0 \wedge 1 < y < 3$, while $\overrightarrow{Z}[y \mapsto 0]$ is the open half-line defined by: $y = 0 \wedge x > 2$.

The construction of the zone graph $Zone(\mathscr{A})$ associated with a timed automaton $\mathscr{A}$ is based on the following observation: starting from a zone Z in some state q, the set of valuations reached in state q' after transition $q \xrightarrow{g,a,r} q'$ is obtained by taking the future of Z, intersecting it with the invariant of state q and with the guard g, and applying on the resulting zone the reset r. The new set of valuations obtained this way:

$$Z' = (\overrightarrow{Z} \cap [[Inv(q)]] \cap [[g]])[r \leftarrow 0]$$

is also a zone from Proposition 9.1. Then, shifting the application of the future operator, we can define the zone graph as follows.

Definition 9.8. *Let $\mathscr{A} = (X, Q, q_0, \Delta, Inv)$ be a timed automaton. The zone graph $Zone(\mathscr{A})$ associated with $\mathscr{A}$ is the transition system whose configurations are pairs (q, Z) where $q \in Q$ and $Z \in \mathscr{Z}(X)$, with:*

- *the initial configuration (q_0, Z_0), where $Z_0 = \overrightarrow{\{\mathbf{0}\}} \cap Inv(q_0)$,*
- *transition $(q, Z) \xrightarrow{a} (q', Z')$ is in $Zone(\mathscr{A})$ if there is a transition $q \xrightarrow{g,a,r} q'$ in Δ with $Z' = \overrightarrow{(Z \cap [[g]])[r \leftarrow 0]} \cap [[Inv(q')]]$.*

To illustrate reachability analysis with this technique, consider again the timed automaton of Fig. 9.4 for the dinner preparation.

- We first want to know if the state (V_3, L_2) can be reached from (V_0, L_0) along the path corresponding to the sequence of actions *startL*, *startV*, *fry*. We start from the initial zone $Z_0 = \overrightarrow{\{\mathbf{0}\}} = [[x = y \geq 0]]$ in state (V_0, L_0), which has no invariant (hence a value true). Then, computing the successive zones, we obtain in state (V_2, L_1), just after transition *fry*, the zone $Z_1 = [[y = 0 \ \wedge \ 1 < x < 3]]$. Then, $Z_2 = \overrightarrow{Z_1} \cap [[x < 3]]$ is defined by $0 \leq y < x - 1 \ \wedge x < 3$, which has an empty intersection with the zone associated with the guard $g: \ 2 < x < 3 \ \wedge \ y > 3$ of transition *end*. Moreover, applying the transition *add* in state (V_2, L_1) will not change the emptiness result. As a consequence, this path must be eliminated to prepare the dinner, implying that Bob must start to cook the vegetable first.

- On the other hand, consistently with intuition, if we consider the path corresponding to the sequence of actions $startV, fry, add, startL$, we obtain in (V_2, L_1) the zone $Z_3 = [[y \geq x \geq 0 \ \wedge x < 3]]$. Intersecting Z_3 with $[[g]]$ yields exactly the non empty zone $[[g]] = [[2 < x < 3 \ \wedge \ y > 3]]$. Hence, this is a possible way of preparing the dinner.
- The third possible path to be investigated is the one corresponding to the sequence $startV, startL, fry$, possibly followed by applications of action add.

Finally note that, as shown in Fig. 9.8, the set of configurations of the zone graph can be infinite. In this example, all zones defined by: $0 \leq y \leq 1 \ \wedge \ y = x - k$, for $k \in \mathbb{N}$ can be reached in state q_0.

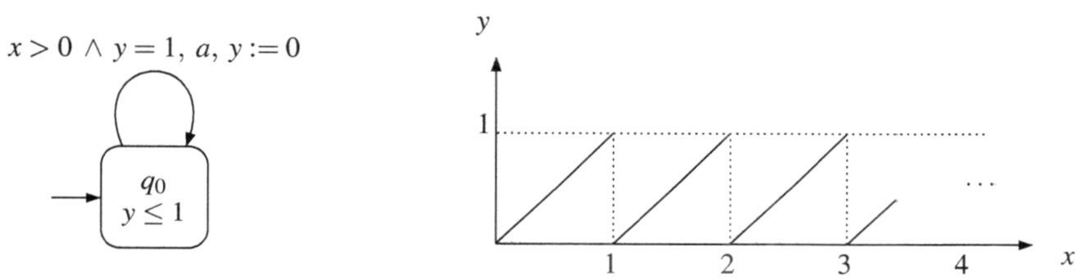

Fig. 9.8 A timed automaton generating an infinite number of reachable zones

Therefore, to perform forward analysis with zones, approximations must be computed to obtain a finite zone graph. Different approximations are studied in details in [12]. On the other hand, dual operations involving the past of a zone or the reverse of a reset can be defined, leading to backward computation of zones. This computation always terminate and can thus be used for reachability analysis.

9.5 Region Graph of a Timed Automaton

The region graph is obtained by applying the classical method of building a quotient graph from the transition system associated with a timed automaton. This quotient should be finite, while retaining enough properties of the original transition system. Since the set of control states is finite, it can be kept unchanged, but a partition of the set of valuations must be built, such that the corresponding equivalence $\sim$ is consistent with time elapsing, reset operations and the satisfaction of clock constraints.

More precisely, for a timed automaton $\mathscr{A} = (X, Q, q_0, \Delta, Inv)$ over aphabet Σ, the consistency properties are expressed by the following conditions:
(C) for two valuations $v, v' \in \mathbb{R}^X_{\geq 0}$ such that $v \sim v'$,

(i) for each clock constraint g of the form $x \bowtie c$, $v \models g$ if and only if $v' \models g$;
(ii) if $v \leq v_1$ for some valuation v_1, then there exists a valuation v'_1 such that $v' \leq v'_1$ and $v_1 \sim v'_1$;
(iii) for each subset r of clocks, $v[r \mapsto 0] \sim v'[r \mapsto 0]$.

Assuming such a quotient $\mathscr{R} = (\mathbb{R}^X_{>0})_{/\sim}$ can be built, we call *region* an element of $\mathscr{R}$. Hence, any region $R \in \mathscr{R}$ is the equivalence class of some valuation v, which is denoted by $R = [v]$. From conditions (C), operations $\leq$ and reset can be defined, as well as the satisfaction relation $R \models g$ for a region $R \in \mathscr{R}$ and a guard $g \in \mathscr{C}(X)$. A synchronized product of $\mathscr{A}$ and $\mathscr{R}$ again yields a transition system $Reg(\mathscr{A})$ which is time abstract bisimilar (in the sense of conditions (C)) to the transition system $\mathscr{T}_{\mathscr{A}}$ of $\mathscr{A}$. More precisely:

Definition 9.9. *Let* $\mathscr{A} = (X, Q, q_0, \Delta, Inv)$ *be a timed automaton and let* $\sim$ *be an equivalence relation satisfying (C). The region graph* $Reg(\mathscr{A})$ *is the transition system whose configurations are pairs* (q,R) *where* $q \in Q$ *and* $R \in \mathscr{R}$. *The initial configuration is* $(q_0, [\mathbf{0}])$ *and transitions are:*

- $(q,R) \xrightarrow{a} (q',R')$ *if there exists a transition* $q \xrightarrow{g,a,r} q' \in \Delta$ *with* $R \models g$ *and* $R' = R[r \mapsto 0]$;
- $(q,R) \xrightarrow{\varepsilon} (q,R')$ *if* $R \leq R'$.

In [3], Alur and Dill built such a relation $\sim$ for timed automata, with a finite number of equivalence classes. If m is the maximal constant appearing in the constraints of the automaton $\mathscr{A}$, the equivalence relation is defined by:
(D) $v \sim v'$ if

1. for each clock x, either the integral parts of $v(x)$ and $v'(x)$ are equal, or $v(x) > m$ and $v'(x) > m$;
2. for each clock x such that $v(x) \leq m$, $frac(v(x)) = 0$ if and only if $frac(v'(x)) = 0$, where $frac(t)$ is the fractional part of the real number t;
3. for each pair (x,y) of clocks such that $v(x) \leq m$ et $v'(x) \leq m$, $frac(v(x)) \leq frac(v(y))$ if and only if $frac(v'(x)) \leq frac(v'(y))$.

It results from the conditions above that regions are particular zones, sometimes called *elementary zones*. Moreover, any zone contained in the part bounded by m is a union of regions.

Consider the two partitions depicted on Fig. 9.9 for a two-clock automaton with $m = 3$. The one on the left satisfies only points 1. and 2. of definition (D) and does not respect point (ii) of condition (C): starting from valuation v and letting time elapse produces the region defined by $x = 2 \wedge 0 < y < 1$ while the same operation applied to v' yields the region defined by $y = 1 \wedge 1 < x < 2$. Hence, these valuations cannot be equivalent. On the other hand, the partition on the right is consistent with all the required conditions, thanks to the comparisons of the fractional parts of the clocks. For instance region R_1 in light gray is defined by

$$2 < x < 3 \wedge 1 < y < 2 \wedge y < x - 1,$$

while region R_2 in darker gray is defined by $x > 3 \wedge 1 < y < 2$. On this figure, regions are of several types:

- either points with integer coordinates between 0 and 3, for instance the three corners of R_1,

- or open segments, for instance the three borders of R_1 defined, respectively by $x=3 \wedge 1<y<2$, $y=1 \wedge 2<x<3$ and $y=x-1 \wedge 2<x<3$,
- or open triangles like R_1,
- or half lines like the low and upper borders of R_2,
- or unbounded rectangles like R_2.

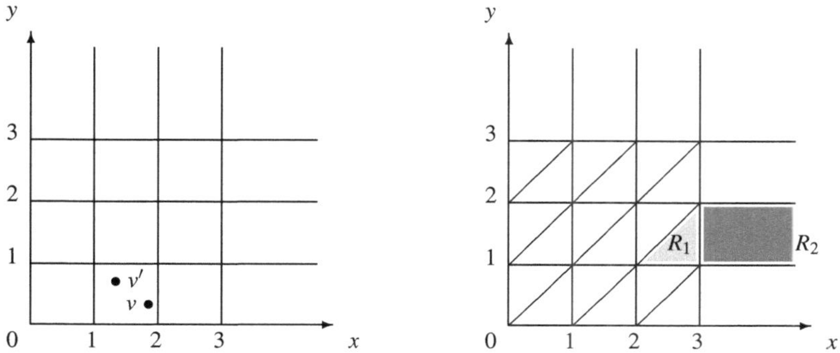

Fig. 9.9 Partitions for two clocks and $m=3$, satisfying only 1. and 2. on the left, satisfying 1., 2. and 3. on the right

In the general case of a set X of n clocks and maximal constant m, a region is defined by:

- for each clock $x \in X$, an interval among $[0,0]$, $]0,1[$, $[1,1]$, $\ldots [m,m]$, $]m,+\infty[$. When the interval is different from $]m,+\infty[$, its lower bound gives the integral part of the clock;
- for the clocks with value less than m, an ordering of the fractional parts of these clocks.

The representation above shows that the total number of regions is bounded by $n! \cdot 2^n \cdot (2m+2)^n$. Therefore, even if the construction is finite, the exponential blow up makes analysis rather inefficient with this method.

For the timed automaton of Fig. 9.8, the maximal constant is $m=1$ and the corresponding region automaton is depicted in Fig. 9.10, with the geometrical view of the regions on the right. Note that, the finiteness of the graph comes from including all zones except the first one in the three unbounded regions defined, respectively by $x>1 \wedge y=0, x>1 \wedge 0<y<1$ and $x>1 \wedge y=1$.

The main result proved by Alur and Dill states that $Untime(\mathscr{L}(\mathscr{A}))$ is a regular language, accepted by the finite automaton $Reg(\mathscr{A})$. More precisely:

Proposition 9.2. *Let $\mathscr{A}$ be a timed automaton and let (AC) be an accepting condition (either a final state condition for finite timed words, or a Muller or Büchi condition for infinite timed words).*

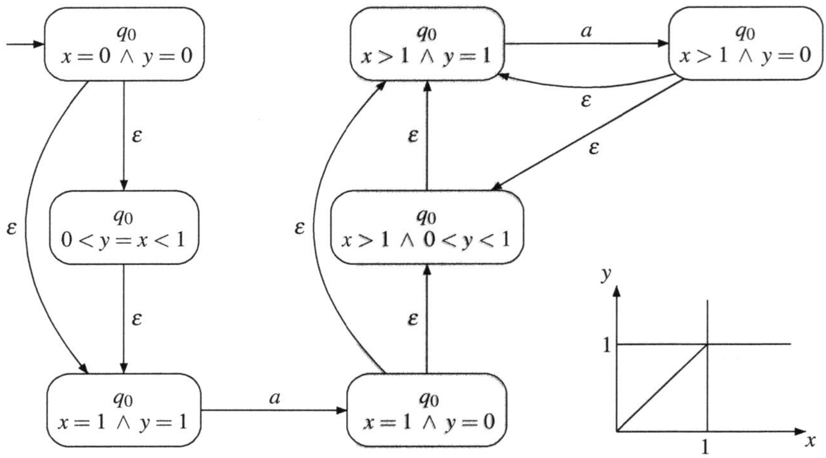

Fig. 9.10 Regions and region automaton for the timed automaton of Fig. 9.8

Then, if L is the timed language accepted by $\mathscr{A}$ with this condition, there exists a condition (AC') such that $Untime(L)$ is the language accepted by $Reg(\mathscr{A})$ for (AC').

Since a timed language L is empty if and only if $Untime(L)$ is empty, a consequence of the proposition above is the decidability of emptiness for a language accepted by a timed automaton $\mathscr{A}$. Hence, reachability of a state of $\mathscr{A}$ is also decidable by considering finite state acceptance. In fact, the following result is proved in [5]:

Theorem 9.1. *The problems of:*

- *emptiness of a timed regular language,*
- *reachability of a state in a timed automaton*

are PSPACE-complete. They are already PSPACE-hard for a fixed number n of clocks, with $n \geq 3$.

This very important result was a breakthrough for the analysis of timed models and initiated a large research field on real-time systems. Model checking timed extensions of CTL was proposed in [4, 17], as well as other analysis methods (for instance [1, 15]) and several tools were developed, based on specific algorithms using zones and zone approximations, like in UPPAAL [10] and KRONOS [24], or composition, like in CMC [18].

9.6 Language Properties

Returning to the initial problem of language inclusion, we observe that the presence of clocks clearly makes the model of timed automata strictly more powerful than the one of finite automata, in particular because the set of configurations is infinite and noncountable. This main difference has significant consequences regarding the properties of the corresponding languages.

In this section, we restrict the scope to families of timed regular languages (on finite timed words): TL_ε denotes the class of timed languages accepted by timed automata (which may contain silent transitions) and TL the subclass of TL_ε where no ε-transition is permitted. It is well known [11] that TL is a strict subclass of TL_ε. In this context, we present several additional results linked to the problem above.

9.6.1 Closure Properties

Like for finite automata, we have:

Proposition 9.3. *The family TL_ε is closed under (finite) union and intersection. It is also closed under projection and concatenation.*

Proof. Since silent transitions are permitted in the model, closure under union, projection and concatenation is trivial.

Concerning the closure under intersection, the proof (see [5]) imitates the one for finite automata. The underlying mechanism is exactly the one used to build a synchronized product: the intersection is obtained by considering a synchronization function $f : (\Sigma_1 \cup \{\varepsilon\}) \times (\Sigma_2 \cup \{\varepsilon\}) \to (\Sigma_1 \cap \Sigma_2) \cup \{\varepsilon\}$ such that $f(a,a) = a$ for $a \in (\Sigma_1 \cap \Sigma_2) \cup \{\varepsilon\}$ and f is undefined otherwise. □

Note that, the first three closure properties (union, intersection and projection) also hold for languages of infinite words accepted with Büchi or Muller conditions. The subclass TL is also closed under union, intersection and concatenation, but not under projection.

However, in contrast to the case of finite automata, the complement of a language in TL_ε is not necessarily in TL_ε. The timed language L accepted by the automaton on the left in Fig. 9.11 consists of all timed words on alphabet $\{a,b\}$ such that there is some a, occurring at time t, such that no event occurs at time $t+1$.

Proposition 9.4. [7] *The language $\overline{L}$ does not belong to TL_ε.*

Proof. Let M be the timed language accepted by the timed automaton on the right of Fig. 9.11. It consists of the words w such that $Untime(w) \in a^*b^*$, with all a events occurring strictly before time 1 and no two a events occur simultaneously. It is easy to see that $Untime(\overline{L} \cap M) = \{a^n b^p \mid p \geq n\}$, which is well known not to be a regular language. Hence, from Proposition 9.2, $\overline{L} \cap M$ itself is not timed regular. Since the class of timed languages is closed under intersection by Proposition 9.3, we can conclude that $\overline{L}$ is not timed regular. □

This nonclosure result also holds for a one-letter alphabet. The timed language N defined by:

$$N = \{(a,t_1)\dots(a,t_n) \mid t_j - t_i = 1 \text{ for some } i < j\}$$

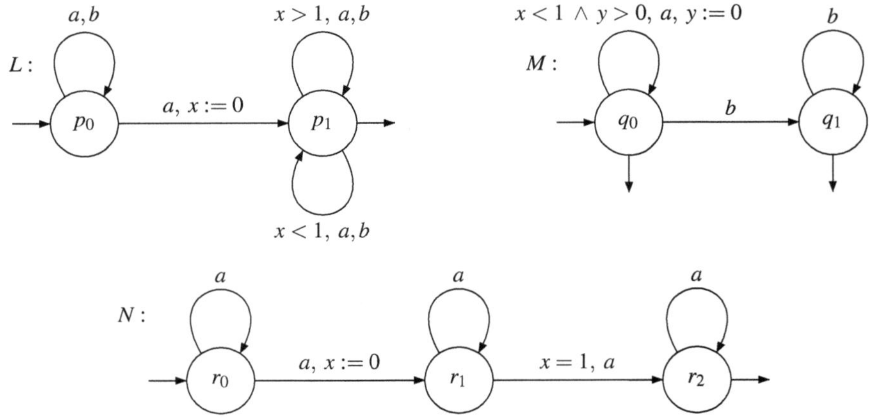

Fig. 9.11 Counter examples for the non closure under complement: languages L and N

is accepted by the automaton at the bottom of Fig. 9.11, hence it belongs to TL. But:

Proposition 9.5. [13] *The language* $\overline{N}$ *does not belong to* TL_ε.

Although the language was proposed in [5], with an intuitive argument, a complete proof only appears in [13].

A consequence of this result concerns the determinization procedure, which is always possible for finite automata on finite words, although with an exponential blow up of the set of states.

Definition 9.10. *A timed automaton* $\mathscr{A} = (X, Q, q_0, \Delta, Inv, F)$ *on alphabet* Σ *is* deterministic *if the two following conditions hold:*

- *no transition is labeled by* ε,
- *two transitions with same source and same label have disjoint guards.*

A key property of deterministic timed automata is that, given a timed word, there is at most one run accepting it. As in the untimed case, complementing such an automaton is easy: it suffices to complete the automaton with respect to Σ and take $Q \setminus F$ as subset of final states.

Let DTL be the subclass of TL containing timed languages accepted by deterministic timed automata. Proposition 9.4 implies that, given a timed automaton $\mathscr{A}$, a deterministic timed automaton $\mathscr{D}$ such that $\mathscr{L}_{fin}(\mathscr{A}) = \mathscr{L}_{fin}(\mathscr{D})$ does not always exists, hence the containment of DTL in TL is strict. Even finding an equivalent timed automaton without silent transitions, which can be done for finite automata, is not possible for timed automata, as soon as these transitions appear in loops. A detailed study on the expressive power of silent transitions can be found in [11].

The good properties of the class DTL makes deterministic timed automata well suited to specification purposes (recall that it is the part that must be complemented) and some work (see for instance [6, 9, 22]) has been devoted to deterministic or determinizable timed automata.

9.6.2 Undecidability Results

Decidability issues concerning the questions of the previous paragraph were investigated in details in [7, 8, 13, 16, 23] and the following problems (with variants) were proved undecidable:

- given a timed automaton, is it determinizable ?
- given a language in TL_ε, does it belong to TL ? (which amounts to asking if the silent transitions can be removed),
- given a language in TL_ε, is its complement in TL_ε ?

For the third question, the subcase of languages in TL was handled in [16] and the more involved general case with silent transitions was solved in [13]. Indeed, when silent transitions are permitted, an infinite number of configurations may be reached while reading a timed word. The case of infinite timed words was also studied in [13].

We finally return to the inclusion problem, which is closely related to the universality problem:

given an automaton $\mathscr{A}$, is $\mathscr{L}(\mathscr{A})$ equal to the set of all words ?

While both problems are decidable for finite automata, it is not the case for timed automata:

Theorem 9.2. [5] *The problems: given two timed automata $\mathscr{A}$ and $\mathscr{B}$,*

- *is $\mathscr{L}_{fin}(\mathscr{A})$ equal to the set of all timed words ?*
- *is $\mathscr{L}_{fin}(\mathscr{A})$ contained in $\mathscr{L}_{fin}(\mathscr{B})$?*

are undecidable.

The inclusion problem was proved to remain undecidable even for rather restricted classes [2]. However, some subclasses where the problems are decidable were identified [20, 22] and a generic construction was proposed in [9], which applies to several classes of timed automata.

9.7 Further Reading

Variants of the timed automata model are proposed in [21, 11, 14] and verification methods are investigated in [15, 4, 17, 1, 12]. Moreover, (un)decidability results about determinization and the inclusion problem can be found in [6, 23, 7, 8, 20, 2, 16, 22, 9, 13].

References

1. Aceto, L., Bouyer, P., Burgueño, A., Larsen, K.G.: The power of reachability testing for timed automata. Theoretical Computer Science 300(1-3), 411–475 (2003)
2. Adams, S., Ouaknine, J., Worrell, J.B.: Undecidability of Universality for Timed Automata with Minimal Resources. In: Raskin, J.-F., Thiagarajan, P.S. (eds.) FORMATS 2007. LNCS, vol. 4763, pp. 25–37. Springer, Heidelberg (2007)
3. Alur, R., Dill, D.L.: Automata for Modeling Real-Time Systems. In: Paterson, M. (ed.) ICALP 1990. LNCS, vol. 443, pp. 322–335. Springer, Heidelberg (1990)
4. Alur, R., Courcoubetis, C., Dill, D.L.: Model-checking in dense real-time. Information and Computation 104(1), 2–34 (1993)
5. Alur, R., Dill, D.L.: A theory of timed automata. Theoretical Computer Science 126, 183–235 (1994)
6. Alur, R., Fix, L., Henzinger, T.A.: A Determinizable Class of Timed Automata. In: Dill, D.L. (ed.) CAV 1994. LNCS, vol. 818, pp. 1–13. Springer, Heidelberg (1994)
7. Alur, R., Madhusudan, P.: Decision Problems for Timed Automata: A Survey. In: Bernardo, M., Corradini, F. (eds.) SFM-RT 2004. LNCS, vol. 3185, pp. 1–24. Springer, Heidelberg (2004)
8. Asarin, E.: Challenges in timed languages from applied theory to basic theory. Bulletin of the European Association for Theoretical Computer Science 83, 106–120 (2004)
9. Baier, C., Bertrand, N., Bouyer, P., Brihaye, T.: When Are Timed Automata Determinizable? In: Albers, S., Marchetti-Spaccamela, A., Matias, Y., Nikoletseas, S., Thomas, W. (eds.) ICALP 2009. Part II. LNCS, vol. 5556, pp. 43–54. Springer, Heidelberg (2009)
10. Behrmann, G., David, A., Larsen, K.G.: A Tutorial on UPPAAL. In: Bernardo, M., Corradini, F. (eds.) SFM-RT 2004. LNCS, vol. 3185, pp. 200–236. Springer, Heidelberg (2004)
11. Bérard, B., Diekert, V., Gastin, P., Petit, A.: Characterization of the expressive power of silent transitions in timed automata. Fundamenta Informaticae 36, 145–182 (1998)
12. Bouyer, P.: Forward analysis of updatable timed automata. Formal Methods in System Design 24(3), 281–320 (2004)
13. Bouyer, P., Haddad, S., Reynier, P.-A.: Undecidability results for timed automata with silent transitions. Fundamenta Informaticae 92(1-2), 1–25 (2009)
14. Bouyer, P., Chevalier, F.: On conciseness of extensions of timed automata. Journal of Automata, Languages and Combinatorics 10(4), 393–405 (2005)
15. Courcoubetis, C., Yannakakis, M.: Minimum and maximum delay problems in real-time systems. Formal Methods in System Design 1(4), 385–415 (1992)
16. Finkel, O.: Undecidable problems about timed automata. CoRR, abs/0712.1363 (2007)
17. Henzinger, T., Nicollin, X., Sifakis, J., Yovine, S.: Symbolic model checking for real-time systems. Information and Computation 111(2), 193–244 (1994)
18. Laroussinie, F., Larsen, K.G.: CMC: a tool for compositional model-checking of real-time systems. In: Proc. IFIP Joint Int. Conf. Formal Description Techniques and Protocol Specification, Testing and Verification, Paris, France (1998)
19. Lasota, S., Walukiewicz, I.: Alternating timed automata. ACM Transactions on Computational Logic 386(3), 169–187 (2007)
20. Ouaknine, J., Worrell, J.: On the language inclusion problem for timed automata: Closing a decidability gap. In: Proceedings of LICS, pp. 54–63. IEEE Computer Society Press (2004)

21. Sifakis, J., Yovine, S.: Compositional Specifications of Timed Systems. In: Puech, C., Reischuk, R. (eds.) STACS 1996. LNCS, vol. 1046, pp. 347–359. Springer, Heidelberg (1996)
22. Suman, P.V., Pandya, P.K., Krishna, S.N., Manasa, L.: Timed Automata with Integer Resets: Language Inclusion and Expressiveness. In: Cassez, F., Jard, C. (eds.) FORMATS 2008. LNCS, vol. 5215, pp. 78–92. Springer, Heidelberg (2008)
23. Tripakis, S.: Folk theorems on the determinization and minimization of timed automata. In: Proc. FORMATS 2003. LNCS, vol. 2791, pp. 182–188. Springer (2004)
24. Yovine, S.: Kronos: A verification tool for real-time systems. International Journal of Software Tools for Technology Transfer 1(1-2), 123–133 (1997)

Part II
Petri Nets

Chapter 10
Introduction to Petri Nets

Maria Paola Cabasino, Alessandro Giua, and Carla Seatzu

10.1 Introduction

Petri nets (PNs) are a discrete event system model first introduced in the early 1960s by Carl Adam Petri in his Ph.D dissertation [14]. In this chapter we focus on the most common class of PNs, called *place/transition* (or *P/T*) *net*. It is a purely *logic model* that does not aim to represent the occurrence time of events, but only the order in which events occur.

Petri nets have been specifically designed to model systems with interacting components and as such are able to capture many characteristics of an event driven system, namely concurrency, asynchronous operations, deadlocks, conflicts, etc. Furthermore, the PN formalism may be used to describe several classes of logical models (e.g., P/T nets, Colored PNs, nets with inhibitor arcs), performance models (e.g., Timed PNs, Time PNs, Stochastic PNs), continuous and hybrid models (continuous PNs, hybrid PNs). Some of these models are considered in this book: timed PNs are studied in Chapters 16 and 17 while continuous PNs are the object of Chapters 18, 19 and 20.

The main features of PNs can be summarized in the following items.

- PNs are both a *graphical* and *mathematical* formalism. Being a graphical formalism, they are easy to interpret and provide a useful visual tool both in the design and analysis phase.
- They provide a *compact representation* of systems with a very large state space. Indeed they do not require to explicitly represent all states of a dynamical system but only an initial one: the rest of the state space can be determined from the rules that govern the system evolution. Thus a finite structure may be used to describe systems with an infinite number of states.

Maria Paola Cabasino · Alessandro Giua · Carla Seatzu
Department of Electrical and Electronic Engineering, University of Cagliari, Italy
e-mail: {cabasino,giua,seatzu}@diee.unica.it

C. Seatzu et al. (Eds.): Control of Discrete-Event Systems, LNCIS 433, pp. 191–211.
springerlink.com

- They permit a *modular representation*, i.e., if a system is composed by several subsystems that interact among them, it is usually possible to represent each subsystem with a simple subnet and then, through appropriate net operators, combine the subnets to obtain a model of the whole system.

Several PN analysis techniques have been presented in the literature. In this chapter we focus on *analysis by enumeration* that requires the construction of the *reachability graph* of the net representing the set of reachable markings and transition firings. If this set is not finite, a finite *coverability graph* may be constructed. Techniques based on *structural analysis*, on the contrary, permit the analysis of several properties based on the net structure, e.g., focusing on the state equation of the net or on the net graph; they are described in the next chapter.

The chapter is structured as follows. P/T nets and the rules that govern their evolution are introduced in Section 10.2. In Section 10.3 elementary PN structures are described and a physical modeling example is presented. In Section 10.4 the reachability and coverability graphs are presented. Behavioral properties of interest are also defined and characterized. Finally, Section 10.5 points out some further interesting reading.

10.2 Petri Nets and Net Systems

We will first define the algebraic and graphical structure of P/T nets. Adding a *marking* to such a structure, a *marked net* (or *net system*), i.e., a discrete event system, is obtained. The laws that govern its dynamical evolution are also studied.

10.2.1 Place/Transition Net Structure

A P/T net is a bipartite weighted directed graph. The two types of vertices are called *places* (represented by circles) and *transitions* (represented by bars or rectangles).

Definition 10.1. *A* place/transition *(or* P/T*)* net *is a structure* $N = (P, T, \boldsymbol{Pre}, \boldsymbol{Post})$ *where:*

- $P = \{p_1, p_2, \cdots p_m\}$ *is the set of m* places.
- $T = \{t_1, t_2, \cdots t_n\}$ *is the set of n* transitions.
- $\boldsymbol{Pre} : P \times T \longrightarrow \mathbb{N}$ *is the* pre-incidence function *that specifies the number of arcs directed from places to transitions (called "pre" arcs) and is represented as* $m \times n$ *matrix.*
- $\boldsymbol{Post} : P \times T \longrightarrow \mathbb{N}$ *is the* post-incidence function *that specifies the number of arcs directed from transitions to places (called "post" arcs) and is represented as* $m \times n$ *matrix.*

Example 10.1. In Fig. 10.1 it is represented the net $N = (P,T,\boldsymbol{Pre},\boldsymbol{Post})$ with set of places $P = \{p_1,p_2,p_3,p_4\}$ and set of transitions $T = \{t_1,t_2,t_3,t_4,t_5\}$. Here:

$$\boldsymbol{Pre} = \begin{bmatrix} 1 & 1 & 0 & 0 & 0 \\ 0 & 0 & 1 & 1 & 0 \\ 0 & 0 & 0 & 0 & 1 \\ 0 & 0 & 0 & 0 & 1 \end{bmatrix} \begin{matrix} p_1 \\ p_2 \\ p_3 \\ p_4 \end{matrix} \qquad \boldsymbol{Post} = \begin{bmatrix} 1 & 0 & 0 & 0 & 1 \\ 0 & 2 & 0 & 0 & 0 \\ 0 & 0 & 1 & 0 & 0 \\ 0 & 0 & 0 & 1 & 0 \end{bmatrix} \begin{matrix} p_1 \\ p_2 \\ p_3 \\ p_4 \end{matrix}$$

$$\quad t_1\ t_2\ t_3\ t_4\ t_5 \qquad\qquad\qquad t_1\ t_2\ t_3\ t_4\ t_5$$

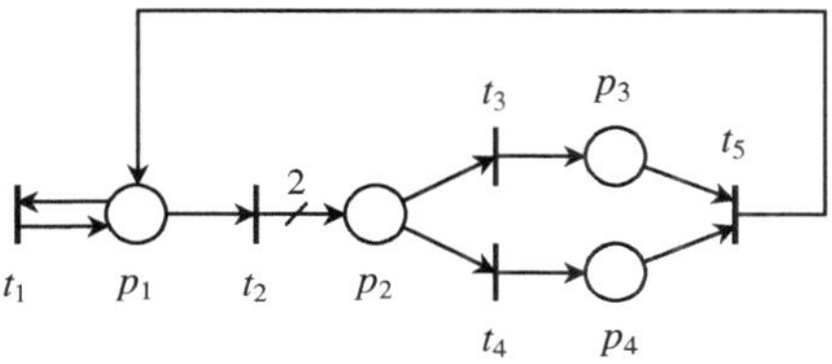

Fig. 10.1 A place/transition net

The element Post$[p_2,t_2] = 2$ denotes that there are two arcs from transition t_2 to place p_2. This is represented in the figure by means of a single barred arc with weight (or multiplicity) 2. ∎

We denote by $\boldsymbol{Pre}[\cdot,t]$ the column of $\boldsymbol{Pre}$ relative to t, and by $\boldsymbol{Pre}[p,\cdot]$ the row of $\boldsymbol{Pre}$ relative to p. The same notation is used for matrix $\boldsymbol{Post}$.

The *incidence matrix* of a net defined as

$$\boldsymbol{C} = \boldsymbol{Post} - \boldsymbol{Pre}, \tag{10.1}$$

is represented by an $m \times n$ matrix of integers where a negative element is associated with a "pre" arc (from place to transition), while a positive element is associated with a "post" arc (from transition to place).

Note that the incidence matrix does not contain, in general, sufficient information to reconstruct the net structure. As an example, in the net in Fig. 10.1 it holds:

$$\boldsymbol{C} = \begin{bmatrix} 0 & -1 & 0 & 0 & 1 \\ 0 & 2 & -1 & -1 & 0 \\ 0 & 0 & 1 & 0 & -1 \\ 0 & 0 & 0 & 1 & -1 \end{bmatrix}.$$

In this net there exist both a "pre" and a "post" arc between place p_1 and transition t_1; we say that p_1 and t_1 form a *self-loop*, i.e., a directed cycle in the graph of the net only involving one place and one transition. In such a case, the algebraic sum of $\boldsymbol{Pre}$ and $\boldsymbol{Post}$ determines an element $C[p_1,t_1] = 0$, hiding the existence of arcs between these two vertices. A net without self-loops is called *pure*.

Finally, given a transition $t \in T$ we denote its set of *input* and *output* places as:

$$^\bullet t = \{p \in P \mid \text{Pre}[p,t] > 0\} \qquad \text{and} \qquad t^\bullet = \{p \in P \mid \text{Post}[p,t] > 0\},$$

while given a place $p \in P$ we denote its set of *input* and *output* transitions as:

$$^\bullet p = \{t \in T \mid \text{Post}[p,t] > 0\} \qquad \text{and} \qquad p^\bullet = \{t \in T \mid \text{Pre}[p,t] > 0\}.$$

As an example, in the net in Fig. 10.1 it holds $^\bullet t_2 = \{p_1\}$, $t_2^\bullet = \{p_2\}$, $^\bullet p_2 = \{t_2\}$ and $p_2^\bullet = \{t_3, t_4\}$.

10.2.2 Marking and Net System

The *marking* of a P/T net defines its state.

Definition 10.2. *A* marking *is a function* $\boldsymbol{m} : P \to \mathbb{N}$ *that assigns to each place a nonnegative integer number of* tokens.

As an example, in the net in Fig. 10.1, a possible marking $\boldsymbol{m}$ is $m[p_1] = 1$, $m[p_2] = m[p_3] = m[p_4] = 0$. Other possible markings are: $\boldsymbol{m}'$ with $m'[p_2] = 2$, $m'[p_1] = m'[p_3] = m'[p_4] = 0$; $\boldsymbol{m}''$ with $m''[p_2] = m''[p_4] = 1$, $m''[p_1] = m''[p_3] = 0$; etc. A marking is usually denoted as a column vector with as many entries as the number m of places. Thus $\boldsymbol{m} = [1\ 0\ 0\ 0]^T$, $\boldsymbol{m}' = [0\ 2\ 0\ 0]^T$, $\boldsymbol{m}'' = [0\ 1\ 0\ 1]^T$.

Graphically, tokens are represented as black bullets inside places. See as an example Fig. 10.4.

Definition 10.3. *A net N with an initial marking $\boldsymbol{m}_0$ is called* marked net *or* net system, *and is denoted* $\langle N, \boldsymbol{m}_0 \rangle$.

A marked net is a discrete event system with a dynamical behavior as discussed in the following section.

10.2.3 Enabling and Firing

Definition 10.4. *A transition t is* enabled *at a marking $\boldsymbol{m}$ if*

$$\boldsymbol{m} \geq \boldsymbol{Pre}[\cdot, t] \tag{10.2}$$

i.e., if each place $p \in P$ contains a number of tokens greater than or equal to $Pre[p,t]$. To denote that t is enabled at $\boldsymbol{m}$ we write $\boldsymbol{m}[t\rangle$. To denote that t' is not enabled at $\boldsymbol{m}$ we write $\neg \boldsymbol{m}[t'\rangle$.

In the net in Fig. 10.1 the set of enabled transitions at $\mathbf{m} = [1\ 0\ 0\ 0]^T$ is $\{t_1, t_2\}$; the set of enabled transitions at $\mathbf{m}' = [0\ 2\ 0\ 0]^T$ is $\{t_3, t_4\}$; $\{t_3, t_4\}$ is also the set of transitions enabled at $\mathbf{m}'' = [0\ 1\ 0\ 1]^T$ since t_5 is not enabled, even if p_4 is marked, because p_3 is not marked.

A transition with no input arcs, such as t in Fig. 10.2, is called a *source transition*. A source transition t is always enabled, since, being in such a case $\mathbf{Pre}[\cdot, t] = \mathbf{0}$, the condition in equation (10.2) is satisfied for all markings $\mathbf{m}$.

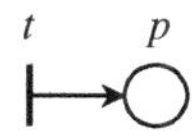

Fig. 10.2 A transition with no input arcs

Definition 10.5. *A transition t enabled at a marking $\mathbf{m}$ can* fire. *The* firing *of t removes $Pre[p,t]$ tokens from each place $p \in P$ and adds $Post[p,t]$ tokens in each place $p \in P$, yielding a new marking*

$$\mathbf{m}' = \mathbf{m} - \mathbf{Pre}[\cdot, t] + \mathbf{Post}[\cdot, t] = \mathbf{m} + \mathbf{C}[\cdot, t]. \tag{10.3}$$

To denote that the firing of t from $\mathbf{m}$ leads to $\mathbf{m}'$ we write $\mathbf{m}[t\rangle\mathbf{m}'$.

Note that the firing of a transition is an *atomic* operation since the removal of tokens from input places and their addition in output places occurs in an indivisible way. Consider the net in Fig. 10.1 at marking $\mathbf{m} = [1\ 0\ 0\ 0]^T$. If t_2 fires, $\mathbf{m}' = [0\ 2\ 0\ 0]^T$ is reached. Note that at marking $\mathbf{m} = [1\ 0\ 0\ 0]^T$, t_1 may also fire; the firing of such a transition does not modify the marking being $\mathbf{C}[\cdot, t_1] = \mathbf{0}$, thus it holds $\mathbf{m}[t_1\rangle\mathbf{m}$. If the marking of the net in Fig. 10.1 is equal to $\mathbf{m}' = [0\ 2\ 0\ 0]^T$, t_4 may fire leading to $\mathbf{m}'' = [0\ 1\ 0\ 1]^T$; note that t_3 is also enabled at $\mathbf{m}' = [0\ 2\ 0\ 0]^T$ and may fire instead of t_4.

Finally, in the marked net in Fig. 10.2 t is always enabled and can repeatedly fire, leading the initial marking $\mathbf{m}_0 = [0]$ to markings $[1]$, $[2]$ etc.

Definition 10.6. *A* firing sequence *at marking $\mathbf{m}_0$ is a string of transitions $\sigma = t_{j_1} t_{j_2} \cdots t_{j_r} \in T^*$, where T^* denotes the Kleene closure of T, such that*

$$\mathbf{m}_0 [t_{j_1}\rangle \mathbf{m}_1 [t_{j_2}\rangle \mathbf{m}_2 \cdots [t_{j_r}\rangle \mathbf{m}_r,$$

i.e., for all $k \in \{1, \ldots, r\}$ transition t_{j_k} is enabled at $\mathbf{m}_{k-1}$ and its firing leads to $\mathbf{m}_k = \mathbf{m}_{k-1} + \mathbf{C}[\cdot, t_{j_k}]$. To denote that the sequence σ is enabled at $\mathbf{m}$ we write $\mathbf{m}[\sigma\rangle$. To denote that the firing of σ at $\mathbf{m}$ leads to the marking $\mathbf{m}'$ we write $\mathbf{m}[\sigma\rangle\mathbf{m}'$.

The empty sequence *ε (i.e., the sequence of zero length) is enabled at all markings $\mathbf{m}$ and is such that $\mathbf{m}[\varepsilon\rangle\mathbf{m}$.*

In the net in Fig. 10.1 a possible sequence of transitions enabled at marking $\mathbf{m} = [1\ 0\ 0\ 0]^T$ is $\sigma = t_1 t_1 t_2 t_3$, whose firing leads to $\mathbf{m}''' = [0\ 1\ 1\ 0]^T$.

Let us now introduce the notion of conflict.

Definition 10.7. *Two transitions t and t' are in* structural conflict *if* $^\bullet t \cap {}^\bullet t' \neq \emptyset$, *i.e., if there exists a place p with a pre arc to both t and t'.*
Given a marking $\boldsymbol{m}$, we say that transitions t and t' are in behavioral conflict *(or in* conflict *for short) if $\boldsymbol{m} \geq \boldsymbol{Pre}[\cdot,t]$ and $\boldsymbol{m} \geq \boldsymbol{Pre}[\cdot,t']$ but $\boldsymbol{m} \ngeq \boldsymbol{Pre}[\cdot,t] + \boldsymbol{Pre}[\cdot,t']$, i.e., they are both enabled at $\boldsymbol{m}$, but $\boldsymbol{m}$ does not contain enough tokens to allow the firing of both transitions.*

In the net in Fig. 10.1 transitions t_3 and t_4 are in structural conflict. Such conflict is also behavioral at marking $\boldsymbol{m}'' = [0\ 1\ 0\ 1]^T$ since $p_2 \in {}^\bullet t_3 \cap {}^\bullet t_4$ only contains one token that can be used for the firing of only one of the two transitions. On the contrary, the conflict is not behavioral at marking $\boldsymbol{m} = [1\ 0\ 0\ 0]^T$ since the two transitions are not enabled. Analogously, the conflict is not behavioral at marking $\boldsymbol{m}' = [0\ 2\ 0\ 0]^T$, since p_2 contains enough tokens to allow the firing of both transitions.

To a marked net $\langle N, \boldsymbol{m}_0 \rangle$ it is possible to associate a well precise dynamics, given by the set of all sequences of transitions that can fire at the initial marking.

Definition 10.8. *The* language *of a marked net $\langle N, \boldsymbol{m}_0 \rangle$ is the set of firing sequences enabled at the initial marking, i.e., the set*

$$L(N, \boldsymbol{m}_0) = \{\sigma \in T^* \mid \boldsymbol{m}_0[\sigma\rangle\}.$$

Finally, it is also possible to define the state space of a marked net.

Definition 10.9. *A marking $\boldsymbol{m}$ is* reachable *in $\langle N, \boldsymbol{m}_0 \rangle$ if there exists a firing sequence σ such that $\boldsymbol{m}_0[\sigma\rangle \boldsymbol{m}$. The* reachability set *of a marked net $\langle N, \boldsymbol{m}_0 \rangle$ is the set of markings that can be reached from the initial marking, i.e., the set*

$$R(N, \boldsymbol{m}_0) = \{\boldsymbol{m} \in \mathbb{N}^m \mid \exists \sigma \in L(N, \boldsymbol{m}_0) : \ \boldsymbol{m}_0[\sigma\rangle \boldsymbol{m}\}.$$

Note that in the previous definition the empty sequence, that contains no transition, is also considered. Indeed, since $\boldsymbol{m}_0[\varepsilon\rangle \boldsymbol{m}_0$, it holds $\boldsymbol{m}_0 \in R(N, \boldsymbol{m}_0)$.

As an example, let us consider the marked net in Fig. 10.3(a), where the initial marking assigns a number r of tokens to p_1. The reachability set is $R(N, \boldsymbol{m}_0) = \{[i\ j\ k]^T \in \mathbb{N}^3 \mid i + j + k = r\}$ and it is thus finite. On the contrary, the language $L(N, \boldsymbol{m}_0)$ of such a net system is infinite since sequences of arbitrary length can fire.

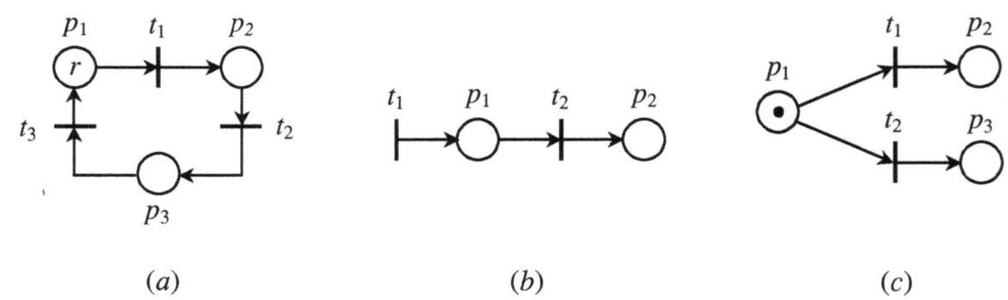

Fig. 10.3 Some examples of marked nets

In the net in Fig. 10.3(*b*), the reachability set is $R(N, \boldsymbol{m}_0) = \mathbb{N}^2$ and such a set is infinite. A generic marking $\boldsymbol{m}$ with $m[p_1] = i$ and $m[p_2] = j$ can be reached from $\boldsymbol{m}_0$ firing the sequence $t_1{}^{i+j}\ t_2{}^j$. Also the language is infinite. Finally, in the net in Fig. 10.3(*c*), it is $R(N, \boldsymbol{m}_0) = \{[1\ 0\ 0]^T, [0\ 1\ 0]^T, [0\ 0\ 1]^T\}$ and $L(N, \boldsymbol{m}_0) = \{\varepsilon, t_1, t_2\}$.

Summarizing, a double meaning is associated with the marking of a net: on one side the marking denotes the current state of the system; on the other side it specifies which activities can be executed, i.e., which transitions can fire. The transition firing determines the dynamical behavior of the marked net.

10.3 Modeling with Petri Nets

In this section we present some elementary P/T structures and the semantics associated with them.

In a discrete event system, the order in which events occur can be subject to constraints of different nature. In a PN model this corresponds to impose some constraints on the order in which transitions fire. In the following we present four main structures.

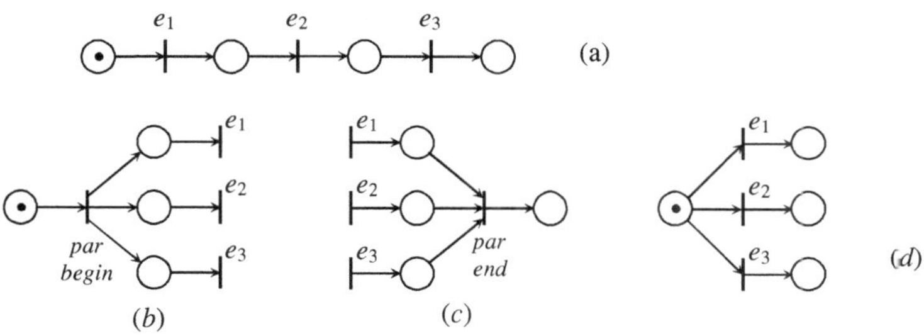

Fig. 10.4 Elementary structures of PNs: (*a*) sequentiality; (*b*) parallelism; (*c*) synchronization; (*d*) choice

Sequentiality. Events occur in a sequential order.
In Fig. 10.4(*a*) event e_2 can only occur after the occurrence of e_1; e_3 can occur only after the occurrence of e_2.

Parallelism (or **structural concurrency**). Events may occur with no fixed order.
In Fig. 10.4(*b*), after the firing of transition *par begin* (parallel begin) events e_1, e_2, and e_3 are simultaneously enabled. Parallelism implies that the three events are not in structural conflict and can occur in any order since the occurrence of any event does not modify the enabling condition of the others. Transition *par begin* creates a fork in the flow of events.

Synchronization. Several parallel events must have occurred before proceeding.

In Fig. 10.4(c), events e_1, e_2 and e_3 can occur in parallel but transition *par end* (parallel end) cannot fire until all of them have occurred. Transition *par end* creates a join in the flow of events.

Choice (or **structural conflict**). Only one event among many possible ones can occur.
In Fig. 10.4(d), only one event among e_1, e_2 and e_3 can occur, because the firing of any transition disables the others. Note that the choice is characterized by two or more transitions sharing an input place that determines a structural conflict.

To the above elementary structures it is often possible to associate a dual semantics, that takes into account the variation of markings, rather than the order in which transitions fire. In such a case, tokens represent available resources.

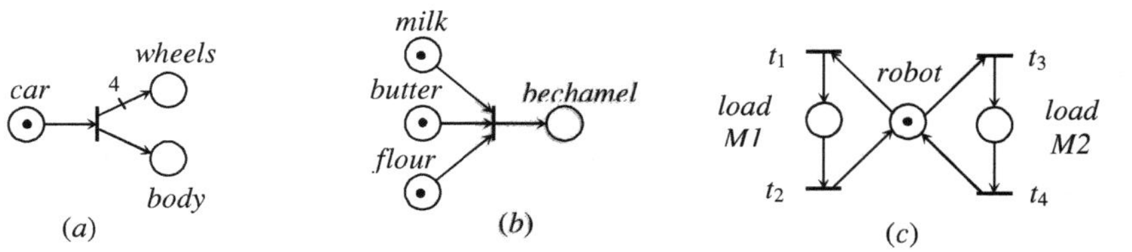

Fig. 10.5 Elementary structures of PNs: (a) disassembly; (b) assembly; (c) mutual exclusion

Disassembly. A composite element is separated into elementary parts.
In Fig. 10.5(a) the marked net represents the disassembly of a car, obtaining four wheels and a body. The transition is similar to the transition *par begin* previously introduced.

Assembly. Several parts are combined to produce a composite element.
In Fig. 10.5(b), the marked net describes the recipe to prepare bechamel sauce. The transition is similar to *par end* introduced above.

Mutual exclusion. A resource (or a set of resources) can be employed in several operations. However, while it has been acquired for a given operation, it is not available for other operations until it is released.
In Fig. 10.5(c), a single robot is available to load parts in two machines. When the place *robot* is marked the robot is available, while if either place *load* $M1$ or *load* $M2$ is marked the robot is acquired for the corresponding operation. From the situation in figure, if t_1 fires, the loading of the first machine starts and place *robot* gets empty: thus t_3, whose firing corresponds to the reservation of the robot for the loading of the second machine, is disabled until the firing of t_2 that moves the token again in place *robot*. Analogously, from the situation in figure the firing of t_3 disables t_1 until the firing of t_4. The structure is similar to "choice" in Fig. 10.4(d).

We conclude this section presenting an example taken from the manufacturing domain. Note that *manufacturing* is one of the application areas where PNs have been more extensively used since the early 1990s [4, 6].

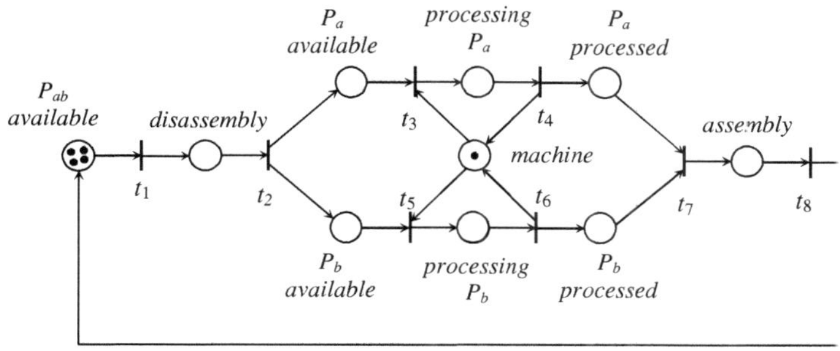

Fig. 10.6 Petri net model of a manufacturing cell

Figure 10.6 presents the Petri net model of a manufacturing cell where composite parts P_{ab} are processed. The cell consists of a single machine. A composite part is initially disassembled in two elementary parts P_a and P_b, that are, one at a time, processed by the machine. Finally, the two processed parts are assembled again and removed from the cell. The PN in Fig. 10.6 describes such a system. Places associated with resources are: P_{ab} *available*, P_a *available*, P_a *processed*, P_b *available*, P_b *processed*, *machine*. In figure, four tokens are initially assigned to place P_{ab} *available*: this denotes the presence of four parts P_{ab} that are available to be disassembled. The firing of t_1 represents the withdrawal of a part P_{ab} to be disassembled. The disassembly operation is modeled by transition t_2. After such an operation one part P_a and one part P_b are available to be processed. A single machine is available to process parts of both types. When transition t_3 fires, the machine starts processing a part of type P_a and no other part can be processed until t_4 fires, i.e., the machine is released. Analogously, transition t_5 represents the acquisition of the machine for the processing of a part of type P_b, while t_6 represents its release. Transitions t_7 models the assembly operation, that can only occur when a part of each type is available. At the end of the assembly operation, the processed part P_{ab} exits the cell and a new part to be processed enters the system. This is modeled by transition t_8. Note that this operation mode is typical of those processes where parts move on pallets, that are available in a finite number.

10.4 Analysis by Enumeration

In this section we present an important technique for the analysis of PNs based on the enumeration of the reachability set of the net and of the transition function between markings. If the reachability set is finite, an exhaustive enumeration is possible and the *reachability graph* of the net is constructed. If the reachability set is not finite, a finite *coverability graph* can still be constructed using the notion of *ω-marking*; the coverability graph provides a larger approximation of the reachability set and of the net language.

10.4.1 Reachability Graph

The main steps for the construction of the reachability graph of a marked net $\langle N, \boldsymbol{m}_0 \rangle$ are summarized in the following algorithm.

Algorithm 10.2. (Reachability Graph).

1. *The initial node of the graph is the initial marking* $\boldsymbol{m}_0$. *This node is initially unlabeled.*
2. *Consider an unlabeled node* $\boldsymbol{m}$ *of the graph.*
   a *For each transition t enabled at* $\boldsymbol{m}$, *i.e., such that* $\boldsymbol{m} \geq \boldsymbol{Pre}[\cdot, t]$*:*
      i. *Compute the marking* $\boldsymbol{m}' = \boldsymbol{m} + \boldsymbol{C}[\cdot, t]$ *reached from* $\boldsymbol{m}$ *firing t.*
      ii. *If no node* $\boldsymbol{m}'$ *is on the graph, add a new node* $\boldsymbol{m}'$ *to the graph.*
      iii. *Add an arc t from* $\boldsymbol{m}$ *to node* $\boldsymbol{m}'$.
   b *Label node* $\boldsymbol{m}$ *"old".*
3. *If there exist nodes with no label, goto Step 2.*

In the case of nets with an infinite reachability set the algorithm does not terminate. However, a simple test to detect this case can be added at Step 2.a: if there exists a marking $\boldsymbol{m}''$, computed previously, such that the new marking $\boldsymbol{m}'$ is greater than and different from $\boldsymbol{m}''$, then stop the computation because the reachability graph is infinite.

An example of reachability graph is given in Fig. 10.7.

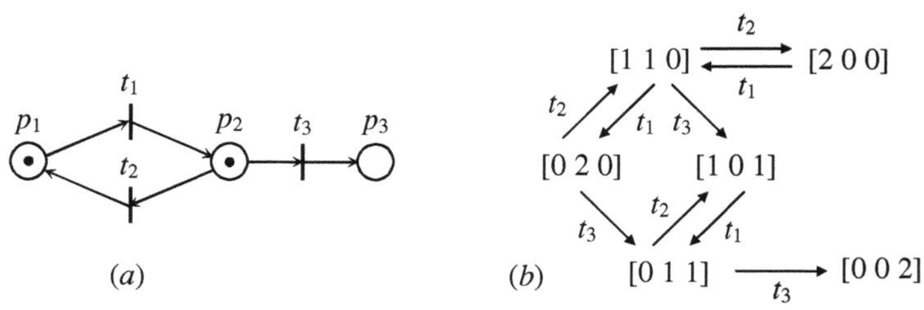

Fig. 10.7 (*a*) A bounded PN system and its reachability graph

The following proposition holds, whose proof immediately follows from the definition of reachability graph.

Proposition 10.1. *Consider a bounded marked net* $\langle N, \boldsymbol{m}_0 \rangle$ *and its reachability graph.*

(a) *Marking* $\boldsymbol{m}$ *belongs to the reachability set* $R(N,\boldsymbol{m}_0) \iff \boldsymbol{m}$ *is a node of the graph.*

(b) *Given* $\boldsymbol{m} \in R(N,\boldsymbol{m}_0)$, *sequence* $\sigma = t_{j_1}\, t_{j_2} \cdots$, *belongs to* $L(N,\boldsymbol{m})$ *and can be generated with the trajectory* $\boldsymbol{m}[t_{j_1}\rangle \boldsymbol{m}'[t_{j_2}\rangle \boldsymbol{m}'' \cdots \iff$ *there exists in the graph a directed path* $\gamma = \boldsymbol{m}\, t_{j_1}\, \boldsymbol{m}'\, t_{j_2}\, \boldsymbol{m}'' \cdots$.

As shown in [5], given a bounded PN, the problem of construction of the reachability graph is not primitive recursive. This implies that every method based on the reachability graph construction has an unpredictable complexity. This explains the importance of structural analysis which is the object of the following Chapter 11.

10.4.2 Coverability Graph

The procedure used for the construction of the reachability graph obviously does not terminate if the net is unbounded. Indeed in such a case, a situation like the following would surely occur. There exists a directed path that starts from $\boldsymbol{m}_0$ to $\tilde{\boldsymbol{m}}$, and from such a node there exists a directed path leading to $\boldsymbol{m}' \gneq \tilde{\boldsymbol{m}}$. To characterize the existence of sequences of transitions whose firing indefinitely increase the marking of some places, we assign a special symbol ω to all entries of $\boldsymbol{m}'$ that are strictly greater than the corresponding entries of $\tilde{\boldsymbol{m}}$.

Definition 10.10. *An* $\omega-$marking *of a net N with m places is a vector* $\boldsymbol{m}_\omega \in (\mathbb{N} \cup \{\omega\})^m$, *where one or more components may be equal to* ω.

Thus ω should be thought as "arbitrarily large" and we assume that $\forall n \in \mathbb{N}$ it holds $\omega > n$ and $\omega \pm n = \omega$.
Using the notion of $\omega-$marking, a finite approximation of the reachability graph, called *coverability graph*, can be constructed. The construction of the coverability graph first requires the construction of the *coverability tree*, a graph with no loops where duplicated nodes may exist. The following algorithm summarizes the main steps for the computation of the coverability tree of a marked net $\langle N, \boldsymbol{m}_0 \rangle$ with incidence matrix $\boldsymbol{C}$.

Algorithm 10.3. (Coverability tree).

1 *The root node of the tree is the initial marking* $\boldsymbol{m}_0$. *This node is initially unlabeled.*

2 *Consider an unlabeled node* $\boldsymbol{m}$ *of the tree.*

  a *For each transition t enabled at* $\boldsymbol{m}$, *i.e., such that* $\boldsymbol{m} \geq \boldsymbol{Pre}[\cdot,t]$:

    i. *Compute the marking* $\boldsymbol{m}' = \boldsymbol{m} + \boldsymbol{C}[\cdot,t]$ *reached from* $\boldsymbol{m}$ *firing t.*

    ii. *For all markings* $\tilde{\boldsymbol{m}} \lneq \boldsymbol{m}'$ *on the path from the root node* $\boldsymbol{m}_0$ *to node* $\boldsymbol{m}$ *and for all* $p \in P$,
    if $\tilde{m}[p] < m'[p]$ *then let* $m'[p] = \omega$.

    iii. *Add a new node* $\boldsymbol{m}'$ *to the tree.*

   iv. Add an arc t from **m** *to the new node* **m**$'$.
   v. If there already exists a node **m**$'$ *in the tree, label the new node* **m**$'$ *"duplicated".*

 b Label node **m** *"old".*

3 If there exist nodes with no label, goto Step 2.

Karp and Miller [10] proved that Algorithm 10.3 always terminates in a finite number of steps even if the net has an infinite state space.

Consider as an example, the marked net in Fig. 10.8(*a*). The coverability tree is shown in Fig. 10.8(*b*) where labels "old" in the internal nodes have been omitted to make the figure more readable.

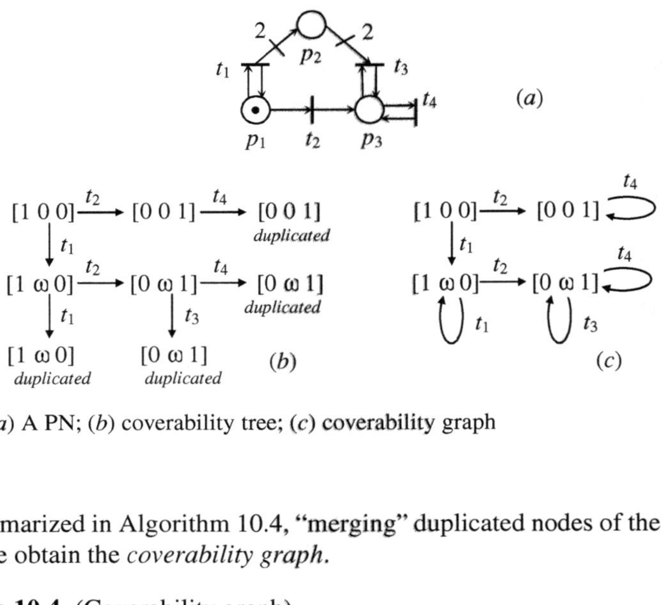

Fig. 10.8 (*a*) A PN; (*b*) coverability tree; (*c*) coverability graph

As summarized in Algorithm 10.4, "merging" duplicated nodes of the coverability tree, we obtain the *coverability graph*.

Algorithm 10.4. (Coverability graph).

1 If the tree contains no nodes with label "duplicated" goto Step 4.

2 Consider a node **m** *of the graph with label "duplicated".*
Such a node has no output arcs but an input arc t from node **m**$'$.
Moreover, there surely exists in the graph another node **m** *with label "old".*

 a Remove arc t from node **m**$'$ *to node* **m** *"duplicated".*
 b Add an arc t from node **m**$'$ *to node* **m** *"old".*
 c Remove node **m** *"duplicated".*

3 If there still exist nodes with label "duplicated" goto Step 2.

4 Remove labels from nodes.

The coverability graph of the marked net in Fig. 10.8(a) is shown in Fig. 10.8(c).

In the case of nets with an infinite reachability set, the coverability graph provides a finite description that approximates this infinite set.

Definition 10.11. *A marking* $\boldsymbol{m} \in \mathbb{N}^m$ *is said to be* ω-covered *by a vector* $\boldsymbol{m}_\omega \in (\mathbb{N} \cup \{\omega\})^m$ *if* $m_\omega[p] = m[p]$ *for all places* p *such that* $m_\omega[p] \neq \omega$*; this relation is denoted* $\boldsymbol{m}_\omega \geq_\omega \boldsymbol{m}$.

Thus a node $\boldsymbol{m}_\omega$ in the graph represents all markings that are ω-covered by it.

Due to the presence of ω-markings, the coverability graph does not always provide necessary and sufficient conditions to decide the reachability of a marking or the existence of a firing sequence. Such results are summarized in the following proposition, where a node that can contain ω components is denoted with the notation $\boldsymbol{m}_\omega$.

Proposition 10.2. *Consider a marked net* $\langle N, \boldsymbol{m}_0 \rangle$ *and its coverability graph.*

(a) Marking $\boldsymbol{m}$ *is reachable* $\Longrightarrow$ *there exists a node* $\boldsymbol{m}_\omega \geq_\omega \boldsymbol{m}$ *in the graph.*
(b) Given a marking $\boldsymbol{m} \in R(N, \boldsymbol{m}_0)$*, sequence* $\sigma = t_{j_1} t_{j_2} \cdots$*, belongs to the language* $L(N, \boldsymbol{m})$ *and can be generated with a trajectory* $\boldsymbol{m}[t_{j_1}\rangle \boldsymbol{m}'[t_{j_2}\rangle \boldsymbol{m}'' \cdots \Longrightarrow$ *there exists in the graph a directed path* $\gamma = \boldsymbol{m}_\omega\ t_{j1}\ \boldsymbol{m}'_\omega\ t_{j2}\ \boldsymbol{m}''_\omega \cdots$*, with* $\boldsymbol{m}_\omega \geq_\omega \boldsymbol{m}$*,* $\boldsymbol{m}'_\omega \geq_\omega \boldsymbol{m}'$ *etc.*

The main feature of the coverability graph is that of not providing a general algorithm, valid in all cases, to determine the reachability of a marking.

Example 10.5. Consider the marked net in Fig. 10.8 and its coverability graph. Based on Proposition 10.2(a) we conclude that marking $[0\ 0\ 1]^T$ is reachable, because it appears in the graph. On the contrary, based on Proposition 10.2(a) marking $[1\ 1\ 1]^T$ is not reachable since it is covered by no node in the graph. Finally, if we consider a marking $[0\ k\ 1]^T$ for a given value $k > 0$, it is not possible to draw a conclusion concerning its reachability, being it covered by node $[0\ \omega\ 1]^T$: as an example, $[0\ 2\ 1]^T$ is a reachable marking, while $[0\ 3\ 1]^T$ is not reachable.

Let us also observe that by Proposition 10.2(b) a coverability graph may contain directed paths associated with sequences that are not enabled. As an example, in the net in Fig. 10.8, $\sigma = t_1 t_2 t_3 t_3$ cannot fire at the initial marking: indeed in an admissible sequence, t_3 can fire at most as many times as t_1, due to the constraint imposed by place p_2 that is initially empty. However, starting from $\boldsymbol{m}_0$ there is in the graph a path whose arcs form sequence σ. ■

We conclude this section introducing the notion of *covering set*, that is a (not necessarily strict) superset of $R(N, \boldsymbol{m}_0)$.

Definition 10.12. *Given a marked net* $\langle N, \boldsymbol{m}_0 \rangle$*, let* $V \subseteq (\{\mathbb{N} \cup \{\omega\})^m$ *be the set of nodes of its coverability graph. The* covering set *of* $\langle N, \boldsymbol{m}_0 \rangle$ *is*

$$CS(N, \boldsymbol{m}_0) = \{\boldsymbol{m} \in \mathbb{N}^m \mid \exists \boldsymbol{m}_\omega \in V,\ m[p] = m_\omega[p] \text{ if } m_\omega[p] \neq \omega\}.$$

By Proposition 10.2, we can state the following result.

Proposition 10.3. *Given a marked net* $\langle N, \boldsymbol{m}_0 \rangle$, *it holds* $R(N, \boldsymbol{m}_0) \subseteq CS(N, \boldsymbol{m}_0)$.

As an example, in the case of the marked net in Fig. 10.8, it holds $CS(N, \boldsymbol{m}_0) = \{[1\ 0\ 0]^T, [0\ 0\ 1]^T\} \cup \{[1\ k\ 0]^T,\ k \in \mathbb{N}\} \cup \{[0\ k\ 1]^T,\ k \in \mathbb{N}\} \subset R(N, \boldsymbol{m}_0)$. However, if N' is a new net obtained from the net in Fig. 10.8 changing the multiplicity of the arcs incident on place p_2 from 2 to 1, then $CS(N', \boldsymbol{m}_0) = R(N', \boldsymbol{m}_0)$.

Other approximations of the reachability set will be given in the following Chapter 11.

10.4.3 Behavioral Properties

In this section we define the main *behavioral properties* of a marked net, i.e., those properties that depend both on the net structure and on the initial marking.

10.4.3.1 Reachability

A fundamental problem in the PN net setting is the following, known as the *reachability problem*.

- Given a marked net $\langle N, \boldsymbol{m}_0 \rangle$ and a generic marking $\boldsymbol{m}$, is $\boldsymbol{m} \in R(N, \boldsymbol{m}_0)$?

As already discussed in the previous section, if the net has a finite state space, such a problem can be solved constructing the reachability graph. However, in the case of nets with an infinite state space, the coverability graph does not provide necessary and sufficient conditions to test if a given marking is reachable.

It is easy to show that the reachability problem is at least semi-decidable[1]. Indeed, if we consider a marked net $\langle N, \boldsymbol{m}_0 \rangle$ and a marking $\boldsymbol{m}$ whose reachability has to be verified, we can generate in an orderly fashion all sequences in $L(N, \boldsymbol{m}_0)$, starting first with those of length 1, then with those of length 2, etc., and compute the markings reached with each of these sequences. If $\boldsymbol{m}$ is reachable with a sequence of length k, at the kth step the algorithm terminates with a positive answer. However, if $\boldsymbol{m}$ is not reachable, this algorithm never halts.

In the 1980s it has been proved that the reachability problem is also *decidable*, even if the corresponding algorithm has a very high complexity [16].

[1] A problem whose solution may either be YES or NO is said to be:

- *decidable* if there exists an algorithm that, for each possible formulation of the problem, halts in a finite number of steps providing the correct solution;
- *semi-decidable* if there exists an algorithm that, for each possible formulation of the problem, halts in a finite number of steps providing the correct solution in one of the two cases (e.g., if the answer is YES), while it may not halt in the other case (e.g., if the answer is NO).

10.4.3.2 Boundedness

The *boundedness* property, associated with a place or with a net, implies that the number of tokens in the place or in the net, never exceeds a given amount. As an example, this property may imply that no overflow occurs in a buffer, or can be used to dimension the number of resources required by a process.

Definition 10.13. *A place p is $k-$*bounded *in $\langle N, \boldsymbol{m}_0 \rangle$ if for all reachable markings $\boldsymbol{m} \in R(N, \boldsymbol{m}_0)$ it holds $m[p] \leq k$. A place 1-bounded is* safe *(or* binary*). A marked net $\langle N, \boldsymbol{m}_0 \rangle$ is $k-$*bounded *if all places are $k-$bounded. A marked net that is $1-$bounded is called* safe *(or* binary*)* .

When it is not important to specify the value of k, the place (net) is simply called *bounded.*

Proposition 10.4. [12] *A marked net $\langle N, \boldsymbol{m}_0 \rangle$ is bounded if and only if it has a finite reachability set.*

Proposition 10.5. *Consider a marked net $\langle N, \boldsymbol{m}_0 \rangle$ and its coverability graph.*

- *A place p is $k-$bounded $\Longleftrightarrow$ for each node $\boldsymbol{m}_\omega$ of the graph it holds $\boldsymbol{m}_\omega[p] \leq k \neq \omega$.*
- *The marked net is bounded $\Longleftrightarrow$ no node of the graph contains the symbol ω.*

The net in Fig. 10.8 is unbounded. Places p_1 and p_3 are safe, while place p_2 is unbounded. The net in Fig. 10.9 is safe.

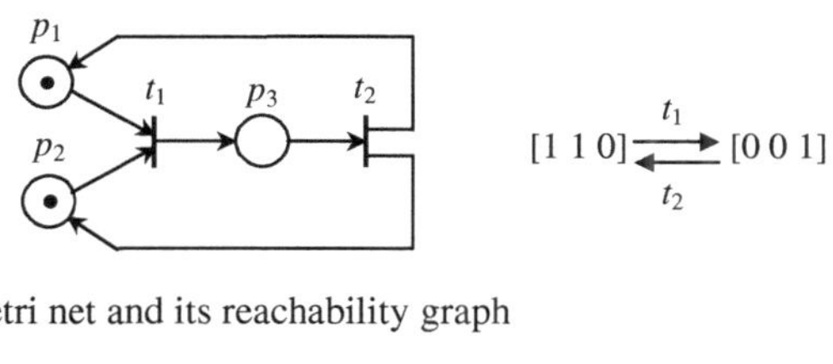

Fig. 10.9 A safe Petri net and its reachability graph

10.4.3.3 Conservativeness

A property strictly related to boundedness is *conservativeness* implying that the weighted sum of tokens in a net remains constant. Such a property ensures that resources are preserved.

Definition 10.14. *A marked net $\langle N, \boldsymbol{m}_0 \rangle$ is* strictly conservative *if for all reachable markings $\boldsymbol{m} \in R(N, \boldsymbol{m}_0)$ the number of tokens that the net can contain does not vary, i.e., if:*

$$\mathbf{1}^T \cdot \boldsymbol{m} = \sum_{p \in P} m[p] = \sum_{p \in P} m_0[p] = \mathbf{1}^T \cdot \boldsymbol{m}_0$$

It is easy to verify graphically if a marked net is strictly conservative. Indeed all transitions should have a number of "pre" arcs equal to the number of "post" arcs.

Note however, that such a condition is not necessary for strict conservativeness: there may exist a transition with a different number of "pre" and "post" arcs that never fires. The net in Fig. 10.7 is strictly conservative since the total number of tokens is always equal to two. The net in Fig. 10.8 is not strictly conservative.

A generalization of strict conservativeness is the following.

Definition 10.15. *A marked net* $\langle N, \boldsymbol{m}_0 \rangle$ *is* conservative *if there exists a vector of* positive *integers* $\boldsymbol{x} \in \mathbb{N}_+^m$ *such that for all reachable markings* $\boldsymbol{m} \in R(N, \boldsymbol{m}_0)$ *it is:*

$$\boldsymbol{x}^T \cdot \boldsymbol{m} = \boldsymbol{x}^T \cdot \boldsymbol{m}_0$$

i.e. the number of tokens weighted through $\boldsymbol{x}$ *does not vary.*

The net in Fig. 10.9 is not strictly conservative, but it is conservative. Indeed, consider the vector $\boldsymbol{x} = [1\ 1\ 2]^T$. It is easy to verify that for all reachable markings $\boldsymbol{m}$ it is $\boldsymbol{x}^T \cdot \boldsymbol{m} = \boldsymbol{x}^T \cdot \boldsymbol{m}_0 = 2$.

Conservativeness is related to boundedness.

Proposition 10.6. *If a marked net* $\langle N, \boldsymbol{m}_0 \rangle$ *is conservative then it is bounded.*

Note however, that there may also exist bounded nets that are not conservative. An example is given in Fig. 10.10 that shows a safe net that is not conservative: indeed its reachability set is $\{[1],[0]\}$, thus chosen an arbitrary positive integer x it holds $0 = x \cdot 0 \neq x \cdot 1 = x$.

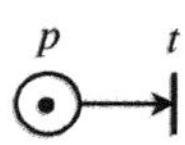

Fig. 10.10 A bounded Petri net that is not conservative

In the following Chapter 11 it will be shown how, using the incidence matrix, it is possible to compute a vector $\boldsymbol{x}$ with respect to whom the net is conservative.

10.4.3.4 Repetitiveness

Repetitiveness of a sequence of transitions ensures that the sequence can occur indefinitely.

Definition 10.16. *Given a marked net* $\langle N, \boldsymbol{m}_0 \rangle$*, let* σ *be a non empty sequence of transitions and* $\boldsymbol{m} \in R(N, \boldsymbol{m}_0)$ *a marking enabling it. Sequence* σ *is called* repetitive *if it can fire an infinite number of times at* $\boldsymbol{m}$*, i.e., it holds*

$$\boldsymbol{m}[\sigma\rangle \boldsymbol{m}_1 [\sigma\rangle \boldsymbol{m}_2 [\sigma\rangle \boldsymbol{m}_3 \cdots$$

A marked net $\langle N, \boldsymbol{m}_0 \rangle$ *is* repetitive *if there exists a repetitive sequence in* $L(N, \boldsymbol{m}_0)$.

The following proposition characterizes repetitive sequences.

Proposition 10.7. *Let σ be a non empty sequence of transitions such that $\boldsymbol{m}[\sigma\rangle\boldsymbol{m}'$. Sequence σ is repetitive if and only if $\boldsymbol{m} \leq \boldsymbol{m}'$.*

We distinguish two types of repetitive sequences.

Definition 10.17. *A repetitive sequence σ enabled at $\boldsymbol{m}$ is called:*

- stationary *if $\boldsymbol{m}[\sigma\rangle\boldsymbol{m}$,*
- increasing *if $\boldsymbol{m}[\sigma\rangle\boldsymbol{m}'$ with $\boldsymbol{m}' \gneq \boldsymbol{m}$.*

As an example, the net in Fig. 10.3(c) does not contain repetitive sequences. In the net in Fig. 10.8 repetitive sequences are t_1^k and t_4^k, with $k \in \mathbb{N}_+$: sequences t_1^k are increasing, while sequences t_4^k are stationary.

Increasing sequences exist only on unbounded nets.

Proposition 10.8. [12] *A marked net $\langle N, \boldsymbol{m}_0\rangle$ is bounded if and only if it does not admit increasing repetitive sequences.*

As discussed in Chapter 11 it is immediate to verify if a given sequence σ is repetitive (either stationary or increasing) using the incidence matrix of the net. Here we only consider the information given by the analysis of the reachability graph.

Proposition 10.9. *Consider a marked bounded net $\langle N, \boldsymbol{m}_0\rangle$ and its reachability graph. A sequence σ is stationary $\Longleftrightarrow$ there exists a directed cycle in the graph whose arcs form σ.*

In the net in Fig. 10.7 each stationary sequence corresponds to a cycle in the reachability graph. Sequences that correspond to elementary cycles are called *elementary*. As an example, $t_1 t_2$ is an elementary sequence, while $t_1 t_2 t_1 t_2$ is not elementary.

Proposition 10.10. *Consider a marked net $\langle N, \boldsymbol{m}_0\rangle$ and its coverability graph.*

- *A sequence σ is repetitive $\Longrightarrow$ there exists a directed cycle in the graph whose arcs form σ.*
- *A sequence σ is stationary $\Longleftarrow$ there exists a directed cycle in the graph that does not pass through markings containing ω and whose arcs form σ.*

Note that a coverability graph has always at least one cycle associated with an increasing sequence. Such is the case of sequence t_1 in the net in Fig. 10.8. Moreover, there can be cycles associated with non repetitive sequences. Such is the case of sequence t_3 in the net in Fig. 10.8: t_3 is not repetitive because its firing leads to decreasing of two units the number of tokens in p_2; however, this is hidden when t_3 fires from $[0\ \omega\ 1]^T$.

10.4.3.5 Reversibility

Reversibility implies that a system can always be reinitialized to its initial state. This is a desirable feature in many man-made systems.

Definition 10.18. *A marked net* $\langle N, \boldsymbol{m}_0 \rangle$ *is* reversible *if for all reachable markings* $\boldsymbol{m} \in R(N, \boldsymbol{m}_0)$ *it holds* $\boldsymbol{m}_0 \in R(N, \boldsymbol{m})$, *i.e., if from any reachable marking it is possible to reach back the initial marking* $\boldsymbol{m}_0$.

As an example, the net in Fig. 10.11(a) is reversible because from any reachable marking $m = [k]$, transition t_2 can fire k times leading the net back to the initial marking $m_0 = [0]$ in figure. On the contrary, the net in Fig. 10.11(b) is not reversible because any token that enters in p_2 can never be removed coming back to the initial marking.

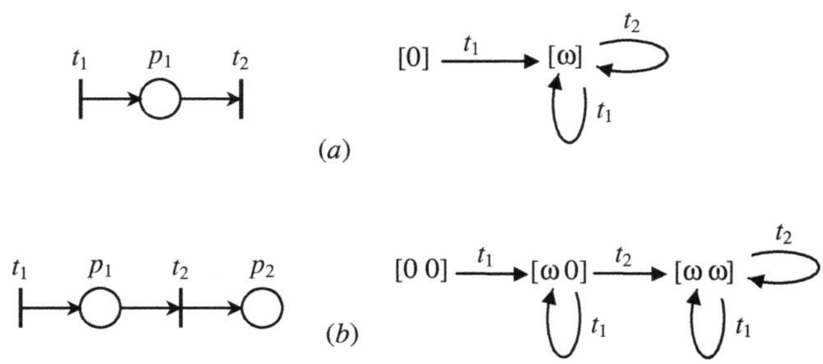

Fig. 10.11 (a) A reversible unbounded marked net and its coverability graph; (b) a non reversible unbounded marked net and its coverability graph

The reachability graph provides necessary and sufficient conditions for reversibiliy. On the contrary, the coverability graph only provides necessary conditions. This is formalized by the following two propositions whose validity derives from Propositions 10.1 and 10.2, respectively.

Proposition 10.11. *Consider a bounded marked net* $\langle N, \boldsymbol{m}_0 \rangle$ *and its reachability graph. The marked net is reversible* $\Longleftrightarrow$ *the graph is strongly connected.*

The reachability graph of the net in Fig. 10.7 is not strongly connected: as an example, there exists no directed path from marking $[1\ 0\ 1]^T$ to the initial marking.

Proposition 10.12. *Consider a marked net* $\langle N, \boldsymbol{m}_0 \rangle$ *and its coverability graph. The net is reversible* $\Longrightarrow$ *each ergodic*[2] *component of the graph contains a node* $\boldsymbol{m}_\omega \geq_\omega \boldsymbol{m}_0$.

As an example, the only ergodic component of the coverability graph of the reversible net in Fig. 10.11(a) contains marking $[\omega] \geq_\omega [0] = m_0$. Note however, that also the net in Fig. 10.11(b) has only one ergodic component that contains the marking $[\omega\ \omega]^T \geq_\omega [0\ 0]^T = \boldsymbol{m}_0$ and it is not reversible. Finally, it is possible to conclude

[2] Consider a maximal strongly connected component of a graph. Such a component is called *ergodic* if there are no edges leading from a node that belongs to the component to a node that does not belong to it. Otherwise the component is called *transient* [2].

by the only analysis of the coverability graph, that the net in Fig. 10.8 is not reversible: indeed it has two ergodic components, each one with a single marking $[0\ 0\ 1]^T$ and $[0\ \omega\ 1]^T$, respectively, none of them covering the initial marking.

Note, finally, that even if the coverability graph does not provide necessary and sufficient conditions for checking reversibility, such a property is decidable. In fact checking for reversibility reduces to checking if the initial marking $\boldsymbol{m}_0$ is a *home marking*, a problem that is known to be decidable [7] (see also Chapter 12, Definition 12.8).

10.4.3.6 Liveness and Deadlock

Liveness of a transition implies the possibility that it can always eventually fire, regardless of the current state of the net.

Definition 10.19. *Given a marked net $\langle N, \boldsymbol{m}_0 \rangle$, we say that a transitions t is:*

- dead *if no reachable marking enables it, i.e.,* $\forall \boldsymbol{m} \in R(N, \boldsymbol{m}_0)\ \neg \boldsymbol{m}[t\rangle$;
- quasi-live *if it is enabled by some reachable marking, i.e.,* $\exists \boldsymbol{m} \in R(N, \boldsymbol{m}_0)\ :\ \boldsymbol{m}[t\rangle$;
- live *if for all reachable markings* $\boldsymbol{m} \in R(N, \boldsymbol{m}_0)$, *t is quasi-live in* $\langle N, \boldsymbol{m} \rangle$.

In the net in Fig. 10.12 transition t_4 is dead, transitions t_1 and t_2 are quasi-live, transition t_3 is live. Note a fundamental difference between quasi-live transitions t_1 and t_2: t_1 can fire an infinite number of times, while t_2 may only fire once.

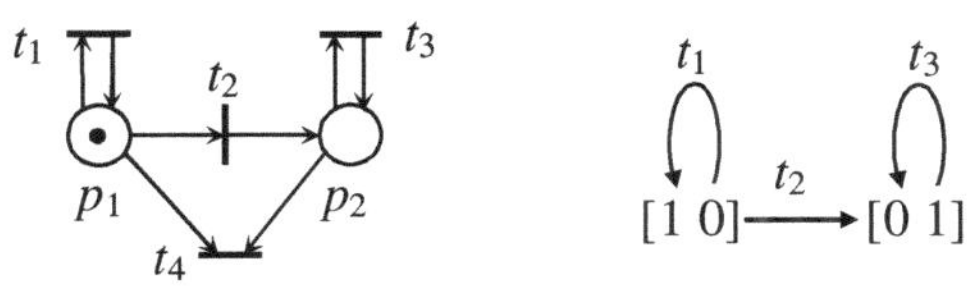

Fig. 10.12 A PN for the study of liveness

It is also possible to define the liveness property for a marked net.

Definition 10.20. *A marked net $\langle N, \boldsymbol{m}_0 \rangle$ is:*

- dead, *if all its transitions are dead;*
- not quasi-live, *if some of its transitions are dead and some are quasi-live;*
- quasi-live, *if all its transitions are quasi-live;*
- live, *if all its transitions are live.*

The net in Fig. 10.12 is not quasi-live because it contains both dead and quasi-live transitions. The two nets in Fig. 10.11 are both live.

Another important concept related to the notion of liveness is *deadlock* that denotes an anomalous state from which no further evolution is possible.

Definition 10.21. *Given a marked net* $\langle N, \boldsymbol{m}_0 \rangle$ *let* $\boldsymbol{m} \in R(N, \boldsymbol{m}_0)$ *be a reachable marking. We say that* $\boldsymbol{m}$ *is a* dead marking *if no transition is enabled at* $\boldsymbol{m}$, *i.e., if* $\langle N, \boldsymbol{m} \rangle$ *is dead. A marked net* $\langle N, \boldsymbol{m}_0 \rangle$ *is* deadlocking *if there exists a dead reachable marking.*

The net in Fig. 10.7 is deadlocking: marking $[0\ 0\ 2]^T$ is dead.

Once again the reachability graph provides necessary and sufficient conditions for the verification of liveness and deadlock.

Proposition 10.13. *Consider a bounded marked net* $\langle N, \boldsymbol{m}_0 \rangle$ *and its reachability graph.*

- *Transition* t *is dead* $\Longleftrightarrow$ *no arc labeled* t *belongs the graph.*
- *Transition* t *is quasi-live* $\Longleftrightarrow$ *an arc labeled* t *belongs the graph.*
- *Transition* t *is live* $\Longleftrightarrow$ *an arc labeled* t *belongs to each ergodic component of the graph.*
- *Reachable marking* $\boldsymbol{m}$ *is dead* $\Longleftrightarrow$ *node* $\boldsymbol{m}$ *in the graph has no output arc.*

The coverability graph provides necessary and sufficient conditions for the analysis of quasi-liveness, but only necessary conditions for the analysis of liveness.

Proposition 10.14. *Consider a marked net* $\langle N, \boldsymbol{m}_0 \rangle$ *and its coverability graph.*

- *Transition* t *is dead* $\Longleftrightarrow$ *no arc labeled* t *belongs the graph.*
- *Transition* t *is quasi-live* $\Longleftrightarrow$ *an arc labeled* t *belongs to the graph.*
- *Transition* t *is live* $\Longrightarrow$ *an arc labeled* t *belongs each ergodic component of the graph.*
- *Reachable marking* $\boldsymbol{m}$ *is dead* $\Longleftarrow$ *node* $\boldsymbol{m}_\omega$ *in the graph has no output arc and* $\boldsymbol{m}_\omega \geq_\omega \boldsymbol{m}$.

Note, finally, that even if the coverability graph does not provide necessary and sufficient conditions for checking liveness, such a property is decidable. In fact checking for liveness can be reduced to a reachability problem [13]. Thus liveness of a net is a decidable property.

10.5 Further Reading

Further details on the proposed topics can be found in the survey paper by Murata [12] and on the books of Peterson [13] and David and Alla [1].

Finally, we address to the book of Girault and Valk [8] for a discussion on the effectiveness of model checking in the verification of the properties introduced in Section 10.4.

References

1. David, R., Alla, H.: Discrete, Continuous and Hybrid Petri Nets. Springer (2005)
2. Diestel, R.: Graph Theory, 4th edn. Springer (2010)
3. Desel, J., Esparza, J.: Free Choice Petri Nets. Cambridge University Press (1995)
4. Desrochers, A.A., Al-Jaar, R.Y.: Applications of Petri Nets in Manufacturing Systems. IEEE Press (1995)
5. Diaz, M. (ed.): Petri nets. Fundamental Models, Verification and Applications. John Wiley and Sons, Inc. (2009)
6. Di Cesare, F., Harhalakis, G., Proth, J.M., Silva, M., Vernadat, F.B.: Petri Net Synthesis for Discrete Event Control of Manufacturing Systems. Kluwer (1993)
7. Finkel, A., Johnen, C.: The home state problem in transition systems. In: Rapport de Recherche, vol. 471, Univ. de Paris-Sud., Centre d'Orsay (1989)
8. Girault, C., Valk, R.: Petri Nets for Systems Engineering. Springer (2003)
9. Hopcroft, J.E., Ullman, J.D.: Introduction to Automata Theory, Languages and Computation. Addison-Wesley (1979)
10. Karp, R., Miller, R.: Parallel program schemata. Journal of Computer and System Sciences 3(2), 147–195 (1969)
11. Martinez, J., Silva, M.: A simple and fast algorithm to obtain all invariants of a generalized Petri net. In: Informatik-Fachberichte: Application and Theory of Petri Nets, vol. 52. Springer (1982)
12. Murata, T.: Petri nets: properties, analysis and applications. Proceedings IEEE 77(4), 541–580 (1989)
13. Peterson, J.L.: Petri Net Theory and the Modeling of Systems. Prentice-Hall (1981)
14. Petri, C.A.: Kommunication Mit Automaten. Institut für Instrumentelle, Mathematik (Bonn, Germany), Schriften des IIM 3 (1962)
15. Reisig, W.: Petri Nets: An Introduction. EATCS Monographs on Theoretical Computer Science (1985)
16. Reutenauer, C.: Aspects Mathématiques des Réseaux de Petri. Prentice-Hall International (1990)

Chapter 11
Structural Analysis of Petri Nets

Maria Paola Cabasino, Alessandro Giua, and Carla Seatzu

11.1 Introduction

This chapter is devoted, as the previous one, to the presentation of background material on P/T nets. In particular, here the main focus is on *structural analysis* that consists in a set of algebraic tools that do not require the enumeration of the reachability set of a marked net but are based on the analysis of the state equation, on the incidence matrix, etc.

It is shown how the existence of some vectors, called P- and T-vectors, characterize the behavior of the net. In particular P-vectors enable to express some important constraints on the number of places in certain subsets of places, e.g., they can be constant, increasing or decreasing during the net evolution. On the contrary, T-vectors are related to transitions and express the effect of some sequences of transitions on the marking of the net, e.g., they can keep it unaltered, or make it increase or decrease.

The structural counterpart of several behavioral properties, that we had defined in the previous chapter, will also be defined. The properties we will consider are boundedness, conservativeness, repetitiveness, reversibility and liveness: in their structural form they are related to the net structure regardless of the initial marking. These properties will be characterized in terms of P- and T-vectors.

Subclasses of Petri nets are finally defined. Some of these classes pose some restrictions on the nature of physical systems they can model. However, this restricted modeling power often leads to simplified analysis criteria and this motivates the interest in these subclasses.

Maria Paola Cabasino · Alessandro Giua · Carla Seatzu
Department of Electrical and Electronic Engineering, University of Cagliari, Italy
e-mail: {cabasino,giua,seatzu}@diee.unica.it

C. Seatzu et al. (Eds.): Control of Discrete-Event Systems, LNCIS 433, pp. 213–233.
springerlink.com

11.2 Analysis via State Equation

In this section we present an approach to characterize marking reachability by means of the state equation of a net, that can be solved using integer programming techniques. The main limitation of this approach consists in the fact that in general it only provides necessary (but not sufficient) conditions for reachability. However, as discussed at the end of this chapter, there exist classes of nets (such as acyclic nets, state machines and marked graphs) for which the analysis based on the state equation provides necessary and sufficient conditions for reachability. Unfortunately, both marked graphs and state machines are very restricted classes of models.

Definition 10.5 introduced the notion of transition firing and allows one to compute the marking reached after the firing of an enabled transition through a simple matrix equation. Such a condition can be generalized to a sequence of transitions σ.

Definition 11.1. *Given a net N with set of transitions $T = \{t_1, t_2, \cdots, t_n\}$ and a sequence of transitions $\sigma \in T^*$, we call* **firing vector** *(or* **firing count vector***) of σ the vector*

$$\boldsymbol{\sigma} = [\sigma[t_1]\ \sigma[t_2]\ \cdots\ \sigma[t_n]]^T \in \mathbb{N}^n$$

whose entry $\sigma[t]$ denotes how many times transition t appears in sequence σ.

Consider the sequence $\sigma = t_1 t_2 t_2$ in the net in Fig. 11.1(a). Since t_1 appears once in σ, while t_2 appears twice, the firing vector of σ is $\boldsymbol{\sigma} = [1\ 2]^T$.

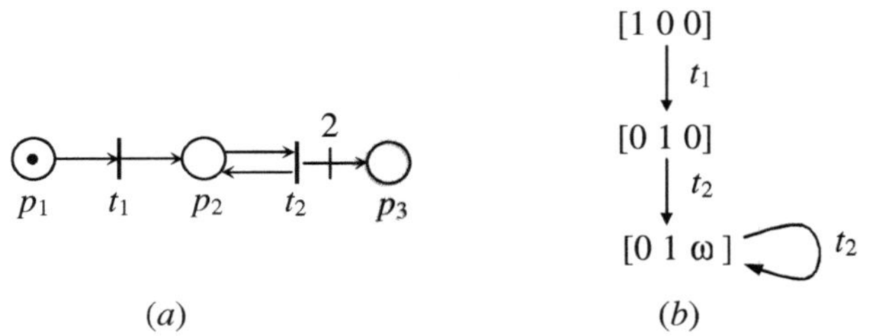

Fig. 11.1 (a) A marked PN; (b) its coverability graph

Proposition 11.1 (State equation). *Let $\langle N, \boldsymbol{m}_0 \rangle$ be a marked net and $\boldsymbol{C}$ its incidence matrix. If $\boldsymbol{m}$ is reachable from $\boldsymbol{m}_0$ firing σ it holds that*

$$\boldsymbol{m} = \boldsymbol{m}_0 + \boldsymbol{C} \cdot \boldsymbol{\sigma}. \tag{11.1}$$

Proof. Assume that $\boldsymbol{m}_0[\sigma\rangle\boldsymbol{m}$ with $\sigma = t_{j_1} t_{j_2} \cdots t_{j_r}$, i.e., $\boldsymbol{m}_0[t_{j_1}\rangle\boldsymbol{m}_1[t_{j_2}\rangle\boldsymbol{m}_2 \cdots [t_{j_r}\rangle\boldsymbol{m}_r$, with $\boldsymbol{m}_r = \boldsymbol{m}$. It holds that:

$$\boldsymbol{m} = \boldsymbol{m}_r = \boldsymbol{m}_{r-1} + \boldsymbol{C}[\cdot, t_{j_r}] = \cdots = \boldsymbol{m}_0 + \sum_{k=1}^{r} \boldsymbol{C}[\cdot, t_{j_k}] = \boldsymbol{m}_0 + \boldsymbol{C} \cdot \boldsymbol{\sigma}$$

and the state equation is satisfied. □

Consider the net system in Fig. 11.1(*a*). It is easy to verify that $\boldsymbol{m}' = [0\ 1\ 4]^T$, reachable from the initial marking $\boldsymbol{m}_0 = [1\ 0\ 0]^T$ firing $\sigma = t_1 t_2 t_2$, satisfies the equation $\boldsymbol{m}' = \boldsymbol{m}_0 + \boldsymbol{C} \cdot \boldsymbol{\sigma}$.

Definition 11.2. *Given a marked net $\langle N, \boldsymbol{m}_0 \rangle$ with m places and n transitions, let $\boldsymbol{C}$ be its incidence matrix. The* potentially reachable set *of $\langle N, \boldsymbol{m}_0 \rangle$ is the set*

$$PR(N, \boldsymbol{m}_0) = \{\boldsymbol{m} \in \mathbb{N}^m \mid \exists \boldsymbol{y} \in \mathbb{N}^n : \ \ \boldsymbol{m} = \boldsymbol{m}_0 + \boldsymbol{C} \cdot \boldsymbol{y}\},$$

i.e., the set of vectors $\boldsymbol{m} \in \mathbb{N}^m$ such that there exists a vector $\boldsymbol{y} \in \mathbb{N}^n$ that satisfies the state equation.

Proposition 11.2. [1] *Let $\langle N, \boldsymbol{m}_0 \rangle$ be a marked net. It is: $R(N, \boldsymbol{m}_0) \subseteq PR(N, \boldsymbol{m}_0)$.*

Proof. We simply need to prove that if $\boldsymbol{m}$ is reachable, then $\boldsymbol{m}$ is also potentially reachable. Indeed, if $\boldsymbol{m}$ is reachable, there exists a sequence σ such that $\boldsymbol{m}_0[\sigma\rangle\boldsymbol{m}$. Thus $\boldsymbol{m} = \boldsymbol{m}_0 + \boldsymbol{C} \cdot \boldsymbol{y}$ with $\boldsymbol{y} = \boldsymbol{\sigma}$, i.e., $\boldsymbol{m} \in PR(N, \boldsymbol{m}_0)$. □

We show by means of an example that the converse of Proposition 11.2 does not hold, i.e., it may happen that $R(N, \boldsymbol{m}_0) \subsetneq PR(N, \boldsymbol{m}_0)$.

Consider the marked net in Fig. 11.1(*a*) whose incidence matrix is $\boldsymbol{C} = [\ -1\ 0;\ 1\ 0;\ 0\ 2\ ]$. The initial marking is $\boldsymbol{m}_0 = [1\ 0\ 0]^T$. Let $\boldsymbol{m} = [1\ 0\ 2]^T$. Equation $\boldsymbol{m} = \boldsymbol{m}_0 + \boldsymbol{C} \cdot \boldsymbol{y}$ is verified by $\boldsymbol{y} = [0\ 1]^T$. However, $\sigma = t_2$ is not enabled at the initial marking and $\boldsymbol{m}$ is not reachable.

Definition 11.3. *Given a marked net $\langle N, \boldsymbol{m}_0 \rangle$, potentially reachable but not reachable markings are said to be* spurious markings.

The presence of spurious markings implies that in general the state equation analysis provides necessary, but not sufficient, conditions for reachability. Note however, that the necessary condition in Proposition 11.2 often allows to verify that a marking is not reachable. In the net in Fig. 11.1(*a*), consider the marking $\boldsymbol{m} = [0\ 2\ 0]^T$ and a generic vector $\boldsymbol{y} = [y_1\ y_2]^T \in \mathbb{N}^2$. The equation $\boldsymbol{m} = \boldsymbol{m}_0 + \boldsymbol{C} \cdot \boldsymbol{y}$ implies

$$\begin{array}{ccc} \boldsymbol{m} - \boldsymbol{m}_0 & = & \boldsymbol{C} \cdot \boldsymbol{y} \\ \begin{bmatrix} -1 \\ 2 \\ 0 \end{bmatrix} & = & \begin{bmatrix} -y_1 \\ y_1 \\ 2y_2 \end{bmatrix} \end{array}.$$

The first and second of such equalities require that: $y_1 = 1$ and $y_1 = 2$. Thus the equation does not admit solution and $\boldsymbol{m}$ is not reachable because it does not satisfy Proposition 11.2.

In the previous Chapter 10 we introduced the notion of *covering set* (cfr. Definition 10.12) that provides an approximation of the reachability set. In particular, by Proposition 10.3, it is $R(N, \boldsymbol{m}_0) \subseteq CS(N, \boldsymbol{m}_0)$. It is natural to wonder if an inclusion relationship exists between $PR(N, \boldsymbol{m}_0)$ and $CS(N, \boldsymbol{m}_0)$. The PN system in Fig. 11.1(*a*) shows that no relationship exists. Indeed, it is

$$CS(N, \boldsymbol{m}_0) = \{[1\ 0\ 0]^T\} \cup \{[0\ 1\ k]^T,\ k \in \mathbb{N}\}$$

as it can be easily verified looking at the coverability graph in Fig. 11.1(*b*). Moreover, it is

$$PR(N,\boldsymbol{m}_0) = \{[1\ 0\ 2k]^T,\ k \in \mathbb{N}\} \cup \{[0\ 1\ 2k]^T,\ k \in \mathbb{N}\}.$$

Therefore there exist markings that belong to $PR(N,\boldsymbol{m}_0)$, but do not belong to $CS(N,\boldsymbol{m}_0)$, and viz. As an example, given $\boldsymbol{m} = [0\ 1\ 3]^T$, it is $\boldsymbol{m} \in CS(N,\boldsymbol{m}_0)$ but $\boldsymbol{m} \notin PR(N,\boldsymbol{m}_0)$. On the contrary, if we consider $\boldsymbol{m}' = [1\ 0\ 2]^T$, it is $\boldsymbol{m}' \in PR(N,\boldsymbol{m}_0)$, but $\boldsymbol{m}' \notin CS(N,\boldsymbol{m}_0)$.

11.3 Analysis Based on the Incidence Matrix

11.3.1 Invariant Vectors

Definition 11.4. *Given a net N with m places and n transitions, let* $\boldsymbol{C}$ *be its incidence matrix. A P-vector*[1] $\boldsymbol{x} \in \mathbb{N}^m$ *with* $\boldsymbol{x} \neq \boldsymbol{0}$ *is called:*

- P-invariant*: if* $\boldsymbol{x}^T \cdot \boldsymbol{C} = \boldsymbol{0}^T$*;*
- P-increasing*: if* $\boldsymbol{x}^T \cdot \boldsymbol{C} \gneq \boldsymbol{0}^T$*;*
- P-decreasing*: if* $\boldsymbol{x}^T \cdot \boldsymbol{C} \lneq \boldsymbol{0}^T$*.*

A T-vector[2] $\boldsymbol{y} \in \mathbb{N}^n$ *with* $\boldsymbol{y} \neq \boldsymbol{0}$ *is called:*

- T-invariant*: if* $\boldsymbol{C} \cdot \boldsymbol{y} = \boldsymbol{0}$*;*
- T-increasing*: if* $\boldsymbol{C} \cdot \boldsymbol{y} \gneq \boldsymbol{0}$*;*
- T-decreasing*: if* $\boldsymbol{C} \cdot \boldsymbol{y} \lneq \boldsymbol{0}$.

Consider the nets in Fig. 11.2 (*a*) and (*b*) whose incidence matrices are respectively

$$\boldsymbol{C}_a = \begin{bmatrix} -1 & 2 \\ 1 & -1 \\ 0 & -1 \end{bmatrix} \quad \text{and} \quad \boldsymbol{C}_b = \boldsymbol{C}_a^T = \begin{bmatrix} -1 & 1 & 0 \\ 2 & -1 & -1 \end{bmatrix}.$$

For the net in figure (*a*) one readily verifies that vector $\boldsymbol{x}_I = [1\ 1\ 1]^T$ is a P-invariant, vector $\boldsymbol{x}_C = [1\ 1\ 0]^T$ is a P-increasing and vector $\boldsymbol{x}_D = [0\ 0\ 1]^T$ is a P-decreasing. For the net in figure (*b*) one readily verifies that vector $\boldsymbol{y}_I = [1\ 1\ 1]^T$ is a T-invariant, vector $\boldsymbol{y}_C = [1\ 1\ 0]^T$ is a T-increasing and vector $\boldsymbol{y}_D = [0\ 0\ 1]^T$ is a T-decreasing.

The following definition holds.

Definition 11.5. *The* support of a P-vector $\boldsymbol{x} \in \mathbb{N}^m$*, denoted* $\|\boldsymbol{x}\|$*, is the set of places* $p \in P$ *such that* $x[p] > 0$*. The* support of a T-vector $\boldsymbol{y} \in \mathbb{N}^n$*, denoted* $\|\boldsymbol{y}\|$*, is the set of transitions* $t \in T$ *such that* $y[t] > 0$.

As an example, for the net in Fig. 11.2(*a*) it is $\|\boldsymbol{x}_I\| = \{p_1, p_2, p_3\}$, $\|\boldsymbol{x}_C\| = \{p_1, p_2\}$ and $\|\boldsymbol{x}_D\| = \{p_3\}$.

[1] A P-vector can also be represented by a function $\boldsymbol{x} : P \rightarrow \mathbb{N}$.

[2] A T-vector can also be represented by a function $\boldsymbol{y} : T \rightarrow \mathbb{N}$.

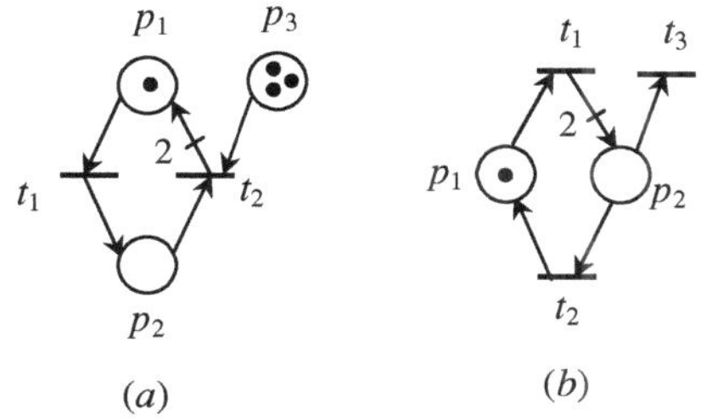

Fig. 11.2 Two PNs for the analysis via invariants

The following proposition allows to give a physical interpretation of P-vectors.

Proposition 11.3. *Given a net N, let $\boldsymbol{x}_I$ be a P-invariant, $\boldsymbol{x}_C$ a P-increasing, $\boldsymbol{x}_D$ a P-decreasing. For all markings $\boldsymbol{m} \in \mathbb{N}^m$ and for all enabled sequences $\sigma \in L(N,\boldsymbol{m})$ such that $\boldsymbol{m}[\sigma\rangle \boldsymbol{m}'$ it holds that*

$$\boldsymbol{x}_I^T \cdot \boldsymbol{m}' = \boldsymbol{x}_I^T \cdot \boldsymbol{m}; \qquad \boldsymbol{x}_C^T \cdot \boldsymbol{m}' \gneq \boldsymbol{x}_C^T \cdot \boldsymbol{m}; \qquad \boldsymbol{x}_D^T \cdot \boldsymbol{m}' \lneq \boldsymbol{x}_D^T \cdot \boldsymbol{m}. \tag{11.2}$$

Proof. If $\boldsymbol{m}'$ is reachable from $\boldsymbol{m}$ with the firing of sequence σ it holds that: $\boldsymbol{m}' = \boldsymbol{m} + \boldsymbol{C} \cdot \sigma$ and for any P-vector $\boldsymbol{x}$ it holds that $\boldsymbol{x}^T \cdot \boldsymbol{m}' = \boldsymbol{x}^T \cdot \boldsymbol{m} + \boldsymbol{x}^T \cdot \boldsymbol{C} \cdot \sigma$.

By definition of P-invariant $\boldsymbol{x}_I^T \cdot \boldsymbol{C} = \boldsymbol{0}^T$, i.e., $\boldsymbol{x}^T \cdot \boldsymbol{C} \cdot \sigma = 0$ hence the first result in (11.2) follows. By definition of P-increasing $\boldsymbol{x}_I^T \cdot \boldsymbol{C} \gneq \boldsymbol{0}^T$ and since $\sigma \gneq \boldsymbol{0}$ it holds that $\boldsymbol{x}^T \cdot \boldsymbol{C} \cdot \sigma \geq 0$, hence the second result in (11.2) follows. The last result can be proved in a similar fashion. □

Applying the result of the previous proposition to the net in Fig. 11.2 (a), where $\boldsymbol{m}_0 = [1\ 0\ 3]^T$, one concludes that for all reachable markings $\boldsymbol{m} \in R(N,\boldsymbol{m}_0)$:

- the sum of the tokens in the net remains constant and equal to 4 because the P-invariant $\boldsymbol{x}_I = [1\ 1\ 1]^T$ ensures that $m[p_1]+m[p_2]+m[p_3] = \boldsymbol{x}_I^T \cdot \boldsymbol{m} = \boldsymbol{x}_I^T \cdot \boldsymbol{m}_0 = 4$;
- the sum of the tokens in places p_1 and p_2 may increase starting from the initial value 1 but never decreases, because the P-increasing $\boldsymbol{x}_C = [1\ 1\ 0]^T$ ensures that $m[p_1]+m[p_2] = \boldsymbol{x}_C^T \cdot \boldsymbol{m} \geq \boldsymbol{x}_C^T \cdot \boldsymbol{m}_0 = 1$;
- the number of tokens in place p_3 may decrease starting from the initial value 3 but never increases, because the P-decreasing $\boldsymbol{x}_D = [0\ 0\ 1]^T$ ensures that $m[p_3] = \boldsymbol{x}_D^T \cdot \boldsymbol{m} \leq \boldsymbol{x}_D^T \cdot \boldsymbol{m}_0 = 3$.

The following proposition provides a physical interpretation of T-vectors.

Proposition 11.4. *Given a net N, let $\boldsymbol{m} \in \mathbb{N}^m$ be a marking and $\sigma \in L(N,\boldsymbol{m})$ be a firing sequence such that $\boldsymbol{m}[\sigma\rangle \boldsymbol{m}'$. The following properties hold:*

- *the firing vector σ is a T-invariant $\Longleftrightarrow \boldsymbol{m}' = \boldsymbol{m}$, i.e., sequence σ is repetitive and stationary;*
- *the firing vector σ is a T-increasing $\Longleftrightarrow \boldsymbol{m}' \gneq \boldsymbol{m}$, i.e., sequence σ is repetitive increasing;*
- *the firing vector σ is a T-decreasing $\Longleftrightarrow \boldsymbol{m}' \lneq \boldsymbol{m}$.*

Proof. The three properties can be easily proved considering the state equation $\boldsymbol{m}' = \boldsymbol{m} + \boldsymbol{C} \cdot \sigma$ and recalling the properties of T-vectors given in Definition 11.4. □

As an example, consider the net in Fig. 11.2 (*b*). Given the marking $\boldsymbol{m} = [1\ 0]^T$ shown in the figure, sequence $\sigma' = t_1 t_2$ is enabled and its firing yields marking $\boldsymbol{m}' = [1\ 1]^T \gneq \boldsymbol{m}$: the firing vector of σ' is the T-increasing $\boldsymbol{y}_C = [1\ 1\ 0]^T$. From the same marking $\boldsymbol{m}$, the firing of sequence $\sigma'' = t_1 t_2 t_3$ yields $\boldsymbol{m}'' = \boldsymbol{m}$: the firing vector of σ'' is the T-invariant $\boldsymbol{y}_I = [1\ 1\ 1]^T$.

Remark 11.1. *Many authors use a different terminology for* P-invariants *and* T-invariants, *and call them, respectively,* P-semiflow *and* T-semiflow. *In particular, this is the terminology used in the following Chapters 18 and 20. Moreover, in these chapters the term* invariant *is used to denote the token conservation law* $\boldsymbol{x}_I^T \cdot \boldsymbol{m}' = \boldsymbol{x}_I^T \cdot \boldsymbol{m}$ *in Proposition 11.3.* ■

11.3.2 P-Invariants Computation

Definition 11.6. *A P-invariant* $\boldsymbol{x} \in \mathbb{N}^m$ *is called:*

- minimal *if there does not exist a P-invariant* $\boldsymbol{x}'$ *such that* $\boldsymbol{x}' \lneq \boldsymbol{x}$*;*
- of minimal support *if there does not exist a P-invariant* $\boldsymbol{x}'$ *such that* $\| \boldsymbol{x}' \| \subsetneq \| \boldsymbol{x} \|$.

Analogous definitions hold for T-invariants.

As an example, vector $\boldsymbol{x}_I = [1\ 1\ 1]^T$ for the net in Fig. 11.2(*a*) is a minimal P-invariant that has also minimal support. Note however, that there may exist minimal invariants that do not have minimal support. As an example, in the net in Fig. 11.3, P-invariants $\boldsymbol{x}' = [1\ 2\ 0]^T$ and $\boldsymbol{x}'' = [1\ 0\ 2]^T$ are minimal and have minimal support; P-invariant $\boldsymbol{x}''' = 0.5(\boldsymbol{x}' + \boldsymbol{x}'') = [1\ 1\ 1]^T$ is minimal but does not have minimal support.

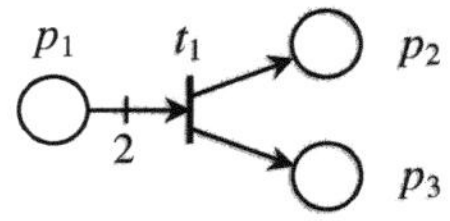

Fig. 11.3 A net with a minimal P-invariant that is not of minimal support

A non minimal or non minimal support P-invariant can always be obtained as the linear combination, with positive coefficients, of one or more minimal and minimal support P-invariants. As an example, given the net in Fig. 11.2(*a*) it holds that $\boldsymbol{x}''' = 0.5(\boldsymbol{x}' + \boldsymbol{x}'')$.

The following algorithm determines a set of P-invariants of a net. In particular, it computes all minimal P-invariants, but also many others that are not minimal (in general an exponential number of non minimal). Solutions for this issue can be found in [12].

Algorithm 11.2. (P-invariants computation). *Consider a net N with m places and n transitions and let $\boldsymbol{C}$ be its incidence matrix.*

1. *Compute the table $\boldsymbol{A} := \mid \boldsymbol{C} \mid \boldsymbol{I}_{m\times m} \mid$, where $\boldsymbol{I}_{m\times m}$ is the $m \times m$ identity matrix.*
2. *For $j := 1, \ldots, n$ (column index associated with transitions):*
   a. *let $J_+ := \{i \mid A[i,j] > 0\}$ be the set of row indices that correspond to positive entries in column j;*
   b. *let $J_- := \{i \mid A[i,j] < 0\}$ be the set of row indices that correspond to negative entries in column j;*
   c. *for each pair $(i_+, i_-) \in J_+ \times J_-$:*
      i. *let $d := lcm\{A[i_+,j], -A[i_-,j]\}$ be the least common multiplier of entries $A[i_+,j]$ and $-A[i_-,j]$;*
      ii. *let $d_+ := d/A[i_+,j]$ and $d_- := -d/A[i_-,j]$;*
      iii. *add the new row $d_+\ \boldsymbol{A}[i_+,\cdot] + d_-\ \boldsymbol{A}[i_-,\cdot]$ to the table (the new row has the $j-$th entry equal to zero by construction);*
   d. *remove from $\boldsymbol{A}$ all rows with index $J_+ \cup J_-$, corresponding to non-null elements along the $j-$th column.*
3. *The resulting table $\boldsymbol{A}$ is in the form $\boldsymbol{A} = \mid \boldsymbol{0}_{r\times m} \mid \boldsymbol{X}^T \mid$, where $\boldsymbol{0}_{r\times m}$ is a null $r \times n$ matrix, while $\boldsymbol{X}$ is a matrix with m rows and r columns. Each column of $\boldsymbol{X}$ is a P-invariant.*

Note that in the previous algorithm if any of the two sets J_+ or J_- is empty, at Step 2(c) no row is added. Moreover, the resulting table will be empty, i.e. $r = 0$, if the net N has no P-invariant.

Finally, note that it could be necessary to divide a column of $\boldsymbol{X}$ for the largest common divisor of its entries to obtain a P-invariant that is minimal.

Let us now present a simple example of application of such algorithm. Consider the net in Fig. 11.4.

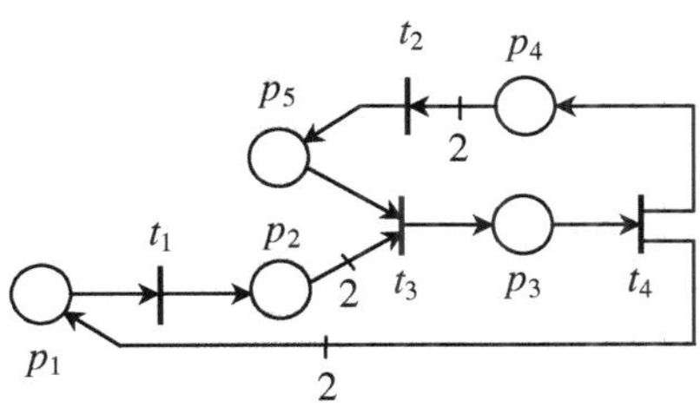

Fig. 11.4 A net for the computation of P-invariants

Initially construct the table

$$\left|\begin{array}{rrrr|ccccc|l} -1 & 0 & 0 & 2 & 1 & 0 & 0 & 0 & 0 & p_1 \\ 1 & 0 & -2 & 0 & 0 & 1 & 0 & 0 & 0 & p_2 \\ 0 & 0 & 1 & -1 & 0 & 0 & 1 & 0 & 0 & p_3 \\ 0 & -2 & 0 & 1 & 0 & 0 & 0 & 1 & 0 & p_4 \\ 0 & 1 & -1 & 0 & 0 & 0 & 0 & 0 & 1 & p_5 \end{array}\right.$$

where for a better understanding each row has been labeled with the corresponding place. At Step $j = 1$ the sum of rows p_1 and p_2 is computed and added to the table, while the two rows are removed, thus obtaining the table

$$\left|\begin{array}{rrrr|ccccc|l} 0 & 0 & 1 & -1 & 0 & 0 & 1 & 0 & 0 & p_3 \\ 0 & -2 & 0 & 1 & 0 & 0 & 0 & 1 & 0 & p_4 \\ 0 & 1 & -1 & 0 & 0 & 0 & 0 & 0 & 1 & p_5 \\ 0 & 0 & -2 & 2 & 1 & 1 & 0 & 0 & 0 & p_1+p_2 \end{array}\right.$$

At step $j = 2$ the linear combination of row p_4 with row p_5 multiplied by 2 is executed, and the two rows are removed, obtaining the table

$$\left|\begin{array}{rrrr|ccccc|l} 0 & 0 & 1 & -1 & 0 & 0 & 1 & 0 & 0 & p_3 \\ 0 & 0 & -2 & 2 & 1 & 1 & 0 & 0 & 0 & p_1+p_2 \\ 0 & 0 & -2 & 1 & 0 & 0 & 0 & 1 & 2 & p_4+2p_5 \end{array}\right.$$

At step $j = 3$ two combinations are computed and added to the table: the sum of row p_3 multiplied by 2 and either row $p_1 + p_2$ or row $p_4 + 2p_5$. Removing the three rows, we get the table

$$\left|\begin{array}{rrrr|ccccc|l} 0 & 0 & 0 & 0 & 1 & 1 & 2 & 0 & 0 & p_1+p_2+2p_3 \\ 0 & 0 & 0 & -1 & 0 & 0 & 2 & 1 & 2 & 2p_3+p_4+2p_5 \end{array}\right.$$

At Step $j = 4$ there are no possible combinations and we simply remove row $2p_3 + p_4 + 2p_5$ that has a non-null entry in the fourth column. The resulting table is

$$\left|\begin{array}{rrrr|ccccc|l} 0 & 0 & 0 & 0 & 1 & 1 & 2 & 0 & 0 & p_1+p_2+2p_3 \end{array}\right.$$

Thus the net has a single minimal and minimum support P-invariant $\boldsymbol{x} = [1\ 1\ 2\ 0\ 0]^T$.

From the definition of P-invariants and T-invariants, one can readily see that the T-invariants of a net N with incidence matrix $\boldsymbol{C}$ are the P-invariants of the *dual net*[3] with incidence matrix $\boldsymbol{C}^T$, and viz. Thus Algorithm 11.2 can also be used to compute the T-invariants initializing the table as $\boldsymbol{A} := \mid \boldsymbol{C}^T \mid \boldsymbol{I}_{n\times n} \mid$. The i–th row in such a case should be labeled by transition t_i.

It is also important to observe that the previous algorithm can be modified to determine increasing (or decreasing) P-vectors: at Step 2(d) rather than eliminating all the rows in $\mathscr{I}_+ \cup \mathscr{I}_-$ only the rows with index $\mathscr{I}_-$ (or $\mathscr{I}_+$) should be removed

[3] See Definition 11.19 for a formal definition of duality.

since if $\boldsymbol{x}$ is an increasing (decreasing) vector positive (negative) entries in the product $\boldsymbol{z}^T = \boldsymbol{x} \cdot \boldsymbol{C}$ are allowed. However, in general, when the algorithm is applied for the computation of increasing or decreasing vectors, the resulting vectors are not only the minimal ones or those of minimum support, but many others as well that can be obtained as a linear combination of them.

11.3.3 Reachability Analysis Using P-Invariants

In this section we discuss how P-invariants can be used to approximate the reachability set of a marked net.

Definition 11.7. [1] *Let $\langle N, \boldsymbol{m}_0 \rangle$ be a net with m places and $\boldsymbol{X} = [\boldsymbol{x}_1\ \boldsymbol{x}_2\ \cdots\ \boldsymbol{x}_k]$ a matrix $m \times k$, whose generic column $\boldsymbol{x}_i$ is a P-invariant of N.*

*The $X-$*invariant set *of $\langle N, \boldsymbol{m}_0 \rangle$ is the set*

$$I_X(N, \boldsymbol{m}_0) = \{\boldsymbol{m} \in \mathbb{N}^m \mid \boldsymbol{X}^T \cdot \boldsymbol{m} = \boldsymbol{X}^T \cdot \boldsymbol{m}_0\},$$

i.e., the set of vectors $\boldsymbol{m} \in \mathbb{N}^m$ such that $\boldsymbol{x}_i^T \cdot \boldsymbol{m} = \boldsymbol{x}_i^T \cdot \boldsymbol{m}_0$ for all $i \in \{1, \ldots, k\}$.

It is easy to prove that the potentially reachable set of a marked net is contained in the $X-$invariant set for any matrix of P-invariants $\boldsymbol{X}$. The following proposition extends the results of Proposition 11.2.

Proposition 11.5. [1] *Let $\langle N, \boldsymbol{m}_0 \rangle$ be a marked net and $\boldsymbol{X}$ a matrix whose columns are P-invariants of N. It holds: $R(N, \boldsymbol{m}_0) \subseteq PR(N, \boldsymbol{m}_0) \subseteq I_X(N, \boldsymbol{m}_0)$.*

Proof. The first inequality derives from Proposition 11.2. It is sufficient to prove that if $\boldsymbol{m}$ is potentially reachable, then $\boldsymbol{m}$ is also $X-$invariant. In fact, if $\boldsymbol{m}$ is potentially reachable, there exists a vector $\boldsymbol{y} \in \mathbb{N}^n$ such that $\boldsymbol{m} = \boldsymbol{m}_0 + \boldsymbol{C} \cdot \boldsymbol{y}$. Thus $\boldsymbol{X}^T \cdot \boldsymbol{m} = \boldsymbol{X}^T \cdot \boldsymbol{m}_0 + \boldsymbol{X}^T \cdot \boldsymbol{C} \cdot \boldsymbol{y}$. Being $\boldsymbol{X}$ a matrix whose columns are P-invariants of N it holds that $\boldsymbol{X}^T \cdot \boldsymbol{C} \cdot \boldsymbol{y} = \boldsymbol{0}$, thus $\boldsymbol{m} \in I_X(N, \boldsymbol{m}_0)$. □

Let us now present an example where it is $R(N, \boldsymbol{m}_0) \subset PR(N, \boldsymbol{m}_0) \subset I_X(N, \boldsymbol{m}_0)$, i.e., where strict inclusion holds. Consider again the net in Fig. 11.1(a). There exists only a minimal P-invariant $\boldsymbol{x} = [1\ 1\ 0]^T$, thus let $\boldsymbol{X} = \boldsymbol{x}$. Moreover,

$$R(N, \boldsymbol{m}_0) = \{\ [1\ 0\ 0]^T\} \cup \{[0\ 1\ 2k]^T,\ k \in \mathbb{N}\ \},$$

and, as already discussed in Section 11.2,

$$PR(N, \boldsymbol{m}_0) = \{\ [1\ 0\ 2k]^T,\ k \in \mathbb{N}\ \} \cup \{\ [0\ 1\ 2k]^T,\ k \in \mathbb{N}\ \} \subset R(N, \boldsymbol{m}_0).$$

A generic $\boldsymbol{m}$ is an $X-$invariant only if:

$$\begin{array}{ccccccc} \boldsymbol{X}^T & \cdot & \boldsymbol{m} & = & \boldsymbol{X}^T & \cdot & \boldsymbol{m}_0 \\ [1\ 1\ 0] & \cdot & \begin{bmatrix} m[p_1] \\ m[p_2] \\ m[p_3] \end{bmatrix} & = & [1\ 1\ 0] & \cdot & \begin{bmatrix} 1 \\ 0 \\ 0 \end{bmatrix} \end{array}$$

or equivalently $m[p_1] + m[p_2] = 1$. Moreover, since each $m[p_i]$ should be a non-negative integer, it holds that:

$$I_X(N,\boldsymbol{m}_0) = \{[1\ 0\ k]^T,\ k \in \mathbb{N}\} \cup \{\ [0\ 1\ k]^T,\ k \in \mathbb{N}\},$$

i.e., it is $PR(N,\boldsymbol{m}_0) \subset I_X(N,\boldsymbol{m}_0)$.

We conclude this section extending the result in Proposition 10.3.

Proposition 11.6. *Let $\langle N,\boldsymbol{m}_0\rangle$ be a marked net and $\boldsymbol{X}$ a matrix whose columns are P-invariants of N. It holds that: $R(N,\boldsymbol{m}_0) \subseteq CS(N,\boldsymbol{m}_0) \subseteq I_X(N,\boldsymbol{m}_0)$.*

Proof. The first inequality derives from Proposition 10.3. It is sufficient to prove that $\boldsymbol{m} \in CS(N,\boldsymbol{m}_0)$ implies that $\boldsymbol{m}$ is also $X-$invariant. To this aim we first observe that the support of no P-invariant contains unbounded places. Therefore, if p is an unbounded place, for all marking $\boldsymbol{m} \in I_x(N,\boldsymbol{m}_0)$, $m[p]$ may take any value in $\mathbb{N}$. This implies that the statement $\boldsymbol{m} \in CS(N,\boldsymbol{m}_0) \Rightarrow \boldsymbol{m} \in I_X(N,\boldsymbol{m}_0)$ may be at most violated by bounded places. However, either of this cannot occur since bounded places in the coverability graph only contain reachable markings being by Proposition 11.5, $R(N,\boldsymbol{m}_0) \subseteq I_X(N,\boldsymbol{m}_0)$. □

11.4 Structural Properties

In this section we define meaningful *structural properties* of a P/T net. In Section 10.4.3 of the previous chapter we defined several behavioral properties of a marked net $\langle N,\boldsymbol{m}_0\rangle$, such as boundedness, conservativeness, repetitiveness, reversibility and liveness. In this section we define the structural counterpart of these properties, relating them to the net structure independent of a particular initial marking. We also show how such properties can be verified using P-vectors and T-vectors.

Most of the proofs of the results presented in this section are omitted; the interested reader is addressed to [13].

11.4.1 Structural Boundedness

This property implies boundedness for all initial markings.

Definition 11.8. *Consider a P/T net N and one of its places $p \in P$.*

- *Place p is* structurally bounded *if it is bounded in $\langle N,\boldsymbol{m}_0\rangle$ for all initial markings $\boldsymbol{m}_0$.*
- *Net N is* structurally bounded *if the marked net $\langle N,\boldsymbol{m}_0\rangle$ is bounded for all initial markings $\boldsymbol{m}_0$.*

This property can be characterized in terms of P-vectors .

Proposition 11.7. *Consider a P/T net N and a place $p \in P$.*

1. *Place p is structurally bounded if and only if there exists a P-invariant or a P-decreasing $\boldsymbol{x}$ with $x[p] > 0$, i.e., $p \in \|\boldsymbol{x}\|$.*
2. *The net N is structurally bounded if and only if there exists a P-invariant or a P-decreasing of positive integers $\boldsymbol{x} \in \mathbb{N}_+^m$, i.e., $P = \|\boldsymbol{x}\|$.*

As an example, the net in Fig. 11.1(a) has P-invariant $[1\ 1\ 0]^T$. Thus places p_1 and p_2 are structurally bounded while place p_3 is not structurally bounded. Therefore the net is not structurally bounded.

Structural boundedness implies (behavioral) boundedness, since it holds for all initial markings. The opposite is not true: consider the net in Fig. 11.1(a) with initial marking $\boldsymbol{m}_0 = [0\ 0\ r]^T$ (for all $r \in \mathbb{N}$). The resulting marked net is bounded, even if it is not structurally bounded: indeed in such a net only places p_1 and p_2 are structurally bounded.

11.4.2 Structural Conservativeness

Definition 11.9. *Consider a P/T net N and one of its places $p \in P$.*

- *Net N is* structurally strictly conservative *if the marked net $\langle N, \boldsymbol{m}_0 \rangle$ is strictly conservative for all initial markings $\boldsymbol{m}_0$.*
- *Net N is* structurally conservative *if the marked net $\langle N, \boldsymbol{m}_0 \rangle$ is conservative for all initial markings $\boldsymbol{m}_0$.*

This property can be characterized in terms of P-invariants.

Proposition 11.8. *Consider a P/T net N.*

- *N is structurally strictly conservative if and only if vector $\mathbf{1} = \{1\}^m$ is a P-invariant.*
- *N is structurally conservative if and only if there exists a P-invariant of positive integers $\boldsymbol{x} \in \mathbb{N}_+^m$, i.e., a P-invariant whose support contains all places.*

Structural conservativeness implies (behavioral) conservativeness, since it holds for all initial markings. The opposite is not true. As an example, the net in Fig. 10.12(*a*) in Chapter 10 is conservative with respect to the vector $[1\ 1]^T$ for the given initial marking, but such a net admits no P-invariant, thus it is not structurally conservative. It is easy to see that the net system $\langle N, \boldsymbol{m}_0 \rangle$ is not conservative if $m_0[p_1] \geq 1$ and $m_0[p_1] + m_0[p_2] \geq 2$, since in such a case transition t_4 would be almost-live and its firing would decrease the number of tokens in the net.

Let us finally observe, as already discussed for the corresponding behavioral properties, that structural conservativeness implies structural boundedness while the opposite is not true. The net in Fig. 10.10 (Chapter 10) is an example of a structurally bounded net being vector $[1]$ a P-decreasing, while it is not structurally conservative since it admits no P-invariant.

11.4.3 Structural Repetitiveness and Consistency

These properties are the structural counterpart of repetitiveness and stationarity introduced for sequences of transitions.

Proposition 11.9. *Consider a P/T net N.*

- *N is* repetitive *if there exists an initial marking* $\boldsymbol{m}_0$ *such that* $\langle N, \boldsymbol{m}_0 \rangle$ *admits a repetitive sequence containing all transitions.*
- *N is* consistent *if there exists an initial marking* $\boldsymbol{m}_0$ *such that* $\langle N, \boldsymbol{m}_0 \rangle$ *admits a repetitive stationary sequence containing all transitions.*

These properties can be characterized in terms of T-vectors.

Proposition 11.10. *Let N be a P/T net.*

- *N is repetitive if and only if it admits either a T-invariant or a T-increasing of strictly positive integers* $\boldsymbol{y} \in \mathbb{N}^n_+$, *i.e., a T-increasing whose support contains all transitions.*
- *N is consistent if and only if it admits a T-invariant of strictly positive integers* $\boldsymbol{y} \in \mathbb{N}^n_+$, *i.e., a T-invariant whose support contains all transitions.*

As an example the net in Fig. 11.8(b) is repetitive and consistent. If such a net is modified assuming that the multiplicity of the arc from t' to p' is equal to 2, the resulting net is repetitive but not consistent.

11.4.4 Structural Liveness

Definition 11.10. *A P/T net N is* structurally live *if there exists an initial marking* $\boldsymbol{m}_0$ *such that the marked net* $\langle N, \boldsymbol{m}_0 \rangle$ *is live.*

It is possible to give a necessary condition for such property.

Proposition 11.11. *A P/T net N with incidence matrix* $\boldsymbol{C}$ *is structurally live only if it does not admit a P-decreasing .*

As an example, the net in Fig. 11.2(*a*) has a P-decreasing $\boldsymbol{x}_D = [0\ 0\ 1]^T$, i.e., the number of tokens in p_3 can never increase. This net is structurally dead: regardless of the initial marking, each time t_2 fires the number of tokens in p_3 decreases and when the place gets empty, t_2 becomes dead.

11.5 Implicit Places

We now introduce the notion of *implicit place* that is useful in a large variety of problems, such as simulation, performance evaluation and deadlock analysis.

Definition 11.11. [21] *Let* $\langle N, \boldsymbol{m}_0 \rangle$ *be a PN system with* $N = (P \cup \{p\}, T, \boldsymbol{Pre}, \boldsymbol{Post})$. *Place* p *is* implicit *if and only if* $L(N, \boldsymbol{m}_0) = L(N', \boldsymbol{m}'_0)$ *where* $N' = (P, T, \boldsymbol{Pre}[P,T], \boldsymbol{Post}[P,T])$ *is the restriction of* N *to* P, *i.e., the net obtained from* N *removing* p *and its input and output arcs, and* $\boldsymbol{m}'_0 = \boldsymbol{m}_0[P]$ *is the projection of* $\boldsymbol{m}_0$ *on* P.

In simple words, a place p of a marked PN is said to be *implicit* if deleting it does not change the "behavior" of the marked net, i.e., the language it can generate. Therefore a place p is implicit if there does not exist a marking $\boldsymbol{m} \in R(N, \boldsymbol{m}_0)$ and a transition $t \in T$ such that $\boldsymbol{m}[P] \geq \boldsymbol{Pre}[P,t]$ and $m[p] < Pre[p,t]$.

As an example, both places p_1 and p_2 in Fig. 11.5(a) are implicit places.

The following theorem provides a sufficient condition for a place p to be implicit.

Theorem 11.1. [21] *Let* $\langle N, \boldsymbol{m}_0 \rangle$ *be a PN system with* $N = (P \cup \{p\}, T, \boldsymbol{Pre}, \boldsymbol{Post})$. *Let*

$$\begin{aligned} \gamma^* = \min \{ \boldsymbol{y}^T \cdot \boldsymbol{m}_0[P] + \mu \mid\ & \boldsymbol{y}^T \cdot \boldsymbol{C}[P,T] \leq \boldsymbol{C}[p,T] \\ & \boldsymbol{y}^T \cdot \boldsymbol{Pre}[P, p^\bullet] + \mu \cdot \boldsymbol{1}^T \geq \boldsymbol{Pre}[p, p^\bullet] \\ & \boldsymbol{y} \geq \boldsymbol{0},\ \mu \geq 0 \}. \end{aligned} \tag{11.3}$$

If $m_0[p] \geq \gamma^*$, *then* p *is implicit.*

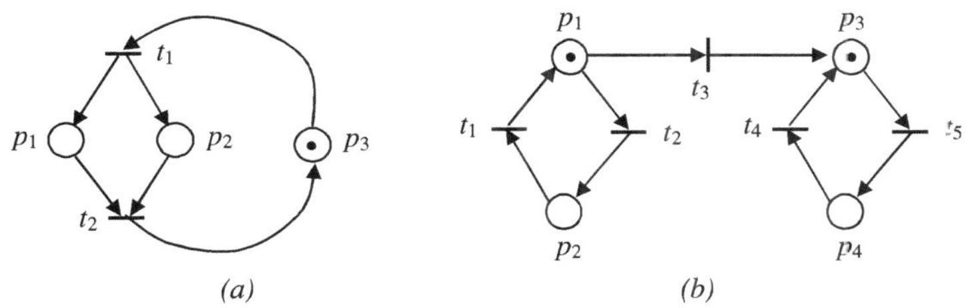

Fig. 11.5 (a) A PN with two implicit places; (b) a PN with a siphon and a trap

If a place p can be made implicit for every possible initial marking then it is called *structurally implicit.*

Definition 11.12. [21] *Let* $N = (P \cup \{p\}, T, \boldsymbol{Pre}, \boldsymbol{Post})$ *be a PN. Place* p *is* structurally implicit *if for every* $\boldsymbol{m}_0[P]$, *there exists a* $m_0[p]$ *such that* p *is implicit in* $\langle N, \boldsymbol{m}_0[P \cup \{p\}] \rangle$.

Both places p_1 and p_2 in Fig. 11.5(a) are also structurally implicit. An example of a place that is implicit but not structurally implicit is given in [21].

A characterization of structurally implicit places is the following:

Theorem 11.2. [21] *Let* $N = \langle P \cup \{p\}, T, \boldsymbol{Pre}, \boldsymbol{Post} \rangle$. *Place* p *is* structurally implicit *iff (equivalently):*

- $\exists\, \boldsymbol{y} > \boldsymbol{1}$ *such that* $\boldsymbol{C}[p,T] \geq \boldsymbol{y}^T \cdot \boldsymbol{C}[P,T]$.
- $\nexists\, \boldsymbol{x} \geq \boldsymbol{1}$ *such that* $\boldsymbol{C}[P,T] \cdot \boldsymbol{x} \geq \boldsymbol{0}$ *and* $\boldsymbol{C}[p,T] < \boldsymbol{0}$.

Structurally implicit places are very important when performing structural analysis. Many approaches in this framework are based on either the addition or the removal of structurally implicit places. Indeed, the *addition* of structurally implicit places can: (i) increase the Hamming distance useful for error/fault detecting and error/fault correcting codes [19]; (ii) decompose the system for computing performance evaluation (divide and conquer techniques) [22]; (iii) decompose the system for decentralized control [25]; (iv) cut spurious (deadlock) solutions improving the characterization of the state equation [1]. On the contrary, the *removal* of structurally implicit places can: (i) simplify the implementation having less nodes and (ii) improve the simulation of continuous PNs under infinite server semantics [16] .

Note that other restrictions of the concept of implicit place have been proposed in the literature [21], e.g., that of *concurrent implicit place* that is particularly useful in performance evaluation or control of timed models. For a detailed study on this, we address to [1].

11.6 Siphons and Traps

Let us now introduce two structural complementary objects, namely *siphons* and *traps*. Siphons and traps are usually introduced for ordinary nets and most of the literature dealing with them refers to such a restricted class of nets. However, some authors extended their definitions to non ordinary nets. See e.g. [23] and the references therein.

Definition 11.13. *A* siphon *of an ordinary PN is a set of places $S \subseteq P$ such that the set of input transitions of S is included in the set of output transitions of S, i.e.,*

$$\bigcup_{p \in S} {}^\bullet p \subseteq \bigcup_{p \in S} p^\bullet .$$

A siphon is minimal *if it is not the superset of any other siphon.*

Definition 11.14. *A* trap *of an ordinary PN is a set of places $S \subseteq P$ such that the set of output transitions of S is included in the set of input transitions of S, i.e.,*

$$\bigcup_{p \in S} p^\bullet \subseteq \bigcup_{p \in S} {}^\bullet p .$$

A trap is minimal *if it is not the superset of any other trap.*

The main interest on siphons and traps derives from the following two considerations. Once a siphon becomes empty, it remains empty during all the future evolutions of the net. Once a trap becomes marked, it remains marked during all the future evolutions of the net.

Consider the net in Fig. 11.5(b). The set $S = \{p_1, p_2\}$ is a siphon. The token initially in p_2 may move to p_1 and again to p_2, through the firing of t_1 and t_2, respectively. However, once t_3 fires, S becomes empty and remains empty during all future evolutions of the net. On the contrary, $S' = \{p_3, p_4\}$ is a trap. Once a token enters in p_3, it can only move to p_4 and to p_3 again, but it will never leave S'.

Traps and siphons have been extensively used for the structural analysis of PNs. Following [21], here we limit to mention three of the most significant results in this context.

- In an ordinary deadlocked system, the subset of unmarked places is a siphon, otherwise one of its input transitions would be enabled.
- Taking into account that traps remain marked, if every siphon of an ordinary net contains an initially marked trap, then the system is deadlock-free.
- If $\boldsymbol{m}$ is a home state of a live system, then every trap must be marked, otherwise once the trap becomes marked — and it will eventually do by liveness — $\boldsymbol{m}$ cannot be reached any more.

Linear algebraic characterizations of siphons and traps have been given since the nineties [21] and can be considered as the starting point for extensive and fruitful theories on liveness analysis and deadlock prevention, particularly in the case of some net subclasses [4, 5, 6, 11, 24].

11.7 Classes of P/T Nets

The P/T net definition given in the previous chapter corresponds to a model sometimes called *general P/T net*. In this section we present some classes of P/T nets that satisfy particular structural conditions. In particular, the considered classes are summarized in the Venn diagram in Fig. 11.6.

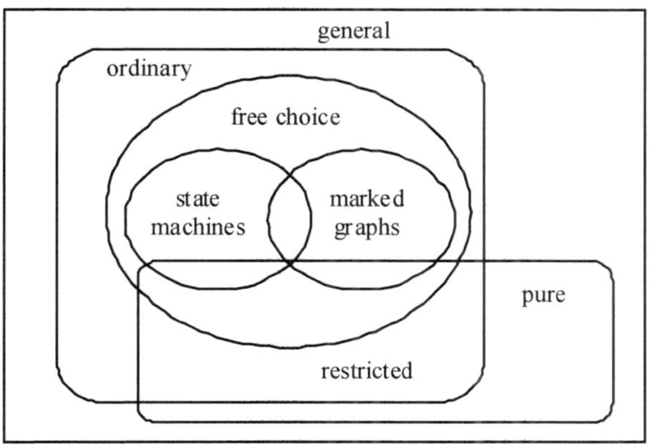

Fig. 11.6 Venn diagram of the different classes of P/T nets

11.7.1 Ordinary and Pure Nets

Definition 11.15. *A P/T net* $N = (P, T, \boldsymbol{Pre}, \boldsymbol{Post})$ *is called:*

- ordinary *if* $\boldsymbol{Pre} : P \times T \longrightarrow \{0,1\}$ *and* $\boldsymbol{Post} : P \times T \longrightarrow \{0,1\}$, *i.e., if all arcs have unitary multiplicity;*
- pure *if for each place p and transition t it holds that* $Pre[p,t] \cdot Post[p,t] = 0$, *i.e., if the net has no self-loop;*
- restricted *if it is ordinary and pure.*

Even if the definition of restricted net seems more restrictive than that of a general P/T net, it is possible to prove that the two formalisms have the same modeling power, in the sense that they can describe the same class of systems.

In the following we present a rather intuitive construction to convert a general net into an equivalent[4] restricted one, by removing arcs with multiplicity greater than one and self-loops.

The first construction, shown in Fig. 11.7, removes arcs with a multiplicity greater than one. Let r be the maximum multiplicity among all "pre" and "post" arcs incident on place p. We replace p with a cycle of r places $p^{(i)}$ and transitions $t^{(i)}$, $i = 1, \ldots, r$, as in the figure. Each arc "post" ("pre") with multiplicity $k \leq r$ is replaced by k arcs, each one directed (coming from) k different places. Disregarding the firing of transitions $t^{(i)}$ and keeping into account that for each marking $\boldsymbol{m}$ of the original net and a corresponding marking $\boldsymbol{m}'$ of the transformed net it holds that $m[p] = \Sigma_{i=1}^{r} m'[p^{(i)}]$, the two nets have the same behavior.

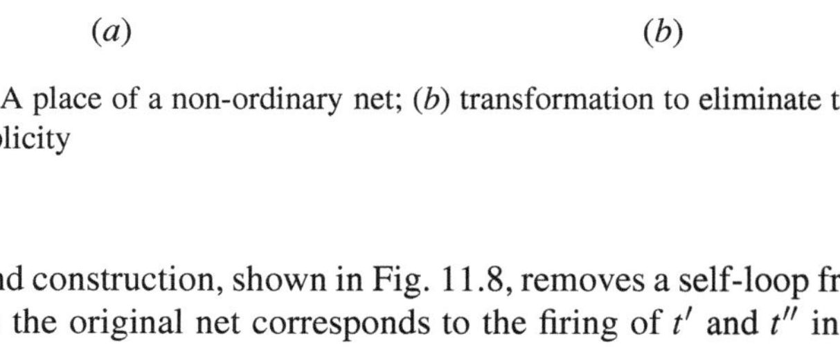

Fig. 11.7 (*a*) A place of a non-ordinary net; (*b*) transformation to eliminate the arcs of non-unitary multiplicity

The second construction, shown in Fig. 11.8, removes a self-loop from t to p: the firing of t in the original net corresponds to the firing of t' and t'' in the modified net.

---

[4] Here the term *equivalent* is used in a purely qualitative fashion: a formal discussion of model equivalence (e.g., in terms of languages, bisimulation, etc.) goes beyond the scope of this book.

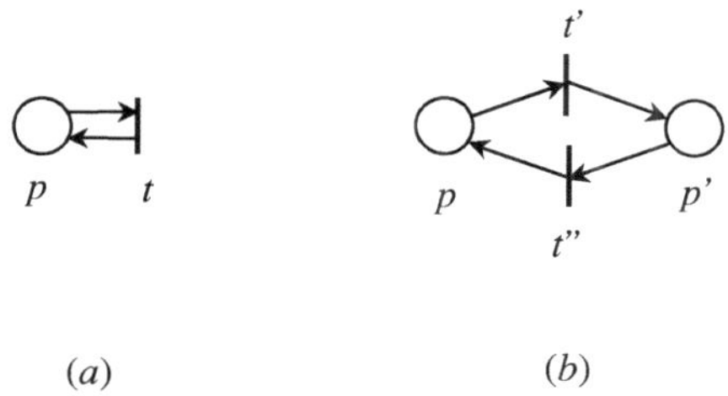

Fig. 11.8 (*a*) A self-loop; (*b*) transformation to remove the self-loop

11.7.2 Acyclic Nets

Definition 11.16. *A P/T net is* acyclic *if its underlying graph is acyclic, i.e., it does not contain directed cycles.*

The net in Fig. 11.3 is acyclic. On the contrary, the net in Fig. 11.1 is not acyclic because it contains the self-loop $p_2 t_2 p_2$. Analogously, the nets in Fig. 11.2 are not acyclic because they both contain cycle $p_1 t_1 p_2 t_2 p_1$.

The main feature of acyclic nets is that their state equation has no spurious solutions.

Proposition 11.12. [13] *Let N be an acyclic net. For all initial markings $\boldsymbol{m}_0$, it holds that $R(N,\boldsymbol{m}_0) = PR(N,\boldsymbol{m}_0)$.*

Therefore for this class of nets, the analysis based on the state equation provides necessary and sufficient conditions to solve the reachability problem.

11.7.3 State Machines

Definition 11.17. *A* state machine *is an ordinary net whose transitions have exactly one input and one output arc, i.e., it holds that $\sum_{p \in P} Pre[p,t] = \sum_{p \in P} Post[p,t] = 1$ for all transitions $t \in T$.*

The net in Fig. 11.9(*a*) is a state machine, while all the other nets in the same figure are not.

A state machine with a single token is analogous to a finite state automaton: each place of the net corresponds to a state of the automaton and the position of the token denotes the current state of the automaton. Since each place can have more than one output transition, as place p_1 in Fig. 11.9(*a*), state machines can model a "choice".

The initial marking can also assign to a state machine a number of tokens greater than one. In such a case it is possible to represent a limited form of "parallelism", that originates from the firing of transitions enabled by different tokens. On the

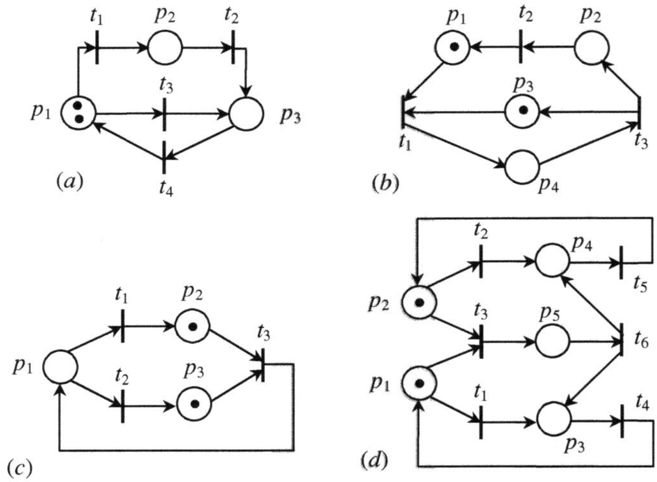

Fig. 11.9 Ordinary PNs: (*a*) a state machine; (*b*) a marked graph; (*c*) a free-choice net; (*d*) a non free-choice net

contrary, it is not possible to model a "synchronization" since the enabling of a transition depends on a single place.

The particular structure of state machines leads to a more restrictive model than ordinary nets, in the sense that it is not possible in general to find a state machine that is equivalent to an arbitrary ordinary net. Nevertheless, such restriction allows to significantly simplify the study of their properties [14, 17]. In particular, the following two important results can be proved.

- A state machine is always bounded.
- If a state machine is connected (but not necessarily strictly connected), then for all initial markings $\boldsymbol{m}_0$, it holds that $R(N,\boldsymbol{m}_0) = PR(N,\boldsymbol{m}_0)$, i.e., a marking is reachable if and only if it is potentially reachable.

11.7.4 Marked Graphs

Definition 11.18. *A* marked graph, *also called* marked event *(or* synchronization*)* graph, *is an ordinary net whose places have exactly one input and one output transition, i.e.,* $\sum_{t\in T} Pre[p,t] = \sum_{t\in T} Post[p,t] = 1$ *for all places* $p \in P$.

The net in Fig. 11.9(*b*) is a marked graph, while all the other nets in the same figure are not.

Since each place of a marked graph has a single output transition, this structure cannot model a "choice". However, it can model "parallelism" because a transition may have more than one output place. Moreover, it can model a "synchronization" since the enabling state of a transition can depend on several places; such is the case of transition t_1 in Fig. 11.9(b).

There exists a dual relation between state machines and marked graphs.

Definition 11.19. *Given a net $N = (P,T,\boldsymbol{Pre},\boldsymbol{Post})$ with m places and n transitions, the* dual *net of N is the net $N^T = (T,P,\boldsymbol{Pre}^T,\boldsymbol{Post}^T)$ with n places and m transitions.*

The dual net can be obtained from the original one simply replacing each node "place" with a node "transition" and viz., and inverting the directions of arcs. If the original net has incidence matrix $\boldsymbol{C}$, the dual net has incidence matrix $\boldsymbol{C}^T$. Moreover, if N' is the dual net of N, then N is the dual net of N'.

The two nets in Fig. 11.9(a)-(b) are one the dual net of the other.

Proposition 11.13. *If N is a state machine (resp., marked graph) its dual net N^T is a marked graph (resp., state machine).*

Proof. The construction of the dual net transforms each node "place" into a node "transition" and viz., and does not change the multiplicity of the arcs but only their orientation. Thus, if N is a state machine each transition has a single input place and a single output place and in N^T each place has a single input transition and a single output transition, thus the resulting net is a marked graph. A similar reasoning applies if N is a marked graph. □

As in the case of state machines, also for marked graphs some important properties can be proved. For a detailed discussion on this we address to [14].

11.7.5 Choice-Free Nets

A generalization of state machines and marked graphs is the following.

Definition 11.20. *A* free-choice net *is an ordinary net such that from $p \in P$ to $t \in T$ is either the single output arc from p or the single input arc to t, i.e., for all places $p \in P$ if $Pre[p,t] = 1$ it holds that:*

$$[\forall t' \neq t : Pre[p,t'] = 0] \vee [\forall p' \neq p : Pre[p',t] = 0].$$

The admissible structures in a free-choice net are shown in Fig. 11.10(a)-(b) while a non-admissible structure is shown in Fig. 11.10(c).

A free-choice net thus can model "choice", "parallelism" and "synchronization". As an example, in the free-choice net in Fig. 11.9(c) transitions t_1 and t_2 are in structural conflict and model a choice in place p_1; t_3 models a synchronization.

However, a free-choice net cannot represent "choice" and "synchronization" relatively to the same transition. As an example, the net in Fig. 11.9(d) is not a free-choice net since transition t_3 models a synchronization and at the same time one of the admissible choices for place p_1.

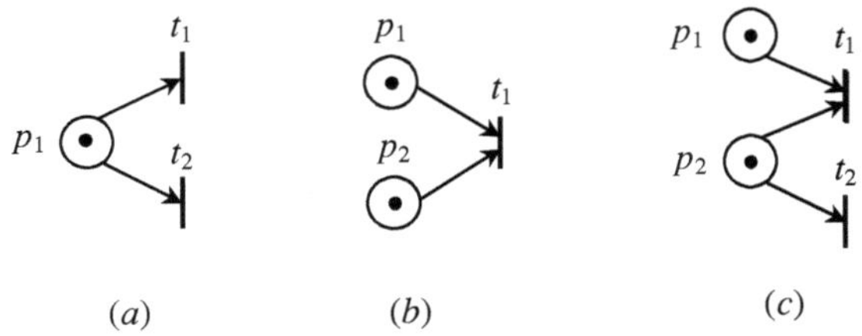

Fig. 11.10 Elementary structures: (a)-(b) free-choice; (c) not free-choice

Note that each state machine and each marked graph are also free-choice net. However, the class of free-choice net is larger than the union of these two classes. The net in Fig. 11.9(c) is a free-choice net even if it is neither a state machine nor a marked graph. Also, free-choice nets satisfy particular conditions that allow to reduce the computational complexity of the analysis of their properties with respect to the case of ordinary nets. A rich literature exists on this topic. See e.g. [3].

11.8 Further Reading

As in the case of Chapter 10, further details on the proposed topics can be found in the survey paper by Murata [13] and on the books of Peterson [14] and David and Alla [2]. Moreover, for more details on methods to improve the state equation based on implicit places, we address to the paper by Silva *et al.* [21]. Significant results related to the analysis of structural boundedness and structural liveness can be found in [20, 21], most of which are based on rank theorems. Finally, very interesting results on deadlock analysis and prevention are summarized in the book of Li and Zhou [10] and in the book edited by Zhou and Fanti [7].

References

1. Colom, J.M., Silva, M.: Improving the Linearly Based Characterization of P/T Nets. In: Rozenberg, G. (ed.) APN 1990. LNCS, vol. 483, pp. 113–145. Springer, Heidelberg (1991)
2. David, R., Alla, H.: Discrete, Continuous and Hybrid Petri Nets. Springer (2005)
3. Desel, J., Esparza, J.: Free Choice Petri Nets. Cambridge University Press (1995)
4. Ezpeleta, J., Colom, J.M., Martinez, J.: A Petri net based deadlock prevention policy for flexible manufacturing systems. IEEE Transactions on Robotics and Automation 11(2), 173–184 (1995)
5. Ezpeleta, J., Recalde, L.: A deadlock avoidance approach for nonsequential resource allocation systems. IEEE Transactions on Systems, Man and Cybernetics, Part A: Systems and Humans 34(1), 93–101 (2004)

6. Ezpeleta, J., Valk, R.: A polynomial deadlock avoidance method for a class of nonsequential resource allocation systems. IEEE Transactions on Systems, Man and Cybernetics, Part A: Systems and Humans 36(6), 1234–1243 (2006)
7. Fanti, M.P., Zhou, M.: Deadlock Resolution in Computer-Integrated Systems. Marcel Dekker/CRC Press (2005)
8. Finkel, A., Johnen, C.: The home state problem in transition systems. In: Rapport de Recherche, vol. 471. Univ. de Paris-Sud., Centre d'Orsay (1989)
9. Karp, R., Miller, C.: Parallel program schemata. Journal of Computer and System Sciences 3(2), 147–195 (1969)
10. Li, Z., Zhou, M.: Deadlock Resolution in Automated Manufacturing Systems. Springer (2009)
11. Liao, H., Lafortune, S., Reveliotis, S., Wang, Y., Mahlke, S.: Synthesis of maximally-permissive liveness-enforcing control policies for Gadara Petri nets. In: 49th IEEE Conference on Decision and Control, Atlanta, USA (2010)
12. Martinez, J., Silva, M.: A simple and fast algorithm to obtain all invariants of a generalized Petri net. In: Informatik-Fachberichte: Application and Theory of Petri Nets, vol. 52. Springer (1982)
13. Murata, T.: Petri nets: properties, analysis and applications. Proceedings IEEE 77(4), 541–580 (1989)
14. Peterson, J.L.: Petri Net Theory and the Modeling of Systems. Prentice-Hall (1981)
15. Petri, C.A.: Kommunication Mit Automaten. Institut für Instrumentelle, Mathematik (Bonn, Germany) Schriften des IIM 3 (1962)
16. Recalde, L., Mahulea, C., Silva, M.: Improving analysis and simulation of continuous Petri nets. In: 2nd IEEE Conf. on Automation Science and Engineering, Shanghai, China (2006)
17. Reisig, W.: Petri Nets: an Introduction. EATCS Monographs on Theoretical Computer Science (1985)
18. Reutenauer, C.: Aspects Mathématiques des Réseaux de Petri. Prentice-Hall International (1990)
19. Silva, M., Velilla, S.: Error detection and correction on Petri net models of discrete control systems. In: IEEE Int. Symp. on Circuits and Systems, Kyoto, Japan (1985)
20. Silva, M., Recalde, L., Teruel, E.: On linear algebraic techniques for liveness analysis of P/T systems. Journal of Circuits, Systems and Computers 8(11), 223–265 (1998)
21. Silva, M., Teruel, E., Colom, J.M.: Linear Algebraic and Linear Programming Techniques for the Analysis of Net Systems. In: Reisig, W., Rozenberg, G. (eds.) APN 1998. LNCS, vol. 1491, pp. 309–373. Springer, Heidelberg (1998)
22. Silva, M., Campos, J.: Performance evaluation of DEDS with conflicts and synchronizations: net-driven decomposition techniques. In: 4th Int. Workshop on Discrete Event Systems, Cagliari, Italy (1998)
23. Tricas, F., Ezpeleta, J.: Some results on siphon computation for deadlock prevention in resource allocation systems modeled with Petri nets. In: IEEE Conf. on Emerging Technologies and Factory Automation, Lisbon, Portugal (2003)
24. Tricas, F., Ezpeleta, J.: Computing minimal siphons in Petri net models of resource allocation systems: a parallel solution. IEEE Transactions on Systems, Man and Cybernetics, Part A: Systems and Humans 36(3), 532–539 (2006)
25. Wang, L., Mahulea, C., Julvez, J., Silva, M.: Decentralized control of large scale systems modeled with continuous marked graphs. In: 18th IFAC World Congress, Milano, Italy (2011)

Chapter 12
Supervisory Control of Petri Nets with Language Specifications

Alessandro Giua

12.1 Introduction

In this chapter, we study Petri nets (PNs) as language generators and we show how PNs can be used for supervisory control of discrete event systems under language specifications.

Supervisory control, originated by the work of Ramadge and Wonham [13], is a system theory approach that has been gaining increasing importance because it provides a unifying framework for the control of Discrete Event Systems (DESs). A general overview of Supervisory Control has been presented in Chapter 3.

In the original work of Ramadge and Wonham finite state machines (FSMs) were used to model plants and specifications. FSMs provide a general framework for establishing fundamental properties of DES control problems. They are not convenient models to describe complex systems, however, because of the large number of states that have to be introduced to represent several interacting subsystems, and because of the lack of structure. More efficient models have been proposed in the DES literature. Here the attention will be drawn to Petri net models.

PNs have several advantages over FSMs. First, PNs have a higher language complexity than FSM, since Petri net languages are a proper superset of regular languages. Second, the states of a PN are represented by the possible markings and not by the places: thus they give a compact description, i.e., the structure of the net may be maintained small in size even if the number of the markings grows[1]. Third, PNs can be used in modular synthesis, i.e., the net can be considered as composed

Alessandro Giua
Department of Electrical and Electronic Engineering, University of Cagliari,
Piazza d'Armi, 09123 Cagliari, Italy
e-mail: giua@diee.unica.it

[1] However, we should point out that many analysis techniques for Petri nets are based on the construction of the reachability graph, that suffers from the same state explosion problem typical of automata. To take advantage of the compact PN representation, other analysis techniques (e.g. structural) should be used.

C. Seatzu et al. (Eds.): Control of Discrete-Event Systems, LNCIS 433, pp. 235–255.
springerlink.com

of interrelated subnets, in the same way as a complex system can be regarded as composed of interacting subsystems.

Although PNs have a greater modeling power than FSMs, computability theory shows that the increase of modeling power often leads to an increase in the computation required to solve problems. This is why a section of this paper focuses on the decidability properties of Petri nets by studying the corresponding languages: note that some of these results are original and will be presented with formal proofs. It will be shown that Petri nets represent a good tradeoff between modeling power and analysis capabilities

The chapter is structured as follows. In Section 1 Petri net generators and languages are defined. In Section 2 the concurrent composition operator on languages is defined and extended to an operator on generators. In Section 3 it is shown how the classical monolithic supervisory design can be carried out using Petri net models. Finally, in Section 4 some issues arising from the use of unbounded PNs in supervisory control are discussed.

12.2 Petri Nets and Formal Languages

This section provides a short but self-standing introduction to Petri net languages. PN languages represent an interesting topic within the broader domain of formal language theory but there are few books devoted to this topic and the relevant material is scattered in several journal publications. In this section and in the following we focus on the definition of Petri net generators and operators that will later be used to solve a supervisory control problem.

12.2.1 Petri Net Generators

Definition 12.1. *A* labeled Petri net system *(or* Petri net generator*) [7, 12] is a quadruple* $G = (N, \ell, \boldsymbol{m}_0, F)$ *where:*

- $N = (P, T, \boldsymbol{Pre}, \boldsymbol{Post})$ *is a* Petri net structure *with* $|P| = m$ *and* $|T| = n$*;*
- $\ell : T \to E \cup \{\lambda\}$ *is a* labeling function *that assigns to each transition a label from the alphabet of events E or assigns the empty word*[2] λ *as a label;*
- $\boldsymbol{m}_0 \in \mathbb{N}^n$ *is an* initial marking*;*
- $F \subset \mathbb{N}^n$ *is a finite set of* final markings.

Three different types of *labeling functions* are usually considered.

- *Free labeling:* all transitions are labeled distinctly and none is labeled λ, i.e., $(\forall t, t' \in T)\ [t \neq t' \Longrightarrow \ell(t) \neq \ell(t')]$ and $(\forall t \in T)\ [\ell(t) \neq \lambda]$.

[2] While in other parts of this book the empty string is denoted ε, in this section we have chosen to use the symbol λ for consistency with the literature on PN languages.

- *λ-free labeling:* no transition is labeled λ.
- *Arbitrary labeling:* no restriction is posed on ℓ.

The labeling function may be extended to a function $\ell : T^* \to E^*$ defining: $\ell(\lambda) = \lambda$ and $(\forall t \in T, \forall \sigma \in T^*)$ $\ell(\sigma t) = \ell(\sigma)\ell(t)$.

Example 12.1. Consider the nets in Fig. 12.1 where the label of each transition is shown below the transition itself. Net (a) is a free-labeled generator on alphabet $E = \{a, b\}$. Nets (b) and (c) are λ-free generators on alphabet $E = \{a\}$. Net (d) is an arbitrary labeled generator on alphabet $E = \{a\}$. ■

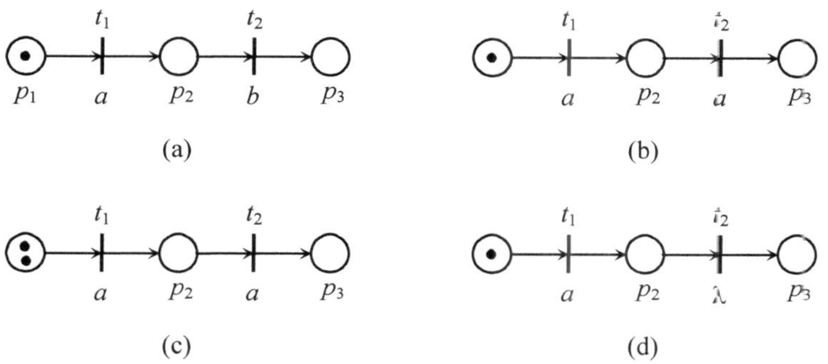

Fig. 12.1 PN generators of Example 12.1

Three languages are associated with a generator G depending on the different notions of terminal strings.

- *L-type* or *terminal language:*[3] the set of strings generated by firing sequences that reach a final marking, i.e.,

$$L_L(G) = \{\ell(\sigma) \mid \boldsymbol{m}_0 \, [\sigma\rangle \, \boldsymbol{m}_f \in F\}.$$

- *G-type* or *covering language* or *weak language:* the set of strings generated by firing sequences that reach a marking $\boldsymbol{m}$ covering a final marking, i.e.,

$$L_G(G) = \{\ell(\sigma) \mid \boldsymbol{m}_0 \, [\sigma\rangle \, \boldsymbol{m} \geq \boldsymbol{m}_f \in F\}.$$

- *P-type* or *prefix language:*[4] the set of strings generated by any firing sequence, i.e.,

$$L_P(G) = \{\ell(\sigma) \mid \boldsymbol{m}_0 \, [\sigma\rangle \, \}.$$

[3] This language is called marked behavior in the framework of Supervisory Control and is denoted $L_m(G)$.

[4] This language is called closed behavior in the framework of Supervisory Control and is denoted $L(G)$.

Example 12.2. Consider the free-labeled generator G in Fig. 12.2. The initial marking, also shown in the figure, is $\boldsymbol{m}_0 = [1\ 0\ 0]^T$. Assume the set of final markings is $F = \{[0\ 0\ 1]^T\}$. The languages of this generator are:

$$L_L(G) = \{a^m c b^m \mid m \geq 0\};$$

$$L_G(G) = \{a^m c b^n \mid m \geq n \geq 0\};$$

$$L_P(G) = \{a^m \mid m \geq 0\} \cup \{a^m c b^n \mid m \geq n \geq 0\}.$$ ■

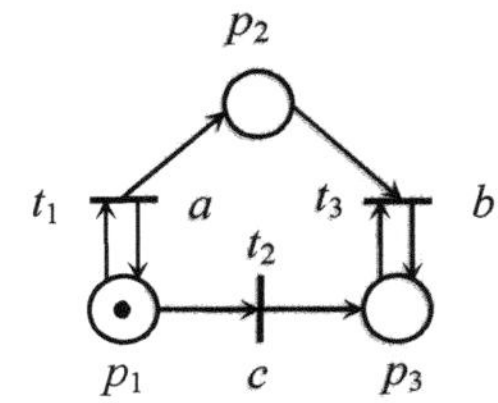

Fig. 12.2 Free-labeled generator G of Example 12.2

12.2.2 Deterministic Generators

A *deterministic* PN generator [7] is such that the word of events generated from the initial marking uniquely determines the marking reached.

Definition 12.2. *A λ-free generator G is* deterministic *iff for all $t, t' \in T$, with $t \neq t'$, and for all $\boldsymbol{m} \in R(N, \boldsymbol{m}_0)$: $\boldsymbol{m}\,[t\rangle \wedge \boldsymbol{m}\,[t'\rangle \Longrightarrow \ell(t) \neq \ell(t')$.*

According to the previous definition, in a deterministic generator two transitions sharing the same label may never be simultaneously enabled and no transition may be labeled by the empty string. Note that a free-labeled generator is also deterministic. On the contrary, a λ-free (but not free labeled) generator may be deterministic or not depending on its structure and also on its initial marking.

Example 12.3. Consider generators (b) and (c) in Fig. 12.1: they have the same net structure and the same λ-free labeling, but different initial marking. The first one is deterministic, because transitions t_1 and t_2, sharing label a can never be simultaneously enabled. On the contrary, the second one is not deterministic, because reachable marking $[1\ 1\ 0]^T$ enables both transitions t_1 and t_2: as an example, the observed word aa may be produced by two different sequences yielding two different markings

$$\begin{bmatrix}2\\0\\0\end{bmatrix} [t_1\rangle \begin{bmatrix}1\\1\\0\end{bmatrix} [t_1\rangle \begin{bmatrix}0\\2\\0\end{bmatrix} \quad or \quad \begin{bmatrix}2\\0\\0\end{bmatrix} [t_1\rangle \begin{bmatrix}1\\1\\0\end{bmatrix} [t_2\rangle \begin{bmatrix}1\\0\\1\end{bmatrix}.$$ ■

The previous definition of determinism was introduced in [18] and used in [7, 12]. It may be possible to extend it as follows.

Definition 12.3. *A λ-free generator G is* deterministic *iff for all $t,t' \in T$, with $t \neq t'$, and for all $\boldsymbol{m} \in R(N,\boldsymbol{m}_0)$: $\boldsymbol{m}\,[t\rangle \wedge \boldsymbol{m}\,[t'\rangle \implies [\ell(t) \neq \ell(t')] \vee [\boldsymbol{Post}[\cdot,t] - \boldsymbol{Pre}[\cdot,t] = \boldsymbol{Post}[\cdot,t'] - \boldsymbol{Pre}[\cdot,t']]$.*

With this extended definition, we accept as deterministic a generator in which two transitions with the same label may be simultaneously enabled at a marking $\boldsymbol{m}$, provided that the two markings reached from $\boldsymbol{m}$ by firing t and t' are the same. Note that with this extended definition, while the word of events generated from the initial marking uniquely determines the marking reached it does not necessarily uniquely determine the sequences that have fired.

12.2.3 Classes of Petri Net Languages

The classes of Petri net languages are denoted as follows:

- $\mathscr{L}^f$ (resp. $\mathscr{G}^f$, $\mathscr{P}^f$) denotes the class of terminal (resp. covering, prefix) languages generated by free-labeled PN generators.
- $\mathscr{L}_d$ (resp. $\mathscr{G}_d$, $\mathscr{P}_d$) denotes the class of terminal (resp. covering, prefix) languages generated by deterministic PN generators.
- $\mathscr{L}$ (resp. $\mathscr{G}$, $\mathscr{P}$) denotes the class of terminal (resp. covering, prefix) languages generated by λ-free PN generators.
- $\mathscr{L}^\lambda$ (resp. $\mathscr{G}^\lambda$, $\mathscr{D}^\lambda$, $\mathscr{P}^\lambda$) denotes the class of terminal (resp. covering, prefix) languages generated by arbitrary labeled PN generators.

Table 12.1 shows the relationship among these classes. Here $A \to B$ represents a strict set inclusion $A \subsetneq B$.

Table 12.1 Known relations among classes of Petri net languages. An arc $\to$ represents the set inclusion

$$\begin{array}{ccccccc}
\mathscr{L}^f & \to & \mathscr{L}_d & \to & \mathscr{L} & \to & \mathscr{L}^\lambda \\
 & & & & \uparrow & & \uparrow \\
\mathscr{G}^f & \to & \mathscr{G}_d & \to & \mathscr{G} & \to & \mathscr{G}^\lambda \\
\uparrow & & \uparrow & & \uparrow & & \uparrow \\
\mathscr{P}^f & \to & \mathscr{P}_d & \to & \mathscr{P} & \to & \mathscr{P}^\lambda
\end{array}$$

While a formal proof of all these relations can be found in [1], we point out that the relations on each line — that compare the same type of languages of nets with different labeling — are rather intuitive. Additionally, one readily understands that any *P-type* language of a generator G may also be obtained as a *G-type* language defining as a set of final markings $F = \{\boldsymbol{0}\}$.

Parigot and Peltz [10] have defined PN languages as regular languages with the additional capability of determining if a string of parenthesis is well formed.

If we consider the class $\mathscr{L}$ of PN languages, it is possible to prove [12] that $\mathscr{L}$ is a strict superset of regular languages and a strict subset of context-sensitive languages. Furthermore, $\mathscr{L}$ and the class of context-free languages are not comparable. An example of a language in $\mathscr{L}$ that is not context-free is: $L = \{a^m b^m c^m \mid m \geq 0\}$. An example of a language that is context-free but is not in $\mathscr{L}$ is: $L = \{ww^R \mid w \in E^*\}$[5] if $|E| > 1$.

All these results are summarized in Fig. 12.3. Note that the class $\mathscr{L}_d$, although contained in $\mathscr{L}$, occupies the same position as $\mathscr{L}$ in the hierarchy shown in the figure.

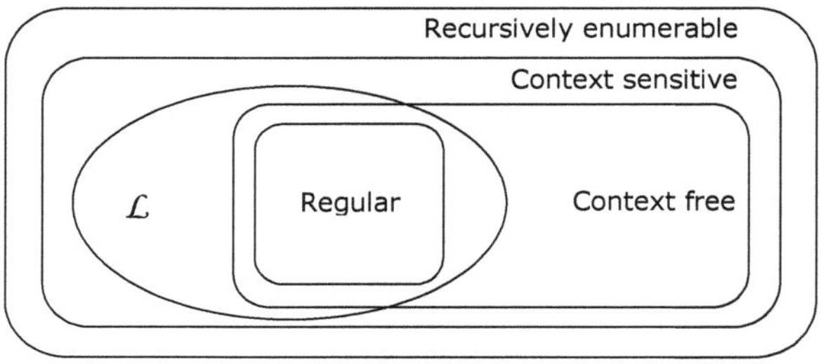

Fig. 12.3 Relations among the class $\mathscr{L}$ and other classes of formal languages

In the framework of Supervisory Control, we will assume that the generators considered are deterministic. In particular, class $\mathscr{L}_d$ (or possibly $\mathscr{G}_d$ for unbounded nets) will be used to describe marked languages, while class $\mathscr{P}_d$ will be used to describe closed languages. There are several reasons for this choice.

- Systems of interest in supervisory control theory are deterministic.
- Although each class of deterministic languages defined here is strictly included in the corresponding class of λ-free languages, it is appropriate to restrict our analysis to deterministic generators. In fact, several properties of interest are decidable for deterministic nets while they are not for λ-free nets [1, 11, 18].
- In [1] it was shown that the classes $\mathscr{G}_d$ and $\mathscr{L}_d$ are incomparable, and furthermore $\mathscr{G}_d \cap \mathscr{L}_d = \mathscr{R}$, where $\mathscr{R}$ is the class of regular languages. Hence taking also into account the G-type language (in addition to the L-type language) one extends the class of control problems that can be modeled by deterministic unbounded PNs.

[5] The string w^R is the reversal of string w.

12.2.4 Other Classes of Petri Net Languages

Gaubert and Giua [1] have explored the use of infinite sets of final markings in the definition of the marked behavior of a net. With each more or less classical subclass of subsets of $\mathbb{N}^m$ — finite, ideal (or upper), semi-cylindrical, star-free, recognizable, rational (or semilinear) subsets — it is possible to associate the class of Petri net languages whose set of accepting states belongs to the class.

When comparing the related Petri net languages, it was shown that for arbitrary or λ-free PN generators, the above hierarchy collapses: one does not increase the generality by considering semilinear accepting sets instead of the usual finite ones. However, for free-labeled and deterministic PN generators, it is shown that one gets new distinct subclasses of Petri net languages, for which several decidability problems become solvable.

12.3 Concurrent Composition and System Structure

In this section we recall the definition of the concurrent composition operator on languages and introduce the corresponding operator on nets.

Definition 12.4 (Concurrent composition of languages). *Given two languages $L_1 \subseteq E_1^*$ and $L_2 \subseteq E_2^*$, their* concurrent composition *is the language L on alphabet $E = E_1 \cup E_2$ defined as follows:*

$$L = L_1 \parallel L_2 = \{\, w \in E^* \mid w \uparrow_{E_1} \in L_1,\ w \uparrow_{E_2} \in L_2 \,\}$$

where $w \uparrow_{E_i}$ denotes the projection of word w on alphabet E_i, for $i = 1,2$.

We now consider the counterpart of this language operator on a net structure.

Definition 12.5 (Concurrent composition of PN generators)
Let $G_1 = (N_1, \ell_1, \boldsymbol{m}_{0,1}, F_1)$ and $G_2 = (N_2, \ell_2, \boldsymbol{m}_{0,2}, F_2)$ be two PN generators. Their concurrent composition, *denoted also $G = G_1 \parallel G_2$, is the generator $G = (N, \ell, \boldsymbol{m}_0, F)$ that generates $L_L(G) = L_L(G_1) \parallel L_L(G_2)$ and $L_P(G) = L_P(G_1) \parallel L_P(G_2)$.*

The structure of G may be determined with the following procedure.

Algorithm 12.4. *Let P_i, T_i and E_i $(i = 1,2)$ be the place set, transition set, and the alphabet of G_i.*

- *The place set P of N is the union of the place sets of N_1 and N_2, i.e., $P = P_1 \cup P_2$.*
- *The transition set T of N and the corresponding labels are computed as follows:*
  - *For each transition $t \in T_1 \cup T_2$ labeled λ, a transition with the same input and output bag of t and labeled λ belongs to T.*

- *For each transition $t \in T_1 \cup T_2$ labeled $e \in (E_1 \setminus E_2) \cup (E_2 \setminus E_1)$, a transition with the same input and output bag of t and labeled e belongs to T.*
- *Consider a symbol $e \in E_1 \cap E_2$ and assume it labels m_1 transitions $T_{e,1} \subseteq T_1$ and m_2 transitions $T_{e,2} \subseteq T_2$. Then $m_1 \times m_2$ transitions labeled e belong to T. The input (output) bag of each of these transitions is the sum of the input (output) bags of one transition in $T_{e,1}$ and of one transition in $T_{e,2}$.*

- $\boldsymbol{m}_0 = [\boldsymbol{m}_{0,1}^T \ \boldsymbol{m}_{0,2}^T]^T$.
- *F is the cartesian product of F_1 and F_2, i.e., $F = \{[\boldsymbol{m}_1^T \ \boldsymbol{m}_2^T]^T \mid \boldsymbol{m}_1 \in F_1, \boldsymbol{m}_2 \in F_2\}$.*

The composition of more than two generators can be computed by repeated application of the procedure. Note that while the set of places grows linearly with the number of composed systems, the set of transitions and of final markings may grow faster.

Example 12.5. Let $G_1 = (N_1, \ell_1, \boldsymbol{m}_{0,1}, F_1)$ and $\mathbf{G_2} = (N_2, \ell_2, \boldsymbol{m}_{0,2}, F_2)$ be the two generators shown in Fig. 12.4. Here $F_1 = \{[1\ 0]^T\}$ and $F_2 = \{[1\ 0]^T, [0\ 1]^T\}$. Their concurrent composition $G = G_1 \parallel G_2$ is also shown in Fig. 12.4. The initial marking of G is $\boldsymbol{m}_0 = [1\ 0\ 1\ 0]^T$ and its set of final markings is $F = \{[1\ 0\ 1\ 0]^T, [1\ 0\ 0\ 1]^T\}$. ■

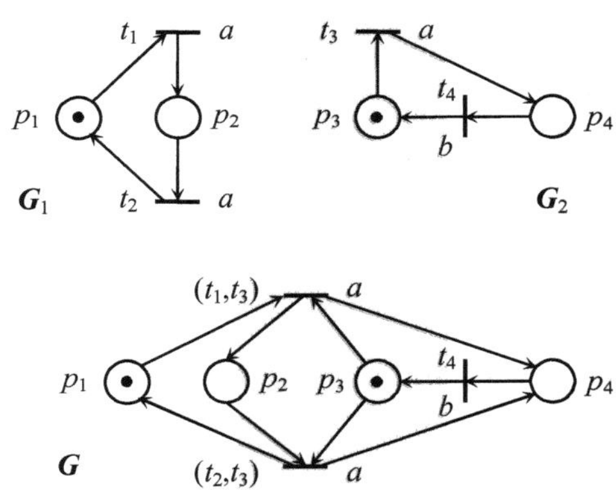

Fig. 12.4 Two generators G_1, G_2 and their concurrent composition G of Example 12.5

12.4 Supervisory Design Using Petri Nets

In this section we discuss how Petri net models may be used to design supervisors for language specifications within the framework of Supervisory Control. The design of a supervisor in the framework of automata was presented in Chapter 3 and we assume the reader is already familiar with this material.

12.4.1 Plant, Specification and Supervisor

Here we comment on some of the assumptions that are peculiar to the PN setting.

- The **plant** is described by a deterministic PN generator G on alphabet E. Its closed language is $L(G) = L_P(G)$ while its marked[6] language is $L_m(G) = L_L(G)$. We assume such a generator is *nonblocking*, i.e., $\overline{L_L(G)} = L_P(G)$.
  The transition set of G is partitioned as follows: $T = T_c \cup T_{uc}$, where T_c are the *controllable transitions* that can be disabled by a control agent, while T_{uc} are the *uncontrollable transitions*. Note that this allows a generalization of the automata settings where the notion of controllability and uncontrollability is associated to the events. In fact, it is possible that two transitions, say t' and t'', have the same event label $\ell(t') = \ell(t'') = e \in E$ but one of them is controllable while the other is not. In the rest of the chapter, however, we will not consider this case and assume that the event alphabet may be partitioned as $E = E_c \cup E_{uc}$ where
  $$E_c = \bigcup_{t \in T_c} \ell(t), \quad E_{uc} = \bigcup_{t \in T_{uc}} \ell(t) \quad \text{and} \quad E_c \cap E_{uc} = \emptyset.$$
  It is also common to consider plants composed by m PN generators $G_1, \ldots, G_m$ working concurrently. The alphabets of these generators are $E_1, \ldots, E_m$. The overall plant is a PN generator $G = G_1 \parallel \cdots \parallel G_m$ on alphabet $E = E_1 \cup \cdots \cup E_m$.
- The **specification** is a language $K \subset \hat{E}^*$, where $\hat{E} \subset E$ is a subset of the plant alphabet. Such a specification defines a set of *legal words* on E given by $\{w \in E^* \mid w \uparrow_{\hat{E}} \in \text{prefix}(K)\}$.
  The specification K is represented by a deterministic nonblocking PN generator H on alphabet $\hat{E}$ whose marked language is $L_m(H) = L_L(H) = K$. As for the plant, other choices for the marked language are possible.
- The **supervisor**[7] is described by a nonblocking PN generator S on alphabet E. It runs in parallel with the plant, i.e., each time the plant generates an event e a transition with the same label is executed on the supervisor. The control law computed by S when its marking is $\boldsymbol{m}$ is given by $g(\boldsymbol{m}) = E_{uc} \cup \{e \in E_c \mid (\exists t \in T_c) \boldsymbol{m}[t\rangle, \ell(t) = e\}$.

12.4.2 Monolithic Supervisor Design

The *monolithic supervisory design* requires three steps. In the first step, a coarse structure for a supervisor is synthesized by means of concurrent composition of the plant and specification. In the second step, the structure is analyzed to check

[6] While in the case of bounded nets the L-type language can describe any marked language, in the case of unbounded generators other choices for the marked language are possible considering the G-type language of the generator or even any other type of terminal languages as mentioned in § 12.2.4. This will be discussed in Section 12.5.

[7] See also Definition 3.9 in Chapter 3.9.

if properties of interest (namely, the absence of uncontrollable and blocking states) hold. In the third step, if the properties do not hold, this structure is trimmed to avoid reaching undesirable states.

Algorithm 12.6. (Monolithic supervisory design). *We are given a plant G and a specification H.*

1. *Construct by concurrent composition the generator $J = G \parallel H$.*
2. *Determine if the generator J satisfies the following properties:*
   - nonblockingness, *i.e., it does not contain blocking markings from which a final marking cannot be reached;*
   - controllability, *i.e., it does not contain uncontrollable markings such that when G and H run in parallel an uncontrollable event is enabled in G but is not enabled in H.*

   If J satisfies both properties, then both H and J are suitable supervisors.
3. *If J contains blocking or uncontrollable markings, we have to trim it to obtain a nonblocking and controllable generator S. The generator S obtained through this procedure is at the same time a suitable maximally permissive supervisor and the corresponding closed-loop system.*

In the previous algorithm, the generator J constructed in step 1 represents the largest behavior of the plant that satisfies all the constraints imposed by the specifications. More precisely, its closed language

$$L(J) = \{w \in E \mid w \in L(G), w \uparrow_{\hat{E}} \in L(H)\}$$

represents the behavior of the plant restricted to the set of legal words, while its marked behavior

$$L_m(J) = \{w \in E \mid w \in L_m(G), w \uparrow_{\hat{E}} \in L_m(H)\}$$

represents the marked behavior of the plant restricted to the set of legal words marked by the specification.

In step 2 we have used informally the term "blocking marking" and "uncontrollable marking". We will formally define these notions in the following.

We first define some useful notation. The structure of the generators is $J = (N, \ell, \boldsymbol{m}_0, F)$, $G = (N_1, \ell_1, \boldsymbol{m}_{0,1}, F_1)$, and $H = (N_2, \ell_2, \boldsymbol{m}_{0,2}, F_2)$, where $N = (P, T, \boldsymbol{Pre}, \boldsymbol{Post})$ and $N_i = (P_i, T_i, \boldsymbol{Pre}_i, \boldsymbol{Post}_i)$, $(i = 1,2)$. We define the *projection of a marking $\boldsymbol{m}$ of N on net N_i*, $(i = 1,2)$, denoted $\boldsymbol{m} \uparrow_i$, is the vector obtained from $\boldsymbol{m}$ by removing all the components associated to places not present in N_i.

We first present the notion of a blocking marking.

Definition 12.6. *A marking $\boldsymbol{m} \in R(N, \boldsymbol{m}_0)$ of generator J is a* blocking marking *if no final marking may be reached from it, i.e., $R(N, \boldsymbol{m}) \cap F = \emptyset$. The generator J is* nonblocking *if no blocking marking is reachable.*

We now present the notion of an uncontrollable marking.

Definition 12.7. *Let $T_u \subseteq T$ be the set of uncontrollable transitions of J. A marking $\boldsymbol{m} \in R(N, \boldsymbol{m}_0)$ of generator J is* uncontrollable *if there exists an uncontrollable transition $t \in T_u$ that is enabled by $\boldsymbol{m} \uparrow_1$ in G but that is not enabled by $\boldsymbol{m} \uparrow_2$ in H. The generator J is* controllable *if no uncontrollable marking is reachable.*

Determining if a generator J is nonblocking and controllable is always possible, as we will show in the next section. We also point out that for bounded nets this test can be done by construction of the reachability graph[8] as in the following example of supervisory design.

Example 12.7. Consider the generators G_1 and G_2, and the specification H in Fig. 12.5 (left). Note that all nets are free-labeled, hence we have an isomorphism between the set of transitions T and the set of events E: in the following each transitions will be denoted by the corresponding event.

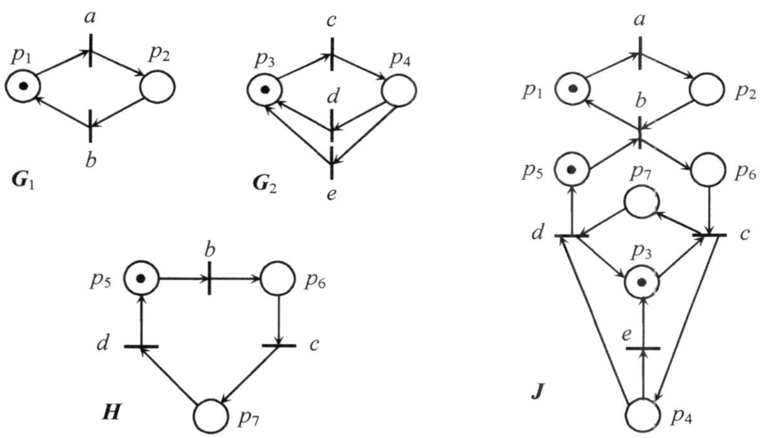

Fig. 12.5 Left: Systems G_1, G_2 and specification H for the control problem of Example 12.7. Right: System $J = G_1 \parallel G_2 \parallel H$

G_1 describes a conveyor that brings in a manufacturing cell a raw part (event a) that is eventually picked-up by a robot (event b) so that a new part can enter. G_2 describes a machine that is loaded with a raw part (event c) and, depending on the operation it performs, may produce parts of type A or type B (events d or e) before returning to the idle state. The set of final states of both generators consists of the initial marking shown in the figure.

The specification we consider, represented by the generator H, describes a cyclic operation process where a robot picks-up a raw part from the conveyor, loads it on

[8] As we have already pointed out, the construction of the reachability graph suffers from the state explosion problem. An open area for future research is the use of more efficient analysis techniques (e.g., structural) to check nonblockingness and controllability for language specification.

the machine and after recognizing that a part of type A has been produced repeats the process. The set of final states consists of the initial marking shown in the figure.

The overall process is $G = G_1 \parallel G_2$ and the generator $J = G \parallel H$, is shown in Fig. 12.5 (right). Its set of final states consists of the initial marking shown in the figure.

Assume now that the controllable transition/event set is $E_c = \{a, c, d, e\}$ and the uncontrollable transition/event set is $E_u = \{b\}$.

It is immediate to show that generator J is blocking and uncontrollable. To show this we have constructed the reachability graph of J in Fig. 12.6. The two markings shown in thick boxes are blocking because from them it is impossible to reach the initial marking (that is also the unique final marking). The three markings shaded in gray are uncontrollable: in fact, in all these markings $m[p_2] = 1$, i.e., uncontrollable transition b is enabled in the plant G, while $m[p_5] = 0$, i.e., b is not enabled in H. ■

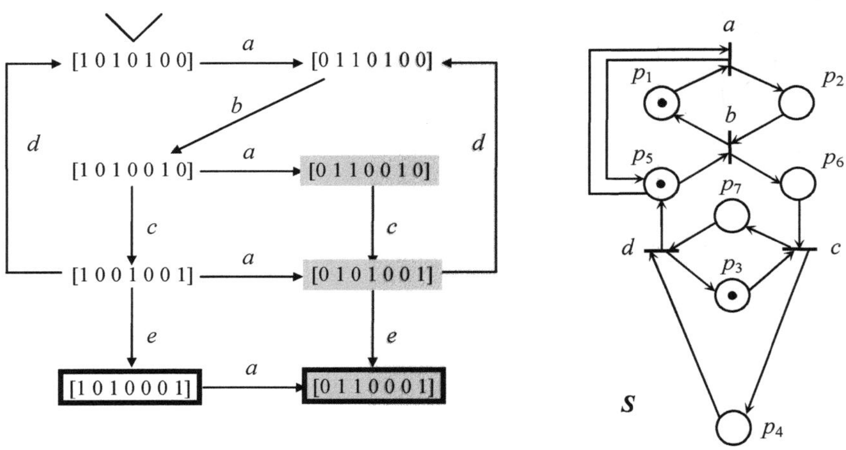

Fig. 12.6 Left: Reachability graph of generator J of Example 12.7. Right: the structure of the trim generator **S** of Example 12.8

12.4.3 *Trimming*

Once the coarse structure of a candidate supervisor is constructed by means of concurrent composition, we need to trim it to obtain a nonblocking and controllable generator.

The next example shows the problems involved in the trimming of a net.

Example 12.8. Let us consider the generator J constructed in Example 12.7.

Refining the PN to avoid reaching the undesirable markings shown in Fig. 12.6 is complex. First, we could certainly remove the transition labeled by e since its firing always leads to an undesirable state and it is controllable. After removal of this transition, the transition labeled by a will be enabled by the following reachable markings: $\boldsymbol{m}' = [1\ 0\ 1\ 0\ 1\ 0\ 0]^T, \boldsymbol{m}'' = [1\ 0\ 1\ 0\ 0\ 1\ 0]^T, \boldsymbol{m}''' = [1\ 0\ 0\ 1\ 0\ 0\ 1]^T$. We want to block the transition labeled a when the markings $\boldsymbol{m}''$ and $\boldsymbol{m}'''$ are reached. Since

$$m'[p_5] = 1 > m''[p_5] = m'''[p_5] = 0,$$

we can add an arc from p_5 to a and from a to p_5 as in Fig. 12.6. ■

The following algorithm can be given for the trimming of a net.

Algorithm 12.9. *Let t be a transition to be controlled, i.e., a transition leading from an admissible marking to an undesirable marking. Let e be its label.*

1. *Determine the set of admissible reachable markings that enable t, and partition this set into the disjoint subsets $\mathcal{M}_a$ (the markings from which t should be allowed to fire), and $\mathcal{M}_{na}$ (the markings from which t should not be allowed to fire, to avoid reaching an undesirable marking). If $\mathcal{M}_a = \emptyset$ remove t and stop, else continue.*
2. *Determine a construct in the form:*

$$\begin{aligned}\mathcal{U}(\boldsymbol{m}) = {} & [(m[p_1^1] \geq n_1^1) \wedge \ldots \wedge (m[p_{k1}^1] \geq n_{k1}^1)] \vee \\ & \ldots \\ & \vee[(m[p_1^l] \geq n_1^l) \wedge \ldots \wedge (m[p_{kl}^l] \geq n_{kl}^l)],\end{aligned}$$

such that $\mathcal{U}(\boldsymbol{m}) = TRUE$ if $\boldsymbol{m} \in \mathcal{M}_a$, and $\mathcal{U}(\boldsymbol{m}) = FALSE$ if $\boldsymbol{m} \in \mathcal{M}_{na}$.

3. *Replace transition t with l transitions $t^1, \ldots, t^l$ labeled a. The input (output) arcs of transition t^j, $j = 1, \ldots, l$, will be those of transition t plus n_i^j arcs inputting from (outputting to) place p_i^j, $i = 1, \ldots, k_j$.*

It is clear that following this construction there is an enabled transition labeled e for any marking in $\mathcal{M}_a$, while none of these transitions are enabled by a marking in $\mathcal{M}_{na}$. We also note that in general several constructs of this form may be determined. The one which requires the minimal number of transitions, i.e., the one with the smallest l, is preferable.

The following theorem gives a sufficient condition for the applicability of the algorithm.

Theorem 12.1. *The construct of Algorithm 12.9 can always be determined if the net is bounded.*

Proof. For sake of brevity, we prove this result for the more restricted class of conservative nets. One should keep in mind, however, that given a bounded non conservative net, one can make the net conservative adding dummy sink places that do not modify its behavior.

A net is conservative if there exists an integer vector $Y > \mathbf{0}$ such that for any two markings $\boldsymbol{m}$ and $\boldsymbol{m}'$ reachable from the initial marking $Y^T\boldsymbol{m} = Y^T\boldsymbol{m}'$. Hence if $\boldsymbol{m} \neq \boldsymbol{m}'$ there exists a place p such that $m[p] > m'[p]$. Also the set of reachable markings is finite.

On a conservative net, consider $\boldsymbol{m}_i \in \mathscr{M}_a$, $\boldsymbol{m}_j \in \mathscr{M}_{na}$. We have that $\mathscr{M}_a$ and $\mathscr{M}_{na}$ are finite sets and also there exists a place p_{ij} such that $m_i[p_{ij}] = n_{ij} > m_j[p_{ij}]$. Hence

$$\mathscr{U}(\boldsymbol{m}) = \bigvee_{i \in \mathscr{M}_a} \left[\bigwedge_{j \in \mathscr{M}_{na}} (m[p_{ij}] \geq n_{ij}) \right]$$

is a construct for Algorithm 12.9. □

Unfortunately, the construct may contain up to $|\mathscr{M}_a|$ *OR* clauses, i.e., up to $|\mathscr{M}_a|$ transitions may be substituted for a single transition to control. Note, however, that it is often possible to determine a simpler construct as in Example 12.8, where the construct for the transition labeled a was $\mathscr{U}(\boldsymbol{m}) = [m[p_5] \geq 1]$.

12.5 Supervisory Control of Unbounded PN Generators

As we have seen in the previous section, the monolithic supervisory design presented in Algorithm 12.6 can always be applied when the plant G and the specification H are bounded PN generators. Here we consider the case of general, possibly unbounded, generators.

In step 1 of the monolithic supervisory design algorithm the unboundedness of the G or H does not require any special consideration, since the procedure to construct the concurrent composition $J = G \parallel H$ is purely structural in the PN setting. Thus we need to focus on the last two steps, and discuss how it is possible to check if an unbounded generator G is nonblocking and controllable, and eventually how it can be trimmed.

We have previously remarked that in the case of bounded nets the L-type language can describe any marked language. In the case of unbounded generators other choices for the marked language are possible considering the G-type language of the generator or even any other type of terminal language mentioned in § 12.2.4.

In the rest of this section we will only consider two types of marked languages for a PN generator G.

- *L-type language*, i.e., $L_m(G) = L_L(G)$. This implies that the set of *marked markings* reached by words in $L_m(G)$ is $\mathscr{F} = F$, i.e., it coincides with the finite set of final markings associated to the generator.
- *G-type marked language*, i.e., $L_m(G) = L_G(G)$. This implies that the set of *marked markings* reached by words in $L_m(G)$ is

$$\mathscr{F} = \bigcup_{\boldsymbol{m}_f \in F} \{\boldsymbol{m} \in \mathbb{N}^m \mid \boldsymbol{m} \geq \boldsymbol{m}_f\},$$

i.e., it is the infinite covering set of F.

12.5.1 Checking Nonblockingness

We will show in this subsection that checking a generator for nonblockingness is always possible.

Let us first recall the notion of home space.

Definition 12.8. *A marking* $\boldsymbol{m} \in \mathbb{N}^m$ *of a Petri net is a* home-marking *if it is reachable from all reachable markings.*

A set of markings $\mathscr{M} \subseteq \mathbb{N}^m$ *of a Petri net is a* home space *if for all reachable marking* $\boldsymbol{m}$ *a marking in* $\mathscr{M}$ *is reachable from* $\boldsymbol{m}$.

The following result is due to Johnen and Frutos Escrig.

Proposition 12.1. [8] *The property of being a home space for finite unions of linear sets*[9] *having the same periods is decidable.*

We can finally state the following original result.

Theorem 12.2. *Given a generator J constructed as in step 1 of Algorithm 12.6 it is decidable if it is nonblocking when its marked language is the L-type or G-type language.*

Proof. Let $\mathscr{F}$ be the set of marked markings of the generator. According to Definition 12.6 generator J is nonblocking iff from every reachable markings $\boldsymbol{m}$ a marked marking in $\mathscr{F}$ is reachable. Thus checking for nonblockingness is equivalent to checking if the set of marked markings $\mathscr{F}$ is a home space.

When the marked language is the L-type language, $\mathscr{F} = F$ and we observe that each marking $\boldsymbol{m}_f$ can be considered as a linear set with base $\boldsymbol{m}_f$ and empty set of generators.

When the marked language is the G-type language,

$$\mathscr{F} = \bigcup_{\boldsymbol{m}_f \in F} \{\boldsymbol{m} \in \mathbb{N}^m \mid \boldsymbol{m} \geq \boldsymbol{m}_f\} = \{\boldsymbol{m}_f + \sum_{i=1}^{m} k_i \boldsymbol{e}_i \mid k_i \in \mathbb{N}\}$$

where vectors $\boldsymbol{e}_i$ are the canonical basis vectors, i.e., $\boldsymbol{e}_i \in \{0,1\}^n$, with $e_i[i] = 1$ and $e_i[j] = 0$ if $i \neq j$.

In both cases $\mathscr{F}$ is the finite unions of linear sets having the same periods, hence checking if it is a home space is decidable by Proposition 12.1. □

12.5.2 Checking Controllability

We will show in this subsection that checking a generator for controllability is always possible. The material presented in this subsection is original and proofs of all results will be given.

[9] We say that $\mathscr{E} \subseteq \mathbb{N}^m$ is a *linear set* if there exists some $\boldsymbol{v} \in \mathbb{N}^m$ and a finite set $\{\boldsymbol{v}_1, \cdots, \boldsymbol{v}_n\} \subseteq \mathbb{N}^m$ such that $\mathscr{E} = \{\boldsymbol{v}' \in \mathbb{N}^m \mid \boldsymbol{v}' = \boldsymbol{v} + \sum_{i=1}^{n} k_i \boldsymbol{v}_i$ with $k_i \in \mathbb{N}\}$. The vector $\boldsymbol{v}$ is called the *base* of $\mathscr{E}$, and $\boldsymbol{v}_1, \cdots, \boldsymbol{v}_n$ are called its *periods*.

We first present some intermediate result.

Lemma 12.1. *Let $\langle N, \boldsymbol{m}_0 \rangle$ be a marked net with $N = (P, T, \boldsymbol{Pre}, \boldsymbol{Post})$ and $|P| = m$. Given a marking $\bar{\boldsymbol{m}} \in \mathbb{N}^m$ and a place $\bar{p} \in P$, we define the set*

$$\mathscr{S}(\bar{\boldsymbol{m}}, \bar{p}) = \{\boldsymbol{m} \in \mathbb{N}^m \mid m[\bar{p}] = \bar{m}[\bar{p}], \ (\forall p \in P \setminus \{\bar{p}\})\, m[p] \geq \bar{m}[p]\}$$

of those markings that are equal to $\bar{\boldsymbol{m}}$ in component $\bar{p}$ and greater than or equal to $\bar{\boldsymbol{m}}$ in all other components.

Checking if a marking in this set is reachable in $\langle N, \boldsymbol{m}_0 \rangle$ is decidable.

Proof. To prove this result, we reduce the problem of determining if a marking in $\mathscr{S}(\bar{\boldsymbol{m}}, \bar{p})$ is reachable to the standard marking reachability problem (see Chapter 10) of a modified net.

Consider in fact net $N' = (P', T', \boldsymbol{Pre}', \boldsymbol{Post}')$ obtained from N as follows. $P' = P \cup \{p_s, p_f\}$; $T' = T \cup \{t_f\} \cup \{t_p \mid \forall p \in P \setminus \{\bar{p}\}\}$. For $p \in P$ and $t \in T$ it holds $\boldsymbol{Pre}'[p,t] = \boldsymbol{Pre}[p,t]$ and $\boldsymbol{Post}'[p,t] = \boldsymbol{Post}[p,t]$, while the arcs incident on the newly added places and transitions are described in the following. Place p_s is self-looped with all transitions in T, i.e., $\boldsymbol{Pre}'[p_s,t] = \boldsymbol{Post}'[p_s,t] = 1$ for all $t \in T$. Place p_f is self-looped with all new transitions t_p, for all $p \in P \setminus \{\bar{p}\}$. Transition t_f has an input arc from place p_s and an output arc to place p_f; furthermore it has $\bar{m}[p]$ input arcs from any place $p \in P \setminus \{\bar{p}\}$. Finally, for all $p \in P \setminus \{\bar{p}\}$ transition t_p is a sink transition with a single input arc from place p.

We associate to N' an initial marking $\boldsymbol{m}'_0$ defined as follows: for all $p \in P$, $m'_0[p] = m_0[p]$, while $m'_0[p_s] = 1$ and $m'_0[p_f] = 0$. Such a construction is shown in Fig. 12.7 where the original net N with set of places $P = \{\bar{p}, p', \ldots, p''\}$ and set of transitions $T = \{t_1, ldots, t_n\}$ is shown in a dashed box. Arcs with starting and ending arrows represent self-loops.

We claim that a marking in the set $\mathscr{S}(\bar{\boldsymbol{m}}, \bar{p})$ is reachable in the original net if and only if marking $\boldsymbol{m}'_f$ is reachable in $\langle N', \boldsymbol{m}'_0 \rangle$, where $m'_f[\bar{p}] = \bar{m}[\bar{p}]$, $m'_f[p_f] = 1$ and $m'_f[\bar{p}] = 0$ for $p \in P' \setminus \{\bar{p}, p_f\}$.

This can be proved by the following reasoning. The evolution of net N' before the firing of t_f mimics that of N. Transition t_f may only fire from a marking greater than or equal to $\bar{\boldsymbol{m}}$ in all components but eventually $\bar{p}$. After the firing of t_f, the transitions of the original net are blocked (p_s is empty) and only the sink transitions t_p, for all $p \in P \setminus \{\bar{p}\}$, may fire thus emptying the corresponding places. The only place whose markings cannot change after the firing of t_f is $\bar{p}$. □

Theorem 12.3. *Given a generator $J = G \parallel H$ constructed as in step 1 of Algorithm 12.6 it is decidable if it is controllable.*

Proof. We will show that the set of uncontrollable markings to be checked can be written as the finite union of sets of the form $\mathscr{S}(\bar{\boldsymbol{m}}, \bar{p})$.

Given an uncontrollable transition $t \in T_{uc}$ let $P_G(t)$ (resp., $P_H(t)$) be the set of input places of t that belong to generator G (resp., H). Consider now a place $p \in P_H(t)$ and an integer $k \in \{0, 1, \ldots, Pre[p,t] - 1\}$ and define the following marking $\boldsymbol{m}_{t,p,k}$ such that $m_{t,p,k}[p] = k$, $m_{t,p,k}[p'] = Pre[p',t]$ if $p' \in P_G(t)$, else $m_{t,p,k}[p'] = 0$. Clearly,

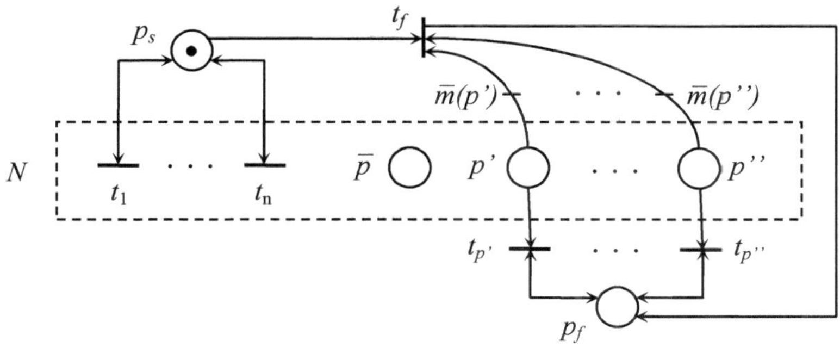

Fig. 12.7 Construction of Lemma 12.1

such a marking is uncontrollable because the places in G contain enough tokens to enable uncontrollable transition t while place p in H does not contain enough tokens to enable it. All other markings in $\mathscr{S}(\boldsymbol{m}_{t,p,k}, p)$ are equally uncontrollable.

Thus the overall set of uncontrollable markings to be checked can be written as the finite union

$$\bigcup_{t\in T_{uc}} \bigcup_{p\in P_H(t)} \bigcup_{k\in\{0,1,\ldots,Pre[p,t]-1\}} \mathscr{S}(\boldsymbol{m}_{t,p,k}, p)$$

and by Lemma 12.1 checking if an uncontrollable marking is reachable is decidable. □

12.5.3 *Trimming a Blocking Generator*

The problem of trimming a blocking net is the following: given a deterministic PN generator G with languages $L_m(G)$ and $L(G) \supset \overline{L_m(G)}$ one wants to modify the structure of the net to obtain a new DES G' such that $L_m(G') = L_m(G)$ and $L(G') = \overline{L_m(G')} = \overline{L_m(G)}$.

On a simple model such as a state machine this may be done, trivially, by removing all states that are reachable but not coreachable (i.e., no final state may be reached from them) and all their input and output edges.

On Petri net models the trimming may be more complex. If the Petri net is bounded, it was shown in the previous section how the trimming may be done without major changes of the net structure, in the sense that one has to add new arcs and eventually duplicate transitions without introducing new places. Here we discuss the general case of possibly unbounded nets.

When the marked language of a net is its L-type Petri net language, the trimming of the net is not always possible as will be shown by means of the following example.

Example 12.10. Let G be the deterministic PN generator in Fig. 12.8 (left), with $\boldsymbol{m}_0 = [1\,0\,0\,0]^T$ and set of final markings $F = \{[0\,0\,0\,1]^T\}$. The marked (L-type) and closed behaviors of this net are: $L_m(G) = \{a^m ba^m b \mid m \geq 0\}$ and $L(G) = \overline{\{a^m ba^n b \mid m \geq n \geq 0\}}$. The infinite reachability graph of this net is partially shown in Fig. 12.8 (right): here the unique final marking is shown in a box.

One sees that all markings of the form $[0\ k\ 0\ 1]^T$ with $k \geq 1$ are blocking. To avoid reaching a blocking marking one requires that p_2 be empty before firing the transition inputting into p_4. However, since p_2 is unbounded this cannot be done with a simple place/transition structure. ■

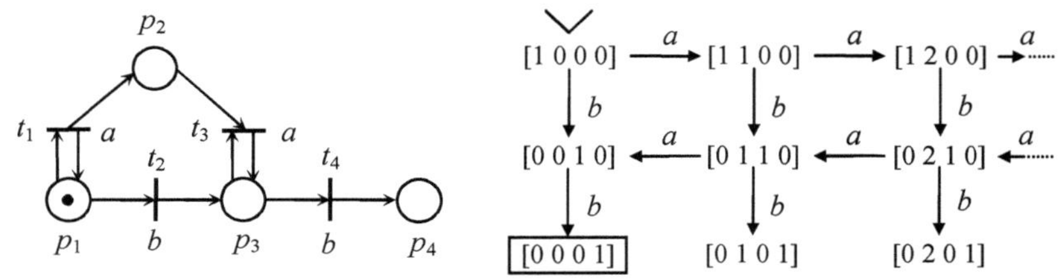

Fig. 12.8 Left: Blocking net of Example 12.10: Right: Its labeled reachability graph

It is possible to prove formally that the prefix closure of the marked language of the net discussed in Example 12.10 is not a P-type Petri net language. The proof is based on the pumping lemma for P-type PN languages, given in [7].

Lemma 12.2. (Pumping lemma). *Consider a PN language $L \in \mathscr{P}$. Then there exist numbers k, l such that any word $w \in L$, with $\mid w \mid \geq k$, has a decomposition $w = xyz$ with $1 \leq \mid y \mid \leq l$ such that $xy^iz \in L, \forall i \geq 1$.*

Proposition 12.2. *Consider the L-type PN language $L' = \{a^m ba^m b \mid m \geq 0\}$. Its prefix closure $L = \overline{L'}$ is not a P-type Petri net language.*

Proof. Given k according to the pumping lemma, consider the word $w = a^k ba^k b \in L$. Obviously, there is no decomposition of this word that can satisfy the pumping lemma. □

When the marked language of a net is its G-type Petri net language, the trimming of the net is always possible because the prefix closure of such a language is a deterministic P-type Petri net language. This follows from the next theorem, that provides an even stronger result.

Theorem 12.4. [4] *Given a deterministic PN generator $G = (N, \ell, \boldsymbol{m}_0, F)$ with $\overline{L_G(G)} \subsetneq L_P(G)$, there exists a finite procedure to construct a new deterministic PN generator G' such that $L_G(G') = L_G(G)$ and $L_P(G') = \overline{L_G(G')}$.*

12.5.4 Trimming an Uncontrollable Generator

In this section we show by means of an example that given a PN generator $J = G \parallel H$ obtained by concurrent composition of a plant and of a specification, it is not always possible to trim it removing the uncontrollable markings.

Example 12.11. Consider a plant G described by the PN generator on the left of Fig. 12.11 (including the dashed transition and arcs). We are interested in the closed language of the net, so we will not specify a set of final markings F: all reachable markings are also final. We assume $T_{uc} = \{t_1, t_3, t_5\}$, i.e., $E_{uc} = \{a\}$.

Consider a specification H described by the PN generator on the left of Fig. 12.9 (excluding the dashed transition and arcs).

On the right of Fig. 12.9 we have represented the labeled reachability graph of G (including the dashed arcs labeled a on the bottom of the graph) and the labeled reachability graph of H (excluding the dashed arcs labeled a on the bottom of the graph). Now if we consider the concurrent composition $J = G \parallel H$ and construct its labeled reachability graph, we obtain a graph isomorphic to the labeled graph of generator H (only the labeling of the nodes changes).

All markings of the form $[0\ k\ 0\ 1]^T$ with $k \geq 1$ are uncontrollable: in fact, when the plant is in such a marking the uncontrollable transition t_5 labeled a is enabled, while no event labeled a is enabled on J. If we remove all uncontrollable markings, we have a generator whose closed language is $L = \{\overline{a^m b a^m b} \mid m \geq 0\}$ that, however, as shown in Proposition 12.2, is not a P-type language. ■

Based on these results in [3] the following result was proven.

Theorem 12.5. The classes $\mathscr{P}_d$, $\mathscr{G}_d$, and $\mathscr{L}_d$ of PN languages are not closed under the supremal controllable sublanguage operator[10].

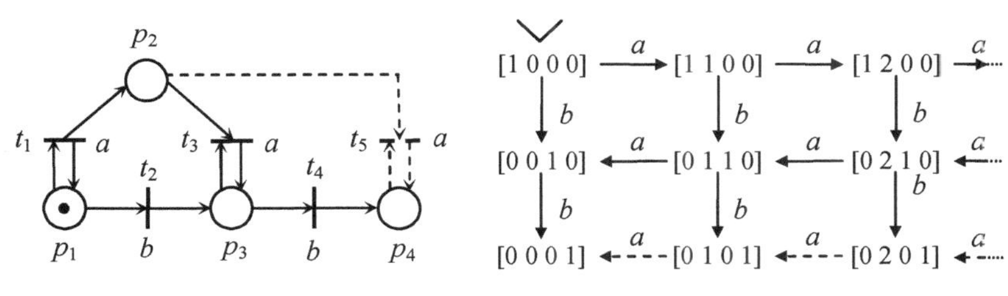

Fig. 12.9 Left: Generators of Example 12.11. Right: Their labeled reachability graphs

[10] See Chapter 3 for a formal definition of this operator.

12.5.5 Final Remarks

The results we have presented in this section showed that in the case of unbounded PN generators a supervisor may not always be represented as a PN. In fact, while it is always possible to check a given specification for nonblockingness and controllability — even in the case of generators with an infinite state space — when these properties are not satisfied the trim behavior of the closed loop system may not be represented as a net. A characterization of those supervisory control problems that admit PN supervisors is an area still open to future research.

12.6 Further Reading

The book by Peterson [12] contains a good introduction to PN languages, while other relevant results can be found in [1, 7, 10, 11, 16, 18].

Many issues related to PNs as discrete event models for supervisory control have been discussed in the survey by Holloway *et al.* [5] and in the works of Giua and DiCesare [3, 4]. The existence of supervisory control policies that enforce liveness have been discussed by Sreenivas in [15, 17].

Finally, an interesting topic that has received much attention in recent years is the supervisory control of PNs under a special class of state specifications called *Generalized Mutual Exclusion Constraints* (GMECs) that can be enforced by controllers called *monitor places* [2]. Several monitor-based techniques have been developed for the control of Petri nets with uncontrollable and unobservable transitions and good surveys can be found in [6, 9].

References

1. Gaubert, S., Giua, A.: Petri net languages and infinite subsets of $\mathbb{N}^m$. Journal of Computer and System Sciences 59, 373–391 (1999)
2. Giua, A., DiCesare, F., Silva, M.: Generalized Mutual Exclusion Constraints for Nets with Uncontrollable Transitions. In: Proc. IEEE Int. Conf. on Systems, Man and Cybernetics, Chicago, USA, pp. 974–799 (1992)
3. Giua, A., DiCesare, F.: Blocking and controllability of Petri nets in supervisory control. IEEE Transactions on Automatic Control 39(4), 818–823 (1994)
4. Giua, A., DiCesare, F.: Decidability and closure properties of weak Petri net languages in supervisory control. IEEE Transactions on Automatic Control 40(5), 906–910 (1995)
5. Holloway, L.E., Krogh, B.H., Giua, A.: A Survey of Petri Net Methods for Controlled Discrete Event Systems. Discrete Event Dynamic Systems 7, 151–190 (1997)
6. Iordache, M.V., Antsaklis, P.J.: Supervision Based on Place Invariants: A Survey. Discrete Event Dynamic Systems 16, 451–492 (2006)
7. Jantzen, M.: Language Theory of Petri Nets. In: Brauer, W., Reisig, W., Rozenberg, G. (eds.) APN 1986. LNCS, vol. 254, pp. 397–412. Springer, Heidelberg (1987)

8. Johnen, C., Frutos Escrig, D.: Decidability of home space property. In: LRI 503. Univ. d'Orsay (1989)
9. Moody, J.O., Antsaklis, P.J.: Supervisory Control of Discrete Event Systems Using Petri Nets. Kluwer (1998)
10. Parigot, M., Pelz, E.: A Logical Formalism for the Study of Finite Behaviour of Petri Nets. In: Rozenberg, G. (ed.) APN 1985. LNCS, vol. 222, pp. 346–361. Springer, Heidelberg (1986)
11. Pelz, E.: Closure Properties of Deterministic Petri Net Languages. In: Brandenburg, F.J., Wirsing, M., Vidal-Naquet, G. (eds.) STACS 1987. LNCS, vol. 247, pp. 373–382. Springer, Heidelberg (1987)
12. Peterson, J.L.: Petri Net Theory and the Modeling of Systems. Prentice-Hall, Englewood Cliffs (1981)
13. Ramadge, P.J., Wonham, W.M.: The control of discrete event systems. Proceedings of the IEEE 77(1), 81–98 (1989)
14. Reutenauer, C.: The Mathematics of Petri Nets. Masson and Prentice-Hall (1990)
15. Sreenivas, R.S.: On the existence of supervisory policies that enforce liveness in discrete-event dynamic systems modeled by controlled Petri nets. IEEE Transactions on Automatic Control 42(7), 928–945 (1997)
16. Sreenivas, R.S.: On minimal representations of Petri net languages. IEEE Transactions on Automatic Control 51(5), 799–804 (2006)
17. Sreenivas, R.S.: On the Existence of Supervisory Policies That Enforce Liveness in Partially Controlled Free-Choice Petri Nets. IEEE Transactions on Automatic Control 57(2), 435–449 (2012)
18. Vidal-Naquet, G.: Deterministic Petri net languages. In: Girault, C., Reisig, W. (eds.) Application and Theory of Petri Net. Informatick-Fachberichte, vol. 52. Springer, New York (1982)

Chapter 13
Structural Methods for the Control of Discrete Event Dynamic Systems – The Case of the Resource Allocation Problem

Juan-Pablo López-Grao and José-Manuel Colom

13.1 Introduction

Resource scarceness is a traditional scenario in many systems engineering disciplines. In such cases, available resources may be shared among concurrent processes, which must compete in order to be granted their allocation. Discrete Event Dynamic Systems (DEDSs) of this kind are named Resource Allocation Systems (RASs). In this paper, we revisit a family of Petri net-based deadlock handling methodologies which have been successfully applied in many application domains, such as Flexible Manufacturing Systems (FMSs) [6], multicomputer interconnection networks [25] or multithreaded software engineering [20]. These methodologies are based in the attention to the RAS view of the system and are basically deployed in three stages.

The first stage is that of abstraction and modelling, in which physical details of the system not relevant to the Resource Allocation Problem (RAP) are discarded, obtaining a Petri net as outcome. The Petri net is processed in the analysis stage, in which potential deadlock situations due to diseased resource allocation patterns are inspected. At the third stage, that of synthesis and implementation, the Petri net is corrected in order to obtain a deadlock-free system, usually by the addition of virtual resources. This correction is finally unfolded in the real system by the correction of the processes involved and the inclusion of control mechanisms. Often structural results exist which enable powerful structure-based analysis and synthesis techniques for identifying and fixing potential or factual deadlocks [6, 22, 28].

Juan-Pablo López-Grao
Department of Computer Science and Systems Engineering, University of Zaragoza, Spain
e-mail: jpablo@unizar.es

José-Manuel Colom
Aragon Institute of Engineering Research (I3A), University of Zaragoza, Spain
e-mail: jm@unizar.es

C. Seatzu et al. (Eds.): Control of Discrete-Event Systems, LNCIS 433, pp. 257–278.
springerlink.com

Regarding the first stage of abstraction, and with a view to obtain a useful, descriptive but tractable model, it is necessary to have into account the different characteristics and physical restrictions derived from the application domain. Consequently, the syntax of the RAS models obtained for different domains can differ notably. On the following, we address a classification of the diverse models developed in the literature, and their specific features with regard to RAS modelling.

Basically, restrictions on Petri net models for RASs can be classified within two categories: (a) on the processes structure, and (b) on the way the processes use the resources. As far as (b) is concerned, resources can be serially reusable (i.e., they are used in a conservative way by all processes), or consumable (i.e., they are consumed and not regenerated). This work focuses on RAS with serially reusable resources.

Regarding the processes structure, most of the current works focus on Sequential RASs (S-RASs), as opposed to Non-Sequential RASs (NS-RASs), in which assembly/disassembly operations are allowed within the processes. Some works ([29, 9]), however, have attempted to approach NS-RASs from the Petri nets perspective, despite that finding effective solutions for them is, in general, much more complicated.

In the field of S-RASs (the scope of this paper), different Petri net models have successively emerged, frequently extending previous results and hence widening the subclass of systems that can be modelled and studied.

One of the first classes aimed to deal with the resource allocation problem in S-RASs is the class of Cooperating Sequential Systems (CSSs) [14]. In CSS, concurrent processes share both the routing pattern and the way the resources are used in the routes (i.e., the process type is unique). These processes may compete for several resource types, allowing multiple instances of each type.

In more recent works, different process types with multiple concurrent instances are allowed, sometimes allowing alternative paths per process. In [10], the path (i.e., route) a process will follow is selected at the beginning of the process execution. Other works consider on-line routing decisions; in particular, this is dealt with in [6], where the seminal System of Simple Sequential Processes with Resources (S^3PR) class is introduced. However, processes in an S^3PR can use at most a single resource unit at a given state. A subclass of S^3PR, called Linear S^3PR (L-S^3PR), was presented in [7], which featured some useful properties.

The mentioned restriction over resources usage is eliminated by the (more general) S^4PR class [27]. This allows processes to simultaneously reserve several resources belonging to distinct types. Many others [13, 29] were defined for this aim, with specific attributes for modelling different configurations.

In Section 13.2, the most general Petri net classes for RASs are reviewed and categorized. In Section 13.3, the concept of insufficiently marked siphon is introduced as an artifact to approach the liveness problem in RASs from the system structure. It will be shown, however, that this artifact falls short on characterizing liveness in the context of multithreaded control software. In Section 13.4, an iterative control policy for RASs is presented. Section 13.5 summarizes the conclusions.

13.2 Abstraction and Modelling: Class Definitions and Relations

S^{4}PR nets are modular models composed of state machines with no internal cycles plus shared resources. One of the most interesting features of this kind of models is their composability. Two S^{4}PR nets can be composed into a new S^{4}PR model via fusion of the common resources. Since multiple resource reservation is allowed, S^{4}PR nets are not ordinary, i.e., the weight of the arcs from the resources to the state machines (or vice versa) is not necessarily equal to one, in contrast to S^{3}PR nets.

Definition 13.1. [28] *Let I_N be a finite set of indices. An S^{4}PR is a connected generalized pure P/T net $N = (P, T, \boldsymbol{Pre}, \boldsymbol{Post})$ where:*

1. $P = P_0 \cup P_S \cup P_R$ *is a partition such that:*
   a. [idle places] $P_0 = \bigcup_{i \in I_N} \{p_{0_i}\}$.
   b. [process places] $P_S = \bigcup_{i \in I_N} P_{S_i}$, *where:*
   $\forall i \in I_N : P_{S_i} \neq \emptyset$, *and* $\forall i, j \in I_N, i \neq j : P_{S_i} \cap P_{S_j} = \emptyset$.
   c. [resource places] $P_R = \{r_1, r_2, r_3, ..., r_n\}, n > 0$.
2. $T = \bigcup_{i \in I_N} T_i$, *where* $\forall i \in I_N : T_i \neq \emptyset$, *and* $\forall i, j \in I_N, i \neq j : T_i \cap T_j = \emptyset$.
3. [i-th process subnet] *For each $i \in I_N$ the subnet generated by $\{p_{0_i}\} \cup P_{S_i}, T_i$ is a strongly connected state machine such that every cycle contains p_{0_i}.*
4. *For each $r \in P_R$ there exists a unique minimal p-semiflow associated to r, $\boldsymbol{y}_r \in \mathbb{N}^{|P|}$, fulfilling:* $\{r\} = \|\boldsymbol{y}_r\| \cap P_R, P_0 \cap \|\boldsymbol{y}_R\| = \emptyset, P_S \cap \|\boldsymbol{y}_r\| \neq \emptyset$, *and* $y_r[r] = 1$.[1]
5. $P_S = \bigcup_{r \in P_R} (\|\boldsymbol{y}_r\| \setminus \{r\})$.

Fig. 13.1 depicts a net system belonging to the S^{4}PR class. Places $R1$ and $R2$ are the *resource places*. A resource place represents a resource type, and the number of tokens in it represents the quantity of free instances of that resource type. If we remove these places, we get two isolated state machines. These state machines represent the different patterns of resource reservation that a process can follow. In the context of FMSs, these two state machines model two different production plans.

Consequently, tokens in a state machine represent parts which are being processed in stages of the same production plan. At the initial state, the unique tokens in each machine are located at the so-called *idle place* (here: $A0$, $B0$). In general, the idle place can be seen as a mechanism which limits the maximum number of concurrent parts being processed in the same production plan. The rest of places model the various stages of the production plan as far as resource reservation is concerned.

Meanwhile, the transitions represent the acquisition or release of resources by the processes along their evolution through the production plan. Every time a transition fires, the total amount of resources available is altered while the part advances to the next stage. The weight of an arc connecting a resource with a transition models the number of instances which are allocated or released when a part advances.

[1] The support of a p-semiflow $\boldsymbol{y}$ (marking $\boldsymbol{m}$), denoted $\|\boldsymbol{y}\|$ ($\|\boldsymbol{m}\|$), is the set of places such that their corresponding components in the vector $\boldsymbol{y}$ ($\boldsymbol{m}$) are non-null.

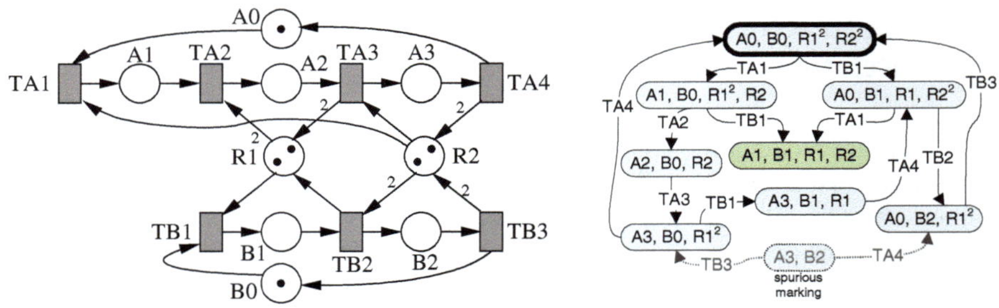

Fig. 13.1 A (non-live) S^4PR with an acceptable initial marking

For instance, place $R1$ could model a set of free robotic arms used to process parts in the stage $A2$ of the first production plan (two arms are needed per each part processed there) as well as in the stage $B1$ of the second production plan (only one arm needed per part processed). Consequently, if transition $TB1$ is fired from the initial marking then one robotic arm will be allocated and one part will visit stage $B1$. Still, there will remain one robotic arm to be used freely by other processes.

Finally, it is worth noting that moving one isolated token of a state machine (by firing its transitions) until the token reaches back the idle state, leaves the resource places marking unaltered. Thus, the resource usage is conservative.

The next definition formalizes the fact that there should exist enough free resource instances in the initial state so as that every production plan is realizable:

Definition 13.2. [28] *Let* $N = (P, T, \boldsymbol{Pre}, \boldsymbol{Post})$ *be an* S^4PR*. An initial marking* $\boldsymbol{m}_0$ *is acceptable for N iff* $\|\boldsymbol{m}_0\| = P_0 \cup P_R$ *and* $\forall p \in P_S, r \in P_R : m_0[r] \geq y_r[p]$.

Nowadays, the most general class of the S^nPR family is the S^*PR class [8], in which processes are ordinary state machines with internal cycles. Many interesting works from different authors present and study other classes in the same vein. For a more detailed revision of these, we refer the reader to [5].

All the aforementioned Petri net classes are frequently presented in the context of FMS modelling, and make sense as artifacts conceived for properly modelling significant physical aspects of this kind of systems. For all of these, except for S^*PR, a siphon-based liveness characterization is known. Due to its structural nature, it opens a door to an efficient detection and correction of deadlocks, by implementing controllers (usually by the addition of places) that restrain the behaviour of the net and avoid the bad markings to be reached.

Although there exist obvious resemblances between the RAP in FMSs and that in parallel or concurrent software, previous attempts to bring these well-known RAS techniques into the analysis of multithreaded control software has been, to the best of our knowledge, either too limiting or unsuccessful. Gadara nets [30] constitute the most recent attempt, yet they fall in the overrestrictive side in the way the resources can be used [19]. Presumably, this is a consequence of being conceived with a primary focus on inheriting the powerful structural liveness results which were fruitful in the context of FMSs. Such a bias works against obtaining a model class capable of properly abstracting RASs in many multithreaded systems [19]. In [20], it is basically analyzed why the net classes and results introduced in the context of FMSs can fail when brought to the field of concurrent programming.

In [20], the class of Processes Competing for Conservative Resources (PC^2R) is introduced. This class is aimed to overcome the deficiencies identified in finding models which properly capture the RAS view of multithreaded software systems. Furthermore, it generalizes other subclasses of the S^nPR family while respecting the design philosophy on these. Hence, previous results are still valid in the new framework. However, PC^2R nets can deal with more complex scenarios which were not yet addressed from the domain of S^nPR nets. The generalization is also useful in the context of FMS configuration but especially in other scenarios where the following elements are more frequent: (1) Internal iterations (e.g., recirculating circuits in manufacturing, nested loops in software); (2) Initial states in which there are resources that are already allocated.

Definition 13.3 presents a subclass of state machines used for modelling the control flow of the processes of a PC^2R net in isolation. Iterations are allowed, as well as decisions within internal cycles, in such a way that the control flow of structured programs can be fully supported. Non-structured processes can be refactored into structured ones as discussed in [20].

Definition 13.3. [20] *An* iterative state machine $N = (P,T,\boldsymbol{Pre},\boldsymbol{Post})$ *is a strongly connected state machine such that: (i) P can be partitioned into three subsets:* $\{p_k\}$, P_1 *and* P_2, *(ii)* $P_1 \neq \emptyset$, *(iii) The subnet generated by* $\{p_k\} \cup P_1, {}^\bullet P_1 \cup P_1{}^\bullet$ *is a strongly connected state machine in which every cycle contains* p_k, *and (iv) If* $P_2 \neq \emptyset$, *the subnet generated by* $\{p_k\} \cup P_2, {}^\bullet P_2 \cup P_2{}^\bullet$ *is an iterative state machine.*

As Fig. 13.2 shows, P_1 contains the set of places of an outermost iteration block, while P_2 is the set of places of the rest of the state machine (the inner structure, which may contain multiple loops within). Consequently, the subnet generated by $\{p_k\} \cup P_1, {}^\bullet P_1 \cup P_1{}^\bullet$ is a strongly connected state machine in which every cycle contains p_k. Meanwhile, inner iteration blocks can be identified in the iterative state machine generated by $\{p_k\} \cup P_2, {}^\bullet P_2 \cup P_2{}^\bullet$. The place p_0 represents the place "p_k" that we choose after removing every iteration block.

The definition of iterative state machine is instrumental for introducing the class of PC^2R nets. PC^2R nets are modular models composed by iterative state machines and shared resources. Two PC^2R nets can be composed into a new PC^2R model via

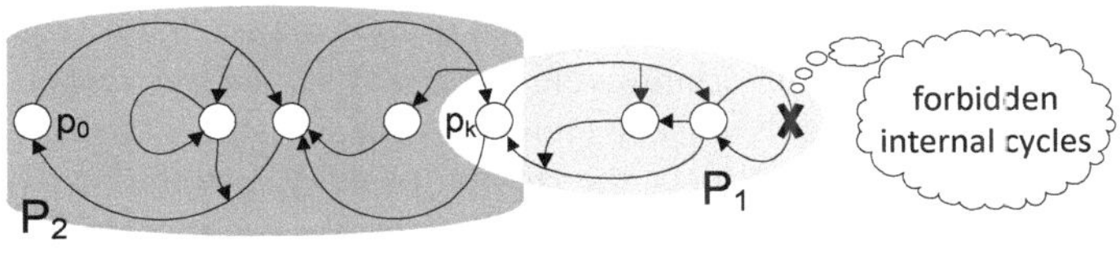

Fig. 13.2 Schematic diagram of an iterative state machine. For simplicity, no transition is drawn

fusion of the common resources. Note that, a PC^{2}R net can simply be one process modelled by an iterative state machine along with the set of resources it uses.

The class supports iterative processes, multiple resource acquisitions, non-blocking wait operations and resource lending. Inhibition mechanisms are not natively supported (although some cases can still be modelled with PC^{2}R nets).

Definition 13.4. [20] *Let I_N be a finite set of indices. A PC^{2}R net is a connected generalized self-loop free P/T net $N = (P, T, \mathbf{Pre}, \mathbf{Post})$ where:*

1. $P = P_0 \cup P_S \cup P_R$ *is a partition such that:*
   a. [idle places] $P_0 = \bigcup_{i \in I_N} \{p_{0_i}\}$.
   b. [process places] $P_S = \bigcup_{i \in I_N} P_{S_i}$, *where:*
   $\forall i \in I_N : P_{S_i} \neq \emptyset$, *and* $\forall i, j \in I_N, i \neq j : P_{S_i} \cap P_{S_j} = \emptyset$.
   c. [resource places] $P_R = \{r_1, r_2, r_3, ..., r_n\}, n > 0$.
2. $T = \bigcup_{i \in I_N} T_i$, *where* $\forall i \in I_N : T_i \neq \emptyset$, *and* $\forall i, j \in I_N, i \neq j : T_i \cap T_j = \emptyset$.
3. [i-th process subnet] *For each $i \in I_N$ the subnet generated by $\{p_{0_i}\} \cup P_{S_i}, T_i$ is an iterative state machine.*
4. *For each $r \in P_R$, there exists a unique minimal p-semiflow associated to r, $\mathbf{y}_r \in \mathbb{N}^{|P|}$, fulfilling:* $\{r\} = \|\mathbf{y}_r\| \cap P_R$, $(P_0 \cup P_S) \cap \|\mathbf{y}_r\| \neq \emptyset$, *and* $y_r[r] = 1$.
5. $P_S \subseteq \bigcup_{r \in P_R} (\|\mathbf{y}_r\| \setminus \{r\})$.

Figure 13.3 depicts a PC^{2}R net which models a special version of the classic philosophers problem introduced in [20]. Since this net is modelling software, the state machines represent the control flow for each type of philosopher (thread). Tokens in a state machine represent concurrent processes/threads which share the same control flow. At the initial state, every philosopher is thinking, i.e., the unique token in each machine is located at the idle place. Note that, in order to have a concise model, we considered the simplest case in which there exist only two philosophers.

The resources (here: the bowl of spaghetti and two forks) are shared among both philosophers. From a real-world point of view, the resources in this context are not necessarily physical (e.g., a file) but can also be logical (e.g., a semaphore). The fact that resources in software engineering do not always have a physical counterpart is a peculiar characteristic with consequences. In this context, processes do not only consume resources but also can *create* them. A process will destroy the newly created resources before its termination. For instance, a process can create a shared memory variable (or a service!) which can be allocated to other processes/threads. Hence, the resource allocation scheme is no longer *first-acquire-later-release*, but it can be the other way round too. Still, all the resources are used conservatively by the processes (either by a create–destroy sequence or by a wait–release sequence). As a side effect, and perhaps counterintuitively, there may not be free resources during the system startup (as they still must be created), yet the system is live.

As a result, and unlike S^{4}PR nets, the support of the $\mathbf{y}_r$ p-semiflows (point 4 of Definition 13.4) may include P_0. For such a resource place r, there exists at least a process which creates (*lends*) instances of r. As a consequence, there might exist additional minimal p-semiflows containing more than one resource place [20].

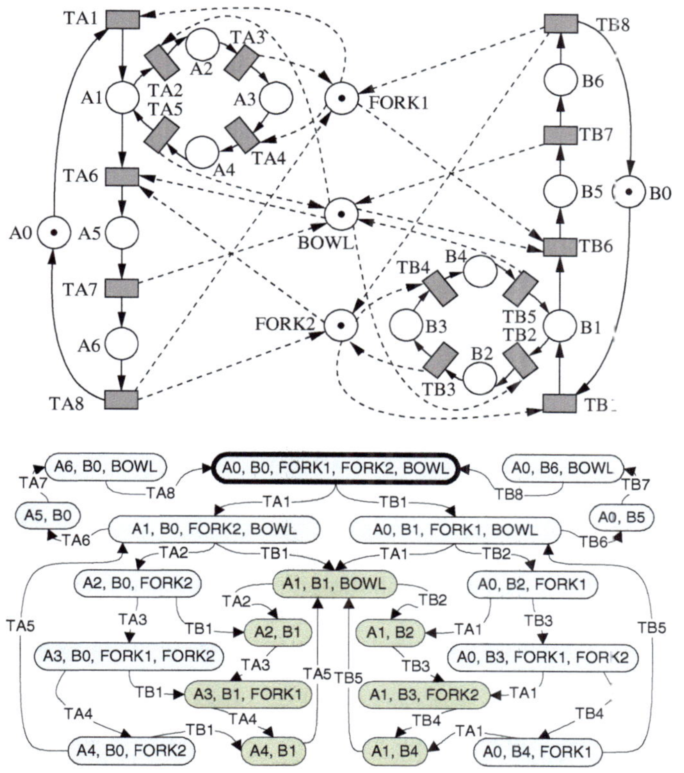

Fig. 13.3 A (non-live) PC^{2}R with a potentially acceptable initial marking

The next definition is strongly related to the notion of acceptable initial marking introduced for the S^{4}PR class. In software systems, all processes/threads are initially inactive and start from the same point (the `begin` statement). Hence, all of the corresponding tokens are in the idle place at the initial marking (the process places being therefore empty). The definition takes this into account and establishes a lower bound for the marking of the resource places.

Definition 13.5. [20] *Let* $N = (P, T, \boldsymbol{Pre}, \boldsymbol{Post})$ *be a PC*^{2}*R. An initial marking* $\boldsymbol{m}_0$ *is potentially acceptable for N iff* $\|\boldsymbol{m}_0\| \setminus P_R = P_0$*, and* $\forall p \in P_S, r \in P_R : \boldsymbol{y}_r^T \cdot \boldsymbol{m}_0 \geq y_r[p]$.

The above definition is syntactically a generalization of the concept of acceptable initial marking for S^{4}PR nets. If the initial marking of some resource place r is lesser than the lower bound established for $m_0[r]$ by Definition 13.5, then there exists at least one dead transition at the initial marking.

At this point, it is worth stressing that an S^{4}PR net with an acceptable initial marking cannot have any dead transition at the initial marking (since every minimal t-semiflow is firable in isolation from it). For PC^{2}R nets, however, having a marking which is greater than that lower bound does not guarantee, in the general case, that there do not exist dead transitions, as discussed in Section 13.3. Accordingly, the preceding adverb *potentially* stresses this fact, in contrast to S^{4}PR nets.

Furthermore, according to Definition 13.5, the initial marking of some resource place r may be empty if some idle place is covered by $\|\mathbf{y}_r\|$, as seen later.

Since the PC^{2}R class generalizes previous classes in the S^{n}PR family, these can be redefined in the new framework as follows.

Definition 13.6. [20] *Previous classes of the S^{n}PR family are defined as follows:*

- *An S^{5}PR [18] is a PC^{2}R where $\forall r \in P_R : \|\mathbf{y}_r\| \cap P_0 = \emptyset$.*
- *An S^{4}PR [28] is an S^{5}PR where $\forall i \in I_N$ the subnet generated by $\{p_{0_i}\} \cup P_i, T_i$ is a strongly connected state machine in which every cycle contains p_{0_i} (i.e., a iterative state machine with no internal cycles).*
- *An S^{3}PR [6] is an S^{4}PR where $\forall p \in P_S : |^{\bullet\bullet}p \cap P_R| = 1, (^{\bullet\bullet}p \cap P_R = p^{\bullet\bullet} \cap P_R)$.*
- *An L-S^{3}PR [7] is an S^{3}PR where $\forall p \in P_S : |^{\bullet}p| = |p^{\bullet}| = 1$.*

Remark 13.1. L-S^{3}PR $\subseteq$ S^{3}PR $\subseteq$ S^{4}PR $\subseteq$ S^{5}PR $\subseteq$ PC^{2}R. ■

The preceding remark is straightforward from Definition 13.6. It is worth remarking that Definition 13.5 collapses with the definition of acceptable initial markings respectively provided for those subclasses [6, 7, 28, 20]. For all of these, the same properties than for S^{4}PR (i.e., no dead transitions at $\boldsymbol{m}_0$, non-empty resources) apply.

Finally, there exists another class for S-RASs, called System of Processes Quarrelling over Resources (SPQR) [18], which does not strictly contain or is contained by the PC^{2}R class. Yet, there exist transformation rules to travel between PC^{2}Rs and Structurally Bounded (SB) SPQRs. Note that, by construction, PC^{2}R nets are conservative, and hence SB, but this is not true for general SPQRs. The SPQR class seems interesting from an analytical point of view thanks to its syntactic simplicity, as discussed in [18].

In the following, we will make clear that the approach followed to date does not work with the new net classes for modelling multithreaded control applications (PC^{2}R, SPQR). In this sense, there does not exist any analogous non-liveness characterization, and their inherent properties are much more complex. In particular, we would like to stress the fact that siphons do not longer work, in general, with the aforementioned superclasses.

Figure 13.4 summarizes the inclusion relations between the reviewed Petri net classes for S-RASs. The lefter on the x-axis, the more complex the process structure can be (i.e., linear state machines are on the right while general state machines are on the left). The upper on the y-axis, the higher degree of freedom in the way the processes use the resources (resource lending is on top). The figure also illustrates the fact that every model of the S^{n}PR family can be transformed into a PC^{2}R net.

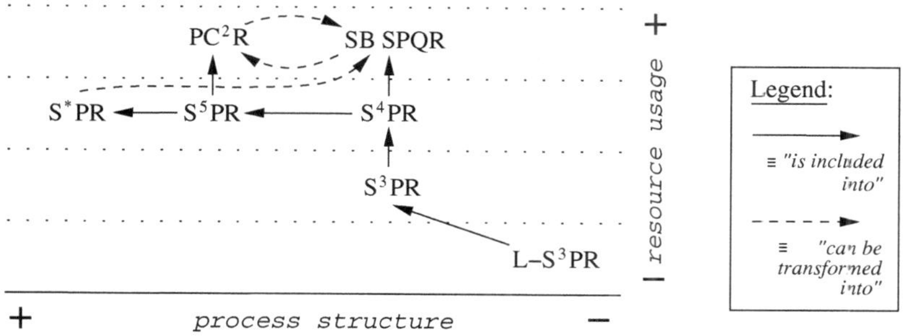

Fig. 13.4 Inclusion relations between Petri net classes for RASs

13.3 Liveness Analysis and Related Properties

Traditionally, empty or insufficiently marked siphons have been a fruitful structural element for characterizing non-live RASs. The more general the net class, however, the more complex the siphon-based characterization is. The following results can be easily obtained from previously published works. The originality here is to point out the strict conditions that the siphons must fulfil.

Theorem 13.1. [6, 20] *Let* $\langle N, \boldsymbol{m}_0 \rangle$ *be a marked S*^{3}*PR with an acceptable initial marking.* $\langle N, \boldsymbol{m}_0 \rangle$ *is non-live iff* $\exists \boldsymbol{m} \in R(N, \boldsymbol{m}_0)$ *and a* <u>*minimal*</u> *siphon* $D : \boldsymbol{m}[D] = \boldsymbol{0}$.

For instance, the marked S^{3}PR in Fig. 13.5 is non-live with $K_0 = K_1 = 1$, $K_3 = 2$. From this acceptable initial marking, the marking $(A4 + B4 + R2 + 2 \cdot R3)$ can be reached by firing σ, where $\sigma = TB1\ TA1\ TB2\ TA2\ TB3\ TA3\ TB4\ TA4$. This firing sequence empties the minimal siphon $\{A1, B1, A5, B5, R1, R4\}$.

However, this characterization is sufficient, but not necessary, in general, for S^{4}PR nets. Hence, the concept of *empty siphon* had to be generalized. Given a

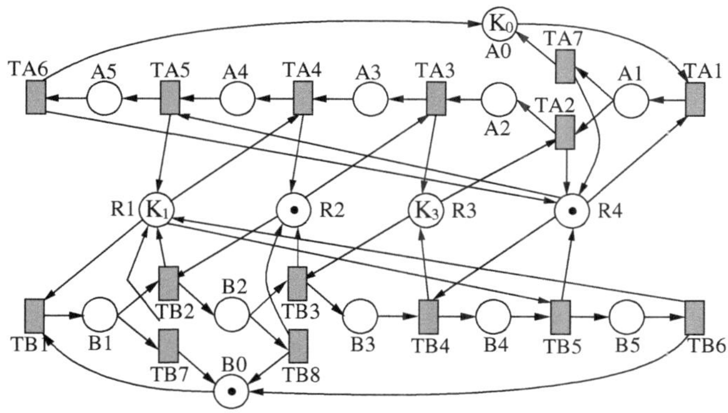

Fig. 13.5 An S^{3}PR which is non-live iff $(K_0 \geq K_1, K_3 \geq 2) \vee (K_0 \cdot K_1 \cdot K_3 = 0)$. Note that $\boldsymbol{m}_0$ is an acceptable initial marking iff $(K_0 \cdot K_1 \cdot K_3 \neq 0)$

marking $\boldsymbol{m}$ in an S^{4}PR net, a transition t is said to be $\boldsymbol{m}$-process-enabled ($\boldsymbol{m}$-process-disabled) iff it has (not) a marked input process place, and $\boldsymbol{m}$-resource-enabled ($\boldsymbol{m}$-resource-disabled) iff its input resource places have (not) enough tokens to fire it, i.e., $\boldsymbol{m}[P_R,t] \geq \boldsymbol{Pre}[P_R,t]$ ($\boldsymbol{m}[P_R,t] \not\geq \boldsymbol{Pre}[P_R,t]$).

Theorem 13.2. [28] *Let $\langle N, \boldsymbol{m}_0 \rangle$ be a marked S^{4}PR with an acceptable initial marking. $\langle N, \boldsymbol{m}_0 \rangle$ is non-live iff $\exists \boldsymbol{m} \in R(N, \boldsymbol{m}_0)$ and a siphon D such that:* i) *There exists at least one* $\boldsymbol{m}$*-process-enabled transition;* ii) *Every* $\boldsymbol{m}$*-process-enabled transition is* $\boldsymbol{m}$*-resource-disabled by some resource place in D;* iii) *All process places in D are empty at* $\boldsymbol{m}$.

Such a siphon D is said to be insufficiently marked at $\boldsymbol{m}$ (also: *bad* siphon). In Theorems 13.1 and 13.2, the siphon captures the concept of circular wait, revealing it from the underlying net structure. In contrast to the S^{3}PR class, it is worth noting the following fact about *minimal* siphons in S^{4}PR nets, which emerges because of their minimal p-semiflows not being strictly binary.

Property 1. [20] *There exists a S^{4}PR net with an acceptable initial marking which is non-live but every siphon characterizing the non-liveness is non-minimal, i.e., minimal siphons are insufficient to characterize non-liveness.*

For instance, the S^{4}PR net in Fig. 13.1 is non-live, but there is no minimal siphon containing both resource places $R1$ and $R2$. Note that, the siphon $D = \{R1, R2, A3, B2\}$ becomes insufficiently marked at $\boldsymbol{m}$, where $\boldsymbol{m} = A1 + B1 + R1 + R2$, but it contains the minimal siphon $D' = \{R2, A3, B2\}$. D' is not insufficiently marked for any reachable marking. It is also worth noting that no siphon is ever emptied.

Thus, non-minimal siphons must be considered in order to deal with deadlocks in systems more complex than S^{3}PR.

On the other hand, insufficiently marked siphons (even considering those non-minimal) are not enough for characterizing liveness for more complex systems such as S^{5}PR models. This means that siphon-based control techniques for RASs do not work in general for concurrent software, even in the 'good' case in which every `wait`-like operation precedes its complementary `signal`-like operation.

Property 2. [20] *There exists an S^{5}PR with an acceptable initial marking $\langle N, \boldsymbol{m}_0 \rangle$ which is non-live but insufficiently marked siphons do not characterize non-liveness (dead markings).*

The S^{5}PR net in Fig. 13.3 evidences the claim stated above. The figure depicts a non-live system with three possibly bad siphons. These siphons are $D_1 = \{A2, A3, A4, A5, A6, B2, B4, B5, B6, FORK2, BOWL\}$, $D_2 = \{A2, A4, A5, A6, B2, B3, B4, B5, B6, FORK1, BOWL\}$ and $D_3 = \{A2, A4, A5, A6, B2, B4, B5, B6, FORK1, FORK2, BOWL\}$. Besides, every transition in the set $\Omega = \{TA2, TA3, TA4, TA5, TB2, TB3, TB4, TB5\}$ is an output transition of D_1, D_2 and D_3. After firing transitions $TA1$ and $TB1$ starting from $\boldsymbol{m}_0$, the state $A1 + B1 + BOWL$ is reached. This marking belongs to a livelock with other six markings. The reader can check that there exists a firable transition in Ω for every marking in the livelock, and in any case there is no insufficiently marked siphon.

In other words, for every reachable marking in the livelock, there exist output transitions of the siphons which are firable. As a result, the siphon-based non-liveness characterization for earlier net classes (such as S^4PR [28]) is not sufficient in the new framework.

This is an a priori unexpected result since these nets fulfil some strong structural properties which also S^4PR nets hold. PC^2R nets are well-formed, i.e., SB and Structurally Live (SL), and therefore also conservative and consistent [20]. Structural boundedness is straightforward, since PC^2R nets are conservative by construction (every place is covered by at least one minimal p-semiflow), and a well-known general result of Petri nets is that conservativeness implies structural boundedness [26]. Structural liveness is derived from the fact that every resource place is a structurally implicit place, and therefore its initial marking can be increased enough so as to make it implicit. When every resource place is implicit, the net system behaves like a set of isolated and marked strongly-connected state machines, and therefore the system is live. Obviously, all the subclasses (S^3PR, S^4PR, etc.) are also well-formed. However, SPQR nets are not necessarily SL [18].

By carefully observing the net in Fig. 13.3, it might seem that the difficulty in finding a liveness characterization for PC^2R nets lies in the appearance of certain types of livelocks. In general, livelocks with dead transitions are not a new phenomenon in the context of Petri net models for RASs. Fig. 13.6 shows that, even for L-S^3PR nets, deadlock-freeness does not imply liveness.

Property 3. [7] *There exists a marked L-S^3PR with an acceptable initial marking such that it is deadlock-free but not live.*

This L-S^3PR net system has no deadlock but two reachable livelocks: (i) $\{(A0+B2+C0+D1+R1),(A1+B2+C0+D1)\}$, and (ii) $\{(A0+B1+C0+D2+R3),(A0+B1+C1+D2)\}$. Nevertheless, these livelocks are captured by insufficiently marked siphons. Unfortunately, this no longer holds for some kind of livelocks in S^5PR or more complex systems. Indeed, PC^2R nets feature some complex properties which complicate the finding of a liveness characterization.

Another relevant property for studying liveness is its monotonicity. In this sense, the negative results emerge further back than probably expected.

Property 4. [20] *There exists an S^3PR such that liveness is not monotonic, neither with respect to the marking of the idle/process places, nor that of the resource places, i.e., liveness is not always preserved when those are increased.*

The net depicted in Fig. 13.5 illustrates this fact:

- With respect to P_R: The system is live with $K_0 = K_1 = K_3 = 1$ and non-live with $K_0 = K_1 = 1$, $K_3 = 2$ (however, it becomes live again if the marking of $R1$, $R2$ and $R4$ is increased enough so as to make every resource place an implicit place).
- With respect to P_0: The system is live with $K_0 = 1$, $K_1 = K_3 = 2$ and non-live with $K_0 = K_1 = K_3 = 2$.

Note that, liveness is monotonic for every net belonging to the L-S^3PR class [7] with respect to the resource places. But, from S^3PR nets upwards, there is a discontinuity zone between the point where the resource places are empty enough so that

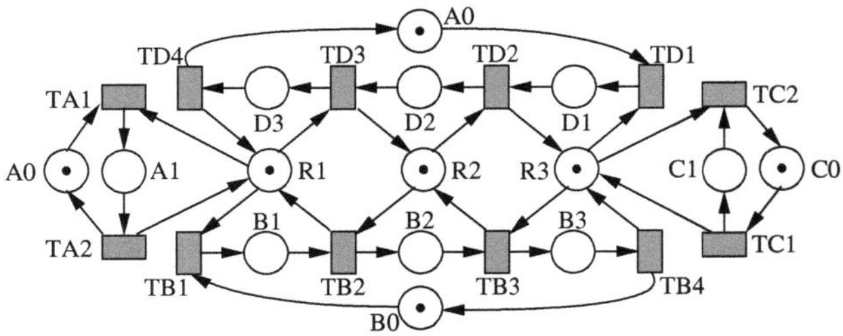

Fig. 13.6 A non-live L-S^{3}PR which is deadlock-free

every transition is dead (also held for lower markings), and the point where every resource place is implicit (liveness is preserved if their marking is increased). Markings within these bounds fluctuate between liveness and non-liveness. The location of those points also depends on the marking of the idle/process places: the more tokens in them, the farther the saturation point.

Nevertheless, an interesting property of S^{4}PR nets is that liveness equals reversibility. This, along with the fact that the idle place does not belong to any p-semiflow $\boldsymbol{y}_r$, is a powerful feature. If every token in a process net can be moved to the idle place, then the net is not dead (yet).

Theorem 13.3. [20] *Let* $\langle N, \boldsymbol{m}_0 \rangle$ *be an S^{4}PR with an acceptable initial marking.* $\langle N, \boldsymbol{m}_0 \rangle$ *is live iff* $\boldsymbol{m}_0$ *is a home state (i.e., the system is reversible).*

However, Theorem 13.3 is false in general for S^{5}PR nets. In fact, the directedness property [1] does not even hold: a live S^{5}PR may have no home state.

Property 5. [20] *There exists an S^{5}PR with an acceptable initial marking* $\langle N, \boldsymbol{m}_0 \rangle$ *such that the system is live but there is no home state.*

The net system in Fig. 13.7 has no home state in spite of being live. It is worth noting that this net is ordinary. Yet S^{5}PR nets still retain an interesting property: its minimal t-semiflows are eventually realizable from an acceptable initial marking.

Theorem 13.4. [20] *Let* $\langle N, \boldsymbol{m}_0 \rangle$ *be an S^{5}PR with an acceptable initial marking. For every (minimal) t-semiflow* $\boldsymbol{x}$*, there exists a reachable marking* $\boldsymbol{m} \in R(N, \boldsymbol{m}_0)$ *such that* $\boldsymbol{x}$ *is realizable from* $\boldsymbol{m}$*, i.e.,* $\exists \sigma$ *such that* $\boldsymbol{m} \xrightarrow{\sigma}$, $\boldsymbol{\sigma} = \boldsymbol{x}$*, where* $\boldsymbol{\sigma}$ *is the firing count vector of* σ.

However, for PC^{2}R nets there may not exist minimal t-semiflows being eventually realizable; even for live systems.

Property 6. [20] *There exists a PC^{2}R with a potentially acceptable initial marking* $\langle N, \boldsymbol{m}_0 \rangle$ *such that the system is live and there exists a minimal t-semiflow* $\boldsymbol{x}$ *such that* $\forall \boldsymbol{m} \in R(N, \boldsymbol{m}_0)$, $\nexists \sigma$ *such that* $\boldsymbol{m} \xrightarrow{\sigma}$ *and* $\boldsymbol{\sigma} = \boldsymbol{x}$*, i.e.* $\boldsymbol{x}$ *is not realizable from* $\boldsymbol{m}$.

The reader can check that the PC^{2}R net system in Fig. 13.8 has no home state in spite of being live. Depending on which transition is fired first (either $T1$ or $T8$)

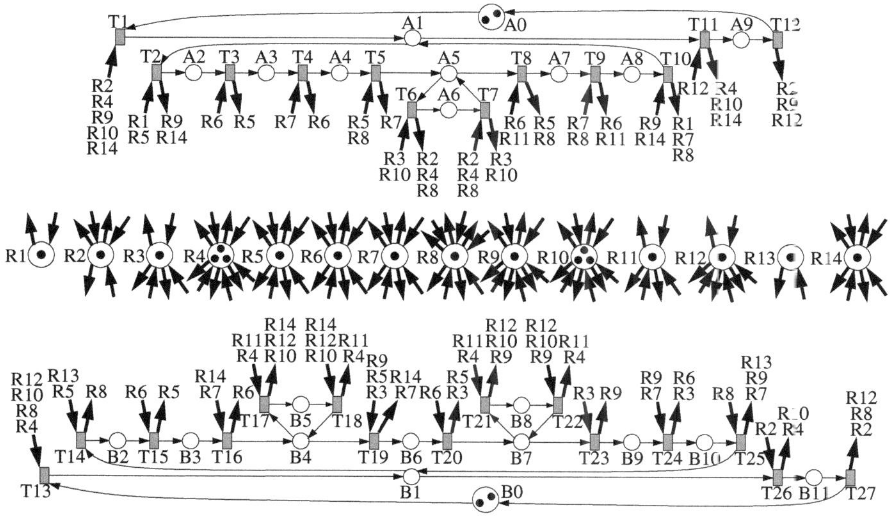

Fig. 13.7 A live S^5PR which has no home state. The arcs from/to P_R are omitted for clarity. Instead, the set of input and output resource places are listed next to each transition

a different livelock is reached. Besides, for every reachable marking, there is no minimal t-semiflow such that it is realizable, i.e., firable in isolation. Instead, both state machines need each other to progress from the very beginning.

Counterintuitively, the impossibility of realizing every t-semiflow in a live PC^2R net cannot be directly linked to the system reversibility. The net system in Fig. 13.8 has no home state. However, the net system in Fig. 13.9 is reversible, live, but no minimal t-semiflow is realizable. In fact, these two properties (reversibility and t-semiflow realizability) are usually strongly linked to the property of liveness for many Petri net classes. Particularly, reversibility is powerful since its fulfilment implies that the net is live iff there are no dead transitions at the initial marking. Both properties together imply that the net is live, as the next theorem states.

Theorem 13.5. [21] *Let* $\langle N, \boldsymbol{m}_0 \rangle$ *be an* PC^2R *with a potentially acceptable initial marking. If the net system is reversible and every (minimal) t-semiflow* $\boldsymbol{x}$ *is eventually realizable (i.e., there exist a reachable marking* $\boldsymbol{m} \in R(N, \boldsymbol{m}_0)$ *and a firing sequence* σ *such that* $\boldsymbol{m} \xrightarrow{\sigma}$, $\boldsymbol{\sigma} = \boldsymbol{x}$*) then the net is live.*

Table 13.1 illustrates in a concise way the relation between those three properties (liveness, reversibility, and eventual firability of all t-semiflows) in the context of general PC^2R nets with potentially acceptable initial markings. The table highlights the fact that those properties are not totally independent because of PC^2R nets being consistent, as proved by Theorem 13.5. It also reveals that the simpler the subclass, the less combinations of the three properties are possible (up to the point that liveness equals reversibility for S^4PR and simpler subclasses). Fig. 13.10 and 13.11, which have not been introduced before, are used to complete the table.

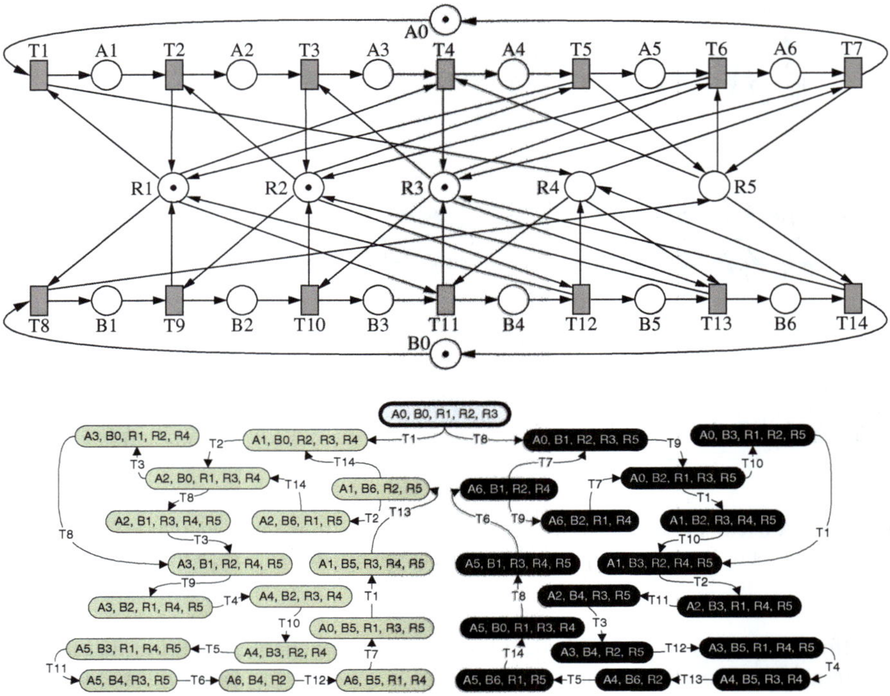

Fig. 13.8 A live PC^{2}R for which no minimal t-semiflow is ever realizable

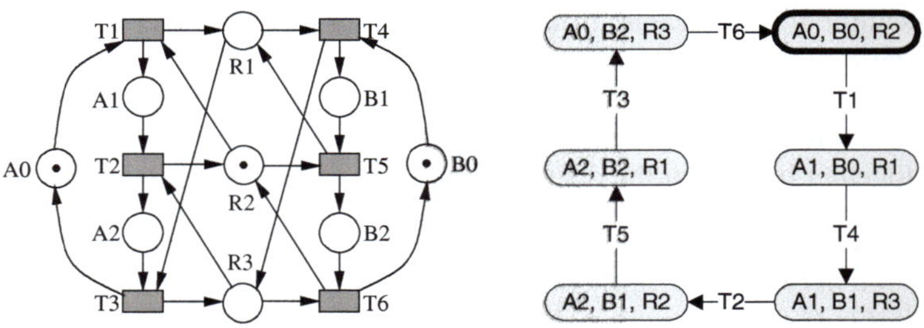

Fig. 13.9 A live and reversible PC^{2}R for which no minimal t-semiflow is ever realizable

Table 13.1 Summary of the relationship between liveness (L), reversibility (R), and eventual realizability of every t-semiflow (T) for PC^{2}R nets with a potentially acceptable initial marking. For each cell, the first line indicates which (sub)class such combination of properties is possible from. The second line references a proof of such behaviour

$L\,R\,T$ From L-S^{3}PR upwards Fig. 13.5 with $K_0 = K_1 = K_3 = 1$	$L\,R\,\overline{T}$ PC^{2}R only Fig. 13.9	$L\,\overline{R}\,T$ From S^{5}PR upwards Figs. 13.7
$L\,\overline{R}\,\overline{T}$ PC^{2}R only Fig. 13.8	$\overline{L}\,R\,T$ IMPOSSIBLE Theorem 13.5	$\overline{L}\,R\,\overline{T}$ PC^{2}R only Fig. 13.10
$\overline{L}\,\overline{R}\,T$ From L-S^{3}PR upwards Fig. 13.5 with $K_0 = K_1 = 1$ and $K_3 = 2$		$\overline{L}\,\overline{R}\,\overline{T}$ PC^{2}R only Fig. 13.11

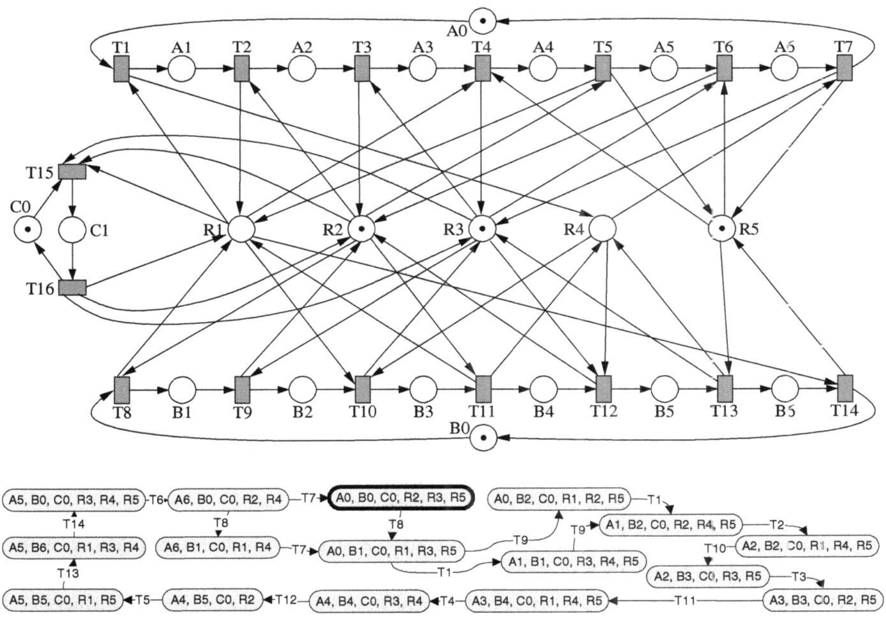

Fig. 13.10 A non-live but reversible PC^{2}R with no realizable minimal t-semiflow. It is worth noting that transitions $T15$ and $T16$ are dead at $\boldsymbol{m}_0$, despite being a *potentially* acceptable initial marking

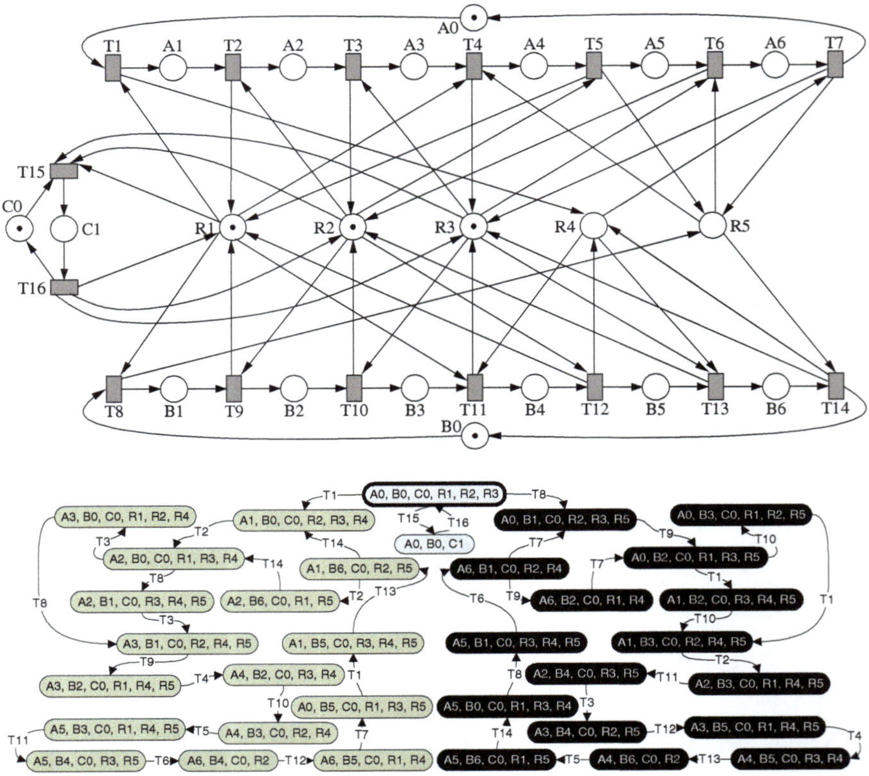

Fig. 13.11 A non-live, non-reversible PC^{2}R with no realizable minimal t-semiflow

13.4 Structure-Based Synthesis Methods: An Iterative Control Policy

In Section 13.3, we have seen that there exists a structural characterization of the deadlock problem for Petri net classes usually explored in the context of FMSs, i.e., S^{4}PR nets and subclasses. Unfortunately, only necessary or sufficient siphon-based conditions have been found for more complex superclasses such as PC^{2}R or SPQR, in the context of static analysis of multithreaded control software.

In the present section, we review an algorithm for computing the system control for S^{4}PR nets [28] which is based in that structural characterization and is successfully deployed in the context of FMSs. With the help of the net state equation, a set of Integer Linear Programming Problems (ILPPs) is constructed which prevents the costly exploration of the state space. The foundation for such approach relies on bad siphons fully capturing non-live behaviours. Since bad siphons do no longer chacterize non-liveness for models beyond the S^{4}PR frontier (Property 2) this also delimits the applicability of such kind of structural techniques in more complex scenarios.

Prior to the introduction of the algorithm, some basic notation must be settled.

In the following, for a given insufficiently marked siphon D, $D_R = D \cap P_R$ and $\mathbf{y}_{D_R} = \sum_{r \in D_R} \mathbf{y}_r$. Notice that $\mathbf{y}_{D_R}$ is the total amount of resource units belonging to D (in fact, to D_R) used by each active process in their process places. Also

Definition 13.7. *Let $\langle N, \boldsymbol{m}_0 \rangle$ be a marked S^4PR. Let D be a siphon of N. Then, $\mathcal{T}h_D = \|\mathbf{y}_{D_R}\| \setminus D$ is the* set of thieves *of D, i.e., the set of process places of the net that use resources of the siphon and do not belong to that siphon.*

In each iteration, the algorithm searches for a bad siphon. If found, a control place is suggested to prevent that siphon from ever becoming insufficiently marked. Such control place will be a virtual resource, in such a way that the resulting Petri net remains into the S^4PR class. Thanks to this, a new iteration of the algorithm can be executed. The algorithm terminates as soon as there do not exist more siphons to be controlled, i.e., the system is live.

The next system of restrictions relates the liveness characterization introduced in Theorem 13.2 with the ILPPs which are used in the forthcoming algorithm. Essentially, the structural characterization is reformulated into a set of linear restrictions given a reachable marking and a related bad siphon. It is worth noting that in order to compact the system of linear restrictions, three sets of variables have been parameterized: v_p, e_t and e_{rt} (the latter being doubly parameterized, both by resource places r and transitions t). Obviously, the ILPP would have appeared considerably larger if it had not been compacted in this way for the sake of concision.

Proposition 13.1. [28] *Let $\langle N, \boldsymbol{m}_0 \rangle$ be a marked S^4PR. The net is non-live if and only if there exist a siphon D and a marking $\boldsymbol{m} \in R(N, \boldsymbol{m}_0)$ such that the following set of inequalities has, at least, one solution ($D = \{p \in P_S \cup P_R | v_p = 0\}$):*

$$\left\{\begin{array}{lll}
\boldsymbol{m}[P_S] \geq \mathbf{0} \wedge \boldsymbol{m}[P_S] \neq \mathbf{0} & & \\
\forall t \in T \setminus P_0{}^\bullet : being\{p\} = {}^\bullet t \cap P_S, & & \\
 & m[p] \geq e_t & \\
 & e_t \geq \frac{m[p]}{sb(p)} & \\
\forall r \in P_R : & \forall t \in r^\bullet \setminus P_0{}^\bullet : & \frac{m[r]}{Pre[r,t]} + v_r \geq e_{rt} \\
 & & e_{rt} \geq \frac{m[r] - Pre[r,t] + 1}{m_0[r] - Pre[r,t] + 1} \\
 & & e_{rt} \geq v_r \\
 & \forall t \in T \setminus P_0{}^\bullet : & \sum_{r \in {}^\bullet t \cap P_R} e_{rt} < |{}^\bullet t \cap P_R| + 1 - e_t \\
 & \forall p \in P \setminus P_0 : & v_p \in \{0,1\} \\
 & \forall t \in T \setminus P_0{}^\bullet : & e_t \in \{0,1\} \\
 & \forall t \in r^\bullet \setminus P_0{}^\bullet : & e_{rt} \in \{0,1\}.
\end{array}\right\} \tag{13.1}$$

where $sb(p)$ denotes the structural bound[2] *of p [4].*

Thanks to the addition of the net state equation as another linear restriction, the following theorem constructs an ILPP which can compute a marking and a bad

[2] $sb(p)$ is the max. of the following ILPP: $sb(p) = \max\ m[p]$ s.t. $\boldsymbol{m} = \boldsymbol{m}_0 + \boldsymbol{C} \cdot \boldsymbol{\sigma}, \boldsymbol{m} \geq \mathbf{0}$, $\boldsymbol{\sigma} \in \mathbb{N}^{|T|}$.

siphon holding System (13.1). Nevertheless, that marking can be a spurious solution of the state equation. Since this kind of nets can have killing spurious solutions (i.e, spurious solutions which are non-live when the original net system is live) then the theorem establishes a necessary but not sufficient condition. This is usually not a problem when the objective is to obtain a live system: the only consequence can be that some harmless, unnecessary control places are added. These control places would forbid some markings which are not really reachable.

Since one siphon must be selected, the ILPP selects that with a minimal number of places, hoping that controlling the smallest siphons first may prevent controlling the bigger ones. Other works present analogous techniques with a different objective function for this ILPP [12].

Theorem 13.6. [28] *Let $\langle N, \boldsymbol{m}_0 \rangle$ be a marked S^4PR. If the net is non-live, then there exist a siphon D and a marking $\boldsymbol{m} \in PR(N, \boldsymbol{m}_0)$ such that the following set of inequalities has, at least, one solution with $D = \{p \in P_S \cup P_R \mid v_p = 0\}$:*

$$\begin{array}{ll}\max & \sum_{p \in P \setminus P_0} v_p \\ s.t. & \boldsymbol{m} = \boldsymbol{m}_0 + \boldsymbol{C} \cdot \boldsymbol{\sigma} \\ & \boldsymbol{m} \geq \boldsymbol{0}, \boldsymbol{\sigma} \in \mathbb{N}^{|T|} \\ & \forall p \in P \setminus P_0, \forall t \in {}^{\bullet}p : v_p \geq \sum_{q \in {}^{\bullet}t} v_q - |{}^{\bullet}t| + 1 \\ & \sum_{p \in P \setminus P_0} v_p < |P \setminus P_0| \\ & \text{System (13.1)}\end{array}$$

The previous theorem can compute a marking $\boldsymbol{m}$ and a related bad siphon D. However, siphon D can be related with a high number of deadlocks, and not only with that represented with $\boldsymbol{m}$. For that reason, the aim is to compute a control place able to cut every *unwanted* marking which the siphon D is related to. Consequently, two different strategies are raised from the observation of the set of unwanted markings: (i) adding a place that introduces a lower bound of the number of available resources in the siphon for every reachable marking (*D-resource-place*), or (ii) adding a place that introduces an upper bound of the number of active processes which are retaining tokens from the siphon (*D-process-place*).

In order to define the initial marking of such places, two constants must be computed which are the result of two ILPPs. These ILPPs evaluate every unwanted marking that a bad siphon is related to:

Definition 13.8. [28] *Let $\langle N, \boldsymbol{m}_0 \rangle$ be a marked S^4PR. Let D be an insufficiently marked siphon, m_D^{max} and m_D^{min} are defined as follows, with $v_p = 0$ iff $p \in D$:*

$$\begin{array}{rll}m_D^{max} = & \max & \sum_{r \in D_R} m[r] \\ & s.t. & \boldsymbol{m} = \boldsymbol{m}_0 + \boldsymbol{C} \cdot \boldsymbol{\sigma} \\ & & \boldsymbol{m} \geq \boldsymbol{0}, \boldsymbol{\sigma} \in \mathbb{N}^{|T|} \\ & & \boldsymbol{m}[P_S \setminus \mathscr{Th}_D] = \boldsymbol{0} \\ & & \text{System (13.1)}\end{array} \qquad \begin{array}{rll}m_D^{min} = & \min & \sum_{p \in \mathscr{Th}_D} m[p] \\ & s.t. & \boldsymbol{m} = \boldsymbol{m}_0 + \boldsymbol{C} \cdot \boldsymbol{\sigma} \\ & & \boldsymbol{m} \geq \boldsymbol{0}, \boldsymbol{\sigma} \in \mathbb{N}^{|T|} \\ & & \boldsymbol{m}[P_S \setminus \mathscr{Th}_D] = \boldsymbol{0} \\ & & \text{System (13.1)}\end{array}$$

The next definition establishes the connectivity and the initial marking of the control place proposed for a given bad siphon D, both whether that place is a D-process-place or a D-resource place.

Definition 13.9. [28] *Let $\langle N, \boldsymbol{m}_0 \rangle$ be a non-live marked S^4PR. Let D be an insufficiently marked siphon, and m_D^{max} and m_D^{min} as in Definition 13.8. Then, the associated D-resource-place, p_D, is defined by means of the addition of the following incidence matrix row and initial marking: $\boldsymbol{C}^{p_D}[p_D, T] = -\sum_{p \in \mathcal{T}h_D} y_{D_R}[p] \cdot \boldsymbol{C}[p, T]$, and $m_0^{p_D}[p_D] = \boldsymbol{m}_0[D] - (m_D^{max} + 1)$. The associated D-process-place, p_D, is defined by means of the addition of the following incidence matrix row and initial marking: $\boldsymbol{C}^{p_D}[p_D, T] = -\sum_{p \in \mathcal{T}h_D} \boldsymbol{C}[p, T]$, and $m_0^{p_D}[p_D] = m_D^{min} - 1$.*

Finally, we can enunciate the algorithm that computes the control places for a given S^4PR net. In those cases in which a D-resource-place with an admissible initial marking cannot be computed, the algorithm proposes the corresponding D-process-place, which always has an admissible initial marking [28].

Algorithm 1. [28] (Synthesis of live S^4PR net systems).

1. Compute an insufficiently marked siphon using the ILPP of Theorem 13.6.
2. Compute m_D^{max} (Definition 13.8).
   a. If the associated D-resource-place (Definition 13.9) has an acceptable initial marking according to Definition 13.2, then let p_D be that place, and go to step 3.
   b. Else, compute m_D^{min} (Definition 13.8). Let p_D be the associated D-process-place (Definition 13.9).
3. Add the control place p_D.
4. Go to step 1, taking as input the partially controlled system, until no insufficiently marked siphons exist.

Theorem 13.7. [28] *Let $\langle N, \boldsymbol{m}_0 \rangle$ be a marked S^4PR. Algorithm 1 applied to $\langle N, \boldsymbol{m}_0 \rangle$ terminates. The resulting controlled system, $\langle N^C, \boldsymbol{m}_0^C \rangle$, is a live marked S^4PR such that $R(N^C, \boldsymbol{m}_0^C) \subseteq R(N, \boldsymbol{m}_0)$.*

Let us apply Algorithm 1 to the net depicted in Fig. 13.1. There exists one deadlock ($\boldsymbol{m} = A1 + B1 + R1 + R2$) and two insufficiently marked siphons at $\boldsymbol{m}$, $D_1 = \{R1, R2, A3, B2\}$ and $D_2 = D_1 \cup \{A2\}$. None of these is minimal. When applied step 1 of Algorithm 1, the ILPP of Theorem 13.6 returns D_1. In step 2, we compute $m_D^{max} = 2$. Since the associated D-resource-place has not an acceptable initial marking, then we compute $m_D^{min} = 2$. In step 3, we add the associated D-process-place p_D to the net. And finally, we go back to step 1. But now the net is live and the ILPP of Theorem 13.6 has no solution, so the algorithm terminates. The resulting controlled system is depicted in Fig. 13.12.

Let us now apply the algorithm to the S^3PR net depicted in Fig. 13.5 with $K_0 = K_1 = 1$, $K_3 = 2$. In this case, there exists one deadlock ($\boldsymbol{m} = A4 + B4 + R2 + 2 \cdot R3$) which is reachable by firing the sequence $\sigma = TB1\, TA1\, TB2\, TA2\, TB3\, TA3\, TB4\, TA4$. This sequence empties the minimal siphon $D = \{A1, B1, A5, B5, R1, R4\}$, which is also the siphon returned by the computation in step 1 of the algorithm. In step 2, we obtain $m_D^{max} = 0$. Then the associated D-resource-place p_D has an acceptable initial marking ($m_0^{p_D}[p_D] = 1$), with

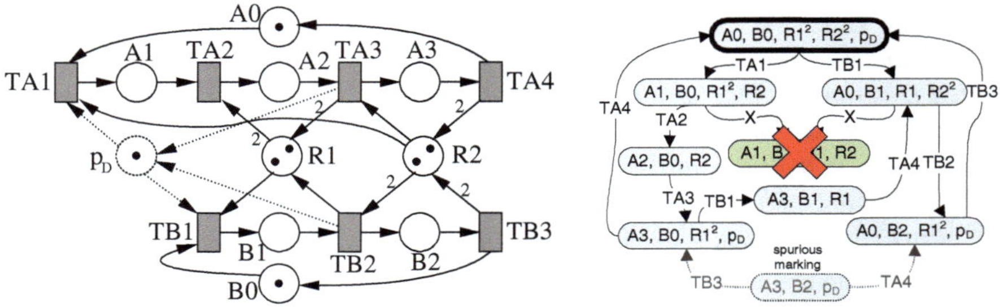

Fig. 13.12 The controlled system after applying Algorithm 1 on the net depicted in Fig. 13.1

$\mathbf{y}_r \in \{0,1\}^{|P|}$ and $\|\mathbf{y}_r\| = \{A4, B4\}$. Therefore, p_D can be aggregated to the net (step 3), and we go back to step 1. Since the resulting net is live and the ILPP of Theorem 13.6 has no solution, the algorithm terminates.

13.5 Further Reading

The material of this chapter is essentially related to structural techniques for dealing with deadlocks in S-RASs with online routing decisions. The reader can find some reference works on the family of Petri net models for tackling this kind of systems in the following references: [6, 7, 8, 20, 27, 28, 22]

From a modelling point of view, the problem of integrating assembly/dissassembly operations has been approached in works such as [29, 13, 8]. However, structural liveness enforcing approaches can be computationally demanding in this scenario, as evidenced for augmented marked graphs in [3]. An insight on the computational complexity of such approaches on the S-RAS context is driven in [17].

Most works on modelling RASs by way of Petri nets are focused on FMSs. The article [5] can serve as a succinct introduction to the discipline. The books [16, 24] also focus on the RAP from this perspective. Besides, the latter book also features some approximation to Automated Guided Vehicle (AGV) Transportation Systems from the point of view of RAS modelling through Petri nets. Recently, AGVs have been comprehensively tackled in [25]. Other application domains in which similar methodological approaches have been deployed include multiprocessor interconnection networks [25] and multithreaded software [30, 19, 20].

The most significant proliferation of works can be found in the context of synthesis. Most of this papers are related to siphon computation (two recent works on this issue can be found in [2, 15]) as well as to applying Mixed Integer Programming to liveness enforcing [12, 28]. Another family of works focuses on synthesis based on reachability state analysis and on the theory of regions [11, 23].

References

1. Best, E., Voss, K.: Free Choice Systems Have Home States. Acta Informatica 21, 89–100 (1984)
2. Cano, E.-E., Rovetto, C.-A., Colom, J.-M.: On the Computation of the Minimal Siphons of S^4PR nets from a Generating Family of Siphons. In: Proc. 15th IEEE International Conf. on Emerging Technologies and Factory Automation, Bilbao, Spain (2010)
3. Chu, F., Xie, X.: Deadlock Analysis of Petri Nets using Siphons and Mathematical Programming. IEEE Transactions on Robotics and Automation 13(6), 793–804 (1997)
4. Colom, J.M., Silva, M.: Improving the Linearly Based Characterization of P/T Nets. In: Rozenberg, G. (ed.) APN 1990. LNCS, vol. 483, pp. 113–145. Springer, Heidelberg (1991)
5. Colom, J.-M.: The Resource Allocation Problem in Flexible Manufacturing Systems. In: van der Aalst, W.M.P., Best, E. (eds.) ICATPN 2003. LNCS, vol. 2679, pp. 23–35. Springer, Heidelberg (2003)
6. Ezpeleta, J., Colom, J.-M., Martínez, J.: A Petri net Based Deadlock Prevention Policy for Flexible Manufacturing Systems. IEEE Transactions on Robotics and Automation 2(11), 173–184 (1995)
7. Ezpeleta, J., Garcia-Vallés, F., Colom, J.-M.: A Class of Well Structured Petri Nets for Flexible Manufacturing Systems. In: Desel, J., Silva, M. (eds.) ICATPN 1998. LNCS, vol. 1420, pp. 64–83. Springer, Heidelberg (1998)
8. Ezpeleta, J., Tricas, F., García–Vallés, F., Colom, J.-M.: A Banker's Solution for Deadlock Avoidance in FMS with Flexible Routing and Multiresource States. IEEE Transactions on Robotics and Automation 18(4), 621–625 (2002)
9. Ezpeleta, J., Recalde, L.: A Deadlock Avoidance Approach for Non–Sequential Resource Allocation Systems. IEEE Transactions on Systems, Man and Cybernetics, Part A: Systems and Humans 34(1), 93–101 (2004)
10. Fanti, M.P., Maione, B., Mascolo, S., Turchiano, B.: Event-Based Feedback Control for Deadlock Avoidance in Flexible Production Systems. IEEE Transactions on Robotics and Automation 13(3), 347–363 (1997)
11. Ghaffari, A., Rezg, N., Xie, X.: Design of a Live and Maximally Permissive Petri Net Controller using the Theory of Regions. IEEE Transactions on Robotics 19(1), 137–141 (2003)
12. Hu, H., Zhou, M.C., Li, Z.W.: Supervisor Optimization for Deadlock Resolution in Automated Manufacturing Systems With Petri Nets. IEEE Transactions on Automation Science and Engineering 8(4), 794–804 (2011)
13. Jeng, M.-D., Xie, X.-L., Peng, M.-Y.: Process Nets with Resources for Manufacturing Modeling and their Analysis. IEEE Transactions on Robotics 18(6), 875–889 (2002)
14. Lautenbach, K., Thiagarajan, P.S.: Analysis of a Resource Allocation Problem Using Petri Nets. In: Proc. 1st European Conf. on Parallel and Distributed Processing, France (1979)
15. Li, Z.-W., Zhou, M.-C.: Control of Elementary and Dependent Siphons in Petri Nets and Their Application. IEEE Transactions on Systems, Man and Cybernetics, Part A: Systems and Humans 38(1), 133–148 (2008)
16. Li, Z.-W., Zhou, M.-C.: Deadlock Resolution in Automated Manufacturing Systems: A Novel Petri Net Approach. Springer, New York (2009)
17. López-Grao, J.-P., Colom, J.-M.: Resource Allocation Systems: Some Complexity Results on the S^4PR Class. In: Najm, E., Pradat-Peyre, J.-F., Donzeau-Gouge, V.V. (eds.) FORTE 2006. LNCS, vol. 4229, pp. 323–338. Springer, Heidelberg (2006)

18. López–Grao, J.-P., Colom, J.-M.: Lender Processes Competing for Shared Resources: Beyond the S^4PR Paradigm. In: Proc. 2006 IEEE Int. Conf. on Systems, Man and Cybernetics, Taipei, Taiwan (2006)
19. López–Grao, J.-P., Colom, J.-M.: On the Deadlock Analysis of Multithreaded Control Software. In: Proc. 16th IEEE Int. Conf. on Emerging Technologies and Factory Automation, Toulouse, France (2006)
20. López-Grao, J.-P., Colom, J.-M.: A Petri Net Perspective on the Resource Allocation Problem in Software Engineering. In: Jensen, K., Donatelli, S., Kleijn, J. (eds.) Transactions on Petri Nets and Other Models of Concurrency V. LNCS, vol. 6900, pp. 181–200. Springer, Heidelberg (2012)
21. López–Grao, J.-P.: Contributions to the Deadlock Problem in Multithreaded Software Applications Observed as Resource Allocation Systems. PhD Thesis. University of Zaragoza, Spain (2012)
22. Park, J., Reveliotis, S.A.: Deadlock Avoidance in Sequential Resource Allocation Systems with Multiple Resource Acquisitions and Flexible Routings. IEEE Transactions on Automatic Control 46(10), 1572–1583 (2001)
23. Piroddi, L., Cordone, R., Fumagalli, I.: Combined Siphon and Marking Generation for Deadlock Prevention in Petri Nets. IEEE Transactions on Systems, Man and Cybernetics, Part A: Systems and Humans 39(3), 650–661 (2009)
24. Reveliotis, S.A.: Real-Time Management of Resource Allocation Systems: A Discrete Event Systems Approach. International Series in Operations Research & Management Science (2004)
25. Rovetto, C.-A.: Métodos Basados en Redes de Petri para el Diseño de Algoritmos de Encaminamiento Adaptativos Mínimos Libres de Bloqueos. PhD Thesis. University of Zaragoza, Spain (2011)
26. Silva, M., Colom, J.-M.: On the Computation of Structural Synchronic Invariants in P/T Nets. In: Rozenberg, G. (ed.) APN 1988. LNCS, vol. 340, pp. 386–417. Springer, Heidelberg (1988)
27. Tricas, F.: Deadlock Analysis, Prevention and Avoidance in Sequential Resource Allocation Systems. PhD Thesis. University of Zaragoza, Spain (2003)
28. Tricas, F., García–Valles, F., Colom, J.-M., Ezpeleta, J.: A Petri net Structure-Based Deadlock Prevention Solution for Sequential Resource Allocation Systems. In: Proc. 2005 Int. Conf. on Robotics and Automation, Barcelona, Spain (2005)
29. Xie, X., Jeng, M.-D.: ERCN-merged Nets and Their Analysis using Siphons. IEEE Transactions on Robotics and Automation 29(4), 692–703 (1999)
30. Wang, Y., Liao, H., Reveliotis, S.A., Kelly, T., Mahlke, S., Lafortune, S.: Gadara Nets: Modeling and Analyzing Lock Allocation for Deadlock Avoidance in Multithreaded Software. In: Proc. 49th IEEE Conf. on Decision and Control, Atlanta, Georgia, USA (2009)

Chapter 14
Diagnosis of Petri Nets

Maria Paola Cabasino, Alessandro Giua, and Carla Seatzu

14.1 Introduction

Failure detection and isolation in industrial systems is a subject that has received a lot of attention in the past few decades. A failure is defined to be any deviation of a system from its normal or intended behavior. Diagnosis is the process of detecting an abnormality in the system behavior and isolating the cause or the source of this abnormality.

As usual the first significant contributions in this framework have been presented in the area of time driven systems. However, a quite rich literature on diagnosis of discrete event systems has also been produced in the last two decades. Faults may be described by discrete events that represent abnormal conditions. As an example, in a telecommunication system, a fault may correspond to a message that is lost or not sent to the appropriate receiver. Similarly, in a transportation system, a fault may be a traffic light that does not switch from red to green according to the given schedule. In a manufacturing system, it may be the failure of a certain operation, e.g., a wrong assembly, or a part put in a wrong buffer, and so on.

In the first part of this chapter we recall an approach proposed in [16], that is a generalization of [11]. The main feature of the diagnosis approach here presented is the concept of basis marking. This concept allows us to represent the reachability space in a compact manner, i.e., it requires to enumerate only a subset of the reachability space. In particular, arbitrary labeled Petri nets (PNs) are considered where there is an association between sensors and observable events, while no sensor is available for certain activities — such as faults or other unobservable but regular transitions — due to budget constraints or technology limitations.

It is also assumed that several different transitions might share the same sensor in order to reduce cost or power consumption; in such a case if two transitions

Maria Paola Cabasino · Alessandro Giua · Carla Seatzu
Department of Electrical and Electronic Engineering, University of Cagliari, Italy
e-mail: {cabasino,giua,seatzu}@diee.unica.it

C. Seatzu et al. (Eds.): Control of Discrete-Event Systems, LNCIS 433, pp. 279–300.
springerlink.com

are simultaneously enabled and one of them fires, it is impossible to distinguish which one has fired, thus they are called *undistinguishable*. Four diagnosis states are defined that model different degrees of alarm. The only limitation on the system is that the unobservable subnet should be acyclic. Such an assumption is common to all discrete event systems approaches, not only those based on PNs, but those based on finite state automata as well.

Note that the approach here presented, as most of the approaches dealing with diagnosis of discrete event systems [21, 28, 37, 39], assumes that the faulty behavior is completely known, thus a fault model is available. Such an assumption is applicable to interesting classes of problems, e.g., this is the case of many man-made systems where the set of possible faults is often predictable and finite in number [2, 24, 31, 44].

In the second part of this chapter, several extensions of this basic approach are presented. The first result originates from the requirement of relaxing the assumption of acyclicity of the unobservable subnet. To address this issue fluidification is proposed [7, 8, 14] (see Chapters 18-20 of this book for a comprehensive presentation of fluidification in Petri nets). Second, the diagnosability problem is considered: a system is said to be diagnosable if when a fault occurs, it is possible to detect its occurrence after a finite number of events occurrences. Obviously, this is a major requirement when performing online fault diagnosis. Finally, the problem of designing a decentralized diagnoser is discussed. Indeed, due to the everincreasing complexity of nowadays systems and the intrinsic distributed nature of many realistic systems, performing online centralized diagnosis often reveals not convenient, or even unfeasible.

14.2 Basic Definitions and Notations

As already discussed in the Introduction, the goal of a diagnosis problem consists in reconstructing the occurrence of a fault event based on the observation of the output of some sensors. The association between sensors and transitions can be captured by a *labeling function* $\mathscr{L} : T \to L \cup \{\varepsilon\}$ assigns to each transition $t \in T$ either a symbol from a given alphabet L or the empty string ε.

The set of transitions whose label is ε is denoted as T_u, i.e., $T_u = \{t \in T \mid \mathscr{L}(t) = \varepsilon\}$. Transitions in T_u are called *unobservable* or *silent*. T_o denotes the set of transitions labeled with a symbol in L. Transitions in T_o are called *observable* because when they fire their label can be observed. In this chapter, it is assumed that the same label $l \in L$ can be associated with more than one transition. In particular, two transitions $t_1, t_2 \in T_o$ are called *undistinguishable* if they share the same label, i.e., $\mathscr{L}(t_1) = \mathscr{L}(t_2)$. The set of transitions sharing the same label l is denoted T_l.

In the following, let $\boldsymbol{C}_u$ ($\boldsymbol{C}_o$) be the restriction of the incidence matrix to T_u (T_o) and n_u and n_o, respectively, be the cardinality of these two sets. Moreover, given a sequence $\sigma \in T^*$, $P_u(\sigma)$, resp., $P_o(\sigma)$, denotes the projection of σ over T_u, resp., T_o.

The word w of events associated with sequence σ is $w = P_o(\sigma)$. Note that, the length of a sequence σ (denoted $|\sigma|$) is always greater than or equal to the length of the corresponding word w (denoted $|w|$). In fact, if σ contains k' transitions in T_u then $|\sigma| = k' + |w|$.

Definition 14.1. [16] *Let $\langle N, \boldsymbol{m}_0\rangle$ be a labeled net system with labeling function $\mathscr{L} : T \to L \cup \{\varepsilon\}$, where $N = (P, T, \boldsymbol{Pre}, \boldsymbol{Post})$ and $T = T_o \cup T_u$. Let $w \in L^*$ be an observed word. Let*

$$\mathscr{S}(w) = \{\sigma \in L(N, \boldsymbol{m}_0) \mid P_o(\sigma) = w\}$$

be the set of firing sequences consistent *with $w \in L^*$, and*

$$\mathscr{C}(w) = \{\boldsymbol{m} \in \mathbb{N}^m \mid \exists \sigma \in T^* : P_o(\sigma) = w \wedge \boldsymbol{m}_0[\sigma\rangle \boldsymbol{m}\}$$

be the set of reachable markings consistent *with $w \in L^*$.*

In plain words, given an observation w, $\mathscr{S}(w)$ is the set of sequences that may have fired, while $\mathscr{C}(w)$ is the set of markings in which the system may actually be.

Example 14.1. Consider the PN in Fig. 14.1. Assume $T_o = \{t_1, t_2, t_3, t_4, t_5, t_6, t_7\}$ and $T_u = \{\varepsilon_8, \varepsilon_9, \varepsilon_{10}, \varepsilon_{11}, \varepsilon_{12}, \varepsilon_{13}\}$, where for a better understanding unobservable transitions have been denoted ε_i rather than t_i. The labeling function is defined as follows: $\mathscr{L}(t_1) = a$, $\mathscr{L}(t_2) = \mathscr{L}(t_3) = b$, $\mathscr{L}(t_4) = \mathscr{L}(t_5) = c$, $\mathscr{L}(t_6) = \mathscr{L}(t_7) = d$.

First consider $w = ab$. The set of firing sequences that is consistent with w is $\mathscr{S}(w) = \{t_1t_2,\ t_1t_2\varepsilon_8, t_1t_2\varepsilon_8\varepsilon_9, t_1t_2\varepsilon_8\varepsilon_9\varepsilon_{10}, t_1t_2\varepsilon_8\varepsilon_{11}\}$, and the set of markings consistent with w is $\mathscr{C}(w) = \{[0\ 0\ 1\ 0\ 0\ 0\ 0\ 1\ 0\ 0\ 0]^T, [0\ 0\ 0\ 1\ 0\ 0\ 0\ 1\ 0\ 0\ 0]^T,$ $[0\ 0\ 0\ 0\ 1\ 0\ 0\ 1\ 0\ 0\ 0]^T, [0\ 1\ 0\ 0\ 0\ 0\ 0\ 1\ 0\ 0\ 0]^T, [0\ 0\ 0\ 0\ 0\ 1\ 0\ 1\ 0\ 0\ 0]^T\}$.

If $w = acd$ is considered it holds $\mathscr{S}(w) = \{t_1t_5t_6, t_1t_5\varepsilon_{12}\varepsilon_{13}t_7\}$ and $\mathscr{C}(w) = \{[0\ 1\ 0\ 0\ 0\ 0\ 0\ 1\ 0\ 0\ 0]^T\}$. Thus, two different firing sequences may have fired (the second one also involving silent transitions), but they both lead to the same marking. ■

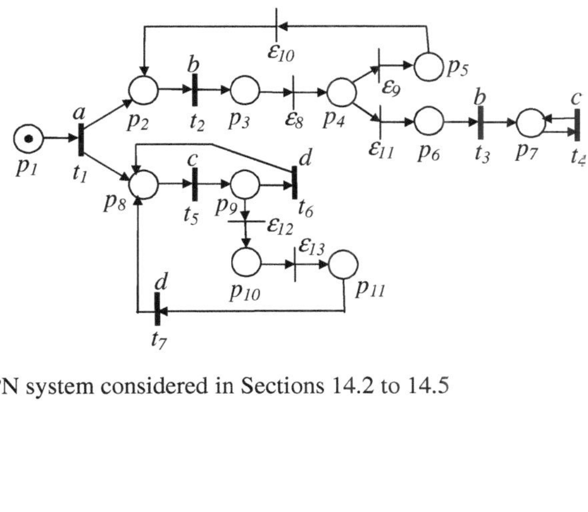

Fig. 14.1 The PN system considered in Sections 14.2 to 14.5

Definition 14.2. *Given a net $N = (P,T,\boldsymbol{Pre},\boldsymbol{Post})$, and a subset $T' \subseteq T$ of its transitions, let us define the T'-*induced subnet *of N as the new net $N' = (P,T',\boldsymbol{Pre}',\boldsymbol{Post}')$ where $\boldsymbol{Pre}',\boldsymbol{Post}'$ are the restrictions of $\boldsymbol{Pre},\boldsymbol{Post}$ to T'. The net N' can be thought as obtained from N removing all transitions in $T \setminus T'$.*

Finally, given a sequence $\sigma \in T^*$, we denote as $\pi : T^* \to \mathbb{N}^{|T|}$ the function that associates with σ a vector $\boldsymbol{y} \in \mathbb{N}^{|T|}$, called the firing vector of σ. In particular, $y[t] = k$ if the transition t is contained k times in σ.

14.3 Characterization of the Set of Consistent Markings

To solve a diagnosis problem, it is essential to be able to compute the set of sequences and markings consistent with a given observation w. In this section a formalism that allows one to characterize these sets without resorting to explicit enumeration is provided.

14.3.1 Minimal Explanations and Minimal e-Vectors

Let us now first recall the notion of minimal explanation for unlabeled PNs. Then, it is shown how such a notion can be extended to labeled PNs.

Definition 14.3. [16] *Given a marking $\boldsymbol{m}$ and an observable transition $t \in T_o$, let*

$$\Sigma(\boldsymbol{m},t) = \{\sigma \in T_u^* \mid \boldsymbol{m}[\sigma\rangle \boldsymbol{m}',\ \boldsymbol{m}' \geq \boldsymbol{Pre}[\cdot,t]\}$$

be the set of explanations *of t at $\boldsymbol{m}$, and let*

$$Y(\boldsymbol{m},t) = \pi(\Sigma(\boldsymbol{m},t))$$

be the e-vectors *(or* explanation vectors*), i.e., firing vectors associated with the explanations.*

Thus $\Sigma(\boldsymbol{m},t)$ is the set of unobservable sequences whose firing at $\boldsymbol{m}$ enables t. Among the above sequences select those whose firing vector is minimal. The firing vector of these sequences are called *minimal e-vectors*.

Definition 14.4. [16] *Given a marking $\boldsymbol{m}$ and a transition $t \in T_o$, let us define*[1]

$$\Sigma_{\min}(\boldsymbol{m},t) = \{\sigma \in \Sigma(\boldsymbol{m},t) \mid \nexists\, \sigma' \in \Sigma(\boldsymbol{m},t) \ :\ \pi(\sigma') \lneq \pi(\sigma)\}$$

[1] Given two vectors $\boldsymbol{x}$ and $\boldsymbol{y}$, $\boldsymbol{x} \lneq \boldsymbol{y}$ denotes that all components of $\boldsymbol{x}$ are less than or equal to the corresponding component of $\boldsymbol{y}$ and there exists at least one component of $\boldsymbol{x}$ that is strictly less than the corresponding component of $\boldsymbol{y}$.

the set of minimal explanations *of t at $\boldsymbol{m}$, and let us define*

$$Y_{\min}(\boldsymbol{m},t) = \pi(\Sigma_{\min}(\boldsymbol{m},t))$$

the corresponding set of minimal e-vectors.

Example 14.2. Consider the PN in Fig. 14.1 introduced in Example 14.1. It holds that $\Sigma(\boldsymbol{m}_0,t_1) = \{\varepsilon\}$. Then $\Sigma(\boldsymbol{m}_0,t_2) = \emptyset$. Finally, let $\boldsymbol{m} = [\,0\,0\,1\,0\,0\,0\,0\,1\,0\,0\,0\,]^T$, it holds that $\Sigma(\boldsymbol{m},t_5) = \{\varepsilon,\varepsilon_8,\varepsilon_8\varepsilon_9,\varepsilon_8\varepsilon_{11},\varepsilon_8\varepsilon_9\varepsilon_{10}\}$, while $\Sigma_{min}(\boldsymbol{m},t_5) = \{\varepsilon\}$. It follows that $Y(\boldsymbol{m},t_5) = \{[0\ 0\ 0\ 0\ 0\ 0]^T,\ [1\ 0\ 0\ 0\ 0\ 0]^T,\ [1\ 1\ 0\ 0\ 0\ 0]^T,\ [1\ 0\ 0\ 1\ 0\ 0]^T,\ [1\ 1\ 1\ 0\ 0\ 0]^T\}$, and $Y_{min}(\boldsymbol{m},t_5) = \{[0\,0\,0\,0\,0\,0]^T\}$. ■

In [19] it was shown that, if the unobservable subnet is acyclic and backward conflict-free, then $|Y_{\min}(\boldsymbol{m},t)| = 1$. Different approaches can be used to compute $Y_{\min}(\boldsymbol{m},t)$, e.g., [6, 29]. In [16] it is suggested an approach that simply requires algebraic manipulations, and is inspired by the procedure proposed by [32] for the computation of minimal P-invariants.

14.3.2 Basis Markings and j-Vectors

Given a sequence of observed events $w \in L^*$, a basis marking $\boldsymbol{m}_b$ is a marking reached from $\boldsymbol{m}_0$ with the firing of a sequence of transitions whose observable projection is w and whose unobservable transitions interleaved with the observable ones are strictly necessary to enable it. Such a sequence of unobservable transitions is called *justification*. Note that, in general several sequences $\sigma_o \in T_o^*$ may correspond to the same w, i.e., there are several sequences of observable transitions such that $\mathscr{L}(\sigma_o) = w$ that may have actually fired. Moreover, in general, to any of such sequences a different sequence of unobservable transitions interleaved with it is necessary to make it firable at the initial marking. Thus, the introduction of the following definition of pairs (sequence of transitions in T_o labeled w; corresponding justification) is needed.

Definition 14.5. [16] *Let $\langle N,\boldsymbol{m}_0\rangle$ be a net system with labeling function $\mathscr{L} : T \to L \cup \{\varepsilon\}$, where $N = (P,T,\boldsymbol{Pre},\boldsymbol{Post})$ and $T = T_o \cup T_u$. Let $w \in L^*$ be a given observation. Let*

$$\begin{aligned}\hat{\mathscr{J}}(w) = \{\, & (\sigma_o,\sigma_u),\ \sigma_o \in T_o^*,\ \mathscr{L}(\sigma_o) = w,\ \sigma_u \in T_u^* \mid \\ & [\exists\, \sigma \in \mathscr{S}(w)\ :\ \sigma_o = P_o(\sigma),\ \sigma_u = P_u(\sigma)] \wedge \\ & [\nexists \sigma' \in \mathscr{S}(w) :\ \sigma_o = P_o(\sigma'),\ \sigma_u' = P_u(\sigma') \wedge\ \pi(\sigma_u') \lneq \pi(\sigma_u)\,]\,\}\end{aligned}$$

be the set of pairs (sequence $\sigma_o \in T_o^*$ with $\mathscr{L}(\sigma_o) = w$, corresponding justification *of w). Moreover, let*

$$\begin{aligned}\hat{Y}_{\min}(\boldsymbol{m}_0,w) = \{ & (\sigma_o,\boldsymbol{y})\sigma_o \in T_o^*,\ \mathscr{L}(\sigma_o) = w,\ \boldsymbol{y} \in \mathbb{N}^{n_u} \mid \\ & \exists(\sigma_o,\sigma_u) \in \hat{\mathscr{J}}(w) : \pi(\sigma_u) = \boldsymbol{y}\}\end{aligned}$$

be the set of pairs (sequence $\sigma_o \in T_o^*$ with $\mathscr{L}(\sigma_o) = w$, corresponding j-vector*).*

In simple words, $\hat{\mathscr{J}}(w)$ is the set of pairs whose first element is the sequence $\sigma_o \in T_o^*$ labeled w and whose second element is the corresponding sequence of unobservable transitions interleaved with σ_o whose firing enables σ_o and whose firing vector is minimal. The firing vectors of these sequences are called *j-vectors*.

Example 14.3. Consider the PN in Fig. 14.1 previously introduced in Example 14.1. Assume $w = ab$. In this case $\hat{\mathscr{J}}(w) = \{(t_1t_2, \varepsilon)\}$ and $\hat{Y}_{min}(\boldsymbol{m}_0, w) = \{(t_1t_2, \boldsymbol{0})\}$. Now, consider $w = acd$. The set $\hat{\mathscr{J}}(w) = \{(t_1t_5t_6, \varepsilon), (t_1t_5t_7, \varepsilon_{12}\varepsilon_{13})\}$ and $\hat{Y}_{min}(\boldsymbol{m}_0, w) = \{(t_1t_5t_6, \boldsymbol{0}), (t_1t_5t_7, [0\,0\,0\,0\,1\,1]^T)\}$. ■

The main difference among minimal explanations and justifications is that the first ones are functions of a generic marking $\boldsymbol{m}$ and transition t, while justifications are functions of the initial marking $\boldsymbol{m}_0$ and w. Moreover, as will be claimed in the following Proposition 14.1, in the case of acyclic unobservable subnets, justifications can be computed recursively summing up minimal explanations.

Definition 14.6. [16] *Let $\langle N, \boldsymbol{m}_0 \rangle$ be a net system with labeling function $\mathscr{L} : T \to L \cup \{\varepsilon\}$, where $N = (P, T, \boldsymbol{Pre}, \boldsymbol{Post})$ and $T = T_o \cup T_u$. Let w be a given observation and $(\sigma_o, \sigma_u) \in \hat{\mathscr{J}}(w)$ be a generic pair (sequence of observable transitions labeled w; corresponding justification). The marking*

$$\boldsymbol{m}_b = \boldsymbol{m}_0 + \boldsymbol{C}_u \cdot \boldsymbol{y} + \boldsymbol{C}_o \cdot \boldsymbol{y}', \qquad \boldsymbol{y} = \pi(\sigma_u), \;\; \boldsymbol{y}' = \pi(\sigma_o),$$

i.e., the marking reached firing σ_o interleaved with the justification σ_u, is called basis marking *and $\boldsymbol{y}$ is called its* j-vector *(or* justification-vector*).*

Obviously, because in general more than one justification exists for a word w (the set $\hat{\mathscr{J}}(w)$ is generally not a singleton), the basis marking may be not unique as well.

Definition 14.7. [16] *Let $\langle N, \boldsymbol{m}_0 \rangle$ be a net system with labeling function $\mathscr{L} : T \to L \cup \{\varepsilon\}$, where $N = (P, T, \boldsymbol{Pre}, \boldsymbol{Post})$ and $T = T_o \cup T_u$. Let $w \in L^*$ be an observed word. Let*

$$\begin{aligned} \mathscr{M}(w) = \{(\boldsymbol{m}, \boldsymbol{y}) \mid\; & (\exists\, \sigma \in \mathscr{S}(w) \;:\; \boldsymbol{m}_0[\sigma\rangle \boldsymbol{m}) \wedge \\ & (\exists (\sigma_o, \sigma_u) \in \hat{\mathscr{J}}(w) \;:\; \sigma_o = P_o(\sigma), \;\; \sigma_u = P_u(\sigma), \;\; \boldsymbol{y} = \pi(\sigma_u))\} \end{aligned}$$

be the set of pairs (basis marking, relative j-vector) that are consistent *with $w \in L^*$.*

Note that, the set $\mathscr{M}(w)$ does not keep into account the sequences of observable transitions that may have actually fired. It only keeps track of the basis markings that can be reached and of the firing vectors relative to sequences of unobservable transitions that have fired to reach them. Indeed, this is the information really significant when performing diagnosis. The notion of $\mathscr{M}(w)$ is fundamental to provide a recursive way to compute the set of minimal explanations.

Proposition 14.1. [16] *Given a net system* $\langle N, \boldsymbol{m}_0 \rangle$ *with labeling function* $\mathscr{L} : T \rightarrow L \cup \{\varepsilon\}$, *where* $N = (P, T, \boldsymbol{Pre}, \boldsymbol{Post})$ *and* $T = T_o \cup T_u$. *Assume that the unobservable subnet is acyclic*[2]. *Let* $w = w'l$ *be a given observation. It holds:*

$$\hat{Y}_{\min}(\boldsymbol{m}_0, w'l) = \{(\sigma_o, \boldsymbol{y}) \mid \sigma_o = \sigma_o' t \wedge \boldsymbol{y} = \boldsymbol{y}' + \boldsymbol{e} : \\ (\sigma_o', \boldsymbol{y}') \in \hat{Y}_{\min}(\boldsymbol{m}_0, w'),\ (t, \boldsymbol{e}) \in \hat{Y}_{\min}(\boldsymbol{m}_b', l),\ \mathscr{L}(t) = l\},$$

where $\boldsymbol{m}_b' = \boldsymbol{m}_0 + \boldsymbol{C}_u \cdot \boldsymbol{y}' + \boldsymbol{C}_o \cdot \pi(\sigma_o')$ *and* $\hat{Y}_{\min}(\boldsymbol{m}_b', l)$ *is the set of pairs (transition labeled* l *that may have fired at* $\boldsymbol{m}_b'$, *corresponding j-vector) introduced in Definition 14.5.*

Example 14.4. Consider the PN in Fig. 14.1 previously introduced in Example 14.1. Assume $w = ab$. As shown in Example 14.3 $\hat{\mathscr{J}}(w) = \{(t_1 t_2, \varepsilon)\}$, thus the basis marking is $\boldsymbol{m}_b = [0\,0\,1\,0\,0\,0\,0\,1\,0\,0\,0]^T$, and $\mathscr{M}(w) = \{(\boldsymbol{m}_b, \boldsymbol{0})\}$.

Now, consider $w = acd$. As computed in Example 14.3, the set $\hat{\mathscr{J}}(w) = \{(t_1 t_5 t_6, \varepsilon),\ (t_1 t_5 t_7, \varepsilon_{12}\varepsilon_{13})\}$. All the above j-vectors lead to the same basis marking $\boldsymbol{m}_b'$=$[0\,1\,0\,0\,0\,0\,0\,1\,0\,0\,0]^T$ thus $\mathscr{M}(w)$=$\{(\boldsymbol{m}_b', \boldsymbol{0}), (\boldsymbol{m}_b', [0\,0\,0\,0\,1\,1]^T)\}$. ∎

By Proposition 14.1, under the assumption of acyclicity of the unobservable subnet, the set $\mathscr{M}(w)$ can be easily constructed as follows.

Algorithm 14.5. (Computation of the basis markings and j-vectors). [16]

1. Let $w = \varepsilon$.
2. Let $\mathscr{M}(w) = \{(\boldsymbol{m}_0, \boldsymbol{0})\}$.
3. Wait until a new label l *is observed.*
4. Let $w' = w$ *and* $w = w'l$.
5. Let $\mathscr{M}(w) = \emptyset$.
6. For all $\boldsymbol{m}'$ *such that* $(\boldsymbol{m}', \boldsymbol{y}') \in \mathscr{M}(w')$ *, do*

6.1. for all $t \in T_l$, *do*
6.1.1. for all $\boldsymbol{e} \in Y_{\min}(\boldsymbol{m}', t)$, *do*
6.1.1.1. let $\boldsymbol{m} = \boldsymbol{m}' + \boldsymbol{C}_u \cdot \boldsymbol{e} + \boldsymbol{C}[\cdot, t]$,
6.1.1.2. for all $\boldsymbol{y}'$ *such that* $(\boldsymbol{m}', \boldsymbol{y}') \in \mathscr{M}(w')$, *do*
6.1.1.2.1 let $\boldsymbol{y} = \boldsymbol{y}' + \boldsymbol{e}$,
6.1.1.2.2. let $\mathscr{M}(w) = \mathscr{M}(w) \cup \{(\boldsymbol{m}, \boldsymbol{y})\}$.

7. Goto Step 3.

In simple words, the above algorithm can be explained as follows. Assume that, after a certain word w' has been observed, a new observable t fires and its label $l = \mathscr{L}(t)$ is observed. Consider all basis markings at the observation w' and select among them those that may have allowed the firing of at least one transition $t \in T_l$, also taking into account that this may have required the firing of appropriate sequences

[2] As discussed in Chapter 11 (see Proposition 11.12), the main feature of acyclic nets is that their state equation has no spurious solutions. This implies that the reachability set coincides with the potential reachability set.

of unobservable transitions. In particular, let us focus on the minimal explanations, and thus on the corresponding minimal e-vectors (Step 6.1.1). Finally, update the set $\mathcal{M}(w't)$ including all pairs of new basis markings and j-vectors, taking into account that for each basis marking at w' it may correspond more than one j-vector.

Definition 14.8. [16] *Let* $\langle N, \boldsymbol{m}_0 \rangle$ *be a net system where* $N = (P, T, \boldsymbol{Pre}, \boldsymbol{Post})$ *and* $T = T_o \cup T_u$. *Assume that the unobservable subnet is acyclic. Let* $w \in T_o^*$ *be an observed word. Let*

$$\mathcal{M}_{basis}(w) = \{\boldsymbol{m} \in \mathbb{N}^m \mid \exists \boldsymbol{y} \in \mathbb{N}^{n_u} \text{ and } (\boldsymbol{m}, \boldsymbol{y}) \in \mathcal{M}(w)\}$$

be the set of basis markings at w. *Moreover, denote as* $\mathcal{M}_{basis} = \bigcup_{w \in T_o^*} \mathcal{M}_{basis}(w)$ *the set of all basis markings for any observation* w.

Note that if the net system is bounded then the set $\mathcal{M}_{basis}$ is *finite* being the set of basis markings a subset of the reachability set.

Theorem 14.1. [16] *Consider a net system* $\langle N, \boldsymbol{m}_0 \rangle$ *whose unobservable subnet is acyclic. For any* $w \in L^*$ *it holds that*

$$\mathcal{C}(w) = \{\boldsymbol{m} \in \mathbb{N}^m \mid \boldsymbol{m} = \boldsymbol{m}_b + \boldsymbol{C}_u \cdot \boldsymbol{y} \; : \boldsymbol{y} \geq \boldsymbol{0} \text{ and } \boldsymbol{m}_b \in \mathcal{M}_{basis}(w)\}.$$

The above result shows that the set $\mathcal{C}(w)$ can be characterized in linear algebraic terms given the set $\mathcal{M}_{basis}(w)$, thus not requiring exhaustive enumeration. This is the main advantage of the approach here presented.

14.4 Diagnosis Using Petri Nets

Assume that the set of unobservable transitions is partitioned into two subsets: $T_u = T_f \cup T_{reg}$ where T_f includes all fault transitions (modeling anomalous or fault behavior), while T_{reg} includes all transitions relative to unobservable but regular events. The set T_f is further partitioned into r different subsets T_f^i, where $i = 1, \ldots, r$, that model the different fault classes. Usually, fault transitions that belong to the same fault class are transitions that represent similar physical faulty behavior.

Definition 14.9. [16] *A* diagnoser *is a function* $\Delta : L^* \times \{T_f^1, T_f^2, \ldots, T_f^r\} \rightarrow \{0, 1, 2, 3\}$ *that associates with each observation* $w \in L^*$ *and with each fault class* T_f^i, $i = 1, \ldots, r$, *a* diagnosis state.

- $\Delta(w, T_f^i) = 0$ *if for all* $\sigma \in \mathcal{S}(w)$ *and for all* $t_f \in T_f^i$ *it holds* $t_f \notin \sigma$.
  In such a case the ith fault cannot have occurred, because none of the firing sequences consistent with the observation contains fault transitions of class i.
- $\Delta(w, T_f^i) = 1$ *if:*
  (i) there exist $\sigma \in \mathcal{S}(w)$ *and* $t_f \in T_f^i$ *such that* $t_f \in \sigma$ *but*
  (ii) for all $(\sigma_o, \sigma_u) \in \hat{\mathcal{J}}(w)$ *and for all* $t_f \in T_f^i$ *it holds that* $t_f \notin \sigma_u$.

In such a case a fault transition of class i may have occurred but is not contained in any justification of w.

- $\Delta(w, T_f^i) = 2$ *if there exist* $(\sigma_o, \sigma_u), (\sigma_o', \sigma_u') \in \hat{\mathscr{J}}(w)$ *such that*
  (i) there exists $t_f \in T_f^i$ *such that* $t_f \in \sigma_u$*;*
  (ii) for all $t_f \in T_f^i$, $t_f \notin \sigma_u'$.
  In such a case a fault transition of class i is contained in one (but not in all) justification of w.
- $\Delta(w, T_f^i) = 3$ *if for all* $\sigma \in \mathscr{S}(w)$ *there exists* $t_f \in T_f^i$ *such that* $t_f \in \sigma$.
  In such a case the ith fault must have occurred, because all firable sequences consistent with the observation contain at least one fault in T_f^i.

Example 14.6. Consider the PN in Fig. 14.1 previously introduced in Example 14.1. Let $T_f = \{\varepsilon_{11}, \varepsilon_{12}\}$. Assume that the two fault transitions belong to different fault classes, i.e., $T_f^1 = \{\varepsilon_{11}\}$ and $T_f^2 = \{\varepsilon_{12}\}$.

Let us observe $w = a$. Then $\Delta(w, T_f^1) = \Delta(w, T_f^2) = 0$, being $\hat{\mathscr{J}}(w) = \{(t_1, \varepsilon)\}$ and $\mathscr{S}(w) = \{t_1\}$. In simple words, no fault of both fault classes may have occurred.

Let us observe $w = ab$. Then $\Delta(w, T_f^1) = 1$ and $\Delta(w, T_f^2) = 0$, being $\hat{\mathscr{J}}(w) = \{(t_1t_2, \varepsilon)\}$ and $\mathscr{S}(w) = \{t_1t_2, t_1t_2\varepsilon_8, t_1t_2\varepsilon_8\varepsilon_9, t_1t_2\varepsilon_8\varepsilon_9\varepsilon_{10}, t_1t_2\varepsilon_8\varepsilon_{11}\}$. This means that a fault of the first fault class may have occurred (firing the sequence $t_1t_2\varepsilon_8\varepsilon_{11}$) but it is not contained in any justification of *ab*, while no fault of the second fault class can have occurred.

Now, consider $w = abb$. In this case $\Delta(w, T_f^1) = 2$ and $\Delta(w, T_f^2) = 0$, being $\hat{\mathscr{J}}(w) = \{(t_1t_2t_2, \varepsilon_8\varepsilon_9\varepsilon_{10}), (t_1t_2t_3, \varepsilon_8\varepsilon_{11})\}$ and $\mathscr{S}(w) = \{t_1t_2\varepsilon_8\varepsilon_9\varepsilon_{10}t_2,\ t_1t_2\varepsilon_8\varepsilon_9\varepsilon_{10}t_2\varepsilon_8,\ t_1t_2\varepsilon_8\varepsilon_9\varepsilon_{10}t_2\varepsilon_8\varepsilon_9,\ t_1t_2\varepsilon_8\varepsilon_9\varepsilon_{10}t_2\varepsilon_8\varepsilon_9\varepsilon_{10},\ t_1t_2\varepsilon_8\varepsilon_9\varepsilon_{10}t_2\varepsilon_8\varepsilon_{11},\ t_1t_2\varepsilon_8\varepsilon_{11}t_3\}$. This means that no fault of the second fault class can have occurred, while a fault of the first fault class may have occurred since one justification does not contain ε_{11} and one justification contains it.

Finally, consider $w = abbccc$. In this case $\Delta(w, T_f^1) = 3$ and $\Delta(w, T_f^2) = 1$. In fact since $\hat{\mathscr{J}}(w) = \{(t_1t_2t_3t_5t_4t_4, \varepsilon_8\varepsilon_{11}), (t_1t_2t_3t_4t_5t_4, \varepsilon_8\varepsilon_{11}), (t_1t_2t_3t_4t_4t_5, \varepsilon_8\varepsilon_{11}), (t_1t_2t_3t_4t_4t_4, \varepsilon_8\varepsilon_{11})\}$ a fault of the first fault class must have occurred, while a fault of the second fault class may have occurred (e.g. $t_1t_2\varepsilon_8\varepsilon_{11}t_3t_4t_4t_5\varepsilon_{12}$) but it is not contained in any justification of w. ■

The following two results proved in [11] for unlabeled PNs still hold in the case of labeled PNs [16]. In particular, the following proposition presents how the diagnosis states can be characterized analyzing basis markings and justifications.

Proposition 14.2. [16] *Consider an observed word* $w \in L^*$.

- $\Delta(w, T_f^i) \in \{0, 1\}$ *iff for all* $(\boldsymbol{m}, \boldsymbol{y}) \in \mathscr{M}(w)$ *and for all* $t_f \in T_f^i$ *it holds* $y[t_f] = 0$.
- $\Delta(w, T_f^i) = 2$ *iff there exist* $(\boldsymbol{m}, \boldsymbol{y}) \in \mathscr{M}(w)$ *and* $(\boldsymbol{m}', \boldsymbol{y}') \in \mathscr{M}(w)$ *such that:*
  (i) there exists $t_f \in T_f^i$ *such that* $y[t_f] > 0$,
  (ii) for all $t_f \in T_f^i$, $\boldsymbol{y}'[t_f] = 0$.
- $\Delta(w, T_f^i) = 3$ *iff for all* $(\boldsymbol{m}, \boldsymbol{y}) \in \mathscr{M}(w)$ *there exists* $t_f \in T_f^i$ *such that* $y[t_f] > 0$.

Proposition 14.3. [16] *For a PN whose unobservable subnet is acyclic, let $w \in L^*$ be an observed word such that for all $(\boldsymbol{m}, \boldsymbol{y}) \in \mathscr{M}(w)$ it holds $y[t_f] = 0 \; \forall \; t_f \in T_f^i$. Consider the constraint set*

$$\mathscr{T}(\boldsymbol{m}, T_f^i) = \begin{cases} \boldsymbol{m} + \boldsymbol{C}_u \cdot \boldsymbol{z} \geq \boldsymbol{0}, \\ \sum_{t_f \in T_f^i} z[t_f] > 0, \\ \boldsymbol{z} \in \mathbb{N}^{n_u}. \end{cases} \tag{14.1}$$

- $\Delta(w, T_f^i) = 0$ *if* $\forall \; (\boldsymbol{m}, \boldsymbol{y}) \in \mathscr{M}(w)$ *the constraint set* (14.1) *is not feasible.*
- $\Delta(w, T_f^i) = 1$ *if* $\exists \; (\boldsymbol{m}, \boldsymbol{y}) \in \mathscr{M}(w)$ *such that the constraint set* (14.1) *is feasible.*

On the basis of the above two results, if the unobservable subnet is acyclic, diagnosis may be carried out by simply looking at the set $\mathscr{M}(w)$ for any observed word w and, should the diagnosis state be either 0 or 1, by additionally evaluating whether the corresponding integer constraint set (14.1) admits a solution.

Example 14.7. Consider the PN in Fig. 14.1 where $T_f^1 = \{\varepsilon_{11}\}$ and $T_f^2 = \{\varepsilon_{12}\}$.

Let $w = a$. In this case $\mathscr{M}(w) = \{(\boldsymbol{m}_1, \boldsymbol{0})\}$, where $\boldsymbol{m}_1 = [0\,1\,0\,0\,0\,0\,0\,0\,1\,0\,0\,0]^T$. Being $\mathscr{T}(\boldsymbol{m}_1, T_f^i)$ not feasible both for $i = 1$ and $i = 2$ it holds $\Delta(w, T_f^1) = \Delta(w, T_f^2) = 0$.

Let $w = ab$. In this case $\mathscr{M}(w) = \{(\boldsymbol{m}_2, \boldsymbol{0})\}$, where $\boldsymbol{m}_2 = [0\,0\,1\,0\,0\,0\,0\,0\,1\,0\,0\,0]^T$. Being $\mathscr{T}(\boldsymbol{m}_2, T_f^i)$ feasible only for $i = 1$ it holds $\Delta(w, T_f^1) = 1$ and $\Delta(w, T_f^2) = 0$.

Let $w = abb$. It is $\mathscr{M}(w) = \{(\boldsymbol{m}_2, [1\;1\;1\;0\;0\;0]^T),\; (\boldsymbol{m}_3, [1\;0\;0\;1\;0\;0]^T)\}$, where $\boldsymbol{m}_3 = [0\;0\;0\;0\;0\;0\;1\;1\;0\;0\;0]^T$. It is $\Delta(w, T_f^1) = 2$ and $\Delta(w, T_f^2) = 0$ being both $\mathscr{T}(\boldsymbol{m}_2, T_f^2)$ and $\mathscr{T}(\boldsymbol{m}_3, T_f^2)$ not feasible.

Let $w = abbccc$. In this case $\mathscr{M}(w) = \{(\boldsymbol{m}_3, [1\,0\,0\,1\,0\,0]^T), (\boldsymbol{m}_4, [1\,0\,0\,1\,0\,0]^T)\}$, where $\boldsymbol{m}_4 = [0\,0\,0\,0\,0\,0\,1\,0\,1\,0\,0]^T$. It is $\Delta(w, T_f^1) = 3$ and being $\mathscr{T}(\boldsymbol{m}_4, T_f^2)$ feasible it holds $\Delta(w, T_f^2) = 1$. ■

The approach described above requires to compute for each observed word w and for each fault class i a diagnosis state $\Delta(w, T_f^i)$. Let us conclude this section with a brief discussion on the definition of diagnosis states $\Delta = 1$ and $\Delta = 2$. First, observe that both the diagnosis states correspond to *uncertain states* even if a higher degree of alarm is associated with $\Delta = 2$ with respect to $\Delta = 1$. Second, observe that an advantage in terms of computational complexity can be obtained by splitting the uncertain condition in two diagnosis states, namely $\Delta = 1$ and $\Delta = 2$. In fact, the diagnosis approach is based on the preliminary computation of the set $\mathscr{M}(w)$. If $\Delta = 2$ or $\Delta = 3$ no additional computation is required. On the contrary, to distinguish among $\Delta = 0$ and $\Delta = 1$ an integer programing problem should be solved.

14.5 Basis Reachability Graph

Diagnosis approach described in the previous section can be applied both to bounded and unbounded PNs. The proposed approach is an online approach that for each new observed event updates the diagnosis state for each fault class computing the set of basis markings and j-vectors. Moreover, if for the fault class T_f^i is necessary to distinguish between diagnosis states 0 and 1, it is also necessary to solve for each basis marking $\boldsymbol{m}_b$ the constraint set $\mathscr{T}(\boldsymbol{m}_b, T_f^i)$.

In this section, it is shown that if the considered net system is bounded, the most burdensome part of the procedure can be moved offline defining a graph called *Basis Reachability Graph* (BRG).

Definition 14.10. [16] *The BRG is a deterministic graph that has as many nodes as the number of possible basis markings.*

To each node is associated a different basis marking $\boldsymbol{m}$ *and a row vector with as many entries as the number of fault classes. The entries of this vector may only take binary values:* 1 *if* $\mathscr{T}(\boldsymbol{m}, T_f^i)$ *is feasible,* 0 *otherwise.*

Arcs are labeled with observable events in L and e-vectors: an arc exists from a node containing the basis marking $\boldsymbol{m}$ *to a node containing the basis marking* $\boldsymbol{m}'$ *if and only if there exists a transition t for which an explanation exists at* $\boldsymbol{m}$ *and the firing of t and one of its minimal explanations leads to* $\boldsymbol{m}'$*. The arc going from* $\boldsymbol{m}$ *to* $\boldsymbol{m}'$ *is labeled* $(\mathscr{L}(t), \boldsymbol{e})$*, where* $\boldsymbol{e} \in Y_{\min}(\boldsymbol{m}, t)$ *and* $\boldsymbol{m}' = \boldsymbol{m} + \boldsymbol{C}_u \cdot \boldsymbol{e} + \boldsymbol{C}[\cdot, t]$.

Note that, the number of nodes of the BRG is always finite being the set of basis markings a subset of the set of reachable markings, that is finite being the net bounded by assumption. Moreover, the row vector of binary values associated with the nodes of the BRG allows us to distinguish between the diagnosis state 1 or 0.

The main steps for the computation of the BRG in the case of labeled PNs are summarized in the following algorithm.

Algorithm 14.8. (Computation of the BRG). [16]

1. Label the initial node $(\boldsymbol{m}_0, \boldsymbol{x}_0)$ *where* $\forall i = 1, \ldots, r$,

$$\boldsymbol{x}_0[T_f^i] = \begin{cases} 1 \text{ if } \mathscr{T}(\boldsymbol{m}_0, T_f^i) \text{ is feasible,} \\ 0 \text{ otherwise.} \end{cases}$$

Assign no tag to it.

2. While nodes with no tag exist, select a node with no tag and do

2.1. let $\boldsymbol{m}$ *be the marking in the node* $(\boldsymbol{m}, \boldsymbol{x})$*,*

2.2. for all $l \in L$

2.2.1. for all $t : L(t) = l \wedge Y_{\min}(\boldsymbol{m}, t) \neq \emptyset$*, do*

- *for all* $\boldsymbol{e} \in Y_{\min}(\boldsymbol{m}, t)$*, do*
  - *let* $\boldsymbol{m}' = \boldsymbol{m} + \boldsymbol{C}_u \cdot \boldsymbol{e} + \boldsymbol{C}[\cdot, t]$*,*
  - *if* $\nexists$ *a node* $(\boldsymbol{m}, \boldsymbol{x})$ *with* $\boldsymbol{m} = \boldsymbol{m}'$*, do*
    - *add a new node to the graph containing*

$(\boldsymbol{m}',\boldsymbol{x}')$ *where* $\forall i = 1,\ldots,r$,
$\boldsymbol{x}'[T_f^i] = \{\, 1$ *if* $\mathscr{T}(\boldsymbol{m}',T_f^i)$ *is feasible,* 0 *otherwise.*
and arc $(l,\boldsymbol{e})$ *from* $(\boldsymbol{m},\boldsymbol{x})$ *to* $(\boldsymbol{m}',\boldsymbol{x}')$

- *else*
  - *add arc* $(l,\boldsymbol{e})$ *from* $(\boldsymbol{m},\boldsymbol{x})$ *to* $(\boldsymbol{m}',\boldsymbol{x}')$ *if it does not exist yet*

2.3. tag the node "old".

3. Remove all tags.

The algorithm constructs the BRG starting from the initial node to which it corresponds the initial marking and a binary vector that specifies for all fault classes, if some transition in the class may occur at $\boldsymbol{m}_0$. Now, consider all the labels $l \in L$ such that there exists a transition t with $\mathscr{L}(t) = l$ for which a minimal explanation at $\boldsymbol{m}_0$ exists. For each of these transitions compute the marking resulting from firing t at $\boldsymbol{m}_0 + \boldsymbol{C}_u \cdot \boldsymbol{e}$, for any $\boldsymbol{e} \in Y_{\min}(\boldsymbol{m}_0,t)$. If a pair (marking, binary vector) not contained in the previous nodes is obtained, a new node is added to the graph. The arc going from the initial node to the new node is labeled $(l,\boldsymbol{e})$. The procedure is iterated until all basis markings have been considered. Note that, the approach here presented always requires to enumerate a state space that is a subset (usually a strict subset) of the reachability space. However, as in general for diagnosis approaches, the combinatory explosion cannot be avoided.

Example 14.9. Consider again the PN in Fig. 14.1, where $T_o = \{t_1,t_2,t_3,t_4,t_5,t_6,t_7\}$, $T_u = \{\varepsilon_8,\varepsilon_9,\varepsilon_{10},\varepsilon_{11},\varepsilon_{12},\varepsilon_{13}\}$, $T_f^1 = \{\varepsilon_{11}\}$ and $T_f^2 = \{\varepsilon_{12}\}$. The labeling function is defined as follows: $\mathscr{L}(t_1) = a$, $\mathscr{L}(t_2) = \mathscr{L}(t_3) = b$, $\mathscr{L}(t_4) = \mathscr{L}(t_5) = c$, $\mathscr{L}(t_6) = \mathscr{L}(t_7) = d$.

The BRG is shown in Fig. 14.2. The notation used in this figure is detailed in Tables 14.1 and 14.2. Each node contains a different basis marking and a binary row vector of dimension two, being two the number of fault classes. As an example, the binary vector [0 0] is associated with $\boldsymbol{m}_0$ because $\mathscr{T}(\boldsymbol{m}_0,T_f^i)$ is not feasible for $i = 1$ and $i = 2$. From node $\boldsymbol{m}_0$ to node $\boldsymbol{m}_1$ there is one arc labeled a with the null vector as minimal explanation. The node containing the basis marking $\boldsymbol{m}_2$ has binary vector [1 0], because $\mathscr{T}(\boldsymbol{m}_2,T_f^i)$ is feasible only for $i = 1$. Node $(\boldsymbol{m}_5,[0\ 1])$ has two output arcs both labeled with d and both directed to node $(\boldsymbol{m}_1,[0\ 0])$ with

Table 14.1 The markings of the BRG in Fig. 14.2.

$\boldsymbol{m}_0$	$[\,1\,0\,0\,0\,0\,0\,0\,0\,0\,0\,0\,]^T$
$\boldsymbol{m}_1$	$[\,0\,1\,0\,0\,0\,0\,0\,1\,0\,0\,0\,]^T$
$\boldsymbol{m}_2$	$[\,0\,0\,1\,0\,0\,0\,0\,1\,0\,0\,0\,]^T$
$\boldsymbol{m}_3$	$[\,0\,0\,0\,0\,0\,0\,1\,1\,0\,0\,0\,]^T$
$\boldsymbol{m}_4$	$[\,0\,0\,0\,0\,0\,0\,1\,0\,1\,0\,0\,]^T$
$\boldsymbol{m}_5$	$[\,0\,1\,0\,0\,0\,0\,0\,0\,1\,0\,0\,]^T$
$\boldsymbol{m}_6$	$[\,0\,0\,1\,0\,0\,0\,0\,0\,1\,0\,0\,]^T$

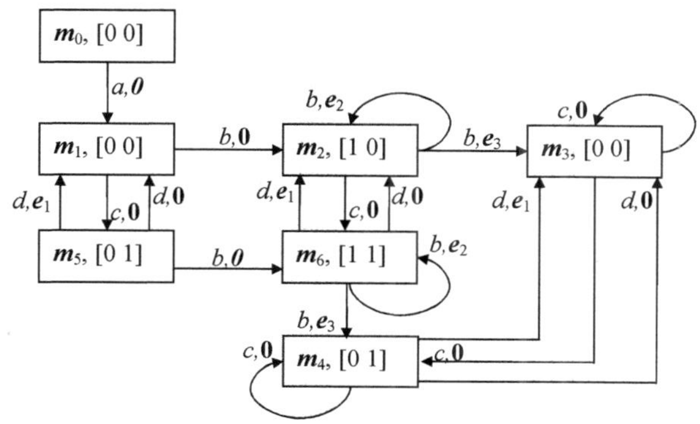

Fig. 14.2 The BRG of the PN in Fig. 14.1

two different minimal explanations $\mathbf{0}$ and $\boldsymbol{e}_1$, respectively, plus another output arc $(b,\mathbf{0})$ directed to node $(\boldsymbol{m}_6,[1\ 1])$. ■

The following algorithm summarizes the main steps of the online diagnosis carried out by looking at the BRG.

Algorithm 14.10. (Diagnosis using the BRG). [16]

1. Let $w=\varepsilon$.
2. Let $\mathcal{M}(w)=\{(\boldsymbol{m}_0,\mathbf{0})\}$.
3. Wait until a new observable transition fires.
Let l be the observed event.
4. Let $w'=w$ and $w=w'l$.
5. Let $\mathcal{M}(w)=\emptyset$, **[Computation of $\mathcal{M}(w)$]**
6. For all nodes containing $\boldsymbol{m}'$: $(\boldsymbol{m}',\boldsymbol{y}')\in\mathcal{M}(w')$, do

6.1 for all arcs exiting from the node with $\boldsymbol{m}'$, do
6.1.1. let $\boldsymbol{m}$ be the marking of the output node and $\boldsymbol{e}$ be the minimal e-vector on the edge from $\boldsymbol{m}'$ to $\boldsymbol{m}$,
6.1.2. for all $\boldsymbol{y}'$ such that $(\boldsymbol{m}',\boldsymbol{y}')\in\mathcal{M}(w')$, do
6.1.2.1. let $\boldsymbol{y}=\boldsymbol{y}'+\boldsymbol{e}$,
6.1.2.2. let $\mathcal{M}(w)=\mathcal{M}(w)\cup\{(\boldsymbol{m},\boldsymbol{y})\}$,

7. for all $i=1,\ldots,r$, do **[Computation of the diagnosis state]**

7.1. if $\forall\ (\boldsymbol{m},\boldsymbol{y})\in\mathcal{M}(w)\wedge\forall t_f\in T_f^i$ it is $y[t_f]=0$, do
7.1.1. if $\forall\ (\boldsymbol{m},\boldsymbol{y})\in\mathcal{M}(w)$ it holds $\boldsymbol{x}_{=}0$, where $\boldsymbol{x}$ is the binary vector in node $\boldsymbol{m}$, do
7.1.1.1. let $\Delta(w,T_f^i)=0$,
7.1.2. else
7.1.2.1. let $\Delta(w,T_f^i)=1$,

Table 14.2 The e-vectors of the BRG in Fig. 14.2.

	ε_8	ε_9	ε_{10}	ε_{11}	ε_{12}	ε_{13}
$\boldsymbol{e}_1$	0	0	0	0	1	1
$\boldsymbol{e}_2$	1	1	1	0	0	0
$\boldsymbol{e}_3$	1	0	0	1	0	0

7.2. if $\exists$ $(\boldsymbol{m},\boldsymbol{y}) \in \mathcal{M}(w)$ *and* $(\boldsymbol{m}',\boldsymbol{y}') \in \mathcal{M}(w)$ *s.t.:*
(i) $\exists t_f \in T_f^i$ *such that* $y[t_f] > 0$,
(ii) $\forall t_f \in T_f^i$, $\boldsymbol{y}'[t_f] = 0$, *do*
7.2.1. let $\Delta(w,T_f^i) = 2$,
7.3. if $\forall$ $(\boldsymbol{m},\boldsymbol{y}) \in \mathcal{M}(w)$ $\exists t_f \in T_f^i : y[t_f] > 0$, *do*
7.3.1. let $\Delta(w,T_f^i) = 3$.

8. Goto Step 3.

Steps 1 to 6 of Algorithm 14.10 enable us to compute the set $\mathcal{M}(w)$. When no event is observed, namely $w = \varepsilon$, then $\mathcal{M}(w) = \{(\boldsymbol{m}_0, \boldsymbol{0})\}$. Now, assume that a label l is observed. All couples $(\boldsymbol{m},\boldsymbol{y})$ such that an arc labeled l exits from the initial node and ends in a node containing the basis marking $\boldsymbol{m}$ are included in the set $\mathcal{M}(l)$. The corresponding value of $\boldsymbol{y}$ is equal to the e-vector in the arc going from $\boldsymbol{m}_0$ to $\boldsymbol{m}$, being $\boldsymbol{0}$ the j-vector relative to $\boldsymbol{m}_0$. In general, if w' is the actual observation, and a new event labeled l fires, one has to consider all couples $(\boldsymbol{m}',\boldsymbol{y}') \in \mathcal{M}(w')$ and all nodes that can be reached from $\boldsymbol{m}'$ with an arc labeled l. Let $\boldsymbol{m}$ be the basis marking of the generic resulting node. Include in $\mathcal{M}(w) = \mathcal{M}(w'l)$ all couples $(\boldsymbol{m},\boldsymbol{y})$, where for any $\boldsymbol{m}$, $\boldsymbol{y}$ is equal to the sum of $\boldsymbol{y}'$ plus the e-vector labeling the arc from $\boldsymbol{m}'$ to $\boldsymbol{m}$.

Step 7 of Algorithm 14.10 computes the diagnosis state. Consider the generic ith fault class. If $\forall(\boldsymbol{m},\boldsymbol{y}) \in \mathcal{M}(w)$ and $\forall t_f \in T_f^i$ it holds $y[t_f] = 0$, the ith entry of all the binary row vectors associated with the basis markings $\boldsymbol{m}$ has to be checked, such that $(\boldsymbol{m},\boldsymbol{y}) \in \mathcal{M}(w)$. If these entries are all equal to 0, it holds $\Delta(w,T_f^i) = 0$, otherwise it holds $\Delta(w,T_f^i) = 1$. On the other hand, if there exists at least one pair $(\boldsymbol{m},\boldsymbol{y}) \in \mathcal{M}(w)$ with $y[t_f] > 0$ for any $t_f \in T_f^i$, and there exists at least one pair $(\boldsymbol{m}',\boldsymbol{y}') \in \mathcal{M}(w)$ with $y[t_f] = 0$ for all $t_f \in T_f^i$, then $\Delta(w,T_f^i) = 2$. Finally, if for all pairs $(\boldsymbol{m},\boldsymbol{y}) \in \mathcal{M}(w)$, $y[t_f] > 0$ for any $t_f \in T_f^i$, then $\Delta(w,T_f^i) = 3$.

Example 14.11. Consider the PN in Fig. 14.1 and its BRG in Fig. 14.2. Let $w = \varepsilon$. By looking at the BRG it holds that $\Delta(\varepsilon,T_f^1) = \Delta(\varepsilon,T_f^2) = 0$ being both entries of the row vector associated with $\boldsymbol{m}_0$ equal to 0.

Now, consider $w = ab$. In such a case $\mathcal{M}(w) = \{(\boldsymbol{m}_2, \boldsymbol{0})\}$. It holds $\Delta(ab,T_f^1) = 1$ and $\Delta(ab,T_f^2) = 0$ being the row vector in the node equal to $[1\ 0]$.

Finally, for $w = abbc$ it holds $\Delta(abbc,T_f^1) = 2$ and $\Delta(abbc,T_f^2) = 1$. In fact $\mathcal{M}(w) = \{(\boldsymbol{m}_6,\boldsymbol{y}_1),(\boldsymbol{m}_3,\boldsymbol{y}_2),(\boldsymbol{m}_4,\boldsymbol{y}_3)\}$, where $\boldsymbol{y}_1 = \boldsymbol{e}_2$, $\boldsymbol{y}_2 = \boldsymbol{y}_3 = \boldsymbol{e}_3$, and the row vectors associated with $\boldsymbol{m}_6, \boldsymbol{m}_3$ and $\boldsymbol{m}_4$ are respectively $[1\ 1]$, $[0\ 0]$ and $[0\ 1]$. ■

14.6 Some Important Problems Strictly Related to Online Diagnosis

Let us discuss some important problems strictly related to online diagnosis.

14.6.1 Online Diagnosis via Fluidification

In the case of discrete event systems with a large number of reachable states the problem of fault detection, as well as many others, becomes computationally prohibitive because of the state explosion. As discussed in detail in Chapters 18 to 20, a common technique to overcome this is *fluidification*. Several discrete-event based fluid models have been proposed in the literature, some of them derived by the fluidification of queuing networks [4, 18, 43] or Petri nets [41, 20]. The main idea of the fluidification of PNs is the relaxation of the transitions firings allowing them to fire in positive real amounts. Therefore, the content of the places is no more restricted to take natural values, but it may be expressed by non-negative real numbers. This implies a series of significant properties. As an example, the reachability set is convex [36]. Moreover, as proved in [7, 8, 14], in the case of partial observation of the transitions firings (namely in the presence of silent transitions), the set of markings that are consistent with a given observation is convex. Using this convexity property, the fault detection problem for untimed continuous Petri nets has been studied [14]. In particular, as in the previous section, it is assumed that certain transitions are not observable, including fault transitions and transitions modeling a regular behavior. Thus, faults should only be detected on the basis of the observation of a subset of transitions. Note, however, that the case of undistinguishable transitions has not been dealt in [14], i.e., all transitions that are observable are also distinguishable. Fault transitions are partitioned into different fault classes and three different diagnosis states are defined, each one representing a different degree of alarm: N means that *no* fault of a given class has surely occurred; U means that a fault of a given class may have occurred or not (*uncertain* state); F means that a *fault* of a given class has surely occurred. A criterion to define, for each fault class, the value of the diagnosis state, given the observation of a sequence of transitions firings has been derived. Thus, another difference with the case in the previous section is that only three diagnosis states, rather than four, are considered. This is a major requirement originating from a technical assumption that is not discussed here for the sake of brevity.

In [14] general PN structures are considered under the assumption, common to all works dealing with fault diagnosis, that the unobservable subnet has no spurious markings, i.e., all solutions of the state equation are reachable markings. Since in continuous case this assumption is not very restrictive, this allows one to also deal with unobservable subnets that are *cyclic*, making the procedure more general with respect to the approach developed for discrete PNs [11]. To the best of our knowledge, the approach for fault diagnosis in [14] is the only one that can be applied to systems whose unobservable part contains cycles, regardless of the discrete-event formalism used to model the plant: automata, Petri nets, etc.

On the other hand, in the case of acyclic unobservable subnets, it is not so easy to evaluate the effectiveness of fluidification. In particular, the computational complexity of the proposed approach is not a priori comparable with that of the diagnosis approach for discrete nets in [11], based on the notion of basis markings. As it can be shown via some numerical example, the computational complexity of both procedures depends on the particular net structure, on the observed word and on the initial marking as well, thus no general conclusion can be drawn. Nevertheless, there exist cases in which the proposed approach provides a considerable improvement on the computational costs.

As a result of the above considerations, we believe that fluidification is an appropriate technique when the unobservable subnet is cyclic, being in such a case the only viable approach, or when the advantages in terms of computational complexity are really significant such as in the case of systems with a very large number of reachable states.

Note that, Section 19.7 of Chapter 19 is fully devoted to on-line diagnosis via fluidification.

14.6.2 Diagnosability Analysis

In the fault analysis framework two different problems can be solved: the problem of diagnosis and the problem of diagnosability. As explained in detail in the above sections, solving a problem of diagnosis means that to each observed string of events is associated a diagnosis state, such as "normal" or "faulty" or "uncertain". Solving a problem of diagnosability is equivalent to determine if the system is diagnosable, i.e., to determine if, once a fault has occurred, the system can detect its occurrence in a finite number of steps. This problem has been addressed in two different settings, depending on the boundedness of the net system. In particular, in [9] a solution that only applies to bounded nets has been proposed, where the major feature is to allow to perform, using the same framework, both diagnosis and diagnosability analysis. The computational complexity of such an approach is actually under investigation as well as a comparison among the approach in [9] and the automata-based approach by Sampath *et al.* [37].

Moreover, in [10, 17] an approach that also applies to unbounded PNs has been presented. In particular in [17] two different notions of diagnosability have been considered: diagnosability and diagnosability in K steps, or K-diagnosability. Diagnosability in K steps is stronger than diagnosability and implies not only that the system is diagnosable, i.e., when the fault occurs we are able to detect it in a finite number of transition firings, but also that if the fault occurs we are able to detect it in at most K steps. Necessary and sufficient conditions for both notions of diagnosability have been given together with a test to study both diagnosability and K-diagnosability based on the analysis of the coverability graph of a special PN, called *Verifier Net*, that is built starting from the initial system. Moreover, a procedure to compute the bound K in the case of K-diagnosable systems is presented. Then, sufficient conditions on diagnosability are derived using linear programing techniques. To the best of our knowledge, this is the first time that necessary and sufficient conditions for diagnosability and K-diagnosability of labeled unbounded Petri nets are presented.

14.6.3 Decentralized Diagnosis

The extension of the approach presented in the previous sections to decentralized fault diagnosis has been investigated in [12, 13, 15]. In particular, exploiting the classical decentralized diagnosis architecture, it is assumed that the system is monitored by a set of sites. Each site knows the structure of the net and the initial marking but observes the evolution of the system with a different mask, i.e., the set of observable transitions is different for each site. Diagnosis is locally performed using the approach founded on basis markings introduced in [11, 16]. Using its own observation, each site performs diagnosis and, according to a given protocol, communicates it, eventually with some other information, to the coordinator who calculates global diagnosis states. In particular, three different protocols have been defined that differ for the amount of information exchanged between the coordinator and the local sites, and vice versa. In all cases, an important property has been proved, namely that the coordinator never produces false alarms.

Finally, the diagnosability property under decentralization has been investigated, i.e., it is investigated if it is possible to guarantee that any fault occurrence can be detected after a finite number of observations. To this aim, the definition of failure ambiguous strings is first introduced. Second, it is shown that the absence of such kind of sequences is a sufficient condition for the diagnosability of a given net system in a decentralized framework, regardless of the considered protocol. It is also shown that the absence of failure ambiguous strings is also a necessary condition for codiagnosability, i.e., diagnosability in the case in which there is no communication between the sites and the coordinator. Moreover, a procedure to detect the presence of failure ambiguous strings is presented. Such a procedure is based on the construction of a particular net called *Modified Verifier Net* (MVN), that is an extension of the *Verifier Net* (VN), introduced in [10] to analyze diagnosability in a centralized framework.

Note that, the following Chapter 15 presents approaches for online diagnosis and diagnosability analysis of large distributed systems modeled with Petri nets.

14.7 Further Reading

The first contributions in the framework of fault diagnosis of discrete event systems have been proposed in the context of automata by Sampath *et al.* [37, 38] who proposed an approach to failure diagnosis where the system is modeled as a nondeterministic automaton in which the failures are treated as unobservable events. In [37], they provided a definition of diagnosability in the framework of formal languages and established necessary and sufficient conditions for diagnosability. Moreover, in [39] Sampath *et al.* presented an integrated approach to control and diagnosis. More specifically, the authors presented an approach for the design of diagnosable systems by appropriate design of the system controller and this approach is called active diagnosis. They formulated the active diagnosis problem as a supervisory control problem. In [21], Debouk *et al.* proposed a coordinated decentralized architecture consisting of two local sites communicating with a coordinator that is responsible for diagnosing the failures occurring in the system. In [5], Boel and van Schuppen addressed the problem of synthesizing communication protocols and failure diagnosis algorithms for decentralized failure diagnosis of discrete event systems with costly communication between diagnosers. In [28], a state-based approach for on-line passive fault diagnosis is presented.

More recently, PN models have been used in the context of diagnosis. Indeed, the use of PNs offers significant advantages because of their twofold representation: graphical and mathematical. Moreover, the intrinsically distributed nature of PNs where the notion of state (i.e., marking) and action (i.e., transition) is local reduces the computational complexity involved in solving a diagnosis problem.

Among the first pioneer works dealing with PNs, let us recall the approach of Prock [34]. He proposes an online technique whose goal is the identification of sensor or process errors which are manifested in signals related to physical conservation quantities. After a fault is detected a prognosis of the future system's behavior can be provided. Note that, preliminary results in this respect have already been presented in the 80s by Silva and Velilla [40].

Sreenivas and Jafari [42] employ time PNs to model the discrete event system (DES) controller and backfiring transitions to determine whether a given state is invalid. Later on, time PNs have been employed by Ghazel *et al.* [26] who propose a monitoring approach for DES with unobservable events and to represent the "a priori" known behavior of the system, and track online its state to identify the events that occur.

Hadjicostis and Veghese [27] use PN models to introduce redundancy into the system and additional P-invariants allow the detection and isolation of faulty markings.

Wu and Hadjicostis [45] use redundancy into a given PN to enable fault detection and identification using algebraic decoding techniques. In this paper, the authors consider two types of faults: place faults that corrupt the net marking, and transition faults that cause a not correct update of the marking after event occurrence. Although this approach is general, the net marking has to be periodically observable even if unobservable events occur. Analogously, Lefebvre and Delherm [30] investigate on the determination of the set of places that must be observed for the exact and immediate estimation of faults occurrence.

Miyagi and Riascos [33] introduce a methodology, based on the hierarchical and modular integration of PNs, for modeling and analyzing fault-tolerant manufacturing systems that not only optimizes normal productive processes, but also performs detection and treatment of faults.

Ramirez-Treviño *et al.* [35] employ Interpreted PNs to model the system behavior that includes both events and states partially observable. Based on the Interpreted PN model derived from an online methodology, a scheme utilizing a solution of a programing problem is proposed to solve the problem of diagnosis.

Genc and Lafortune [25] propose a diagnoser on the basis of a modular approach that performs the diagnosis of faults in each module. Subsequently, the diagnosers recover the monolithic diagnosis information obtained when all the modules are combined into a single module that preserves the behavior of the underlying modular system. A communication system connects the different modules and updates the diagnosis information. Even if the approach does not avoid the state explosion problem, an improvement is obtained when the system can be modeled as a collection of PN modules coupled through common places.

The main advantage of the approaches in [25] consists in the fact that, if the net is bounded, the diagnoser may be constructed offline, thus moving offline the most burdensome part of the procedure. Nevertheless, a characterization of the set of markings consistent with the actual observation is needed. Thus, large memory may be required.

An improvement in this respect has been given in [1, 3, 22].

In particular, Benveniste *et al.* [3] use a net unfolding approach for designing an online asynchronous diagnoser. The state explosion is avoided but the online computation can be high due to the online building of the PN structures by means of the unfolding.

Basile *et al.* [1] build the diagnoser online by defining and solving Integer Linear Programing (ILP) problems. Assuming that the fault transitions are not observable, the net marking is computed by the state equation and, if the marking has negative components, an unobservable sequence is occurred. The linear programing solution provides the sequence and detects the fault occurrences. Moreover, an offline analysis of the PN structure reduces the computational complexity of the ILP problem.

Dotoli *et al.* [22] propose a diagnoser that works online in order to avoid the redesign and the redefinition of the diagnoser when the structure of the system changes. In particular, the diagnoser waits for an observable event and an algorithm decides whether the system behavior is normal or may exhibit some possible faults. To this aim, some ILP problems are defined and provide eventually the minimal sequences of unobservable transitions containing the faults that may have occurred. The proposed approach is a general technique since no assumption is imposed on the reachable state set that can be unlimited, and only few properties must be fulfilled by the structure of the PN modeling the system fault behavior. A problem strictly related to diagnosis has been recently studied by Dotoli *et al.* [23]. They address the problem of identifying the model of the unobservable behavior of PN systems in the industrial automation framework. Assuming that the fault-free system structure and dynamics are known, the paper proposes an algorithm that monitors the system online, storing the occurred observable event sequence and the corresponding reached states.

References

1. Basile, F., Chiacchio, P., De Tommasi, G.: An efficient approach for online diagnosis of discrete event systems. IEEE Transactions on Automatic Control 54(4), 748–759 (2009)
2. Baviehi, S., Chong, E.K.P.: Automated fault diagnosis using a discrete event systems. In: Proc. 1994 IEEE Int. Symposium on Intelligent Control, Ohio, USA (1994)
3. Benveniste, A., Fabre, E., Haar, S., Jard, C.: Diagnosis of asynchronous discrete event systems: A net unfolding approach. IEEE Transactions on Automatic Control 48(5), 714–727 (2003)
4. Bertsimas, D., Gamarnik, D., Tsitsiklis, J.N.: Stability conditions for multiclass fluid queueing networks. IEEE Transaction on Automatic Control 41(11), 1618–1631 (2002)
5. Boel, R.K., van Schuppen, J.H.: Decentralized failure diagnosis for discrete-event systems with costly communication between diagnosers. In: Proc. 6th Int. Workshop on Discrete Event Systems, Zaragoza, Spain (2002)
6. Boel, R.K., Jiroveanu, G.: Distributed contextual diagnosis for very large systems. In: Proc. 7th Int. Workshop on Discrete Event Systems, Reims, France (2004)
7. Cabasino, M.P., Mahulea, C., Seatzu, C., Silva, M.: Fault diagnoser design for untimed continuous Petri nets. In: Proc. Int. Symposium on Intelligent Control, Saint Petersburg, Russia (2009)
8. Cabasino, M.P., Mahulea, C., Seatzu, C., Silva, M.: New Results for Fault Detection of untimed Continuous Petri nets. In: Proc. 48th IEEE Conf. on Decision and Control, Shanghai, China (2009)
9. Cabasino, M.P., Giua, A., Seatzu, C.: Diagnosability of bounded Petri nets. In: Proc. 48th IEEE Conf. on Decision and Control, Shanghai, China (2009)
10. Cabasino, M.P., Giua, A., Lafortune, S., Seatzu, C.: Diagnosability analysis of unbounded Petri nets. In: Proc. 48th IEEE Conf. on Decision and Control, Shanghai, China (2009)
11. Cabasino, M.P., Giua, A., Seatzu, C.: Fault detection for discrete event systems using Petri nets with unobservable transitions. Automatica 46(9), 1531–1539 (2010)
12. Cabasino, M.P., Giua, A., Paoli, A., Seatzu, C.: A new protocol for the decentralized diagnosis of labeled Petri nets. In: Proc. 10th IFAC Int. Workshop on Discrete Event Systems, Berlin, Germany (2010)
13. Cabasino, M.P., Giua, A., Paoli, A., Seatzu, C.: Decentralized diagnosis of Petri nets. In: Proc. 2010 American Control Conference, Baltimore, USA (2010)
14. Cabasino, M.P., Mahulea, C., Seatzu, C., Silva, M.: Fault diagnosis of discrete-event systems using continuous Petri nets. IEEE Transactions on Systems, Man and Cybernetics, Part A (to appear, 2012)
15. Cabasino, M.P., Giua, A., Paoli, A., Seatzu, C.: Decentralized diagnosability analysis of discrete event systems using Petri nets. In: Proc. 18th IFAC World Congress, Milan, Italy (2011)
16. Cabasino, M.P., Giua, A., Pocci, M., Seatzu, C.: Discrete event diagnosis using labeled Petri nets. An application to manufacturing systems. Control Engineering Practice 19(9), 989–1001 (2011)

17. Cabasino, M.P., Giua, A., Lafortune, S., Seatzu, C.: A new approach for diagnosability analysis of Petri nets using Verifier Nets. Transactions on Automatic Control (to appear, 2012)
18. Chen, H., Yao, D.D.: Fundamentals of queueing networks: Performance, asymptotics and optimization. Springer (2011)
19. Corona, D., Giua, A., Seatzu, C.: Marking estimation of Petri nets with silent transitions. IEEE Transactions on Automatic Control 52(9), 1695–1699 (2007)
20. David, R., Alla, H.: Discrete, Continuous and Hybrid Petri Nets. Springer (2004)
21. Debouk, R., Lafortune, S., Teneketzis, D.: Coordinated decentralized protocols for failure diagnosis of discrete-event systems. Discrete Event Dynamic Systems 10(1), 33–86 (2000)
22. Dotoli, M., Fanti, M.P., Mangini, A.M.: Fault detection of discrete event systems using Petri nets and integer linear programming. In: Proc. 17th IFAC World Congress, Seoul, Korea (2008)
23. Dotoli, M., Fanti, M.P., Mangini, A.M., Ukovich, W.: Identification of the unobservable behaviour of industrial automation systems by Petri nets. Control Engineering Practice 19(9), 958–966 (2011)
24. Garcia, E., Correcher, A., Morant, F., Quiles, E., Blasco, R.: Modular fault diagnosis based on discrete event systems. Discrete Event Dynamic Systems 15(3), 237–256 (2005)
25. Genc, S., Lafortune, S.: Distributed Diagnosis of Place-Bordered Petri Nets. IEEE Transactions on Automation Science and Engineering 4(2), 206–219 (2007)
26. Ghazel, M., Togueni, A., Bigang, M.: A monitoring approach for discrete events systems based on a time Petri net model. In: Proc. 6th IFAC World Congress, Prague, Czech Republic (2005)
27. Hadjicostis, C.N., Veghese, G.C.: Monitoring Discrete Event Systems Using Petri Net Embeddings. In: Donatelli, S., Kleijn, J. (eds.) ICATPN 1999. LNCS, vol. 1639, pp. 188–207. Springer, Heidelberg (1999)
28. Zad, H., Kwong, R.H., Wonham, W.M.: Fault diagnosis in discrete-event systems: framework and model reduction. IEEE Transactions on Automatic Control 48(7), 1199–1212 (2003)
29. Jiroveanu, G., Boel, R.K.: Contextual analysis of Petri nets for distributed applications. In: Proc. 16th Int. Symp. on Mathematical Theory of Networks and Systems, Leuven, Belgium (2004)
30. Lefebvre, D., Delherm, C.: Diagnosis of DES with Petri net models. IEEE Transactions on Automation Science and Engineering 4(1), 114–118 (2007)
31. Lunze, J., Schroder, J.: Sensor and actuator fault diagnosis of systems with discrete inputs and outputs. IEEE Transactions on Systems, Man and Cybernetics, Part B: Cybernetics 34(3), 1096–1107 (2004)
32. Martinez, J., Silva, M.: A simple and fast algorithm to obtain all invariants of a generalized Petri net. In: Informatik-Fachberichte: Application and Theory of Petri Nets, vol. 52, pp. 301–310. Springer (1982)
33. Miyagi, P.E., Riascos, L.A.M.: Modeling and analysis of fault-tolerant systems for machining operations based on Petri nets. Control Engineering Practice 14(4), 397–408 (2010)
34. Prock, J.: A new tecnique for fault detection using Petri nets. Automatica 27(2), 239–245 (1991)
35. Ramirez-Treviño, A., Ruiz-Beltrán, E., Rivera-Rangel, I., Lopez-Mellado, E.: Online fault diagnosis of discrete event systems. A Petri net-based approach. IEEE Transactions on Automation Science and Engineering 4(1), 31–39 (2007)

36. Recalde, L., Teruel, E., Silva, M.: Autonomous Continuous P/T Systems. In: Donatelli, S., Kleijn, J. (eds.) ICATPN 1999. LNCS, vol. 1639. Springer, Heidelberg (1999)
37. Sampath, M., Sengupta, R., Lafortune, S., Sinnamohideen, K., Teneketzis, D.: Diagnosability of discrete-event systems. IEEE Transactions on Automatic Control 40(9), 1555–1575 (1995)
38. Sampath, M., Sengupta, R., Lafortune, S., Sinnamohideen, K., Teneketzis, D.: Failure diagnosis using discrete-event models. IEEE Transactions on Control Systems Technology 4(2), 105–124 (1996)
39. Sampath, M., Lafortune, S.: Active diagnosis of discrete-event systems. IEEE Transactions on Automatic Control 43(7), 908–929 (1998)
40. Silva, M., Velilla, S.: Error detection and correction on Petri net models of discrete control systems. In: Proc. IEEE Int. Symp. on Circuits and Systems, Kyoto, Japan (1985)
41. Silva, M., Recalde, L.: On fluidification of Petri net models: from discrete to hybrid and continuous models. Annual Reviews in Control 28(2), 253–266 (2004)
42. Sreenivas, V.S., Jafari, M.A.: Fault detection and monitoring using time Petri nets. IEEE Transactions on Systems, Man and Cybernetics 23(4), 1155–1162 (1993)
43. Sun, G., Cassandras, C.G., Panayiotou, C.G.: Perturbation analysis of multiclass stochastic fluid models. Discrete Event Dynamic Systems 14(3), 267–307 (2004)
44. Viswanadham, N., Johnson, T.L.: Fault detection and diagnosis of automated manufacturing systems. In: Proc. 2th IEEE Conf. on Decision and Control, Austin, Texas (1988)
45. Wu, Y., Hadjicostis, C.N.: Algebraic approaches for fault identification in discrete-event systems. IEEE Transactions on Robotics and Automation 50(12), 2048–2053 (2005)

Chapter 15
Diagnosis with Petri Net Unfoldings

Stefan Haar and Eric Fabre

15.1 Motivation

Large systems or softwares are generally obtained by designing independent modules or functions, and by assembling them through appropriate interfaces to obtain more elaborate functions and modules. The latter can in turn be assembled, up to forming huge systems providing sophisticated services. Consider for instance the various components of a computer, telecommunication networks, plane ticket reservation sofwares for a company, etc. Such systems are not only modular in their design, but often multithreaded, in the sense that many events may occur in parallel.

From a discrete event system perspective, such modular or distributed systems can be modeled in a similar manner, by first designing component models and then assembling them through an adequate composition operation. A first approach to this design principle has been presented in Chapter 5 (see Section 5.5): composition can be defined as the synchronous product of automata. The transitions of each component carry labels, and the product proceeds by synchronizing transitions with identical labels, while all the other transitions remain private. This construction is recalled in Fig. 15.1 on the simple case of three tiny automata. The size of the resulting system is rather surprising, given the simplicity of the three components! And this deserves a detailed study.

One first notices the classical state space explosion phenomenon: the number of states in the global system is the product of the number of states of their components (here $2 \times 2 \times 3 = 12$). So, the number of states augments exponentially fast with the

Stefan Haar
INRIA/LSV, CNRS & ENS de Cachan, 61, avenue du Président Wilson,
94235 CACHAN Cedex, France
e-mail: Stefan.Haar@inria.fr

Eric Fabre
INRIA/IRISA, Campus de Beaulieu, F-35042 Rennes Cedex, France
e-mail: fabre@irisa.fr

C. Seatzu et al. (Eds.): Control of Discrete-Event Systems, LNCIS 433, pp. 301–318
springerlink.com

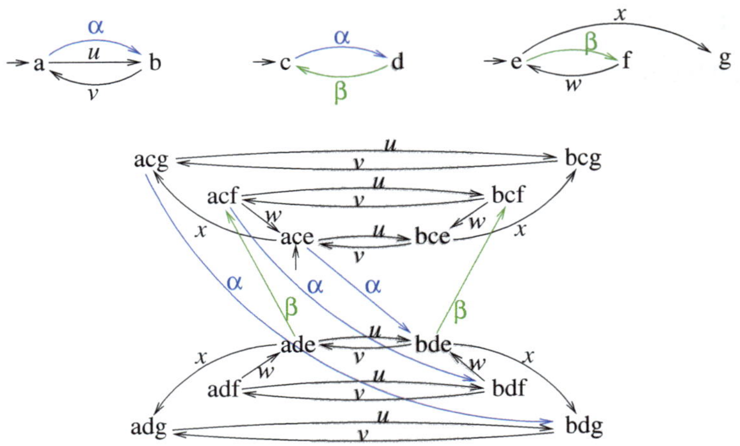

Fig. 15.1 Three components (top) as labeled automata, and their synchronous product (bottom)

number of components. Second, the number of transitions explodes as well: private transitions of a component are cloned many times (see transition (a,u,b) of the first component for example), and creates the so-called concurrency diamonds, representing different possible orderings of transitions (from state ade, one can reach bcf by firing either $u\beta$ or βu).

These phenomena motivate alternate methods of assembling components in order to make explicit the concurrency of transitions. Petri nets are a natural tool toward this objective. Reconsidering the above example under the form of Petri nets, one gets Fig. 15.2. Components are recast into simple PNs with a single token, and their assembling amounts to gluing transitions with identical labels. The explosion both in states and in transitions is now kept under control, and the token semantics of PN makes explicit the fact that several transitions are simultaneously firable. Observe

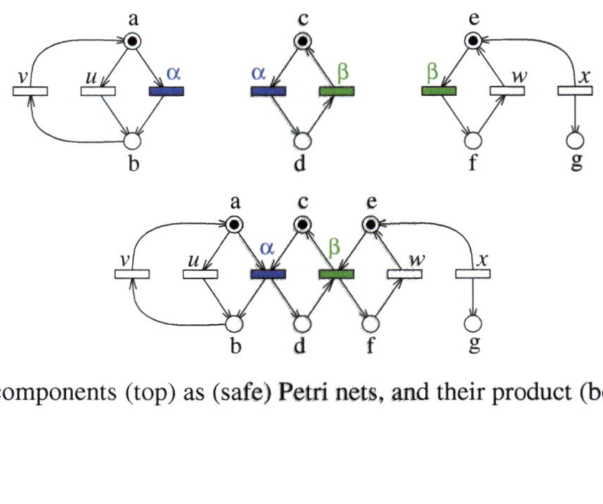

Fig. 15.2 Three components (top) as (safe) Petri nets, and their product (bottom)

in passing that the PNs obtained in this manner are *safe*: each place contains at most one token. Moreover, here the number of tokens is constant and characterizes the number of automata that were assembled.

Did all difficulties of large systems vanish with this simple modeling trick? Not really. When considering runs of (safe) Petri nets, one may still face explosive phenomena. In the usual sequential semantics, trajectories are modeled as sequences of events. So, one recovers the difficulty that different interleaving of concurrent events correspond to different trajectories. For example, trajectories ux and xu both lead from the inital state ace to state bcf, but correspond to two distinct trajectories. While it is clear that the exact ordering in which the private events u and x of the first and third component respectively does not really matter.

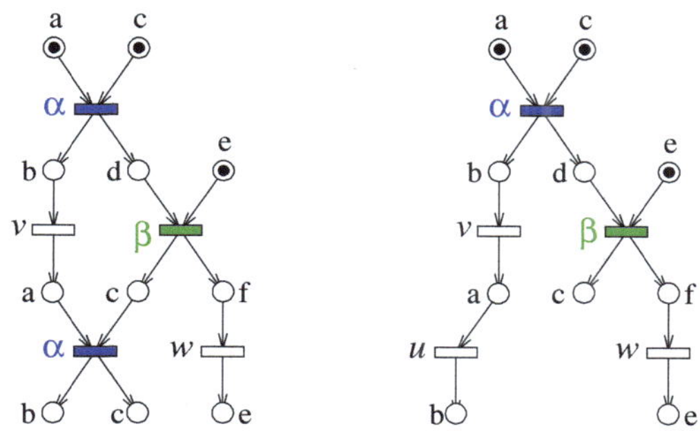

Fig. 15.3 Two trajectories of the PN in Fig. 15.2, as partial orders of events. 'Time' (or precedence, or causality) is oriented from top to bottom

To avoid the explosion in the number of possible runs of a large concurrent system, people soon realized the the parallelism, or the concurrency, also had to be handled in the description of trajectories. The first ideas in this direction came from Mazurkiewicz traces (not treated here), which consist in establishing an equivalence relation between sequences of events that only differ by the ordering of their concurrent events. A related idea is to directly represent runs as *partial orders* of events, rather than sequences. Fig. 15.3 illustrates this idea for the PN of Fig. 15.2. This representation encodes the causality of events, as derived by the use of resources (tokens), but discards any unnecessary timing information. The run on the left, for example, encodes that events v and β occur after the first α, but their order is unspecified. These simple five events partial order stands for five possible sequences, obtained as different interleaving of concurrent events (exercise).

Such *true concurrency semantics,* which handle time as partially ordered, offer many advantages. The first one being to keep under control the explosion due to the intrinsic parallelism of events in large distributed systems. But it also allows one to model that some global knowledge on the system may be unaccessible. It is a common place that the knowledge of global time, or of global state, may be unreachable in distributed asynchronous systems. This idea was already illustrated in

Chapter 5 in the case of distributed observations: if a sensor is placed on each component, the events observed locally by each sensor may be totally ordered, but the exact interleaving of observations collected on different sensors can not (always) be recovered. Therefore, one should also be able to represent distributed observations as partial orders of observations, rather than sequences.

This chapter aims at introducing the main concepts that make possible working with partial orders of event, or true concurrency semantics. It first recalls the notion of (safe) Petri net, and then introduces occurrence nets, a compact data structure to handle sets of runs, where runs are partial orders of events. It then examines how diagnosis can be performed with such semantics, by relating partially ordered observations to possible runs of a Petri net. As for automata, the approach extends to distributed systems. The chapter closes on the notion of diagnosability, which takes a new meaning in this context.

15.2 Asynchronous Diagnosis with Petri Net Unfoldings

Nets and homomorphisms. A *net* is a triple $N = (P,T,F)$, where P and T are disjoint sets of *places* and *transitions*, respectively, and $F \subset (P \times T) \cup (T \times P)$ is the *flow relation.* [1] In figures, places are represented by circles, rectangular boxes represent transitions, and arrows represent F; Fig. 15.4 shows two nets. For node $x \in P \cup T$, call ${}^\bullet x \triangleq \{x' \mid F(x',x)\}$ the *preset*, and $x^\bullet \triangleq \{x' \mid F(x,x')\}$ the *postset* of x. Let $<$ be the transitive closure of F and $\leqslant$ the reflexive closure of $<$; further, let $[x] \triangleq \{x' \mid x' \leqslant x\}$ be the *prime configuration* or *cone* of x, and $[x[\triangleq [x] \setminus \{x\}$ the *pre-cone* of x.

A *net homomorphism*[2] from N to N' is a map $\pi : P \cup T \longmapsto P' \cup T'$ such that *(i)* $\pi(P) \subseteq P'$, $\pi(T) \subseteq T'$, and *(ii)* $\pi_{|{}^\bullet t} : {}^\bullet t \to {}^\bullet \pi(t)$ and $\pi_{|t^\bullet} : t^\bullet \to \pi(t)^\bullet$ induce bijections, for every $t \in T$.

Petri Nets. Let $N = (P,T,F)$ be a finite net. A *marking* of net N is a multi-set $m : P \to \mathbb{N}$. A *Petri net* (PN) is a pair $\mathscr{N} = (N, m_0)$, where $m_0 : P \to \mathbb{N}$ is an *initial* marking. Transition $t \in T$ is *enabled* at marking m, written $m \stackrel{t}{\longrightarrow}$, iff ${}^\bullet t \leqslant m$, where we interpret ${}^\bullet t$ as the multi set whose value is 1 on all preplaces of t, and 0 otherwise. If $m \stackrel{t}{\longrightarrow}$, then t can *fire*, leading to $m' = (m - {}^\bullet t) + t^\bullet$ (in the multi-set interpretation); write in that case $m \stackrel{t}{\longrightarrow} m'$. The set $\mathbf{R}(m_0)$ contains m_0 and the markings of $\mathscr{N}$ *reachable* through the transitive closure $\stackrel{+}{\longrightarrow}$ of $\longrightarrow$.

Only safe nets are considered in this article; the net on the left-hand side of Fig. 15.4 is safe. If $m(p) > 0$, we will draw $m(p)$ black *tokens* in the circle representing p. A Petri net $\mathscr{N} = (N, m_0)$ is *safe* if for all $m \in \mathbf{R}(m_0)$ and $p \in P$, $m(p) \in \{0,1\}$.

---

[1] Only *ordinary* nets are considered here, i.e. with arc weights 0 or 1.

[2] There exist several notions of morphisms for nets and for Petri nets, which are needed e.g. to formalize composition of nets; see [27, 19, 18, 6, 17] and the references therein.

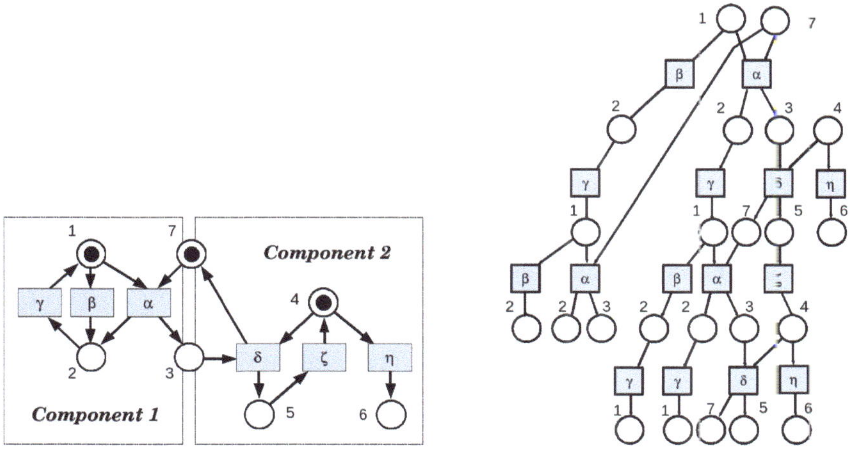

Fig. 15.4 A Petri net (left) and a prefix of its unfolding (right)

Semantics. The behavior of Petri nets can be recorded in either an *interleaved* or a *concurrent* fashion. To formalize this, we introduce *Occurrence Nets* (due to [28]) and *Branching Processes*. Occurrence nets are characterized by a particular structure. In a net $N = (P,T,F)$, let $<_N$ the transitive closure of F, and $\leqslant_N$ the reflexive closure of $<_N$. Further, set $t_1 \#_{im} t_2$ for transitions t_1 and t_2 if and only if $t_1 \neq t_2$ and ${}^\bullet t_1 \cap {}^\bullet t_2 \neq \emptyset$, and define $\# = \#_N$ by

$$a \# b \Leftrightarrow \exists t_a, t_b \in T : [(t_a \#_{im} t_b) \wedge (t_a \leqslant_N a) \wedge (t_b \leqslant_N b)].$$

Finally, define concurrency relation $\mathbf{co} = \mathbf{co}_{\mathscr{N}}$ by setting, for any nodes $a, b \in P \cup T$,

$$a \,\mathbf{co}\, b \iff \neg(a \leqslant b) \wedge \neg(a \# b) \wedge \neg(b < a).$$

Definition 15.1. *A net $ON = (B,E,G)$ is an* **occurrence net** *if and only if it satisfies*

1. *$\leqslant_{ON}$ is a partial order;*
2. *for all $b \in B$, $|{}^\bullet b| \in \{0,1\}$;*
3. *for all $x \in B \cup E$, the set $[x] = \{y \in B \cup E \mid y \leqslant_{ON} x\}$ is finite;*
4. *no self-conflict,i.e. there is no $x \in B \cup E$ such that $x \#_{ON} x$;*
5. *the set cut_0 of $\leqslant_{ON}$-minimal nodes is contained in B and finite.*

In occurrence nets, the nodes of E are called *events*, and the elements of B are denoted *conditions*. The right-hand side of Fig. 15.4 shows an occurrence net.

Occurrence nets are the mathematical form of the *partial order unfolding semantics* for Petri nets [30]; although more general applications are possible, we will focus here on unfoldings of *safe* Petri nets only.

A *branching process* of safe Petri net $\mathcal{N} = (N, m_0)$ is a pair $\beta = (ON, \pi)$, where $ON = (B, E, G)$ is an occurrence net, and π is a homomorphism from ON to N such that:

1. The restriction of π to cut_0 is a bijection from cut_0 to the set $m_0 \triangleq \{p \in P : m_0(p) = 1\}$, and
2. for every $e_1, e_2 \in E$, if ${}^\bullet e_1 = {}^\bullet e_2$ and $h(e_1) = h(e_2)$ then $e_1 = e_2$.

Branching processes $\beta_1 = (ON_1, \pi_1)$ and $\beta_2 = (ON_2, \pi_2)$ for $\mathcal{N}$ are isomorphic iff there exists a bijective homomorphism $h : ON_1 \to ON_2$ such that $\pi_1 = \pi_2 \circ h$. The unique (up to isomorphism) maximal branching process $\beta_{\mathcal{U}} = (ON_{\mathcal{U}}, \pi_{\mathcal{U}})$ of $\mathcal{N}$ is called the *unfolding* of $\mathcal{N}$. A canonical algorithm (see [30] for details) for constructing the unfolding of $\mathcal{N} = (P, T, F, m_0)$ proceeds as follows: For any branching process $\beta = (ON_\beta, \pi_\beta)$ of $\mathcal{N} = (P, T, F)$, with $ON_\beta = (B_\beta, E_\beta, G_\beta)$, denote as $\mathbf{PE}(\beta) \subseteq T \times \text{Pwrset}(B)$ the set of *possible extensions* of β, i.e. the set of the pairs (t, W) such that

- W is a co-set of ON_β , i.e. for $a, b \in W$, either $a = b$ or a **co** b,
- ${}^\bullet t = \pi_\beta(W)$,
- E_β contains no event e such that $\pi_\beta(e) = t$ and ${}^\bullet e = W$.

Now, let $cut_0 \triangleq m_0 \times \{\emptyset\}$ and initialize $\beta = (cut_0, \emptyset, \emptyset)$; recursively, for given $\beta = (ON_\beta, \pi_\beta)$ with $ON_\beta = (B_\beta, E_\beta, G_\beta)$, compute $\mathbf{PE}(ON_\beta)$ and replace:

$$
\begin{aligned}
&E_\beta \text{ by } E_\beta \cup \mathbf{PE}(ON_\beta),\\
&B_\beta \text{ by } B_\beta \cup \{(P, e) \mid e \in \mathbf{PE}(ON_\beta), p \in \pi_\beta(e)^\bullet\}, \text{ and}\\
&G_\beta \text{ by } G_\beta \cup \{(b, (t, W)) \mid (t, W) \in \mathbf{PE}(ON_\beta), b \in W\}\\
&\qquad \cup \{(e, (P, e)) \mid e \in \mathbf{PE}(ON_\beta), p \in \pi_\beta(e)^\bullet\}.
\end{aligned}
$$

We will assume that all transitions $t \in T$ have at least one output place, i.e. $t^\bullet$ is not empty.

Occurrence nets give rise to a specific kind of partially ordered set with *conflict* relation that is known in computer science as *event structure*.

Definition 15.2 (compare [28]). *A prime event structure is a tuple* $\mathcal{E} = (E, \leqslant, \#, \lambda)$, *where* $E = \|\mathcal{E}\|$ *is the* support, *or* set of events *of* $\mathcal{E}$, and such that

1. $\leqslant \subseteq E \times E$ is a partial order satisfying the property of *finite causes* i.e. setting $[e] \triangleq \{e' \in E \mid e' \leqslant e\}$, one has for all $e \in E$, $|[e]| < \infty$;
2. $\# \subseteq E \times E$ an irreflexive symmetric *conflict* relation satisfying the property of *conflict heredity*, i.e.

$$
\forall\, e, e', e'' \in E : \; e \# e' \wedge e' \leqslant e'' \Rightarrow e \# e''. \tag{15.1}
$$

Events $e, e' \in E$ are *concurrent*, written e **co** e', iff neither $e \leqslant e'$ nor $e' < e$ nor $e \# e'$ hold. If **co** is the empty relation, we call $\mathcal{E}$ *sequential*.

One notices quickly that occurrence nets form particular cases of event structures. The canonical association of an event structure to an occurrence net ON is by restricting $\leqslant$ and # to the event set E, "forgetting" conditions.

Sequential and Nonsequential behavior. In the net on the left-hand side of Fig. 15.4, the transition sequence $\alpha\delta\gamma\zeta$ is enabled; so is the sequence $\alpha\gamma\delta\zeta$, and it is immaterial to us which of the two sequences actually occurs; both lead to the same final marking (which is identical with the initial marking), and the same actions are performed, only in different order.

We therefore would like to use a unifying way to reason about such collections of firing sequences without having to examine each individual one. One way of capturing the equivalence up to permutation of independent events is developed in the theory of *Mazurkiewicz traces*, see [15, 26]. We will use another relation, which includes also the marking equivalence and which is provided by the concept of *configuration*: a unique partially ordered set that represents in a unique and compact way all enabled *interleavings* of a set of events. Let us formalize this.

Prefixes and Configurations. The *set of causes* or *prime configuration* of $e \in E$ is $[e] \triangleq \{e' \mid e' \leqslant e\}$, as defined above. A *prefix* of $\mathscr{E}$ is any downward closed subset $D \subseteq E$, i.e. such that for every $e \in D$, $[e] \subseteq D$. Prefixes of $\mathscr{E}$ induce, in the obvious way, subevent structures of $\mathscr{E}$ in the sense of the above definition. Denote the set of $\mathscr{E}$'s prefixes as $\mathscr{D}(\mathscr{E})$. Prefix $\mathbf{c} \in \mathscr{D}(\mathscr{E})$ is a *configuration* if and only if it is conflict-free, i.e. if $e \in \mathbf{c}$ and $e\#e'$ imply $e' \notin \mathbf{c}$. Denote as $\mathscr{C}(\mathscr{E})$ the set of $\mathscr{E}$'s configurations. Call any $\subseteq$-maximal element of $\mathscr{C}(\mathscr{E})$ a *run* of $\mathscr{E}$; denote the set of $\mathscr{E}$'s runs as $\Omega(\mathscr{E})$, or simply Ω if no confusion can arise.

In Fig. 15.4, the leftmost branch, with events labeled β, γ, β, is an example of a configuration.

Every *finite* configuration $\mathbf{c}$ terminates at a cut, i.e. a $\subseteq$-maximal co-set, which we denote $cut_{\mathbf{c}}$. The mapping $\mathbf{c} \mapsto cut_{\mathbf{c}}$ is bijective; for each cut cut, the union of the cones of all conditions in cut yield the unique configuration $\mathbf{c}$ such that $cut = \mathbf{c}_{cut}$. Moreover, one has the following two correspondences:

If $\mathbf{c}$ is a configuration of $\mathscr{U}_{\mathscr{N}}$ with $\mathscr{N} = (N, m_0)$, then every occurrence sequence σ obtained as a linear order extension, i.e. an *interleaving*, of the partial order $\leqslant_{\mathbf{c}}$ yields a firable transition sequence of $\mathscr{N}$. Conversely, every firable transition sequence of $\mathscr{N}$ corresponds to a linear order extension of some configuration of $\mathscr{U}_{\mathscr{N}}$. To sum up: the nonsequential executions of $\mathscr{N}$ are in one-to-one correspondence with the configurations of $\mathscr{U}(\mathscr{N})$. We will therefore speak of *$\mathscr{N}$'s configurations* and write $\mathscr{C}(\mathscr{N}) \triangleq \mathscr{C}(\mathscr{U}_{\mathscr{N}})$ and $\Omega(\mathscr{N}) \triangleq \Omega(\mathscr{U}_{\mathscr{N}})$.

• For every reachable marking m of $\mathscr{N}$, there exists at least one cut cut of $\mathscr{U}(\mathscr{N})$ such that $\|\pi(cut)(p) = m(p)\|$ for all $p \in P$, and for the unique configuration $\mathbf{c}$ such that $cut_{\mathbf{c}} = cut$, execution of $\mathbf{c}$ takes m_0 to m; write $m_0 \xrightarrow{\mathbf{c}} m$ for this. Conversely, every finite configuration $\mathbf{c}$ corresponds to a unique reachable marking $m(\mathbf{c})$ given by $m(\mathbf{c}) \triangleq \pi(cut_{\mathbf{c}})$. We call configurations such that $m(\mathbf{c}) = m(\mathbf{c}')$ *marking equivalent*, and denote this by $\mathbf{c} \equiv_m \mathbf{c}'$.

15.3 Asynchronous Diagnosis

The fundamental challenge is the same as in the case of finite state machines: correlation of the observation $\lambda \in \mathbb{A}1^*$ with the system model, and thus extract those runs that are compatible with the observation, i.e. whose image under the observation mask λ agrees with λ. To this end, one lets the observation steer the evolution of the system model, by synchronizing with observable transitions. Formally, this is ensured via a *synchronous* or **Labeled Product:** Let $\mathscr{N}_1 = (P_1, T_1, F_1, m_0^1)$ and $\mathscr{N}_2 = (P_2, T_2, F_2, m_0^2)$ be two Petri nets, with associated labellings $\lambda_1 : T_1 \to \mathbb{A}1$ and $\lambda_2 : T_2 \to \mathbb{A}1$ into the same label alphabet $\mathbb{A}1$. The *λ-synchronized product of $\mathscr{N}_1$ and $\mathscr{N}_2$* is the Petri net $\mathscr{N}_1 \times \mathscr{N}_2 \triangleq (P, T, F, m_0)$, where

1. $P_{\mathscr{V}} = P_1 \uplus P_2$,
2. for $i \in \{1,2\}$, $T_i^{\varepsilon} \triangleq \{t \in T_i \mid \lambda(t) = \varepsilon\}$,
3. $T_{12} \triangleq t\{t \in T_1 \mid \lambda(t) \neq \varepsilon\}$,
4. $F_{\varepsilon} \triangleq \bigcup_{i=1}^{2}(F_i \cap P_i \times T_i^{\varepsilon}) \cup \bigcup_{i=1}^{2}(F_i \cap T_i^{\varepsilon} \times P_i)$,
5. $F_{12} \triangleq \bigcup_{i=1}^{2}(F_i \cap P_i \times T_{12}) \cup \bigcup_{i=1}^{2}(F_i \cap T_{12} \times P_i)$,
6. $T_{\mathscr{V}} \triangleq T_1^{\varepsilon} \uplus T_2^{\varepsilon} \uplus T_{12}$ and $F_{\mathscr{V}} \triangleq F_{\varepsilon} \uplus F_{12}$,
7. $m_0 \triangleq m_0^1 + m_0^2$

Figure 15.5 shows the product of a system model $\mathscr{N}$ and a Petri net model of a partially ordered alarm pattern $\mathscr{A}$. The unfolding of this product is shown on the right-hand side; it exhibits the behavior of $\mathscr{N}$ *steered* by the observation $\mathscr{A}$. Unfolding $\mathscr{U}_{\mathscr{N} \times \mathscr{A}}$ thus contains exactly those behaviors that explain at least a prefix of $\mathscr{A}$; the *full* explanations are highlighted as κ_1 and κ_2 in the figure.

Note that, the product of two 1-safe Petri nets is a 1-safe Petri net. Moreover, in perfect analogy with the synchronous product of finite automata, the semantics of the product projects into the semantics of the two factors; i.e., if $\Pi_{\mathscr{N}_i}$ denotes the operation of erasing, from any prefix of $\mathscr{U}_{\mathscr{N}}$, all arcs and conditions that are not mapped to parts of $\mathscr{N}_i$, one has (see [10]):

$$\forall \mathbf{c} \in \mathscr{C}(\mathscr{N}) : \begin{cases} \Pi_{\mathscr{N}_1} \in \mathscr{C}(\mathscr{N}_1) \\ \Pi_{\mathscr{N}_1} \in \mathscr{C}(\mathscr{N}_1) \end{cases} \tag{15.2}$$

Remark: The above construction can be formalized as a *pullback* in appropriate categories.

The advantage is that results such as 15.2 can be derived from much stronger results which imply that the unfolding of the pullback of two safe nets is isomorphic to the pullback of the two unfoldings. The theory necessary to detail these algebraic tools is beyond the scope of this chapter; see [6, 7, 18] . In the asynchronous diagnosis setting, observations are *partially ordered*. The representation of alarm patterns thus generalizes with respect to the linear automaton model above. Figure 15.5 shows , in the center, an alarm pattern represented as an occurrence net $\mathfrak{A}$ (without conflict), with concurrently observed alarm labels; for instance, the β-labeled event on top is concurrent with the α-labeled one to its right. Time flows top-down; we

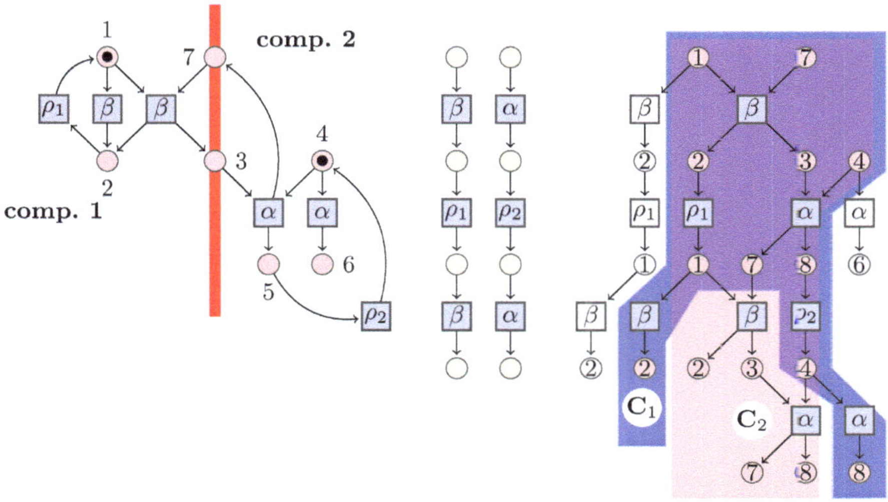

Fig. 15.5 The methodology for unfolding-based Diagnosis. From left to right: A Petri net model $\mathcal{N}$ of a system, taken from Fig. 15.4; the marked places are represented by thick-lined circles. Then, a partially ordered observation (alarm pattern) $\mathscr{A}$ consisting of two disjoint totally ordered chains of event labels $\beta \to \rho 1 \to \beta$ and $\alpha \to \rho 2 \to \alpha$, is represented as a small Petri net (arrows point downward) with added places between successive events. Then, one forms the product $\mathcal{N} \times \mathscr{A}$ by synchronizing transitions that bear the same label. Finally (right), the unfolding $\mathscr{U}_{\mathcal{N}\times\mathscr{A}}$ contains exactly two explanations for $\mathscr{A}$, namely the configurations κ_1 and κ_2. The events not belonging to either of these explanations are shown in gray; they cannot be extended into any explanation of $\mathscr{A}$ and can be *pruned* away

have sometimes omitted arrows. The central step in the diagnosis procedure now consists in computing computing $\mathscr{U}_{\mathcal{N}\times\mathfrak{A}}$, shown on the right hand side.

Remark: In the example, this unfolding is finite, as opposed to the unfolding of the system net $\mathcal{N}$. This is an important feature, which requires a strong *observability* property (compare Section 15.5):

- There must not exist any *cyclic* firing sequence $m_1 \xrightarrow{t_1} m_2 \xrightarrow{t_2} \ldots m_n \xrightarrow{t_n} m_1$ between reachable markings in $\mathcal{N}$ such that $\lambda(t_i) = \varepsilon$ for all $i \in \{1,\ldots\ n\}$.

In fact, otherwise $\mathscr{U}_{\mathcal{N}\times\mathfrak{A}}$ will contain at least one infinite branch, since the transitions of any such loop can fire indefinitely, unrestrained by $\mathfrak{A}$. Conversely, if any loop in the behavior of $\mathcal{N}$ must contain at least one observable transition, then it can be performed only a finite number of times since $\mathfrak{A}$ is finite. If this requirement cannot be met, another remedy consists in truncating branches that produce two nested marking-equivalent configuration that are observation-equivalent; such a pair $\kappa \subseteq \kappa'$ with $\kappa' \backslash \kappa \neq \emptyset$ need not be explored further. *Cutoff* criteria like this, have been exploited by [29] and others to ensure all analysis can be carried out on a *complete finite prefix*; 1-safeness of the system model ensures that for any fixed

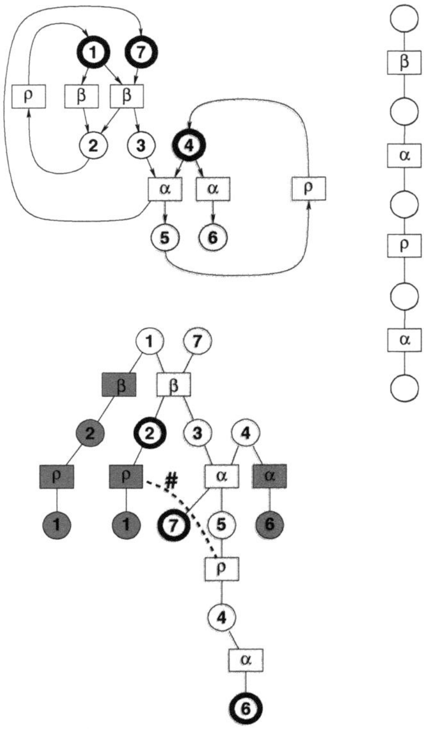

Fig. 15.6 Illustrating the correlation of an alarm pattern $\mathscr{A}$ on the right with a *linearly ordered* alarm pattern (right-hand side)

observation $\mathscr{A}$, a prefix of finite size of $\mathscr{U}_{\mathscr{N}\times\mathscr{A}}$ is sufficient to find essentially all explanations — up to the above surgery to remove unobservable loops — for $\mathscr{A}$.

By relation (15.2), we deduce that the set of configurations that *explain* $\mathfrak{A}$ is obtained as the following prefix of $\mathscr{U}_{\mathfrak{A}\times\mathscr{N}}$:

$$D_{\mathfrak{A}} \triangleq \Pi_{\mathscr{N}}\left(\Pi_{\mathfrak{A}}^{-1}(\mathfrak{A})\right), \tag{15.3}$$

where as above $\Pi_{\mathfrak{A}}$ is the operation of removing all non-$\mathfrak{A}$ parts from $\mathscr{U}_{\mathfrak{A}\times\mathscr{N}}$. Therefore, we have as diagnosis set

$$\mathbf{diag}(\mathfrak{A}) = \{\mathbf{c} \in \mathscr{C}(\mathscr{N}) : \exists\, \bar{\mathbf{c}} \in \mathscr{C}(\mathscr{U}_{\mathfrak{A}\times\mathscr{N}}) : \mathbf{c} \subseteq \Pi_{\mathscr{N}}(\bar{\mathbf{c}})\}. \tag{15.4}$$

Notice that $\mathbf{diag}(\mathfrak{A})$ is in general a proper *superset* of $\Omega(\mathscr{U}_{\mathfrak{A}\times\mathscr{N}})$. For the final *diagnosis* task it remains, once $\mathbf{diag}(\mathfrak{A})$ has been computed, to inspect all its configurations for the presence of an occurrence of ϕ.

It should be noted that the computation of (a sufficient prefix of) $\mathscr{U}_{\mathscr{N}\times\mathscr{A}}$ can be "sequentialized" and considerably simplified if $\mathscr{A}$ is *linearly ordered*, i.e. an observation sequence $\mathscr{A} = a_1 a_2 \ldots a_n \in \mathbb{A}1^*$ (see [10, 29] for more details). In fact, letting $\mathscr{A}_i$ be the ith prefix of $\mathscr{A}$, one obtains $\mathbf{diag}(\mathfrak{A}_{i+1})$ from $\mathbf{diag}(\mathfrak{A}_i)$ in the following way:

1. Compute the extension with new events following the unfolding algorithm,
2. Stop each branch either after after the first occurrence e of an observable transition.
3. Then, remove all such e whose label is *not* a_{i+1}, and ..
4. ... prune away all those events that do not allow to explain $\mathfrak{A}_{i+1}$, i.e. that are in conflict with *all* occurrences of a_{i+1} computed in the previous steps.

In Fig. 15.6, we see that only the configuration shown in white in the unfolding prefix at the bottom is capable of explaining *entirely* the alarm pattern on the right-hand side. In fact, the ρ-labeled event shown in gray is part of an explanation for an observation sequence $\beta\alpha\rho$ (formed by its own prime configuration plus the α-event on the right hand side), but this configuration cannot be extended to explain the subsequent occurrence of α in the pattern. As a result, it is pruned away in the fourth round, since it is in conflict with the second α in white, a conflict inherited from the one indicated in the figure. Similarly, the other events shown in gray are pruned away since they cannot provide explanations for the present alarm observation, nor — a fortiori — for any of its extensions.

15.4 Taking the Methodology Further

The use of unfoldings brings conceptual and technical gains since it allows to abstract away from interleavings of concurrent events. Still, the *computation* of the diagnosis sets can still be hampered by the size of the necessary unfolding prefixes. One notices that the main factor that leads to high *widths* of branching processeses is the number of conflicting branches. Two approaches have launched for improving the data structures used, and both tackle the impact of branching:

1. First, suppose that the supervised system $\mathscr{N}$ is decomposed into subcomponents $\mathscr{N}_1, \ldots, \mathscr{N}_n$ that are supervised separately and locally; the observation $\mathfrak{A}$ is therefor also fragmented into local portions $\mathfrak{A}_1, \ldots, \mathfrak{A}_n$. The *global* diagnosis prefix $D_{\mathfrak{A}}$ is in general too big to be computed directly; by contrast, *local* prefixes $D_{\mathfrak{A}_i}$ obtained by unfolding $\mathscr{N}_i \times \mathfrak{A}_i$ are of more manageable size, and can be computed locally. The number $\mathbf{B}(D_{\mathfrak{A}})$ of branches in $D_{\mathfrak{A}}$ is bounded by $K \triangleq \prod_{i=1}^{n} \mathbf{B}(D_{\mathfrak{A}_i})$, so we can expect an exponential gain in the storage space required. However, care must be taken to compute the right local diagnosis: since not all combinations of local branches match into a global run, $D_{\mathfrak{A}_i}$ is an *over-approximation* of the local diagnosis obtained as the projection $\Pi_i(D_{\mathfrak{A}})$ of the global diagnosis $D_{\mathfrak{A}_i}$ to the ith component. The nontrivial task is thus to orchestrate correctly the distributed computation of the unfolding of an n-component net; see Fig. 15.7 for an illustration of the communication between two "unfolders" in the context of the running example. The work [17] carries out this task

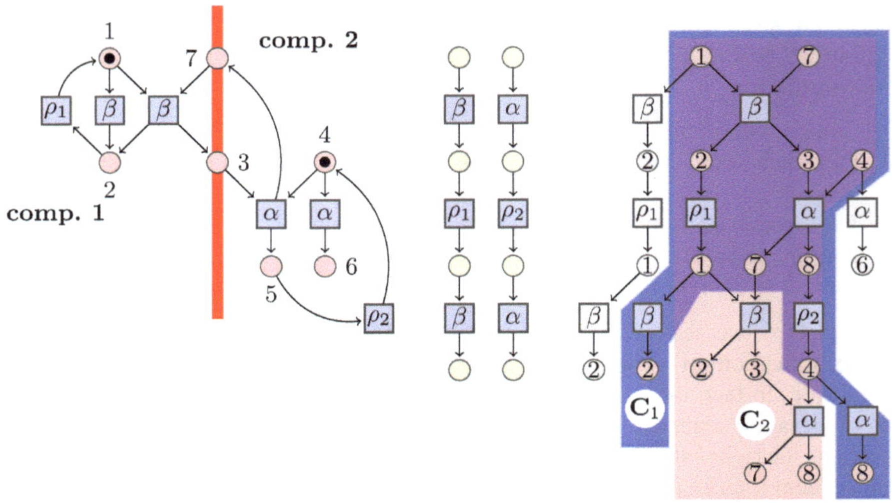

Fig. 15.7 The example from Fig. 15.5, with the distributed computation of the diagnosis net on the right, in two components

in the context of composition *via shared places*, as in Fig. 15.7. Another lead was followed in [6] where the composition was required to form a *pullback* in a suitable Petri net category (see also [18]), which allows to use powerful algebraic tools to characterize the data exchange between two components. While the details of this research are beyond the scope of this presentation, one should note that all results point to the fact that best theoretical (and practical ?) results can be obtained if the interfaces — i.e. the net parts that are shared between two components — should be *concurrency*-free. This is the assumption made, e.g. in [27].

2. In [16], the distributed approach is combined with a methodology for reducing the width of unfoldings by using *trellis* structures: when a state is reached on two or more different branches, the branches are fused on that state, and share the different future extensions. This avoids the width explosion of the stored data structures for long observations, by quotienting the occurrence net structure that forces the inheritance of conflict and thus the separation of branches even if they differ only on an initial segment. The technical challenges of this approach are tackled and solved in [16, 19].

Note also that the approach presented here for the case of *static* system topologies has been extended, in [21] and [7] to graph transformation systems (GTS) for modeling dynamically evolving system topologies; GTS are a proper generalization of Petri nets, yet share with them many properties and techniques, such as partial order unfoldings.

15.5 Asynchronous Diagnosability: Weak versus Strong

Let us turn now to analyzing the power of the unfolding-based diagnosis approach, and ask under which circumstances a fault ϕ is diagnosable. Recall the classical definition of diagnosability given by [31], which we give in the equivalent presentation of [14]. Write $s \sim_\eta s'$ iff $s, s' \in T^*$ are mapped to the same observable word in o^*, and call any sequence s such that ϕ occurs in s a *faulty* sequence, and all other sequences *healthy*. Then:

Definition 15.3 (Strong Diagnosability). *Language* $\mathscr{L}$ *is* not (strongly) diagnosable *iff there exist sequences* $s_N, s_Y \in \mathscr{L}$ *such that:*

1. s_Y *is faulty,* s_N *is healthy, and* $s_N \sim_\eta s_Y$*;*
2. *moreover,* s_Y *with the above is arbitrarily long after the first fault, i. e. for every* $k \in \mathbb{N}$ *there exists a choice of* $s_N, s_Y \in \mathscr{L}$ *with the above properties and such that the suffix* $s_{Y/\phi}$ *of* s_Y *after the first occurrence of fault* ϕ *in* s_Y *satisfies* $|s_Y| \geq k$*.*

Note that, verification of this property is possible *without* using unfoldings, see [12, 13] and this book. The verification of strong diagnosability *with* the use of unfoldings is studied in [29], via the construction of a *verifier net*: the verifier $\mathscr{V}$ is obtained as the product of two isomorphic copies $\mathscr{N}_1$ and $\mathscr{N}_2$ of the diagnosed system $\mathscr{N}$, with synchronization only on *observable* transitions. Therefore, two unsynchronized copies ϕ_1 and ϕ_2 of the unobservable fault event exist; the verification then consists in checking (on a suitable *complete* finite prefix of the unfolding) whether $\mathscr{V}$ allows some infinite run ω on which ϕ_1 occurs and ϕ_2 does not.

Weak Diagnosability. However, it was shown in [24, 25] that for Petri nets, this property is not the only relevant one; a net may violate strong diagnosability and still be *weakly* diagnosable, in the following sense : on any faulty execution, bounded observation is sufficient to detect that on all *maximal concurrent runs* are compatible with the observation, ϕ must have occurred *or is inevitable*, possibly in the future. The presence of these weak and strong properties reflects the choice of *semantics* that produces the event structure model of behavior for the system that is investigated.

What Interleavings do and do not see. Figure 15.8 illustrates that choosing a partial order versus an interleaving semantics has important consequences. Assume that a is the only observable transition. In *sequential semantics*, the net is *not* observable: Consider the run $\omega_s \in \Omega(\mathscr{E}_{seq})$ which consists only of occurrences of v and u; it contains no observable event. Further, when choosing fault $\phi = y$, the net is not diagnosable, since all runs without an occurrence y are observationally indiscernible from the run ω' formed only by occurrences of b and a; this $\sim_\eta$-class therefore contains both faulty and healthy runs.

By contrast, when we consider the partial order semantics of the same net $\mathscr{N}$, the above ω_s is not a run. Its only extension $\bar{\omega}$ into a maximal configuration contains also an infinite number of occurrences of a and b; $\bar{\omega}$ is also the only run with this observation pattern. In fact, all runs $\omega \in \Omega(\mathscr{E}_{\mathscr{U}})$ are fault-definite, i.e. *every* run must

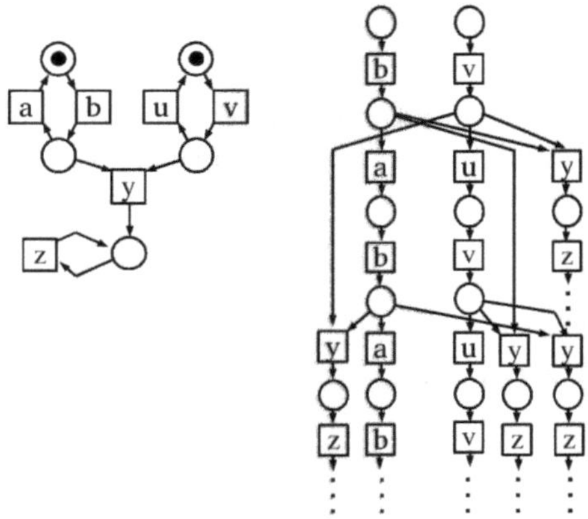

Fig. 15.8 A Petri Net (left) with a prefix of its unfolding (right) to illustrate the difference of strong and weak diagnosability

contain an occurrence of v. The example allows to observe several important phenomena. In fact, it illustrates that decentralized systems with weak synchronization between subsystems may elude diagnosis under the interleaved viewpoint, while being well captured under partial order semantics. In the example, consider now b the fault event, instead of v, and a observable. Then, the new system is neither classically observable nor classically diagnosable. However, removing the loop $u-v$ from the system leaves a classically diagnosable system. In other words, it is the presence of the second loop, running in parallel and without influence on the fault occurrence, that blocks diagnosis of the fault.[3] Thus, the partial order approach actually increases precision for partial observation of highly concurrent systems.

Following [25], we argue that systems like this which allow to derive from the observation that the fault inevitably occurs are *diagnosable* as well, albeit in a weaker sense: *weakly* diagnosable. The formalization given in [25] develops a topological description which we will not follow here.

Defining Weak Diagnosability. Let $\Phi \subseteq E$ be a set of *invisible fault events*; in particular, no event in Φ is observable, i.e. $\lambda(\Phi) \cup \mathrm{Dom}(\eta) = \emptyset$. A configuration $\mathbf{c} \in \mathscr{C}(\mathscr{E})$ is called *faulty* iff $\mathbf{c} \cap \Phi \neq \emptyset$, and *healthy* otherwise. Denote as Ω_F ($\mathscr{C}_F$) the set of faulty runs (configurations), and Ω_{NF} the set of healthy runs. Finally, set, for $\omega \in \Omega$:

$$[[\omega]]_\eta \triangleq \left\{\omega' \in \Omega \mid \omega \sim_\eta \omega'\right\}.$$

Then *weak diagnosability* for a Petri net means that for all *maximal configurations*, observation equivalence implies fault equivalence:

[3] Thanks to A. Giua who made the first author discover this aspect by a remark in a DISC workshop discussion.

Definition 15.4. *Safe Petri net $\mathcal{N} = (P, T, F, m_0)$ is* weakly F-diagnosable *iff for every $\omega \in \Omega(\mathcal{N})$,*

$$\omega \in \Omega_{NF} \Rightarrow [[\omega]]_\eta \subseteq \Omega_{NF}, \tag{15.5}$$

and weakly N-diagnosable *iff for every $\omega \in \Omega(\mathcal{N})$,*

$$\omega \in \Omega_F \Rightarrow [[\omega]]_\eta \subseteq \Omega_F \tag{15.6}$$

It is interesting to note that both notions are equivalent, i.e. $\mathcal{N}$ is weakly F-diagnosable iff it is N-diagnosable. This property — obtained in [25] from the symmetry of pseudometrics — confirms a result in [32] for strong diagnosability, and shows that the symmetry is intrinsic to the concept of diagnosability, rather than a property of the semantic framework.

15.6 Conclusion and Outlook

The use of unfoldings in diagnosis constitutes an important tool in managing large and highly distributed systems, since it allows to avoid the explosion of state space size and the associated huge number of interleaved sequences that would otherwise have to be dealt with. The exploitation of the partial order semantics allows to exhibit concurrency and, dually, causal precedence exactly, thus permitting to focus on essential dependencies in the system. Techniques of correlation via event synchronizations, verifier construction, etc. that had been known in the sequential framework carry over in a natural way to the concurrent case. Also, one notices that systems that are highly distributed in space — both for the execution of their processes *and* their observation — may necessitate a *distributed* multisupervisor approach, to factorize the branching structure of the set of processes and curb the number of such processes to be handled by any *one* diagnoser. This field still leaves room for developments, both concerning diagnosis procedures and verification methods for diagnosability.

Effective verification of *weak* diagnosability is work in progress. The verification of strong diagnosability has been shown to **PSPACE**-complete for the sequential case in [8]. This theoretical bound is a fortiori true for the non-sequential case. It is therefore important now to develop efficient algorithms for verification of *weak* diagnosability.

Another approach to partial observation in concurrent systems, introduced in [22, 23, 24], consists in looking for *inevitable* occurrences that are revealed by observation, regardless of the possible time for occurrence (which may be concurrent with the observation, with no synchronization). Knowledge of such relations in the system allows to raise alarms and start countermeasures as soon as the threat becomes apparent, without waiting for evidence of its actual occurrence.

Finally, let us point out that probabilistic measures for concurrent runs of Petri net unfoldings have been studied in [1, 2, 3, 4, 5, 9, 20, 11]. It remains to develop, in the concurrency setting, probabilistic diagnosis methods on the one hand, and characterizations and verification methods of probabilistic diagnosability on the other, generalizing the existing works for the sequential case.

15.7 Further Reading

The central reference for asynchronous diagnosis with Petri nets is [10]; for the extension to graph grammar models of systems with evolving topology see [7]. Diagnosability for the unfolding-based approaches is treated in the references [22, 23, 24, 25, 29]. Readers that which to better understand occurrence nets and the partial order semantics in general may wish to read [28] and/or compare with Mazurkiewicz Traces [15, 26]. The practical computation of (complete prefixes of) unfoldings are very well described by [30].

References

1. Abbes, S.: The (True) Concurrent Markov Property and Some Applications to Markov Nets. In: Ciardo, G., Darondeau, P. (eds.) ICATPN 2005. LNCS, vol. 3536, pp. 70–89. Springer, Heidelberg (2005)
2. Abbes, S., Benveniste, A.: Branching Cells as Local States for Event Structures and Nets: Probabilistic Applications. In: Sassone, V. (ed.) FOSSACS 2005. LNCS, vol. 3441, pp. 95–109. Springer, Heidelberg (2005)
3. Abbes, S., Benveniste, A.: Probabilistic true-concurrency models: branching cells and distributed probabilities for event structures. Information & Computation 204(2), 231–274 (2006)
4. Abbes, S., Benveniste, A.: Probabilistic true-concurrency models: Markov nets and a law of large numbers. Theoretical Computer Science 390(2-3), 129–170 (2008)
5. Abbes, S., Benveniste, A.: Concurrency, σ-Algebras, and Probabilistic Fairness. In: de Alfaro, L. (ed.) FOSSACS 2009. LNCS, vol. 5504, pp. 380–394. Springer, Heidelberg (2009)
6. Baldan, P., Haar, S., König, B.: Distributed Unfolding of Petri Nets. In: Aceto, L., Ingólfsdóttir, A. (eds.) FOSSACS 2006. LNCS, vol. 3921, pp. 126–141. Springer, Heidelberg (2006)
7. Baldan, P., Chatain, T., Haar, S., König, B.: Unfolding-based diagnosis of systems with an evolving topology. Information and Computation 208(10), 1169–1192 (2010)
8. Bauer, A., Pinchinat, S.: A topological perspective on diagnosis. In: 9th International Workshop on Discrete Event Systems, Gothenburg, Sweden (2008)
9. Benveniste, A., Fabre, E., Haar, S.: Markov nets: Probabilistic models for distributed and concurrent systems. IEEE Transactions on Automatic Control 48(11), 1936–1950 (2003)
10. Benveniste, A., Fabre, E., Haar, S., Jard, C.: Diagnosis of asynchronous discrete event systems: A net unfolding approach. IEEE Transactions on Automatic Control 48(5), 714–727 (2003)

11. Bouillard, A., Haar, S., Rosario, S.: Critical Paths in the Partial Order Unfolding of a Stochastic Petri Net. In: Ouaknine, J., Vaandrager, F.W. (eds.) FORMATS 2009. LNCS, vol. 5813, pp. 43–57. Springer, Heidelberg (2009)
12. Cabasino, M.P., Giua, A., Seatzu, C.: Diagnosability of bounded Petri nets. In: Proc. 48th IEEE Conference on Decision and Control, Shangai, China (2009)
13. Cabasino, M.P., Giua, A., Lafortune, S., Seatzu, C.: Diagnosability analysis of unbounded Petri nets. In: Proc. 48th IEEE Conference on Decision and Control, Shangai, China (2009)
14. Cassandras, C.G., Lafortune, S.: Introduction to Discrete Event Systems, 2nd edn. Springer (2008)
15. Diekert, V., Rozenberg, G. (eds.): The Book of Traces. World Scientific (1995)
16. Fabre, E.: Distributed diagnosis based on trellis processes. In: 44th Conference on Decision and Control, Seville, Spain (2005)
17. Fabre, E., Benveniste, A., Haar, S., Jard, C.: Distributed monitoring of concurrent and asynchronous systems. Discrete Event Dynamic Systems: Theory and Applications 15(1), 33–84 (2005)
18. Fabre, E.: On the construction of pullbacks for safe Petri nets. In: Applications and Theory of Petri Nets and other Models of Concurrency, Turku, Finland (2006)
19. Fabre, E.: Trellis processes: A compact representation for runs of concurrent systems. Discrete Event Dynamic Systems 17, 267–306 (2007)
20. Haar, S.: Probabilistic cluster unfoldings. Fundamenta Informaticae 53(3-4), 281–314 (2002)
21. Haar, S., Benveniste, A., Fabre, A., Jard, C.: Fault diagnosis for distributed asynchronous dynamically reconfigured discrete event systems. In: Proc. 16th IFAC World Congress, Prague, Czech Republic (2005)
22. Haar, S.: Unfold and cover: Qualitative diagnosability for Petri nets. In: Proc. 46th IEEE Conference on Decision and Control, New Orleans, LA, USA (2007)
23. Haar, S.: Qualitative diagnosability of labeled Petri nets revisited. In: Proc. 48th IEEE Conference on Decision and Control and 28th Chinese Control Conference (CDC 2009), Shanghai, China (2009)
24. Haar, S.: Types of asynchronous diagnosability and the *reveals*-relation in occurrence nets. IEEE Transactions on Automatic Control 55(10), 2310–2320 (2010)
25. Haar, S.: What topology tells us about diagnosability in partial order semantics In: Proc. 10th Workshop on Discrete Event Systems, Berlin (2010)
26. Kummetz, R., Kuske, D.: The topology of Mazurkiewicz Traces. Theoretical Computer Science 305, 237–258 (2003)
27. Madalinski, A., Fabre, E.: Modular Construction of Finite and Complete Prefixes of Petri Net Unfoldings. Fundamenta Informaticae 95(1), 219–244 (2009)
28. Nielsen, M., Plotkin, G., Winskel, G.: Petri nets, event structures and domains (I). Theoretical Computer Science 13, 85–108 (1981)
29. Nouioua, F., Madalinski, A., Dague, P.: Diagnosability verification with Petri net unfoldings. KES Journal 14(2), 49–55 (2010); Long version: Rapport de recherche No. 1516, UMR 8623, CNRS. Université Paris-Sud (March 2009)
30. Römer, S., Esparza, J., Vogler, W.: An improvement of Mcmillan's unfolding algorithm. Formal Methods in System Design 20(3), 285–310 (2002)
31. Sampath, M., Sengupta, R., Lafortune, S., Sinnamohideen, K., Teneketzis, D.: Diagnosability of discrete-event systems. IEEE Transactions on Automatic Control 40(9), 1555–1575 (1995)
32. Wang, Y., Lafortune, S., Yoo, T.-S.: Decentralized diagnosis of discrete event systems using unconditional and conditional decisions. In: Proc. 44th IEEE Conference on Decision and Control, Seville, Spain (2005)

Chapter 16
Petri Nets with Time

Béatrice Bérard, Maria Paola Cabasino, Angela Di Febbraro, Alessandro Giua, and Carla Seatzu

16.1 Introduction and Motivation

Place/Transition nets have been used in previous chapters to model Discrete Event Systems (DESs) with the aim of analyzing logical properties. However, since they do not consider the duration associated with the activities occurring in a system, they cannot be used for performance analysis of a DES, i.e., for computing the execution time of a given process, identifying bottlenecks, optimizing the use of resources, and so on. Petri nets with time are an extension of Place/Transition (P/T) nets endowed with a timing structure and can be used as performance models.

When defining Petri nets with time, three main elements should be specified: *topological structure*, *timing structure*, and *transition firing rules*. While the topological structure is generally that of a P/T net, the definition of the timing structure is a crucial problem: several timing structures have been proposed in the literature to extend P/T nets and firing rules are also based on them.

The chapter is structured as follows. Next section briefly describes the timing structures and other basic concepts related to Petri nets with time. Starting with Section 16.3 we focus on T-Timed Petri nets, the most commonly used class of Petri nets with time, and discuss different firing rules that can be used in this context. In Sections 16.4 and 16.5 we present several results related to deterministic

Béatrice Bérard
LIP6, Univ. P. et M. Curie, France
e-mail: Beatrice.Berard@lip6.fr

Maria Paola Cabasino · Alessandro Giua · Carla Seatzu
Department of Electrical and Electronic Engineering, University of Cagliari, Italy
e-mail: {cabasino,giua,seatzu}@diee.unica.it

Angela Di Febbraro
Department of Mechanical Engineering, Energetics, Production, Transportation and Mathematical Models, University of Genova, Italy
e-mail: angela.difebbraro@unige.it

C. Seatzu et al. (Eds.): Control of Discrete-Event Systems, LNCIS 433, pp. 319–341.
springerlink.com

and stochastic T-Timed Petri nets. Section 16.6 deals with a different class of Petri nets with time, called T-Time Petri nets. Further readings are finally suggested in Section 16.7.

16.2 Timing Structure and Basic Concepts

In this section, we point out a few general issues associated with a timing structure that can be associated with Petri nets.

A P/T net is a *logical* DES model, and a possible evolution of a net is described by a sequence

$$\boldsymbol{m}_0 \left[t_{j_1}\right\rangle \boldsymbol{m}_1 \left[t_{j_2}\right\rangle \boldsymbol{m}_2 \left[t_{j_3}\right\rangle \boldsymbol{m}_3 \ \dots \ \boldsymbol{m}_{k-1} \left[t_{j_k}\right\rangle \boldsymbol{m}_k \ \dots$$

of markings (i.e., states) $\boldsymbol{m}_k$ (for $k = 0, 1, 2 \dots$) and transitions (i.e., events) t_{j_k} (for $k = 1, 2, 3 \dots$). Marking $\boldsymbol{m}_0$ is the initial marking and the firing of transition t_{j_k} changes the marking from $\boldsymbol{m}_{k-1}$ to $\boldsymbol{m}_k$.

In a PN with time the evolution of a system initialized at time τ_0 is described by a sequence

$$\boldsymbol{m}_0 \left[t_{j_1}, \tau_1\right\rangle \boldsymbol{m}_1 \left[t_{j_2}, \tau_2\right\rangle \boldsymbol{m}_2 \left[t_{j_3}, \tau_3\right\rangle \boldsymbol{m}_3 \ \dots \ \boldsymbol{m}_{k-1} \left[t_{j_k}, \tau_k\right\rangle \boldsymbol{m}_k \ \dots$$

where $\tau_k \geq \tau_{k-1}$ and τ_k denotes the *firing time* of transition t_{j_k} (for $k = 1, 2, 3 \dots$) or equivalently

$$\boldsymbol{m}_0 \left[t_{j_1}, \theta_1\right\rangle \boldsymbol{m}_1 \left[t_{j_2}, \theta_2\right\rangle \boldsymbol{m}_2 \left[t_{j_3}, \theta_3\right\rangle \boldsymbol{m}_3 \ \dots \ \boldsymbol{m}_{k-1} \left[t_{j_k}, \theta_k\right\rangle \boldsymbol{m}_k \ \dots$$

where $\theta_k = \tau_k - \tau_{k-1}$ (for $i = 2, 3, 4 \dots$) denotes the *delay* between the firing of transition $t_{j_{k-1}}$ and t_{j_k} and $\theta_1 = \tau_1 - \tau_0$ denotes the *delay* between the firing of transition t_{j_1} and the initial time.

A *timing structure* specifies the value that these delays may take.

16.2.1 Timed Elements

Although, in a timed evolution the delays denote the time elapsed between the firing of two transitions, from a structural point of view a delay can be associated with different elements of a net, such as places, transitions, or arcs.

As an example, consider the simple net in Fig. 16.1. Assume the systems behavior is such that the firing of transition t should occur $\theta = 3$ seconds after the initial time. This can be done associating a delay θ with one of the following elements.

- *Place* p_1: this denotes that the token in the place becomes available for transition firings only after it has been in the place for θ seconds.

- *Transition t*: this denotes that the transition will fire only after it has been enabled for θ seconds.
- *Arc* (p_1,t): this denotes that the token in the place becomes available for this arc only after the token has reached an age of θ seconds (assuming its age at the initial time was 0).

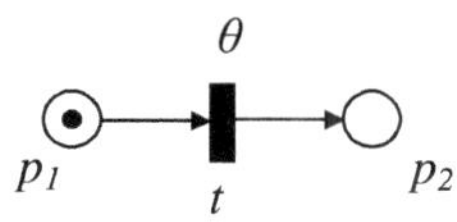

Fig. 16.1 A simple Petri net

In the rest of the chapter, we will assume that the timing structure associates delays to transitions, and we denote θ_i the delay associated with transition t_i.

16.2.2 Timed Petri Nets and Time Petri Nets

Another significant difference is among *Timed Petri nets* and *Time Petri nets*.

In *Timed Petri nets* (TdPNs) a delay is represented by a single value θ. As an example, consider a net with delays associated with transitions: if a transition becomes enabled at time τ and remains enabled henceforth, it must fire at time $\tau+\theta$.

In *Time Petri nets* (TPNs) a delay is represented by a time interval of the form $[l,u]$, where $l \in \mathbb{R}_{\geq 0}$, $u \in \mathbb{R}_{\geq 0} \cup \{+\infty\}$, and $l \leq u$. As an example, consider a net where interval $[l,u]$ is associated with a transition: if the transition becomes enabled at time τ and remains enabled henceforth, it cannot fire before time $\tau+l$ and it must fire at latest at time $\tau+u$.

When it is required to specify that delays are associated with transitions, one speaks of *T-Timed Petri nets* and *T-Time Petri nets*. On the contrary, when delays are associated with places one speaks of *P-Timed Petri nets* and *P-Time Petri nets*.

16.2.3 Deterministic and Stochastic Nets

The timing structure of a net can be: *deterministic*, when the delays are known *a priori*, or *stochastic*, when the delays are random variables.

Consider, as an example, the class of T-Timed Petri nets to which most of this chapter is dedicated. According to the nature of the associated delay, timed transitions can be classified as follows.

Definition 16.1. *A transition t_i of a T-Timed Petri net is called:*

- Immediate, *if it fires as soon as it is enabled, or equivalently, if its time delay is null.*
- Deterministic, *if the delay θ is chosen deterministically. Note that the deterministic delay may be a* constant *value θ_i, may be* variable *according to a sequence $\{\theta_{i,1}, \theta_{i,2}, \theta_{i,3}, \ldots\}$ of delay times known a priori, and finally may also be* marking dependent.
- Stochastic, *if the delay time θ_i is a random variable with a known probability distribution.*
  If the delay θ_i has an exponential distribution $f_i(\tau) = \lambda_i e^{-\lambda_i \tau}$ (with $\lambda_i > 0$) transition t_i is called stochastic exponential. *If the delay is a random variable with a distribution different from the exponential one the transition is called* generalized stochastic. *Finally if the parameters of the distribution depend on current marking of the net, the transition is called* stochastic marking dependent.

In this chapter only immediate, deterministic constant, and exponential stochastic are considered. Therefore, in the following, the last two types of transitions are briefly called deterministic and stochastic, respectively. In the most general case, the same Petri net may contain transitions of all three types mentioned above (immediate, deterministic, and stochastic); however, this increases the analysis complexity and very few analysis results exist for such a general net.

In Fig. 16.2 different PNs are shown. A deterministic transition t_i is represented by a black rectangle and is labeled with the value of its constant delay θ_i. A stochastic timed transition t_i is represented by a white rectangle and is labeled with the value of its parameter λ_i. An immediate transition is represented by a black bar with no label.

For a detailed comparison of the various timing mechanisms we refer to [6].

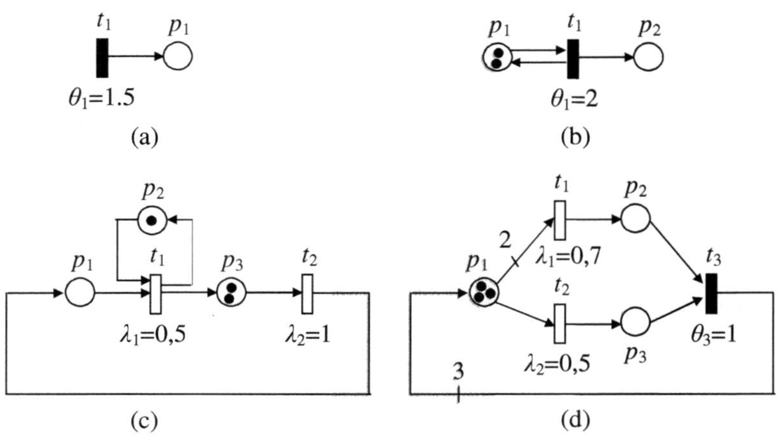

Fig. 16.2 T-Timed Petri nets

16.3 T-Timed Petri Nets and Firing Rules

In this section we focus on T-Timed Petri nets, that in the following we will simply call Timed Petri nets (TdPNs).

A series of "rules" or "conventions" should be specified in order to clarify the behavior of a given timed net. In the following subsections the most significant ones are discussed, with some comments on their expressive power.

16.3.1 Atomic vs. Non Atomic Firing

In a P/T the firing of a transition is assumed to be an *atomic* event, i.e., in the same instant the tokens that enable the transition are removed from the input places and new tokens are produced in the output places.

In the case of TdPNs this notion depends on the semantics given to the delay θ associated with a transition t.

- *The delay represents the time that must pass between the enabling and firing of a transition.* In this case, if transition t is enabled at time τ and remains enabled henceforth, it fires atomically at time $\tau+\theta$. If $\boldsymbol{m}$ is the marking before the firing, then the firing yields the new marking $\boldsymbol{m}' = \boldsymbol{m} - \boldsymbol{Pre}[\cdot,t] + \boldsymbol{Post}[\cdot,t]$.
- *The delay represents the time required to fire a transition.* In this case, assume marking $\boldsymbol{m}$ enabling transition t is reached at time τ. The transition starts its firing at τ and all tokens from the input places are removed, yielding marking $\hat{\boldsymbol{m}} = \boldsymbol{m} - \boldsymbol{Pre}[\cdot,t]$. At time $\tau+\theta$, the firing is completed producing the tokens in the output places thus yielding marking $\boldsymbol{m}' = \hat{\boldsymbol{m}} + \boldsymbol{Post}[\cdot,t]$. Such a firing policy is called *non atomic firing*.

Note that following the non atomic firing rule, the intermediate marking $\hat{\boldsymbol{m}}$ may not represent a reachable marking in the underlying P/T net and many of the analysis techniques for P/T nets, such as those based on invariants, do not apply. For this reason, we will only consider atomic firings in the rest of this chapter.

16.3.2 Enabling Semantics

Another important "rule" concerns the different strategies for the enabling of a transition.

- *Reserved marking:* as soon as a transition is enabled, the tokens of the input places of such a transition that are necessary to enable it, are reserved becoming completely invisible to all the other transitions. Moreover, in the case of actual conflict, tokens are immediately assigned to transitions, with a criterion that is in general independent of the length of their time delays.

- *Concurrent enabling:* tokens are always visible to all places and priority is given to the transition that first finishes of being enabled for a time period equal to its time delay. So, even if a transition starts being enabled later, but its time delay is small, it may happen that it fires before a transition that was enabled earlier but whose time delay is longer.

The strategy of concurrent enabling is more general than the one of reserved marking. In fact, it is always possible to transform the structure of a net following the strategy of reserved marking to an equivalent net based on concurrent enabling. This can be done using the simple scheme illustrated via the example in Fig. 16.3.

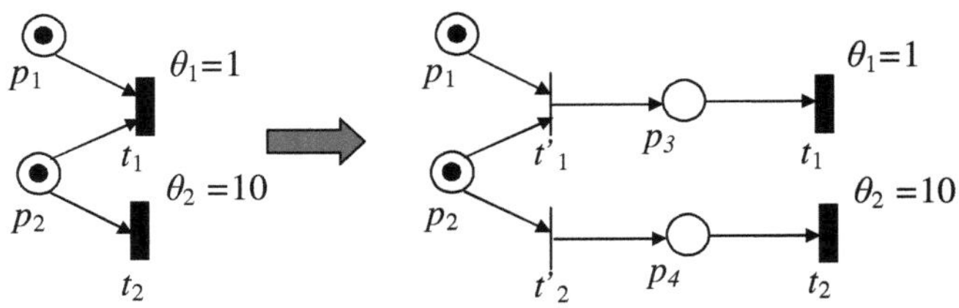

Fig. 16.3 Reserved marking strategy versus concurrent enabling strategy

16.3.3 Server Semantics

Another fundamental notion that needs to be specified when defining a TdPN is the so-called *server semantics*. Possible choices are described in the following:

- *infinite server semantics:* each transition represents an operation that can be executed by an infinite number of operation units that work in parallel; as an example, this is the case of the net in Fig. 16.4.*a*, where transition t_1 fires three times at time θ_1 because the operation units can use (process) all tokens simultaneously;
- *single server:* each transition represents an operation that can be executed by a single operation unit; an example of this is given in Fig. 16.4.*b*, where transition t_1 fires at time instants θ_1, $2\theta_1$, and $3\theta_1$ since the single operation unit can only consume (process) one token at a time;
- *multiple servers:* each transition represents an operation that can be executed by a finite number k of operation units; this is the case of the net in Fig. 16.4.*c*, where, assuming $k = 2$, transition t_1 fires twice at time θ_1 and once at time $2\theta_1$, since the two operation units can process only two tokens at a time.

In the rest of this chapter, we always assume infinite server semantics. In fact, starting from such a notion, it is possible to also represent the other two via appropriate places (as place p in Fig. 16.4.*b* and in Fig. 16.4.*c*), that limit the maximum enabling degree of the generic transition, as it will be explained in the following section.

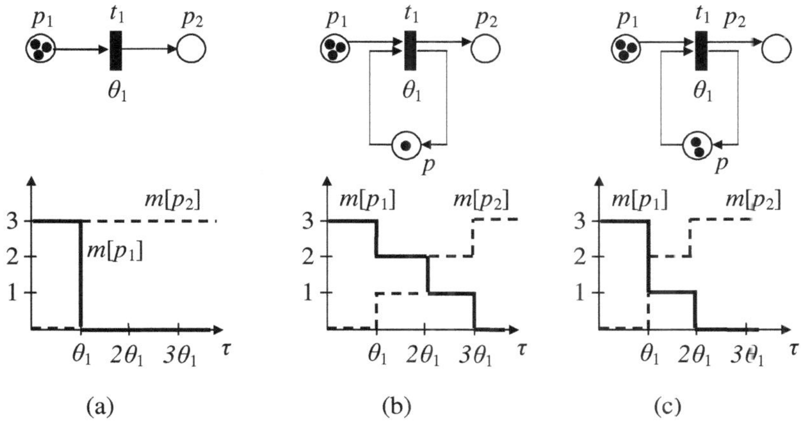

Fig. 16.4 Transitions with different server semantics

16.3.4 Memory Policy

Another notion that has to be specified concerns the *memory* associated with transitions. We have seen that a transition t_i can fire only if a time θ_i has elapsed since its enabling. Now, let us observe the net in Fig. 16.5; assuming $\theta_2 < \theta_1 < 2\theta_2$, from the initial marking (at time $\tau_0 = 0$) transitions t_1 and t_2 are enabled, thus at time $\tau_1 = \theta_2$ transition t_2 fires and yields to the marking $[0\ 0\ 1]^T$. After a delay equal to θ_3, i.e., at time $\tau_2 = \tau_1 + \theta_3$, transition t_3 fires and the net reaches again the initial marking. Two different notions of memory can be introduced.

1. *Total memory*: transition t_1 "remembers" being already enabled for a time interval equal to θ_2 and fires after a delay equal to $\theta_1 - \theta_2$, i.e., at time $\tau_3 = \tau_2 + (\theta_1 - \theta_2)$;
2. *enabling memory*: transition t_1 has only memory of the current enabling and can only fire after a delay equal to θ_1, i.e., at time $\tau_4 = \tau_2 + \theta_1$.

In the rest of this chapter, we consider as basic notion the enabling memory policy.

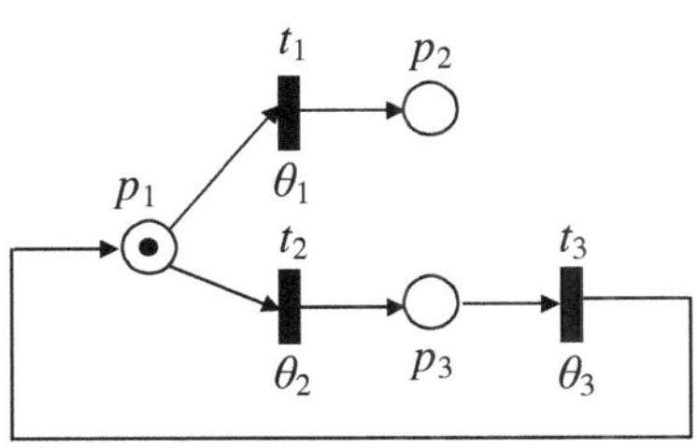

Fig. 16.5 Timed net with conflict

16.4 Deterministic Timed Petri Nets

The first extension of the P/T nets via deterministic delays has been presented in [21]. This approach uses timed transitions to address the idea of modeling the duration of activities of the represented DES, being in general the actions associated with the transitions. These nets are called *Deterministic Timed Transitions Petri Nets* or *Deterministic T-Timed Petri Nets*.

As discussed in Subsection 16.2.1, another timing structure is the one that assigns the time to places [11] that are seen as processes that require a given execution time. These nets, that are called *Deterministic Timed Places Petri Nets* or *Deterministic P-Timed Petri Nets*, represent an excellent applicative field of DES modeling approach based on *max-plus algebra* or *minimax* [2].

Finally there have been proposed also nets where the time is associated with arcs. As an example, Zhu and Denton [24] showed that such Petri nets are more general than those where the time is associated either with transitions or places.

In the rest of this section we focus on *Deterministic T-Timed Petri Nets* that are most commonly used in the literature. As a result of this, they are often called *Deterministic Timed Petri Nets* (DTdPN), without making explicit that delays are associated with transitions. Delays can be either constant or variable as clarified in the following definition.

Definition 16.2. *A deterministic timed Petri net is characterized by the algebraic structure* $N_d = (N, \Theta)$ *where:*

- $N = (P, T, \boldsymbol{Pre}, \boldsymbol{Post})$ *is a P/T net defined as in Definition 10.1 in Chapter 10;*
- $\Theta = \{\Theta_i : t_i \in T\}$, *with* $\Theta_i = \{\theta_{i,1}, \theta_{i,2}, \ldots\}$, $t_i \in T$, $\theta_{i,k} \in \mathbb{R}_+ \cup \{0\}$, $k \in \mathbb{N}_+$ *is a deterministic timing structure; if the time delays are constant, the generic element* $\theta_{i,k}$ *is denoted* θ_i, $\forall k \in \mathbb{N}_+$.

Even for timed Petri nets it is possible to define a marked Petri net. In general, the marking vector at the time instant τ_j is denoted $\boldsymbol{m}_j$.

Definition 16.3. *A deterministic timed Petri net* N_d *with a marking* $\boldsymbol{m}_0$ *at the initial time instant* τ_0 *is said a* marked deterministic timed Petri net, *and is denoted* $\langle N_d, \boldsymbol{m}_0 \rangle$.

16.4.1 Dynamical Evolution

The state of a DTdPN is determined not only by the marking, as for P/T nets, but also by the clocks associated with transitions.

Definition 16.4. *A transition* t_i *is* enabled *at a marking* $\boldsymbol{m}_j$ *if each place* $p \in P$ *of the net contains a number of tokens equal to or greater than* $\boldsymbol{Pre}[p, t_i]$, *i.e.,* $\boldsymbol{m}_j \geq \boldsymbol{Pre}[\cdot, t_i]$.

The enabling degree *of a transition t_i enabled at a marking $\boldsymbol{m}_j$ is the biggest integer number k such that $\boldsymbol{m}_j \geq k\ \boldsymbol{Pre}[\cdot,t_i]$. The enabling degree of t_i at $\boldsymbol{m}_j$ is denoted $\alpha_i(j)$.*

In the net in Fig. 16.2.*a*, transition t_1 has an infinite enabling degree; in the marked net in Fig. 16.2.*b*, transition t_1 has enabling degree equal to 2; in the marked net Fig. 16.2.*c*, transition t_1 is not enabled because p_1 is empty, while transition t_2 has enabling degree equal to 2; in the marked net in Fig. 16.2.*d*, transition t_1 has enabling degree equal to 1, because its firing needs 2 tokens and in its pre place p_1 there are only 3 tokens, transition t_2 has enabling degree equal to 3, while transition t_3 is not enabled. The case in which the enabling degree of a transition is infinite, as the case of Fig. 16.2.*a*, is a degenerate case. In the following, we assume that the enabling degree of a transition is finite (but not necessarily bounded).

At each time instant the number of clocks associated with a transition t_i is equal to its current enabling degree; this number changes with the enabling degree, thus it can change each time the net evolves from one marking to another one, namely each time that a transition fires.

The net evolution occurs in an asynchronous way on the basis of the events occurrence that is regulated by the clocks associated with the events according to the algorithm of evolution here reported.

Algorithm 16.1. (Temporal evolution of a DTdPN). *Assume that the DTdPN at the time instant τ_j is in the marking $\boldsymbol{m}_j$ and that the minimum values of the clocks associated with the transitions, $o_i = min\{o_{i,1},\ldots,o_{i,\alpha_i(j)}\}$, $\forall t_i \in T$, are known; the marking evolution of the DTdPN follows the repetition of these steps:*

1. *Let o^**

$$o^* = \min_{i:t_i \in T} \{o_i\} \tag{16.1}$$

   be the minimum among the values of the clocks o_i associated with the transitions enabled at marking $\boldsymbol{m}_j$; if o^* is not unique more than one transition could fire at the same time according to a sequence that should be specified a priori.
2. *At the time instant $\tau_{j+1} = \tau_j + o^*$ transition t^* fires yielding the system from marking $\boldsymbol{m}_j$ to the marking $\boldsymbol{m}_{j+1} = \boldsymbol{m}_j + C[\cdot,t]$.*
3. *Once marking $\boldsymbol{m}_{j+1}$ is reached, the clock associated with t^* is discarded. Clocks associated with each transition $t_i \in T$ are updated as follows:*
   - *if the enabling degree $\alpha_i(j+1)$ at marking $\boldsymbol{m}_{j+1}$ is less than the enabling degree $\alpha_i(j)$ that transition t_i had at previous marking $\boldsymbol{m}_j$, then $[\alpha_i(j) - \alpha_i(j+1)]$ clocks associated with t_i have to be discarded: clocks that are discarded from set $\{o_{i,1},\ldots,o_{i,\alpha_i(j)}\}$ are those having the higher values;*
   - *if $\alpha_i(j+1) > \alpha_i(j)$, $[\alpha_i(j+1) - \alpha_i(j)]$ new clocks are associated with t_i and initialized to the values specified by the timing structure Θ;*
   - *if $\alpha_i(j+1) = \alpha_i(j)$, do nothing;*
   - *reduce to an amount equal to o^* the values of all old clocks.*
4. *Repeat from step 1, setting $j+1 \rightarrow j$.*

Note that if transition t_i is not enabled at a marking, it has no clocks associated with it, i.e., it has no *active* clocks. If at a marking $\boldsymbol{m}_j$ the minimum value of the clock o_i of a transition t_i corresponds to more than one clock, as an example k, in the set $\{o_{i,1},\dots,o_{i,\alpha_i(j)}\}$, this means that if the transition will be the next one to fire, it will fire k times at the same time.

The algorithm is based on the assumption of *enabling memory* and *infinite server semantics*. If *total memory* is used, step 2 of the algorithm should be modified. Note that the memory chosen depends on the kind of study one wants to do on the system while does not depend on the system itself. For the server semantics we have chosen the most general: in fact one can always lead the system to single or multiple servers adding self loops to transitions, as shown in Fig. 16.4.*b* and 16.4.*c*. Obviously, the algorithm is simplified if all transitions have single server semantics, because in such a case each transition has associated one single clock.

Finally, at step 1 it is said that a politic for the *resolution of conflicts* has to be applied if more than one transition can fire at the same time, to decide the sequence in which these transitions will fire. This is needful only when the firing of one transition can disable other transitions, namely when the net is not *persistent*. Either a firing priority or a firing probability can be associated with transitions.

Example 16.2. Let us consider the net in Fig. 16.2.*b* whose temporal evolution is shown in Fig. 16.6.

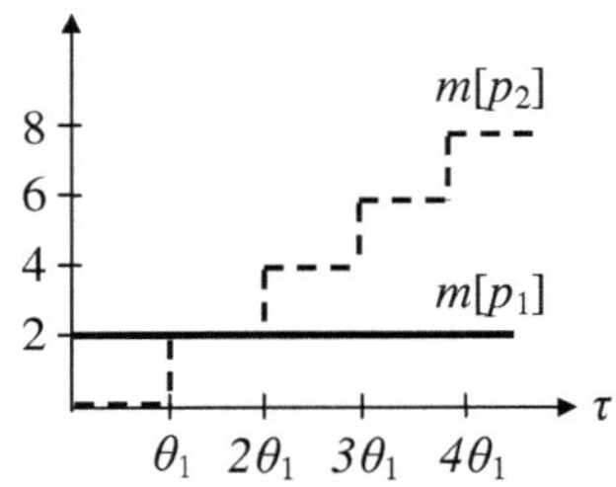

Fig. 16.6 Evolution of the net in Fig. 16.2.*b*

Transition t_1 (with time delay $\theta_1 = 2$) has enabling degree $\alpha_1(0)$ equal to 2 at the initial time instant $\tau_0 = 0$; it has initially two active clocks $o_{1,1}$, and $o_{1,2}$. After two time instants, i.e., at $\tau_1 = \tau_0 + \theta_1$, both clocks are deleted, transitions t_1 fires twice and the two clocks are again set to $\theta_1 = 2$. At the reached marking $\boldsymbol{m}_1 = [2\ 2]^T$, transition t_1 still has an enabling degree equal to 2 and the two clocks $o_{1,1}$, and $o_{1,2}$ are again active. The net continues to evolve following Algorithm 16.1. ■

Example 16.3. A production line is composed by two machines M_1 and M_2, two robotics arms R_1 and R_2, and two conveyor belts. Each machine uses one robotic arm that loads and unloads parts that the machine has to process. One of the conveyor belt can carry only two parts, while the other one carries empty pallets. Pallets in the system are three.

Each part is processed by machine M_1 first and machine M_2 later, with process times respectively equal to 10 and 20 time units. The loading and unloading processes require 1 time unit, while the time spent in the conveyor belts is assumed negligible.

The DTdPN modeling this production system is shown in Fig. 16.7, while Table 16.1 contains the meaning of places and transitions and the transitions delay.

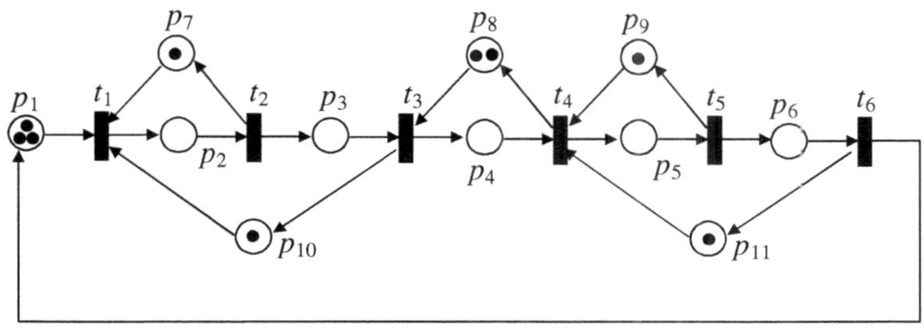

Fig. 16.7 DTdPN modeling a production line formed by two machines

At the initial marking $\boldsymbol{m}_0 = [3\ 0\ 0\ 0\ 0\ 0\ 1\ 2\ 1\ 1\ 1]^T$ transition t_1 is enabled and after one time unit fires yielding to the marking $\boldsymbol{m}_1 = [2\ 1\ 0\ 0\ 0\ 0\ 0\ 2\ 1\ 0\ 1]^T$, where transition t_2 is enabled. After 10 time units t_2 fires and yields the net to the new marking $\boldsymbol{m}_2 = [2\ 0\ 1\ 0\ 0\ 0\ 1\ 21\ 0\ 1]^T$. The net continues to evolve following the procedure indicated above. ■

16.4.2 *Timed Marked Graphs*

A *Timed Marked Graph* (TdMG) is a DTdPN where each place has only one input transition and one output transition and all arcs have unitary weight. A more restricted class of such nets are the *strongly connected timed marked graphs* whose importance is due to the fact that there exist some criteria to analyze the performance of the system in an easy way.

Definition 16.5. *A* (deterministic) Strongly Connected Timed Marked Graph *(SCTdMG) is a DTdPN N_d satisfying the following properties:*

- *the net structure N_d is a timed marked graph;*
- *the net is strongly connected, namely there exists an oriented path from each node to any node: this implies that each place and each transition of the net belongs to an oriented cycle; the set of the oriented elementary cycles of N_d is denoted $\Gamma = \{\gamma_1, \ldots, \gamma_r\}$;*
- *the timing structure Θ associated with transitions is deterministic and has constant time delays.*

Table 16.1 Description of places and transitions in Fig. 16.7

Place	Description
p_1	availability of parts and pallets
p_2	M_1 is working
p_3	part ready to be unloaded by M_1
p_4	part ready to be processed by M_2
p_5	m_2 is working
p_6	part ready to be unloaded by M_2
p_7	M_1 is available
p_8	availability on the conveyor belts
p_9	M_2 is available
p_{10}	R_1 is available
p_{11}	R_2 is available

Transition	Description	Delay
t_1	R_1 loads a part on M_1	1
t_2	M_1 ends the processing	10
t_3	R_1 ends the processing M_1 to the conveyor belt	1
t_4	R_2 loads a part on M_2	1
t_5	M_2 ends the processing	20
t_6	R_2 removes from M_2 a processed part	1

Although, these nets could seem too much restrictive, they can model important classes of discrete event systems. As an example two important classes of production systems, such as *job-shop* systems and systems based on the *Kanban* philosophy, can be modeled using SCTdMGs [14].

16.4.2.1 Performance Analysis

Let us now present some results that allow to perform the analysis, in steady conditions, in the case of TdMGs and SCTdMGs.

Theorem 16.1. *In a TdMG, the number of tokens in a cycle remains constant for any firing sequence.*

The proof of this theorem is based on the structural characteristics of a TdMG, where each place has one single input and output transition. In fact, each time a transition in a cycle fires, it removes a token from the input place that belongs to the cycle and put a token in the output place that belongs to the same cycle, thus the number of tokens in the cycle remains unchanged.

Let us now introduce the notion of *cycle time* that can be a performance index in a system modeled as a SCTdMG.

Definition 16.6. *The* cycle time C_i *of a transition* t_i *of a SCTdMG is defined on the basis of its generic kth time firing* $\tau_{i,k}$

$$C(t_i) = \lim_{k \to \infty} \frac{\tau_{i,k}}{k} \tag{16.2}$$

where $\tau_{i,k}$ *is the time instant at which transition* t_i *fires for the kth time, starting from the initial time instant* τ_0.

The above definition allows one to give two important results.

Theorem 16.2. [7, 12, 23] *In a SCTdMG, all transitions belonging to a cycle* $\gamma_j \in \Gamma$ *have the same cycle time* C_{γ_j}*, defined as the ratio between the sum of the delay times of transitions that form* γ_j *and the number of tokens circulating in it, i.e.,*

$$C_{\gamma_j} = \frac{\sum_{t_i \in \gamma_j} \theta_i}{\sum_{p_k \in \gamma_j} m[p_k]} \tag{16.3}$$

Theorem 16.3. [7, 10] *In a SCTdMG in steady conditions, all transitions in a cycle have the same cycle time* C*, equivalent to:*

$$C = \max_{\gamma_j \in \Gamma} C_{\gamma_j} = \max_{\gamma_j \in \Gamma} \left\{ \frac{\sum_{t_i \in \gamma_j} \theta_i}{\sum_{p_k \in \gamma_j} m[p_k]} \right\} \tag{16.4}$$

that identifies the maximum among the cycle times of all elementary cycles of a SCTdMG. This means that in steady conditions all transitions have the same firing frequency equal to $\lambda_r = 1/C$.

The result presented above is intuitive, in fact due to the structural characteristics of a SCTdMG, in steady conditions all cycles are synchronized on the "slower" cycle.

Example 16.4. Let us consider again the DTdPN shown in Fig. 16.7 already introduced in Example 16.3. The elementary cycles that form the set Γ are 6: γ_1: $p_7t_1p_2t_2p_7$; γ_2: $p_2t_2p_3t_3p_{10}t_1p_2$; γ_3: $p_8t_3p_4t_4p_8$; γ_4: $p_9t_4p_5t_5p_9$; γ_5: $p_5t_5p_6t_5p_{11}t_4p_5$; γ_6: $p_1t_1p_2t_2p_3t_3p_4t_4p_5t_5p_6t_6p_1$. By Theorem 16.2 the time cycles are respectively: $C_{\gamma_1} = 11$; $C_{\gamma_2} = 12$; $C_{\gamma_3} = 1$; $C_{\gamma_4} = 21$; $C_{\gamma_5} = 22$; $C_{\gamma_6} = 11{,}3$. Thus, by Theorem 16.3, the firing frequency of all transitions of this DTdPN in steady condition is $\lambda_r = 1/C = 1/\max_{\gamma_j \in \Gamma} C_{\gamma_j} = 1/22 = 0{,}045$. ■

16.5 Stochastic Timed Petri Nets

In this section, we present *Stochastic Timed Petri Nets* (STdPNs), i.e., P/T nets where the delays associated with transitions are random variables. As a result of this randomness STdPN can be considered stochastic processes.

We consider STdPNs with atomic firing and time delay described by a random variable with negative exponential distribution function. At each stochastic transition t_i is associated a parameter λ_i that characterizes its distribution, called *firing rate* or *firing frequency* of the transition. Note that if we denote $\bar{\theta}_i$ the average firing delay of transition t_i, it holds $\bar{\theta}_i = 1/\lambda_i$.

Definition 16.7. *A stochastic timed Petri net (STdPN) is a triple $N_s = (N, \Psi, \boldsymbol{\lambda})$ where:*

- *$N = (P, T, \boldsymbol{Pre}, \boldsymbol{Post})$ is a Petri net defined as in Definition 10.1 in Chapter 10;*
- *$\Psi = \{\Psi_i : t_i \in T\}$ is a timing stochastic structure; Ψ_i is a probability distribution function defined in $\mathbb{R}_+ \cup \{0\}$, from which are extracted the values of the random variables that form the delay firing $\theta_{i,k}$ of transition t_i, $t_i \in T$, $k \in \mathbb{N}_+$; in particular in this section we consider all Ψ_i as negative exponential distribution functions;*
- *$\boldsymbol{\lambda} = [\lambda_1\ \lambda_2\ \cdots]$ is the vector of the firing frequencies of transitions; elements λ_i can depend on the marking, namely it can be $\lambda_i = \lambda_i(\boldsymbol{m}_k)$, $k \in \mathbb{N}_+$.*

For a stochastic Petri net the firing of a transition follows the same rule of a P/T net, except for the fact that the choice of the next transition to fire is made on the basis of the firing probabilities of single transitions. The probability that transition t_i, enabled at marking $\boldsymbol{m}_k$, fires is equal to

$$Pr\{t_i \mid \boldsymbol{m}_k\} = \frac{\lambda_i(\boldsymbol{m}_k)}{\sum\limits_{t_j \in \mathscr{A}(\boldsymbol{m}_k)} \lambda_j(\boldsymbol{m}_k)} \tag{16.5}$$

where $\mathscr{A}(\boldsymbol{m}_k)$ is the set of transitions enabled at marking $\boldsymbol{m}_k$.

To describe the behavior of a STdPN, as for DTdPNs, clocks are associated with transitions. For simplicity, it is assumed that each transition is associated with a single clock, which is initialized to the value of the delay when the transition is enabled for the first time after a firing. Clocks operate as in the case of DTdPNs. In more detail, each time a new marking is reached, each enabled transition t_i resamples a new instance θ_i from the probability density function associated with its delay.

Many researchers formally demonstrated the potentiality of STdPNs as a tool for performance analysis of real systems, particularly showing that, from the point of view of the dynamic behavior, a STdPN is equivalent to a *Continuous Time Markov Chain* (CTMC). This connection has been proved by the following results that can be found in many classical books, such as [23].

Theorem 16.4. *In a STdPN, times spent in each marking are exponentially distributed.*

Theorem 16.5. *The time evolution of a STdPN can be described by a CTMC where each state corresponds to a different marking reachable by the STdPN.*

As a result of the above two theorems and of the STdPN evolution rules, it is possible to compute the distribution of times spent in a given marking $\boldsymbol{m}_j$ since the delay of all the enabled transitions is a random variable with exponential distribution. Therefore, the time spent in a marking is the smallest of the delays before next transition firing. As a consequence, the parameter that characterizes its exponential distribution is $\alpha_j = \sum_{t_i \in \mathscr{A}(\boldsymbol{m}_j)} \lambda_i(\boldsymbol{m}_j)$; this element also identifies the negative component $-q_{jj}$ along the diagonal frequency matrix $\boldsymbol{Q}$ of the CTMC equivalent to the considered STdPN.

Example 16.5. Consider the STdPN in Fig. 16.8 whose reachability graph is reported in the same figure, where $\boldsymbol{m}_0 = [1\ 0\ 0\ 0\ 0]^T$, $\boldsymbol{m}_1 = [0\ 1\ 1\ 0\ 0]^T$, $\boldsymbol{m}_2 = [0\ 0\ 1\ 1\ 0]^T$, $\boldsymbol{m}_3 = [0\ 1\ 0\ 0\ 1]^T$, e $\boldsymbol{m}_4 = [0\ 0\ 0\ 1\ 1]^T$.

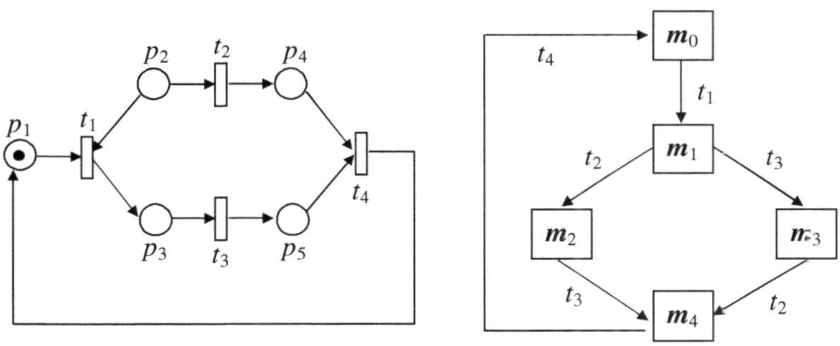

Fig. 16.8 A stochastic timed Petri net and its reachability graph

Let $\boldsymbol{\pi}$ be the row vector with as many components as the number of reachable markings, where each component represents the steady state probability associated with the corresponding marking. Since the graph is finite and strongly connected, we can compute vector $\boldsymbol{\pi}$ solving the linear system

$$\begin{cases} \boldsymbol{\pi Q} = \boldsymbol{0} \\ \sum_{x_j \in X} \pi_j = 1 \end{cases}$$

that in this example is equal to

$$\begin{cases} \boldsymbol{\pi} \begin{bmatrix} -\lambda_1 & \lambda_1 & 0 & 0 & 0 \\ 0 & -(\lambda_2+\lambda_3) & \lambda_2 & \lambda_3 & 0 \\ 0 & 0 & -\lambda_3 & 0 & \lambda_3 \\ 0 & 0 & 0 & -\lambda_2 & \lambda_2 \\ \lambda_4 & 0 & 0 & 0 & -\lambda_4 \end{bmatrix} = \begin{bmatrix} 0\,0\,0\,0\,0 \end{bmatrix} \\ \sum_{j=0}^{4} \pi_j = 1 \end{cases}$$

Supposing that $\lambda_j = 1$, $j = 0,\ldots,4$, the system solution is $\pi_0 = \pi_4 = 2/7$, $\pi_1 = \pi_2 = \pi_3 = 1/7$. ■

The following section formalizes the rules of construction of the CTMC equivalent to a given STdPN.

16.5.1 Construction of the Markov Chain Equivalent to the STdPN

The CTMC equivalent to a given STdPN can be easily generated using the following algorithm:

Algorithm 16.6. (CTMC equivalent to a STdPN).

1. *Create a bijective correspondence between states X of the Markov chain and the reachability set $R(N_s, \boldsymbol{m}_0)$, such that to each marking $\boldsymbol{m}_k$ corresponds the state $x_k \in X$.*
2. *Let $\pi_0(0) = 1$ be the initial state probability vector, i.e., associate the maximum probability with state x_0 corresponding to $\boldsymbol{m}_0$.*
3. *Let the transition frequencies of the Markov chain, namely the elements of matrix $\boldsymbol{Q}$, equal to*

$$-q_{kk} = \sum_{t_i \in \mathscr{A}(\boldsymbol{m}_k)} \lambda_i(\boldsymbol{m}_k) \tag{16.6}$$

$$q_{kj} = \sum_{t_i \in \mathscr{A}_j(\boldsymbol{m}_k)} \lambda_i(\boldsymbol{m}_k) \tag{16.7}$$

where $\mathscr{A}_j(\boldsymbol{m}_k)$ is the subset of $\mathscr{A}(\boldsymbol{m}_k)$ that includes transitions whose firing yields to $\boldsymbol{m}_j$, i.e., $\mathscr{A}_j(\boldsymbol{m}_k) = \{t_i \in \mathscr{A}(\boldsymbol{m}_k) \mid \boldsymbol{m}_k[t_i\rangle \boldsymbol{m}_j\}$; in general there exists only one transition whose firing yields the state from marking $\boldsymbol{m}_k$ to marking $\boldsymbol{m}_j$.

Example 16.7. Consider the behavior of a machine: when it is available, it can load one part and starts its processing. The end of the process makes the machine available to process another part. The machine, however, may fail while working and therefore needs to be repaired. After the repair, the machine is ready to work again. The behavior of this machine is modeled by the STdPN in Fig. 16.9. The meaning of places and transitions is described in Table 16.2, where are also described the characteristic parameters of exponential distributions that determine the firing time of transitions.

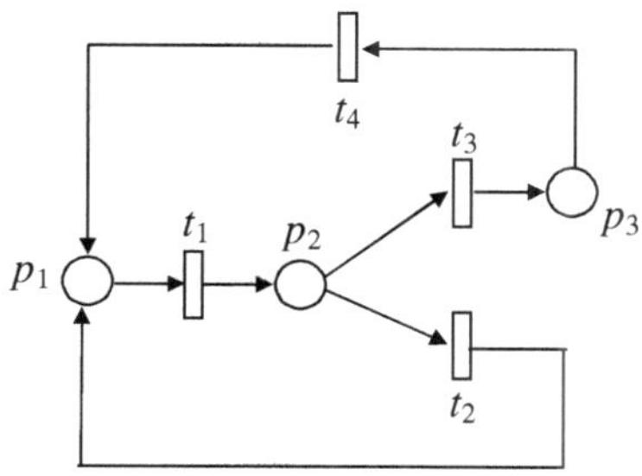

Fig. 16.9 STdPN of a machine that may fail

Table 16.2 Description of places and transitions in Fig. 16.9

Place	Description
p_1	the machine is available
p_2	the machine is working
p_3	the machine is being repaired

Transition	Description	Firing frequency
t_1	beginning of a process	α
t_2	end of a process	β
t_3	failure of the machine	μ
t_4	completed repairing	λ

States in which the machine can be are: x_0 = machine available, corresponding to marking $\boldsymbol{m}_0 = [1\ 0\ 0]^T$; x_1 = machine working, corresponding to marking $\boldsymbol{m}_1 = [0\ 1\ 0]^T$; x_2 = faulty machine, corresponding to marking $\boldsymbol{m}_2 = [0\ 0\ 1]^T$. The Markov chain equivalent to the STdPN is characterized by the frequency transition matrix $\boldsymbol{Q}$

$$Q = \begin{bmatrix} -\alpha & \alpha & 0 \\ \beta & -(\beta+\mu) & \mu \\ \lambda & 0 & -\lambda \end{bmatrix}$$

and the corresponding transitions diagram is shown in Fig. 16.10.

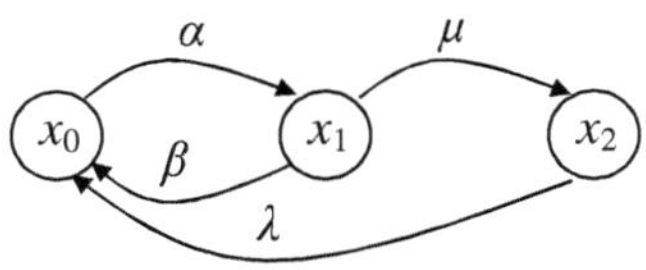

Fig. 16.10 State transition frequency diagram of the CTMC equivalent to the STdPN in Fig. 16.9

Note that such a diagram can be obtained by the reachability graph of the STdPN, represented in Fig. 16.11, substituting to each marking the corresponding state of the equivalent Markov chain and to each transition of the STdPN the parameter that characterizes the exponential distribution of the firing time. ■

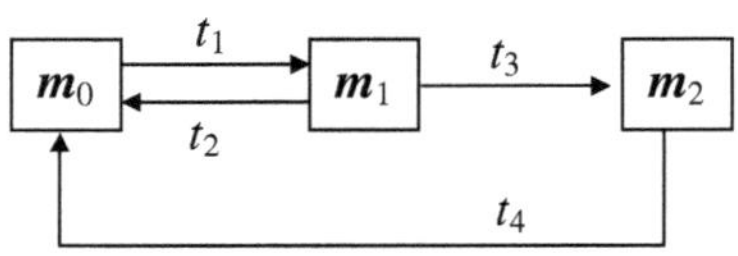

Fig. 16.11 Reachability graph of the STdPN in Fig. 16.9

16.5.2 Performance Analysis

An homogeneous CTMC that is finite and irreducible is always ergodic[1] [9]. This implies that a bounded STdPN whose reachability graph is strongly connected always corresponds to an ergodic CTMC. The reachability graph of a STdPN is equivalent to the one of a P/T net obtained removing time delays. This happens because time delays associated with transitions have probability density defined in $\mathbb{R}_+$. As a result, the criteria and methodologies introduced in Chapter 11 for the structural analysis of P/T nets also apply to STdPNs.

[1] A Markov chain is *ergodic* if and only if, for any initial probability distribution, there exists a limit probability distribution, i.e., there exists $\lim_{t\to\infty} \pi(t)$, and such a distribution is independent of the initial marking. Note that for ergodicity it is not necessary that the graph is irreducible. It is sufficient that there exists a unique ergodic component, i.e., a strongly connected component with no output arcs.

The analysis of a STdPN is usually targeted to measure aggregate performance indices that are more significant than the steady state probabilities $\boldsymbol{\pi}$ of the markings. In the following items are reported the most common performance indices [1].

- The probability of event e defined as a function of the marking (e.g. no token in a given set of places or at least one token in a place when another one is empty, etc.) can be computed summing up the probabilities of all markings that satisfy the condition expressed by the event; thus, the steady probability of event e is

$$Pr\{e\} = \sum_{\boldsymbol{m}_k \in \mathscr{M}_e} \pi_k$$

  where $\mathscr{M}_e$ is the set of markings satisfying the condition expressed by e; note that we can sum up the probabilities of the single markings because they are mutually exclusive.
- The probability of having a certain number of tokens in a place p_i can be computed as a special event; then, if $e_{i,j}$ denotes the event of having j tokens in place p_i, the average number of tokens in p_i can be computed as

$$\bar{n}_i = \sum_j jPr\{e_{i,j}\} \tag{16.8}$$

- The firing frequency f_j of a transition t_j, i.e., the average number of times that the transition fires in the time unit, under steady conditions can be computed as the weighted sum of the firing rates of transitions enabled at each marking $\boldsymbol{m}_i$

$$f_j = \sum_{i:t_j \in \mathscr{A}(\boldsymbol{m}_i)} \lambda_j(\boldsymbol{m}_i)\pi_i \tag{16.9}$$

- The average time $\bar{\theta}$ that a token spend to pass through a subnet in steady conditions can be computed applying the Little's law [9], that can be written for such a case as

$$\bar{\theta} = \frac{\bar{n}}{\lambda} \tag{16.10}$$

  where $\bar{n}$ is the average number of tokens passing through the subnet and λ is the average speed of the tokens that are entering in the subnet.

Finally, note that the main problem of the evaluation of the performance indices for a STdPN is the necessity of working with the equilibrium equations based on the reachability graph. In fact, the dimension of the reachability graph grows exponentially both with the number of tokens of the initial marking $\boldsymbol{m}_0$ and with the number of places; thus, except for some particular classes of nets, the dimension of the reachability graph and the computational complexity of the procedure do not allow to have exact analytical solutions.

16.6 Time Petri Nets

In this section, we focus on *T-Time Petri Nets*, i.e., P/T nets where a timing interval is associated with each transition. As in the case of T-Timed Petri nets, "T" (for transition) is often assumed as implicit and the model is more briefly called *Time Petri nets* (TPNs). This interval-based variant was first proposed in [20] and applied later to other timed models (see [8, 16, 17, 19]). The basic principle is the following. When an interval $[l_i, u_i]$ from the time domain is associated with a transition t_i in a P/T net, the bounds of the interval represent respectively the minimal and maximal delay for firing the transition. In this case, an implicit clock can be associated with the transition and the transition can be fired only if the clock value belongs to the interval.

We give a formal definition for this model and its semantics, described by timed transition systems.

We denote here by $\mathscr{I}$ the set of closed intervals with a lower bound in $\mathbb{Q}_+$ and an upper bound in $\mathbb{Q}_+ \cup \{\infty\}$, associated with transitions. In particular, $I(t_i) = [l_i, u_i]$ denotes the interval associated with transition t_i.

For an interval I, the backward closure of I is defined by: $I^{\downarrow} = \{x \mid \exists y \in I,\ x \leq y\}$.

Definition 16.8. *A* Time Petri Net *is characterized by the algebraic structure* $N_T = (N, I)$ *where:*

- *N is a P/T net defined as in Definition 10.1 in Chapter 10;*
- *$I : T \to \mathscr{I}$ associates with each transition a* firing interval.

Fig. 16.12 depicts a Time Petri net. Each transition is equipped with its firing interval. For instance, transition t_1 has interval $I(t_1) = [1, \infty[$. The initial marking has two tokens in place p_1 and one token in place p_2.

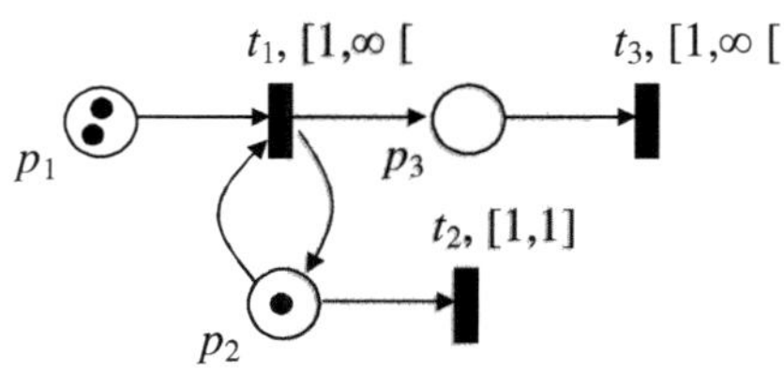

Fig. 16.12 A time Petri net, with time intervals on transitions

We now explain the timing conditions, with $\mathbb{R}_+$ as dense time domain. A transition t can be fired if the time elapsed since the last update belongs to its interval $I(t)$. Moreover, for all enabled transitions, time cannot progress when one of the upper bounds is reached, thus enforcing urgency.

A configuration of N_T is a pair $(\boldsymbol{m}, \nu)$, for a marking $\boldsymbol{m}$ and a valuation $\nu \in (\mathbb{R}_+ \cup \{\bot\})^T$. Relevant values of ν are those for which t belongs to $\mathscr{A}(\boldsymbol{m})$, and $\nu(t)$

contains the time elapsed since the last update in this case. We write $\nu(t) = \perp$ otherwise. For a real number d, the valuation $\nu + d$ is defined by $(\nu + d)(t) = \nu(t) + d$ for any t, with adequate conventions for $\perp$ values. A configuration is *admissible* if for all enabled transitions, $\nu(t) \in I(t)^{\downarrow}$. We denote by $ADM(N_T)$ the set of admissible configurations of N_T.

The important point that remains to be defined is the update of timing information upon transition firing. In other words, we should precise when the implicit clock associated with the transition is reset: transition t' is said to be *newly enabled* after firing t from marking $\boldsymbol{m}$ if the predicate $\uparrow enabled(t', \boldsymbol{m}, t)$ defined by:

$$\uparrow enabled(t', \boldsymbol{m}, t) = (t' \in \mathscr{A}(\boldsymbol{m} - \boldsymbol{Pre}[\cdot, t] + \boldsymbol{Post}[\cdot, t])) \wedge (t' \notin \mathscr{A}(\boldsymbol{m}))$$

evaluates to *true*.

Thus, t' is newly enabled if it was not enabled before firing t but becomes enabled by this firing. This corresponds to the so-called *persistent atomic* semantics, which is not the most frequently used but is easier to explain and equivalent to the other ones for safe time Petri nets. Discussions and comparisons with *atomic* and *intermediate* semantics can be found in [3, 22].

Definition 16.9. *The semantics of a time Petri net N_T is the timed transition system $\mathscr{T} = (S, s_0, E)$ where:*

- $S = ADM(N_T)$*;*
- $s_0 = (\boldsymbol{m}_0, \boldsymbol{0})$*, where* $\boldsymbol{0}$ *denotes the valuation with null values for all transitions enabled in* $\boldsymbol{m}_0$ *and* $\perp$ *otherwise;*
- $E \subseteq S \times (T \cup \mathbb{R}_+) \times S$ *contains the two following types of transitions, from an admissible configuration* $(\boldsymbol{m}, \nu)$*:*
  - *For each transition t enabled in $\boldsymbol{m}$ such that $\nu(t) \in I(t)$, a discrete transition $(\boldsymbol{m}, \nu) \xrightarrow{t} (\boldsymbol{m} - \boldsymbol{Pre}[\cdot, t] + \boldsymbol{Post}[\cdot, t]), \nu')$ such that for all $t' \in \mathscr{A}(\boldsymbol{m} - \boldsymbol{Pre}[\cdot, t] + \boldsymbol{Post}[\cdot, t])$,*
  $$\nu'(t') = \begin{cases} 0 & \text{if } \uparrow enabled(t', \boldsymbol{m}, t), \\ \nu(t') & \text{otherwise.} \end{cases}$$
  - *For each $d \in \mathbb{R}_+$, such that for each t in $\mathscr{A}(M), \nu(t) + d \in I(t)^{\downarrow}$, a delay transition $(\boldsymbol{m}, \nu) \xrightarrow{d} (\boldsymbol{m}, \nu + d)$.*

For instance, a possible run of the net in Fig. 16.12 is the following:

$$(\boldsymbol{m}_0, [0,0,\perp]) \xrightarrow{1} (\boldsymbol{m}_0, [1,1,\perp]) \xrightarrow{t_1} (\boldsymbol{m}_1, [1,1,0]) \xrightarrow{t_1} (\boldsymbol{m}_2, [\perp,1,0])$$
$$\xrightarrow{t_2} (\boldsymbol{m}_3, [\perp,\perp,0]) \xrightarrow{1.5} (\boldsymbol{m}_3, [\perp,\perp,1.5]) \xrightarrow{t_3} (\boldsymbol{m}_4, [\perp,\perp,1.5]) \cdots$$

with markings $\boldsymbol{m}_0 = [2\ 1\ 0]^T$, $\boldsymbol{m}_1 = [1\ 1\ 1]^T$, $\boldsymbol{m}_2 = [0\ 1\ 2]^T$, $\boldsymbol{m}_3 = [0\ 0\ 2]^T$, and $\boldsymbol{m}_4 = [0\ 0\ 1]^T$.

This definition corresponds to what is called *strong* semantics, and implies "*urgency*" for transition firing, because time delays cannot disable transitions. This can be used to enforce priorities between transitions, which can lead to time locks, i.e., deadlocks due to conflicting timing constraints. In contrast, in the definition of *weak* semantics [22], time elapsing is permitted beyond the upper bound of the interval of a transition, by removing the condition: for each t in $\mathscr{A}(M), \nu(t)+d \in I(t)^{\downarrow}$ for a transition with delay d. When this occurs, the transition is disabled, in a mechanism similar to what happens in ordinary P/T nets.

The class of time Petri nets with strong intermediate semantics has been largely studied (see for instance [4, 18]), and tools like ROMÉO [15] and TINA [5] have been developed for the analysis of bounded nets in this class.

16.7 Further Reading

This chapter is based on the Italian texbook on discrete event systems by Di Febbraro and Giua [13].

Many references are already cited along the chapter. Among them, particular attention should be devoted to [1, 23] dealing with Timed nets, and to [4, 6] devoted to Time nets. Interesting comparisons among different semantics for Time Petri nets are given in [3, 22].

References

1. Ajmone Marsan, M.: Stochastic Petri Nets: an Elementary Introduction. In: Rozenberg, G. (ed.) APN 1989. LNCS, vol. 424. Springer, Heidelberg (1990)
2. Baccelli, F., Cohen, G., Olsder, G.J., Quadrat, J.P.: Synchronization and linearity. Wiley (1992)
3. Bérard, B., Cassez, F., Haddad, S., Lime, D., Roux, O.H.: Comparison of Different Semantics for Time Petri Nets. In: Peled, D.A., Tsay, Y.-K. (eds.) ATVA 2005. LNCS, vol. 3707, pp. 293–307. Springer, Heidelberg (2005)
4. Berthomieu, B., Diaz, M.: Modeling and verification of time dependent systems using time Petri nets. IEEE Transactions on Software Engineering 17(3), 259–273 (1991)
5. Berthomieu, B., Vernadat, F.: Time Petri nets analysis with TINA. In: Proc. 3rd Int. Conf. on the Quantitative Evaluation of Systems, Riverside, California, USA (2006)
6. Boyer, M., Roux, O.H.: On the compared expressiveness of arc, place and transition time Petri nets. Fundamenta Informaticae 88(3), 225–249 (2008)
7. Campos, J., Chiola, G., Colom, J.M., Silva, M.: Properties and performance bounds for timed marked graphs. IEEE Transactions on Circuits and Systems I: Fundamental Theory and Applications 39(5), 386–401 (1992)
8. Campos, S., Clarke, E.M.: Theories and experiences for real-time system development. AMAST Series in Computing, vol. 2, pp. 129–145. World Scientific (1995)
9. Cassandras, C.G., Lafortune, S.: Introduction to Discrete Event Systems, 2nd edn. Springer (2008)
10. Chretienne, P.: Les reseaux de Petri Temporises. Master Thesis. Univ. of Paris VI, Paris (1983)
11. Coolahan, J., Roussopoulos, N.: Timing requirements for time-driven systems using augmented Petri nets. IEEE Transactions on Software Engineering 9(5), 603–616 (1983)

12. Desrochers, A.A., Al-Jaar, R.Y.: Applications of Petri nets in manufacturing systems. IEEE Press (1995)
13. Di Febbraro, A., Giua, A.: Sistemi ad eventi discreti. McGraw Hill (2002) (in Italian)
14. Di Cesare, F., Harhalakis, G., Proth, J.M., Silva, M., Vernadat, F.B.: Practice of Petri Nets in Manufacturing. Chapman and Hall (1993)
15. Gardey, G., Lime, D., Magnin, M., Roux, O.H.: Romeo: A Tool for Analyzing Time Petri Nets. In: Etessami, K., Rajamani, S.K. (eds.) CAV 2005. LNCS, vol. 3576, pp. 418–423. Springer, Heidelberg (2005)
16. Henzinger, T.A., Manna, Z., Pnueli, A.: What Good are Digital Clocks? In: Kuich, W. (ed.) ICALP 1992. LNCS, vol. 623, pp. 545–558. Springer, Heidelberg (1992)
17. Henzinger, T.A., Manna, Z., Pnueli, A.: Temporal proof methodologies for timed transition systems. Information and Computation 112(2), 273–337 (1994)
18. Jones, N.D., Landweber, L.H., Lien, Y.E.: Complexity of some problems in Petri nets. Theoretical Computer Scienc 4, 277–299 (1977)
19. Laroussinie, F., Markey, N., Schnoebelen, P.: On Model Checking Durational Kripke Structures. In: Nielsen, M., Engberg, U. (eds.) FOSSACS 2002. LNCS, vol. 2303, pp. 264–279. Springer, Heidelberg (2002)
20. Merlin, P.M.: A Study of the Recoverability of Computing Systems. UCI, Univ. California, Irvine (1974)
21. Ramchandani, C.: Analysis of asynchronous concurrent systems by Petri nets. Project MAC, TR-120. M.I.T, Cambridge (1974)
22. Reynier, P.-A., Sangnier, A.: Weak Time Petri Nets Strike Back! In: Bravetti, M., Zavattaro, G. (eds.) CONCUR 2009. LNCS, vol. 5710, pp. 557–571. Springer, Heidelberg (2009)
23. Wang, J.: Timed Petri Nets: Theory and Applications. CRC Press (1998)
24. Zhu, J.J., Denton, R.T.: Timed Petri nets and their application to communication protocol specification. In: Proc. Military Communications Conference, San Diego, CA (1988)

Chapter 17
The On-Line Diagnosis of Time Petri Nets

René K. Boel and George Jiroveanu

17.1 Introduction

The Petri Net models that we analyse in the following consider the time as a quantifiable and continuous parameter whereas in an untimed Petri Net model the time is taken into account only via the partial order relation between the transitions that are executed in the plant.

As presented in Chapter 16 there are two main time extensions of Petri Nets (PNs) namely Time Petri Nets [21] and Timed Petri Nets [22]. The difference between the two is that in Time Petri Nets (TPNs) a transition can be fired after its enabling with a delay that belongs to a given time-interval and the execution takes no time to complete, while in Timed Petri Nets a transition fires as soon as possible (without delay) but its execution requires a certain amount of time to complete. Among the two we have chosen Time Petri Nets for modelling our plant since this formalism is convenient for expressing most of the temporal constraints regarding the execution and the duration of the events.

For the diagnosis problem we adopt the setting proposed in [13] and [23]. Thus, we consider the plant observation given via a subset of transitions (observable transitions) and the faults that should be detected are modelled by a subset of the unobservable (silent) transitions. We assume that the model is correct and there are no delays in receiving the plant observation. Moreover, the execution time of an observed transition is measured with perfect accuracy w.r.t. a global clock. In this

René K. Boel
EESA - SYSTeMS Research Group, Ghent University, Technologiepark 914, Zwijnaarde 9052, Belgium
e-mail: rene.boel@ugent.be

George Jiroveanu
Transelectrica SA - Romanian National Power Grid Company, Brestei 5, 200581 Craiova, Romania
e-mail: george.jiroveanu@transelectrica.ro

C. Seatzu et al. (Eds.): Control of Discrete-Event Systems, LNCIS 433, pp. 343–364.
springerlink.com

setting, the goal of this chapter is to design an on-line algorithm that derives the plant diagnosis at time ξ based on the known plant model and on the observation received up to time ξ.

As for untimed PNs all the interesting problems for the analysis of TPN models can be reduced to reachability analysis (e.g. for the diagnosis problem one must calculate all the feasible time-traces that obey the received observation to check whether or not they contain fault transitions). In a TPN, a transition t can be executed in a given marking at a certain time if t is enabled as in an untimed PN and additionally some timing constrains are satisfied. There are timing constrains that relate the delay between the time when t has become enabled and the time when t is allowed to be executed, as well as timing constrains regarding the execution time of t before some other transitions that are also enabled in that marking are forced to be executed by reaching the maximum delay allowed by the model for their execution. In general to decide if a time-trace τ^θ is legal (valid) or not it is necessary to check if its untimed support trace τ is legal in the untimed PN support of the TPN model and additionally to check if the times when the transitions in τ are assumed to be executed provide a solution for a system of linear inequalities (called the characteristic system of τ). To calculate all the legal time-traces in the TPN model requires to derive for each legal trace τ of the untimed PN model the entire set of solutions of its characteristic system. Since a transition in a TPN can fire at any time in its firing domain, TPN models have in general infinite state spaces because a state may have an infinite number of successor states. Methods based on grouping states that are equivalent under a certain equivalence relation into so-called state classes were proposed in [4, 5, 12, 29, 30] where it was shown in [4] that for bounded TPNs the state class graph is finite and it compactly represents the set of all sequences of transitions that can be executed. Thus, the potentially infinite state spaces of TPNs can be finitely represented and the analysis can be reduced to a decidable problem.

The on-line algorithms that we design can be briefly described as follows. First, we construct the state class graph up to the time ξ. Then for each observable transition considered in the state class graph we derive the earlier time and the latest time when it can be executed. If an observable transition is either executed sooner than allowed or it is not observed prior to the latest time it could have been executed, then the arc labelled by that transition is deleted. The arc is also deleted if some other observable transition has been executed in the plant. Otherwise an equality relation is added to the characteristic system to express the fact that the observable transition occurred at the time given by the received observation. Hence, there could be cases when one can infer that a fault has happened for sure in the plant even though no observation is received from the plant, since traces can be deleted from the state class graph when the latest execution time of an observable transition has elapsed.

It is known that the analysis of PNs is a computationally hard problem because of the state space explosion due to the interleaving of concurrent transitions (e.g. the time computational complexity to construct the reachability graph is exponential in the number of places of the PN model [9]). The same problem remains also for TPN models where additionally the computation requires to solve for each untimed trace a system of linear inequalities (that can be solved by a standard algorithm in polynomial time). To deal with this difficulty methods based on partial orders were proposed for the analysis of untimed PNs [3, 11, 20], as well as for the analysis of TPNs [1, 8, 19, 24].

As for the analysis based on state classes, the set of all legal time-traces can be obtained using partial orders by computing for each configuration $\boldsymbol{c}$ of the unfolding of the untimed PN support of the TPN model the entire solution set of a $(max,+)$-system of linear inequalities called the characteristic system of configuration $\boldsymbol{c}$. The characteristic system of a configuration $\boldsymbol{c}$ comprises $(max,+)$ inequalities relating the execution times of the events within $\boldsymbol{c}$ and $(max,+)$ inequalities that assure that a conflicting event (an event that is not considered in $\boldsymbol{c}$ but that has its pre-set of conditions in the set of conditions of $\boldsymbol{c}$) was not forced to be executed. One way to derive the entire set of solutions of the characteristic system of a configuration $\boldsymbol{c}$ is by deciding each *max*-term (i.e. by choosing nondeterministically which is the maximum element of the *max*-term) and then solving systems of linear inequalities [1]. However, by deciding each *max*-term basically the concurrent events whose occurrence times are included in a *max*-term are interleaved, which is exactly what we tried to avoid by using partial orders. Alternatively it has been shown in [17] that the system of $(max,+)$ inequalities can be put in the form of an Extended Linear Complementarity Problem (ELCP) and then the solution set of the characteristic system of a configuration is provided more efficiently as a union of faces of a polyhedron that satisfy a cross-complementarity condition [10].

The chapter is organized as follows. In Section 17.2 we introduce Time Petri Nets and then in Section 17.3 we present the setting and the diagnosis problem. The analysis of TPN models based on construction of the state class graph and based on time processes is presented in Section 17.4. In Section 17.5 we discuss issues regarding the on-line implementation and then conclude the chapter in Section 17.6 with some suggestions for further readings.

17.2 Time Petri Nets

A Time Petri Net structure $N_T = (N, \mathscr{I})$ consists of an (untimed) Petri Net $N = (P, T, \boldsymbol{Pre}, \boldsymbol{Post})$ called the untimed support of N_T and the static firing time interval function $I : T \to \mathscr{I}(\mathbb{Q}^+)$, where $EFT_t, LFT_t \in \mathbb{Q}^+$ are the earliest static firing time respectively the latest static firing time and $I(t) = [EFT_t, LFT_t]$ represents all possible time delays associated to transition $t \in T$. A TPN system consists of a pair $\langle N_T, \boldsymbol{m}_0^\theta \rangle$, where N_T is a TPN structure and $\boldsymbol{m}_0^\theta$ is its initial marking.

We make the assumption that there is a start-up transition that fires only once at the time zero producing the tokens considered by the initial marking. If this is not suitable for the model under investigation then consider for each token in the initial marking a fictitious transition that can fire only once and its static interval represents the date of birth of the token in the initial marking [8]. The time the analysis starts is translated then to the time when all the fictitious transitions become enabled (in each input place of the fictitious transitions simultaneously one token is produced).

The following definitions are borrowed from [4] with the difference that we consider as in [29] that the time is counted according to a global clock relative to the time the analysis starts (assumed as 0 for simplicity). This is motivated by our purpose to perform on-line diagnosis where we assume that the plant observation includes also the time stamp when a transition is executed.

In a TPN $\langle N_T, \boldsymbol{m}_0^\theta \rangle$ we say that a transition t becomes enabled at the time θ_t^{en} (according to a global clock) then the clock attached to t is started and transition t can and must fire at some time $\theta_t \in [\theta_t^{en} + EFT_t, \theta_t^{en} + LFT_t]$, provided t did not become disabled because of the firing of another transition. Notice that t is forced to fire if it is still enabled at the time $\theta_t^{en} + LFT_t$.

Definition 17.1. *A state at the time θ (according to a global clock) of a TPN $\langle N_T, \boldsymbol{m}_0^\theta \rangle$ is a pair $S = (\boldsymbol{m}, FI)$, where $\boldsymbol{m}$ is a marking and $FI : T \to \mathscr{I}(\mathbb{Q}^+)$ is a firing interval function that associates an interval $FI(t) = [l_t, u_t]$ to each enabled transition t in $\boldsymbol{m}$.*

We write $(\boldsymbol{m}, FI) \xrightarrow{\langle t, \theta_t \rangle} (\boldsymbol{m}', FI')$ or simply $S \xrightarrow{\langle t, \theta_t \rangle} S'$ if $\theta_t \in \mathbb{Q}^+$ and:

1. $(\boldsymbol{m} \geq \boldsymbol{Pre}[\cdot, t] \wedge \theta_t \geq \theta_t^{en} + EFT_t) \wedge (\forall t' \in T, \boldsymbol{m} \geq \boldsymbol{Pre}[\cdot, t'] \Rightarrow \theta_t \leq \theta_{t'}^{en} + LFT_{t'})$
2. $\boldsymbol{m}' = \boldsymbol{m} - \boldsymbol{Pre}[\cdot, t] + \boldsymbol{Post}[\cdot, t]$
3. $\forall t'' \in T$ *s.t.* $\boldsymbol{m}' \geq \boldsymbol{Pre}[\cdot, t'']$ *we have:*

   a. *if* $t'' \neq t$ *and* $\boldsymbol{m} \geq \boldsymbol{Pre}[\cdot, t'']$ *then* $FI(t'') = [\max(\theta_{t''}^{en} + EFT_{t''}, \theta_t), \theta_{t''}^{en} + LFT_{t''}]$
   b. *else* $\theta_{t''}^{en} = \theta_t$ *and* $FI(t'') = [\theta_{t''}^{en} + EFT_{t''}, \theta_{t''}^{en} + LFT_{t''}]$.

Condition (1) above assures that a transition t that fires at the time θ_t is enabled by the marking $\boldsymbol{m} \geq \boldsymbol{Pre}[\cdot, t]$ and that it fires in its temporal interval $FI(t)$ unless it is disabled by another enabled transition that is executed sooner. Condition (2) represents the marking transformation rule while (3) gives the rule how the firing intervals are modified according to:

(3.a) if a transition t'' remains enabled after firing t then its lower limit remains unaffected if $\theta_t \leq \theta_{t''}^{en} + EFT_{t''}$ while if $\theta_t \geq \theta_{t''}^{en} + EFT_{t''}$ then the earliest firing time becomes θ_t

(3.b) for a newly enabled transition t'' its firing interval $FI(t'')$ is obtained by adding the time $\theta_{t''}^{en}$ when transition t'' has become enabled ($\theta_t = \theta_{t''}^{en}$) to its static firing interval $I(t'') = [EFT_{t''}, LFT_{t''}]$.

Definition 17.2. *A time-trace τ^θ in $\langle N_T, \boldsymbol{m}_0^\theta \rangle$ is defined as follows:*

i) $\tau^\theta = S_0 \xrightarrow{\langle t_1, \theta_{t_1} \rangle} S_1 \ldots S_{n-1} \xrightarrow{\langle t_n, \theta_{t_n} \rangle} S_n$
ii) $\exists \theta_{t_{i+1}}$ *s.t.* $S_i \xrightarrow{\langle t_{i+1}, \theta_{t_{i+1}} \rangle} S_{i+1}$ *according to Definition 17.1 for* $i = 0, \ldots, n-1$
iii) $\tau = \boldsymbol{m}_0 \xrightarrow{t_1} \boldsymbol{m}_1 \ldots \boldsymbol{m}_{n-1} \xrightarrow{t_n} \boldsymbol{m}_n$ *is the untimed support of* τ^θ.

Definition 17.3. *Denote the reflexive and transitive closure of $\to$ as $\xrightarrow{*}$. The state graph of $\langle N_T, \boldsymbol{m}_0^\theta \rangle$ is $SG = (\mathscr{S}, \xrightarrow{*}, S_0)$, where:*

i) $\mathscr{S} = \left\{S \mid S_0 \xrightarrow{*} S\right\}$ *is the set of reachable states from the initial state* S_0

ii) $S_0 = (\boldsymbol{m}_0, FI_0)$, *where* $FI_0(t) = I(t)$ *for all transitions t s.t.* $\boldsymbol{m}_0 \geq \boldsymbol{Pre}[\cdot\ t]$ *otherwise* $FI_0(t)$ *is not defined.*

In the following we use also the notation $\boldsymbol{m}_0^\theta$ for the initial state S_0 . Let $L_{N_T}^\theta(\boldsymbol{m}_0^\theta)$ be the set of all possible time-traces that can be executed in $\langle N_T, \boldsymbol{m}_0^\theta\rangle$. Then denote by $L_{N_T}(\boldsymbol{m}_0^\theta) = \left\{\tau \mid \exists \tau^\theta \in L_{N_T}^\theta(\boldsymbol{m}_0^\theta)\right\}$ the untimed support language of $L_{N_T}^\theta(\boldsymbol{m}_0^\theta)$.

For a TPN $\langle N_T, \boldsymbol{m}_0^\theta\rangle$ we have that $L_{N_T}(\boldsymbol{m}_0^\theta) \subseteq L_N(\boldsymbol{m}_0)$ i.e. the untimed support language of a TPN is included in the language of its untimed support PN. In other words the timing information eliminates some untimed traces because some transitions that are concurrent in the untimed model are forced to be executed in a certain order in the timed model. The timing constraints of the time model may also eliminate some choices that would be possible in the untimed model but are impossible in the timed model.

Given a global time $\xi \in \mathbb{Q}^+$ denoted by $L_{N_T}^\theta(\boldsymbol{m}_0^\theta, \xi)$ the set of all time-traces that can be executed up to the time ξ, that is $\tau_\xi^\theta \in L_{N_T}^\theta(\boldsymbol{m}_0^\theta, \xi)$ iff:

i) $\tau_\xi^\theta \in L_{N_T}^\theta(\boldsymbol{m}_0^\theta)$ and $S_0 \xrightarrow{\tau_\xi^\theta} S_\xi$

ii) the last transition in τ_ξ^θ is executed at a time smaller than ξ or equal

iii) any transition enabled in S_ξ can be executed only at times larger than ξ.

Then the set of untimed support traces that can be executed up to the time ξ is:

$$L_{N_T}(\boldsymbol{m}_0^\theta, \xi) = \left\{\tau_\xi \mid \tau_\xi^\theta \in L_{N_T}^\theta(\boldsymbol{m}_0^\theta, \xi)\right\}.$$

17.3 The Diagnosis of TPNs – The Setting and the Problem Description

We assume throughout this chapter that the following assumptions on the structure and the dynamics of the plant are satisfied:

i) the TPN model $\langle N_T, \boldsymbol{m}_0^\theta\rangle$ is untimed 1-safe, i.e. the untimed PN support $\langle N, \boldsymbol{m}_0\rangle$ is 1-safe

ii) $T = T_o \cup T_{uo}$ and $T_o \cap T_{uo} = \emptyset$, where T_o is the set of observable transitions and T_{uo} is the set of unobservable (silent) transitions

iii) $\ell : T \to \mathbb{A} \cup \{\varepsilon\}$ is the observation labeling function, where $\mathbb{A}$ is a set of labels and ε is the empty label; $\ell(t) = \varepsilon$ iff $t \in T_{uo}$ and $\ell(t) \in \mathbb{A}$ iff $t \in T_o$. ℓ is not necessarily injective in $\mathbb{A}$, that is $\exists (t_1, t_2) \in T_o \times T_o$ s.t. $t_1 \neq t_2$ and $\ell(t_1) = \ell(t_2)$

iv) each time when an observable transition $t^o \in T_o$ is executed in the plant the label $\ell(t^o)$ is emitted together with the global time $\theta_{\ell(t^o)} \in \mathbb{Q}^+$ when this execution of t^o took place

v) the observation is always correct, it is always received (there is no loss of observation) and there are no delays in receiving the observation *vi*) the execution of an unobservable transition does not emit anything (is silent)

vii) $T_f \subseteq T_{uo}$ is the set of faulty transitions; T_f is partitioned as $T_f = T_{\mathtt{F}_1} \cup \ldots \cup T_{\mathtt{F}_k}$, where $T_{\mathtt{F}_i}$ is the subset of fault transitions that models a fault of class $\mathtt{F}_i$ for $i = 1, \ldots, k$

viii) the time diverges when an infinite number of transitions are executed (e.g. any cycle that can be executed infinitely often contains at least one transition that has a nonzero lower limit of its static firing time interval).

Given the plant model as described above the goal of this chapter is to design an on-line algorithm that derives at time ξ the diagnosis of the plant based on the plant model and on the received observation up to time ξ. The exact meaning of diagnosis is defined as follows.

Denote the observation received up to time ξ by $w^\theta_\xi = \langle obs_1, \theta_{obs_1}\rangle \ldots \langle obs_n, \theta_{obs_n}\rangle$, where $obs_1, \ldots, obs_n \in \mathbb{A}$ are the labels that are emitted and $\theta_{obs_1} \leq \theta_{obs_2} \ldots \leq \theta_{obs_n} \leq \xi$ are the global times at which the observable transitions occurred.

We say that a time trace $\tau^\theta_\xi \in L^\theta_{N_T}(\boldsymbol{m}^\theta_0, \xi)$ obeys the observation w^θ_ξ if $\ell(\tau_\xi) = w_\xi$ and the kth observable transition in τ^θ_ξ is assumed to be executed at the time $\theta_{t_k} = \theta_{obs_k}$ for $k = 1, \ldots, n$. Then let $L^\theta_{N_T}(\boldsymbol{m}^\theta_0, \xi, w^\theta_\xi)$ denotes the set of all time-traces that are feasible in $\langle N_T, \boldsymbol{m}^\theta_0\rangle$ up to the time ξ and that obey the received observation w^θ_ξ.

Denote by $\mathscr{D}(w^\theta_\xi)$ the set of untimed strings that are obtained by projecting onto the set of fault transitions T_f the untimed supports of the time-traces contained in $L^\theta_{N_T}(\boldsymbol{m}^\theta_0, \xi, w^\theta_\xi)$:

$$\mathscr{D}(w^\theta_\xi) = \left\{ \tau_f \mid \tau_f = \Pi_{T_f}(\tau) \text{ and } \tau^\theta \in L^\theta_{N_T}(\boldsymbol{m}^\theta_0, \xi, w^\theta_\xi) \right\} \tag{17.1}$$

where Π is the standard projection function that erases from a string τ all its elements that do not belong to a given set (in our case T_f).

If we have the set of fault transitions partitioned into different fault classes $T_f = T_{\mathtt{F}_1} \cup \ldots \cup T_{\mathtt{F}_k}$ then the detection of a fault of class $\mathtt{F}_i$ is obtained by further projecting the strings of $\mathscr{D}(w^\theta_\xi)$ onto $T_{\mathtt{F}_i}$:

$$\mathscr{D}_{\mathtt{F}_i}(w^\theta_\xi) = \left\{ \tau_{\mathtt{F}_i} \mid \tau_{\mathtt{F}_i} = \Pi_{T_{\mathtt{F}_i}}(\tau_f) \text{ and } \tau_f \in \mathscr{D}(w^\theta_\xi) \right\} \tag{17.2}$$

Definition 17.4. *For a given fault class, say* $\mathtt{F}_i$*, we have that the diagnosis result of the plant w.r.t. fault class* $\mathtt{F}_i$ *at time* ξ *with received observation* w^θ_ξ *is:*

$$\Delta_{\mathtt{F}_i}(w^\theta_\xi) = \begin{cases} \mathtt{F}_{\mathtt{F}_i} & \textit{iff } \varepsilon \notin \mathscr{D}_{\mathtt{F}_i}(w^\theta_\xi) \\ \mathtt{N}_{\mathtt{F}_i} & \textit{iff } \{\varepsilon\} = \mathscr{D}_{\mathtt{F}_i}(w^\theta_\xi) \\ \mathtt{UF}_{\mathtt{F}_i} & \textit{iff } \{\varepsilon\} \subsetneq \mathscr{D}_{\mathtt{F}_i}(w^\theta_\xi) \end{cases} \tag{17.3}$$

where as in [23]: $\mathtt{F}_{\mathtt{F_i}}$ *means that a fault of kind* $\mathtt{F_i}$ *did necessarily happen in the plant* ($\forall \tau_f \in \mathscr{D}_{\mathtt{F_i}}(w_\xi^\theta) \Rightarrow \Pi_{T_{\mathtt{F_i}}}(\tau_f) \neq \varepsilon$)*;* $\mathtt{N}_{\mathtt{F_i}}$ *means that a fault of kind* $\mathtt{F_i}$ *did not happen in the plant* ($\forall \tau_f \in \mathscr{D}_{\mathtt{F_i}}(w_\xi^\theta) \Rightarrow \Pi_{T_{\mathtt{F_i}}}(\tau_f) = \varepsilon$)*; and* $\mathtt{UF}_{\mathtt{F_i}}$ *means that it is uncertain whether a fault of kind* $\mathtt{F_i}$ *happened or not in the plant (there exist two legal time-strings* $\tau_f, \tau_f' \in \mathscr{D}_{\mathtt{F_i}}(w_\xi^\theta)$ *s.t.* $\Pi_{T_{\mathtt{F_i}}}(\tau_f) \neq \varepsilon$ *and* $\Pi_{T_{\mathtt{F_i}}}(\tau_f') = \varepsilon$*).*

17.4 The Analysis of TPNs

In this section we consider first the analysis of TPN models based on state class graph construction [4, 5, 29, 30] while the last part is devoted to the use of partial order methods for the analysis of TPNs [1, 8, 19, 24].

17.4.1 Analysis of TPNs Based on State Classes

Consider a time-trace $\tau^\theta = S_0 \xrightarrow{\langle t_1, \theta_{t_1}\rangle} S_1 \dots S_{n-1} \xrightarrow{\langle t_n, \theta_{t_n}\rangle} S_n$. The times at which $\{t_1, t_2, \dots, t_n\}$ are executed $\{\theta_{t_1}, \theta_{t_2}, \dots, \theta_{t_n}\}$ are called the path variables and are denoted by the vector Θ. Given a state $S = (\boldsymbol{m}, FI)$, let $\vartheta_t \in FI(t)$ denote the possible firing time of an enabled transition $t \in T$ in S. Then the vector $\boldsymbol{v}$ represents the firing times of the transitions enabled in S.

For a time-trace τ^θ the path variables Θ and the state variables $\boldsymbol{v}$ are related by a system of inequalities (called the characteristic system of τ) denoted K_τ having the shape:

$$\begin{cases} \boldsymbol{A} \cdot \Theta \leq \boldsymbol{a} \\ \Theta^{en} = \boldsymbol{B} \cdot \Theta \\ \Theta^{en} + \boldsymbol{l} \leq \boldsymbol{v} \leq \Theta^{en} + \boldsymbol{u} \end{cases} \tag{17.4}$$

where: $\boldsymbol{A}$ is an $n \times n$ matrix (with n the number of transitions in τ^θ) that relates the execution times of the transitions in τ; $\boldsymbol{a}$ is an n-vector of constant rational numbers. $\boldsymbol{B}$ is a $k \times n$ matrix (with k the number of enabled transitions in S_n) that determines the enabling times of the enabled transitions in S_n; Θ^{en} is a vector of dimension k having the component i equal to the global time $\theta_{t_i}^{en}$ when t_i has become enabled; $\boldsymbol{l}, \boldsymbol{u}$ are vectors of dimension k with the components specifying the lower limit (EFT_t) respectively the upper limit (LFT_t) of the static firing intervals of the transitions enabled in S_n.

Notice that by the assumption that $\langle N, \boldsymbol{m}_0\rangle$ is 1-safe we have that the parents of a transition t_i (the transitions in ${}^{\bullet\bullet}t_i$ whose occurrence made t_i enabled) are uniquely defined for any trace τ^θ. We have that $\theta_{t_i}^{en} = \max_{t_j \in {}^{\bullet\bullet}t_i}(\theta_{t_j})$ if the input places of t_i

are not marked by tokens from the initial marking. If a transition t_i is enabled by the tokens from the initial marking then $\theta_{t_i}^{en} = 0$ since the marking is assumed produced by a fictitious transition that fires at the time 0.

In order to have a finite representation of the set of reachable states [4] proposed the analysis of the TPN models based on state classes (called also linear state classes to distinguish them from some other state classes that are proposed in the literature [5, 30]).

However, these methods do not consider a global clock (absolute time) but only local clocks that measure the relative time between the enabling of a transition and its execution. This could be simply understood by considering for each transition t the new variable $\vartheta_t^r = \vartheta_t - \theta_t^{en}$ (i.e. the state variables w.r.t the relative time). Let $\boldsymbol{v}^r$ and $FI^r(t)$ be the vector of state variables respectively the firing time interval of an enabled transition defined w.r.t to the relative time. The states w.r.t. the relative time are then defined as pairs $S^r = (\boldsymbol{m}, FI^r)$ and are represented by the state graph SG^r.

Linear state classes are constructed based on the following fact: consider two time-traces τ_i^θ and τ_j^θ that lead from the initial state S_0^r into S_i^r respectively S_j^r such that: *i)* $\boldsymbol{m}_i = \boldsymbol{m}_j$ and *ii)* $FI_i^r(t) = FI_j^r(t)$ for each transition t enabled in $\boldsymbol{m}_i$. Then the subgraphs of the state graph SG^r rooted in S_i^r and S_j^r are isomorphic. The states are grouped in state classes such that all the states contained in a state class have the same marking while the firing domain of a state class is the union of the firing domains of each state included in the state class. It is proven in [4] that for a bounded TPN the linear state class graph $LSCG^r$ is finite and it explicitly represents the untimed support language of $\langle N_T, \boldsymbol{m}_0^\theta \rangle$.

If we consider the state variables w.r.t a global clock we have also that the subgraphs of the state graph SG rooted in $S_i = (\boldsymbol{m}_i, FI_i)$ and $S_j = (\boldsymbol{m}_j, FI_j)$ are isomorphic if *i)* $\boldsymbol{m}_i = \boldsymbol{m}_j$ and *ii)* $FI_i(t) = FI_j(t)$ for each transition t enabled in $\boldsymbol{m}_i$. Obviously, in this case the linear state class graph $LSCG$ will not have in general a finite number of state classes because the state variables ϑ_t are defined w.r.t. the global clock and the global clock progress is infinite. Consequently, $LSCG$ is a directed acyclic graph (obtained by unrolling $LSCG^r$).

We give the following motivation for choosing to define the states in the TPN model w.r.t. a global clock. In the first place, our aim is different than the verification of some properties of the behaviour of the time models as in [5, 30]. We perform on-line diagnosis so that we are not interested in a compact representation of the full plant behaviour (i.e. to construct $LSCG^r$) but only to make some calculations in advance (depending on the computational capabilities and the diagnosis requirements). Second, for most of the current man-made systems the plant observation includes also the time stamp according to a global clock when an observable transition is executed. Hence, it would be necessary anyway to have some path variables that give the global time along a trace to be able to calculate the earliest time and the latest time when an observable transition can fire. Moreover, the plant observation is taken into account by adding extra timing constraints to the characteristic system K_τ of each trace τ and adding these constraints implies that $LSCG^r$ should be refined.

Definition 17.5. *The linear state class graph of a TPN $\langle N_T, \boldsymbol{m}_0^\theta \rangle$ is:*

$$LSCG = (\mathscr{K}_{/\cong}, \xrightarrow{*}, [\{S_0\}]_{\cong})$$

where $\mathscr{K} = \bigcup_{\tau \in L_N(\boldsymbol{m}_0)} SC_\tau$ is a cover of $\mathscr{S}$ (a set of subsets of $\mathscr{S}$ whose union includes $\mathscr{S}$) inductively defined as:

1. $SC_\varepsilon = \{S_0\}$ *and* $SC_{\tau t} = \left\{ S' \mid \exists S \in SC_\tau \text{ s.t. } S \xrightarrow{\langle t, \theta_t \rangle} S' \right\}$
2. $SC \cong SC'$ *iff* $(\exists S \in SC)(\exists S' \in SC')$ *s.t.* $\boldsymbol{m}(S) = \boldsymbol{m}(S')$ *and*

$$\bigcup_{S \in SC} FI(S) = \bigcup_{S' \in SC'} FI(S')$$

3. $SC \xrightarrow{t} SC''$ *iff* $\exists S \in SC$ *and* $\exists S'' \in SC''$ *s.t.* $S \xrightarrow{\langle t, \theta_t \rangle} S''$.

Denote by $LSCG_\xi$ the initial part of $LSCG$ developed up to the time ξ, that is $LSCG_\xi$ contains all the state classes SC_i of $LSCG$ that have at least one enabled transition t such that the lower bound of its firing time interval l_t is smaller than ξ. Each state class SC_i that has no successors in $LSCG_\xi$ either has no successors in $LSCG$ or all the enabled transitions in SC_i can be executed only at times larger than the time ξ. Let $L(LSCG_\xi)$ be the language generated by $LSCG_\xi$. We have then the following results.

Proposition 17.1. *If ξ is finite and $\langle N_T, \boldsymbol{m}_0^\theta \rangle$ is bounded then $LSCG_\xi$ has a finite number of states.*

Proof. By assumption we have that the time diverges when an infinite number of transitions are executed and that $\langle N_T, \boldsymbol{m}_0^\theta \rangle$ is bounded. Thus, in a finite time ξ only a finite number of untimed traces can be executed. For each untimed trace τ there exists only one state class in $LSCG$ such that $SC_0 \xrightarrow{\tau} SC$. Hence, $LSCG_\xi$ has a finite number of state classes. □

Theorem 17.1. *Given a TPN $\langle N_T, \boldsymbol{m}_0^\theta \rangle$ and a finite time ξ we have that $L(LSCG_\xi) = L_{N_T}(\boldsymbol{m}_0^\theta, \xi)$.*

Proof. The proof is straightforward from Theorem 3 in [4], Definition 17.5 and the way $LSCG_\xi$ is defined. □

Based on Proposition 17.1 and Theorem 17.1 we have that the plant diagnosis is computable and provides the correct diagnosis result. K_τ is constructed in a recursive manner for $\tau_n = t_1 \dots t_n$ by running the following algorithm.

Algorithm (Computing characteristic systems)

1 $K_\varepsilon = \{EFT_t \leq \vartheta_t \leq LFT_t \mid t \in T \wedge \boldsymbol{m}_0 \geq \boldsymbol{Pre}[\cdot,t]\}$

2 Assume $SC_0 \xrightarrow{\tau_i} SC_i$ *(where* $\tau_i = t_1 \ldots t_i$*). Then for* $i = 0,\ldots,n-1$ *we have that* t_{i+1} *is fireable from* SC_i *iff:*

2.1 $\boldsymbol{m}_i \geq \boldsymbol{Pre}[\cdot,t_{i+1}]$

2.2 $K_{\tau_i} \wedge \{\vartheta_{t_{i+1}} \leq \vartheta_{t_j} \mid t_{i+1} \neq t_j \wedge \boldsymbol{m}_i \geq \boldsymbol{Pre}[\cdot,t_j]\}$ *is consistent (the solution set is not empty)*

3 if t_{i+1} *is fireable then* $K_{\tau_{i+1}}$ *is computable from* K_{τ_i} *as:*

3.1 the fireability constraints for t_{i+1} *given by* 2.2 *above are added to the characteristic system* K_{τ_i}

3.2 the state variable $\vartheta_{t_{i+1}}$ *is renamed into a path variable* $\theta_{t_{i+1}} = \vartheta_{t_{i+1}}$

3.3 for each newly enabled transition t_k *at* $\boldsymbol{m}_{i+1} = \boldsymbol{m}_i - \boldsymbol{Pre}[\cdot,t_{i+1}] + \boldsymbol{Post}[\cdot,t_{i+1}]$*, a new state variable* ϑ_{t_k} *is introduced and then:*

3.3.1 for all the transitions t_j *that remained enabled after firing* t_{i+1} *we have that:* $\max(\theta^{en}_{t_j} + EFT_{t_j}, \theta_{t_{i+1}}) \leq \vartheta_{t_j} \leq \theta^{en}_{t_j} + LFT_{t_j}$

3.3.2 for the newly enabled transitions t_k *we have that* $\theta^{en}_{t_k} = \theta_{t_{i+1}}$ *and* $\theta^{en}_{t_k} + EFT_{t_k} \leq \vartheta_{t_k} \leq \theta^{en}_{t_k} + LFT_{t_k}$.

The solution set of the characteristic system K_{τ_n} is derived using a standard algorithm to solve systems of linear inequalities, e.g. Floyd's shortest path algorithm whose time computational complexity is $O(n^3)$. Then the projection of the solution set onto $\boldsymbol{v}$ provides the firing time interval $FI(t)$ for each transition enabled in the marking $\boldsymbol{m}_n$ of SC_n. To construct *LSCG* we need to take into consideration all the possible untimed traces $\tau \in L(\boldsymbol{m}_0)$. We illustrate this by the following example.

Example 17.1. Consider the TPN in Fig. 17.1 were the static firing time intervals are: $I(t_i) = [3,10]$ for $i = 1,\ldots,4$; $I(t_j) = [4,8]$ for $j = 6,\ldots,9$ while the synchronizing transition t_5 has its static firing time interval $I(t_5) = [7,8]$.

The linear state class graph is constructed as follows. In the initial marking $\boldsymbol{m}_0 = [1,0,0,0,1,0,0,0]^T$ we have t_1 and t_6 as enabled transitions, and we assume that the tokens arrived in p_1, respectively p_5 at global time 0 (when the analysis starts).

$$SC_0 \quad K_\varepsilon = \begin{cases} 3 \leq \vartheta_{t_1} \leq 10 \\ 4 \leq \vartheta_{t_6} \leq 8 \end{cases} \tag{17.5}$$

Consider the case that t_1 fires first. The new marking is $\boldsymbol{m}_1 = [0,1,0,0,1,0,0,0]^T$, the set of newly enabled transition is $\{t_2,t_3\}$ while t_6 remains enabled after firing t_1. The characteristic system K_{t_1} of the state class SC_1 is:

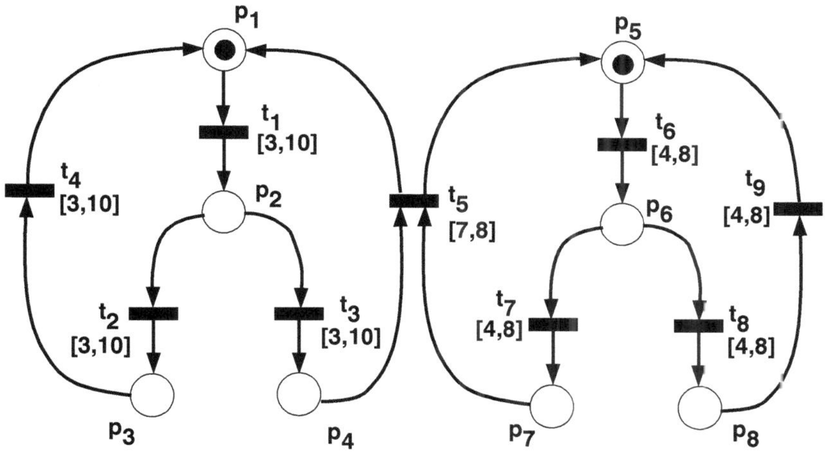

Fig. 17.1 The TPN model of Example 17.1

$$SC_1 \quad K_{t_1} = \begin{cases} 3 \le \theta_{t_1} \le 10 \\ 4 \le \vartheta_{t_6} \le 8 \\ \theta_{t_1} \le \vartheta_{t_6} \\ \theta_{t_1} + 3 \le \vartheta_{t_2} \le \theta_{t_1} + 10 \\ \theta_{t_1} + 3 \le \vartheta_{t_3} \le \theta_{t_1} + 10 \end{cases} \qquad FI(SC_1) = \begin{cases} 6 \le \vartheta_{t_2} \le 18 \\ 6 \le \vartheta_{t_3} \le 18 \\ 4 \le \vartheta_{t_6} \le 8 \end{cases} \tag{17.6}$$

Assume that t_2 is the first transition to fire after t_1. The new marking is $\boldsymbol{m}_3 = [0,0,1,0,1,0,0,0]^T$ and the set of newly enabled transitions is $\{t_4\}$ while t_6 remains enabled after firing t_2. The characteristic system $K_{t_1 t_2}$ of the state class SC_2 is:

$$SC_2 \quad K_{t_1 t_2} = \begin{cases} 3 \le \theta_{t_1} \le 8 \\ \theta_{t_1} + 3 \le \theta_{t_2} \le \theta_{t_1} + 10 \\ 4 \le \vartheta_6 \le 8 \\ \theta_{t_2} \le \vartheta_{t6} \\ \theta_{t_2} + 3 \le \vartheta_{t4} \le \theta_{t_2} + 10 \end{cases} \qquad FI(SC_2) = \begin{cases} 9 \le \vartheta_{t_4} \le 18 \\ 6 \le \vartheta_{t_5} \le 8 \end{cases} \tag{17.7}$$

If t_6 is the first transition to fire after t_1 we obtain the new marking $\boldsymbol{m}_4 = [0,1,0,0,0,1,0,0]^T$ and the set of newly enabled transition is $\{t_7, t_8\}$ while t_2, t_3 remain enabled after firing t_6. The characteristic system $K_{t_1 t_6}$ of SC_3 is:

$$SC_3 \quad K_{t_1t_6} = \begin{cases} 3 \le \theta_{t_1} \le 8 \\ 4 \le \theta_{t_6} \le 8 \\ \theta_{t_1} \le \theta_{t_6} \\ \theta_{t_6} \le \vartheta_{t_2} \\ \theta_{t_6} \le \vartheta_{t_3} \\ \theta_{t_1} + 3 \le \vartheta_{t_2} \le \theta_{t_1} + 10 \\ \theta_{t_1} + 3 \le \vartheta_{t_3} \le \theta_{t_1} + 10 \\ \theta_{t_6} + 4 \le \vartheta_{t_7} \le \theta_{t_6} + 8 \\ \theta_{t_6} + 4 \le \vartheta_{t_8} \le \theta_{t_6} + 8 \end{cases} \qquad FI(SC_3) = \begin{cases} 6 \le \vartheta_{t_2} \le 18 \\ 6 \le \vartheta_{t_3} \le 18 \\ 8 \le \vartheta_{t_7} \le 16 \\ 8 \le \vartheta_{t_8} \le 16 \end{cases} \tag{17.8}$$

Now consider that t_1 fires at the time $\theta_{t_1} = 3$ leading the plant into the state $S_1 \in SC_1$ ($S_0 \xrightarrow{\langle t_1,3\rangle} S_1$). It is easy to check that S_1 has successors in SC_2 and SC_3 while if t_1 fires at $\theta_{t_1} = 6$ the state $S_1' \in SC_1$ has successors in SC_3 but not in SC_2 because t_2 can fire in this case at the earliest time 9 which is larger than the latest time when t_6 is forced to fire at time 8. ■

Thus, we have that in general not all the states within a linear state class have successors in a successor state class. In order to assure the atomicity property that all the states within a state class have successors in all the successor state classes the linear state classes have to be split. Although, this remark is not important for the centralized diagnosis since *LSCG* preserves all the untimed support traces of the TPN model, for a distributed setting the local computations and the consistency check may not be performed properly unless the atomicity property is maintained for the local calculations of each local site. The procedure to refine the linear state class graph *LSCG* such that the atomicity property is satisfied was proposed in [30] and then improved in [5].

As for untimed PNs the approach based on state graph construction suffers from the state space explosion due to the interleaving of the concurrent transitions. Even for TPN models having a reasonable size the plant computation becomes practically impossible. In order to overcome the state space explosion when the plant exhibits a high degree of concurrency we present in the next section the analysis of Time Petri Nets based on partial orders.

17.4.2 Analysis of TPNs Based on Time Processes

When applied to partially observable TPNs the state class graph method has the drawback that the interleavings of the unobservable transitions must be enumerated and this can make the analysis of TPNs of reasonable size sometimes impossible. We illustrate this via the following example.

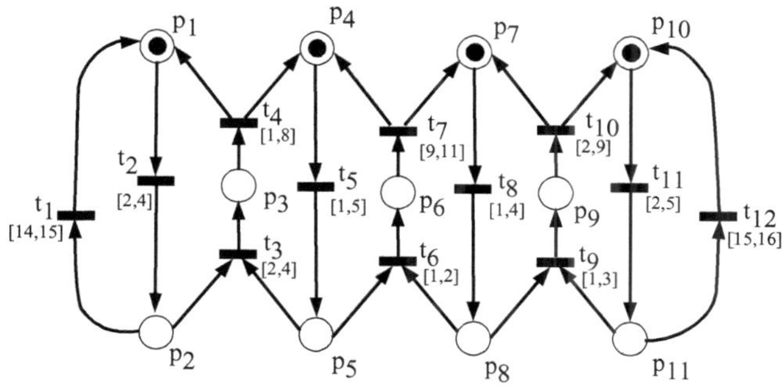

Fig. 17.2 The TPN of the Example 17.2

Example 17.2. Consider the TPN displayed in Fig. 17.2. Static firing time intervals are attached to each transition. The observable transitions are t_4, t_7 and t_{10} and they emit the same label. t_1 and t_{12} are faulty transitions.

For instance, if the plant analysis is based on the state class graph construction one should consider all the possible interleavings of the unobservable concurrent transitions t_2, t_5, t_8, and t_{11}. ■

Hence, in this example the timing information does not reduce the number of the interleavings of the concurrent (unobservable) transitions that are considered. The partial order reduction techniques developed for untimed PNs [3, 11, 20], are shown in [1, 8, 15, 19, 24] to be applicable for TPNs.

In what follows we assume the reader is familiar with the partial order techniques for untimed PNs and we refer to Chapter 15 for the formal definitions of Petri Net Unfoldings. Consider a configuration $\boldsymbol{c}$ in the unfolding $\mathscr{U}_{\mathscr{N}}$ of the untimed PN support of a TPN $\langle N_T, \boldsymbol{m}_0^\theta \rangle$. Then consider a valuation Θ of the execution times at which the events $e \in E_c$ in the configuration $\boldsymbol{c}$ are executed. I.e. for each $e \in E_c$ consider a time value $\theta_e \in \mathbb{Q}^+$ at which e occurs and Θ is an $|E_c|$-tuple comprising all the values at which all the events $e \in E_c$ are executed.

An untimed configuration $\boldsymbol{c}$ together with a valuation Θ of the execution times for its events is called a time configuration (time process in [1]) of the TPN model and is denoted as $\boldsymbol{c}^\theta = (\boldsymbol{c}, \Theta)$.

A time configuration is legal if there is a legal time-trace $\tau^\theta \in L^\theta_{N_T}(\boldsymbol{m}_0^\theta)$ in the TPN $\langle N_T, \boldsymbol{m}_0^\theta \rangle$ whose untimed support τ is a linear order extension (an interleaving) of the partial order $\leq_{\boldsymbol{c}}$, while the execution time θ_t of each transition t considered in the time-trace τ^θ is identical with the valuation θ_e of the event e whose image via π is t.

We compute the entire set of legal time configurations in the following way. Consider an untimed configuration $\boldsymbol{c} \in \mathscr{C}$. Then attach to each event e the static firing interval $I(t)$ that corresponds in the original TPN to transition t s.t. $\pi(e) = t$. We use the notation EFT_e and LFT_e for the lower respectively the upper bound of the static firing time interval of e.

Denote by $\widetilde{K}_c$ the following system of inequalities:

$$\widetilde{K}_c = \left\{ \max_{e' \in {}^{\bullet\bullet}e} (\theta_{e'}) + EFT_e \leq \theta_e \leq \max_{e' \in {}^{\bullet\bullet}e} (\theta_{e'}) + LFT_e \text{ for all } e \in E_c \right. \tag{17.9}$$

where in (17.9) ${}^{\bullet\bullet}e = \emptyset$ implies $\max_{e' \in {}^{\bullet\bullet}e}(\theta_{e'}) = 0$. For each $e \in E_c$ denoted by $\widetilde{FI}(e) = [\widetilde{l}_e, \widetilde{u}_e]$ the time interval obtained solving $\widetilde{K}_c$.

Proposition 17.2. *$\forall \tau^\theta \in L^\theta_{N_T}(\boldsymbol{m}^\theta_0)$ we have that if τ is a linear order extension of $\leq_{\boldsymbol{c}}$ for some $\boldsymbol{c} \in \mathscr{C}$, then Θ is a solution of $\widetilde{K}_{c^\theta}$, where $\Theta = (\theta_{t_1}, \ldots, \theta_{t_{|E_c|}}) = (\theta_{e_1}, \ldots, \theta_{e_{|E_c|}})$ with $\pi(e_i) = t_i$ for $i = 1, \ldots, |E_c|$.*

Proof. The proof is straightforward since for a 1-safe PN there exists an unique configuration $\boldsymbol{c}$ in the net unfolding $\mathscr{U}_{\mathscr{N}}$ s.t. $\tau = \pi(\sigma)$ and σ is obtained by interleaving the events in $\boldsymbol{c}$. Obviously the conditions required for Θ to be a solution of $\widetilde{K}_c$ are satisfied by any legal time-trace. □

Next we should impose the condition that the configuration $\boldsymbol{c}$ is complete w.r.t. the global time (i.e. none of its concurrent parts is left behind). We say that a configuration $\boldsymbol{c}$ is complete w.r.t time ξ if $\forall e \in E_c \Rightarrow \theta_e \leq \xi$ and $\forall e \in \mathit{enabled}(\boldsymbol{c})$, $\max_{e' \in {}^{\bullet\bullet}e}(\theta_{e'}) + EFT_e > \xi$ (where $\mathit{enabled}(\boldsymbol{c})$ is the set of events that can be appended from $\mathit{cut}_{\boldsymbol{c}}$, i.e. $e \in \mathit{enabled}(\boldsymbol{c})$ iff t is enabled in the marking $\boldsymbol{m} = \pi(\mathit{cut}_{\boldsymbol{c}})$ and $\pi(e) = t$).

We cannot claim yet that for a configuration $\boldsymbol{c} \in \mathscr{C}$ there exists at least one legal time configuration $(\boldsymbol{c}, \Theta)$ because for a general TPN the enabling of a transition does not guarantee that it eventually fires because some conflicting transition may be forced to fire before. All we have up to now is that $\widetilde{K}_c$ can be used to explore the future behaviour of the plant to roughly anticipate what kinds of fault events could possibly occur and in which time interval.

To compute the exact plant behaviour we need to take into consideration the set of conflicting events of a configuration $\boldsymbol{c} \in \mathscr{C}$ (denoted by $\check{E}_c$). $\check{E}_c$ comprises the events that could have been executed, but are not included in E_c:

$$\check{E}_c = \{\check{e} \in E \setminus E_c \mid \exists e \in E_c \text{ s.t. } e \,\sharp\, \check{e}\}$$

The characteristic system K_c of configuration $\boldsymbol{c} \in \mathscr{C}$ is obtained adding to $\widetilde{K}_c$ all the inequalities regarding all the conflicting events:

$$K_c = \begin{cases} \max\limits_{e' \in {}^{\bullet\bullet}e} (\theta_{e'}) + EFT_e \leq \theta_e \leq \max\limits_{e' \in {}^{\bullet\bullet}e} (\theta_{e'}) + LFT_e \text{ for all } e \in E_c \\ \min\limits_{e' \sharp \check{e}}(\theta_{e'}) \leq \max\limits_{e'' \in {}^{\bullet\bullet}\check{e}} (\theta_{e''}) + LFT_{\check{e}} \text{ for all } \check{e} \in \check{E}_c \end{cases} \tag{17.10}$$

Theorem 17.2. *Given an arbitrary time ξ we have that $\tau^\theta \in L^\theta_{N_T}(\boldsymbol{m}^\theta_0, \xi)$ iff:*

1. *$\tau = \pi(\sigma)$ and σ is a linear order extension of $\leq_{\boldsymbol{c}}$ for some $\boldsymbol{c} \in \mathscr{C}$*
2. *Θ is a solution of K_c*

3. $\forall e \in E_c \Rightarrow \theta_e \leq \xi$,
4. $\forall e \in enabled(\boldsymbol{c}) \Rightarrow \max_{e' \in {}^{\bullet\bullet}e}(\theta_{e'}) + EFT_e > \xi$.

Proof. $\Rightarrow$ Condition 1, 3 and 4 are trivial and the proof that $\Theta = (t_1, \ldots, t_n)$ is a solution of K_c is by induction.
$\Leftarrow$ The proof is trivial. □

Thus, we need to derive for each untimed configuration the entire solution set of its characteristic system. The naive approach to enumerate all the possible *max*-elements would imply to interleave concurrent events which is exactly what we wanted to avoid by using partial orders to represent the plant behaviour. One way to cope with this difficulty is to put the characteristic system in an ELCP form.

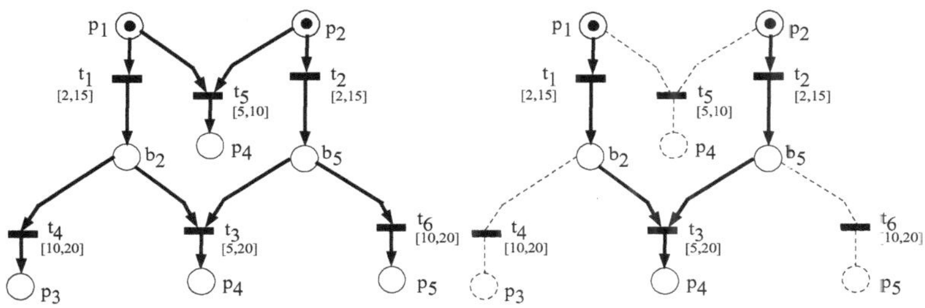

Fig. 17.3 The TPN model of Example 17.3 and one of its configurations

Example 17.3. Consider the TPN $\langle N_T, \boldsymbol{m}_0^{\theta} \rangle$ displayed in Fig. 17.3-left. Static firing time intervals are attached to each transition. Since N_T is acyclic and 1-safe, N_T is isomorphic with its unfolding. Consider in Fig. 17.3-right configuration $\boldsymbol{c}$ that includes t_1, t_2, t_3. The set of conflicting events is $\breve{E}_c = \{t_4, t_5, t_6\}$ (depicted by dotted lines). The following $(max, +)$-system of linear inequalities provides the characteristic system K_c of configuration $\boldsymbol{c}$.

$$\begin{cases} 2 \leq \theta_{t_1} \leq 15 \\ 2 \leq \theta_{t_2} \leq 15 \\ \max(\theta_{t_1}, \theta_{t_2}) + 5 \leq \theta_{t_3} \leq \max(\theta_{t_1}, \theta_{t_2}) + 20 \\ \min(\theta_{t_1}, \theta_{t_2}) \leq 10 \\ \theta_{t_3} \leq \theta_{t_1} + 20 \\ \theta_{t_3} \leq \theta_{t_2} + 20 \end{cases} \tag{17.11}$$

The first three inequalities ($\widetilde{K}_c$) represent the firing conditions of the transitions t_1, t_2 and t_3 within $\boldsymbol{c}$, while the last three inequalities assure that the conflicting events t_4, t_5 and t_6 are not forced to fire by elapsing the maximum delay after their enabling.

We use the notation x_i for the execution time θ_{t_i} of transition t_i, $i = 1,2,3$; $z_1 = \min(x_1,x_2)$, $y_1 = \max(x_1,x_2)$ and $\boldsymbol{A}_k$ for the kth row of matrix $\boldsymbol{A}$. We have the characteristic system K_c in (17.11) expressed in the form of an ELCP [10]:

$$\begin{cases} \boldsymbol{A}\cdot\boldsymbol{x} \geq \boldsymbol{0} \\ (\boldsymbol{A}_1\cdot\boldsymbol{x})(\boldsymbol{A}_2\cdot x) = \boldsymbol{0} \\ (\boldsymbol{A}_3\cdot\boldsymbol{x})(\boldsymbol{A}_4\cdot x) = \boldsymbol{0} \end{cases} \tag{17.12}$$

where $\boldsymbol{x} = [x_1\ x_2\ x_3\ y_1\ z_1\ \delta]^T$ and

$$\boldsymbol{A} = \begin{array}{c} \begin{array}{cccccc} x_1 & x_2 & x_3 & y_1 & z_1 & \delta \end{array} \\ \begin{bmatrix} 1 & 0 & 0 & 0 & -1 & 0 \\ 0 & 1 & 0 & 0 & -1 & 0 \\ -1 & 0 & 0 & 1 & 0 & 0 \\ 0 & -1 & 0 & 1 & 0 & 0 \\ -1 & 0 & 0 & 0 & 0 & 15 \\ 1 & 0 & 0 & 0 & 0 & -2 \\ 0 & -1 & 0 & 0 & 0 & 15 \\ 0 & 1 & 0 & 0 & 0 & -2 \\ 0 & 0 & 1 & -1 & 0 & -5 \\ 0 & 0 & -1 & 1 & 0 & 20 \\ 0 & 0 & 0 & 0 & -1 & 10 \\ 1 & 0 & -1 & 0 & 0 & 20 \\ 0 & 1 & -1 & 0 & 0 & 20 \end{bmatrix} \end{array}$$

Additionally consider that $x_3 = 23$, which simply means that we want to derive the solution set of K_c considering that t_3 is executed at the time $\theta_{t_3} = 23$.

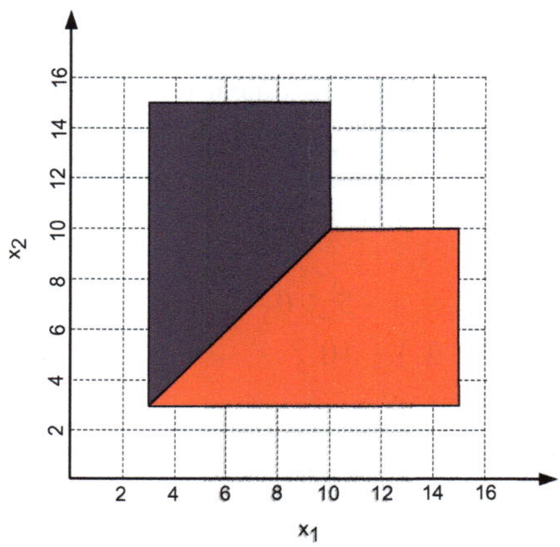

Fig. 17.4 The projection of the solution set of the characteristic system K_c of the Example 17.3 onto the plane $(\theta_{t_1}, \theta_{t_2})$

The solution set of the characteristic system is displayed in Fig. 17.4 as a union of 2 polytopes (trapezia). The first one has as vertices $(3,3)$, $(3,15)$, $(10,15)$, $(10,10)$ and the second one has as vertices $(3,3)$, $(10,10)$, $(15,10)$, $(15,3)$. ■

The plant behaviour up to time ξ is derived using time processes as follows:

1) Construct the branching process $\widetilde{\beta}_\xi$ of the untimed PN support $\langle N, \boldsymbol{m}_0 \rangle$ s.t.:

   a. $\forall e \in \widetilde{\beta}_\xi \Rightarrow \widetilde{l}_e \leq \xi$
   b. $\forall e' \in enabled(\widetilde{\beta}_\xi) \Rightarrow \widetilde{l}_{e'} > \xi$

   where $\widetilde{l}_e$ and $\widetilde{l}_{e'}$ are obtained solving the system of inequalities $\widetilde{K}_c$ given by 17.9.
2) For each final configuration $\boldsymbol{c}$ in $\widetilde{\beta}_\xi$ compute the entire solution set of K_c.
3) Add to $\mathscr{C}_\xi$ all configurations $\boldsymbol{c}$ such that the solution set of K_c is not empty and $\forall e \in E_c$, $l_e \leq \xi$ and $\forall e' \in enabled(\boldsymbol{c})$, $l_{e'} > \xi$
4) $L(\mathscr{C}_\xi) = L_{N_T}(\boldsymbol{m}_0^\theta, \xi)$, where $\tau \in L(\mathscr{C}_\xi)$ iff $\tau = \pi(\sigma)$ and σ is a linear order extension of $\leq_{\boldsymbol{c}}$ for some $\boldsymbol{c} \in \mathscr{C}_\xi$.

Notice that step 3 is needed because if we solve K_c we may have for some events in a configuration $\boldsymbol{c} \in \mathscr{C}_{\widetilde{\beta}_\xi}$ that $l_e > \xi$ although $\widetilde{l}_e \leq \xi$. Since these events can be executed only at times larger than ξ they should not be considered in the configurations that are added to $\mathscr{C}_\xi$.

17.5 The On-Line Implementation

The problem that we should answer next is: *"Up to what time ξ to make the calculations for the on-line diagnosis ?"*.

There are different solutions to answer this question, depending on the computational capability, the plant behaviour and the requirements for the diagnosis result.

A first solution is to make calculations in advance e.g. up to the time when the plant halts in a *"quiescent state"* (a state s.t. no transition is executed until a trigger event is executed). This solution is appropriate for a plant known to have a cyclic behaviour, where each operation cycle is initiated by the plant operator. However, a drawback of this method could be the large amount of calculations that is required each time when an update is performed (i.e. when a new observation is taken into account both the past behaviour and the future behaviour must be refined). Also the size of the characteristic systems of the final traces/configurations maybe too large to handle.

The second solution would be to make calculations up to a discarding time that is given by the observable transition whose latest execution time is the smallest among the observable transitions that are considered on different branches. When the calculations are made based on partial orders one keeps track of the time progress of the entire time process and new unobservable events are appended only if their earliest time of execution is not larger than the smallest among the latest execution times of the observable events already appended.

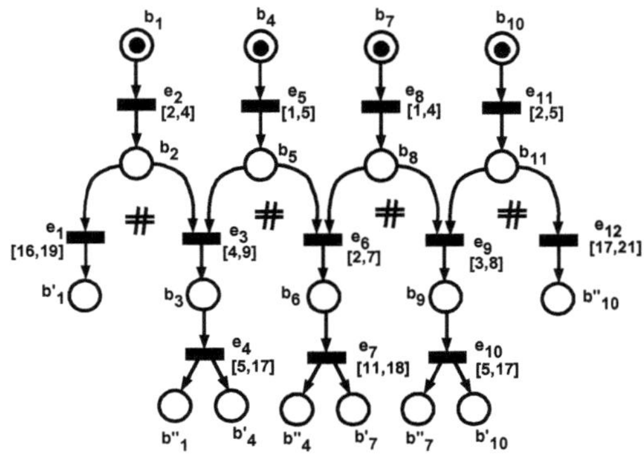

Fig. 17.5 A part of the unfolding of the TPN displayed in Fig. 17.2

Example 17.4. Consider again the TPN displayed in Fig. 17.2. The computation based on partial orders is made as follows (see Fig. 17.5 where we attach to each event $e \in E$ the time interval $\widetilde{FI}(e)$ that results from solving $\widetilde{K}_c$). Each condition b_i has the same lower index i as place $p_i \in P$ if $\pi(b_i) = p_i$ and similarly each event e_j has the same lower index j as transition t_j if $\pi(e_j) = t_j$. First we append the events e_2, e_5, e_8 and e_{11} that have $\widetilde{FI}(e_2) = I(e_2)$, $\widetilde{FI}(e_5) = I(e_5)$, $\widetilde{FI}(e_8) = I(e_8)$ and $\widetilde{FI}(e_{11}) = I(e_{11})$. Then we append e_1 and we obtain $\widetilde{FI}(e_1) = [16,19]$ by adding to $\widetilde{FI}(e_2) = [2,4]$ the static firing interval of t_1, $I(e_1) = [14,15]$. Next we append e_3, e_6, e_9 and e_{12} and after that we append the events e_4, e_7 and e_{10} that correspond to the observable transitions t_4, t_7 and t_{10} respectively.

At this moment the computation stops because we do not append events after the observable events e_4, e_7 and e_{10} and from b_1' and b_{10}'' we can only append unobservable events whose execution times are larger than 17 (which is the latest execution time of e_4 and e_{10}). If it is received only one observation until the time 17, then we have four possible configurations that are listed in Fig. 17.6. For each configuration we calculate the solution set of its characteristic system with the timing constraint taking into account the observation.

If the first observation is received at the time 17.5, then the only feasible configuration is $\boldsymbol{c}_4$. In this case we are sure that either the fault t_1 has already happened or it will happen for sure by the time 19 when e_1 will reach its latest execution time and will be forced to be executed.

Now consider configuration $\boldsymbol{c}_1$. It contains two observable events. If we have observed only one event executed at time 15 we have two cases: either e_4 was executed at time 15 and in this case e_5 can be executed only at a time larger than 15 or vice-versa. ■

Another choice is to make computations at regular times, e.g. compute first the plant behaviour up to time ξ. Then record the plant observations up to time ξ and refine what was already computed. Then extend the calculations up to the time 2ξ and

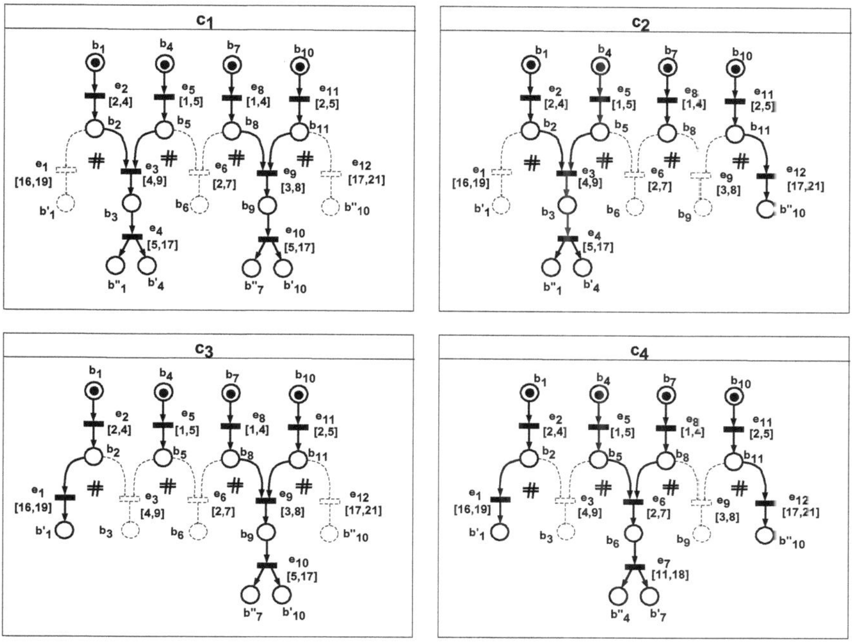

Fig. 17.6 The valid configurations of Example 17.4

so on. The time ξ should be large enough to be sure that the calculation can be performed and small enough to be able to take as fast as possible isolation actions against the faults that are detected.

Notice that we have assumed here that the faults in the plant are unpredictable (i.e. from any state of the plant there is at least one continuation that does not contain a fault transition). This is a reasonable assumption since otherwise one should make the calculations in advance (e.g. up to a *"quiescent state"*) if there is the need of detection a fault occurrence at the earliest possible time. In general the unpredictability of a fault in a time model cannot be checked as it is the case for the untimed models. However, there is a structural assumption on the TPN model that is a sufficient condition for the faults to be unpredictable $\forall t \in T_f$, $\exists t' \in T \setminus T_f$ such that $i)$ ${}^\bullet t' \subseteq {}^\bullet t$ and $ii)$ $EFT_{t'} \leq LFT_t$.

The computational complexity of the two approaches based on the construction of the state class graph and based on time processes cannot be compared because they heavily depend on the TPN structure and on the number of observable transitions. For instance if the untimed PN support is Extended Free Choice the calculations can be performed very easily using partial orders since for such models there are no conflicting events and the solution set is provided simply by $\widetilde{K}_c$.

As a general consideration it should be noted that the use of partial orders reduces the computational effort to make calculations in the untimed PN, because it avoids the interleaving of the concurrent transitions as it is the case when the *LSCG* is constructed. On the other hand, for a system of linear inequalities the computational complexity to derive the solution set is polynomial (it requires $O(n^3)$ time and this can be reduced to $O(n^2)$ if the system of inequalities is constructed in an incremental manner) whereas to derive the solution set of a system of $(max,+)$-linear inequalities is a computationally harder problem. For instance the time required to solve ELCP may grow exponentially with the size of the system. In general its computational complexity depends on many factors and is difficult to be expressed in terms of the number of variables and inequalities.

Although the ELCP is not suited for problems with a large size there are efficient algorithms to decide if an ELCP has at least one solution. In the diagnosis setup that we have considered this decision test would be enough to check the validity of a configuration and thus to calculate the correct diagnosis result. However, the calculation from time to time of the entire set of solutions is required to eliminate inequalities that are redundant, reducing in this way the size of the system.

17.6 Further Reading

Various settings and methods for the monitoring and diagnosis of time Discrete Event Systems were proposed in [2, 8, 14, 17, 18, 25, 26, 31]. They differ in the way the untimed part of plant is modelled: as an automaton in [18, 25, 31], or a Petri Net in [2, 8, 14, 17], in the way the faults are represented: as fault transitions in [2, 8, 14, 17, 18] or fault states in [25, 26, 31], and in how the calculations in the untimed support are performed: via forward reachability as in [2, 8, 14, 17, 18, 31], via backward reachability as in [25, 26] or using algebraic techniques as in [2]. Also there are some other differences regarding how the time and the observations are taken into account. The closest approaches to what has been presented in this chapter can be found in [8, 14] and [17] with the remark that in [8, 14] the plant observation does not include the time stamp when a transition is executed. Also distributed implementations for diagnosis of TPNs are presented in [6] and [16].

The results of this chapter can be easily extended on the following directions. The 1-safe assumption can be relaxed. The only condition needed is the boundedness of the TPN model. The approach can deal with different other representations of the faults e.g. a fault can be also defined as a violation of a certain time separation between the execution of two transitions. Although only the centralized setting has been considered in this chapter, the results can be applied for a distributed implementation that comprises different local sites modelled by TPNs. For instance the distributed implementation is quite simple for the class of Extended Free Choice TPNs.

Notice that here the problem of diagnosability of TPN models was not addressed, i.e. if the model has the property that a fault occurrence is always detected within a finite delay after its occurrence for any plant behaviour. We refer to the work in [7, 27] and [28] where algorithms to test diagnosability are provided for Timed

Automata. These algorithms can be easily adapted for TPN models and require the construction of the state class graph $LSCG^r$ with the state variables defined w.r.t. the relative time.

References

1. Aura, T., Lilius, J.: Time processes of Time Petri Nets. In: Azéma, P., Balbo, G. (eds.) ICATPN 1997. LNCS, vol. 1248, pp. 136–155. Springer, Heidelberg (1997)
2. Basile, F., Chiacchio, P., De Tommasi, G.: Improving on-line fault diagnosis for Discrete Event Systems using time. In: IEEE Int. Conf. on Automation Science and Engineering, Scottsdale, USA (2007)
3. Benveniste, A., Fabre, E., Jard, C., Haar, S.: Diagnosis of asynchronous Discrete Event Systems, a net unfolding approach. IEEE Transactions on Automatic Control 48(5), 714–727 (2003)
4. Berthomieu, B., Menasche, M.: An enumerative approach for analyzing Time Petri Nets. In: Proc. IFIP 9th World Computer Congress, Paris, France (1983)
5. Berthomieu, B., Vernadat, F.: State class constructions for branching analysis of Time Petri Nets. In: Proc. 9th Int. Conf. on Tools and Algorithms for the Construction and Analysis of Systems, Warsaw, Poland (2003)
6. Boufaied, A., Subias, A., Combacau, M.: Distributed time constraints verification modelled with Time Petri Nets. In: Proc. 17th IMACS World Congress, Paris, France (2005)
7. Bouyer, P., Chevalier, F., D'Souza, D.: Fault Diagnosis Using Timed Automata. In: Sassone, V. (ed.) FOSSACS 2005. LNCS, vol. 3441, pp. 219–233. Springer, Heidelberg (2005)
8. Chatain, T., Jard, C.: Time Supervision of Concurrent Systems Using Symbolic Unfoldings of Time Petri Nets. In: Pettersson, P., Yi, W. (eds.) FORMATS 2005. LNCS, vol. 3829, pp. 196–210. Springer, Heidelberg (2005)
9. Ciardo, G., Miner, A.S.: Storage Alternatives for Large Structured State Spaces. In: Marie, R., Plateau, B., Calzarossa, M.C., Rubino, G.J. (eds.) TOOLS 1997. LNCS, vol. 1245. Springer, Heidelberg (1997)
10. De Schutter, B., De Moor, B.: The Extended Linear Complementarity Problem. Mathematical Programming 71(3), 289–325 (1995)
11. Esparza, J.: Model checking using net unfoldings. Science of Computer Programming 23(2-3), 151–195 (1994)
12. Gardey, G., Roux, O.H., Roux, O.F.: State space computation and analysis of Time Petri Nets. Theory and Practice of Logic Programming 6(3), 301–320 (2006)
13. Genc, S., Lafortune, S.: Distributed diagnosis for DES using Petri Nets. In: Proc. Conf. on Applications and Theory of Petri, Eindhoven, The Netherlands (2003)
14. Ghazel, M., Toguyéni, A., Yim, P.: State observer for DES under partial observation with Time Petri Nets. Discrete Event Dynamic Systems 19(2), 137–165 (2009)
15. Hulgaard, H., Burns, S.M.: Efficient Timing Analysis of a Class of Petri Nets. In: Wolper, P. (ed.) CAV 1995. LNCS, vol. 939, pp. 423–436. Springer, Heidelberg (1995)
16. Jiroveanu, G., Boel, R.K.: A distributed approach for fault detection and diagnosis based on Time Petri Nets. Mathematics and Computers in Simulation 70(5-6), 287–313 (2006)
17. Jiroveanu, G., De Schutter, B., Boel, R.K.: In: On-line diagnosis for Time Petri Nets. In: Proc. Int. Workshop on Principles of Diagnosis, Burgos, Spain (2006)

18. Khoumsi, A.: Fault prognosis in real-time Discrete Event Systems. In: Proc. Int. Workshop on Principles of Diagnosis, Stockholm, Sweden (2009)
19. Lilius, J.: Efficient state space search for Time Petri Nets. Electronic Notes in Theoretical Computer Science 18, 113–133 (1998)
20. McMillan, K.L.: Using Unfoldings to Avoid the State Space Explosion Problem in Verification of Asynchronous Circuits. In: Probst, D.K., von Bochmann, G. (eds.) CAV 1992. LNCS, vol. 663, pp. 164–177. Springer, Heidelberg (1993)
21. Merlin, P.: A Study of Recoverability of Computer Science. University of California, Irvine (1974)
22. Ramchandani, C.: Analysis of asynchronous concurrent systems by Timed Petri Nets. Massachusetts Institute of Technology, Project MAC, Technical Report 120 (1974)
23. Sampath, M., Sengupta, R., Lafortune, S., Sinnamohideen, K., Teneketzis, D.: Diagnosability of Discrete Event Systems. IEEE Transactions on Automatic Control 40(9), 1555–1575 (1995)
24. Semenov, A., Yakovlev, A.: Verification of asynchronous circuits using Time Petri Net unfolding. In: Proc. 33rd ACM/IEEE Design Automation Conference, Las Vegas, USA (1996)
25. Simeu-Abazia, Z., Di Mascoloa, M., Knotekb, M.: Fault diagnosis for Discrete Event Systems: modelling and verification. Reliability Engineering and System Safety 95(4), 369–378 (2010)
26. Srinivasan, V.S., Jafari, M.A.: Fault detection/monitoring using Time Petri Nets. IEEE Transactions on Systems Managment and Cybernetics 23(4), 1155–1162 (1994)
27. Tripakis, S.: Fault Diagnosis for Timed Automata. In: Damm, W., Olderog, E.-R. (eds.) FTRTFT 2002. LNCS, vol. 2469, pp. 205–224. Springer, Heidelberg (2002)
28. Xu, S., Jiang, S., Kumar, R.: Diagnosis of dense-time systems using digital clocks. IEEE Transactions on Automation Science and Engineering 7(4), 870–878 (2010)
29. Wang, J., Deng, Y., Xu, G.: Reachability analysis of real-time systems using Time Petri Nets. IEEE Transactions on Systems Managment and Cybernetics 30(5), 725–736 (2000)
30. Yoneda, T., Ryuba, H.: CTL model checking of Time Petri Nets using geometric regions. EICE Transaction on Information & Systems E99-D, 1–11 (1998)
31. Zad, S.H., Zwong, S.H., Wonham, W.M.: Fault diagnosis in Discrete Event Systems: framework and model reduction. IEEE Transactions on Automatic Control 48(7), 1199–1212 (2003)

Chapter 18
Introduction to Fluid Petri Nets

C. Renato Vázquez, Cristian Mahulea, Jorge Júlvez, and Manuel Silva

18.1 Introduction and Motivation

Several results can be found in the literature regarding the analysis of DEDS by using models from the Petri net (PN) paradigm. Applications involve the implementation of sequence controllers, validation in software development, analysis of communication protocols and networks, manufacturing systems, supply chains, etc.

It is well known that one of the most important limitations in the analysis (and synthesis problems) of DEDS is the computational complexity that may appear. In particular, the set of reachable markings of a Petri net frequently increases exponentially w.r.t. the initial marking, what is known as the *state explosion problem*, making prohibitive the application of enumerative techniques even for net systems with a small structure (i.e., small number of places and transitions).

In this context, the *fluidization* or *continuization* (i.e., getting a continuous-state approximation) has been proposed as a relaxation technique in order to overcome the state explosion problem. The idea consists in the analysis of the DEDS via a relaxed continuous approximation, i.e., a continuous-state system if behaves in a "similar" way than the original model (or conserves certain interesting properties), reducing thus the computational efforts. Nevertheless, not all PNs models allow some continuous approximations. In DEDS, fluidization has been explored in queueing networks (e.g., [11, 13, 1]), PNs ([6, 34]) and more recently in Process Algebra [16].

Regarding PNs, David and Alla first introduced fluid PNs with *constant* and *variable speeds* [6, 7]. From another perspective, the relaxation of the fundamental or state equation of the PN system was proposed in [29] (in the same meeting), in order to systematically use Linear Programming for structural analysis. These two approaches lead to continuous state equations (see Fig. 18.1), but the proposal

C. Renato Vázquez · Cristian Mahulea · Jorge Júlvez · Manuel Silva
Instituto de Investigación en Ingeniería de Aragón (I3A), University of Zaragoza, Spain
e-mail: {cvazquez,cmahulea,julvez,silva}@unizar.es

C. Seatzu et al. (Eds.): Control of Discrete-Event Systems, LNCIS 433, pp. 365–386.
springerlink.com

(more conceptual) of David and Alla leads to the possibility of describing the transient behavior of timed models. The resulting fluid PN models can be analyzed as state-continuous systems but behave (quantitatively) as T-timed discrete PNs (examples can be found in [9]). This topic was revisited in [26], making emphasis in the connection with the original discrete models. In fact, there the *infinite server semantics* (which is the same that the variable speed) was derived as the approximation of the average behavior of a Markovian stochastic T-timed PN (a PN whose transitions fire after exponentially distributed random time delays). From another perspective, different authors have proposed hybrid PN systems (some transitions remain discrete while the others are fluidified) that can be used as models *per se*, for instance *fluid stochastic PNs* [36], *differential PNs* [4], *batch hybrid PNs* [10], *first-order hybrid PNs* [2], etc. These hybrid models enjoy a broad expressive power, but the analysis of some of these systems is technically very complex.

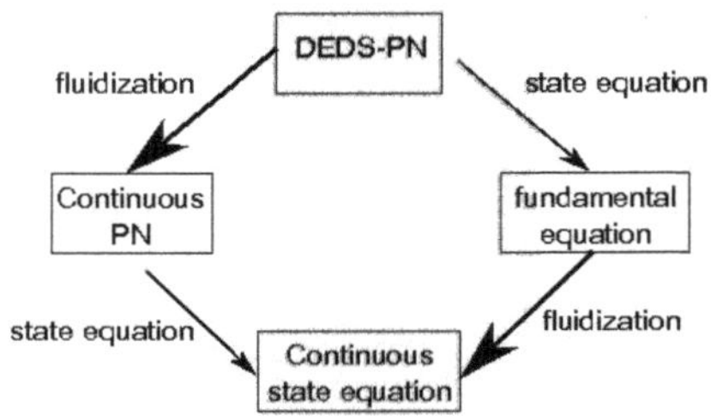

Fig. 18.1 Two ways for the fluidization of Petri nets

In this and the following chapters, the basic model in [9, 26] will be mostly considered in both its autonomous version (untimed) and timed version (mostly under *infinite server semantics* or *variable speed).* Then, from our perspective, a continuous Petri nets model is derived as a potential approximation of an original discrete PN. In this way, the analysis of these continuous models provides coherence, synergy, and economy with respect to the Petri net paradigm [31, 32].

The fluidization seems a promising technique when the initial marking can be assumed as "large enough"(where the relative errors, and their consequences, tend to be small, because the rounding effects are relatively less significant). In fact, increasing the population does not affect the complexity of the analysis via fluid models, since, in the resulting continuous PN, the number of *state variables* is upper bounded by the number of places, being independent of the number of tokens in the net system (thus, of the initial marking).

The risk of fluidization is a loss of fidelity. As linearization of nonlinear models, fluidization does not always lead to relaxed models of similar behavior, i.e., they are not always approximations. The first problem that may arise when using this approach is that the fluid model does not necessarily preserve all the behavioral properties of the original DEDS model. For example, mutex properties (e.g., two places cannot be concurrently marked) are always lost, spurious markings (solutions of the state equation that are not reachable markings in the discrete PN) may appear in the continuous model or liveness is not preserved in the general case. Thus, for certain cases, the analysis through fluidization may be useless. In other cases, the fluidization may provide only an educated guess. Moreover, the resulting fluid model may exhibit an important technical complexity. For instance, a timed continuous PN under infinite server semantics is a *piecewise linear* system (a linear system whose state and input matrices change, among a countable set of matrices, according to the state), from a continuous-state dynamical systems' perspective. The number of embedded linear operation modes (equivalently, sets of linear differential equations) usually increases exponentially w.r.t. the number of transitions representing *rendez-vous* (synchronizations). Thus, even if the behavior of a DEDS is approximately preserved by its corresponding fluid relaxation, this could be still too complex to be properly analyzed. The number of state variables does not depend on the initial marking, but in the number of places. Thus, large net structures may lead to continuous models with a large number of state variables, since one state variable is usually defined per each place. Furthermore, the addition of time in the continuous model brings important additional difficulties for analysis and synthesis. In this sense, the expressive power of timed continuous PNs (under infinite server semantics) is surprisingly high, because they can simulate Turing Machines [27]! This means that certain important properties as the existence of a steady-state are undecidable.

On the other hand, when a system admits a "reasonable" fluidization (in the sense that the fluid model preserves the desired properties of the discrete one), several advantages can be visualized by using continuous models: the first one is obviously the reduction of the complexity related to a large marking population, since in continuous models the state explosion problem does not appear. Furthermore, certain problems can be analyzed by using more efficient algorithms, for instance, the set of reachable markings (including infinite firing sequences) is convex, thus reducing the complexity of optimization problems. Additionally, the ability to fire in isolation minimal T-semiflows makes behavioral and structural synchronic relations [1] to collapse, thus, for example, the bound of a place can be straightforwardly computed in polynomial time [28, 29]. Another interesting advantage is that techniques and concepts developed in the Control Theory for continuous-sate systems can be applied to the timed continuous PN model. For instance, techniques for the analysis and application of performance controllers that reject perturbations, stability, observability, estimation, etc. In this way, fluidization represents a bridge between particular classes of continuous-state and discrete event systems.

In this chapter, *continuous Petri nets* are first introduced as untimed, i.e., fully non-deterministic, and later as timed formalisms. The relationship between the properties of (discrete) PNs and the corresponding properties of their continuous

[1] *Synchrony theory* is a branch of general net theory that deals with the characterization of transition firing dependencies. Two transition subsets are in a given *synchronic relation* if the corresponding quantitative firing dependency is bounded. *Behavioral relations* depend on the initial marking, while *structural relations* hold for any (finite) initial marking.

approximation is considered at several points. Observability and Control of *timed continuous Petri nets* (TCPNs) will be considered in forthcoming chapters.

18.2 Fluidization of Untimed Net Models

This section presents the formalism of continuous Petri nets and its behavior in the untimed framework. It deals with basic concepts, as lim-reachability [25] and desired logical properties, and relates them to those ones of the discrete systems.

18.2.1 The Continuous PN Model

A continuous Petri net system [9, 26] is understood as the fluid relaxation of all the transitions of a *discrete* Petri net one (as a consequence, the marking at all the places becomes continuous). In the sequel, the set of input and output nodes of v will be denoted as $^\bullet v$ and $v^\bullet$, respectively.

Definition 18.1. *A* continuous PN *system is a pair* $\langle N, \boldsymbol{m}_0\rangle$ *where* N *is a P/T net (like in a P/T system) and* $\boldsymbol{m}_0 \in \mathbb{R}^{|P|}_{\geq 0}$ *is the initial marking. The evolution rule is different to the case of discrete P/T systems, since in continuous PNs the firing is not restricted to integer amounts, and so the marking* $\boldsymbol{m} \in \mathbb{R}^{|P|}_{\geq 0}$ *is not forced to be integer. Instead, a transition* t_i *is* enabled *at* $\boldsymbol{m}$ *iff for every* $p_j \in {^\bullet t_i}$, $m[p_i] > 0$*; and its* enabling degree *is* $enab(t_i, \boldsymbol{m}) = \min_{p_j \in {^\bullet t_i}}\{m[p_j]/Pre[p_j, t_i]\}$. *The firing of* t_i *in a certain amount* $0 \leq \alpha \leq enab(t_i, \boldsymbol{m})$ *leads to a new marking* $\boldsymbol{m}' = \boldsymbol{m} + \alpha \cdot \boldsymbol{C}[P, t_i]$, *where* $\boldsymbol{C} = \boldsymbol{Post} - \boldsymbol{Pre}$ *is the* token flow *or* incidence *matrix, and* $\boldsymbol{C}[P, t_i]$ *denotes the column of* $\boldsymbol{C}$ *corresponding to* t_i.

The usual PN system, $\langle N, \boldsymbol{M}_0\rangle$ with $\boldsymbol{M}_0 \in \mathbb{N}^{|P|}$, will be said to be *discrete* so as to distinguish it from a *continuous* PN system $\langle N, \boldsymbol{m}_0\rangle$, in which $\boldsymbol{m}_0 \in \mathbb{R}^{|P|}_{\geq 0}$. In the following, the marking of a continuous PN will be denoted in lower case $\boldsymbol{m}$, while the marking of the corresponding *discrete* one will be denoted in upper case $\boldsymbol{M}$. Observe that $Enab(t_i, \boldsymbol{M}) \in \mathbb{N}$ in discrete PNs, while $enab(t_i, \boldsymbol{m}) \in \mathbb{R}_{\geq 0}$ in continuous PNs. Notice that to decide whether a transition in a continuous system is enabled or not, it is not necessary to consider the weights of the arcs going from the input places to the transition. However, the arc weights are important to compute the enabling degrees. Here no policy for the firing of transitions is imposed, that is, a full non-determinism is assumed for the order and amounts in which transitions are fired.

Right and left natural annullers of the token flow matrix are called T- and P-*semiflows*, respectively (i.e., vectors $\boldsymbol{y}$ and $\boldsymbol{x}$, whose entries belong to $\mathbb{N} \cap \{0\}$, fulfilling $\boldsymbol{y}^T \cdot \boldsymbol{C} = \boldsymbol{0}$ and $\boldsymbol{C} \cdot \boldsymbol{x} = \boldsymbol{0}$, respectively). The existence of P-semiflows induces conservation laws, i.e., if $\exists \boldsymbol{y} \geq \boldsymbol{0}, \boldsymbol{y}^T \cdot \boldsymbol{C} = \boldsymbol{0}$ then by the state equation it holds $\boldsymbol{y}^T \cdot \boldsymbol{m}_0 = \boldsymbol{y}^T \cdot \boldsymbol{m}$ for any initial marking $\boldsymbol{m}_0$ and any reachable marking $\boldsymbol{m}$. On the other hand, T-semiflows represent potential cyclic behaviors, i.e., if $\exists \boldsymbol{x} \geq \boldsymbol{0}$,

$\boldsymbol{C}\cdot\boldsymbol{x}=\boldsymbol{0}$ then $\exists \boldsymbol{m}_0$ such that $\boldsymbol{m}_0 \xrightarrow{\sigma} \boldsymbol{m}_0$ with σ being a firing sequence whose firing count vector equals $\boldsymbol{x}$. As in discrete nets, when $\boldsymbol{y}^T\cdot\boldsymbol{C}=\boldsymbol{0}$ for some $\boldsymbol{y}>\boldsymbol{0}$ the net is said to be *conservative*, and when $\boldsymbol{C}\cdot\boldsymbol{x}=\boldsymbol{0}$ for some $\boldsymbol{x}>\boldsymbol{0}$ the net is said to be *consistent*. Given a vector $\boldsymbol{x}\in\mathbb{R}^{|T|}$, its support is defined as the set of transitions $||\boldsymbol{x}||=\{t_i\in T \mid x[i]\neq 0\}$, where $\boldsymbol{x}[i]$ denotes the ith entry of $\boldsymbol{x}$. Similarly, for a vector $\boldsymbol{y}\in\mathbb{R}^{|P|}$, $||\boldsymbol{y}||=\{p_i\in P \mid y[i]\neq 0\}$.

The definitions of subclasses that depend only on the structure of the net are also generalized to continuous nets. For instance, in *join free* nets (JF) each transition has at most one input place, in *choice free* nets (CF) each place has at most one output transition, and in *equal conflict* nets (EQ) all conflicts are equal, i.e., ${}^\bullet t\cap{}^\bullet t'\neq\emptyset \Rightarrow \boldsymbol{Pre}[P,t]=\boldsymbol{Pre}[P,t']$. Moreover, a net N is said to be *proportional equal conflict* if ${}^\bullet t\cap{}^\bullet t'\neq\emptyset \Rightarrow \exists q\in\mathbb{R}_{>0}$ such that $\boldsymbol{Pre}[P,t]=q\cdot\boldsymbol{Pre}[P,t']$. Finally, a net N is said to be *mono-T-semiflow* (MTS) iff it is conservative and has a unique minimal T-semiflow whose support contains all the transitions (a T-semiflow is minimal if its support does not contain the support of another T-semiflow).

18.2.2 Reachability

Let us now illustrate the firing rule in an untimed continuous Petri net system. For this, consider the system in Fig. 18.2(a). The only enabled transition at the initial marking is t_1, whose enabling degree is 1. Hence, it can be fired in any real quantity going from 0 to 1. For example, by firing it an amount equal to 1, the marking $\boldsymbol{m}_1=[1\ 1]^T$ is reached. At $\boldsymbol{m}_1$ transition t_2 has enabling degree equal to 1; if it is fired in an amount of 0.5, the resulting marking is $\boldsymbol{m}_2=[1.5\ 0.5]^T$. In this way, both $\boldsymbol{m}_1$ and $\boldsymbol{m}_2$ are reachable markings with *finite* firing sequences. On the other hand, at $\boldsymbol{m}_1$ the marking at p_1 is equal to 1, leading to an enabling degree for t_1 of 0.5, i.e., the half amount that at $\boldsymbol{m}_0$. By firing t_1 an amount of 0.5, the marking reached is $\boldsymbol{m}_3=[0.5\ 1.5]^T$. Notice that by successively firing t_1 with an amount equal to its enabling degree, the marking of p_1 will approach to 0. The marking reached in the limit (with an infinite firing sequence), namely $\boldsymbol{m}'=[0\ 2]^T$, corresponds to the emptying of an initially marked trap $\Theta=\{p_1\}$ (a trap is a set of places $\Theta\subseteq P$ s.t. $\Theta^\bullet\subseteq{}^\bullet\Theta$), fact that can not occur in discrete systems. Thus, in continuous systems traps may not *trap*! Nevertheless, such trap cannot be emptied with a finite firing sequence. This leads us to consider two different reachability concepts.

Definition 18.2. *A marking is said to be* reachable *if it is reachable with a finite firing sequence. On the other hand, a marking is said to be* lim-reachable *if it can be reached with either a* finite *or an* infinite *firing sequence. More formally:* $\boldsymbol{m}\in\mathbb{R}^{|P|}_{\geq 0}$ *is* lim-reachable *if a sequence of reachable markings* $\{\boldsymbol{m}_i\}_{i\geq 1}$ *exists such that* $\boldsymbol{m}_0\xrightarrow{\sigma_1}\boldsymbol{m}_1\xrightarrow{\sigma_2}\boldsymbol{m}_2\cdots\cdots\boldsymbol{m}_{i-1}\xrightarrow{\sigma_i}\boldsymbol{m}_i\cdots$ *and* $\lim_{i\to\infty}\boldsymbol{m}_i=\boldsymbol{m}$. *For a given system* $\langle\mathcal{N},\boldsymbol{m}_0\rangle$, *the set of all markings that are reachable in a finite number of firings is denoted as* $\mathbf{R}(N,\boldsymbol{m}_0)$, *while* lim-$\mathbf{R}(N,\boldsymbol{m}_0)$ *denotes the set of* lim-reachable *markings.*

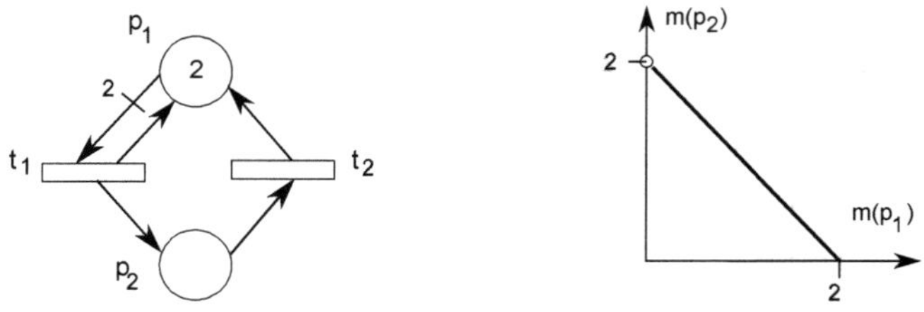

Fig. 18.2 (a) Autonomous continuous system (b) Lim-Reachability space

As an example, Fig. 18.2(b) depicts lim-$\mathbf{R}(N,\boldsymbol{m}_0)$ of the system in Fig. 18.2(a). The set of lim-reachable markings is composed of the points on the line connecting $[2\ 0]^T$ and $[0\ 2]^T$. On the other hand, all the points of that line excepting the circled one $[0\ 2]^T$ belong to $\mathbf{R}(N,\boldsymbol{m}_0)$, i.e., $\mathbf{R}(N,\boldsymbol{m}_0) = \text{lim-}\mathbf{R}(N,\boldsymbol{m}_0) \setminus \{[0\ 2]^T\}$.

For any continuous system $\langle N,\boldsymbol{m}_0\rangle$, the differences between $\mathbf{R}(N,\boldsymbol{m}_0)$ and lim-$\mathbf{R}(N,\boldsymbol{m}_0)$ are just in the border points on their convex spaces. In fact, it holds that $\mathbf{R}(N,\boldsymbol{m}_0) \subseteq \text{lim-}\mathbf{R}(N,\boldsymbol{m}_0)$ and that the closure of $\mathbf{R}(N,\boldsymbol{m}_0)$ is equal to the closure of lim-$\mathbf{R}(N,\boldsymbol{m}_0)$ [17].

18.2.3 Some Advantages

A system can be fluidizable with respect to a given property, i.e., the continuous model preserves that property of the discrete one, but may be not with respect to other properties. Thus, the usefulness of continuous models highly depends on the properties to be analyzed and the systems being studied.

An interesting property is that $\mathbf{R}_D(N,\boldsymbol{m}_0) \subseteq \mathbf{R}(N,\boldsymbol{m}_0)$ (where $\mathbf{R}_D(N,\boldsymbol{m}_0)$ is the set of markings reachable by the system as discrete). This can be explained as follows: for any marking $\boldsymbol{m}$ reached by firing transitions in discrete amounts from $\boldsymbol{m}_0 \in \mathbb{N}^{|P|}_{\geq 0}$, i.e., as if the system were discrete, $\boldsymbol{m}$ is also reachable by the system as continuous just by applying the same firing sequence.

The fact that $\mathbf{R}_D(N,\boldsymbol{m}_0) \subseteq \mathbf{R}(N,\boldsymbol{m}_0)$ might involve a mismatch among some properties of the discrete and continuous systems, e.g., the new reachable markings might make the system live or might deadlock it (examples can be found in [25, 35]). This *non-fluidizability* of discrete net systems with respect to the deadlock-freeness property, that may be surprising at first glance, can be easily accepted if one thinks, for example, on the existence of non-linearizable differential equations systems .

Let us recall the concept of boundedness in discrete Petri nets: a *PN* system is said *bounded* if $k \in \mathbb{N}$ exists such that for every reachable marking $\boldsymbol{m}$, $\boldsymbol{m} \leq k \cdot \mathbf{1}$, with $\mathbf{1}$ is the vector of ones, and it is *structurally bounded* if it is bounded with every initial marking. These concepts can also be applied to continuous Petri net systems. Similarly, a continuous system is said lim-live if for any transition t and any lim-reachable marking $\boldsymbol{m}$, a successor $\boldsymbol{m}'$ exists such that t is enabled [25].

A continuous net N is said structural lim-live if there exists an initial marking $\boldsymbol{m}_0$ such that the continuous system $\langle N, \boldsymbol{m}_0 \rangle$ is lim-live.

The concept of limit-reachability in continuous systems provides an interesting approximation to liveness properties of discrete nets, in the sense that lim-liveness and boundedness of the continuous system is a sufficient condition for structural liveness and boundedness of the discrete one [25]:

Proposition 18.1. *Let $\langle N, \boldsymbol{m}_0 \rangle$ be a bounded lim-live continuous PN system. Then, N is structurally live and structurally bounded as a discrete net.*

From Proposition 18.1 it is clear that any necessary condition for a discrete system to be str. live and str. bounded, is also necessary for it to be str. lim-live and bounded as continuous. In particular rank theorems [24] establish necessary liveness conditions based on consistency, conservativeness and the existence of an upper bound on the rank of the token flow matrix, which is the number of equal conflict sets. Even more, for the EQ subclass, such structural conditions are also sufficient for lim-liveness [25].

There are some interesting advantages when dealing with fluid PN models. An important one is that the reachability set $\mathbf{R}(N, \boldsymbol{m}_0)$ is *convex* [25].

Proposition 18.2. *Let $\langle N, \boldsymbol{m}_0 \rangle$ be a continuous PN system. The set $\mathbf{R}(N, \boldsymbol{m}_0)$ is convex, i.e., if two markings $\boldsymbol{m}_1$ and $\boldsymbol{m}_2$ are reachable, then for any $\alpha \in [0,1]$, $\alpha \boldsymbol{m}_1 + (1-\alpha)\boldsymbol{m}_2$ is also a reachable marking.*

Notice that in a continuous system any enabled transition can be fired in a sufficiently small quantity such that it does not become disabled. This implies that every transition is fireable if and only if a strictly positive marking is reachable. This is equivalent to the non existence of empty *siphons* (a siphon is a set of places Σ s.t. ${}^\bullet\Sigma \subseteq \Sigma^\bullet$, then an unmarked siphon cannot gain marks). From this, realizability of T-semiflows can be deduced [25], and therefore behavioral and structural synchronic relations [28] coincide in continuous systems in which every transition is fireable at least once. In particular, considering boundedness and structural boundedness as in discrete systems, both concepts coincide in continuous systems in which every transition is fireable. And, as in discrete systems, structurally boundedness is equivalent to the existence of $\boldsymbol{y} > \boldsymbol{0}$ such that $\boldsymbol{y} \cdot \boldsymbol{C} \leq \boldsymbol{0}$ (see, for example, [3, 33]).

Another interesting property is that $\mathrm{RS}_D(N, \boldsymbol{m}_0) \subseteq \mathbf{R}(N, \boldsymbol{m}_0)$ implies that boundedness of the continuous system is a sufficient condition for boundedness of the discrete one. Moreover, it is important to stress that the set lim-$\mathbf{R}(N, \boldsymbol{m}_0)$ can be easily characterized if some common conditions are fulfilled [25].

Proposition 18.3. *Let $\langle N, \boldsymbol{m}_0 \rangle$ be consistent and such that each transition can be fired at least once. Then $\boldsymbol{m} \in$ lim-$\mathbf{R}(N, \boldsymbol{m}_0)$ iff there exists $\sigma > 0$ such that $\boldsymbol{m} = \boldsymbol{m}_0 + \boldsymbol{C} \cdot \sigma$.*

Hence, if a net is consistent and all the transitions are fireable, then the set of lim-reachable markings is fully characterized by the state equation . This immediately implies convexity of lim-$\mathbf{R}(N, \boldsymbol{m}_0)$ and the inclusion of every spurious discrete solution in lim-$\mathbf{R}(N, \boldsymbol{m}_0)$. Fortunately, every spurious solution in the border of the convex set lim-$\mathbf{R}(N, \boldsymbol{m}_0)$ can be *cut* by adding some implicit places (more precisely the so-called *cutting implicit places* [5]) what implies clear improvements in the state equation representation. This will be detailed in Subsection 18.4.1.

On the other hand, if $\langle N, \boldsymbol{m}_0 \rangle$ is not consistent or some transitions cannot be fired, lim-$\mathbf{R}(N, \boldsymbol{m}_0)$ can still be characterized by using the state equation plus a simple additional constraint concerning the fireability of the transitions in supportσ. The set $\mathbf{R}(N, \boldsymbol{m}_0)$ can also be fully determined by adding one further constraint related to the fact that a finite firing sequence cannot empty a trap [17] (in contrast to infinite sequences which might empty initially marked traps as shown in the previous section).

18.3 Fluidization of Timed Net Models

This section introduces the notion of time in the continuous Petri net formalism presenting the most used firing semantics. The main focus will be on *infinite server semantics*. Some basic properties will be discussed.

18.3.1 Server Semantics

If a timed interpretation is included in the *continuous* model (time is associated to the transitions, thus they fire with certain speed), the fundamental equation depends on time: $\boldsymbol{m}(\tau) = \boldsymbol{m}_0 + \boldsymbol{C} \cdot \sigma(\tau)$, which, by assuming that $\sigma(\tau)$ is differentiable, leads to $\dot{\boldsymbol{m}}(\tau) = \boldsymbol{C} \cdot \dot{\sigma}(\tau)$. The derivative of the firing count vector of the sequence σ is $\boldsymbol{f}(\tau) = \dot{\sigma}(\tau)$, called the *(firing) flow*, and leads to the following equation for the dynamics of the timed continuous PN (TCPN) system:

$$\dot{\boldsymbol{m}}(\tau) = \boldsymbol{C} \cdot \boldsymbol{f}(\tau). \tag{18.1}$$

Depending on how the flow $\boldsymbol{f}$ is defined, different firing *semantics* can be obtained. For *finite server semantics* (FSS), if the markings of the input places of t_j are strictly greater than zero (*strongly enabled*), its flow will be constant, equal to λ_j, i.e., all servers work at full speed. Otherwise (*weakly enabled*), the flow will be the minimum between its maximal firing speed and the total input flow to the input empty places. Formally,

$$f_j(\tau) = \begin{cases} \lambda_j & \text{if } \forall p_i \in {}^\bullet t_j, m[p_i] > \mathbf{0} \\ min\left\{ \underset{p_i \in {}^\bullet t_j | \boldsymbol{m}_i = 0}{min} \left\{ \sum_{t' \in {}^\bullet p_i} \frac{f[t'] \cdot Post[t', p_i]}{Pre[p_i, t_j]} \right\}, \lambda_j \right\} & \text{otherwise} \end{cases} \tag{18.2}$$

The dynamical system under FSS corresponds to a piecewise constant system and a switch occurs when a marked place becomes empty and the new flow values are computed ensuring that the marking of all places remain positive. Many examples using this semantics are given in [9] while in [21] a model is studied using this and the following semantics.

In the case of *infinite server semantics* (ISS), the flow of transition t_j is given by:

$$f_j(\tau) = \lambda_j \cdot enab(t_j, \boldsymbol{m}(\tau)) = \lambda_j \cdot \min_{p_i \in {}^\bullet t_j} \frac{m_i}{Pre[p_i, t_j]}. \tag{18.3}$$

Under ISS, the flow through a transition is proportional to the marking of the input place determining the enabling degree. As already advanced, TCPNs under ISS have the capability to simulate Turing machines [27], thus they enjoy an important expressive power; nevertheless, certain important properties are *undecidable* (for example, marking coverability, submarking reachability or the existence of a steady-state).

The flow through a transition under *product* (population) semantics (PS) [2] is given by

$$f_j(\tau) = \lambda_j \cdot \prod_{p_i \in {}^\bullet t_j} \frac{m_i}{Pre[p_i, t_j]}. \tag{18.4}$$

Product semantics can lead to *chaotic* models, i.e., models of deterministic dynamical systems that are extremely sensitive to initial conditions. This semantic is frequently adopted in population systems, like predator/pray, biochemistry, etc. (see, for example, [15]).

In the case of manufacturing or logistic systems, it is natural to assume that the transition firing flow is the minimum between the number of clients and servers and, FSS or ISS are mainly used [35, 9]. Since, these two semantics provide two different approximations of the discrete net system, an immediate problem is to decide which semantics will approximate "better" the original system. In [9], the authors observed that most frequently ISS approximates better the marking of the discrete net system. Furthermore, for mono-T-semiflow reducible net systems [18], it has been proved that ISS approximates better the flow in steady state under some (general) conditions [21]. For this reason, ISS will be mainly considered in the rest of this work. Let us recall the formal definition.

Definition 18.3. *A* timed continuous Petri net *(TCPN) is a time-driven continuous-state system described by the tuple $\langle N, \boldsymbol{\lambda}, \boldsymbol{m_0}\rangle$, where $\langle N, \boldsymbol{m_0}\rangle$ is a* continuous *PN and the vector $\boldsymbol{\lambda} \in \mathbb{R}_{>0}^{|T|}$ represents the transitions rates that determine the temporal evolution of the system. Under* ISS *the* flow *(the firing speed) through a transition t_i is defined as the product of the rate λ_i and $enab(t_i, \boldsymbol{m})$, i.e., $f_i(\boldsymbol{m}) = \lambda_i \cdot enab(t_i, \boldsymbol{m}) = \lambda_i \cdot \min_{p \in {}^\bullet t_i}\{m[p]/Pre[p, t_i]\}$.*

Let us recall now some useful concepts.

[2] Through discoloration of colored (discrete) nets, the *minimum* operator of ISS is transformed into a PS [34].

Definition 18.4. *A* configuration *of a TCPN, denoted as $\mathscr{C}_k$, is a set of (p,t) arcs, one per transition,* covering *the set T of transitions. It is said that the system at marking $\boldsymbol{m}$ is at a configuration $\mathscr{C}_k$ if the arcs in $\mathscr{C}_k$ provide the minimum ration in the ISS definition (18.3), or equivalently, $\mathscr{C}_k$ is the active configuration at $\boldsymbol{m}$. A configuration matrix $|T| \times |P|$ is associated to each configuration as follows:*

$$\boldsymbol{\Pi}_k[t_j, p_i] = \begin{cases} \frac{1}{Pre[p_i,t_j]}, & \text{if } (p_i,t_j) \in \mathscr{C}_k \\ 0, & \text{otherwise} \end{cases} \tag{18.5}$$

The reachability set of a TCPN system can be partitioned (except on the borders) according to the configurations, and inside each obtained convex region *$\mathscr{R}_k$ the system dynamics is linear. More formally, $\mathscr{R}_k = \{\boldsymbol{m} \in \text{lim-}\mathbf{R}(N, \boldsymbol{m}_0) | \boldsymbol{\Pi}_k \boldsymbol{m} \le \boldsymbol{\Pi}_j \boldsymbol{m}, \forall \boldsymbol{\Pi}_j \in \{\boldsymbol{\Pi}\}\}$, where $\{\boldsymbol{Pi}\}$ denotes the set of all configuration matrices.*

The number of regions and configurations is upper bounded by $\prod_{t \in |T|} |^{\bullet}t|$ but some of them can be redundant, thus removed [22]. Let us define the firing rate matrix $\boldsymbol{\Lambda} = diag(\{bm\lambda\})$ (i.e., a diagonal $|T| \times |T|$ matrix containing the rates of the transitions). The evolution of a TCPN, under ISS, can be represented as:

$$\dot{\boldsymbol{m}}(\tau) = \boldsymbol{C} \cdot \boldsymbol{f}(\tau) = \boldsymbol{C} \cdot \boldsymbol{\Lambda} \cdot \boldsymbol{\Pi}(\boldsymbol{m}) \cdot \boldsymbol{m}(\tau), \tag{18.6}$$

where $\boldsymbol{\Pi}(\boldsymbol{m})$ is the configuration matrix operator ($\boldsymbol{\Pi}(\boldsymbol{m}) = \boldsymbol{\Pi}_k$ where $\mathscr{C}_k$ is the active configuration at $\boldsymbol{m}$). Notice that, while the system is evolving inside a region $\mathscr{R}_k$, it behaves linearly as $\dot{\boldsymbol{m}} = \boldsymbol{C}\boldsymbol{\Lambda}\boldsymbol{\Pi}_k\boldsymbol{m}$, thus a TCPN under ISS is a piecewise-linear system.

18.3.2 Qualitative Properties under ISS

According to (18.3), it is obvious to remark that being the initial marking of a continuous net system positive, the marking will remain positive during the (unforced or non-controlled) evolution. Hence, it is not necessary to add constraints to ensure the non-negativity of the markings. On the other hand, according to (18.3) as well, two homothetic properties are dynamically satisfied:

- if $\boldsymbol{\lambda}$ is multiplied by a constant $k > 0$ then identical markings will be reached, but the system will evolve k times faster;
- if the initial marking is multiplied by k, the reachable markings are multiplied by k and the flow will also be k times bigger.

Unfortunately, ISS has not only "good" properties and some "paradoxes" appear. For example, it could be thought that, since fluidization relax some restrictions, the throughput (flow at steady-state) of the continuous system should be at least that of the timed discrete one. However, the throughput of a TCPN is not in general an upper bound of the throughput of the discrete PN [35]. Moreover, if only some components of $\boldsymbol{\lambda}$ or only some components of $\boldsymbol{m}_0$ are increased the steady state throughput is not monotone in general.

Two monotonicity results of the steady-state throughput are satisfied under some general conditions [21]:

Proposition 18.4. *Assume $\langle N, \boldsymbol{\lambda}_i, \boldsymbol{m}_i \rangle$, $i = 1,2$ are mono-T-semiflow TCPNs under ISS which reach a steady-state. Assume that the set of places belonging to the arcs of the steady state configuration contains the support of a P-semiflow. If*

1. $\langle N, \boldsymbol{\lambda}_1, \boldsymbol{m}_1 \rangle$ and $\langle N, \boldsymbol{\lambda}_1, \boldsymbol{m}_2 \rangle$ verify $\boldsymbol{m}_1 \leq \boldsymbol{m}_2$ or
2. $\langle N, \boldsymbol{\lambda}_1, \boldsymbol{m}_1 \rangle$ and $\langle N, \boldsymbol{\lambda}_2, \boldsymbol{m}_1 \rangle$ verify $\lambda_1 \leq \boldsymbol{\lambda}_2$,

then the steady state flows satisfy $\boldsymbol{f}_1 \leq \boldsymbol{f}_2$.

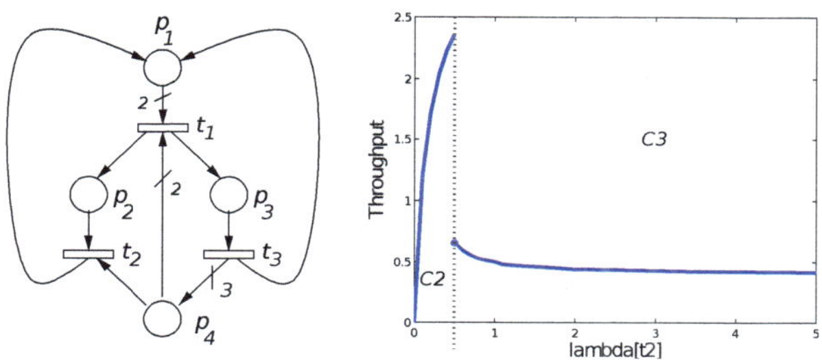

Fig. 18.3 A Mono-T-semiflow net and its "fluid" throughout in steady-state. Observe that it is not smooth, and that increasing $\lambda_2 > 0.5$ the throughput is counterintuitive (faster machine, slower behavior

Let us consider for instance, the mono-T-semiflow TCPN in Fig. 18.3(a) under ISS with $\lambda_1 = \lambda_3 = 1$ and $\boldsymbol{m}_0 = [15\ 1\ 1\ 0]^T$. Different modes can govern the evolution of the system at steady-state. For example, if $0 < \lambda_2 \leq 0.5$, the flow in steady-state is $f_1(\tau) = m_1(\tau)$, $f_2(\tau) = m_4(\tau)$ and $f_3(\tau) = m_3(\tau)$, respectively (i.e., $\frac{m_1(\tau)}{2} < \frac{m_4(\tau)}{2}$ and $m_4(\tau) < m_2(\tau)$ in steady state). Therefore, $\mathscr{C}_2 = \{(p_1,t_1),(p_4,t_2),(p_3,t_3)\}$ is the steady-state configuration and the set of places $\{p_1, p_4, p_3\}$ determines (constrains) the flow. Since this set contains the support of the P-semiflow $\boldsymbol{y} = [1\,0\,4\,1]$, the steady-state flow is monotone (Fig. 18.3(b)). When $\lambda_2 > 0.5$, the steady-state configuration becomes $\mathscr{C}_3 = \{(p_4,t_1),(p_2,t_2),(p_3,t_3)\}$, i.e., the set of places governing the evolution becomes $\{p_4, p_2, p_3\}$, that is the support of the P-flow $\boldsymbol{y} = [0\,1\,-3\,-1]$, not a P-semiflow, and monotonicity may not hold (Fig. 18.3(b)).

The connection between liveness in the autonomous (untimed) continuous model and the timed ones has been investigated. First of all, notice that if a steady-state exists in the timed model, from (18.1) $\dot{\boldsymbol{m}} = \boldsymbol{C} \cdot \boldsymbol{f}_{ss} = \boldsymbol{0}$ is obtained (independently of the firing semantics), where $\boldsymbol{f}_{ss}$ is the flow vector of the timed system in the steady state, $\boldsymbol{f}_{ss} = \lim_{\tau \to \infty} \boldsymbol{f}(\tau)$. Therefore, the flow in the steady state is a T-semiflow of the net. Deadlock-freeness and liveness definitions can be extended to timed continuous systems as follows:

Definition 18.5. *Let $\langle N, \boldsymbol{\lambda}, \boldsymbol{m}_0 \rangle$ be a timed continuous PN system and $\boldsymbol{f}_{ss}$ be the vector of flows of the transitions in the steady state.*

- $\langle N, \boldsymbol{\lambda}, \boldsymbol{m}_0 \rangle$ *is* timed-deadlock-free *if* $\boldsymbol{f}_{ss} \neq \mathbf{0}$;
- $\langle N, \boldsymbol{\lambda}, \boldsymbol{m}_0 \rangle$ *is* timed-live *if* $\boldsymbol{f}_{ss} > \mathbf{0}$;
- $\langle N, \boldsymbol{\lambda} \rangle$ *is* structurally timed-live *if* $\exists\ \boldsymbol{m}_0$ *such that* $\boldsymbol{f}_{ss} > \mathbf{0}$.

Notice that if a timed system is not timed-live (timed-deadlock-free), it can be concluded that, seen as untimed, the system is not lim-live (lim-deadlock-free), since the evolution of the timed system just gives a particular trajectory that can be fired in the untimed system. This fact allows us to establish a one-way bridge from liveness conditions of timed and untimed systems. The reverse is not true, i.e., the untimed system can deadlock, but a given $\boldsymbol{\lambda}$ can *drive* the marking along a trajectory without deadlocks. In other words, the addition of an arbitrary transition-timed semantics to a system imposes constraints on its evolution what might cause the timed system to satisfy some properties, as boundedness and liveness, that are not necessarily satisfied by the untimed system [38, 39]. The relationships among liveness definitions are depicted in Fig. 18.4.

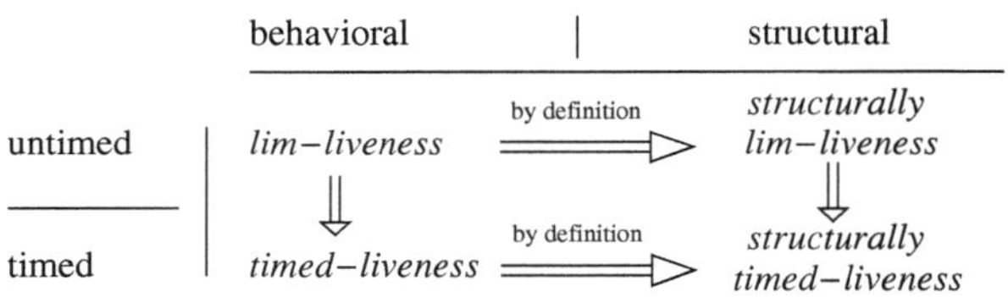

Fig. 18.4 Relationships among liveness definitions for continuous models

As an example, let us show how some conditions initially obtained for timed systems can be applied to untimed ones. It is known that if a MTS timed net $\langle N, \boldsymbol{\lambda} \rangle$ is structurally live for any $\boldsymbol{\lambda} > \mathbf{0}$ then for every transition t there exists $p \in {}^{\bullet}t$ such that $p^{\bullet} = \{t\}$, i.e., p is *structurally persistent* or *conflict-free* [19]. Let $\langle N, \boldsymbol{\lambda} \rangle$ be a MTS timed net containing a transition t such that for every $p \in {}^{\bullet}t$, $|p^{\bullet}| > 1$. According to the mentioned condition, $\boldsymbol{\lambda}$ exists such that $\langle N, \boldsymbol{\lambda} \rangle$ is not structurally timed-live. Therefore N is not structurally lim-live, since structurally timed-liveness is a necessary condition for structurally lim-liveness (see Fig. 18.4).

18.3.3 On the Quantitative Approximation under ISS

Fluid PNs are usually considered as relaxations of original discrete models. In fact, the definitions for the most usual semantics for timed continuous PNs were inspired by the average behavior of high populated timed discrete PNs [9, 26]. Nevertheless, the dynamic behavior of a timed continuous PN model does not always approximate that of the corresponding timed discrete PN. For this reason, it is important to investigate the conditions that lead to a valid relaxation. In some sense, this subsection deals with the *legitimization* of the so called ISS and the consideration of some issues that affect the quality of the approximation.

Let us consider Markovian Petri nets (MPN), i.e., stochastic discrete Petri nets with exponential delays associated to the transitions and conflicts solved by a race policy [23]. In [37] it was shown that, in certain cases, the marking of a TCPN under ISS approximates the average marking of the corresponding MPN (having the same structure, rates and initial marking, under the assumption of ergodicity). The approximation is better when the enabling degrees of the transitions (the number of active servers) is large and the system mainly evolves inside one marking region (according to one configuration or linear mode), i.e., for each synchronization, a single place is *almost always* constraining the throughput. Errors in the approximation may appear due to the existence of *sychronizations*: *arc weights* greater than one and *joins* (*rendez-vous*). The reason is that the enabling degree (thus the flow) definition for the TCPN does not accurately describe the enabling degree (thus the throughput) of the discrete model in these cases. In fact, the approximation is perfect for ordinary Join-Free Petri nets.

Let us provide an intuitive explanation about how the arc weights introduce approximation errors. Assume that the continuous marking of a TCPN approximates the average marking of the corresponding MPN at certain time τ, i.e., $E\{\boldsymbol{M}(\tau)\} \sim \boldsymbol{m}(\tau)$. Given an arc with weight q connecting a place p_j to a transition t_i, the *expected* enabling degree of t_i in the MPN would be $E\{Enab(t_i)\} = E\{\lfloor M[p_j]/q \rfloor\}$, which is different from the enabling degree in the TCPN $enab(t_i) = m[p_j]/q \sim E\{M[p_j]\}/q$, due to the presence of the operator $\lfloor \cdot \rfloor$ (note that in ordinary arcs $q = 1$, thus $\lfloor M[p_j]/q \rfloor = M[p_j]/q$), and the quality of the approximation will be reduced or lost for future time.

As an example, consider the MPN system of Fig. 18.5(a) with timing rates $\lambda_1 = \lambda_2 = 1$, and initial marking $\boldsymbol{M}_0 = [k \cdot q,\ 0]^T$, where $k, q \in \mathbb{N}^+$. This system, and its corresponding TCPN, were evaluated for different values of k and q. The obtained values for the throughput and flow of t_1, at steady state, are shown in Table 18.1. Note that, when $k = 1$ (Table 18.1(a), here the marking is relatively very small), the larger the weight of the input arc of t_1 (i.e., q), the larger the difference (error) between the throughput in the MPN ($\chi[t_1]$) and the flow in the TCPN ($f[t_1]$). Observe that the flow in the continuous model remains unchanged. On the other hand, when the arc weights are fixed but the initial marking (i.e., k) is increased (Table 18.1(b)), the relative approximation error decreases (in such case, $E\{\lfloor M[p_1]/q \rfloor\} \sim E\{M[p_1]\}/q$ for $M[p_1] >> q$). Concluding, the relative errors introduced by arc weights become smaller when the marking in the net is increased w.r.t. those weights.

Let us illustrate now, how joins (rendez-vous) introduce approximation errors. Given a synchronization t_i with two input places $\{p_j, p_k\}$, the expected enabling in the MPN would be $E\{Enab(t_i)\} = E\{min(M[p_j], M[p_k])\}$, which is not equal to the

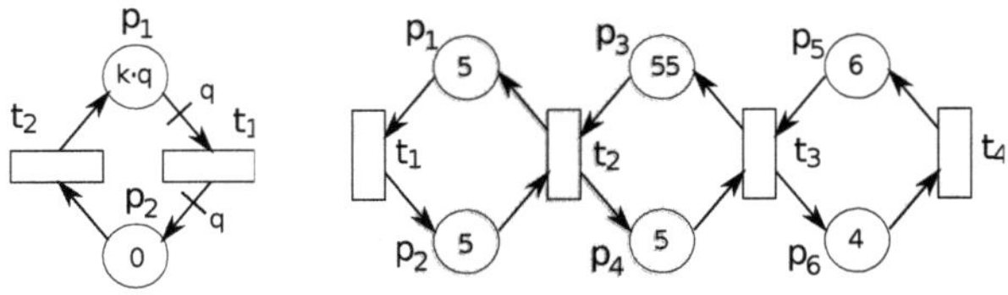

Fig. 18.5 a) Cycle net with arc weights. b) Marked graph which evolves through different regions

Table 18.1 Throughput and its approximation for t_1 of the net of Fig. 18.5(a)

(a)

$k=1 \setminus q=$	1	2	4	8	16
$MPN, \chi[t_1]$	0.50	0.40	0.32	0.26	0.22
$TCPN, f[t_1]$	0.50	0.50	0.50	0.50	0.50

(b)

$q=4 \setminus k=$	1	2	4	8	16
$MPN, \chi[t_1]$	0.32	0.80	1.76	3.78	7.68
$TCPN, f[t_1]$	0.50	1.00	2.00	4.00	8.00

enabling in the TCPN $enab(m[p_j], m[p_k]) \sim min(E\{M[p_j]\}, E\{M[p_k]\})$, because the order in which the expect value and the "min" operator are applied cannot be commuted.

As an example, the MPN system of Fig. 18.5(b) was simulated with timing rates $\lambda_1 = \lambda_2 = \lambda_3 = 1$ and different rates for t_4: $\lambda_4 \in \{2, 1.5, 1.2, 1\}$. The corresponding TCPN model was also simulated. The average markings at the steady state are shown in Table 18.2 (columns MPN and TCPN). The column denoted as $E\{Enab(t_4)\}$ is the average enabling degree of t_4 at the steady state (this represents a lower bound for the number of active servers in the transitions). The value in column $P(\boldsymbol{M} \in \mathscr{R}_{ss})$ is the probability that the marking is inside the region $\mathscr{R}_{ss}$, related to the steady state of the TCPN (equivalently, the fraction of time that $\boldsymbol{M}(\tau)$ evolves according to a single configuration or linear mode). Note that the lower the probability that $\boldsymbol{M}(\tau)$ belongs to $\mathscr{R}_{ss}$, the larger the difference (the error) between the MPN and the TCPN, even if the average enabling degrees increase. On the other hand, a good approximation is provided when the probability that $\boldsymbol{M}(\tau) \in \mathscr{R}_{ss}$ is high, which occurs for $\lambda_4 = 2$. The approximation holds because, in this case, $\boldsymbol{M}(\tau)$ mainly evolves in one region $\mathscr{R}_{ss}$ (in particular, $E\{min(M[p_4], M[p_5])\} \sim E\{M[p_4]\}$ and $E\{min(M[p_2], M[p_3])\} \sim E\{M[p_2]\}$).

Table 18.2 Marking approximation of p_3 for the MPN of Fig. 18.5(b)

λ_4	MPN	TCPN	TnCPN	$E\{Enab(t_4)\}$	$P(\boldsymbol{M} \in \mathscr{R}_{ss})$
2	54.62	55	54.63	2.53	0.8433
1.5	53.87	55	53.88	3.22	0.661
1.2	51.16	55	51.17	3.88	0.413
1	29.97	55	30.73	4.93	0.036

From a continuous-systems perspective, it can be deduced that approximation does not accumulate in time if the steady state marking of the continuous model is *asymptotically stable* (because the deviations of the MPN from its expected behavior, which is similar to that of the TCPN, vanish with the time evolution). Therefore, asymptotic stability is a necessary condition (together with *liveness*, otherwise, the continuous system may die while the discrete is live) for the approximation of the steady state.

18.4 Improving the Approximation: Removing Spurious Solutions, Addition of Noise

Since the approximation provided until now by a fluid PN is not always accurate, a question that may arise is the possibility of improving such approximation by means of modifying the continuous Petri net definition. Through this section, a couple of approaches, for such improvement, will be discussed.

18.4.1 Removing Spurious Solutions

The state equation provides a full characterization of the lim-reachable markings (in the autonomous continuous model) for consistent nets with no empty siphons. This allows one to use the state equation to look for deadlocks, i.e., markings at which every transition has at least one empty input place. In some cases, at such a deadlock $\boldsymbol{m}$ there is an empty trap that was initially marked. Nevertheless, it is well known that initially marked traps cannot be completely emptied in discrete nets. Thus, $\boldsymbol{m}$ is a *spurious solution* of the state equation if we consider the system as discrete, equivalently, deadlock-freeness has not been preserved during fluidization.

Consider for instance the net in Fig. 18.6 with $\boldsymbol{m}_0 = [10\ 11\ 0]^T$. The marking $\boldsymbol{m} = [0\ 1\ 10]^T$ is a deadlock and can be obtained as a solution of the state equation with $\sigma = [10\ \ 0]^T$ as firing count vector. Thus given that the system satisfies the conditions of Proposition 18.3, $\boldsymbol{m}$ is lim-reachable, i.e., the system lim-deadlocks. Notice that p_1 is a trap ($^\bullet p_1 = p_1{}^\bullet$) that was initially marked and can be emptied by an infinite firing sequence. Thus, $\boldsymbol{m}$ is not reachable in the discrete net, thus it is an spurious deadlock.

Fortunately, some spurious deadlocks can be removed from the state equation by adding implicit places [5]. For this, it is first necessary to detect if a deadlock $\boldsymbol{m}$ is spurious or not. Let us define $\boldsymbol{Pre}_\Theta$ and $\boldsymbol{Post}_\Theta$ as $|P| \times |T|$ sized matrices such that:

- $Pre_\Theta[p,t] = 1$ if $Pre[p,t] > 0$, $Pre_\Theta[p,t] = 0$ otherwise
- $Post_\Theta[p,t] = |^\bullet t|$ if $Post[p,t] > 0$, $Post_\Theta[p,t] = 0$ otherwise.

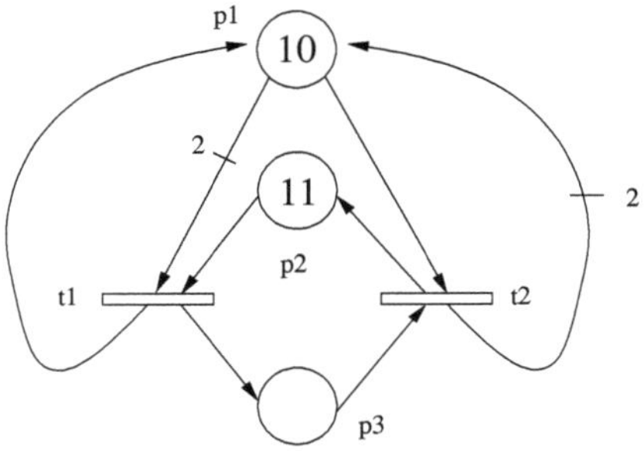

Fig. 18.6 A continuous MTS system that integrates a discrete spurious deadlock $\boldsymbol{m} = [0\ 1\ 10]^T$, reachable through the firing sequence $5t_1, 2.5t_1, 1.25t_1, \ldots$

Equations $\{\boldsymbol{y}^T \cdot \boldsymbol{C}_\Theta = \boldsymbol{0},\ \boldsymbol{y} \geq \boldsymbol{0}\}$ where $\boldsymbol{C}_\Theta = \boldsymbol{Post}_\Theta - \boldsymbol{Pre}_\Theta$ define a generator of traps (Θ is a trap iff $\exists \boldsymbol{y} \geq \boldsymbol{0}$ such that $\Theta =$ support$\boldsymbol{y}$, $\boldsymbol{y}^T \cdot \boldsymbol{C}_\Theta = \boldsymbol{0}$) [14, 33]. Hence:

Proposition 18.5. *Given* $\boldsymbol{m}$*, if the following bilinear system:*

- $\boldsymbol{m} = \boldsymbol{m}_0 + \boldsymbol{C} \cdot \boldsymbol{\sigma},\ \ \boldsymbol{m}, \boldsymbol{\sigma} \geq 0,$ $\{$*state equation*$\}$
- $\boldsymbol{y}^T \cdot \boldsymbol{C}_\Theta = \boldsymbol{0}, \boldsymbol{y} \geq \boldsymbol{0},$ $\{$*trap generator*$\}$
- $\boldsymbol{y}^T \cdot \boldsymbol{m}_0 \geq 1,$ $\{$*initially marked trap*$\}$
- $\boldsymbol{y}^T \cdot \boldsymbol{m} = 0,$ $\{$*trap empty at* $\boldsymbol{m}\}$

has solution, then $\boldsymbol{m}$ *is a spurious solution, and the support of vector* $\boldsymbol{y}$ *defines a trap that becomes empty.*

The result of Proposition 18.5 follows directly from the fact that the support of $\boldsymbol{y}$ is a trap that has been emptied. Let us illustrate now how spurious solutions can be cut by adding an *implicit* place (a place is said to be *implicit* if it is never the unique place that forbids the firing of its output transitions, i.e., it does not constraint the behavior of the sequential net system).

Recalling the example of Fig. 18.6, since p_1 is an initially marked trap, its marking must satisfy $m[p_1] \geq 1$. This equation together with the conservative law $m[p_1] + m[p_3] = 10$ leads to $m[p_3] \leq 9$. This last inequality can be forced by adding a slack variable, i.e., a *cutting implicit place* q_3, such that $m[p_3] + m[q_3] = 9$. Thus q_3 is a place having t_2 as input transition, t_1 as output transition, and 9 as initial marking. The addition of q_3 to the net system renders p_2 implicit (structurally identical but with a higher marking) and therefore p_2 can be removed without affecting the system behavior [5, 33]. In the resulting net system, $\boldsymbol{m} = [0\ 1\ 10]^T$ is not any more a solution of the state equation, i.e., it is not lim-reachable and then the net system does not deadlock as continuous.

Since deadlock markings in continuous systems are always in the borders of the convex set of reachable markings, discrete spurious deadlocks can be cut by the described procedure. Nevertheless, such an addition creates more traps that might be treated similarly in order to improve further the quality of the continuous approximation. It is important to remark that, by eliminating spurious deadlocks, the approximation of the performance of the discrete net system, provided by the timed relaxation, is also improved even if the deadlock is not reached in the timed continuous model. In any case, removing spurious solutions always represents an improvement of the fluidization, being specially important when those are deadlocks or represent non-live steady states.

As an example, consider again the MPN given by the net of Fig. 18.6 with initial marking $\boldsymbol{M}_0 = [10\,11\,0]^T$ and rates $\boldsymbol{\lambda} = [0.4\,1]$. As already shown, this PN has a spurious deadlock, which can be removed by eliminated the two *frozen tokens* from p_2. This is equivalent to consider $\boldsymbol{M}'_0 = [10\ 9\ 0]^T$ as the initial marking. The MPN and the corresponding fluid model TCPN have been simulated for both initial markings $\boldsymbol{M}_0$ (with spurious deadlocks) and $\boldsymbol{M}'_0$, for different rates at t_1 ranging in $\lambda_1 \in [0.4, 4]$. The throughput at t_1, for both models, is shown if Fig. 18.7(a). It can be seen that the MPN is live for any $\lambda_1 \in [0.4, 4]$, furthermore, the throughput seems as a smooth function of λ_1. On the other hand, the continuous model with the original $\boldsymbol{M}_0$ reaches the (spurious) deadlock for any $\lambda_1 \in (2, 4]$. Note the discontinuity

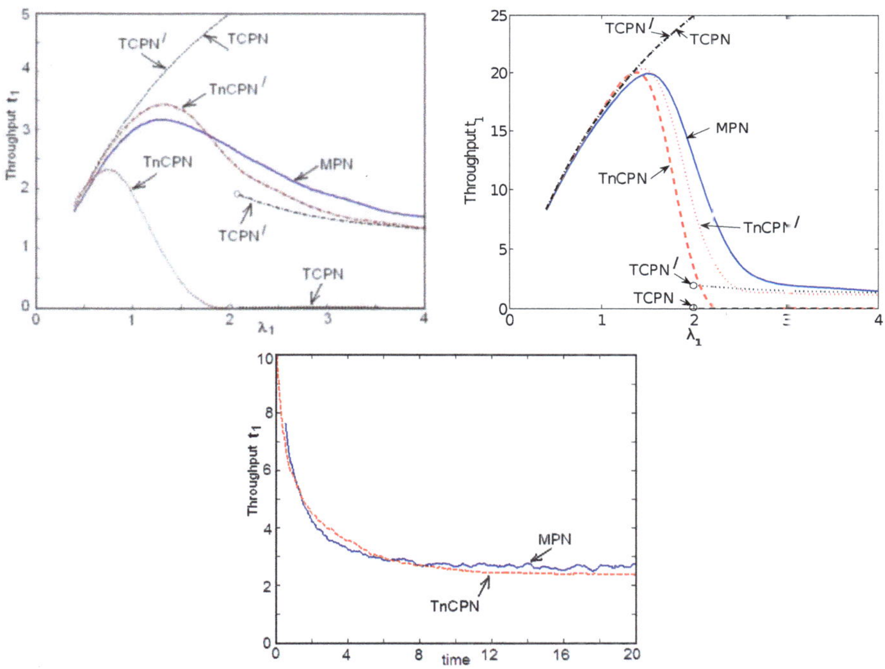

Fig. 18.7 Throughput for the MPN of Fig. 18.6 and its continuous relaxations for $\lambda_1 \in [0.4, 4]$, $\lambda_2 = 1$ and initial marking a) $\boldsymbol{M}_0 = [10\ 11\ 0]^T$ and b) $\boldsymbol{M}_0 = [50\ 55\ 0]^T$. TCPN′ and TnCPN′ represent the models without spurious deadlock. c) Average transient throughput at t_1 for $\boldsymbol{M}_0 = [10\ 11\ 0]^T$ and $\lambda_1 = 2$

at $\lambda_1 = 2$ for the TCPN model with both initial markings, i.e., the continuous model is neither monotonic nor smooth w.r.t the timing. Finally, it can be appreciated that the TCPN provides a much better approximation when the spurious deadlock is removed (TCPN′ with $\boldsymbol{M}'_0$), for any $\lambda_1 > 2$ (for $\lambda_1 \leq 2$ there is no change in the TCPN).

18.4.2 Adding Noise: Stochastic T-Timed Continuous PN

The approximation of the average marking of an ergodic (ergodicity means that the steady state is independent on the initial state, providing the P-flows have the same total marking) Markovian Petri net may be improved by adding white noise to the transitions flow of the TCPN [37]. Intuitively speaking, the transition firings of a MPN are stochastic processes, then, the noise added to the flow in the TCPN may help to reproduce such stochastic behavior, which even at steady state is particularly relevant at the synchronizations. The model thus obtained (here denoted as TnCPN) is represented, in discrete time with a sampling $\Delta\tau$, as: $\boldsymbol{m}_{k+1} = \boldsymbol{m}_k + \boldsymbol{C}(\boldsymbol{\Lambda\Pi}(\boldsymbol{m}_k)\boldsymbol{m}_k\Delta\tau + \boldsymbol{v}_k)$, with $\boldsymbol{v}_k$ being a vector of independent normally distributed random variables with zero mean and covariance matrix $\Sigma_{\boldsymbol{v}_k} = diag(\boldsymbol{\Lambda\Pi}(\boldsymbol{m}_k)\boldsymbol{m}_k\Delta\tau)$. This modification is particularly relevant when the system evolves near to the border between different regions, because in these cases, the continuous flow does not approximate the throughput of the discrete transitions (remember that, in a join $\{p_i^1,..,p_i^k\} = {}^\bullet t_i$, the difference between $E\{Enab(t_i)\} = E\{min(M_i^1,..,M_i^k)\}$ and its continuous approximation $enab(t_i) = min(m[p_i^1],..,m[p_i^k]) \sim min(E\{M[p_i^1]\}\ ,..,E\{M[p_i^k]\})$ may become large). The approximation improves as the enabling degrees of the transitions (the number of active servers) increase, as already mentioned, assuming *asymptotic stability* and *liveness* in the continuous system (thus it is important to previously remove any spurious deadlock).

An interesting issue is that the new continuous stochastic model approximates not only the average value but also the covariance of the marking of the original MPN. Moreover, since the TnCPN model is actually the TCPN one with zero-mean gaussian noise, many of the results known for the deterministic model can be used for analysis and synthesis in the stochastic continuous one. Nevertheless, the addition of noise cannot reduce the error introduced by arc weights.

For instance, consider again the MPN system of Fig. 18.5(b). The corresponding TnCPN was simulated for $\lambda_4 \in \{2, 1.5, 1.2, 1\}$. The average steady state marking is also shown in Table 18.2. As it was pointed out in the previous section, the lower the probability that $\boldsymbol{M}_k$ belongs to $\mathscr{R}_{ss}$, the larger the difference (the error) between the MPN and the deterministic TCPN. On the other hand, the approximation provided by the TnCPN system is good for all of those rates.

Now, consider again the MPN of Fig. 18.6 with $\boldsymbol{M}_0 = [1\,0\,1\,1\,0]^T$. The steady state throughput of the MPN and its different relaxations is shown if Fig. 18.7(a), for different values $\lambda_1 \in [0.4, 4]$. Note that the noise added to the TCPN makes this to reach the spurious deadlock quickly and the approximation to the MPN does not hold since the liveness precondition is not fulfilled. On the other hand, after removing the spurious deadlock with $\boldsymbol{M}'_0 = [10\ 9\ 0]^T$, the TnCPN approximates better the MPN than the TCPN model (curves TnCPN′ and MPN). Fig. 18.7(a) shows the results of the same experiment but with a bigger population. In this case, $\boldsymbol{M}_0 = 5 \cdot [10\ 11\ 0]^T = [50\ 55\ 0]^T$ and the spurious solution is removed by considering the initial marking $\boldsymbol{M}'_0 = [50\ 49\ 0]^T$ (in this case, six frozen tokens are removed from p_2). Note that this marking is not equal to five times the one used in the first case, i.e., $\boldsymbol{M}'_0 \neq 5 \cdot [10\ 9\ 0]^T$, then the curve TCPN′ in Fig. 18.7(b) is not in homothetic relation with that in Fig. 18.7(a) (but the original TCPN does it). It can be observed in Fig. 18.7(b) that now the continuous models provide a better approximation than in the case of Fig. 18.7(a), because the population is bigger. Finally, Fig. 18.7(c) shows the transient trajectory described by the average throughput of t_1, for the case $\boldsymbol{M}_0 = [10\ 11\ 0]^T$ and $\lambda_1 = 2$. It can be observed, that not only the steady state of the MPN is well approximated by the TnCPN′ (after removing the spurious deadlock), but also the transient evolution.

18.5 Steady State: Performance Bounds and Optimization

Product semantics may lead to continuous PN systems with steady orbits or limit cycles [34]. This semantic also allows the existence of chaotic behaviors. Analogously, when ISS are considered, a TCPN system may exhibit stationary oscillations (that can be maintained for ever). An example of an oscillatory behavior, can be found in [20] (Fig. 2 and 3). Usually, a TCPN evolves toward a steady state marking, like in the examples of Section 18.3. The knowledge of this final marking is interesting for performance evaluation purposes, since this represents the number of useful resources and active clients in the long term behavior of the modeled system. This is explored through this section.

A performance measure that is often used in discrete PN systems is the throughput of a transition in the steady state (assuming it exists), i.e., the number of firings per time unit. In the continuous approximation, this corresponds to the firing flow.

In order to study the throughput in discrete systems, the classical concept of "visit ratio" (from the queueing network theory) is frequently used. In Petri net terms, the visit ratio of a transition t_j with respect to t_i, $\boldsymbol{v}^{(i)}[t_j]$, is the average number of times t_j is visited (fired), for each visit to (firing of) the reference transition t_i.

Let us consider, consistent nets without empty siphons at $\boldsymbol{m}_0$ (Proposition 18.3). In order to simplify the presentation, let us assume that the net is MTS. Therefore, for any t_i, $\boldsymbol{f}_{ss} = \chi_i \cdot \boldsymbol{v}^{(i)}$, with χ_i the throughput of t_i. The vector of visit ratios is a right annuler of the incidence matrix $\boldsymbol{C}$, and therefore, proportional to the unique T-semiflow in MTS systems. For this class of systems, a throughput bound can be computed using the following nonlinear programming problem that maximize the flow of a transition (any of them, since all are related by the T-semiflow)

$$
\begin{array}{ll}
\max & \boldsymbol{f}_{ss}[t_1] \\
\text{s.t.} & \boldsymbol{m}_{ss} = \boldsymbol{m}_0 + \boldsymbol{C} \cdot \boldsymbol{\sigma} \\
& \boldsymbol{f}_{ss}[t_j] = \lambda_j \cdot \min_{p_i \in {}^{\bullet}t_j} \left\{ \frac{\boldsymbol{m}_{ss}[p_i]}{Pre[p_i,t_j]} \right\}, \forall t_i \in T \\
& \boldsymbol{C} \cdot \boldsymbol{f}_{ss} = 0 \\
& \boldsymbol{m}_{ss}, \sigma \geq 0
\end{array}
\tag{18.7}
$$

where $\boldsymbol{m}_{ss}$ is the steady-state marking. A way to solve (18.7), that due to the minimum operator is nonlinear, consists in using a branch & bound algorithm [18]. Relaxing the problem to a LPP, an upper bound solution can be obtained in polynomial time, although this may lead to a non-tight bound, i.e., the solution may be not reachable if there exists a transition for which the flow equation is not satisfied. If the net is not MTS, similar developments can be done adapting the equations in [12].

In the case of controlled systems, a LPP transformation of (18.7) can be used to compute an optimal steady-state assuming only flow reduction (the speed of the machines can only be reduced), $\boldsymbol{f} \geq \boldsymbol{0}$ and the steady-state flow should be repetitive $\boldsymbol{C} \cdot \boldsymbol{f} = \boldsymbol{0}$. If all transitions are controllable, it can be solved introducing some *slack* variables in order to transform the inequalities derived from the minimum operator in some equality constraints. These slack variables are used after to compute the optimal steady-state control [20]. For example, let us consider the following LPP:

$$
\begin{array}{ll}
\max & \boldsymbol{k}_1 \cdot \boldsymbol{f} - \boldsymbol{k}_2 \cdot \boldsymbol{m} - \boldsymbol{k}_3 \cdot \boldsymbol{m}_0 \\
\text{s.t.} & \boldsymbol{C} \cdot \boldsymbol{f} = \boldsymbol{0}, \boldsymbol{f} \geq \boldsymbol{0} \\
& \boldsymbol{m} = \boldsymbol{m}_0 + \boldsymbol{C} \cdot \boldsymbol{\sigma}, \boldsymbol{m}, \boldsymbol{\sigma} \geq \boldsymbol{0} \\
& f_i = \lambda_i \cdot \left(\frac{m[p_j]}{Pre[p_j,t_i]} \right) - v[p_j,t_i], \forall p_j \in {}^{\bullet}t_i,\ v[p_j,t_i] \geq 0
\end{array}
\tag{18.8}
$$

where $v[p_j,t_i]$ are *slack* variables. The objective function represents the profit that has to be maximized where $\boldsymbol{k}_1$ is a price vector w.r.t. steady-state flow $\boldsymbol{f}$, $\boldsymbol{k}_2$ is the work in process (WIP) cost vector w.r.t. the average marking $\boldsymbol{m}$ and $\boldsymbol{k}_3$ represents depreciations or amortization of the initial investments over $\boldsymbol{m}_0$. Using the slack variables v the optimal control in steady-state for a transition t_i if it is controllable, i.e., permits a control $u_i > 0$, is just $u_i = \min_{p_j \in {}^{\bullet}t_i} v[p_j,t_i]$. Therefore, this control problem (a synthesis problem) seems easier than the computations of performance (an analysis problem) even if, in general, is the opposite. Controllability issues will be considered from dynamic perspective in chapter 19.

18.6 Further Reading

For a broad perspective on fluidization of DEDS dealing with other modeling paradigms (as queueing networks or process algebra) and further reading on the presented topics, the reader is referred to [30]. Two complementary more basic presentations of fluid or continuous Petri net models can be found in [34, 35]. A comprehensive definition and application examples can be found in [9].

References

1. Altman, E., Jiménez, T., Koole, G.: On the comparison of Queueing Systems with their Fluid limits. Probability in the Engineering and Informational Sciences 15, 65–178 (2001)
2. Balduzzi, F., Giua, A., Menga, G.: First-order hybrid Petri nets: a model for optimization and control. IEEE Transactions on Robotics and Automation 16(4), 382–399 (2000)
3. Brams, G.W.: Réseaux de Petri: Théorie et Pratique, Masson (1983)
4. Champagnat, R., Esteban, P., Pingaud, H., Valette, R.: Modeling and simulation of a hybrid system through Pr/Tr PN-DAE model. In: Proc. ADPM 1998, 3rd International Conference on Automation of Mixed Processes, Reims, France, pp. 131–137 (1998)
5. Colom, J.M., Silva, M.: Improving the Linearly Based Characterization of P/T Nets. In: Rozenberg, G. (ed.) APN 1990. LNCS, vol. 483, pp. 113–145. Springer, Heidelberg (1991)
6. David, R., Alla, H.: Continuous Petri nets. In: Proc. 8th European Workshop on Application and Theory of Petri Nets, Zaragoza, Spain (1987)
7. David, R., Alla, H.: Autonomous and timed continuous Petri nets. In: Proc. 11th Int. Conf. on Application and Theory of Petri Nets, Paris, France (1990)
8. David, R., Alla, H.: Autonomous and Timed Continuous Petri Nets. In: Rozenberg, G. (ed.) APN 1993. LNCS, vol. 674, pp. 72–90. Springer, Heidelberg (1993)
9. David, R., Alla, H.: Discrete, Continuous and Hybrid Petri Nets, 2nd edn. Springer, Berlin (2010)
10. Demongodin, I., Audry, N., Prunet, F.: Batches Petri nets. In: Proc. IEEE Conference on Systems, Man and Cybernetics, vol. 1, pp. 607–617 (1993)
11. Chen, H., Mandelbaum, A.: Discrete flow networks: Bottleneck analysis and fluid approximations. Mathematical Operations Research 16, 408–446 (1991)
12. Chiola, G., Anglano, C., Campos, J., Colom, J.M., Silva, M.: Operational analysis of timed Petri nets and application to the computation of performance bounds. In: Quantitative Methods in Parallel Systems, pp. 161–174. Springer (1995)
13. Dai, J.G.: On positive Harris recurrence of multiclass queuing networks: A unified approach via fluid limit models. The Annals of Applied Probability 5(1), 49–77 (1995)
14. Ezpeleta, J., Couvreur, J.M., Silva, M.: A New Technique for Finding a Generating Family of Siphons, Traps and ST-Components. Application to Coloured Petri Nets. In: Rozenberg, G. (ed.) APN 1993. LNCS, vol. 674, pp. 126–147. Springer, Heidelberg (1993)
15. Heiner, M., Gilbert, D., Donaldson, R.: Petri Nets for Systems and Synthetic Biology. In: Bernardo, M., Degano, P., Zavattaro, G. (eds.) SFM 2008. LNCS, vol. 5016, pp. 215–264. Springer, Heidelberg (2008)
16. Hillston, J.: Fluid Flow Approximation of PEPA models. In: Proc. 2nd Int. Conf. on the Quantitative Evaluation of Systems, pp. 33–42 (2005)
17. Júlvez, J., Recalde, L., Silva, M.: On Reachability in Autonomous Continuous Petri Net Systems. In: van der Aalst, W.M.P., Best, E. (eds.) ICATPN 2003. LNCS, vol. 2679, pp. 221–240. Springer, Heidelberg (2003)
18. Júlvez, J., Recalde, L., Silva, M.: Steady state performance evaluation of continuous mono-T-semiflow Petri nets. Automatica 41(4), 605–616 (2005)
19. Júlvez, J., Recalde, L., Silva, M.: Deadlock-freeness analysis of continuous mono-T-semiflow Petri nets. IEEE Transaction on Automatic Control 51(9), 1472–1481 (2006)
20. Mahulea, C., Ramírez, A., Recalde, L., Silva, M.: Steady state control reference and token conservation laws in continuous Petri net systems. IEEE Transactions on Automation Science and Engineering 5(2), 307–320 (2008)

21. Mahulea, C., Recalde, L., Silva, M.: Basic server semantics and performance monotonicity of continuous Petri nets. Discrete Event Dynamic Systems 19(2), 189–212 (2009)
22. Mahulea, C., Recalde, L., Silva, M.: Observability of continuous Petri nets with infinite server semantics. Nonlinear Analysis: Hybrid Systems 4(2), 219–232 (2010)
23. Marsan, M.A., Balbo, G., Conte, G., Donatelli, S., Franceschinis, G.: Modelling with generalized stochastic Petri Nets. John Wiley and Sons (1995)
24. Recalde, L., Teruel, E., Silva, M.: On linear algebraic techniques for liveness analysis of P/T systems. Journal of Circuits, Systems and Computers 8(1), 223–265 (1998)
25. Recalde, L., Teruel, E., Silva, M.: Autonomous Continuous P/T Systems. In: Donatelli, S., Kleijn, J. (eds.) ICATPN 1999. LNCS, vol. 1639, pp. 107–126. Springer, Heidelberg (1999)
26. Recalde, L., Silva, M.: Petri nets fluidification revisited: Semantics and steady state. European Journal of Automation APII-JESA 35(4), 435–449 (2001)
27. Recalde, L., Haddad, S., Silva, M.: Continuous Petri nets: expressive power and decidability issues. Int. Journal of Foundations of Computer Science 21(2), 235–256 (2010)
28. Silva, M.: Towards a synchrony theory for P/T nets. In: Voss, K., et al. (eds.) Concurrency and Nets, pp. 435–460 (1987)
29. Silva, M., Colom, J.M.: On the structural computation of synchronic invariants in P/T nets. In: Proc. 8th European Workshop on Application and Theory of Petri Nets, Zaragoza, Spain (1987)
30. Silva, M., Júlvez, J., Mahulea, C., Vázquez, C.R.: On fluidization of discrete event models: observation and control of continuous Petri nets. Discrete Event Dynamic Systems 21(4), 427–497 (2011)
31. Silva, M., Teruel, E.: A systems theory perspective of discrete event dynamic systems: the Petri net paradigm. In: Borne, P., Gentina, J.C., Craye, E., El Khattabi, S. (eds.) Symp. on Discrete Events and Manufacturing Systems, pp. 1–12 (1996)
32. Silva, M., Teruel, E.: DEDS along their life cycle. Interpreted extensions of Petri nets. In: Proc. IEEE Int. Conf. on Systems, Man and Cybernetics, San Diego, CA, USA (1998)
33. Silva, M., Teruel, E., Colom, J.M.: Linear Algebraic and Linear Programming Techniques for the Analysis of Net Systems. In: Reisig, W., Rozenberg, G. (eds.) APN 1998. LNCS, vol. 1491, pp. 309–373. Springer, Heidelberg (1998)
34. Silva, M., Recalde, L.: Petri nets and integrality relaxations: A view of continuous Petri nets. IEEE Transactions on Systems, Man and Cybernetics 32(4), 314–327 (2002)
35. Silva, M., Recalde, L.: On fluidification of Petri net models: from discrete to hybrid and continuous models. Annual Reviews in Control 28, 253–266 (2004)
36. Trivedi, K., Kulkarni, V.G.: FSPNs: Fluid Stochastic Petri Nets. In: Ajmone Marsan, M. (ed.) ICATPN 1993. LNCS, vol. 691, pp. 24–31. Springer, Heidelberg (1993)
37. Vázquez, C.R., Recalde, L., Silva, M.: Stochastic–continuous state approximation of Markovian Petri net systems. In: Proc. 47th IEEE Conference on Decision and Control, pp. 901–906 (2008)
38. Vázquez, C.R., Silva, M.: Timing-dependent boundedness and liveness in continuous Petri nets. In: Proc. 10th Int. Workshop on Discrete Event Systems, Berlin, Germany (2010)
39. Vázquez, C.R., Silva, M.: Timing and Liveness in Continuous Petri nets. Automatica 47, 283–290 (2011)

Chapter 19
Continuous Petri Nets: Observability and Diagnosis

Cristian Mahulea, Jorge Júlvez, C. Renato Vázquez, and Manuel Silva

19.1 Introduction and Motivation

The observability problem for CPNs has been studied for both untimed and timed models. In the case of untimed systems, the state estimation is close to the one of discrete event systems since the firing of transitions can be assumed/seen as sequential and the corresponding events are not appearing simultaneously. In the case of timed systems, since the evolution can be characterized by a set of switching differential equations, the state estimation problem is more related to the linear and hybrid systems theory.

In this chapter, we first study different aspects of observability of CPN under infinite server semantics. The notion of *redundant configurations* is presented together with a necessary and sufficient condition for a configuration to be redundant. In some cases, this permits to reduce the system dimension. Three different concepts of observability can be defined for timed CPN based on the firing rate vector $\boldsymbol{\lambda}$. The classical observability problem is the one where the question is to estimate the initial state/marking measuring only a subset of states, while assuming a constant value for the firing vector. In this case, the set of differential equations is time invariant and the concept is called *punctual* or *classical observation*. Observability criteria of *piecewise affine systems* [4] can be applied to CPN since CPN is a subclass of those systems.

As already known, observability of a hybrid system requires not only the estimation of the continuous states but also of the discrete ones. To characterize this, the notion of *distinguishable regions* is introduced and a quadratic programing problem (QPP) is given to check if two regions are or are not distinguishable. Then, an observability criterion is given for general CPN systems. Since the complexity of a

Cristian Mahulea · Jorge Júlvez · C. Renato Vázquez · Manuel Silva
Instituto de Investigación en Ingeniería de Aragón (I3A), University of Zaragoza, Spain
e-mail: {cmahulea,julvez,cvazquez,silva}@unizar.es

C. Seatzu et al. (Eds.): Control of Discrete-Event Systems, LNCIS 433, pp. 387–406.
springerlink.com

potential algorithm based on this criterion may be high, some rules permitting the "deletion" of join transitions are given.

In many real systems it is impossible to have the exact values of the machine speeds. In the extreme case, nothing is known about the firing rate vector and the observability criteria of piecewise affine systems cannot be applied anymore. In this case, *structural observability* is defined and approaches based on the graph theory are used to study it. Only the knowledge of the system structure and the firing count vector (even if not constant) is assumed to be known. The idea is to determine which state variables can be estimated independently of the time values associated to transitions.

Finally, if one wants to estimate the system for "almost all" possible values of firing rate, *generic observability* is defined. In many cases, some punctual values of firing count vector can produce the loss of observability but it is not very important since it is observable outside a proper algebraic variety of the parameter space. Also here, graph based approaches are used. This concept is similar to the works on *linear structured systems* [7].

In the last part of the chapter, we present the effect of fluidization of Petri nets with respect to fault diagnosis. Untimed CPNs are considered and it is assumed that the amount in which some transitions are fired can be observed. It is pointed out that the set of potential marking after a sequence of observed transitions is convex. Based on this convexity, two linear programing problems (LPP) are given that permit us to assign three diagnosis states. The fluidization allows us to relax the assumption, common to all discrete event system diagnosis approaches, that there exists no cycle of unobservable transitions.

This chapter is mainly developed based on the theoretical results presented in [10, 17] and in the survey [22] for the observability of timed CPN under infinite server semantics. Theoretical results for state estimation and fault diagnosis for untimed CPN have been presented in [18].

19.2 A Previous Technicality: Redundant Configurations

We assume that the reader is familiar with the notions and definition given in Chapter 18 where CPNs have been introduced.

The number of configurations of a CPN is exponential and upper bounded by $\prod_{t_j \in T} |^{\bullet}t_j|$. A necessary condition for the observability of a CPN system is the observability of all linear systems. Therefore, if some configurations are "removed", the complexity analysis of the observability may decrease. Notice that the notion of *implicit places* [21] and *time implicit arcs* [14] cannot be used in the context of observability since the implicitness in these cases is proved for a given initial marking and for a given time intepretation. In our case, the initial marking is assumed to be partially known. In this section, we study a stronger concept, only depending on the net structure and valid for all possible initial markings $\boldsymbol{m}_0 \in \mathbb{R}_{\geq 0}^{|P|}$, concept called *redundant configuration*.

Definition 19.1. *Let $\mathscr{R}_i$ be a region associated to a CPN system. If for all $\boldsymbol{m}_0$, $\mathscr{R}_i \subseteq \bigcup_{j \neq i} \mathscr{R}_j$ then $\mathscr{R}_i$ is a* redundant region *and the corresponding configuration a* redundant configuration.

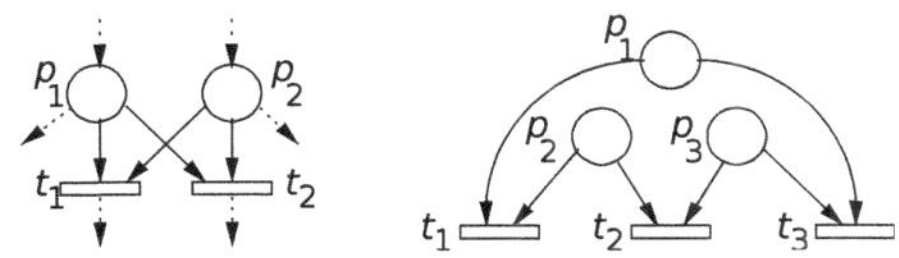

Fig. 19.1 Continuous PN with redundant regions

Example 19.1. Let us consider the left subnet in Fig. 19.1 and assume all $\boldsymbol{m} \in \mathbb{R}^{|P|}_{\geq 0}$ for which the enabling degree of t_1 is given by $m[p_1]$. Therefore, the following inequality is satisfied: $m[p_1] \leq m[p_2]$. Assume also that the enabling degree of t_2 is given by $m[p_2]$. Hence, $m[p_2] \leq m[p_1]$. Finally, let us assume that the enabling degree of all other transitions are given by the same places. Obviously, these markings belong to a region $\mathscr{R}_1$ such that for each marking $\boldsymbol{m} \in \mathscr{R}_1$ the following is true $m[p_1] = m[p_2]$.

Let us consider now all markings $\boldsymbol{m} \in \mathbb{R}^{|P|}_{\geq 0}$ for which the enabling degrees of t_1 and t_2 are given by $m[p_1]$ and the enabling degree of all other transitions is given by the same set of places as for markings belonging to $\mathscr{R}_1$. It is obvious that these markings belong to a region $\mathscr{R}_2$ for which $m[p_1] \leq m[p_2]$.

From the above definition of $\mathscr{R}_1$ and $\mathscr{R}_2$, it is obvious that $\mathscr{R}_1 \subseteq \mathscr{R}_2$ for all $\boldsymbol{m} \in \mathbb{R}^{|P|}_{\geq 0}$. Therefore, $\mathscr{R}_1$ (and the corresponding configuration) can be ignored in the analysis of the CPN system. ■

According to Definition 19.1, a region $\mathscr{R}_i$ is non-redundant if it is a full-dimensional convex polytope in $\mathbb{R}^{|P|}_{\geq 0}$. Therefore, for a given region we need to check if the inequalities composing its definition are strictly satisfied. If for a join t_j with $p_i, p_k \in {}^\bullet t_j$ does not exist $\boldsymbol{m} \in \mathbb{R}^{|P|}_{\geq 0}$ such that $\frac{m[p_i]}{Pre[p_i,t_j]} < \frac{m[p_k]}{Pre[p_k,t_j]}$ then the linear systems of the regions containing in its definition $\frac{m[p_i]}{Pre[p_i,t_j]} \leq \frac{m[p_k]}{Pre[p_k,t_j]}$ are redundant.

Proposition 19.1. *Let N be a timed CPN system. The region $\mathscr{R}_i$ with the corresponding configuration $\mathscr{C}_i$ is redundant iff $\nexists \boldsymbol{m} \in \mathbb{R}^{|P|}_{\geq 0}$ solution of the following system of strict inequalities of the form $\frac{m[p_k]}{Pre[p_k,t_j]} < \frac{m[p_u]}{Pre[p_u,t_j]}$, one for each $\frac{m[p_k]}{Pre[p_k,t_j]} \leq \frac{m[p_u]}{Pre[p_u,t_j]}$ defining $\mathscr{R}_i$.*

The existence of a solution for the system of strict inequalities in Proposition 19.1 can be checked solving a linear programing problem using a variable ε. For each $\frac{m[p_k]}{Pre[p_k,t_j]} < \frac{m[p_u]}{Pre[p_u,t_j]}$, a constraint of the following form is added: $\frac{m[p_k]}{Pre[p_k,t_j]} + \varepsilon \leq \frac{m[p_u]}{Pre[p_u,t_j]}$. The objective function will be to maximize ε. If the resulting LPP is infeasible or has as solution $\varepsilon = 0$ then $\mathscr{R}_i$ is a redundant region.

A pre-arc (an arc connecting a place p_i with a transition t_j) is called *implicit* in an untimed system, if for any reachable marking, the marking of p_i is never constraining/restricting the firing of t_j. If the system is under a timed interpretation, it is called *timed implicit*. It may seem that if a mode is redundant, a set of arcs has to be implicit or timed implicit, since they cannot define the enabling. However, it is not true, since it is not that an arc never defines the enabling, but that a combination of arcs may never define the enabling. For example, in the left net in Fig. 19.1, none of the arcs is implicit, although a region (the one corresponding to $m[p_1] = m[p_2]$) is reduced to its borders. In this example, the redundant mode could also have been avoided by fusing transitions t_1 and t_2 into a single one [14]. However, this kind of transformation cannot always be applied, as shown in the following example.

Example 19.2. Let us consider the right CPN in Fig. 19.1 and let us consider the region $\mathscr{R}_1 = \{m[p_2] \leq m[p_1], m[p_3] \leq m[p_2], m[p_1] \leq m[p_3]\}$ that it is equivalent to assume that the enabling degree of t_1 is given by $m[p_2]$, the one of t_2 by $m[p_3]$ and of t_3 by $m[p_1]$. Applying Proposition 19.1 we want to check if $\mathscr{R}_1$ is redundant. We have to consider the following system:

$$\begin{cases} m[p_2] < m[p_1] & (1) \\ m[p_3] < m[p_2] & (2) \\ m[p_1] < m[p_3] & (3). \end{cases} \tag{19.1}$$

Combining 19.1(2) and 19.1(3) we obtain $m[p_1] < m[p_2]$ that is in contradiction with 19.1(1). Therefore, region $\mathscr{R}_1$ and configuration $\mathscr{C}_1$ are redundant. ∎

The same problem of reducing dimension of a CPN under infinite server semantics has been studied in [19] using the concept of symmetry. It is shown that such a symmetry leads to a permutation of the regions and to equivariant dynamics (dynamical systems that have symmetries). This can be used for reductions to systems of smaller dimension.

19.3 Observability Criteria

Let us assume that the marking of some places $P_o \subseteq P$ can be measured, i.e., the token load at every time instant is known, due to some sensors. The observability problem is to estimate the other marking variables using these measurements. Defining $\boldsymbol{A}_i = \boldsymbol{C} \cdot \boldsymbol{\Lambda} \cdot \boldsymbol{\Pi}_i$ (see Chapter 18 for the definitions of $\boldsymbol{\Lambda}$ and $\boldsymbol{\Pi}_i$), the system dynamic is given by:

$$\begin{cases} \dot{\boldsymbol{m}}(\tau) = \boldsymbol{A}_i \cdot \boldsymbol{m}(\tau), \quad \boldsymbol{m} \in \mathscr{R}_i \\ \boldsymbol{y}(\tau) = \boldsymbol{S} \cdot \boldsymbol{m}(\tau) \end{cases} \tag{19.2}$$

where $\boldsymbol{S}$ is a $|P_o| \times |P|$ matrix, each row of $\boldsymbol{S}$ has all components zero except the one corresponding to the ith measurable place that is 1. Observe that the matrix $\boldsymbol{S}$ is the same for all linear systems since the measured places are characteristic to the CPN system. Here, it is considered that all linear systems are deterministic, i.e., noise-free.

Definition 19.2. *Let* $\langle N, \boldsymbol{\lambda}, \boldsymbol{m}_0 \rangle$ *be a timed CPN system with infinite server semantics and* $P_o \subseteq P$ *the set of measurable places.* $\langle N, \boldsymbol{\lambda}, \boldsymbol{m}_0 \rangle$ *is* observable in infinitesimal time *if it is always possible to compute its initial state* $\boldsymbol{m}_0$ *in any time interval* $[0, \varepsilon)$.

Let us first assume that the system is a Join Free (JF) CPN (a CPN is JF if there is no synchronization, i.e., $\forall t_j \in T, |^{\bullet}t_j| = 1$). Therefore, it is a linear system and let us assume that its dynamical matrix is denoted by $\boldsymbol{A}$. In Systems Theory a very well-known observability criterion exists which allows us to decide whatever a continuous time invariant linear system is observable or not. Besides, several approaches exist to compute the initial state of a continuous time linear system that is observable. The output of the system and the *observability matrix* are:

$$\boldsymbol{y}(\tau) = \boldsymbol{S} \cdot e^{\boldsymbol{A} \cdot \tau} \cdot \boldsymbol{m}(\tau_0) \tag{19.3}$$

$$\vartheta = \left[\boldsymbol{S}^T, (\boldsymbol{SA})^T, \ldots, (\boldsymbol{SA}^{n-1})^T \right]^T. \tag{19.4}$$

Proposition 19.2. [12, 13] *Eq.* (19.3) *is solvable for all* $\mathbf{m}(\tau_0)$ *and for all* $\tau > 0$ *iff the observability matrix* ϑ *has full rank (in our case, rank*$(\vartheta) = |P|$*).*

The initial state can be obtained solving the following system of equations that has a unique solution under the rank condition:

$$\begin{bmatrix} \boldsymbol{y}(0) \\ \frac{d}{dt}\boldsymbol{y}(0) \\ \frac{d^2}{dt^2}\boldsymbol{y}(0) \\ \vdots \\ \frac{d^{n-1}}{dt^{n-1}}\boldsymbol{y}(0) \end{bmatrix} = \vartheta \cdot \boldsymbol{m}(0). \tag{19.5}$$

The observability of a JF CPN systems has been considered in [10], where an interesting interpretation of the observability at the graph level. Let us assume that a place p_i is measured, therefore $m[p_i](\tau)$ and its variation $\dot{m}[p_i](\tau)$ are known at every time moment τ. Because the net has no join, the flow of all its output transitions t_j of p_i is the product of λ_j and $m[p_i](\tau)$ according to the server semantics definition. Assume that p_i is not an attribution (a place $p \in P$ is an attribution if $|^{\bullet}p| \geq 1$). Hence, has at most one input transition t_k. Knowing the derivative and the output flows, the input flow through the input transition t_k is estimated. Applying again the server semantics definition, $f[t_k] = \lambda_k \cdot \boldsymbol{m}[^{\bullet}t_k]$ ($|^{\bullet}t_k| = 1$ since the net is join free). Obviously, the marking of $^{\bullet}t_k$ can be computed immediately. Observe that this is a *backward* procedure: measuring p_i, $^{\bullet}(^{\bullet}p_i)$ is estimated in the absence of joins and attributions.

The problem of state estimation of general CPN systems and not only JF net systems is not so easy. In this case, a very important problem for the observability is the determination of the configuration, also called *discrete state*, i.e, the linear system that governs the system evolution. It may happen that the continuous state estimation fits with different discrete states, i.e., observing some places, it may happen that more than one linear system satisfies the observation. If the continuous states are on the border of some regions, it is not important which linear system is assigned, but it may happen that the solution corresponds to interior points of some regions and it is necessary to distinguish between them.

Example 19.3. Let us consider the left timed CPN in Fig. 19.2. Assume the firing rate of all transitions equal to 1 and $P_o = \{p_3\}$ implying $\boldsymbol{S} = [0\ 0\ 1]^T$. This system has two configurations corresponding to two linear systems:

$$\Sigma_i = \begin{cases} \dot{\boldsymbol{m}}(\tau) = \boldsymbol{A}_i \cdot \boldsymbol{m}(\tau) \\ y(\tau) = [0\ 0\ 1] \cdot \boldsymbol{m}(\tau) \end{cases}, i = 1,2 \tag{19.6}$$

where A_1 is the dynamic matrix corresponding to the configuration in which the marking of p_1 is defining the flow of t_3 while for A_2, the marking of p_2 is giving the flow of t_3.

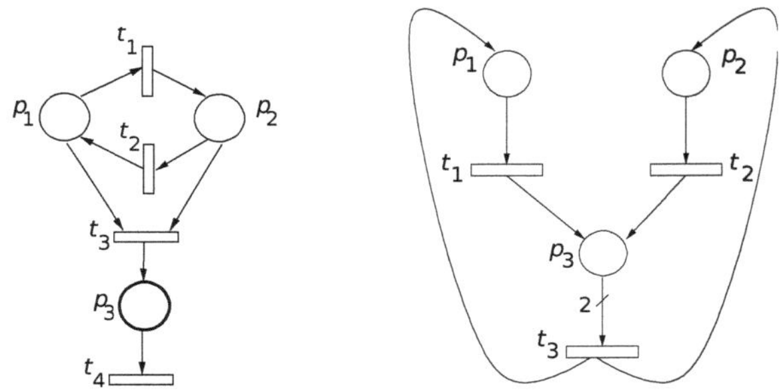

Fig. 19.2 Two CPN

The observability matrices of these two linear systems are:

$$\vartheta_1 = \begin{bmatrix} 0 & 0 & 1 \\ 1 & 0 & -1 \\ -3 & 1 & 1 \end{bmatrix}; \quad \vartheta_2 = \begin{bmatrix} 0 & 0 & 1 \\ 0 & 1 & -1 \\ 1 & -3 & 1 \end{bmatrix}.$$

Both have full rank, meaning that both linear systems are observable. Let us take $\boldsymbol{m}_1 = [1\ 2\ 0]^T \in \mathscr{R}_1 \setminus \mathscr{R}_2$ and $\boldsymbol{m}_2 = [2\ 1\ 0]^T \in \mathscr{R}_2 \setminus \mathscr{R}_1$. As it is well-known, the corresponding observations are $\vartheta_i \cdot \boldsymbol{m}_i(\tau) = [y(\tau)\ \dot{y}(\tau)\ \ldots]^T$. Nevertheless, for the selected markings we have that $\vartheta_1 \cdot \boldsymbol{m}_1 = \vartheta_2 \cdot \boldsymbol{m}_2 = [0\ 1\ \ -1]^T$. Therefore, it is impossible to distinguish between $\boldsymbol{m}_1$ and $\boldsymbol{m}_2$. ∎

Definition 19.3. *Two configurations i and j of a CPN system are* distinguishable *if for any $\boldsymbol{m}_1 \in \mathscr{R}_1 \setminus \mathscr{R}_2$ and any $\boldsymbol{m}_2 \in \mathscr{R}_2 \setminus \mathscr{R}_1$ the observation $\boldsymbol{y}_1(\tau)$ for the trajectory through $\boldsymbol{m}_1$ and the observation $\boldsymbol{y}_2(\tau)$ for the trajectory through $\boldsymbol{m}_2$ are* different *on an interval $[0, \varepsilon)$.*

Remark that we remove the solutions at the border $\mathscr{R}_1 \cap \mathscr{R}_2$ since for those points both linear systems lead to identical behavior, therefore it is not important which one is chosen. If all pairs of modes are distinguishable, it is always possible to uniquely assign a *configuration* (or region) to an observed continuous state. Assuming that the linear systems corresponding to all configurations are observable, a QPP per pair of regions can be proposed to check their distinguishability.

$$\begin{aligned} z = \max\ & \beta^T \cdot \beta \\ \text{s.t.}\quad & \vartheta_1 \cdot \boldsymbol{m}_1 - \vartheta_2 \cdot \boldsymbol{m}_2 = \boldsymbol{0} \\ & \beta = \boldsymbol{m}_1 - \boldsymbol{m}_2 \\ & \boldsymbol{m}_1 \in \mathscr{R}_1 \\ & \boldsymbol{m}_2 \in \mathscr{R}_2 . \end{aligned} \tag{19.7}$$

First, let us observe that if the feasible set of (19.7) is empty (i.e., the problem is infeasible), linear systems are distinguishable. If in QPP (19.7) $z = 0$, using the fact that both systems are observable, i.e., ϑ_1 and ϑ_2 have both full rank, $\boldsymbol{m}_1 = \boldsymbol{m}_2$ is obtained. Therefore, there exist no interior markings $\boldsymbol{m}_1 \in \mathscr{R}_1$ and $\boldsymbol{m}_2 \in \mathscr{R}_2$ with the same observation, i.e., $\vartheta_1 \cdot \boldsymbol{m}_1 = \vartheta_2 \cdot \boldsymbol{m}_2$, and the modes are distinguishable. Finally, if the solution is $z > 0$ the linear systems are undistinguishable being the same evolution in a small interval starting from two markings belonging to different regions. Finally, if the solution is $z > 0$ we cannot say nothing about distinguishability of the linear systems. Moreover, the exact solution of (19.7) is not necessary to be computed and if a feasible solution with $z > \delta$, with δ a small positive number, is found the search can be stopped.

Example 19.4. In Example 19.3, for the left timed CPN in Fig. 19.2 it is shown that $\vartheta_1 \cdot \boldsymbol{m}_1 = \vartheta_2 \cdot \boldsymbol{m}_2 = [0, 1, -1]^T$. Solving QPP (19.7), the problem is found to be unbounded, thus the linear systems Σ_1 and Σ_2 are undistinguishable. For the interpretation of this result, let us consider the equations that govern the evolution of the system:

$$f_3 = \lambda_3 \cdot \min\{m[p_1], m[p_2]\} \tag{19.8}$$

$$\dot{m}[p_1] = \lambda_2 \cdot m[p_2] - \lambda_1 \cdot m[p_1] - f_3 \tag{19.9}$$

$$\dot{m}[p_2] = \lambda_1 \cdot m[p_1] - \lambda_2 \cdot m[p_2] - f_3 . \tag{19.10}$$

Summing (19.9) and (19.10) and integrating, we obtain

$$(m[p_1] + m[p_2])(\tau) = (m[p_1] + m[p_2])(0) - 2\int_0^\tau f_3(\theta) \cdot d\theta \tag{19.11}$$

Obviously, if p_3 is measured, f_3 can be estimated since $f_3(\tau) = \dot{m}[p_3](\tau) + \lambda_4 \cdot m[p_3](\tau)$. Therefore, according to (19.8), the minimum between $m[p_1]$ and $m[p_2]$ is estimated. Moreover, due to (19.11) their sum is also known. Nevertheless, these two equations are not enough to compute the markings, i.e., we have the values but it is impossible to distinguish which one corresponds to which place.

We use the same CPN system to illustrate that if the solution of LPP (19.7) is $z > 0$ or unbounded we cannot decide. Let us take now $\boldsymbol{\lambda} = [2\ 1\ 1\ 1]^T$. In this case, the dynamical matrices are:

$$\boldsymbol{A}_1 = \begin{bmatrix} -3 & 1 & 0 \\ 1 & -1 & 0 \\ 1 & 0 & -1 \end{bmatrix}, \quad \boldsymbol{A}_2 = \begin{bmatrix} -2 & 0 & 0 \\ 2 & -2 & 0 \\ 0 & 1 & -1 \end{bmatrix},$$

and the observability matrices (assuming also $P_o = \{p_3\}$):

$$\vartheta_1 = \begin{bmatrix} 0 & 0 & 1 \\ 1 & 0 & -1 \\ -4 & 1 & 1 \end{bmatrix}; \quad \vartheta_2 = \begin{bmatrix} 0 & 0 & 1 \\ 0 & 1 & -1 \\ 2 & -3 & 1 \end{bmatrix}.$$

Let $\boldsymbol{m}_1 = [1\ 5\ 1]^T \in \mathscr{R}_1 \setminus \mathscr{R}_2$ and $\boldsymbol{m}_2 = [2\ 1\ 1]^T \in \mathscr{R}_2 \setminus \mathscr{R}_1$. Making the computations, we have: $\vartheta_1 \boldsymbol{m}_1 = \vartheta_2 \boldsymbol{m}_2 = [1\ 0\ 2]^T$. So, we have the same observations for these two markings at a time τ but the modes are distinguishable. To see this let us assume the marking at $\tau + \delta$, where δ is a very small value. Being a small time variation, we can consider that the flow of the transitions are constant during the time interval $(\tau, \tau + \delta)$ and we can write:

$$\boldsymbol{m}'_1(\tau+\delta) = \boldsymbol{m}_1(\tau) + \boldsymbol{A}_1 \boldsymbol{m}_1(\tau)\delta = [1+2\delta\ \ 5-4\delta\ \ 1]^T,$$

and

$$\boldsymbol{m}'_2(\tau+\delta) = \boldsymbol{m}_2(\tau) + \boldsymbol{A}_2 \boldsymbol{m}_2(\tau)\delta = [2-4\delta\ \ 1+2\delta\ \ 1]^T.$$

The corresponding observations for these markings are: $\vartheta_1 \boldsymbol{m}'_1 = [1\ 2\delta\ 2-12\delta]^T \neq \vartheta_2 \boldsymbol{m}'_2 = [1\ 2\delta\ 2-14\delta]^T$. Since in all linear systems the set of measured places is the same and the firing rates are also the same can be observed immediately that any $\boldsymbol{m}''_1 \in \mathscr{R}_1, \boldsymbol{m}''_2 \in \mathscr{R}_1$ with $\vartheta_1 \boldsymbol{m}''_1(\tau) = \vartheta_2 \boldsymbol{m}''_2(\tau)$ it holds that $\vartheta_1 \boldsymbol{m}''_1(\tau+\delta) \neq \vartheta_2 \boldsymbol{m}''_2(\tau+\delta)$. Therefore, according to Definition 19.3, the modes are distinguishable. ■

Using the notion of distinguishable modes, an immediate criterion for observability in infinitesimal time is:

Theorem 19.1. *A timed CPN system $\langle N, \boldsymbol{\lambda}, \boldsymbol{m}_0 \rangle$ under infinite server semantics is observable in infinitesimal time iff:*

1. *All pairs of configurations are distinguishable,*
2. *For each region, the associated linear system is observable.*

A complementary observability problem is presented in [11]. For the discrete-time model and measuring some places, the problem is to estimate the firing flow (speed) of the transitions and not the marking of the other places. Since the flow of a transition is the product between its firing rate (constant value) and the enabling degree, in some cases, measuring places or transitions is equivalent. Anyhow, in order to compute the flow through joins it is necessary to measure all of its input places. Moreover, we may also have different markings that have the same firing flow.

19.4 Reducing Complexity

Theorem 19.1 provides a criterion of observability of a CPN system. Observe that the complexity of an algorithm to check this property is not small. The algorithm based on this criterion should be linear in the number of subsystems (for each subsystem the observability matrix and its rank should be computed) but this number is exponential in the number of joins. Moreover, for each pair of subsystems, their distinguishability is necessary to be checked. For this reason, some results have been proposed in order to "delete" the joins without affecting the observable space. After that, observability can be checked using only the observability matrix. This reduction can be done under some general conditions if the net system is attribution free (AF — a net is attribution free if there exists no place $p \in P$ such that $|^\bullet p| \geq 2$) or equal conflict (EQ — a net is equal conflict if for any $t_1, t_2 \in T$ such that $^\bullet t_1 \cap {}^\bullet t_2 \neq \emptyset$ then $\boldsymbol{Pre}[\cdot, t_1] = \boldsymbol{Pre}[\cdot, t_2]$) [17].

Definition 19.4. *Let $N = \langle P, T, \boldsymbol{Pre}, \boldsymbol{Post}\rangle$ be a net and $N' = \langle F, T', \boldsymbol{Pre}', \boldsymbol{Post}'\rangle$ a subnet of N, i.e., $F \subseteq P$, $T' \subseteq T$ and $\boldsymbol{Pre}', \boldsymbol{Post}'$ are the restrictions of $\boldsymbol{Pre}, \boldsymbol{Post}$ to F and T'. N' is a strongly connected p-component of N if for all $p_1, p_2 \in F$ there exists a path from p_1 to p_2 of the form $\langle p_1, t_1, p_i, t_i, \ldots, t_j, p_j, t_2, p_2\rangle$ with $t_1 \in p_1^\bullet$, $p_i \in t_1^\bullet$, $\ldots$, $p_j \in t_j^\bullet$, $t_2 \in p_j^\bullet$, $p_2 \in t_2^\bullet$.*

Further, a strongly connected p-component $N' = \langle F, T', \boldsymbol{Pre}', \boldsymbol{Post}'\rangle$ is called *terminal* if for all $p \in F$ it holds that: there exists a path from p to other place p' implies $p' \in F$.

Proposition 19.3. [17] *Let $\langle N, \boldsymbol{\lambda}\rangle$ be a timed AF CPN and assume that for any join t_i there exists no strongly connected p-component containing all $^\bullet t_i$. Let N' be the net obtained from N by just removing all join transitions together with its input and output arcs. N is observable iff N' is observable.*

Observe that the left net in Fig. 19.2 is not satisfying the conditions of the previous theorem since $^\bullet t_3 = \{p_1, p_2\}$ belongs to a strongly connected p-component. However, if the net has attributions, joins cannot be removed in general.

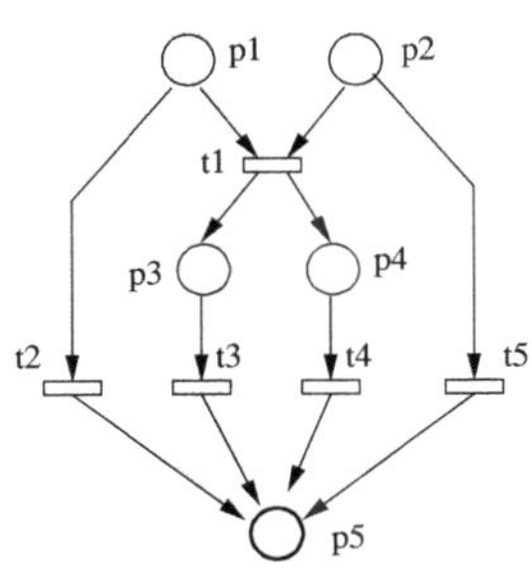

Fig. 19.3 CPN used in Example 19.5

Example 19.5. Let us consider the CPN system in Fig. 19.3 with $\boldsymbol{\lambda} = [a, 1, 2, 3, 4]^T$, $a \in \mathbb{R}_{>0}$ and p_5 the measured place. This net has an attribution in place p_5 and has a join in t_1. The linear system obtained by removing the join t_1 is observable and p_1 and p_2 do not belong to a strongly connected p-component. However, the join transition t_1 cannot be removed without affecting the observability space. The dynamical matrices of the two linear systems are:

$$\boldsymbol{A}_1 = \begin{bmatrix} -1-a & 0 & 0 & 0 & 0 \\ -a & -4 & 0 & 0 & 0 \\ a & 0 & -2 & 0 & 0 \\ a & 0 & 0 & -3 & 0 \\ 1 & 2 & 3 & 4 & 0 \end{bmatrix}, \boldsymbol{A}_2 = \begin{bmatrix} -1 & -a & 0 & 0 & 0 \\ 0 & -4-a & 0 & 0 & 0 \\ 0 & a & -2 & 0 & 0 \\ 0 & a & 0 & -3 & 0 \\ 1 & 2 & 3 & 4 & 0 \end{bmatrix}.$$

Computing the determinants of the corresponding observability matrices, we have:

$$det(\vartheta_1) = 192 \cdot a^3 - 912 \cdot a^2 + 720 \cdot a + 288,$$

which has two positive real roots ($a_1 = 3.5885$ and $a_2 = 1.4498$), and

$$\det(\vartheta_2) = -96 \cdot a^3 - 408 \cdot a^2 - 216 \cdot a + 288,$$

with one positive real root ($a_3 = 0.5885$). Obviously, if λ_1 is equal to one of these roots, the CPN system will not be observable since one of the corresponding linear system will not be observable.

Hence, for some particular values of $\boldsymbol{\lambda}$, the system obtained removing the join is observable but the original system (with join) is not observable. ■

Proposition 19.4. [17] *Let $\langle N, \boldsymbol{\lambda}, \boldsymbol{m}_0 \rangle$ be a timed EQ continuous Petri net system and N' obtained from N by just removing all join transitions together with its input and output arcs. N is observable iff N' is observable.*

The previous two propositions provide necessary and sufficient conditions to "reduce nonlinearity" and study the observability of a nonlinear system on an equivalent, with respect to the observability space, linear system. Hence, it is enough to check the rank of only one observability matrix in order to decide the observability of these CPN net systems.

19.5 Structural and Generic Observability

In this section, the main results of structural and generic observability of CPN are presented. First, we will illustrate by an example that the presence of an attribution may lead to the loss of the observability. For this reason, the main result for structural observability has been given assuming that the net has no attribution while generic observability may be studied easily in the case of nets with attributions.

From the previous graph-based interpretation (the backward strategy) of the observability, it is obvious that the output connectedness is required for a place p to be estimated from an observation. For those places for which there is no path to an output, their marking cannot be estimated. Therefore, the *terminal strongly connected p-components* present a special interest because any place of the net should be connected to those components in order to be able to be estimated.

Definition 19.5. *A strongly connected p-component* $N' = \langle F, T', \boldsymbol{Pre}', \boldsymbol{Post}' \rangle$ *of a net N is said to be* terminal *if there is no path from a place belonging to F to a place not in F.*

Strongly connected p-components of a PN can be computed immediately, adapting the classical polynomial time algorithms (for example the one in [6]) to a bipartite graph.

Definition 19.6. *Let* $\langle N, \boldsymbol{\lambda}, \boldsymbol{m}_0 \rangle$ *be a CPN system and* P_o *the set of measured places. N is* structurally observable *if* $\langle N, \boldsymbol{\lambda}, \boldsymbol{m}_0 \rangle$ *is observable for all values of* $\boldsymbol{\lambda} \in \mathbb{R}^{|T|}_{>0}$.

Proposition 19.5. [17] *Let N be a join and attribution free CPN. N is structurally observable iff at least one place from each terminal strongly connected component is measured.*

Let us consider now attributions and see that this construction can lead to the loss of observability. Assume the right CPN system in Fig. 19.2 where p_3 (an attribution place) is the measured place. Writing down the differential equation we have:

$$\dot{m}[p_3](\tau) = \lambda_1 \cdot m[p_1](\tau) + \lambda_2 \cdot m[p_2](\tau) - \lambda_3 \cdot m[p_3](\tau).$$

From the previous equation, $\lambda_1 \cdot m[p_1](\tau) + \lambda_2 \cdot m[p_2](\tau)$ can be computed since the other variables are known. However, if $\lambda_1 = \lambda_2$, will be impossible to distinguish between $m[p_1](\tau)$ and $m[p_2](\tau)$ and the system is not observable. In general, if there exist two input transitions to an attribution place with the same firing rate, the system is not *observable* [17]. Nevertheless, this is not a general rule since the observability is a global property.

Let us consider the timed CPN is Fig. 19.4 with $\boldsymbol{\lambda} = \mathbf{1}$, assume that p_2 is measured and let us see if the system is observable using the backward strategy presented before. Then $m[p_4]$ and $m[p_5]$ cannot be estimated directly, but their sum (a linear combination of them) is computable (place p_{45} in the figure). Going backwards, $m[p_1]$ is estimated and, even though $m[p_1]$ is an attribution, since $m[p_2]$ is measured, then $m[p_3]$ can also be estimated. Using $m[p_3]$, now $m[p_4]$ is estimated and, through the linear combination of p_{45}, $m[p_5]$ as well. Therefore, by measuring p_2 the system is observable for all $\boldsymbol{\lambda}$, i.e., structurally observable.

Observe that this loss of the observability is due to the presence of attributions happens for very specific values of $\boldsymbol{\lambda}$. If the firing rates of the transitions are chosen randomly in $\mathbb{R}_{>0}$, the probability to have such a loss of observability is almost null. Hence, a concept weaker than structural observability can be studied. It is similar with the concept of "structural observability" defined in [5, 7] for linear systems.

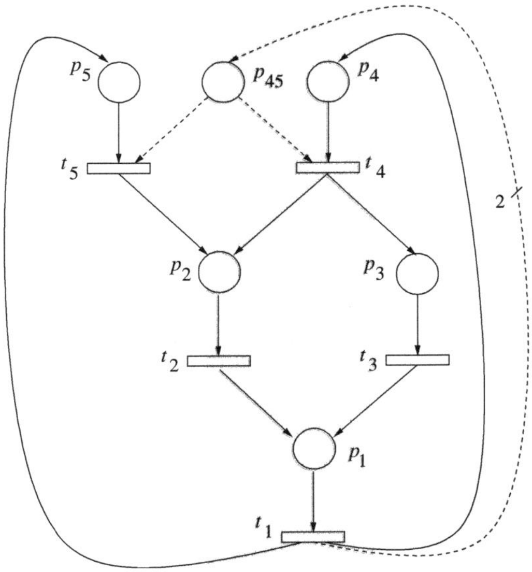

Fig. 19.4 A JF net that is observable measuring the attribution place p_2 even if $\lambda_4 = \lambda_5 = \lambda_2 = \lambda_3$

Definition 19.7. *Let $\langle N, \boldsymbol{\lambda}, \boldsymbol{m}_0 \rangle$ be a CPN system and P_o the set of measured places. N is* generically observable *if $\langle N, \boldsymbol{\lambda}, \boldsymbol{m}_0 \rangle$ is observable for all values of $\boldsymbol{\lambda}$ outside a proper algebraic variety of the parameter space.*

The relation between structural and generic observability is immediate. If N is structurally observable then it is generically observable. In general, the reverse is not true.

In [5], generic observability is studied for structured linear systems using an *associated graph*; observability is guaranteed when there exists a state-output connection for every state variable (the system is said to be *output connected*) and no *contraction* exists. The transformation of a JF net into its corresponding *associated directed graph* is straightforward (see Fig. 19.5 for an example).

Using the associated graph and Proposition 1 in [5], the following result has been obtained to characterize the generic observability.

Corollary 19.1. [17] *Let N be a pure JF CPN. N is generically observable iff at least one place from each terminal strongly connected p-component is measured.*

The previous result can be extended immediately to general CPNs, i.e., it is not true only for JF nets. In Example 19.3, a CPN system is given containing two undistinguishable configurations. Then, changing the firing rates of the transitions in Example 19.4, these modes become distinguishable. Obviously, two configurations are undistinguishable when the path from states (markings) to the outputs are identical in both linear systems. This happens for some particular values of firing rates, e.g., $\lambda_1 = \lambda_2$ in the left CPN of Fig. 19.2. If the firing rates are chosen randomly, the

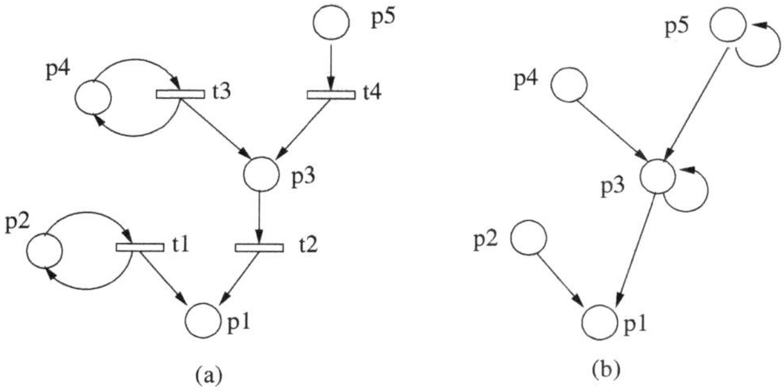

Fig. 19.5 (a) A JF ContPN; (b) Its associated graph

backward paths cannot be identical. Therefore, any pair of subsystems are distinguishable.

Corollary 19.2. [17] *Let N be a pure CPN. N is generically observable iff at least one place from each terminal strongly connected p-component is measured.*

For example, the left net in Fig. 19.2 is not observable (hence neither structurally observable) but it is generically observable.

19.6 Observers Design

JF nets lead to linear systems, for which, Luenberger's observers [12, 13] are frequently used for the estimation of the states. Such an observer for a JF PN, i.e., with a single linear system, can be expressed as:

$$\dot{\tilde{\boldsymbol{m}}}(\tau) = \boldsymbol{A}\cdot\tilde{\boldsymbol{m}}(\tau) + \boldsymbol{K}\cdot(\boldsymbol{y}(\tau) - \boldsymbol{S}\cdot\tilde{\boldsymbol{m}}(\tau)),$$

where $\tilde{\boldsymbol{m}}(\tau)$ is the marking estimation, $\boldsymbol{A}$ and $\boldsymbol{S}$ are the matrices defining the evolution of the marking of the system and its output in continuous time, $\boldsymbol{y}(\tau)$ is the output of the system, and $\boldsymbol{K}$ is a design matrix of parameters.

At a particular time instant, a CPN evolves according to a given linear system. Thus, an online estimation can be performed by designing one (Luenberger) linear observer per potential linear system of the PN (in a similar way to [8] for a class of piecewise linear systems) and selecting the one that accomplishes certain properties. The "goodness" of an estimate can be measured by means of a *residual* [2]. Let us use the 1-norm $||\cdot||_1$, which is defined as $||\boldsymbol{x}||_1 = |\boldsymbol{x}_1| + \ldots + |\boldsymbol{x}_n|$. The residual at a given instant, $r(\tau)$, is the distance between the output of the system and the output that the observer's estimate, $\tilde{\boldsymbol{m}}(\tau)$, yields

$$r(\tau) = ||\boldsymbol{S}\cdot\tilde{\boldsymbol{m}}(\tau) - \boldsymbol{z}(\tau)||_1.$$

In order to be *suitable*, the estimations of the observers must verify the following conditions:

- The residual must tend to zero.
- The estimations of the places in a synchronization have to be *coherent* with the operation mode for which they are computed.

Thus, at a given time instant, only coherent estimations are suitable. Moreover, a criterion must be established to decide which coherent estimation is, at a given time instant, the most appropriate. An adequate *heuristics* is to choose the coherent estimation with minimum residual.

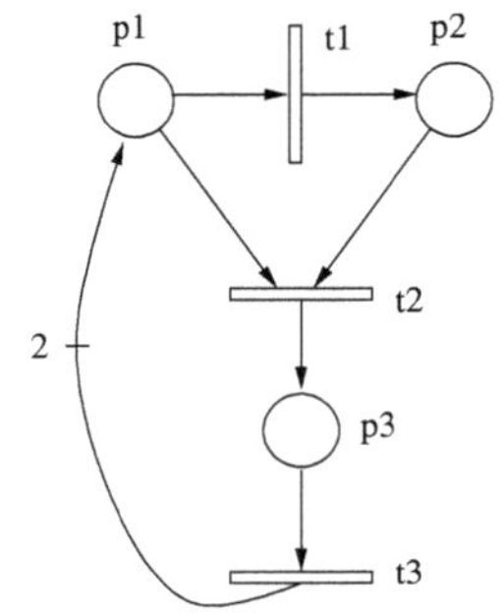

Fig. 19.6 A CPN with two linear systems

Consider the CPN system in Fig. 19.6. Let its output be the marking of place p_1, i.e., $\boldsymbol{S} = [1\ 0\ 0]$. The net has two configurations: $\mathscr{C}_1 = \{(p_1,t_1),(p_1,t_2),(p_3,t_3)\}$ and $\mathscr{C}_2 = \{(p_1,t_1),(p_2,t_2),(p_3,t_3)\}$. For the linear system corresponding to $\mathscr{C}_1$, $m[p_2]$ is not observable. However, for the linear system corresponding to $\mathscr{C}_2$, the marking of all the places can be estimated. Let $\boldsymbol{\lambda} = [0.9\ 1\ 1]^T$ and $\boldsymbol{m}_0 = [3\ 0\ 0]^T$. The marking evolution of this system is depicted in Fig. 19.7(a).

One observer per linear system is designed. Let the initial state of observer 1 be $\boldsymbol{e}_{01} = [1\ \ 2]^T$ and its eigenvalues be $[-12+2\cdot\sqrt{3}\cdot i,\ -12-2\cdot\sqrt{3}\cdot i]$. Since observer 1 can only estimate $m[p_1]$ and $m[p_3]$, the first component of its state vector corresponds to the estimation of $m[p_1]$, and its second component to the estimation of $m[p_3]$. For observer 2, let the initial state be $\boldsymbol{e}_{02} = [1\ 0\ 2]^T$ and its eigenvalues be $[-15,\ -12+2\cdot\sqrt{3}\cdot i,\ -12-2\cdot\sqrt{3}\cdot i]$. The evolution of the coherent estimation with minimum residual is shown in Fig. 19.7(a).

The resulting estimation can be improved by taking into account some considerations. When the first system switch happens, the estimation becomes discontinuous and, what is more undesirable, the estimation for the marking of p_3 becomes worse. A similar effect happens when the second system switch occurs. Another undesirable phenomenon is that, after the first switch, the estimation of $m[p_2]$ just disappears (since it is unobservable in configuration $\mathscr{C}_1$).

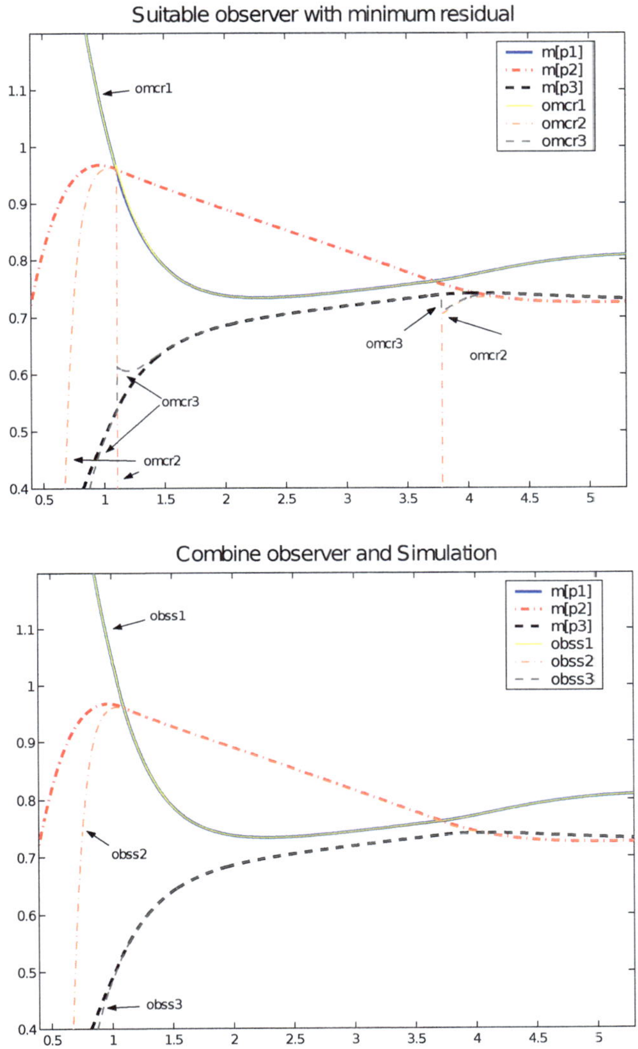

Fig. 19.7 The marking evolution is given by $(m[p_1], m[p_2], m[p_3])$. (a) The estimate of the minimum residual and coherent observer is $(omcr1, omcr2, omcr3)$. (b) The estimate of the observer that makes use of a simulation is $(obss1, obss2, obss3)$

One way to avoid discontinuities in the resulting estimation, is to use the estimation of the observer that is going to be filtered out in order to update the estimation of the observer that is not going to be filtered out. This estimation update must be done when a system switch is detected. In order not to loose the estimation of the marking of a place when it was "almost perfectly" estimated (recall the case of $m[p_2]$ when the first switch happened) a simulation of the system can be launched. The initial marking of this simulation is the estimation of the system just before the observability of the marking is lost. Such a simulation can be seen as an estimation for those markings that are not observable by the observer being considered. The simulation should only be carried out when an estimation for all the places exists and the residual is not significant. Figure 19.7(b) shows the evolution of the estimation obtained by this strategy.

One of the main advantages is that the residual does not increase sharply when the mode of the system changes. Another interesting feature is that the use of a simulation allows one to estimate the marking of places that in some modes are in principle not observable: in Fig. 19.7(b) it can be seen that the marking of p_2 can be estimated, even when it is unobservable due to configuration $\mathscr{C}_1$ being active.

19.7 Diagnosis Using Untimed CPNs

Let us now consider untimed CPN (see Chapter 18 for a short introduction and the differences among the timed and untimed models). Observability and state estimation problems in systems modeled by an *untimed* CPN have also been studied [18]. Nevertheless, in this case it is assumed that the initial marking is known (and not unknown as in previous sections) and the set of transitions is partitioned in two subsets: *observable* ($T_o \subseteq T$) and *unobservable* transitions ($T_u \subseteq T$, $T_o \cap T_u = \emptyset$) (hence transitions are observed and not places). When an observable transition fires, its firing quantity is measured/observed. From the initial marking and given a sequence of observed transitions each one with a given firing amount, it is impossible to uniquely determine the actual marking because the unobservable transitions can fire intercalated with the observable transitions. All markings in which the net may be given the actual observation is called the *set of consistent markings*.

Proposition 19.6. [18] *Let $\langle N, \boldsymbol{m}_0 \rangle$ be a CPN system where $N = \langle P, T, \boldsymbol{Pre}, \boldsymbol{Post} \rangle$ and $T = T_o \cup T_u$. Assume that the net system obtained from N removing all transitions T_o has no spurious solution (solution of the state equation but corresponding to unreachable markings). Given an observed word $t_1(\alpha_1)t_2(\alpha_2)\ldots t_k(\alpha_k)$ with $t_i \in T_o$ $\forall i = 1, \ldots, k$, the set of consistent markings is convex.*

Based on this proposition, an iterative algorithm has be derived [18] in order to characterize the set of consistent markings after an observation word w. The main idea of the algorithm is to start from each vertex of the previous set and compute the vertices of some polytopes. Taking the convex hull of all new vertices, the new set of consistent markings is obtained. The computational complexity of the algorithm is exponential because requires the computation of vertices, but the compact representation as a convex polytope is a real advantage. The fluidization allows us to relax the assumption, common to all the discrete event system diagnosis approaches, that there exist no cycle of unobservable transitions.

Fault diagnosis problem has been considered in Chapter 14 in the case of discrete Petri nets. Similarly, let us assume that a certain number of *anomalous* (or *fault*) behaviors may occur in the system. The occurrence of a fault behavior corresponds to the firing of an unobservable transition, but there may also be other transitions that are unobservable as well, but whose firing corresponds to regular behaviors. Then, assume that fault behaviors may be divided into r main classes (*fault classes*), and we are not interested in distinguishing among fault events in the same class. Usually, fault transitions that belong to the same fault class are transitions that represent similar physical faulty behavior.

This is modeled in PN terms assuming that the set of unobservable transitions is partitioned into two subsets

$$T_u = T_f \cup T_{reg},$$

where T_f includes all *fault* transitions and T_{reg} includes all transitions relative to unobservable but *regular events*. The set T_f is further partitioned into r subsets, namely,

$$T_f = T_f^1 \cup T_f^2 \cup \ldots \cup T_f^r$$

where all transitions in the same subset correspond to the same fault class. We will say that the ith fault has occurred when a transition in T_f^i has fired.

Definition 19.8. *Let $\langle N, \boldsymbol{m}_0 \rangle$ be a CPN system, $T = T_o \cup T_u$ and w an observed word. A* diagnoser *is a function*

$$\Delta : T_o^* \times \{T_f^1, T_f^2, \ldots, T_f^r\} \rightarrow \{N, U, F\}$$

(where T_o^* denotes the possible sequences obtainable combining elements in T_o, where each sequence is characterized by the firing amounts of all the transitions in it) that associates to each observation w and to each fault class T_f^i, $i = 1, \ldots, r$, a diagnosis state.

- $\Delta(w, T_f^i) = N$ *if the ith fault* cannot have occurred. *This is true if* none *of the firing sequences consistent with the observation contains fault transitions of class i.*
- $\Delta(w, T_f^i) = U$ *if a fault transition of class i* may have occurred or not, *i.e., it is uncertain, and we have no criteria to draw a conclusion in this respect.*
- $\Delta(w, T_f^i) = F$ *if the ith* fault has occurred *since all fireable sequences consistent with the observation contain at least one fault transition of class i.* ■

Thus, states N and F correspond to "certain" states: the fault has not occurred or it has occurred for sure; on the contrary state U is an "uncertain" state: the fault may either have occurred or not. Given an observation, the diagnosis state is computed solving two LPPs. Since the set of consistent marking is convex, it can be characterized by a set of vertices. Each vertex of the set of consistent markings is reached from the initial marking by firing the observed word w plus, eventually, some unobservable transitions. Moreover, after the observation w, other unobservable transitions may fire. For a given observed word w, the vectors of unobservable transitions that are fired in order to enable transition in w or after w are called *fireable firing sequences consistent with the observation* w and are denoted by $Y(\boldsymbol{m}_0, w)$.

Proposition 19.7. [18] *Consider an observed word $w \in T_o^*$ and $Y(\boldsymbol{m}_0,w)$ be the polytope containing all fireable sequences consistent with the observation w. Let*

$$\begin{cases} l_i = \min \sum\limits_{t_j \in T_f^i} \rho[t_j] \\ s.t. \\ \rho \in Y(\boldsymbol{m}_0, w) \end{cases} \quad \begin{cases} u_i = \max \sum\limits_{t_j \in T_f^i} \rho[t_j] \\ s.t. \\ \rho \in Y(\boldsymbol{m}_0, w). \end{cases} \tag{19.12}$$

It holds:

$$\begin{aligned} \Delta(w, T_f^i) &= N \Leftrightarrow u_i = 0 \\ \Delta(w, T_f^i) &= U \Leftrightarrow l_i = 0 \wedge u_i > 0 \\ \Delta(w, T_f^i) &= F \Leftrightarrow l_i > 0. \end{aligned}$$

19.8 Further Reading

For timed continuous Petri nets under infinite server semantics the problem of sensor placement has been considered in [15]. It is assumed that each place can be measured using a sensor, each sensor having associated a cost. The problem is to decide the set of places with minimum cost ensuring the observability of the system. Since the observability is a global property, the brute force algorithm has an exponential complexity because has to consider all combinations of places. Some properties permitting to reduce this complexity have been proved in [15]. A similar problem but using a geometrical approach has been considered in [1] where some results in [15] received a different perspective. An observability problem for this firing semantics has been considered also in [11] using a discrete-time model. In this case, the problem was to estimate the firing flow of transitions and not the marking of the places.

In the case of timed CPN under *finite server semantics* the problem has not been considered in the literature. However, for a similar semantics, the continuous part of so called *First-Order Hybrid Petri Nets* [2], a timed reachability problem has been considered in [16]. The observation problem reduces to determining the set of markings, in which the net may be at a given time. It is shown under which conditions the reachability set of the timed net under finite server semantics coincides with that of the untimed one and a procedure to compute the minimum time ensuring that the set of *consistent markings* is equal to the reachability set of untimed system is given for some net classes.

Different problems regarding observability of CPNs deserve a more deep study. For example, to check the distinguishability of two configurations, there exists no necessary and sufficient criterion. Moreover, the concept can be extended to more than two configurations. In the case of redundant regions, the structural symmetry can be considered and, in many cases, such symmetry will conduct to redundant linear systems. In the case of state estimation of untimed CPN, new approaches can be studied in order to decrease the complexity of the actual algorithms.

References

1. Aguayo-Lara, E., Ramirez-Trevino, A., Ruiz-Leon, J.: Invariant subspaces and sensor placement for observability in Continuous Timed Petri Nets. In: Proc. IEEE Conference on Automation Science and Engineering, Trieste, Italy (2011)
2. Balduzzi, F., Giua, A., Menga, G.: First-order hybrid Petri nets: a model for optimization and control. IEEE Trans. on Robotics and Automation 16(4), 382–399 (2000)
3. Balluchi, A., Benvenuti, L., Di Benedetto, M.D., Sangiovanni-Vincentelli, A.L.: Design of Observers for Hybrid Systems. In: Tomlin, C.J., Greenstreet, M.R. (eds.) HSCC 2002. LNCS, vol. 2289, pp. 76–89. Springer, Heidelberg (2002)
4. Collins, P., van Schuppen, J.H.: Observability of Piecewise-Affine Hybrid Systems. In: Alur, R., Pappas, G.J. (eds.) HSCC 2004. LNCS, vol. 2993, pp. 265–279. Springer, Heidelberg (2004)
5. Commault, C., Dion, J.M., Trinh, D.H.: Observability recovering by additional sensor implementation in linear structured systems. In: Proc. 44th IEEE Conference on Decision and Control, Seville, Spain (2005)
6. Cormen, T.H., Leiserson, C.E., Rivest, R.L.: Introduction to Algorithms. MIT Press (1990)
7. Dion, J.M., Commault, C., van der Woude, J.: Generic properties and control of linear structured systems: a survey. Automatica 39(7), 1125–1144 (2003)
8. Juloski, A.L., Heemels, W.P.M.H., Weiland, S.: Observer design for a class of piecewise linear systems. Int. Journal of Robust and Nonlinear Control 17(15), 1387–1404 (2007)
9. Júlvez, J., Recalde, L., Silva, M.: Steady-state performance evaluation of continuous mono-T-semiflow Petri nets. Automatica 41(4), 605–616 (2005)
10. Júlvez, J., Jiménez, E., Recalde, L., Silva, M.: On observability and design of observers in timed continuous Petri net systems. IEEE Transactions on Automation Science and Engineering 5(3), 532–537 (2008)
11. Lefebvre, D.: Estimation of the firing frequencies in discrete and continuous Petri nets models. Int. Journal of Systems Science 32(11), 1321–1332 (2001)
12. Luenberger, D.G.: An introduction to observers. IEEE Transactions on Automatic Control 16(6), 596–602 (1971)
13. Ogata, K.: Discrete-Time Control Systems, 2nd edn. Prentice-Hall (1995)
14. Recalde, L., Mahulea, C., Silva, M.: Improving analysis and simulation of continuous Petri nets. In: Proc. 2nd IEEE Conf. on Automation Science and Engineering, Shanghai, China (2006)
15. Mahulea, C.: Timed Continuous Petri Nets: Quantitative Analysis, Observability and Control. PhD thesis. University of Zaragoza, Spain (2007)

16. Mahulea, C., Cabasino, M.P., Giua, A., Seatzu, C.: A state estimation problem for timed continuous Petri nets. In: Proc. 46th IEEE Conf. on Decision and Control, New Orleans, USA (2008)
17. Mahulea, C., Recalde, L., Silva, M.: Observability of continuous Petri nets with infinite server semantics. Nonlinear Analysis: Hybrid Systems 4(2), 219–232 (2010)
18. Mahulea, C., Seatzu, C., Cabasino, M.P., Silva, M.: Fault diagnosis of discrete-event systems using continuous Petri nets. IEEE Transactions on Systems, Man and Cybernetics, Part A: Systems and Humans (2012), doi:10.1109/TSMCA.2012.2183358
19. Meyer, A., Dellnitz, M., Hessel-von Molo, M.: Symmetries in timed continuous Petri nets. Nonlinear Analysis: Hybrid Systems 5(2), 125–135 (2011)
20. Reisig, W., Rozenberg, G. (eds.): APN 1998. LNCS, vol. 1491. Springer, Heidelberg (1998)
21. Silva, M., Teruel, E., Colom, J.M.: Linear Algebraic and Linear Programming Techniques for the Analysis of Net Systems. In: Reisig, W., Rozenberg, G. (eds.) APN 1998. LNCS, vol. 1491, pp. 309–373. Springer, Heidelberg (1998)
22. Silva, M., Júlvez, J., Mahulea, C., Vázquez, C.R.: On fluidization of discrete event models: observation and control of continuous Petri nets. Discrete Event Dynamic Systems: Theory and Applications 21(4), 1–71 (2011)

Chapter 20
Continuous Petri Nets: Controllability and Control

Jorge Júlvez, C. Renato Vázquez, Cristian Mahulea, and Manuel Silva

20.1 Introduction and Motivation

The previous chapter has dealt with the concept of observability and the design of observers in the framework of continuous Petri nets. *Controllability* and observability can be seen as dual concepts. Similar to observability, a system requirement for a successful design of a control method is controllability, i.e., the possibility to drive the system to "any" desired state. In order to manipulate, i.e., *control*, the system behavior, control actions can be applied on transitions in order to modify (slow-down) their flows.

Example 20.1. As a simple introductory example, consider the net system in 18.2(a) of Chapter 18 and assume that the system works under *infinite* server semantics with $\lambda_1 = \lambda_2 = 1$. Thus, the flow of transitions is $f_1 = m_1/2$ and $f_2 = m_2$. If the system is left to evolve freely from the initial marking $\boldsymbol{m}_0 = [2\ \ 0]^T$, it will tend to the steady-state marking $\boldsymbol{m} = [4/3\ \ 2/3]^T$ at which the flow of both transitions is the same, $f_1 = f_2 = 2/3$. Assume now that control actions can be applied on transitions in order to modify their flow. Let us just apply a constant control action $u_1 = 0.5$ on transition t_1; this means that the original flow of t_1 will be decreased by 0.5, i.e., $f_1 = (m_1/2) - 0.5$ and $f_2 = m_2$ (notice that the flow of transitions cannot be negative, i.e., $f_1 \geq 0$ and $f_2 \geq 0$ must hold). In this particular case, if such a control action is kept indefinitely, the system marking will evolve to the steady state $\boldsymbol{m} = [5/3\ \ 1/3]^T$ at which the flows of transitions are $f_1 = f_2 = 1/3$. It is important to remark that the control actions that will be introduced can only slow down the original flow of transitions, and they are dynamically upper bounded by the enabling degree of the transition, e.g., in this simple example the initial control action for transition t_1 cannot be higher than 1 given that the initial enabling degree of t_1 is 1. Notice that the constraint that the flow of transitions can only be decreased when

Jorge Júlvez · C. Renato Vázquez · Cristian Mahulea · Manuel Silva
Instituto de Investigación en Ingeniería en Aragón (I3A), University of Zaragoza
e-mail: {julvez,cvazquez,cmahulea,silva}@unizar.es

C. Seatzu et al. (Eds.): Control of Discrete-Event Systems, LNCIS 433, pp. 407–428.
springerlink.com

control actions are applied, does not necessarily imply a decrease in the overall system throughput. In fact, due to the non-monotonic behavior that continuous nets can exhibit, its overall throughput can increase. This effect has already been shown in Chapter 18 (Fig. 18.3), where a decrease in the firing speed of a transition can involve a higher global system throughput. ■

Notice that continuous Petri nets are relaxations of discrete Petri nets, but at the same time, they are continuous-state systems (in fact, they are technically hybrid systems in which the discrete state is implicit in the continuous one). That is why it is reasonable to consider at least two different approaches for the controllability and control concepts:

1) the extension of control techniques used in discrete Petri nets, such as the *supervisory-control* theory (for instance, [11, 13, 14]);
2) the application of control techniques developed for continuous-state systems.

Usually, the control objective in the first approach is to meet some safety specifications, like avoiding *forbidden* states, by means of disabling transitions at particular states. The objective of the second approach consists in driving the system, by means of a usually continuous control action, towards a desired steady state, or state trajectory (see, for instance, [10]). Regarding continuous Petri nets, most of the specific works that can be found in the literature deals with the second control approach applied to the *infinite* server semantics model.

Several works in the literature have addressed the study of controllability in the context of continuous Petri nets. For instance, the work in [1] studies controllability for linear nets, i.e., Join-Free nets, pointing out that the classical rank condition is not sufficient (detailed in Subsection 20.3.2). In [15] controllability was studied for Join-Free continuous nets from a different perspective, by characterizing the set of markings that can be reached and maintained. Unfortunately, those results are difficult to extend to general subclasses of nets, where the existence of several regions makes the general reachability problem intractable.

It is important to remark that enforcing a desired target marking in a continuous Petri net is analogous to reaching an average marking in the original discrete model (assuming that the continuous model approximates correctly the discrete one), which may be interesting in several kinds of systems. This idea has been illustrated by different authors. For example, the work in [2] proposes a methodology for the control of open and closed manufacturing lines. The control actions consist in modifying the maximal firing speeds of the controlled transitions. It was also illustrated how the control law can be applied to the original discrete Petri net model (a T-timed model with constant firing delays). This approach has been used in [20] and [19] as well, in the same context of manufacturing lines. A related approach was presented in [27] for a stock-level control problem of an automotive assembly line system originally modeled as a stochastically timed discrete Petri net [9]. The resulting scheme allows to control the average value of the marking at the places that represent the stock-level, by means of applying additional delays to the controllable transitions.

The rest of the chapter is organized as follows: Section 20.2 introduces control actions and the way they modify the flow of transitions in systems working under infinite server semantics. In Section 20.3, the controllability property is discussed and some results are extracted for the case in which all transitions are controllable, and the case in which there are some non-controllable transitions. Section 20.4 describes some control methods for systems in which all transitions are controllable and sketches a couple of methods that can be applied when some transitions are not controllable. Finally, Section 20.5 discusses how a distributed control approach can be designed on a net system composed of several subsystems connected by buffers.

20.2 Control Actions under Infinite Server Semantics

Like in discrete Petri nets, control actions are applied on the transitions. These actions can only consist in the reduction of the flow, because transitions (machines for example) should not work faster than their nominal speed. The set of transitions T is partitioned into two sets T_c and T_{nc}, where T_c is the set of controllable transitions and T_{nc} is the set of uncontrollable transitions. The control vector $\boldsymbol{u} \in \mathbb{R}^{|T|}$ is defined s.t. u_i represents the control action on t_i. In the following *infinite* server semantics will be assumed. Since u_i represents a reduction of the flow, then the following inequality must hold $0 \leq u_i \leq \lambda_i \cdot enab(t_i, \boldsymbol{m})$. The behavior of a *forced* (or controlled) continuous Petri net can be described by the state equation:

$$\begin{array}{c} \dot{\boldsymbol{m}} = \boldsymbol{C}\boldsymbol{\Lambda}\boldsymbol{\Pi}(\boldsymbol{m})\boldsymbol{m} - \boldsymbol{C}\boldsymbol{u} \\ \text{s.t. } \boldsymbol{0} \leq \boldsymbol{u} \leq \boldsymbol{\Lambda}\boldsymbol{\Pi}(\boldsymbol{m})\boldsymbol{m} \text{ and } \forall t_i \in T_{nc}, u_i = 0. \end{array} \tag{20.1}$$

where matrix $\boldsymbol{\Pi}$ is the configuration matrix defined in Chapter 18:

$$\Pi_k[t,p] = \begin{cases} \frac{1}{Pre[p,t]}, & \text{if } (p,t) \in \mathscr{C}_k \\ 0, & \text{otherwise} \end{cases} \tag{20.2}$$

where $\mathscr{C}_k$ denotes a configuration as defined in Chapter 18. Notice that the slack variables introduced in Section 5 of Chapter 18 play a similar role to the one of the control actions. There are, however, important differences, in that case slack variables are associated to places and only the steady state was optimized.

20.3 Controllability

Among the many possible control objectives, we will focus on driving the system, by applying a control law, towards a desired steady state, i.e., a *set-point* control problem, frequently addressed in continuous-state systems. This control objective is related to the classical controllability concept, according to which a system is controllable if for any two states $\boldsymbol{x}_1, \boldsymbol{x}_2$ of the state space it is possible to transfer the system from $\boldsymbol{x}_1$ to $\boldsymbol{x}_2$ in finite time (see, for instance, [10]).

Marking conservation laws frequently exist in most Petri nets with practical significance. Such conservation laws imply that timed continuous Petri net (TCPN) systems are frequently not controllable according to the classical controllability concept [22, 25]. More precisely, if $\boldsymbol{y}$ is a *P-flow* then any reachable marking $\boldsymbol{m}$ must fulfill $\boldsymbol{y}^T\boldsymbol{m} = \boldsymbol{y}^T\boldsymbol{m}_0$, defining thus a *state invariant*. Nevertheless, the study of controllability "over" this invariant is particularly interesting. This set is formally defined as $Class(\boldsymbol{m}_0) = \{\boldsymbol{m} \in \mathbb{R}^{|P|}_{\geq 0} | \boldsymbol{B}_y^T\boldsymbol{m} = \boldsymbol{B}_y^T\boldsymbol{m}_0\}$, where $\boldsymbol{B}_y$ is a basis of P-flows, i.e., $\boldsymbol{B}_y^T\boldsymbol{C} = \boldsymbol{0}$. For a general TCPN system, every reachable marking belongs to $Class(\boldsymbol{m}_0)$.

Another important issue that must be taken into account in TCPN systems is the nonnegativeness and boundedness of the input, i.e., $\boldsymbol{0} \leq \boldsymbol{u} \leq \boldsymbol{\Lambda\Pi}(\boldsymbol{m})\boldsymbol{m}$. An appropriate local controllability concept, once these issues are considered, is [26]:

Definition 20.1. *The TCPN system $\langle N, \boldsymbol{\lambda}, \boldsymbol{m}_0\rangle$* is controllable with bounded input *(BIC)* over $S \subseteq Class(\boldsymbol{m}_0)$ *if for any two markings $\boldsymbol{m}_1, \boldsymbol{m}_2 \in S$ there exists an input $\boldsymbol{u}$ transferring the system from $\boldsymbol{m}_1$ to $\boldsymbol{m}_2$ in finite or infinite time, and it is suitably bounded, i.e., $\boldsymbol{0} \leq \boldsymbol{u} \leq \boldsymbol{\Lambda\Pi}(\boldsymbol{m})\boldsymbol{m}$, and $\forall t_i \in T_{nc}$ $u_i = 0$ along the marking trajectory.*

20.3.1 Controllability When All *the Transitions Are Controllable*

An interesting fact is that when all the transitions are controllable, the controllability of TCPNs, depends exclusively on the structure of the net. Let us give some intuition about this by rewriting the state equation as:

$$\dot{\boldsymbol{m}} = \boldsymbol{C} \cdot \boldsymbol{w} \tag{20.3}$$

where the innovation vector $\boldsymbol{w} = \boldsymbol{\Lambda\Pi}(\boldsymbol{m})\boldsymbol{m} - \boldsymbol{u}$ can be seen as an auxiliary input. The constraints for $\boldsymbol{u}$ are transformed into $\boldsymbol{0} \leq \boldsymbol{w} \leq \boldsymbol{\Lambda\Pi}(\boldsymbol{m})\boldsymbol{m}$. In this way, given a marking $\boldsymbol{m}_1 \in Class(\boldsymbol{m}_0)$, if $\exists \boldsymbol{\sigma} \geq \boldsymbol{0}$ such that $\boldsymbol{C\sigma} = (\boldsymbol{m}_1 - \boldsymbol{m}_0)$ then $\boldsymbol{m}_1$ is reachable from $\boldsymbol{m}_0$. This can be achieved by setting $\boldsymbol{w} = \alpha\boldsymbol{\sigma}$ (with a small enough $\alpha > 0$), so the field vector results $\dot{\boldsymbol{m}} = \boldsymbol{C}\alpha\boldsymbol{\sigma} = \alpha(\boldsymbol{m}_1 - \boldsymbol{m}_0)$ which implies that the system will evolve towards $\boldsymbol{m}_1$ describing a straight trajectory (assuming that the required transitions can be fired from this marking, what always happens if $\boldsymbol{m}$ is a relative interior point of $Class(\boldsymbol{m}_0)$).

Example 20.2. Consider, for instance, the left TCPN in Fig. 20.1 and the markings $\boldsymbol{m}_0 = [2\ 3\ 1\ 1]^T$, $\boldsymbol{m}_1 = [1\ 3\ 2\ 1]^T$ and $\boldsymbol{m}_2 = [2\ 1\ 1\ 3]^T$. Given that this system has 2 P-semiflows (involving $\{p_1, p_3\}$ and $\{p_2, p_4\}$ respectively), the marking of two places is sufficient to represent the whole state. For this system $\exists \boldsymbol{\sigma} \geq \boldsymbol{0}$ such that $\boldsymbol{C\sigma} = (\boldsymbol{m}_1 - \boldsymbol{m}_0)$, but $\nexists \boldsymbol{\sigma} \geq \boldsymbol{0}$ such that $\boldsymbol{C\sigma} = (\boldsymbol{m}_2 - \boldsymbol{m}_0)$, so, $\boldsymbol{m}_1$ is reachable but $\boldsymbol{m}_2$ is not. The shadowed area in right Fig. 20.1 corresponds to the set of reachable markings,

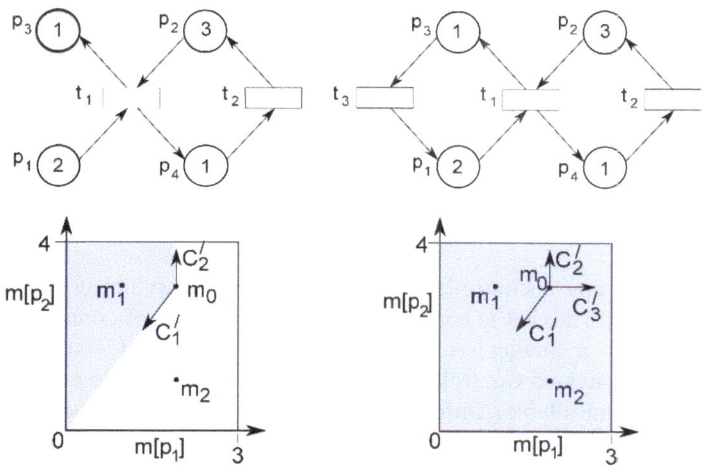

Fig. 20.1 Two TCPN systems with identical P-flows. The shadowed areas correspond to the sets of reachable markings. Only the system on the right is consistent and controllable over $Class(\boldsymbol{m}_0)$.

note that it is the convex cone defined by vectors $\boldsymbol{c}'_1$ and $\boldsymbol{c}'_2$, which represent the columns of $\boldsymbol{C}$ (here restricted to p_1 and p_2). ■

A full characterization of controllability [26] can be obtained from this structural reachability reasoning:

Proposition 20.1. *Let $\langle N, \boldsymbol{\lambda}, \boldsymbol{m}_0 \rangle$ be a TCPN system in which all the transitions are controllable. The system $\langle N, \boldsymbol{\lambda}, \boldsymbol{m}_0 \rangle$ is BIC over the interior of $Class(\boldsymbol{m}_0)$ iff the net is* consistent. *Furthermore, the controllability is extended to the whole $Class(\boldsymbol{m}_0)$ iff (additionally to consistency) there exist no* empty siphon *at any marking in $Class(\boldsymbol{m}_0)$.*

It is important to remark that controllability does not depend on the timing $\boldsymbol{\lambda}$. In fact, the key condition here is consistency, i.e., $\exists \boldsymbol{x} > 0$ such that $\boldsymbol{C} \cdot \boldsymbol{x} = \boldsymbol{0}$. Remember that a reachable marking $\boldsymbol{m} \geq \boldsymbol{0}$ fulfills $\boldsymbol{m} = \boldsymbol{m}_0 + \boldsymbol{C} \cdot \boldsymbol{\sigma}$ with $\boldsymbol{\sigma} \geq \boldsymbol{0}$, which implies $\boldsymbol{B}_y^T \boldsymbol{m} = \boldsymbol{B}_y^T \boldsymbol{m}_0$ (equivalently, $\boldsymbol{m} \in Class(\boldsymbol{m}_0)$). In the opposite sense, if the net is consistent then $\forall \boldsymbol{m} \geq \boldsymbol{0}$ s.t. $\boldsymbol{B}_y^T \boldsymbol{m} = \boldsymbol{B}_y^T \boldsymbol{m}_0$ (i.e., $\boldsymbol{m} \in Class(\boldsymbol{m}_0)$) it exists $\boldsymbol{\sigma} \geq \boldsymbol{0}$ s.t. $\boldsymbol{m} = \boldsymbol{m}_0 + \boldsymbol{C} \cdot \boldsymbol{\sigma}$, thus $\boldsymbol{m}$ is reachable (assuming $\boldsymbol{\sigma}$ is fireable). A very informal and intuitive explanation is that consistency permits movements of marking in any direction inside the reachability space (see right Fig. 20.1), i.e., if there exists $\boldsymbol{\sigma}$ such that $\boldsymbol{m}_1 = \boldsymbol{m}_0 + \boldsymbol{C} \cdot \boldsymbol{\sigma}$, under consistency any $\boldsymbol{\sigma}' = \boldsymbol{\sigma} + k \cdot \boldsymbol{x} \geq 0$, permits the reachability of $\boldsymbol{m}_1$.

Let us consider again the right TCPN system of Fig. 20.1. Given that the net is not consistent, it can be deduced by Proposition 20.1 that it is not controllable over $Class(\boldsymbol{m}_0)$. Let us now consider the system of Fig. 20.1(b). In this case, due to the consistency of the net, it holds that the vector $(\boldsymbol{m} - \boldsymbol{m}_0)$ is in the convex cone defined by the vectors $\boldsymbol{c}'_1$, $\boldsymbol{c}'_2$ and $\boldsymbol{c}'_3$ for any marking $\boldsymbol{m} \in Class(\boldsymbol{m}_0)$. Therefore $\boldsymbol{m}$ is reachable from $\boldsymbol{m}_0$. Furthermore, since at the border markings of $Class(\boldsymbol{m}_0)$ there are not unmarked siphons then, according to Proposition 20.1, the system is *BIC* over $Class(\boldsymbol{m}_0)$.

20.3.2 Controllability When Some Transitions Are Uncontrollable

If a TCPN contains uncontrollable transitions it becomes not controllable over $Class(\boldsymbol{m}_0)$, even if the net is consistent. Thus, the concept of controllability must be constrained to a smaller set of markings. The work in [15] studies this idea by defining a set named Controllability Space (*CS*) for Join-Free nets, over which the system is controllable. Unfortunately, this set depends on the marking, and therefore, its characterization for general subclasses of nets is difficult. The existence of several regions makes the general reachability problem intractable. For practical reasons, the controllability was studied in [26] over sets of *equilibrium markings* : $\boldsymbol{m}^q \in Class(\boldsymbol{m}_0)$ is an equilibrium marking if $\exists \boldsymbol{u}^q$ suitable such that $\boldsymbol{C}(\boldsymbol{\Lambda}\boldsymbol{\Pi}(\boldsymbol{m}^q)\boldsymbol{m}^q - \boldsymbol{u}^q) = \boldsymbol{0}$, i.e., there exists a control action that keeps the system marking constant at $\boldsymbol{m}^q$. They represent the *possible stationary operating points* of the original discrete system. These markings are particularly interesting, since controllers are frequently designed in order to drive the system towards a desired *stationary operating point*.

Given that inside each region $\mathscr{R}_i$ the state equation is linear ($\boldsymbol{\Pi}(\boldsymbol{m})$ is constant), it becomes convenient to study, in a first step, the controllability over equilibrium markings in each region and later over the union of them. This approach is supported by the following proposition:

Proposition 20.2. *Let $\langle N, \boldsymbol{\lambda}, \boldsymbol{m}_0 \rangle$ be a TCPN system. Consider some equilibrium sets S_1, S_2,..., S_j related to different regions $\mathscr{R}_1$, $\mathscr{R}_2$,..., $\mathscr{R}_j$. If the system is BIC (in finite time) over each one and their union $\bigcup_{i=1}^{j} S_i$ is connected, the system is BIC over the union.*

The connectivity of the set of all the equilibrium markings in $Class(\boldsymbol{m}_0)$ has not been demonstrated for the general case. Nevertheless, in every studied system such property holds.

Example 20.3. Let us consider as an example the timed continuous marked graph of Fig. 20.2 with $T_c = \{t_4\}$ and $\boldsymbol{\lambda} = [1\,1\,1\,2]^T$. According to the net structure, there are four possible configurations, but given the initial marking, one of them cannot occur. The polytope in Fig. 20.2 represents the $Class\{\boldsymbol{m}_0\}$. Since the system has 3 P-semiflows, the marking at $\{p_1, p_3, p_5\}$ is enough to represent the whole state. This is divided into the regions $\mathscr{R}_1$, $\mathscr{R}_3$ and $\mathscr{R}_4$, related to the feasible configurations. The segments $E_1 = [\boldsymbol{m}_1, \boldsymbol{m}_2]$, $E_3 = [\boldsymbol{m}_2, \boldsymbol{m}_3]$ and $E_4 = [\boldsymbol{m}_3, \boldsymbol{m}_4]$ are the sets of equilibrium markings in regions $\mathscr{R}_1$, $\mathscr{R}_3$ and $\mathscr{R}_4$, respectively. Since the union of E_1, E_3 and E_4 is connected, if the system was *BIC* over each E_i (this will be explored in a

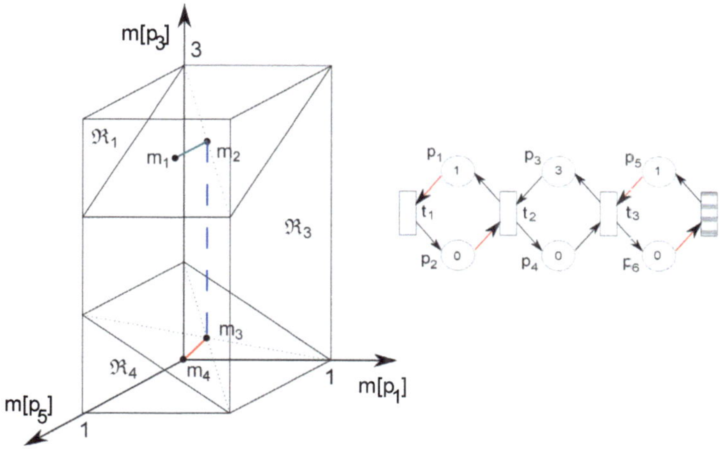

Fig. 20.2 TCPN system with its $\mathbb{E}$. Transition t_4 is the only controllable one. There are four possible configurations: $\mathscr{C}_1 = \{(p_2,t_2), (p_4,t_3)\}$, $\mathscr{C}_2 = \{(p_3,t_2), (p_4,t_3)\}$, $\mathscr{C}_3 = \{(p_2,t_2), (p_5,t_3)\}$ and $\mathscr{C}_4 = \{(p_3,t_2), (p_5,t_3)\}$, however, $\mathscr{C}_2$ cannot occur from the given $\boldsymbol{m}_0$ because p_3 and p_4 cannot concurrently constrain t_2 and t_3, respectively. Equilibrium sets depend on the timing, but regions do not.

forthcoming example) then, according to Proposition 20.2, the system would be *BIC* over $E_1 \cup E_3 \cup E_4$. For instance, the system could be driven from $\boldsymbol{m}_3$ to $\boldsymbol{m}_1$ and in the opposite sense. ■

Notice that the behavior of the TCPN system is linear and time-invariant in a given region $\mathscr{R}_i$, then some of the classical results in control theory can be used for its analysis. Null-controllability (controllability around the origin) of this kind of systems with input constraints was studied in [7]. Recalling from there, if a linear system $\dot{\boldsymbol{x}} = \boldsymbol{A}\boldsymbol{x} + \boldsymbol{B}\boldsymbol{u}$, with input constraint $\boldsymbol{u} \in \Omega$ (called the set of admissible inputs), is controllable then the controllability matrix $Contr(\boldsymbol{A},\boldsymbol{B}) = [\boldsymbol{B}\,\boldsymbol{A}\boldsymbol{B}\,..\boldsymbol{A}^{n-1}\boldsymbol{B}]$ has full rank (equivalently, $\forall \boldsymbol{x}_1, \boldsymbol{x}_2$: $\exists \boldsymbol{z}$ s.t. $(\boldsymbol{x}_2 - \boldsymbol{x}_1) = Contr(\boldsymbol{A},\boldsymbol{B}) \cdot \boldsymbol{z}$). Moreover, if $\boldsymbol{0}$ is in the interior of Ω then the previous rank condition is also sufficient for null-controllability. Otherwise, if there are inputs that can be only settled as positive (or negative) then the controllability depends also on the eigenstructure of the state matrix. These results can be adapted to TCPNs. For this, the state equation of a TCPN is first transformed in order to represent the behavior around an equilibrium marking $\boldsymbol{m}^q$, i.e., the evolution of $\Delta\boldsymbol{m} = \boldsymbol{m} - \boldsymbol{m}^q$. As a consequence, some transformed inputs $\Delta\boldsymbol{u} = (\boldsymbol{u} - \boldsymbol{u}^q)$ can be settled only as nonnegative while others can be settled as either positive or negative. The set of transitions related to this last kind of inputs is denoted as $T^i_{cf} \subseteq T_c$. Let us denote as E^*_i the set of all equilibrium markings in a region $\mathscr{R}_i$ s.t. $\Delta\boldsymbol{u}[T^i_{cf}]$ can be settled as either positive or negative (equivalently, $[\boldsymbol{\Lambda}\boldsymbol{\Pi}_i\boldsymbol{m}^q]_j > u^q_j > 0$ for all $t_j \in T^i_{cf}$). In this way, it can be proved that if a TCPN is controllable over a set E^*_i then $\forall \boldsymbol{m}_2, \boldsymbol{m}_1 \in E^*_i$: $\exists \boldsymbol{z}$ s.t. $(\boldsymbol{m}_2 - \boldsymbol{m}_1) = Contr((\boldsymbol{C}\boldsymbol{\Lambda}\boldsymbol{\Pi}_i),\boldsymbol{C}[T_c]) \cdot \boldsymbol{z}$. This condition is only necessary, as already pointed out in [1], because the existence of input constraints. Furthermore, a system is controllable (in finite time) over E_i^* if $\forall \boldsymbol{m}_2, \boldsymbol{m}_1 \in E_i^*$: $\exists \boldsymbol{z}$ s.t. $(\boldsymbol{m}_2 - \boldsymbol{m}_1) = Contr((\boldsymbol{C\Lambda\Pi}_i), \boldsymbol{C}[T_{cf}^i]) \cdot \boldsymbol{z}$. This sufficient condition is also necessary if $T_{cf}^i = T_c$ (but not if $T_{cf}^i \subset T_c$). Note that now the controllability depends not only on the structure of the net, but also on the timing [26].

As an example, let us consider the region $\mathscr{R}_3$ in the system of Fig. 20.2, where $T_{cf}^3 = \{t_4\}$. Given that $T_{cf}^3 = T_c$ then the span condition introduced above is sufficient and necessary for controllability. In this case, it can be verified that the system is *BIC* over $E_3^* = E_3$. Consider now the same system but with $\lambda_4 = 1$ instead of $\lambda_4 = 2$. In this case, $T_{cf}^3 = \emptyset$ (this set depends on the timing), then we cannot use the same sufficient condition. Nevertheless, it is still fulfilled that $\forall \boldsymbol{m}_2, \boldsymbol{m}_1 \in E_3^* : \exists \boldsymbol{z}$ s.t. $(\boldsymbol{m}_2 - \boldsymbol{m}_1) = Contr((\boldsymbol{C\Lambda\Pi}_3), \boldsymbol{C}[T_c]) \cdot \boldsymbol{z}$. Therefore, the controllability matrices do not provide enough information for deciding whether the system is *BIC* or not over E_3^*. By using other results from [26], it can be proved that the system is not *BIC* with $\lambda_4 = 1$. This implies that controllability is a timing-dependent property.

20.4 Control Techniques under Infinite Server Semantics

This section describes some few techniques proposed in the literature for the control of TCPNs when all transitions are controllable. Similarly to the *set-point* control problem in state-continuous systems, the control objective here consists in driving the system towards a desired target marking here denoted as $\boldsymbol{m}_d$. This desired marking can be selected, in a preliminarily planning stage, according to some optimality criterion [24], e.g., maximizing the flow. Most of the work done on this issue is devoted to centralized dynamic control assuming that *all* the transitions are controllable. We will first present those control techniques that require all the transitions to be controllable, then a basic comparison of such techniques will be performed, and finally a couple of approaches where uncontrollable transitions are allowed will be presented.

20.4.1 Control for a Piecewise-Straight Marking Trajectory

This subsection introduces a control approach that aims at reaching a given target marking by following a piecewise-straight trajectory. A similar approach was studied in [16] for Join-Free nets where the tracking control problem of a mixed ramp-step reference signal was explored, and later extended to general Petri nets in [17]. In such a work, a high & low gain proportional controller is synthesized, while a ramp-step reference trajectory, as a sort of *path-planning* problem at a higher level, is computed. We will discuss the more simple synthesis procedure introduced in [3].

Let us consider the line l connecting $\boldsymbol{m}_0$ and $\boldsymbol{m}_d$, and the markings in the intersection of l with the region's borders, denoted as $\boldsymbol{m}_c^1$, $\boldsymbol{m}_c^2$, ..., $\boldsymbol{m}_c^n$. Define $\boldsymbol{m}_c^0 = \boldsymbol{m}_0$

and $\boldsymbol{m}_c^{n+1} = \boldsymbol{m}_d$. Then, $\forall k \in \{0,n\}$ compute τ_k by solving the linear programming problem (LPP):

$$\begin{array}{ll} \min \tau_k & \\ s.t.: & \boldsymbol{m}_c^{i+1} = \boldsymbol{m}_c^i + \boldsymbol{C} \cdot \boldsymbol{x} \\ & \boldsymbol{0} \le \boldsymbol{x}_j \le \boldsymbol{\lambda}_j \boldsymbol{\Pi}^z_{ji} \min\{\boldsymbol{m}^i_{c,i}, \boldsymbol{m}^{i+1}_{c,i}\} \tau_k \\ & \forall j \in \{1,...,|T|\} \text{ where i satisfies } \boldsymbol{\Pi}^z_{ji} \neq \boldsymbol{0} \end{array} \quad (20.4)$$

where the first constraint is the fundamental state equation and the second constraint ensures the applicability of the input actions. This way, the control law to be applied is $\boldsymbol{w} = \boldsymbol{x}/\tau_k$ (the model is represented as in (20.3)), when the system is between the markings $\boldsymbol{m}_c^k$ and $\boldsymbol{m}_c^{k+1}$. The time required for reaching the desired marking is given by $\tau_f = \sum_{k=0}^n \tau_k$. *Feasibility* and *convergence* to $\boldsymbol{m}_d$ were proved in [3].

If one aims at obtaining faster trajectories, intermediate states, not necessarily on the line connecting the initial and the target marking, can be introduced [17]. According to [3], they can be computed by means of a *bilinear* programming problem (BPP). The idea is to currently compute the intermediate markings $\boldsymbol{m}_c^k$, on the borders of the regions that minimizes the total time $\tau_f = \sum_{k=0}^n \tau_k$ with some additional monotonicity constraints. Finally, the same algorithm can be adapted in order to recursively compute intermediate markings in the interior of the regions, obtaining thus faster trajectories.

20.4.2 Model Predictive Control

Within the Model Predictive Control (MPC) framework, two main solutions can be considered based on the *implicit* and *explicit* methods (see, for instance, [6]). The evolution of the timed continuous Petri net model (20.3), in *discrete-time*, can be represented by the difference equation: $\boldsymbol{m}(k+1) = \boldsymbol{m}(k) + \Theta \cdot \boldsymbol{C} \cdot \boldsymbol{w}(k)$, subject to the constraints $\boldsymbol{0} \le \boldsymbol{w}(k) \le \boldsymbol{f}(k)$ with $\boldsymbol{f}(k)$ being the flow without control, which is equivalent to $\boldsymbol{G} \cdot [\boldsymbol{w}^T(k), \boldsymbol{m}^T(k)]^T \le \boldsymbol{0}$, for a particular matrix $\boldsymbol{G}$. The sampling Θ must be chosen small enough in order to avoid spurious markings, in particular, for ensuring the positiveness of the markings. For that, the following condition is required to be fulfilled $\forall\, p \in P\ :\ \sum_{t_j \in p^\bullet} \lambda_j \Theta < 1$.

A MPC control scheme can be derived [23] by using this representation of the continuous Petri net. The considered goal is to drive the system towards a desired marking $\boldsymbol{m}_d$, while minimizing the quadratic performance index

$$\begin{aligned} J(\boldsymbol{m}(k),N) = & (\boldsymbol{m}(k+N) - \boldsymbol{m}_d)'\boldsymbol{Z}(\boldsymbol{m}(k+N) - \boldsymbol{m}_d) \\ & + \textstyle\sum_{j=0}^{N-1} [(\boldsymbol{m}(k+j) - \boldsymbol{m}_d)'\boldsymbol{Q}(\boldsymbol{m}(k+j) - \boldsymbol{m}_d) \\ & \qquad + (\boldsymbol{w}(k+j) - \boldsymbol{w}_d)'\boldsymbol{R}(\boldsymbol{w}(k+j) - \boldsymbol{w}_d)] \end{aligned}$$

where $\boldsymbol{Z}$, $\boldsymbol{Q}$ and $\boldsymbol{R}$ are positive definite matrices and N is a given time horizon. This leads to the following optimization problem that needs to be solved in each time step:

$$\begin{aligned} &\min J(\boldsymbol{m}(k),N) \\ &\text{s.t.}: \forall j \in \{0,...,N-1\},\ \boldsymbol{m}(k+j+1) = \boldsymbol{m}(k+j) + \Theta \cdot \boldsymbol{C} \cdot \boldsymbol{w}(k+j) \\ &\qquad\qquad\qquad\qquad\quad \boldsymbol{G} \cdot \begin{bmatrix} \boldsymbol{w}(k+j) \\ \boldsymbol{m}(k+j) \end{bmatrix} \leq \boldsymbol{0} \\ &\qquad\qquad\qquad\qquad\quad \boldsymbol{w}(k+j) \geq \boldsymbol{0} \end{aligned} \tag{20.5}$$

Let us show that, in general, the standard MPC approach does not guarantee convergence [23].

Example 20.4. Consider the net system in Fig. 20.3 with $\boldsymbol{\lambda} = [1\ 5]^T$. Let $\Theta = 0.1$, $\boldsymbol{m}_d = [0\ 1]^T$ and $\boldsymbol{w}_d = [0\ 0]^T$. Moreover, let $\boldsymbol{Q} = \boldsymbol{Z} = \boldsymbol{R} = \boldsymbol{I}$ and $N = 1$.

Fig. 20.4 shows the marking evolution of the system controlled with the MPC policy. It can be seen that the desired marking is not reached. Observe that to obtain $\boldsymbol{m}_d$, only t_1 should fire. Given that the timing horizon is too short and $\lambda_2 = 5 >> \lambda_1 = 1$, the optimality of (20.5) implies that it is better to fire at the beginning "a little" t_2 so that m_1 approaches the desired final value $m_{f,1} = 0$. However, once t_2 has fired, $\boldsymbol{m}_d$ cannot be reached because there is not enough marking in p_1 to be transferred to p_2. ■

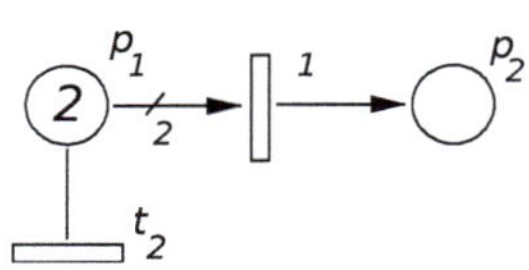

Fig. 20.3 Example of an unstable TCPN system with basic MPC scheme

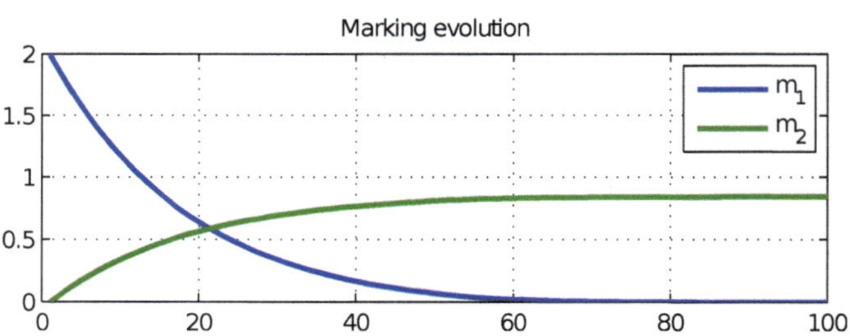

Fig. 20.4 Marking evolution of the TCPN system in Fig. 20.3

The work in [23] shows that the standard techniques used for ensuring converge in linear/hybrid systems (i.e., terminal constraints or terminal cost) cannot be applied in continuous nets if the desired marking has zero components. However, several approaches can be considered to improve convergence. Let us discuss one of them that consists in constraining the system state at time $k+N$ to belong to the straight line $\boldsymbol{m}(k)$ — $\boldsymbol{m}_d$. Roughly, this is equivalent to add a terminal constraint of the form:

$$\begin{cases} \boldsymbol{m}(k+N) = \boldsymbol{m}_k + \alpha \cdot (\boldsymbol{m}_d - \boldsymbol{m}(k)) \\ 0 \leq \alpha \leq 1 \end{cases} \tag{20.6}$$

to the optimization problem (20.5), where α is a new decision variable. As stated in the following proposition, the inclusion of this constraint guarantees asymptotically stability.

Proposition 20.3. *Consider a TCPN system with $\boldsymbol{m}_0$ and $\boldsymbol{m}_d$ the initial and target markings, respectively, being $\boldsymbol{m}_0 > \boldsymbol{0}$ and $\boldsymbol{m}_d$ reachable from $\boldsymbol{m}_0$. Assume that the system is controlled using MPC with a terminal constraint of the form* (20.6) *and prediction horizon $N = 1$. Then, the closed-loop system is asymptotically stable.*

Example 20.5. Let us exemplify this result through the TCPN in Fig. 20.5. Assume $\boldsymbol{m}_0 = [1\ 0.1]^T > \boldsymbol{0}$, $\boldsymbol{m}_d = [0\ 0]^T$, $\boldsymbol{w}_d = [0\ 0\ 0]^T$, $\boldsymbol{\lambda} = [1\ 1\ 1]^T$, $\boldsymbol{Z} = \boldsymbol{R} = \boldsymbol{I}$, $\boldsymbol{Q} = [1\ 0; 0\ 100]$ and $\Theta = 0.1$.

Fig. 20.6 shows the marking evolution after applying MPC with the terminal constraint $\boldsymbol{m}(k+N) = \alpha \cdot \boldsymbol{m}_d + (1-\alpha) \cdot \boldsymbol{m}(k)$, for $N = 1$ and $N = 2$ respectively. It can be observed that if $N = 1$, $\boldsymbol{m}_d$ is reached, but if $N = 2$, $\boldsymbol{m}_d$ is not reached.

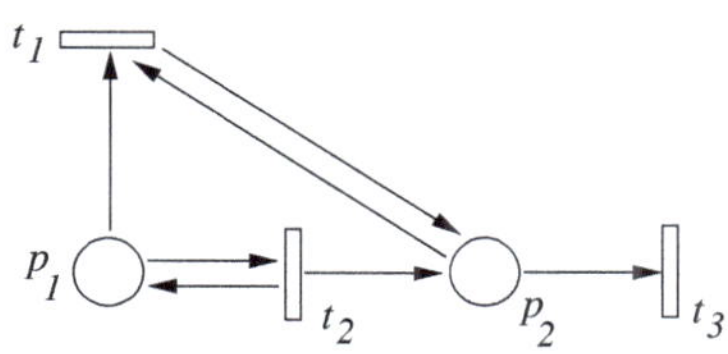

Fig. 20.5 A TCPN showing that the terminal equality constraint may not ensure stability if $N > 1$

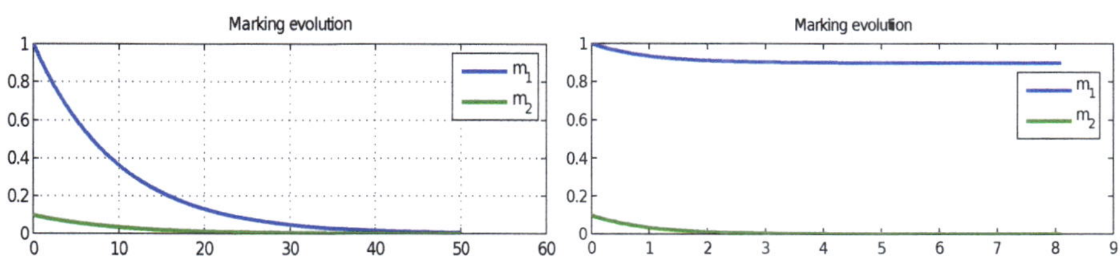

Fig. 20.6 Marking evolution of the net in Fig. 20.5 with $N = 1$ (left) and $N = 2$ (right)

Notice that $\boldsymbol{m}_d = [0\,0]^T$ is on the boundary of the feasible states since in a Petri net, $\boldsymbol{m}(k) \geq 0$ for all reachability markings. ■

An alternative MPC approach for this problem is the so-called *explicit* solution [6], where the set of all states that are controllable is split into polytopes. In each polytope the control command is defined as a piecewise affine function of the state. The closed-loop *stability* is guaranteed with this approach. On the contrary, when either the order of the system or the length of the prediction horizon are not small, the *complexity* of the explicit controller becomes computationally prohibitive. Furthermore, the computation of the polytopes sometimes is unfeasible.

20.4.3 ON-OFF Control

If the control problem is constrained to particular net subclasses, stronger results may be obtained. For instance, for *structurally persistent continuous Petri nets*, i.e., net systems where the enabling of any transition t_j cannot decrease by the firing of any other transition $t_i \neq t_j$ (in continuous nets this corresponds to choice-free nets), the minimum-time control problem has been solved [30].

The solution to this problem can be obtained as follows. First, a minimal firing count vector $\boldsymbol{\sigma}$ s.t. $\boldsymbol{m}_d = \boldsymbol{m}_0 + \boldsymbol{C}\boldsymbol{\sigma}$ is computed ($\boldsymbol{\sigma}$ is minimal if for any T-semiflow $\boldsymbol{x}$, $||\boldsymbol{x}|| \not\subseteq ||\boldsymbol{\sigma}||$, where $||\cdot||$ stands for the support of a vector). Later, the control law is defined, for each transition t_j, as:

$$\boldsymbol{u}[t_j] = \begin{cases} 0 & \text{if } \int_0^{\tau^-} \boldsymbol{w}(t_j,\delta)d\delta < \boldsymbol{\sigma}[t_j] \\ \boldsymbol{f}[t_j] & \text{if } \int_0^{\tau^-} \boldsymbol{w}(t_j,\delta)d\delta = \boldsymbol{\sigma}[t_j] \end{cases}$$

where $\boldsymbol{w}(t_j,\delta)$ is the controlled flow of t_j at time δ. This means that if t_j has not been fired an amount of $\boldsymbol{\sigma}[t_j]$, then t_j is completely ON. Otherwise, t_j is completely OFF (it is blocked).

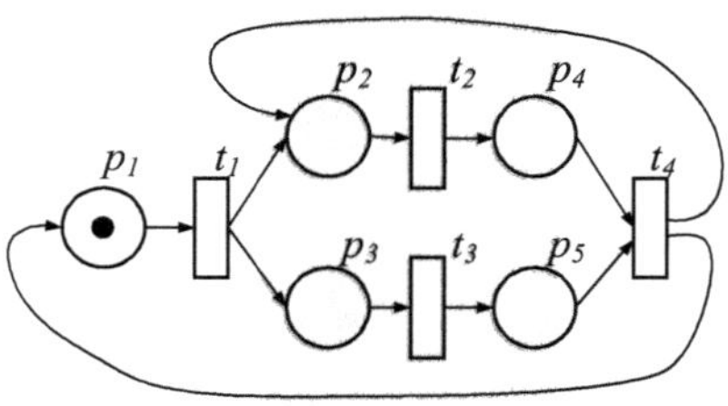

Fig. 20.7 Structurally persistent Petri net system

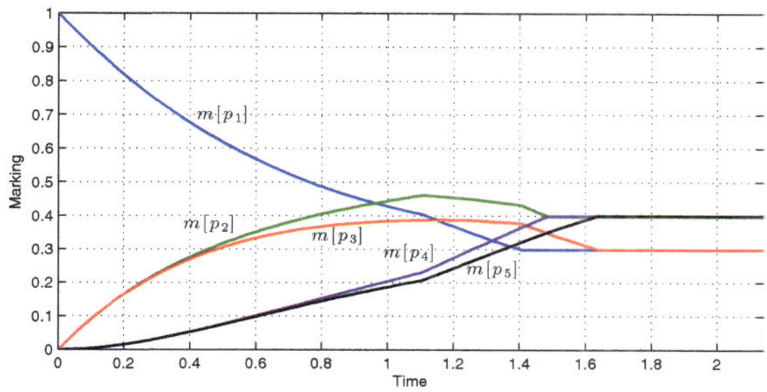

Fig. 20.8 Marking trajectories of the net system in Fig. 20.7 under ON-OFF control

Example 20.6. Let us consider the structurally persistent net in Fig. 20.7 to show the ON-OFF control. Let us assume that the initial marking is $\boldsymbol{m}_0 = [1\ 0\ 0\ 0\ 0]^T$, the target marking is $\boldsymbol{m}_d = [0.3\ 0.4\ 0.3\ 0.4\ 0.4]^T$ and that $\boldsymbol{\lambda} = [1\ 1\ 1\ 1]^T$. Fig. 20.8 shows the marking trajectory after the application of the control. It can be appreciated that the trajectory exhibits *sudden* changes (the first derivative is not continuous) due to the change from ON to OFF in the transitions. The marking $\boldsymbol{m}_d$ is reached in 1.65 time units. ■

According to the ON-OFF approach, once the final marking is reached, all transitions are stopped. This trivially produces a steady state with no flow in which the final marking is kept. As it has been seen in previous sections, steady states with positive flows can be easily maintained as long as the flow vector is a T-semiflow. The system in Fig. 20.7 has no T-semiflows, and therefore, the final marking cannot be kept with a non-null flow vector.

In [30], it is proved that this ON-OFF control policy drives structurally persistent continuous Petri net systems towards $\boldsymbol{m}_d$ in *minimum time*. An intuitive reason for this is that, for persistent nets, the firing order is irrelevant for reaching a marking. Hence, what only matters is the amount of firings required, which is provided by $\boldsymbol{\sigma}$.

20.4.4 Comparison of Control Methods

The availability of several control methods for TCPNs raises concerns about the selection of the most appropriate technique for a given particular system and purpose. In order to make an appropriate choice, several properties may be taken into account, e.g., feasibility, closed-loop stability, robustness, computational complexity (for the synthesis and during the application), etc.

Table 20.1 summarizes a few qualitative properties of some of the control methods described above. According to the presented properties, if the TCPN under consideration is structurally persistent, then the natural choice will be an ON-OFF control law, since it does not exhibit computational problems, ensures convergence and provides the minimum-time transient behavior. For non-persistent nets, MPC

Table 20.1 Qualitative characteristics of control laws (assuming $\boldsymbol{m}_d > 0$). The following abbreviations are used: config. (configuration), min. (minimize), func. (function), compl. (complexity) and poly. (polynomial).

Technique	Computational issues	Optimality criterion	Subclass	Stability
PW-straight trajectory	a LPP for each config.	heuristic for min. time	all	yes
MPC	poly. compl. on $\|T\|,N$	min. quadratic or linear func. of $\boldsymbol{m},\boldsymbol{u}$	all	under suf. conditions
ON-OFF	linear compl. on $\|T\|$	minimum time	structurally persistent	yes

ensures convergence and minimizes a quadratic criterion. Nevertheless, when the number of transitions grows, the complexity may become intractable. In such a case, control synthesis based on other approaches as piecewise-straight trajectories would be more appropriate.

Given a TCPN system with just few configurations and transitions most of the described control laws could be synthesized and applied to it, ensuring convergence. In such a case, the criterion for selecting one of them may be a quantitative one, like minimizing either a quadratic optimization criterion or the time spent for reaching the desired marking.

20.4.5 Control with Uncontrollable Transitions

This subsection briefly discusses two control methods that can be used when the system contains uncontrollable transitions.

Gradient-based control with uncontrollable transitions [21]. This method produces control actions that reduce the rates of the controllable transitions from their nominal maximum values. This is equivalent to reducing the transitions flow, as considered along this chapter. However, the goal of the control problem is slightly different, since it is no longer required to drive the whole marking of the system to a desired value, but only the marking of a subset of places (the *output* of the system). The analysis is achieved in discrete time. Let us provide the basic idea for the case of a single-output system. Firstly, a cost function is defined as $v(k) = 1/2\varepsilon(k)^2$, where $\varepsilon(k)$ denotes the output error. The control proposed has a structure like: $\boldsymbol{u}(k) = \boldsymbol{u}(k-1) - (\boldsymbol{s}(k)\boldsymbol{s}(k)^T + \alpha \boldsymbol{I})^{-1}\boldsymbol{s}(k)\varepsilon(k)$, where the input $\boldsymbol{u}(k)$ is the rate of the controllable transitions and $\boldsymbol{s}(k)$ is the output sensitivity function vector with respect to the input (the gradient vector $\nabla_{\boldsymbol{u}} y$). The factor $\alpha > 0$ is a small term added to avoid ill conditioned matrix computations. The gradient is computed by using a first order approximation method. One of the advantages of this approach is that the change of regions (or configurations) is not explicitly taken into account during the computation of the gradient. Furthermore, a sufficient *condition for stability* is provided.

Pole assignment control with uncontrollable transitions [28]. This technique assumes initially that the initial and desired markings are equilibrium ones and belong to the same region. The control approach considered has the following structure: $\boldsymbol{u} = \boldsymbol{u}'_d + \boldsymbol{K}(\boldsymbol{m} - \boldsymbol{m}'_d)$, where $(\boldsymbol{m}'_d, \boldsymbol{u}'_d)$ is a suitable intermediate equilibrium marking. The gain matrix $\boldsymbol{K}$ is computed, by using any pole-assignment technique, in such a way that the controllable poles are settled as distinct, real and negative. Intermediate markings $\boldsymbol{m}'_d$, with their corresponding input $\boldsymbol{u}'_d$, are computed during the application of the control law (either at each sampling period or just at an arbitrary number of them) by using a given LPP with linear complexity that guarantees that the required input constraints are fulfilled. Later, those results are extended in order to consider several regions. For this, it is required that the initial and desired markings belong to a connected union of equilibrium sets (as defined in Subsection 20.3.2), i.e., $\boldsymbol{m}_0 \in E_1^*$, $\boldsymbol{m}_d \in E_n^*$ and $\cup_{i=1}^n E_i^*$ is connected. Thus, there exist equilibrium markings $\boldsymbol{m}_1^q, ..., \boldsymbol{m}_{n-1}^q$ on the borders of consecutive regions, i.e., $\boldsymbol{m}_j^q \in E_j \cap E_{j+1}, \forall j \in \{1,..,j-1\}$. A gain matrix $\boldsymbol{K}_j$, satisfying the previously mentioned conditions, is computed for each region. Then, inside each jth region, the control action $\boldsymbol{u} = \boldsymbol{u}'_d + \boldsymbol{K}_j(\boldsymbol{m} - \boldsymbol{m}'_d)$ is applied, where $\boldsymbol{m}'_d$ is computed, belonging to the segment $[\boldsymbol{m}_j^q, \boldsymbol{m}_{j+1}^q]$, by using a similar LPP. It was proved that this control law can always be computed and applied (*feasibility*). Furthermore, *convergence* to the desired $\boldsymbol{m}_d$ was also demonstrated, whenever the conditions for controllability are fulfilled and $\cup_{i=1}^n E_i^*$ is connected (see Section 20.3.2). The main drawback of this technique is that a gain matrix and a LPP have to be derived for each region in the marking path.

Similarly to the previous subsection, Table 20.2 summarizes the main features of the two presented methods.

Table 20.2 Qualitative characteristics of control laws (assuming $\boldsymbol{m}_d > 0$) with uncontrollable transitions. The following abbreviations are used: min. (minimize), compl. (complexity) and poly. (polynomial).

Technique	Computational issues	Optimality criterion	Subclass	Stability
Gradient-based	poly. compl. on # outputs	min. quadratic error	all	under a suf. condition
Pole-assignment	a pole-assignment for each config.	none	all	yes

Given that a pole assignment is required for each configuration, if the TCPN has many configurations, the implementation of the pole assignment method becomes tedious although automatizable. This problem does not appear for the gradient based controller. On the contrary, the gradient based controller does not guarantee convergence for the general case, while the pole assignment does it.

20.5 Towards Distributed Control

A natural approach to deal with systems having large net structures is to consider decentralized and distributed control strategies. In a completely distributed approach, the model can be considered as composed of several subsystems that share information through *communication channels*, modeled by places. This problem has been addressed in few works. For instance, [29] proposes the existence of an upper-level controller, named *coordinator*. This coordinator may receive and send information to the local controllers, but it cannot apply control actions directly to the TCPN system. The existence of such coordinator increases the capability of the local controllers, allowing to consider wider classes for the net subsystems (they are assumed to be separately live and consistent, but they are not restricted to particular net subclasses). Affine control laws are proposed for local controllers. Feasibility and concurrent convergence to the required markings are proved.

We will describe in more detail an alternative approach [4] that considers a system composed of *mono-T-semiflow* (MTS) subsystems working under infinite server semantics connected through places (recall that a net is said to be MTS if it is conservative and has a unique minimal T-semiflow whose support contains all the transitions). For each subsystem, a local controller will be designed, being its goal to drive its subsystem from its initial marking to a required one. In order to achieve this goal, it must take into account the interaction with the other subsystems. For this, it is required that neighboring local controllers share information in order to meet a *consensus* that determines the amounts in which transitions must fire in order to reach the target marking. We propose to reach such a consensus by means of an iterative algorithm executed locally at each subsystem.

In order to illustrate the kind of systems that will be handled, consider a simple net modeling a car manufacturing factory composed by two plants A and D in two different cities. The Petri net model is given in Fig. 20.9. The plant A produces the car body (place p_1) and then sends it to the plant D (place p_2). The plant A can produce concurrently a limited number of car bodies (the initial marking of p_3). In plant D, the engine is constructed (p_4) and then it is put in an intermediate buffer p_5. The same plant paints the body received from plant A in p_6 and puts it in p_7 to be assembled together with the engine. The firing of t_8 means the production of a new car. We assume that D can produce concurrently a limited number of engines (initial marking of p_8) and can paint a limited number of car bodies in parallel (initial marking of p_9). Place p_b is the buffer containing the car bodies produced by plant A while p_a is the buffer containing the finished products. Since we do not want to produce more than we sell, the plant A starts to produce a new body (firing of t_1) only when a car is sold.

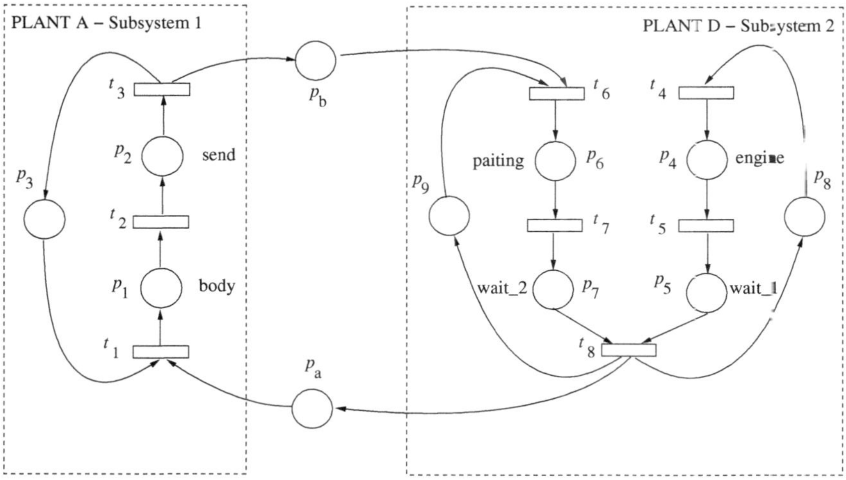

Fig. 20.9 A *distributed* marked graph modeling a car manufacturing plant where p_a and p_b are buffer places

20.5.1 Distributed Continuous Petri Nets

We will focus on distributed continuous Petri nets (DcontPN) which consists of a set of MTS net systems (called subsystems) interconnected through buffers modeled as places. Let K denote the set of subsystems of a given DcontPN. The set of places, transitions and token flow matrix of subsystem $k \in K$ is denoted by P^k, T^k and $\boldsymbol{C}^k \in \mathbb{R}^{|P^k|\times|T^k|}$, respectively. We assume, $P^k \cap P^l = \emptyset$ and $T^k \cap T^l = \emptyset$, $\forall k,l \in K,\ k \neq l$. The directional connection between subsystems is provided by a set of places called *channel* or *buffer places*. In particular, the directional connection from subsystem k to l is provided by a set of places denoted $B^{(k,l)}$, whose input transitions are contained only in subsystem k and output transitions are contained only in subsystem l, i.e., $B^{(k,l)} = \{p \in P| {}^\bullet p \in T^k,\ p^\bullet \in T^l,\ p \notin P^q\ \ \forall q \in K\}$ for every $k,l \in K$, $k \neq l$, and $B^{(l,l)} = \emptyset$ for every $l \in K$.

Note that a place $p \in B^{(k,l)}$ is an *input buffer* of subsystem l and an *output buffer* of subsystem k. The set of all output buffers of subsystem k is denoted by $B^{(k,*)}$, i.e., $B^{(k,*)} = \bigcup_{l \in K} B^{(k,l)}$, and the set of all input channels of subsystem k is denoted by $B^{(*,k)}$, i.e., $B^{(*,k)} = \bigcup_{l \in K} B^{(l,k)}$.

The marking vector of a subsystem k is denoted by $\boldsymbol{m}[P^k] \in \mathbb{R}_{\geq 0}^{|P^k|}$. When designing a controller, it must be taken into account that the controller of a given subsystem only knows its marking and the marking of its input buffers, i.e., the marking of the other subsystems and their input buffers are not observable.

Among the different existing control problems, we will deal with a control problem of DcontPN which aims at reaching a particular target marking $\boldsymbol{m}_f$ at each subsystem. That is, after a finite period of time each subsystem is at its target marking. In contrast to a centralized control, each subsystem is equipped with its own controller that computes the control actions that drive the subsystem to the target marking. Given that the subsystems are interconnected, they may require resources to be available in the communication buffers to reach the target marking. The following example shows this situation.

Example 20.7. Consider the DcontPN in Fig. 20.9 with $\boldsymbol{m}_0[P^1] = [0\,0\,3]^T$, $\boldsymbol{m}_0[P^2] = [0\ 0\ 0\ 0\ 2\ 2]^T$, $m_0[p_a] = 1$, $m_0[p_b] = 0$ and let $\boldsymbol{m}_d[P^1] = [0\ 0\ 3]^T$, $\boldsymbol{m}_d[P^2] = [0\ 0\ 1\ 0\ 2\ 1]^T$ be the target markings of each subsystem. Let the flow integrals of subsystem 1 and 2 be denoted as $\boldsymbol{s}^1$ and $\boldsymbol{s}^2$ respectively.

Let us assume that the controller of the second subsystem computes $s^2[t_6] = 1$, $s^2[t_4] = s^2[t_5] = s^2[t_7] = s^2[t_8] = 0$ so that the subsystem reaches the target marking. Given that the initial marking and target marking of subsystem 1 are the same, a controller for that subsystem could yield: $s^1[t_1] = s^1[t_2] = s^1[t_3] = 0$. Since $m_0[p_b] = 0$, transition t_6 cannot fire unless t_3 fires. Unfortunately, according to the computed controls, t_3 will not fire ($s^1[t_3] = 0$). Hence, these controls are not valid to reach the desired target marking of subsystem 2. In order to solve this situation, subsystem 2 may ask subsystem 1 to put enough tokens in p_b. This can be achieved easily by firing t_3. However, this will imply that subsystem 1 moves away from its desired target marking. ■

Apart from the problem of tokens (resources) required in the buffer places at the initial time, it could happen that the target markings cannot be reached due to the system structure and initial marking (the following example deals with this case). When stating the problem we are implicitly assuming that all target markings of subsystems are reachable, meaning that the final marking of the overall net system, i.e., the net containing all subsystems and buffers, is reachable.

Example 20.8. Consider again the DcontPN in Fig. 20.9. For subsystem 1, let the target marking be $\boldsymbol{m}_d[P^1] = [0\ 0\ 3]^T$ which is reachable from $\boldsymbol{m}_0[P^1] = [0\ 0\ 3]^T$ locally. For subsystem 2, let the target marking be $\boldsymbol{m}_d[P^2] = [0\ 0\ 1\ 0\ 2\ 1]^T$ which is reachable from $\boldsymbol{m}_0[P^2] = [0\ 0\ 0\ 0\ 2\ 2]^T$ locally by firing t_6, i.e., if it is considered isolated from the rest of the system. But when both subsystems are connected through the buffers p_a and p_b with $m_0[p_a] = m_0[p_b] = 0$, the target markings are not reachable. ■

20.5.2 A Control Strategy for DcontPNs

In this subsection, a distributed controller for DcontPN with *tree structure* is proposed (this extends the results in [5] where the problem has been studied for DcontPN with two subsystems). In a system with tree structure cycles are not allowed. The following assumptions will be taken on the considered DcontPNs: (A1) The target marking $\boldsymbol{m}_d$ is strictly positive and reachable at the overall system; (A2) The DcontPN is composed of MTS subsystems. The minimal T-semiflows of the subsystem i is denoted by $\boldsymbol{x}^i$; (A3) The overall system is a MTS net system.

The first assumption is simply a necessary condition for reachability of the target markings. The second assumption reduces the class of DcontPN to those systems composed by MTS subsystems while the third one states that the overall system is MTS. In order to drive the subsystems from their initial states to the target states, Algorithm 20.9 is developed. It represents logic of the rules to be executed in each subsystem to meet a consensus.

Algorithm 20.9. *[Distributed controller of subsystem k]*
Input: $\boldsymbol{C}^k, \boldsymbol{m}_0[P^k], \boldsymbol{m}_d[P^k], B^{(k,*)}, B^{(*,k)}, \boldsymbol{m}_0[B^{(k,*)}]$
Output: *flow integral vector* $\boldsymbol{s}$
1) Solve

$$\begin{aligned} \min\ & \mathbf{1}^T \cdot \bar{\boldsymbol{s}} \\ s.t.\ & \boldsymbol{m}_d[P^k] - \boldsymbol{m}_0[P^k] = \boldsymbol{C}^k \cdot \bar{\boldsymbol{s}}, \\ & \bar{\boldsymbol{s}} \geq 0 \end{aligned} \tag{20.7}$$

2) ***Repeat*** $|K| - 1$ *times*
3) For every $p \in B^{(*,k)}$ *calculate*

$$q_p^{req} = \Big(\sum_{t \in p^\bullet} Pre[p,t] \cdot \bar{s}[t]\Big) - m_0[p] \tag{20.8}$$

4) For all $p \in B^{(*,k)}$ *send* q_p^{req} *to the connected subsystem*
5) For all $p \in B^{(k,*)}$ *receive* r_p^{req} *from the connected subsystem*
6) For all $p \in B^{(k,*)}$ *calculate*

$$h_p = \Big(\sum_{t \in {}^\bullet p} Post[p,t] \cdot \bar{s}[t]\Big) - r_p^{req} \tag{20.9}$$

7) ***If*** $\min_{p \in B^{(k,*)}} \{h_p\} < 0$ ***then*** *solve*

$$\begin{aligned} \min\ & \mathbf{1}^T \cdot \boldsymbol{s} \\ s.t.\ & \boldsymbol{m}_d[P^k] - \boldsymbol{m}_0[P^k] = \boldsymbol{C}^k \cdot \boldsymbol{s}, \\ & \sum_{t \in {}^\bullet p} Post[p,t] \cdot s[t] \geq r_p^{req}, \forall p \in B^{(k,*)} \\ & \boldsymbol{s} \geq 0 \end{aligned} \tag{20.10}$$

Else
$\boldsymbol{s} = \bar{\boldsymbol{s}}$
End If
8) $\bar{\boldsymbol{s}} = \boldsymbol{s}$
9) ***End Repeat***
10) ***return s***

In step 1, each subsystem computes the flow integral $\bar{\boldsymbol{s}}$ required to reach its target marking without taking into account the marking of the buffers. Step 2 computes the amounts of tokens q_p^{req} to be produced in each input buffer p in order to be able to fire $\bar{\boldsymbol{s}}$. The connected subsystems are informed about the amounts of required tokens q_p^{req} in step 4. In step 5, each subsystem receives the amount of tokens it has to produce (if any) in its output buffers. In step 6, it is computed how many tokens would remain in each output buffer if the present control was applied. If this value is negative, more tokens must be produced in the output buffers, and therefore the control law must be recomputed. This re-computation is achieved in step 7 using LPP (20.10). Observe that comparing with LPP (20.7) of step 2 only one extra constraint is added in order to ensure that enough tokens are produced in the output buffers. Steps 4-7 are repeated $|K|-1$ times in order to allow the communication along the longest path connecting a pair of subsystems.

The following Theorem shows that Algorithm 20.9 computes a control law for all subsystems that ensures the reachability of their target markings (see [4] for the proof).

Proposition 20.4. *Let N be a DcontPN with tree structure satisfying assumptions (A1), (A2) and (A3), and let $\boldsymbol{s}^k$ be the flow integral vectors computed by Algorithm 1 for each subsystem for a given initial and target marking. The application of $\boldsymbol{s}^k$ drives the subsystems to their target markings.*

Example 20.10. Consider the net system in Fig. 20.9 used also in Example 20.8. Assume for the first subsystem the same initial and desired markings: $\boldsymbol{m}_0[P^1] = \boldsymbol{m}_d[P^1] = [0\ 0\ 3]^T$ while for the second one: $\boldsymbol{m}_0[P^2] = [0\ 0\ 0\ 0\ 2\ 2]^T$ and $\boldsymbol{m}_d[P^2] = [0\ 0\ 1\ 0\ 2\ 1]^T$. For the buffers, let us assume $m_0[p_a] = 1, m_0[p_b] = 0$. Let us compute local control laws in each subsystem. For the first one, since $\boldsymbol{m}_0[P^1] = \boldsymbol{m}_f[P^1]$, the minimum firing vector is unique equal to $\boldsymbol{s}^1 = [0\ 0\ 0]^T$, i.e., not firing any transition. For the second subsystem, it is easy to observe that the minimum firing vector is $\boldsymbol{s}^2 = [0\ 0\ 1\ 0\ 0]^T$, i.e., firing t_6 in an amount equal to 1. Notice that t_6 cannot fire from the initial marking because $m_0[p_b] = 0$. In order to avoid this, it is possible to fire once the T-semiflow of subsystem 1 (equal to the vector of ones). This is the control action that Algorithm 20.9 computes for subsystem 1 after the first iteration. The algorithm performs just one iteration because, in this example, $|K| = 2$. ■

Once the flow integral vectors $\boldsymbol{s}$ of the evolution from the initial marking to the target marking have been computed by Algorithm 20.9, the value of the control actions $\boldsymbol{u}$ can be derived in several ways (for example applying the procedure in [3]) as long as $\boldsymbol{s} = \int_{\tau_a}^{\tau_b} (\boldsymbol{f} - \boldsymbol{u}) d\tau$ is satisfied where τ_a and τ_b are the initial and final time instants respectively. Remark that $\boldsymbol{s}$ can be seen as a firing count vector in the untimed system and the problem of finding a control law $\boldsymbol{u}$ is equivalent to a reachability problem: if the desired marking is reachable in the untimed net system it is reachable in the timed one with an appropriate control law if all transitions are controllable. This result is proved in [22] (Prop. 14. 3) where a procedure that executes a firing sequence of the untimed system in the timed one is also presented.

20.6 Further Reading

The reader is referred to [24] for a general introduction to fluidization with some advanced topics, and to the the book by R. David and H. Alla [8] for several examples. Further information about some of the presented control topics can be found in the survey paper [25].

Several control approaches have been proposed for continuous nets in which all transitions are controllable. In [12] a fuzzy control approach is presented where it is shown that the flow of a fluid transition, under infinite servers semantics with an implicit self-loop, can be represented as the output of two fuzzy rules under the Sugeno model. A low and high gain control method that generates a piecewise-straight marking trajectory for a tracking control problem was first suggested for Join-Free nets [16] and the extended to general nets [17]. Model predictive control techniques can be consulted in [23]. Details about an approach related to proportional control synthesis with LMI can be found in [18]. Finally, the ON-OFF approach is described in [30].

With respect to approaches that allow uncontrollable transitions, other approaches, as well as model predictive control have been proposed. In [21] a gradient-based control that manipulated the rates of transitions is presented. A description of control method based on pole assignment can be found in [28].

References

1. Amer-Yahiaj, A., Zerhouni, N., El-Moudni, A., Ferney, M.: State variable description and controllability of a class of continuous Petri nets. In: Proc. of the IEEE Int. Symp. on Circuits and Systems, pp. 68–71 (1996)
2. Amrah, A., Zerhouni, N., El-Moudni, A.: On the control of manufacturing lines modelled by controlled continuous Petri nets. Int. Journal of Systems Science 29(2), 127–137 (1998)
3. Apaydin-Özkan, H., Júlvez, J., Mahulea, C., Silva, M.: An efficient heuristics for minimum time control of continuous Petri nets. In: Proc. 3rd IFAC Conf. on Analysis and Design of Hybrid Systems, Zaragoza, Spain (2009)
4. Apaydin-Özkan, H., Mahulea, C., Júlvez, J., Silva, M.: An iterative control method for distributed continuous Petri nets with tree structures. In: Proc. 49th IEEE Conf. on Decision and Control, CDC 2010, Atlanta, USA (2010)
5. Apaydin-Özkan, H., Mahulea, C., Júlvez, J., Silva, M.: A control method for timed distributed continuous Petri nets. In: Proc. American Control Conference 2010, Baltimore, Maryland, USA (2010)
6. Bemporad, A., Morari, M., Dua, V., Pistikopoulos, E.N.: The explicit linear quadratic regulator for constrained systems. Automatica 38(1), 3–20 (2002)
7. Brammer, R.: Controllability in linear autonomous systems with positive controllers. SIAM Journal of Control 10(2), 329–353 (1972)
8. David, R., Alla, H.: Discrete, Continuous and Hybrid Petri Nets, 2nd edn. Springer, Berlin (2004)
9. Dub, M., Pipan, G., Hanzálek, Z.: Stock optimization of a kanban-based assembly line. In: Proc. 12th Int. Conf. on Flexible Automation and Intelligent Manufacturing, pp. 1–10 (2002)
10. Chen, C.: Linear System Theory and Design. Oxford University Press, New York (1984)
11. Giua, A., DiCesare, F., Silva, M.: Petri Net supervisors for generalized mutual exclusion constraints. In: Proc. 12th IFAC World Congress, Sidney, Australia (1993)

12. Hennequin, S., Lefebvre, D., El-Moudni, A.: Fuzzy control of variable speed continuous Petri nets. In: Proc. 38th Conf. on Decisions and Control, CDC 1999 (1999)
13. Holloway, L.E., Krogh, B.H., Giua, A.: A survey of Petri nets methods for controlled discrete event systems. Discrete Event Dynamic Systems 7(2), 151–190 (1997)
14. Iordache, M.V., Antsaklis, P.J.: Supervisory control of concurrent systems. Birkhäuser, Boston (2006)
15. Jiménez, E., Júlvez, J., Recalde, L., Silva, M.: On controllability of timed continuous Petri net systems: the join free case. In: Proc. 44th IEEE Conf. on Decision and Control, Seville, Spain (2005)
16. Jing, X., Recalde, L., Silva, M.: Tracking control of join-free timed continuous Petri net systems under infinite servers semantics. Discrete Event Dynamic Systems 18(2), 263–283 (2008)
17. Jing, X., Recalde, L., Silva, M.: Tracking control of timed continuous Petri net systems under infinite servers semantics. In: Proc. of IFAC World Congress, CDROM (2008)
18. Kara, R., Ahmane, M., Loiseau, J.J., Djennoune, S.: Constrained regulation of continuous Petri nets. Nonlinear Analysis: Hybrid Systems 3(4), 738–748 (2009)
19. Kara, R., Djennoune, S., Loiseau, J.J.: State feedback control for the manufacturing systems modeled by continuous Petri nets. In: Proc. Information Control Problems in Manufacturing, IFAC Papers Online (2006)
20. Lefebvre, D.: Feedback control designs for manufacturing systems modelled by continuous Petri nets. Int. Journal of Systems Science 30(6), 591–600 (1999)
21. Lefebvre, D., Catherine, D., Leclerq, E., Druaux, F.: Some contributions with Petri nets for the modelling, analysis and control of HDS. In: Proc. Int. Conference on Hybrid Systems and Applications, vol. 1(4), pp. 451–465 (2007)
22. Mahulea, C., Ramírez, A., Recalde, L., Silva, M.: Steady state control reference and token conservation laws in continuous Petri net systems. IEEE Transactions on Automation Science and Engineering 5(2), 307–320 (2008)
23. Mahulea, C., Giua, A., Recalde, L., Seatzu, C., Silva, M.: Optimal model predictive control of timed continuous Petri nets. IEEE Transactions on Automatic Control 53(7), 1731–1735 (2008)
24. Silva, M., Recalde, L.: On fluidification of Petri net models: from discrete to hybrid and continuous models. Annual Reviews in Control 28, 253–266 (2004)
25. Silva, M., Júlvez, J., Mahulea, C., Vázquez, C.R.: On fluidization of discrete event models: observation and control of continuous Petri nets. Discrete Event Dynamic Systems 21(4), 427–497 (2011)
26. Vázquez, C.R., Ramírez, A., Recalde, L., Silva, M.: On Controllability of Timed Continuous Petri Nets. In: Egerstedt, M., Mishra, B. (eds.) HSCC 2008. LNCS, vol. 4981, pp. 528–541. Springer, Heidelberg (2008)
27. Vázquez, C.R., Silva, M.: Performance control of Markovian Petri nets via fluid models: a stock-level control example. In: Proc. 5th IEEE Conf. on Automation Science and Engineering, pp. 30–36 (2009)
28. Vázquez, C.R., Silva, M.: Piecewise-linear constrained control for timed continuous Petri nets. In: Proc. 48th IEEE Conf. on Decision and Control, Shangai, China, pp. 5714–5720 (2009)
29. Vázquez, C.R., van Schuppen, J.H., Silva, M.: A modular-coordinated control for continuous Petri nets. In: Proc. 18th IFAC World Congress, Milano, Italy (2011)
30. Wang, L., Mahulea, C., Júlvez, J., Silva, M.: Minimum-time control for structurally persistent continuous Petri nets. In: Proc. 49th IEEE Conf. on Decision and Control (CDC 2010), Atlanta, USA (2010)

Part III
Systems in Dioids

Chapter 21
Discrete-Event Systems in a Dioid Framework: Modeling and Analysis

Thomas Brunsch, Jörg Raisch, Laurent Hardouin, and Olivier Boutin

21.1 Timed Event Graphs

As discussed in the previous chapters, a Petri net graph is a directed bipartite graph $N = (P, T, A, w)$, where $P = \{p_1, \ldots, p_n\}$ is the finite set of places, $T = \{t_1, \ldots, t_m\}$ is the (finite) set of transitions, $A \subseteq (P \times T) \cup (T \times P)$ is the set of directed arcs from places to transitions and from transitions to places, and $w : A \to \mathbb{N}$ is a weight function. Note that this notation differs slightly from the notation introduced in Part II of this book. However, for the description of timed event graphs, i.e., a specific type of Petri nets, our notation is very convenient. In the sequel, the following notation is used for Petri net graphs:

$$^{\bullet}t_j := \{p_i \in P \mid (p_i, t_j) \in A\}$$

Thomas Brunsch · Jörg Raisch
Technische Universität Berlin, Fachgebiet Regelungssysteme, Einsteinufer 17, 10587 Berlin, Germany
e-mail: {brunsch,raisch}@control.tu-berlin.de

Jörg Raisch
Fachgruppe System- und Regelungstheorie, Max-Planck-Institut für Dynamik komplexer technischer Systeme, Sandtorstr. 1, 39106 Magdeburg, Germany

Laurent Hardouin · Thomas Brunsch
LUNAM, University of Angers, LISA, ISTIA, 62 Av. Notre-Dame du Lac, 49000 Angers, France
e-mail: laurent.hardouin@istia.univ-angers.fr

Olivier Boutin
Calle Santiago 2 – 4°C, 11005 Cadiz, Spain
e-mail: olivier.research@gmail.com

C. Seatzu et al. (Eds.): Control of Discrete-Event Systems, LNCIS 433, pp. 431–450.
springerlink.com

is the set of all input places for transition t_j, i.e., the set of places with arcs to t_j.

$$t_j^{\bullet} := \{p_i \in P \,|\, (t_j, p_i) \in A\}$$

denotes the set of all output places for transition t_j, i.e., the set of places with arcs from t_j. Similarly,

$$^{\bullet}p_i := \{t_j \in T \,|\, (t_j, p_i) \in A\}$$

is the set of all input transitions for place p_i, i.e., the set of transitions with arcs to p_i, and

$$p_i^{\bullet} := \{t_j \in T \,|\, (p_i, t_j) \in A\}$$

denotes the set of all output transitions for place p_i, i.e., the set of transitions with arcs from p_i. Obviously, $p_i \in {}^{\bullet}t_j$ if and only if $t_j \in p_i^{\bullet}$, and $t_j \in {}^{\bullet}p_i$ if and only if $p_i \in t_j^{\bullet}$. Graphically, places are shown as circles, transitions as bars, and arcs as arrows. The number attached to an arrow is the weight of the corresponding arc. Usually, weights are only shown explicitly if they are different from one.

A Petri net system (or Petri net) is a pair $(N, \boldsymbol{m}^0)$, where $N = (P, T, A, w)$ is a Petri net graph and $\boldsymbol{m}^0 \in \mathbb{N}_0^n$ with $n = |P|$ is a vector of initial markings. In graphical representations, the vector of initial markings is shown by m_i^0 dots ("tokens") within the circles representing places $p_i, i = 1, \ldots, n$. A Petri net can be interpreted as a dynamical system with a state signal $\boldsymbol{m} : \mathbb{N}_0 \to \mathbb{N}_0^n$ and an initial state $\boldsymbol{m}(0) = \boldsymbol{m}^0$. Its dynamics is governed by two rules, also called firing rules:

(i) In state $\boldsymbol{m}(k)$, a transition t_j can occur (or "fire") if and only if all of its input places contain at least as many tokens as the weight of the arc from the respective place to the transition t_j, i.e., if

$$m_i(k) \geq w(p_i, t_j), \forall p_i \in {}^{\bullet}t_j.$$

(ii) If a transition t_j fires, the number of tokens in all its input places is decreased by the weight of the arc connecting the respective place to the transition t_j, and the number of tokens in all its output places is increased by the weight of the arc connecting t_j to the respective place, i.e., the state changes according to

$$m_i(k+1) = m_i(k) - w(p_i, t_j) + w(t_j, p_i), \; i = 1, \ldots, n,$$

where $m_i(k)$ and $m_i(k+1)$ represent the numbers of tokens in place p_i before and after the firing of transition t_j.

Note that a place can simultaneously be an element of ${}^{\bullet}t_j$ and $t_j^{\bullet}$. Hence the number of tokens in a certain place can appear in the firing condition for a transition whilst being unaffected by the actual firing. It should also be noted that a transition enabled to fire might not actually do so. In fact, it is well possible that, in a certain state, several transitions are enabled simultaneously, and that the firing of one of them will disable the other ones.

A Petri net $(N, \boldsymbol{m}^0)$ is called an event graph (or synchronization graph), if each place has exactly one input transition and one output transition, i.e.,

$$|{}^\bullet p_i| = |p_i{}^\bullet| = 1, \forall p_i \in P,$$

and if all arcs have weight 1. It is obvious that an event graph cannot model conflicts or decisions, but it does model synchronization effects. A small example of an event graph is given in Fig. 21.1.

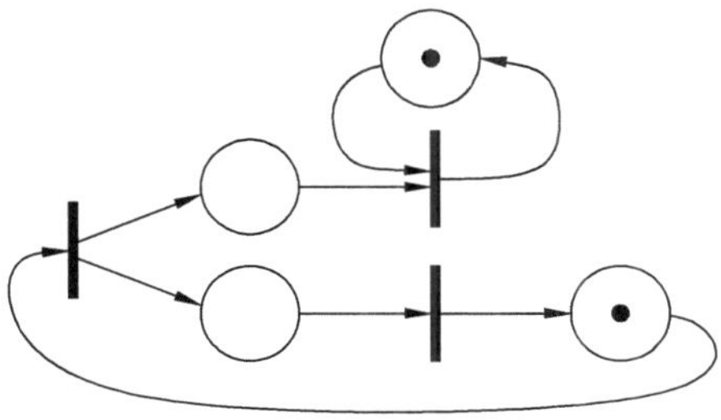

Fig. 21.1 Example of an event graph

A standard Petri net $(N, \boldsymbol{m}^0)$ only models the ordering of firings of transitions, but not the actual firing times. However, it is possible to "attach" timing information to the "logical" DES model $(N, \boldsymbol{m}^0)$. This can be done in two ways: time can be associated with transitions (representing transition delays) or with places (representing holding times). In timed event graphs (TEG), transition delays can always be transposed into holding times by simply shifting each transition delay to all input places of the corresponding transition. However, it is in general not possible to convert holding times into transition delays. Therefore, we will only consider timed event graphs with holding times. In TEG with holding times, tokens in place p_i have to be held for a certain time (called "holding time") before they can contribute to the firing of the output transition of p_i. The holding time for a token in place p_i is denoted v_i.

Figure 21.2 shows a part of a general timed event graph with holding times. In general, the earliest time instant when place p_i receives its kth token is denoted $\pi_i(k)$,

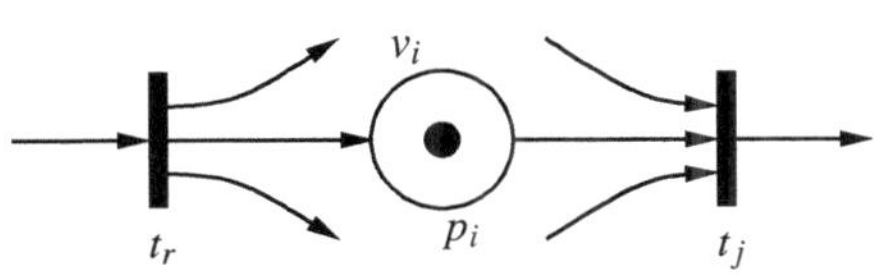

Fig. 21.2 Part of a general timed event graph with holding times

and $\tau_j(k)$ denotes the earliest time instant when transition t_j can fire for the kth time. Then the earliest time of the kth firing of transition t_j can be determined by

$$\tau_j(k) = \max_{p_i \in {}^\bullet t_j} (\pi_i(k) + v_i),$$

i.e., a transition is enabled to fire as soon as all input places of this transition have received their k^{th} token and their corresponding holding times have elapsed. Similarly, the earliest time instant when place p_i receives its $(k+m_i^0)^{\text{th}}$ token can be determined by the kth firing of its input transition, i.e., $\pi_i(k+m_i^0) = \tau_r(k)$, $t_r \in {}^\bullet p_i$. Note that a place in an event graph has exactly one input transition and consequently, the place can only receive tokens when this input transition fires. Therefore, it is possible to eliminate the π_i to give recursive equations for the (earliest) firing times of transitions.

21.2 Motivational Example

From the previous section, it is clear that it is possible to recursively compute the earliest possible firing times for transitions in timed event graphs. In the corresponding equations, two operations were needed: maximization and addition. To illustrate this, a small example (taken from Cassandras, Lafortune & Olsder [4]) is used. Imagine a simple public transportation system consisting of three lines: an inner loop and two outer loops. There are two stations where passengers can change lines, and four rail tracks connecting the stations. The basic structure of the system is given in Fig. 21.3. Initially, it is assumed that the train company operates one train on each track. A train needs 3 units of time to travel from station 1 to station 2, 5 units of time for the track from station 2 to station 1, and 2 and 3 units of time for the outer loops, respectively. The aim is to implement a user-friendly policy where trains wait for each other at the stations to allow passengers to change lines without delay, i.e., the departure times of trains from stations shall be synchronized. This can be easily represented in a timed event graph with holding times (see Fig. 21.4). The tokens

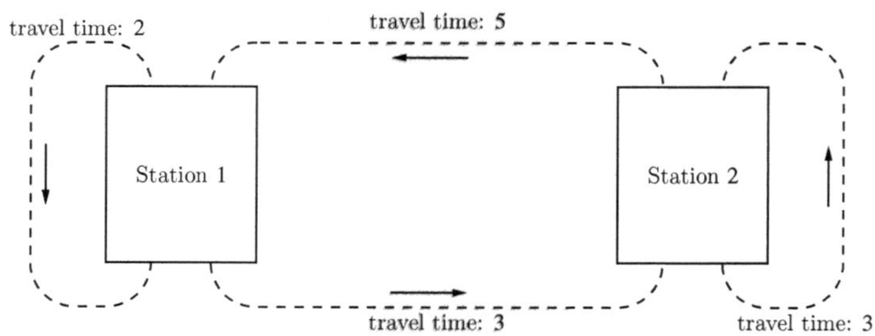

Fig. 21.3 Simple transportation network taken from [4]

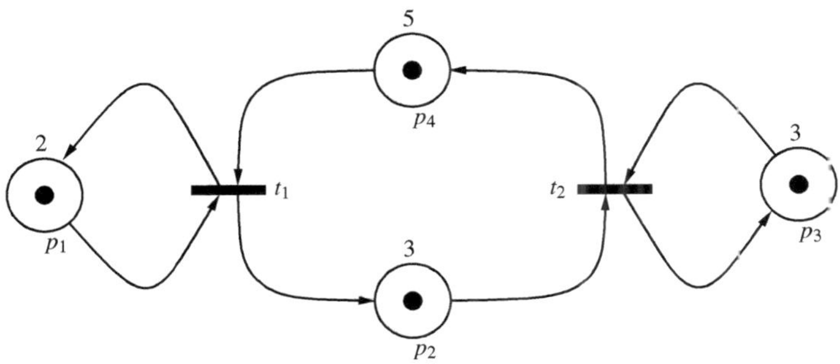

Fig. 21.4 Timed event graph of the transportation network

in places p_1 to p_4 model trains on the tracks and the holding times are the traveling times for trains on the tracks. The firing times of transitions t_1 and t_2 represent the departure times of the trains in stations 1 and 2, respectively. These times can therefore be interpreted as the "time table" for the simple public transportation system. The recursive equations for the firing times of transitions t_1 and t_2 are

$$\begin{aligned} \tau_1(k) &= \max(\pi_1(k)+2, \pi_4(k)+5) \\ \tau_2(k) &= \max(\pi_2(k)+3, \pi_3(k)+3) \end{aligned} \tag{21.1}$$

and

$$\pi_1(k+m_1^0) = \pi_1(k+1) = \tau_1(k) \tag{21.2}$$
$$\pi_2(k+m_2^0) = \pi_2(k+1) = \tau_1(k) \tag{21.3}$$
$$\pi_3(k+m_3^0) = \pi_3(k+1) = \tau_2(k) \tag{21.4}$$
$$\pi_4(k+m_4^0) = \pi_4(k+1) = \tau_2(k). \tag{21.5}$$

Inserting (21.2)–(21.5) into (21.1), it is possible to eliminate $\pi_i(k)$

$$\begin{aligned} \tau_1(k+1) &= \max\big(\tau_1(k)+2, \tau_2(k)+5\big) \\ \tau_2(k+1) &= \max\big(\tau_1(k)+3, \tau_2(k)+3\big). \end{aligned} \tag{21.6}$$

Now, given initial firing times, i.e., the first departures of trains, $\tau_1(1) = \tau_2(1) = 0$, the timetable can be determined as follows:

$$\begin{bmatrix}0\\0\end{bmatrix}, \begin{bmatrix}5\\3\end{bmatrix}, \begin{bmatrix}8\\8\end{bmatrix}, \begin{bmatrix}13\\11\end{bmatrix}, \begin{bmatrix}16\\16\end{bmatrix}, \ldots$$

If the initial departure times are chosen to be $\tau_1(1) = 1$ and $\tau_2(1) = 0$, the sequence is

$$\begin{bmatrix}1\\0\end{bmatrix}, \begin{bmatrix}5\\4\end{bmatrix}, \begin{bmatrix}9\\8\end{bmatrix}, \begin{bmatrix}13\\12\end{bmatrix}, \begin{bmatrix}17\\16\end{bmatrix}, \ldots$$

In both cases the average departure interval is 4 units of time, however, in the second case the departure interval is constant, i.e., the system has a so-called 1-periodic behavior, while in the first case the system shows a 2-periodic behavior.

Eventually, the train company may consider to realize shorter (average) departure intervals. This could be achieved by using additional trains. For example the company provides a second train in the inner loop, e.g., initially on the track from station 1 to station 2. With respect to the timed event graph shown in Fig. 21.4, this means to add a second token in place p_2. In this case the firing times of transitions t_1 and t_2 as functions of π_i (Eq. 21.1) do not change, but the time instants when places p_i receive their tokens change to

$$\pi_1(k+1) = \tau_1(k) \tag{21.7}$$
$$\pi_2(k+2) = \tau_1(k) \tag{21.8}$$
$$\pi_3(k+1) = \tau_2(k) \tag{21.9}$$
$$\pi_4(k+1) = \tau_2(k). \tag{21.10}$$

Therefore the recursive equations for the firing times τ_1 and τ_2 change to

$$\tau_1(k+1) = \max\big(\tau_1(k)+2, \tau_2(k)+5\big) \tag{21.11}$$
$$\tau_2(k+2) = \max\big(\tau_1(k)+3, \tau_2(k+1)+3\big). \tag{21.12}$$

By introducing a new variable τ_3, with $\tau_3(k+1) := \tau_1(k)$, one may transform (21.11) and (21.12) into a system of first-order difference equations

$$\tau_1(k+1) = \max\big(\tau_1(k)+2, \tau_2(k)+5\big) \tag{21.13}$$
$$\tau_2(k+1) = \max\big(\tau_3(k)+3, \tau_2(k)+3\big) \tag{21.14}$$
$$\tau_3(k+1) = \tau_1(k). \tag{21.15}$$

Initializing this system with $\tau_1(1) = \tau_2(1) = \tau_3(1) = 0$, the following evolution can be determined:

$$\begin{bmatrix}0\\0\\0\end{bmatrix}, \begin{bmatrix}5\\3\\0\end{bmatrix}, \begin{bmatrix}8\\6\\5\end{bmatrix}, \begin{bmatrix}11\\9\\8\end{bmatrix}, \begin{bmatrix}14\\12\\11\end{bmatrix}, \cdots$$

After a short transient phase, trains depart from both stations at intervals of three units of time. Obviously, shorter intervals cannot be reached by additional trains in the inner loop of the system, as the outer loop at station 2 now represents the "bottleneck" of the system.

In this simple example, several phenomena have been encountered: 1-periodic solutions (for $\tau_1(1) = 1$ and $\tau_2(1) = 0$), 2-periodic solutions (for $\tau_1(1) = \tau_2(1) = 0$), and a transient phase (for the extended system). These phenomena (and more) can be conveniently analyzed and explained within the formal framework of idempotent semirings or dioids.

21.3 Dioid Algebraic Structures

An idempotent semiring (or dioid) is an algebraic structure based on two monoids (one of which is commutative). A structure $(M, \cdot, e)$ is a monoid if $\cdot$ is an internal law, associative and of which e is the identity element. If the law $\cdot$ is commutative, $(M, \cdot, e)$ is called a commutative monoid. Then, a dioid (or idempotent semiring) is a set, $\mathscr{D}$, endowed with two binary operations, addition ($\oplus$) and multiplication ($\otimes$), such that:

- $(\mathscr{D}, \oplus, \varepsilon)$ is an idempotent commutative monoid, i.e., $\forall a \in \mathscr{D}, a \oplus a = a$
- $(\mathscr{D}, \otimes, e)$ is a monoid
- $\otimes$ is distributive with respect to $\oplus$
- ε is absorbing for $\otimes$, i.e., $\forall a \in \mathscr{D}, \varepsilon \otimes a = a \otimes \varepsilon = \varepsilon$.

The identity element of addition (ε) is also called the zero element of the dioid, while the identity element of multiplication (e) is also called the unit element of the dioid. Often, the multiplication sign is omitted in written equations when they are unambiguous.

Due to the idempotency property, any dioid can be endowed with a natural (partial) order defined by $a \preceq b \Leftrightarrow a \oplus b = b$, i.e., the sum of two elements a and b is the least upper bound of a and b. Thus, any dioid forms a sub-semilattice. A dioid is said to be complete if it is closed for infinite sums, i.e., there exists the greatest element of $\mathscr{D}$ given by $\top = \bigoplus_{x \in \mathscr{D}} x$, and if $\otimes$ distributes over infinite sums. Formally, a complete dioid forms a complete lattice for which the greatest lower bound of a and b is denoted $a \wedge b$ [1].

Example 21.1. [Max-plus algebra] Probably the best known idempotent semiring is the so-called max-plus algebra, often denoted $\mathbb{Z}_{\max}$. It is defined over the set $\mathbb{Z} \cup \{-\infty\}$ and its binary operations are defined as follows:

- addition: $a \oplus b := \max(a, b)$
- multiplication: $a \otimes b := a + b$

and the zero and unit elements are $\varepsilon = -\infty$ and $e = 0$, respectively. Defining the operations over the set $\mathbb{Z} \cup \{-\infty, \infty\}$ defines a complete dioid (often denoted $\overline{\mathbb{Z}}_{\max}$), with the top element $\top = +\infty$. ■

Example 21.2. [Min-plus algebra] The set $\mathbb{Z} \cup \{-\infty, \infty\}$ endowed with addition defined as $a \oplus b := \min(a, b)$, and multiplication ($a \otimes b := a + b$) forms a complete dioid also known as min-plus algebra, often denoted $\overline{\mathbb{Z}}_{\min}$. Its zero, unit, and top elements are $\varepsilon = \infty$, $e = 0$, and $\top = -\infty$, respectively. ■

Note that the (partial) order is a property of a given dioid. For two elements $a = 5$ and $b = 3$ the max-plus addition is defined as $a \oplus b = 5 \oplus 3 = 5$, which indicates that $5 \succeq 3$ in $\overline{\mathbb{Z}}_{\max}$. The same calculation can be done in min-plus algebra, i.e., a and b belonging to $\overline{\mathbb{Z}}_{\min}$. In this case we get $a \oplus b = 5 \oplus 3 = \min(5, 3) = 3$ and, according to the definition of the natural order of a dioid, this means $5 \preceq 3$ in $\overline{\mathbb{Z}}_{\min}$.

As in classical algebra the binary operations can be extended to the matrix case in dioids. For the matrices $\boldsymbol{A},\boldsymbol{B} \in \mathscr{D}^{n\times m}$, and $\boldsymbol{C} \in \mathscr{D}^{m\times p}$ we get

$$\text{Addition:} \quad [\boldsymbol{A}\oplus\boldsymbol{B}]_{ij} = [\boldsymbol{A}]_{ij} \oplus [\boldsymbol{B}]_{ij}$$

$$\text{Multiplication:} \quad [\boldsymbol{A}\otimes\boldsymbol{C}]_{ij} = \bigoplus_{k=1}^{m} \left([\boldsymbol{A}]_{ik} \otimes [\boldsymbol{C}]_{kj}\right).$$

Note that max-plus algebra (and min-plus algebra) may also be defined on other sets, e.g., the set of real numbers. Depending on the set of definition, these idempotent semirings are denoted $\overline{\mathbb{R}}_{\max}$ or $\overline{\mathbb{R}}_{\min}$, respectively. An exhaustive description on idempotent semirings such as max-plus and min-plus algebra can be found e.g. in [1, 8, 11, 14].

21.4 Linear Dynamical Systems in Max-Plus Algebra

The usefulness of max-plus algebra becomes clear, when we take a second look at the previously introduced transportation network. Recall that the recursive equations for the earliest departure times in Fig. 21.4 are

$$\begin{aligned} \tau_1(k+1) &= \max\bigl(\tau_1(k)+2, \tau_2(k)+5\bigr) \\ \tau_2(k+1) &= \max\bigl(\tau_1(k)+3, \tau_2(k)+3\bigr). \end{aligned}$$

Due to the max-operation, these equations are nonlinear in conventional algebra. However, rewriting these equations in max-plus algebra with $\boldsymbol{x} := [\tau_1 \ \ \tau_2]^T$, results in the following linear system:

$$\begin{aligned} \boldsymbol{x}(k+1) &= \begin{bmatrix} 2 & 5 \\ 3 & 3 \end{bmatrix} \otimes \boldsymbol{x}(k) \\ &= \boldsymbol{A} \otimes \boldsymbol{x}(k). \end{aligned}$$

In general, it is possible to convert any timed event graph into a linear system in max-plus algebra, also called a “max-plus linear system”. In max-plus algebra, the variable $x_i(k)$ is the earliest possible time instant that event x_i occurs for the kth time. Therefore, max-plus algebraic functions $\boldsymbol{x}(k)$ are often called dater functions, giving every event a precise earliest date. Note that, besides max-plus algebra, it is also possible to convert any timed event graph into a min-plus linear system. Then, the min-plus variable $x_i(t)$ represents the maximal number of occurrences of event x_i up to time t. Min-plus algebraic functions are, therefore, often called counter functions.

As mentioned in Section 21.2, depending on the vector of initial firing times, a number of different phenomena have been observed: 1- and 2-periodic behaviors, with and without an initial transient phase. For many application scenarios as, e.g., transportation networks, a 1-periodic solution is desirable. It is therefore natural to ask which initial firing vectors will indeed generate 1-periodic solutions and what the duration for one period is. In conventional algebra this means that the following equation shall be satisfied:

$$\tau_i(k+1) = \lambda + \tau_i(k), \quad \begin{array}{l} k = 1,2,\ldots \\ i = 1,2,\ldots,n. \end{array}$$

Rewriting this requirement in max-plus algebra provides

$$x_i(k+1) = \lambda x_i(k), \quad \begin{array}{l} k = 1,2,\ldots \\ i = 1,2,\ldots,n, \end{array}$$

or, equivalently,

$$\boldsymbol{x}(k+1) = \lambda \boldsymbol{x}(k), \quad k = 1,2,\ldots.$$

This amounts to solving the max-plus eigenproblem. If, for a given $\boldsymbol{A} \in \overline{\mathbb{Z}}_{\max}^{n\times n}$, there exists $\boldsymbol{\xi} \in \overline{\mathbb{Z}}_{\max}^{n}$ and $\lambda \in \mathbb{Z}$ such that

$$\boldsymbol{A\xi} = \lambda \boldsymbol{\xi},$$

we call λ *eigenvalue* and $\boldsymbol{\xi}$ *eigenvector* of matrix A. If we choose the vector of initial firing times, $\boldsymbol{x}(1)$, as an eigenvector, we get

$$\boldsymbol{x}(2) = \boldsymbol{Ax}(1) = \lambda \boldsymbol{x}(1)$$

and therefore

$$\boldsymbol{x}(k) = \lambda^{(k-1)}\boldsymbol{x}(1), \quad k = 1,2,\ldots.$$

This is the desired 1-periodic behavior and the period length is the eigenvalue λ. Note that powers in max-plus algebra (and in dioids in general) are defined by $a^i = a \otimes a^{i-1}$, with $a^0 = e$.

To solve the max-plus eigenproblem, we need the notion of matrix (ir)reducibility, i.e, a matrix $\boldsymbol{A} \in \mathscr{D}^{n\times n}$ is called *reducible*, if there exists a permutation matrix $\boldsymbol{P}$, i.e., a square binary matrix that has exactly one 1-element in each column and row and zeros elsewhere, such that $\tilde{\boldsymbol{A}} = \boldsymbol{PAP}^T$ is upper block-triangular. Otherwise, $\boldsymbol{A}$ is called *irreducible*. If $\boldsymbol{A} \in \mathscr{D}^{n\times n}$ is irreducible, there exists precisely one eigenvalue. It is given by

$$\lambda = \bigoplus_{j=1}^{n} \left(tr(\boldsymbol{A}^j)\right)^{1/j}, \tag{21.16}$$

where "trace" and the jth root are defined as in conventional algebra, i.e , for any $\boldsymbol{B} \in \mathscr{D}^{n\times n}$,

$$tr(\boldsymbol{B}) = \bigoplus_{i=1}^{n} b_{ii}$$

and for any $\alpha \in \mathscr{D}$,

$$\left(\alpha^{1/j}\right)^j = \alpha.$$

Whereas an irreducible matrix $\boldsymbol{A} \in \mathscr{D}^{n\times n}$ has a unique eigenvalue λ, it may possess several distinct eigenvectors. We will not go into the details of how to compute the eigenvectors of matrices in dioids. There are several algorithms, e.g., Howard's algorithm and the power algorithm (see [14] for details), to determine the eigenvalue and the corresponding eigenvector(s).

Recalling the transportation network, we have determined the max-plus system matrix

$$\boldsymbol{A} = \begin{bmatrix} 2 & 5 \\ 3 & 3 \end{bmatrix} \quad \text{and one can check that} \quad \boldsymbol{A}^2 = \begin{bmatrix} 8 & 8 \\ 6 & 8 \end{bmatrix}.$$

According to (21.16), the eigenvalue of the system can be calculated by

$$\lambda = \bigoplus_{j=1}^{2} \left(tr\left(\boldsymbol{A}^j\right)\right)^{1/j} = 3^1 \oplus 8^{1/2} = 3 \oplus 4 = 4.$$

It turns out that $\boldsymbol{\xi} = [1\ e]^T$ is an eigenvector of matrix $\boldsymbol{A}$, i.e., it satisfies $\boldsymbol{A} \otimes \boldsymbol{\xi} = \lambda \otimes \boldsymbol{\xi}$. Not surprisingly, this result confirms our observation of the transportation network. When the initial departure times are set to $\boldsymbol{x}(1) = [1\ 0]^T = \boldsymbol{\xi}$, we obtain a 1-periodic schedule with a departure interval of $\lambda = 4$.

21.5 The 2-Dimensional Dioid $\mathscr{M}_{in}^{ax}[\![\gamma,\delta]\!]$

Max-plus algebra is, as shown in the previous section, suitable to model the behavior of timed event graphs. However, for more complex TEG, the corresponding representation in max-plus algebra becomes more complicated. For the transportation network with three trains in the inner loop, we had to introduce a new variable τ_3 to find the state model. Taking a look at another TEG (taken from [1]), this issue becomes even clearer. The linear dynamical system of Fig. 21.5 in max-plus algebra, with $x_i(k), u_j(k), y(k)$ being the earliest time instants that the transitions x_i, u_j, and y fire for the kth time, is

$$\begin{aligned}
x_1(k+1) &= 1u_1(k+1) \oplus 4x_2(k) \\
x_2(k+1) &= 5u_2(k) \oplus 3x_1(k+1) \\
x_3(k+1) &= 3x_1(k+1) \oplus 4x_2(k+1) \oplus 2x_3(k-1) \\
y(k+1) &= x_2(k) \oplus 2x_3(k+1).
\end{aligned}$$

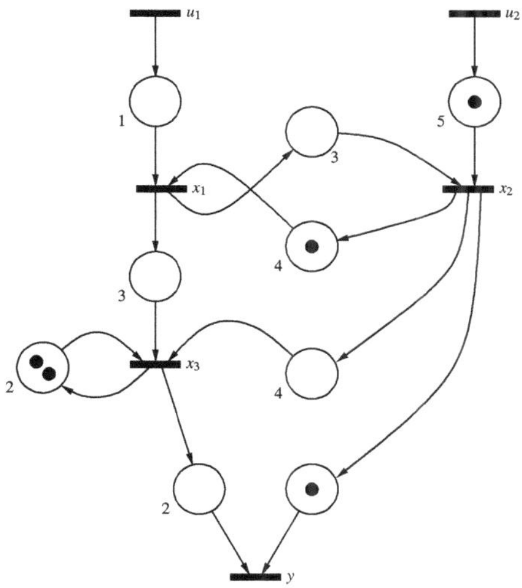

Fig. 21.5 Timed event graph (taken from [1])

Rewriting the linear system in matrix-vector form, we get

$$\begin{aligned} \boldsymbol{x}(k+1) &= \boldsymbol{A}_0\boldsymbol{x}(k+1) \oplus \boldsymbol{A}_1\boldsymbol{x}(k) \oplus \boldsymbol{A}_2\boldsymbol{x}(k-1) \oplus \boldsymbol{B}_0\boldsymbol{u}(k+1) \oplus \boldsymbol{B}_1\boldsymbol{u}(k) \\ y(k) &= \boldsymbol{C}_0\boldsymbol{x}(k) \oplus \boldsymbol{C}_1\boldsymbol{x}(k-1), \end{aligned} \tag{21.17}$$

where

$$\boldsymbol{A}_0 = \begin{bmatrix} \varepsilon & \varepsilon & \varepsilon \\ 3 & \varepsilon & \varepsilon \\ 3 & 4 & \varepsilon \end{bmatrix}, \quad \boldsymbol{A}_1 = \begin{bmatrix} \varepsilon & 4 & \varepsilon \\ \varepsilon & \varepsilon & \varepsilon \\ \varepsilon & \varepsilon & \varepsilon \end{bmatrix}, \quad \boldsymbol{A}_2 = \begin{bmatrix} \varepsilon & \varepsilon & \varepsilon \\ \varepsilon & \varepsilon & \varepsilon \\ \varepsilon & \varepsilon & 2 \end{bmatrix},$$

$$\boldsymbol{B}_0 = \begin{bmatrix} 1 & \varepsilon \\ \varepsilon & \varepsilon \\ \varepsilon & \varepsilon \end{bmatrix}, \quad \boldsymbol{B}_1 = \begin{bmatrix} \varepsilon & \varepsilon \\ \varepsilon & 5 \\ \varepsilon & \varepsilon \end{bmatrix}, \quad \boldsymbol{C}_0 = \begin{bmatrix} \varepsilon & \varepsilon & 2 \end{bmatrix}, \quad \boldsymbol{C}_1 = \begin{bmatrix} \varepsilon & e & \varepsilon \end{bmatrix}.$$

In this example one observes that the equations are not in explicit form, and they are not first-order. Recursively inserting the equation into itself changes the system to

$$\begin{aligned} \boldsymbol{x}(k+1) &= \boldsymbol{A}_0\boldsymbol{x}(k+1) \oplus \boldsymbol{A}_1\boldsymbol{x}(k) \oplus \boldsymbol{A}_2\boldsymbol{x}(k-1) \oplus \boldsymbol{B}_0\boldsymbol{u}(k+1) \oplus \boldsymbol{B}_1\boldsymbol{u}(k) \\ &= \boldsymbol{A}_0 \underbrace{\left(\boldsymbol{A}_0\boldsymbol{x}(k+1) \oplus \boldsymbol{A}_1\boldsymbol{x}(k) \oplus \boldsymbol{A}_2\boldsymbol{x}(k-1) \oplus \boldsymbol{B}_0\boldsymbol{u}(k+1) \oplus \boldsymbol{B}_1\boldsymbol{u}(k)\right)}_{\boldsymbol{x}(k+1)} \oplus \\ &\quad \boldsymbol{A}_1\boldsymbol{x}(k) \oplus \boldsymbol{A}_2\boldsymbol{x}(k-1) \oplus \boldsymbol{B}_0\boldsymbol{u}(k+1) \oplus \boldsymbol{B}_1\boldsymbol{u}(k) \\ &= \boldsymbol{A}_0^2\boldsymbol{x}(k+1) \oplus (\boldsymbol{I} \oplus \boldsymbol{A}_0)\left(\boldsymbol{A}_1\boldsymbol{x}(k) \oplus \boldsymbol{A}_2\boldsymbol{x}(k-1) \oplus \boldsymbol{B}_0\boldsymbol{u}(k+1) \oplus \boldsymbol{B}_1\boldsymbol{u}(k)\right) \\ &= \boldsymbol{A}_0^3\boldsymbol{x}(k+1) \oplus (\boldsymbol{I} \oplus \boldsymbol{A}_0 \oplus \boldsymbol{A}_0^2)\left(\boldsymbol{A}_1\boldsymbol{x}(k) \oplus \boldsymbol{A}_2\boldsymbol{x}(k-1) \oplus \boldsymbol{B}_0\boldsymbol{u}(k+1) \oplus \boldsymbol{B}_1\boldsymbol{u}(k)\right) \end{aligned}$$

Since matrix $\boldsymbol{A}_0$ is not cyclic, $\boldsymbol{A}_0^{\eta}$ contains only ε-entries for $\eta \geq n$. Therefore, the first term of the right-hand side is equal to zero, i.e.,

$$
\begin{aligned}
\boldsymbol{x}(k+1) &= (\boldsymbol{I} \oplus \boldsymbol{A}_0 \oplus \boldsymbol{A}_0^2)\big(\boldsymbol{A}_1\boldsymbol{x}(k) \oplus \boldsymbol{A}_2\boldsymbol{x}(k-1) \oplus \boldsymbol{B}_0\boldsymbol{u}(k+1) \oplus \boldsymbol{B}_1\boldsymbol{u}(k)\big) \\
&= \begin{bmatrix} \varepsilon & 4 & \varepsilon \\ \varepsilon & 7 & \varepsilon \\ \varepsilon & 11 & \varepsilon \end{bmatrix} \boldsymbol{x}(k) \oplus \begin{bmatrix} \varepsilon & \varepsilon & \varepsilon \\ \varepsilon & \varepsilon & \varepsilon \\ \varepsilon & \varepsilon & 2 \end{bmatrix} \boldsymbol{x}(k-1) \\
&\quad \begin{bmatrix} 1 & \varepsilon \\ 4 & \varepsilon \\ 8 & \varepsilon \end{bmatrix} \boldsymbol{u}(k+1) \oplus \begin{bmatrix} \varepsilon & \varepsilon \\ \varepsilon & 5 \\ \varepsilon & 9 \end{bmatrix} \boldsymbol{u}(k), \\
y(k) &= \begin{bmatrix} \varepsilon & \varepsilon & 2 \end{bmatrix} \boldsymbol{x}(k) \oplus \begin{bmatrix} \varepsilon & e & \varepsilon \end{bmatrix} \boldsymbol{x}(k-1).
\end{aligned}
$$

Then, we can transform this system into a first-order system by suitably augmenting the state space, e.g., by defining a new state vector $\tilde{\boldsymbol{x}}(k) = \begin{bmatrix} \boldsymbol{x}(k) & \boldsymbol{x}(k-1) & \boldsymbol{u}(k) \end{bmatrix}^T$. The resulting first-order system is

$$
\begin{aligned}
\tilde{\boldsymbol{x}}(k+1) &= \tilde{\boldsymbol{A}}\tilde{\boldsymbol{x}}(k) \oplus \tilde{\boldsymbol{B}}\boldsymbol{u}(k+1) \\
y(k) &= \tilde{\boldsymbol{C}}\tilde{\boldsymbol{x}}(k),
\end{aligned}
$$

where

$$
\tilde{\boldsymbol{A}} = \begin{bmatrix}
\varepsilon & 4 & \varepsilon & \varepsilon & \varepsilon & \varepsilon & \varepsilon & \varepsilon \\
\varepsilon & 7 & \varepsilon & \varepsilon & \varepsilon & \varepsilon & \varepsilon & 5 \\
\varepsilon & 11 & \varepsilon & \varepsilon & \varepsilon & 2 & \varepsilon & 9 \\
e & \varepsilon & \varepsilon & \varepsilon & \varepsilon & \varepsilon & \varepsilon & \varepsilon \\
\varepsilon & e & \varepsilon & \varepsilon & \varepsilon & \varepsilon & \varepsilon & \varepsilon \\
\varepsilon & \varepsilon & e & \varepsilon & \varepsilon & \varepsilon & \varepsilon & \varepsilon \\
\varepsilon & \varepsilon & \varepsilon & \varepsilon & \varepsilon & \varepsilon & \varepsilon & \varepsilon \\
\varepsilon & \varepsilon & \varepsilon & \varepsilon & \varepsilon & \varepsilon & \varepsilon & \varepsilon
\end{bmatrix}, \quad
\tilde{\boldsymbol{B}} = \begin{bmatrix}
1 & \varepsilon \\ 4 & \varepsilon \\ 8 & \varepsilon \\ \varepsilon & \varepsilon \\ \varepsilon & \varepsilon \\ \varepsilon & \varepsilon \\ e & \varepsilon \\ \varepsilon & e
\end{bmatrix}, \quad
\tilde{\boldsymbol{C}} = \begin{bmatrix} \varepsilon & \varepsilon & 2 & \varepsilon & e & \varepsilon & \varepsilon & \varepsilon \end{bmatrix}. \qquad (21.18)
$$

Clearly, modeling even a simple TEG as given in Fig. 21.5 in max-plus algebra is not really efficient. However, applying the so-called γ-transform in max-plus algebra results in another idempotent semiring. The γ-transform is defined by $\boldsymbol{x}(\gamma) = \bigoplus_{k\in\mathbb{Z}} \boldsymbol{x}(k)\gamma^k$. The resulting dioid is denoted $\overline{\mathbb{Z}}_{\max}[\![\gamma]\!]$ and is the set of formal power series in one variable γ with coefficients in $\overline{\mathbb{Z}}_{\max}$ and exponents in $\mathbb{Z}$ [1]. Addition and multiplication of two formal power series s and s' are defined by

$$
\begin{aligned}
\left(s \oplus s'\right)(k) &= s(k) \oplus s'(k), \\
\left(s \otimes s'\right)(k) &= \bigoplus_{i+j=k} s(i) \otimes s'(j).
\end{aligned}
$$

The neutral element of addition is the series $\varepsilon(\gamma) = \bigoplus_{k\in\mathbb{Z}} \varepsilon\gamma^k$ and the neutral element of multiplication is the series $e(\gamma) = \bigoplus_{k\in\mathbb{N}} e\gamma^k$. Furthermore $\top(\gamma) = \bigoplus_{k\in\mathbb{Z}} \top\gamma^k$.

Applying the γ-transform to (21.17), we get

$$\gamma \boldsymbol{x}(\gamma) = \boldsymbol{A}_0\gamma\boldsymbol{x}(\gamma) \oplus \boldsymbol{A}_1\gamma^2\boldsymbol{x}(\gamma) \oplus \boldsymbol{A}_2\gamma^3\boldsymbol{x}(\gamma) \oplus \boldsymbol{B}_0\gamma\boldsymbol{u}(\gamma) \oplus \boldsymbol{B}_1\gamma^2\boldsymbol{u}(\gamma)$$

or equivalently

$$\begin{aligned} \boldsymbol{x}(\gamma) &= \underbrace{(\boldsymbol{A}_0 \oplus \gamma\boldsymbol{A}_1 \oplus \gamma^2\boldsymbol{A}_2)}_{\boldsymbol{A}(\gamma)}\boldsymbol{x}(\gamma) \oplus \underbrace{(\boldsymbol{B}_0 \oplus \gamma\boldsymbol{B}_1)}_{\boldsymbol{B}(\gamma)}\boldsymbol{u}(\gamma) \\ y(\gamma) &= \underbrace{(\boldsymbol{C}_0 \oplus \gamma\boldsymbol{C}_1)}_{\boldsymbol{C}(\gamma)}\boldsymbol{x}(\gamma) \end{aligned}$$

where

$$\boldsymbol{A}(\gamma) = \begin{bmatrix} \varepsilon & 4\gamma & \varepsilon \\ 3 & \varepsilon & \varepsilon \\ 3 & 4 & 2\gamma^2 \end{bmatrix}, \quad \boldsymbol{B}(\gamma) = \begin{bmatrix} 1 & \varepsilon \\ \varepsilon & 5\gamma \\ \varepsilon & \varepsilon \end{bmatrix}, \quad \boldsymbol{C}(\gamma) = \begin{bmatrix} \varepsilon & \gamma & 2 \end{bmatrix}. \qquad (21.19)$$

Consequently, the TEG given in Fig. 21.5 can be modeled by (21.19), i.e., the γ-transformed system, instead of (21.18), i.e., the system one would obtain by augmenting the state space as described above. Hence, it may be more efficient to use the dioid $\overline{\mathbb{Z}}_{\max}[\![\gamma]\!]$ to model more complex TEG.

The system given in Fig. 21.5 can also be modeled in min-plus algebra (denoted $\overline{\mathbb{Z}}_{\min}$). In this case the state space has to be augmented as well, in order to obtain a first-order model. However, similar to the γ-transformation in max-plus algebra, there is the so-called δ-transformation in min-plus algebra. The transform constitutes an idempotent semiring denoted $\overline{\mathbb{Z}}_{\min}[\![\delta]\!]$. It is the set of formal power series in δ with coefficients in $\overline{\mathbb{Z}}_{\min}$ and exponents in $\mathbb{Z}$.

As mentioned before, it is possible to model any timed event graph in max-plus algebra as well as in min-plus algebra. Which dioid one uses is often based on the specific application which shall be modeled. If for example the model is supposed to be used in combination with a PLC (programmable logic controller) it may be favorable to model the system in min-plus algebra as the min-plus variable $\boldsymbol{x}(t)$ is dependent on time and a PLC also works with a specific cycle time.

It would, however, of course be preferable to combine the assets of max-plus and min-plus algebra. To do so, we first introduce a 2-dimensional dioid denoted $\mathbb{B}[\![\gamma,\delta]\!]$. It is the set of formal power series in two variables (γ,δ) with Boolean coefficients, i.e, $\mathbb{B} = \{\varepsilon, e\}$ and exponents in $\mathbb{Z}$ (see [1, 5, 10] for more information). Thus, a series in this dioid may be, for example, $s = \gamma^1\delta^1 \oplus \gamma^3\delta^2 \oplus \gamma^4\delta^5$. Graphically the series s can be represented as dots in the *event-time plane* (see Fig. 21.6). The interpretation of a monomial $\gamma^k\delta^t$ is that the $(k+1)^{\text{st}}$ occurrence of the corresponding event happens exactly at time t[1]. In terms of timed event graphs, however,

[1] Following the convention defined between Remarks 5.22 and 5.23 in [1, Section 5.4].

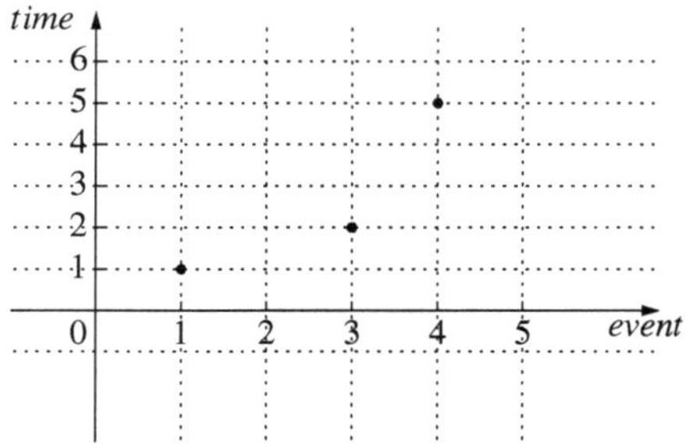

Fig. 21.6 Graphical representation of the series $s = \gamma^1\delta^1 \oplus \gamma^3\delta^2 \oplus \gamma^4\delta^5 \in \mathbb{B}[\![\gamma,\delta]\!]$

the following interpretation of $\gamma^k\delta^t$ would be more useful: *the* $(k+1)^{st}$ *occurrence of the event happens at time t at the earliest* or *at time t, the event has occurred at most* $(k+1)$ *times*. If, in this interpretation, we add two monomials $\gamma^k\delta^t$ and $\gamma^{k+\kappa}\delta^{t-\tau}$, $\kappa,\tau \in \mathbb{N}_0$, we clearly get $\gamma^k\delta^t$. In words, if an event can happen at time t at most $k+1$ times and it can happen at an earlier time $t-\tau$ at most $k+1+\kappa$ times, the resulting statement is that at time t it can happen at most $k+1$ times. We therefore identify all points $\gamma^k\delta^t$ and $\gamma^{k+\kappa}\delta^{t-\tau}$, $\kappa,\tau \in \mathbb{N}_0$, i.e., instead of a single point, we consider the "South-East cone" of this point. This establishes an equivalence relation and the resulting dioid of equivalence classes (quotient dioid) in $\mathbb{B}[\![\gamma,\delta]\!]$ is denoted $\mathcal{M}_{in}^{ax}[\![\gamma,\delta]\!]$ (see [1, 5, 10]). Note that this dioid admits only nondecreasing series and has the following properties:

$$\begin{aligned}
\gamma^k\delta^t \oplus \gamma^l\delta^t &= \gamma^{\min(k,l)}\delta^t \\
\gamma^k\delta^t \oplus \gamma^k\delta^\tau &= \gamma^k\delta^{\max(t,\tau)} \\
\gamma^k\delta^t \otimes \gamma^l\delta^\tau &= \gamma^{(k+l)}\delta^{(t+\tau)}.
\end{aligned}$$

The zero, unit, and top element of $\mathcal{M}_{in}^{ax}[\![\gamma,\delta]\!]$ are $\varepsilon = \gamma^{+\infty}\delta^{-\infty}$, $e = \gamma^0\delta^0$, and $\top = \gamma^{-\infty}\delta^{+\infty}$, respectively. As a consequence of this construction, e.g., the series $\tilde{s} = \gamma^1\delta^1 \oplus \gamma^3\delta^2 \oplus \gamma^4\delta^5 \oplus \gamma^5\delta^2$ is equivalent to the series $s = \gamma^1\delta^1 \oplus \gamma^3\delta^2 \oplus \gamma^4\delta^5$, which is shown in Fig. 21.7.

Using $\mathcal{M}_{in}^{ax}[\![\gamma,\delta]\!]$, we can immediately write down a dioid model for TEG. For example, the TEG in Fig. 21.5 can be represented by

$$\begin{aligned}
x_1 &= \gamma^1\delta^4 x_2 \oplus \gamma^0\delta^1 u_1 \\
x_2 &= \gamma^0\delta^3 x_1 \oplus \gamma^1\delta^5 u_2 \\
x_3 &= \gamma^0\delta^3 x_1 \oplus \gamma^0\delta^4 x_2 \oplus \gamma^2\delta^2 x_3 \\
y &= \gamma^1\delta^0 x_2 \oplus \gamma^0\delta^2 x_3
\end{aligned}$$

or, more compactly,

$$\begin{aligned}
\boldsymbol{x} &= \boldsymbol{A}\boldsymbol{x} \oplus \boldsymbol{B}\boldsymbol{u} \\
y &= \boldsymbol{C}\boldsymbol{x},
\end{aligned}$$

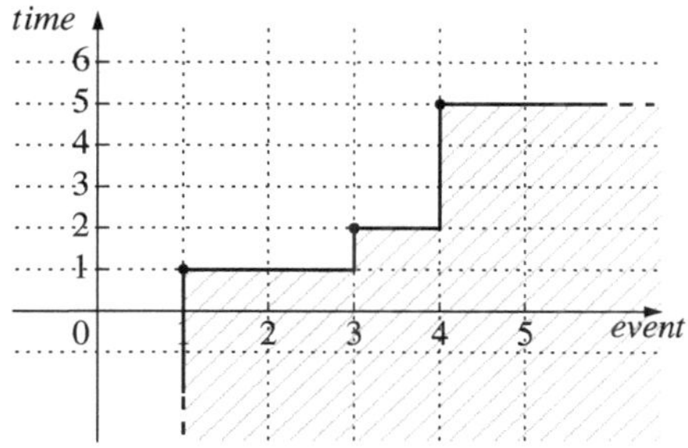

Fig. 21.7 Graphical representation of the series $s = \gamma^1\delta^1 \oplus \gamma^3\delta^2 \oplus \gamma^4\delta^5 \in \mathcal{M}_{in}^{ax}[\![\gamma,\delta]\!]$

where

$$\boldsymbol{A} = \begin{bmatrix} \varepsilon & \gamma^1\delta^4 & \varepsilon \\ \gamma^0\delta^3 & \varepsilon & \varepsilon \\ \gamma^0\delta^3 & \gamma^0\delta^4 & \gamma^2\delta^2 \end{bmatrix}, \quad \boldsymbol{B} = \begin{bmatrix} \gamma^0\delta^1 & \varepsilon \\ \varepsilon & \gamma^1\delta^5 \\ \varepsilon & \varepsilon \end{bmatrix}, \quad \boldsymbol{C} = \begin{bmatrix} \varepsilon \ \gamma^1\delta^0 \ \gamma^0\delta^2 \end{bmatrix}.$$

Clearly, it is possible to model even very complex TEG in a compact form using the dioid $\mathcal{M}_{in}^{ax}[\![\gamma,\delta]\!]$.

21.6 High-Throughput Screening Systems

Among the vast variety of systems that can be modeled as timed event graphs is the operation of so-called *High-Throughput Screening Systems*. High-throughput screening (HTS) has become an important technology to rapidly test thousands of biochemical substances [13, 20]. In pharmaceutical industries, for example, HTS is often used for a first screening in the process of drug discovery. In general, high-throughput screening plants are fully automated systems containing a fixed set of devices performing liquid handling, storage, reading, plate handling, and incubation steps. All operations which have to be conducted to analyze one set of substances are combined in a so-called batch. The testing vessel carrying the biochemical substances in HTS systems is called microplate. It features a grid of up to 3456 wells. The number of wells is historically grown and represents a multiple of 96 reflecting the original microplate with 96 wells [18]. Several microplates may be included in a batch to convey reagents or waste material. While conducting a screening run, more than one batch may be present in the system at the same time, a single batch may pass the same machine more than once, a single batch may occupy two (or more) resources simultaneously, e.g., when being transferred from one resource to another, and there are minimal and maximal processing times defined by the user.

For better understanding we introduce a simple example of an HTS operation. One single batch of this example consists of three activities which are executed on three different resources. The first activity, executed on R_1, represents the filling of some biochemical substance A into the wells of a microplate. After that, the microplate is moved to a second resource R_2, where the second activity is executed. This activity basically mixes substance A with some substance B. Substance B is provided by another activity executed on resource R_3. This resource operates independently of the other resources and may produce substance B even when resources R_1 and R_2 are not working. Due to hardware constraints the user sets the following minimal time durations:

- act_1 :
  - filling compound A into the wells of microplate: 7 units of time
  - post-processing after transfer of microplate to R_2: 1 unit of time
- act_2 :
  - pre-processing before transfer of microplate from R_1: 1 unit of time
  - waiting time before substance B can be added: 1 unit of time
  - mixing of substances: 12 units of time
- act_3 :
  - providing one heap of substance B: 5 units of time
  - post-processing after providing one heap of substance B: 1 unit of time
- The transfer processes are assumed to be possible in zero time.

The corresponding timed event graph is given in Fig. 21.8.

For real HTS systems, further activities, such as incubation steps or reading operations, would be executed on the compound AB. Screening runs of HTS plants may easily involve 150 resource allocations, i.e., activities, per batch. Since all relations

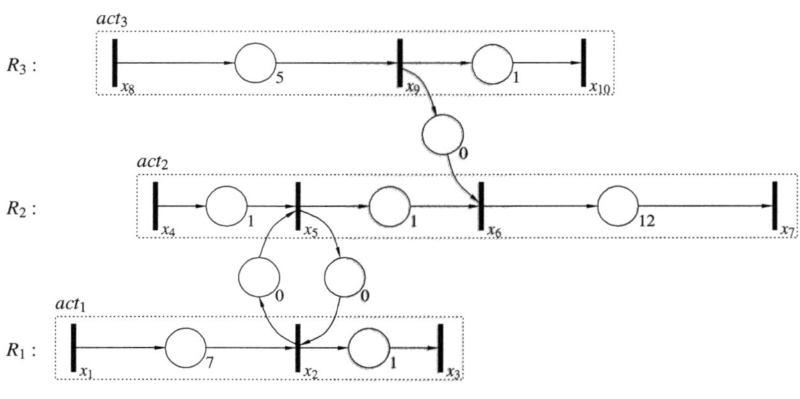

Fig. 21.8 TEG of a single batch of our example HTS operation

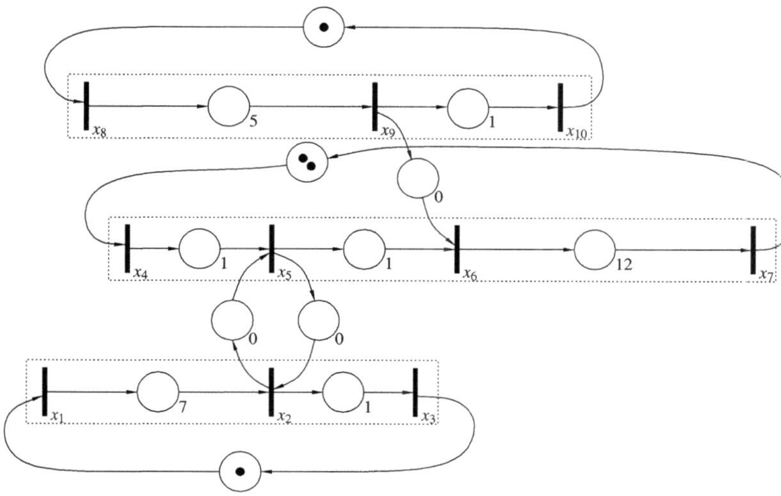

Fig. 21.9 TEG of a single batch and resource capacities

given in Fig. 21.8 belong to one single batch, there are no tokens in the timed event graph. However, additional constraints between events of different batches need to be taken into account. For example, all resources have a specific capacity. If a resource has a capacity of 1, an activity may only occupy this resource if the preceding activity (which may be of a different batch) has been finished. In our small example resources R_1 and R_3 have a capacity of 1 while resource R_2 has a capacity of 2, i.e., this resource may execute two activities at the same time. To model this, the TEG has to be extended by places with 1 and 2 initial tokens. The resulting TEG is shown in Fig. 21.9. Finally, the user has to evaluate which transitions he or she is able to control and what the output of the system is. For HTS plants, it is usually possible to control the starting events of every activity and the output is directly connected to the last transition of a single batch. Note that, depending on the specific system, the controllable transitions and the output transitions may be different. Without loss of generality, we fix all starting events of all activities of our example, i.e., x_1, x_4, and x_8, to be controllable and the output of our system shall be the last event of a single batch, i.e., x_7. Thus the timed event graph modeling the single batch of our HTS operation is extended according to Fig. 21.10. Then the TEG can be written as a $\mathcal{M}_{in}^{ax}[\![\gamma,\delta]\!]$-model:

$$\begin{aligned} \boldsymbol{x} &= \boldsymbol{A}\boldsymbol{x} \oplus \boldsymbol{B}\boldsymbol{u} \\ y &= \boldsymbol{C}\boldsymbol{x}, \end{aligned}$$

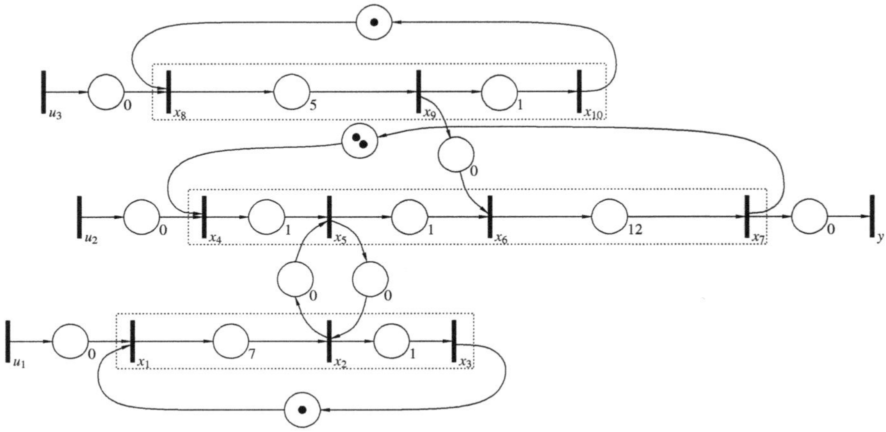

Fig. 21.10 TEG of a single batch, with resource capacities, and input and output transitions

with

$$
\boldsymbol{A}=\begin{bmatrix}
\varepsilon & \varepsilon & \gamma & \varepsilon & \varepsilon & \varepsilon & \varepsilon & \varepsilon & \varepsilon & \varepsilon \\
\delta^7 & \varepsilon & \varepsilon & \varepsilon & e & \varepsilon & \varepsilon & \varepsilon & \varepsilon & \varepsilon \\
\varepsilon & \delta & \varepsilon & \varepsilon & \varepsilon & \varepsilon & \varepsilon & \varepsilon & \varepsilon & \varepsilon \\
\varepsilon & \varepsilon & \varepsilon & \varepsilon & \varepsilon & \varepsilon & \gamma^2 & \varepsilon & \varepsilon & \varepsilon \\
\varepsilon & e & \varepsilon & \delta & \varepsilon & \varepsilon & \varepsilon & \varepsilon & \varepsilon & \varepsilon \\
\varepsilon & \varepsilon & \varepsilon & \varepsilon & \delta & \varepsilon & \varepsilon & \varepsilon & e & \varepsilon \\
\varepsilon & \varepsilon & \varepsilon & \varepsilon & \varepsilon & \delta^{12} & \varepsilon & \varepsilon & \varepsilon & \varepsilon \\
\varepsilon & \varepsilon & \varepsilon & \varepsilon & \varepsilon & \varepsilon & \varepsilon & \varepsilon & \varepsilon & \gamma \\
\varepsilon & \varepsilon & \varepsilon & \varepsilon & \varepsilon & \varepsilon & \varepsilon & \delta^5 & \varepsilon & \varepsilon \\
\varepsilon & \varepsilon & \varepsilon & \varepsilon & \varepsilon & \varepsilon & \varepsilon & \varepsilon & \delta & \varepsilon
\end{bmatrix}, \quad
\boldsymbol{B}=\begin{bmatrix}
e & \varepsilon & \varepsilon \\
\varepsilon & \varepsilon & \varepsilon \\
\varepsilon & \varepsilon & \varepsilon \\
\varepsilon & e & \varepsilon \\
\varepsilon & \varepsilon & \varepsilon \\
\varepsilon & \varepsilon & \varepsilon \\
\varepsilon & \varepsilon & \varepsilon \\
\varepsilon & \varepsilon & e \\
\varepsilon & \varepsilon & \varepsilon \\
\varepsilon & \varepsilon & \varepsilon
\end{bmatrix},
$$

$$
\boldsymbol{C}=\begin{bmatrix}\varepsilon & \varepsilon & \varepsilon & \varepsilon & \varepsilon & \varepsilon & e & \varepsilon & \varepsilon & \varepsilon\end{bmatrix}.
$$

Clearly, the $\mathscr{M}_{in}^{ax}[\![\gamma,\delta]\!]$-model of the HTS operation is very compact. In the next chapter, it is explained how this model can be used to efficiently compute control laws for such systems.

21.7 Further Reading

In this chapter, the modeling and analysis of linear systems in a dioid framework have been presented. However, the chapter only provides a rough overview of this topic. A more exhaustive and mathematical presentation of dioids and systems in dioids can be found in [1, 2]. Many other works on max-plus algebra, dioids in general, and performance evaluation in idempotent semirings have been published, e.g., [5, 8, 10, 11, 14]. Besides that, there are several software packages available to handle max-plus algebraic systems, e.g., the max-plus algebra toolbox for ScicosLab (www.scicoslab.org), or the C++ library *MinMaxGD* to manipulate periodic series in $\mathcal{M}_{in}^{ax}[\![\gamma,\delta]\!]$ [6]. For more information on high-throughput screening systems the reader is referred to [13, 18, 20]. However, please note, that HTS is only one possible application that has a linear representation in dioids. Other applications are, e.g., traffic systems, computer communication systems, production lines, and flows in networks. The reader is also invited to read the next chapter on control theory, developed for discrete-event systems in a dioid framework.

References

1. Baccelli, F., Cohen, G., Olsder, G.J., Quadrat, J.P.: Synchronization and Linearity–An Algebra for Discrete Event Systems. John Wiley and Sons (1992) Web Edition, http://www-rocq.inria.fr/metalau/cohen/SED/SED1-book.html (2012)
2. Blyth, T.S.: Lattices and Ordered Algebraic Structures. Springer (2005)
3. Bosscher, D., Polak, I., Vaandrager, F.: Verification of an Audio Control Protocol. In: Langmaack, H., de Roever, W.-P., Vytopil, J. (eds.) FTRTFT 1994 and ProCoS 1994. LNCS, vol. 863, pp. 170–192. Springer, Heidelberg (1994)
4. Cassandras, C.G., Lafortune, S., Olsder, G.J.: Introduction to the modelling, control and optimization of discrete event systems. In: Isidori, A. (ed.) Trends in Control–A European Perspective, pp. 217–291 (1995)
5. Cohen, G., Moller, P., Quadrat, J.-P., Viot, M.: Algebraic tools for the performance evaluation of discrete event systems. Proceedings of the IEEE 77(1), 39–58 (1989)
6. Cottenceau, B., Hardouin, L., Lhommeau, M., Boimond, J.L.: Data processing tool for calculation in dioid. In: Proc. 5th Int. Workshop on Discrete Event Systems, Ghent, Belgium (2000)
7. Cottenceau, B., Hardouin, L., Boimond, J.L., Ferrier, J.L.: Model reference control for timed event graphs in dioids. Automatica 37(9), 1451–1458 (2001)
8. Cuninghame-Green, R.A.: Minimax algebra. Lecture Notes in Economics and Mathematical Systems. Springer (1979)
9. Cuninghame-Green, R.A.: Minimax algebra and applications. Fuzzy Sets and Systems 41, 251–267 (1989)
10. Gaubert, S., Klimann, C.: Rational computation in dioid algebra and its application to performance evaluation of discrete event systems. In: Gérard, J., Lamnabhi-Lagarrigue, F. (eds.) Algebraic Computing in Control. LNCIS, vol. 165, pp. 241–252. Springer, Heidelberg (1991)
11. Gaubert, S.: Methods and Applications of (max,+) Linear Algebra. In: Reischuk, R., Morvan, M. (eds.) STACS 1997. LNCS, vol. 1200, pp. 261–282. Springer, Heidelberg (1997)
12. Gérard, J., Lamnabhi-Lagarrigue, F. (eds.): Algebraic Computing in Control. LNCIS, vol. 165. Springer (1991)
13. Harding, D., Banks, M., Fogarty, S., Binnie, A.: Development of an automated high-throughput screening system: a case history. Drug Discovery Today 2(9), 385–390 (1997)

14. Heidergott, B., Olsder, G.J., van der Woude, J.: Max Plus at Work–Modeling and Analysis of Synchronized Systems: A Course on Max-Plus Algebra and Its Applications. Princeton University Press, Princeton (2006)
15. Isidori, A. (ed.): Trends in Control – A European Perspective. Springer (1995)
16. Lhommeau, M., Hardouin, L., Cottenceau, B.: Disturbance decoupling of timed event graphs by output feedback controller. In: Proc. 6th Int. Workshop on Discrete Event Systems, Zaragoza, Spain (2002)
17. Maia, C.A., Hardouin, L., Santos-Mendes, R., Cottenceau, B.: Optimal closed-loop control of timed event graphs in dioids. IEEE Transactions on Automatic Control 48(12), 2284–2287 (2003)
18. Mayr, L.M., Fuerst, P.: The future of high-throughput screening. Journal of Biomolecular Screening 13(6), 443–448 (2008)
19. Reischuk, R., Morvan, M. (eds.): STACS 1997. LNCS, vol. 1200. Springer, Heidelberg (1997)
20. Rogers, M.V.: High-throughput screening. Drug Discovery Today 2(11), 503–504 (1997)

Chapter 22
Discrete-Event Systems in a Dioid Framework: Control Theory

Laurent Hardouin, Olivier Boutin, Bertrand Cottenceau, Thomas Brunsch, and Jörg Raisch

22.1 Motivation

This chapter deals with the control of discrete-event systems which admit linear models in dioids (or idempotent semirings), introduced in the previous chapter (see also the books [1, 19]). These systems are characterized by synchronization and delay phenomena. Their modeling is focused on the evaluation of the occurrence dates of events. The outputs correspond to the occurrence dates of events produced by the system and the inputs are the occurrence dates of exogenous events acting on it. The control of the system inputs will lead to a scheduling of these occurrence dates such that a specified objective for the system be achieved. For example, in a manufacturing management context, the controller will have to schedule the input dates of raw materials or the starting of jobs in order to obtain required output dates for the produced parts.

Laurent Hardouin · Bertrand Cottenceau · Thomas Brunsch
LUNAM, University of Angers, LISA, ISTIA, 62 Av. Notre-Dame du Lac, 49000 Angers, France
e-mail: {laurent.hardouin,bertrand.cottenceau}@univ-angers.fr

Olivier Boutin
Calle Santiago 2 – 4°C, 11005 Cadiz, Spain
e-mail: olivier.research@gmail.com

Thomas Brunsch · Jörg Raisch
Technische Universität Berlin, Fachgebiet Regelungssysteme, Einsteinufer 17, 10587 Berlin, Germany
e-mail: {brunsch,raisch}@control.tu-berlin.de

Jörg Raisch
Fachgruppe System- und Regelungstheorie, Max-Planck-Institut für Dynamik komplexer technischer Systeme, Sandtorstr. 1, 39106 Magdeburg, Germany

C. Seatzu et al. (Eds.): Control of Discrete-Event Systems, LNCIS 433, pp. 451–469.
springerlink.com

The nature of the considered systems is such that they cannot be accelerated, indeed the fastest behavior of the system is fixed and known. The only degree of freedom is to delay the inputs. For example, in a manufacturing system, the maximal throughput of a machine is obtained with its fastest behavior, hence the only thing which can be done is to delay the starting of jobs or decrease its production speed. Hence, the control synthesis presented in this part will aim at controlling the input dates in the system in order to get the outputs at the specified date. More precisely, it will be shown that the proposed control strategies will aim at delaying the inputs as much as possible while ensuring that the outputs occur before the specified dates. This kind of strategy is rather popular in an industrial context, since it is optimal with regard to the just-in-time criterion, which means that the jobs will start as late as possible, while ensuring they will be completed before the required date. The optimization of this criterion means that the parts will spend a minimal time in the system, *i.e.*, it will avoid useless waiting time. This strategy is worthwhile from an economic point of view.

This chapter will be organized as follows. In a first step, some theoretical results will be given; they come in addition to the ones introduced in Chapter 21. Then the control strategies will be presented, starting with open-loop strategies and concluding with closed-loop ones. For both, an illustration based on the example introduced in Chapter 21 will be provided.

22.2 Theory and Concepts

The systems considered and the algebraic context were described in the previous chapter. Classically, the control of a system has to do with system inversion. Hence we recall in this section several algebraic results allowing us to invert linear equations in dioids, *i.e.*, we focus on the resolution of a system $\boldsymbol{A} \otimes \boldsymbol{X} = \boldsymbol{B}$, where $\boldsymbol{A}$, $\boldsymbol{X}$ and $\boldsymbol{B}$ are matrices with entries in a given dioid. The product law of semirings does not necessarily admit an inverse, nevertheless it is possible to consider residuation theory to get a pseudo inverse. This theory (see [3]) allows us to deal with inversion of mappings defined over ordered sets. In our algebraic framework, it will be useful to get the greatest solution of the inequality $\boldsymbol{A} \otimes \boldsymbol{X} \preceq \boldsymbol{B}$. Indeed, a dioid is an ordered set, because the idempotency of the addition law defines a canonical order ($a \oplus b = a \Leftrightarrow a \succeq b$). Hence this theory is suitable to invert mappings defined over dioids. The main useful results are recalled in the first part of the chapter. Another useful key point is the characterization of solutions of implicit equations such as $x = a \otimes x \oplus b$. It will be recalled in a second step that the latter admits a smallest solution, namely a^*b, where $*$ is the Kleene star operator. Some useful properties involving this operator will then be recalled.

22.2.1 Mapping Inversion Over a Dioid

Definition 22.1 (Isotone mapping). *An order-preserving mapping $f : \mathscr{D} \to \mathscr{C}$, where $\mathscr{D}$ and $\mathscr{C}$ are ordered sets, is a mapping such that: $x \succeq y \Rightarrow f(x) \succeq f(y)$. It is said to be* ***isotone*** *in the sequel.*

Definition 22.2 (Residuated mapping). *Let $\mathscr{D}$ and $\mathscr{C}$ be two ordered sets. An isotone mapping $\Pi : \mathscr{D} \to \mathscr{C}$ is said to be residuated, if the inequality $\Pi(x) \preceq b$ has a greatest solution in $\mathscr{D}$ for all $b \in \mathscr{C}$.*

The following theorems yield necessary and sufficient conditions to characterize these mappings.

Theorem 22.1. [2, 1] *Let $\mathscr{D}$ and $\mathscr{C}$ be two ordered sets. Let $\Pi : \mathscr{D} \to \mathscr{C}$ be an isotone mapping. The following statements are equivalent.*

(i) Π is residuated.
(ii) There exists an isotone mapping denoted $\Pi^\sharp : \mathscr{C} \to \mathscr{D}$ such that $\Pi \circ \Pi^\sharp \preceq \mathsf{Id}_{\mathscr{C}}$ and $\Pi^\sharp \circ \Pi \succeq \mathsf{Id}_{\mathscr{D}}$.

Theorem 22.2. *Let $\mathscr{D}$ and $\mathscr{C}$ be two ordered sets. Let $\Pi : \mathscr{D} \to \mathscr{C}$ be a residuated mapping; then the following equalities hold:*

$$\begin{aligned} \Pi \circ \Pi^\sharp \circ \Pi &= \Pi \\ \Pi^\sharp \circ \Pi \circ \Pi^\sharp &= \Pi^\sharp \end{aligned} \tag{22.1}$$

Example 22.1. In [1], the following mappings are considered:

$$\begin{aligned} L_a : \mathscr{D} &\to \mathscr{D} \\ x &\mapsto a \otimes x \quad \text{(left product by } a\text{)}, \\ R_a : \mathscr{D} &\to \mathscr{D} \\ x &\mapsto x \otimes a \quad \text{(right product by } a\text{)}, \end{aligned} \tag{22.2}$$

where $\mathscr{D}$ is a dioid. It can be shown that these mappings are residuated. The corresponding residual mappings are denoted:

$$\begin{aligned} L_a^\sharp(x) &= a \backslash\!\!\circ\, x \quad \text{(left division by } a\text{)}, \\ R_a^\sharp(x) &= x \,/\!\!\circ\, a \quad \text{(right division by } a\text{)}. \end{aligned} \tag{22.3}$$

■

Therefore, equation $a \otimes x \preceq b$ has a greatest solution denoted $L_a^\sharp(b) = a \backslash\!\!\circ\, b$. In the same way, $x \otimes a \preceq b$ admits $R_a^\sharp(b) = b \,/\!\!\circ\, a$ as its greatest solution. Let $\boldsymbol{A}, \boldsymbol{D} \in \mathscr{D}^{m \times n}$, $\boldsymbol{B} \in \mathscr{D}^{m \times p}, \boldsymbol{C} \in \mathscr{D}^{n \times p}$ be some matrices. The greatest solution of inequality $\boldsymbol{A} \otimes \boldsymbol{X} \preceq \boldsymbol{B}$ is given by $\boldsymbol{C} = \boldsymbol{A} \backslash\!\!\circ\, \boldsymbol{B}$ and the greatest solution of $\boldsymbol{X} \otimes \boldsymbol{C} \preceq \boldsymbol{B}$ is given by $\boldsymbol{D} = \boldsymbol{B} \,/\!\!\circ\, \boldsymbol{C}$. The entries of matrices $\boldsymbol{C}$ and $\boldsymbol{D}$ are obtained as follows:

$$\begin{aligned} \boldsymbol{C}_{ij} &= \bigwedge_{k=1}^{m} (\boldsymbol{A}_{ki} \diamond\backslash \boldsymbol{B}_{kj}) \\ \boldsymbol{D}_{ij} &= \bigwedge_{k=1}^{p} (\boldsymbol{B}_{ik} \diamond/ \boldsymbol{C}_{jk}), \end{aligned} \tag{22.4}$$

where $\wedge$ is the greatest lower bound. It must be noted that $\forall a \in \mathscr{D}$, $\varepsilon \otimes x \preceq a$ admits $\varepsilon \diamond\backslash a = \top$ as its greatest solution. Indeed, ε is absorbing for the product law. Furthermore, $\top \otimes x \preceq a$ admits $\top \diamond\backslash a = \varepsilon$ as a solution, except if $a = \top$. Indeed, in the latter case $\top \diamond\backslash \top = \top$, that is to say $\top \otimes x \preceq \top$ admits $\top$ as a greatest solution.

Example 22.2. Let us consider the following relation $\boldsymbol{AX} \preceq \boldsymbol{B}$ with $\boldsymbol{A}^t = \begin{bmatrix} 2 & \varepsilon & 0 \\ 5 & 3 & 7 \end{bmatrix}$ and $\boldsymbol{B}^t = \begin{bmatrix} 6 & 4 & 9 \end{bmatrix}$ being matrices with entries in $\overline{\mathbb{Z}}_{\max}$. By recalling that in this dioid the product corresponds to the classical sum, the residuation corresponds to the classical minus. That is to say, $5 \otimes x \preceq 8$ admits the solutions set $\mathscr{X} = \{x | x \preceq 5 \diamond\backslash 8$, with $5 \diamond\backslash 8 = 8 - 5 = 3\}$, with 3 being the greatest solution of this set[1]. Hence, by applying the computation rules of Equation (22.4), the following result is obtained:

$$\boldsymbol{C} = \boldsymbol{A} \diamond\backslash \boldsymbol{B} = \begin{bmatrix} 2 \diamond\backslash 6 \wedge \varepsilon \diamond\backslash 4 \wedge 0 \diamond\backslash 9 \\ 5 \diamond\backslash 6 \wedge 3 \diamond\backslash 4 \wedge 7 \diamond\backslash 9 \end{bmatrix} = \begin{bmatrix} 4 \\ 1 \end{bmatrix}.$$

This matrix $\boldsymbol{C}$ is the greatest such that $\boldsymbol{A} \otimes \boldsymbol{C} \preceq \boldsymbol{B}$. See the verification below.

$$\boldsymbol{A} \otimes (\boldsymbol{A} \diamond\backslash \boldsymbol{B}) = \begin{bmatrix} 2 & 5 \\ \varepsilon & 3 \\ 0 & 7 \end{bmatrix} \begin{bmatrix} 4 \\ 1 \end{bmatrix} = \begin{bmatrix} 6 \\ 4 \\ 8 \end{bmatrix} \preceq \begin{bmatrix} 6 \\ 4 \\ 9 \end{bmatrix} = \boldsymbol{B}.$$

According to Theorem 22.2, the reader is invited to check that:

$$\boldsymbol{A} \diamond\backslash \big(\boldsymbol{A} \otimes (\boldsymbol{A} \diamond\backslash \boldsymbol{B})\big) = \boldsymbol{A} \diamond\backslash \boldsymbol{B}. \tag{22.5}$$

Furthermore, it can be noticed that equation $\boldsymbol{A} \otimes \boldsymbol{X} = \boldsymbol{B}$ admits a greatest solution if matrix $\boldsymbol{B}$ is in the image of matrix $\boldsymbol{A}$, which is equivalent to say that there exists a matrix $\boldsymbol{L}$ such that $\boldsymbol{B} = \boldsymbol{A} \otimes \boldsymbol{L}$. By considering the matrix

$$\boldsymbol{B} = \boldsymbol{A} \otimes \boldsymbol{L} = \begin{bmatrix} 2 & 5 \\ \varepsilon & 3 \\ 0 & 7 \end{bmatrix} \begin{bmatrix} 0 \\ 1 \end{bmatrix} = \begin{bmatrix} 6 \\ 4 \\ 8 \end{bmatrix},$$

the reader is invited to check that $\boldsymbol{A} \otimes (\boldsymbol{A} \diamond\backslash \boldsymbol{B}) = \boldsymbol{B}$. ■

As in the classical algebraic context, in the absence of ambiguity, the operator $\otimes$ will be omitted in the sequel.

Below, some properties about this "division" operator are recalled. Actually, only the ones that are useful in control synthesis are recalled. The reader is invited to consult [1, pp. 182-185] and [9] to get a more comprehensive presentation.

[1] Residuation achieves equality in the scalar case.

Property 22.3

Left product	Right product	
$a(a \diagdown x) \preceq x$	$(x \diagup a)a \preceq x$	(22.6)
$a \diagdown (ax) \succeq x$	$(xa) \diagup a \succeq x$	(22.7)
$a\big(a \diagdown (ax)\big) = ax$	$\big((xa) \diagup a\big)a = xa$	(22.8)
$(ab) \diagdown x = b \diagdown (a \diagdown x)$	$x \diagup (ba) = (x \diagup a) \diagup b$	(22.9)
$(a \diagdown x)b \preceq a \diagdown (xb)$	$b(x \diagup a) \preceq (bx) \diagup a$	(22.10)

Proof. Proofs are given for the left product, the same hold for the right one.
Equations (22.6) and (22.7): Theorem 22.1 leads to $L_a \circ L_a^\sharp \preceq \mathsf{Id}_{\mathscr{D}}$ (*i.e.*, $a(a \diagdown x) \preceq x$) and $L_a^\sharp \circ L_a \succeq \mathsf{Id}_{\mathscr{D}}$ (*i.e.*, $a \diagdown (ax) \succeq x$).
Equation (22.8): Theorem 22.2 leads to $L_a \circ L_a^\sharp \circ L_a = L_a$ (*i.e.*, $a\big(a \diagdown (ax)\big) = ax$).
Equation (22.9): It is a direct consequence of the associativity of the product law.
Equation (22.10): According to associativity of the product law, $a(xb) = (ax)b$ then $a\big((a \diagdown x)b\big) = \big(a(a \diagdown x)\big)b$ and $a \diagdown \Big(a\big((a \diagdown x)b\big)\Big) = a \diagdown \Big(\big(a(a \diagdown x)\big)b\Big)$. According to Equation (22.7), $(a \diagdown x)b \preceq a \diagdown \Big(a\big((a \diagdown x)b\big)\Big)$ and according to Equation (22.6) $a \diagdown \Big(\big(a(a \diagdown x)\big)b\Big) \preceq a \diagdown (xb)$. Then, inequality (22.10) holds. □

22.2.2 *Implicit Equations over Dioids*

Definition 22.3. *Let $\mathscr{D}$ be a dioid and $a \in \mathscr{D}$. The Kleene star operator is defined as follows:*

$$a^* \triangleq e \oplus a \oplus a^2 \oplus \ldots = \bigoplus_{i \in \mathbb{N}_0} a^i,$$

with $a^0 = e$ and $a^n = \underbrace{a \otimes a \otimes \cdots \otimes a}_{n \text{ times}}$.

Theorem 22.3. *a^*b is the least solution of equation $x = ax \oplus b$.*

Proof. By repeatedly inserting equation into itself, one can write:

$$x = ax \oplus b = a(ax \oplus b) \oplus b = a^2x \oplus ab \oplus b = a^n x \oplus a^{n-1}b \oplus \cdots \oplus a^2 b \oplus ab \oplus b.$$

According to Definition 22.3, this leads to $x = a^\infty x \oplus a^*b$. Hence $x \succeq a^*b$. Furthermore, it can be shown that a^*b is a solution. Indeed:

$$a(a^*b) \oplus b = (aa^* \oplus e)b = a^*b,$$

which implies that a^*b is the smallest solution. □

Now we give some useful properties of the Kleene star operator.

Property 22.4 Let $\mathscr{D}$ be a complete dioid. $\forall a, b \in \mathscr{D}$,

$$a^*a^* = a^* \tag{22.11}$$
$$(a^*)^* = a^* \tag{22.12}$$
$$a(ba)^* = (ab)^*a \tag{22.13}$$
$$(a \oplus b)^* = (a^*b)^*a^* = b^*(ab^*)^* = (a \oplus b)^*a^* = b^*(a \oplus b)^* \tag{22.14}$$

Proof. Equation (22.11): $a^* \otimes a^* = (e \oplus a \oplus a^2 \oplus \ldots) \otimes (e \oplus a \oplus a^2 \oplus \ldots) = e \oplus a \oplus a^2 \oplus a^3 \oplus a^4 \oplus \ldots = a^*$.
Equation (22.12): $(a^*)^* = e \oplus a^* \oplus a^*a^* \oplus \ldots = e \oplus a^* \oplus a^*$.
Equation (22.13): $a(ba)^* = a(e \oplus ba \oplus baba \oplus \ldots) = a \oplus aba \oplus ababa \oplus \ldots = (e \oplus ab \oplus abab \oplus \ldots) \otimes a = (ab)^*a$.
Equation (22.14): $x = (a \oplus b)^*$ is the smallest solution of $x = (a \oplus b) \otimes x \oplus e = ax \oplus bx \oplus e$. On the other hand, this equation admits $x = (a^*b)x \oplus a^* = (a^*b)^*a^*$ as its smallest solution. Hence $(a \oplus b)^* = (a^*b)^*a^*$. The other equalities can be obtained by commuting a and b. □

Theorem 22.4. [7, 32] *Let $\mathscr{D}$ be a complete dioid and $a, b \in \mathscr{D}$. Elements $(a \backslash\!\!\circ a)$ and $(b /\!\!\circ b)$ are such that*

$$\begin{aligned} a \backslash\!\!\circ a &= (a \backslash\!\!\circ a)^*, \\ b /\!\!\circ b &= (b /\!\!\circ b)^*. \end{aligned} \tag{22.15}$$

Proof. According to Equation (22.7), $a \backslash\!\!\circ (ae) = a \backslash\!\!\circ a \succeq e$. Furthermore, according to Theorem 22.2, $a \backslash\!\!\circ (a \otimes (a \backslash\!\!\circ a)) = a \backslash\!\!\circ a$. Moreover, according to Inequality (22.10), $a \backslash\!\!\circ (a \otimes (a \backslash\!\!\circ a)) \succeq (a \backslash\!\!\circ a) \otimes (a \backslash\!\!\circ a)$. Hence, $e \preceq (a \backslash\!\!\circ a) \otimes (a \backslash\!\!\circ a) = (a \backslash\!\!\circ a)^2 \preceq (a \backslash\!\!\circ a)$. By recalling that the product law is isotone, these inequalities can be extended to $e \preceq (a \backslash\!\!\circ a)^n \preceq (a \backslash\!\!\circ a)$. And then by considering the definition of the Kleene star operator, $e \preceq (a \backslash\!\!\circ a)^* = (a \backslash\!\!\circ a)$. A similar proof can be developed for $b /\!\!\circ b$. □

Theorem 22.5. *Let $\mathscr{D}$ be a complete dioid and $x, a \in \mathscr{D}$. The greatest solution of*

$$x^* \preceq a^* \tag{22.16}$$

is $x = a^*$.

Proof. First, according to the Kleene star definition (see Definition 22.3), the following equivalence holds:

$$x^* = e \oplus x \oplus x^2 \oplus \ldots \preceq a^* \Leftrightarrow \begin{cases} e \preceq a^* \\ x \preceq a^* \\ x^2 \preceq a^* \\ \ldots \\ x^n \preceq a^* \\ \ldots \end{cases} . \tag{22.17}$$

Let us focus on the right-hand side of this equivalence. From Definition 22.3, the first inequality holds. The second admits a^* as its greatest solution, which is also a solution to the other ones. Indeed, according to Equation (22.11), $a^*a^* = a^*$ and $(a^*)^n = a^*$. Hence $x = a^*$ satisfies all these inequalities and is then the greatest solution. □

22.2.3 Dioid $\mathcal{M}_{in}^{ax}[\![\gamma,\delta]\!]$

In the previous chapter, the semiring of non-decreasing series in two variables with exponents in $\overline{\mathbb{Z}}$ and with Boolean coefficients, denoted $\mathcal{M}_{in}^{ax}[\![\gamma,\delta]\!]$, was considered to model timed event graphs and to obtain transfer relations between transitions. In this section, the results introduced in the previous chapter are extended. Formally, the considered series are defined as follows:

$$s = \bigoplus_{i\in\mathbb{N}_0} s(i)\gamma^{n_i}\delta^{t_i}, \tag{22.18}$$

with $s(i) \in \{e,\varepsilon\}$ and $n_i, t_i \in \mathbb{Z}$. The support of s is defined by $Supp(s) = \{i \in \mathbb{N}_0 | s(i) \neq \varepsilon\}$. The valuation in γ of s is defined as: $val(s) = min\{n_i | i \in Supp(s)\}$. A series $s \in \mathcal{M}_{in}^{ax}[\![\gamma,\delta]\!]$ is said to be a polynomial if $Supp(s)$ is finite. Furthermore, a polynomial is said to be a monomial if there is only one element.

Definition 22.4 (Causality). *A series $s \in \mathcal{M}_{in}^{ax}[\![\gamma,\delta]\!]$ is causal if $s = \varepsilon$ or if both $val(s) \geq 0$ and $s \succeq \gamma^{val(s)}$. The set of causal elements of $\mathcal{M}_{in}^{ax}[\![\gamma,\delta]\!]$ has a complete semiring structure denoted $\mathcal{M}_{in}^{ax+}[\![\gamma,\delta]\!]$.*

Definition 22.5 (Periodicity). *A series $s \in \mathcal{M}_{in}^{ax}[\![\gamma,\delta]\!]$ is said to be a periodic series if it can be written as $s = p \oplus q \otimes r^*$, with $p = \bigoplus_{i=1}^{n} \gamma^{n_i}\delta^{t_i}$, $q = \bigoplus_{j=1}^{m} \gamma^{N_j}\delta^{T_j}$ are polynomials and $r = \gamma^{\nu}\delta^{\tau}$, with $\nu,\tau \in \mathbb{N}$, is a monomial depicting the periodicity and allowing to define the asymptotic slope of the series as $\sigma_\infty(s) = \nu/\tau$. A matrix is said to be periodic if all its entries are periodic series.*

Definition 22.6 (Realizability). *A series $s \in \mathcal{M}_{in}^{ax}[\![\gamma,\delta]\!]$ is said to be realizable if there exists matrices such that $s = \boldsymbol{C}(\gamma\boldsymbol{A}_1 \oplus \delta\boldsymbol{A}_2)^*\boldsymbol{B}$ where $\boldsymbol{A}_1$ and $\boldsymbol{A}_2$ are $n \times n$ matrices with n a finite integer, $\boldsymbol{B}$ and $\boldsymbol{C}$ are $n \times 1$ and $1 \times n$ matrices respectively, with the entries of these matrices are in $\{\varepsilon, e\}$. A matrix $H \in \mathcal{M}_{in}^{ax}[\![\gamma,\delta]\!]^{p\times q}$ is said to be realizable if all its entries are realizable.*

In other words, a matrix H is realizable if it corresponds to the transfer relation of a timed event graph[2].

[2] Timed event graphs constitute a subclass of Petri nets as introduced in the previous chapters.

Theorem 22.6. [4] *The following statements are equivalent:*

(a) A series s is realizable.
(b) A series s is periodic and causal.

Definition 22.7 (Canonical form of polynomials). *A polynomial* $p = \bigoplus_{i=0}^{n} \gamma^{n_i}\delta^{t_i}$ *is in canonical form if* $n_0 < n_1 < \cdots < n_n$ *and* $t_0 < t_1 < \cdots < t_n$.

Definition 22.8 (Canonical form of series). *A series* $s = p \oplus q \otimes r^*$, *with* $p = \bigoplus_{i=0}^{n} \gamma^{n_i}\delta^{t_i}$, $q = \bigoplus_{j=0}^{m} \gamma^{N_j}\delta^{T_j}$ *and* $r = \gamma^{\nu}\delta^{\tau}$ *is in* ***proper form***, *if*

- *p and q are in canonical form*
- $n_n < N_0$ *and* $t_n < T_0$
- $N_m - N_0 < \nu$ *and* $T_m - T_0 < \tau$

furthermore, a series is in ***canonical form*** *if it is in proper form and if*

- (n_n, t_n) *is minimal*
- (ν, τ) *is minimal*

Sum, product, Kleene star, and residuation of periodic series are well-defined (see [12, 20]), and algorithms and software toolboxes are available in order to handle periodic series and to compute transfer relations (see [6]). Below, useful computational rules are recalled

$$\delta^{t_1}\gamma^{n} \oplus \delta^{t_2}\gamma^{n} = \delta^{max(t_1,t_2)}\gamma^{n}, \tag{22.19}$$

$$\delta^{t}\gamma^{n_1} \oplus \delta^{t}\gamma^{n_2} = \delta^{t}\gamma^{min(n_1,n_2)}, \tag{22.20}$$

$$\delta^{t_1}\gamma^{n_1} \otimes \delta^{t_2}\gamma^{n_2} = \delta^{t_1+t_2}\gamma^{n_1+n_2}, \tag{22.21}$$

$$(\delta^{t_1}\gamma^{n_1}) \backslash\!\!\!\circ (\delta^{t_2}\gamma^{n_2}) = \delta^{t_2-t_1}\gamma^{n_2-n_1}, \tag{22.22}$$

$$\delta^{t_1}\gamma^{n_1} \wedge \delta^{t_2}\gamma^{n_2} = \delta^{min(t_1,t_2)}\gamma^{max(n_1,n_2)}. \tag{22.23}$$

Furthermore, we recall that the order relation is such that $\delta^{t_1}\gamma^{n_1} \succeq \delta^{t_2}\gamma^{n_2} \Leftrightarrow n_1 \leq n_2$ and $t_1 \geq t_2$. Let p and p' be two polynomials composed of m and m' monomials respectively, then the following rules hold:

$$p \oplus p' = \bigoplus_{i=1}^{m} \gamma^{n_i}\delta^{t_i} \oplus \bigoplus_{j=1}^{m'} \gamma^{n'_j}\delta^{t'_j}, \tag{22.24}$$

$$p \otimes p' = \bigoplus_{i=1}^{m}\bigoplus_{j=1}^{m'} \gamma^{n_i+n'_j}\delta^{t_i+t'_j}, \tag{22.25}$$

$$p \wedge p' = \bigoplus_{i=1}^{m}(\gamma^{n_i}\delta^{t_i} \wedge p') = \bigoplus_{i=1}^{m}\bigoplus_{j=1}^{m'} \gamma^{max(n_i,n'_j)}\delta^{min(t_i,t'_j)}, \tag{22.26}$$

$$p' \backslash\!\!\!\circ p = (\bigoplus_{j=1}^{m'} \gamma^{n'_j}\delta^{t'_j}) \backslash\!\!\!\circ p = \bigwedge_{j=1}^{m'} (\gamma^{-n'_j}\delta^{-t'_j} \otimes p) = \bigwedge_{j=1}^{m'}\bigoplus_{i=1}^{m} \gamma^{n_i-n'_j}\delta^{t_i-t'_j}. \tag{22.27}$$

Let s and s' be two series, the asymptotic slopes satisfy the following rules

$$\sigma_\infty(s \oplus s') = \min\big(\sigma_\infty(s), \sigma_\infty(s')\big), \tag{22.28}$$
$$\sigma_\infty(s \otimes s') = \min\big(\sigma_\infty(s), \sigma_\infty(s')\big), \tag{22.29}$$
$$\sigma_\infty(s \wedge s') = \max\big(\sigma_\infty(s), \sigma_\infty(s')\big), \tag{22.30}$$
$$\text{if } \sigma_\infty(s) \le \sigma_\infty(s') \text{ then } \sigma_\infty(s' \diagdown\!\!\!\circ\, s) = \sigma_\infty(s), \qquad \text{else } s' \diagdown\!\!\!\circ\, s = \varepsilon. \tag{22.31}$$

Theorem 22.7. [7] *The canonical injection* $\mathsf{Id}_{|\mathcal{M}_{in}^{ax+}[\![\gamma,\delta]\!]} : \mathcal{M}_{in}^{ax+}[\![\gamma,\delta]\!] \to \mathcal{M}_{in}^{ax}[\![\gamma,\delta]\!]$ *is residuated and its residual is denoted* $\mathsf{Pr}_+ : \mathcal{M}_{in}^{ax}[\![\gamma,\delta]\!] \to \mathcal{M}_{in}^{ax+}[\![\gamma,\delta]\!]$.

$\mathsf{Pr}_+(s)$ is the greatest causal series less than or equal to series $s \in \mathcal{M}_{in}^{ax}[\![\gamma,\delta]\!]$. From a practical point of view, for all series $s \in \mathcal{M}_{in}^{ax}[\![\gamma,\delta]\!]$, the computation of $\mathsf{Pr}_+(s)$ is obtained by:

$$\mathsf{Pr}_+\Big(\bigoplus_{i\in I} s(n_i,t_i)\gamma^{n_i}\delta^{t_i}\Big) = \bigoplus_{i\in I} s_+(n_i,t_i)\gamma^{n_i}\delta^{t_i}$$

where

$$s_+(n_i,t_i) = \begin{cases} e & \text{if } (n_i,t_i) \ge (0,0) \\ \varepsilon & \text{otherwise} \end{cases}.$$

22.3 Control

22.3.1 Optimal Open-Loop Control

As in classical linear system theory, the control of systems aims at obtaining control inputs of these systems in order to achieve a behavioral specification. The first result appeared in [4]. The optimal control proposed therein tracks a trajectory *a priori* known in order to minimize the just-in-time criterion. It is an open-loop control strategy, well detailed in [1], and some refinements are proposed later in [29, 30]. The solved problem is the following.

The model of a linear system in a dioid is assumed to be known, which implies that entries of matrices $\boldsymbol{A} \in \mathscr{D}^{n\times n}$, $\boldsymbol{B} \in \mathscr{D}^{n\times p}$, $\boldsymbol{C} \in \mathscr{D}^{q\times n}$ are given and the evolution of the state vector $\boldsymbol{x} \in \mathscr{D}^n$ and the output vector $\boldsymbol{y} \in \mathscr{D}^q$ can be predicted thanks to the model given by:

$$\begin{aligned} \boldsymbol{x} &= \boldsymbol{A}\boldsymbol{x} \oplus \boldsymbol{B}\boldsymbol{u} \\ \boldsymbol{y} &= \boldsymbol{C}\boldsymbol{x} \end{aligned} \tag{22.32}$$

where $\boldsymbol{u} \in \mathscr{D}^p$ is the input vector. Furthermore, a specified output $\boldsymbol{z} \in \mathscr{D}^q$ is assumed to be known. This corresponds to a trajectory to track. It can be shown that there exists a unique greatest input denoted $\boldsymbol{u}_{opt}$ such that the output corresponding to this control, denoted $\boldsymbol{y}_{opt}$, is lower than or equal to the specified output $\boldsymbol{z}$. In practice, the greatest input means that the inputs in the system will occur as late as possible while ensuring that the occurrence dates of the output will be lower than the one given by the specified output $\boldsymbol{z}$. Control $\boldsymbol{u}_{opt}$ is then optimal according to the just-in-time criterion, *i.e.*, the inputs will occur as late as possible while ensuring that the corresponding outputs satisfy the constraints given by $\boldsymbol{z}$. This kind of problem is very popular in the framework of manufacturing systems: the specified output $\boldsymbol{z}$ corresponds to the customer demand, and the optimal control $\boldsymbol{u}_{opt}$ corresponds to the input of raw parts in the manufacturing system. The latter is delayed as much as possible, while ensuring that optimal output $\boldsymbol{y}_{opt}$ of the processed parts occurs before the customer demand. Hence the internal stocks are reduced as much as possible, which is worthwhile from an economic point of view. According to Theorem 22.3, the smallest solution of System (22.32) is given by:

$$\begin{aligned} \boldsymbol{x} &= \boldsymbol{A}^*\boldsymbol{B}\boldsymbol{u} \\ \boldsymbol{y} &= \boldsymbol{C}\boldsymbol{A}^*\boldsymbol{B}\boldsymbol{u}. \end{aligned} \tag{22.33}$$

This smallest solution represents the fastest evolution of the system. Formally, the optimal problem consists in computing the greatest $\boldsymbol{u}$ such that $\boldsymbol{y} \preceq \boldsymbol{z}$. By recalling that the left product is residuated (see Example 22.1), the following equivalence holds:

$$\boldsymbol{y} = \boldsymbol{C}\boldsymbol{A}^*\boldsymbol{B}\boldsymbol{u} \preceq \boldsymbol{z} \Leftrightarrow \boldsymbol{u} \preceq (\boldsymbol{C}\boldsymbol{A}^*\boldsymbol{B}) \circ\!\!\!\!\backslash\, \boldsymbol{z}.$$

In other words the greatest control achieving the objective is:

$$\boldsymbol{u}_{opt} = (\boldsymbol{C}\boldsymbol{A}^*\boldsymbol{B}) \circ\!\!\!\!\backslash\, \boldsymbol{z}. \tag{22.34}$$

22.3.1.1 Illustration

High-Throughput Screening (HTS) systems allow researchers to quickly conduct millions of chemical, genetic, or pharmacological tests. In the previous chapter, an elementary one was considered as an illustrative example. Below, its model in the semiring $\mathcal{M}_{in}^{ax}[\![\gamma,\delta]\!]$ is recalled.

$$\begin{aligned} \boldsymbol{x}(\gamma,\delta) &= \begin{bmatrix} \varepsilon & \varepsilon & \gamma & \varepsilon & \varepsilon & \varepsilon & \varepsilon & \varepsilon & \varepsilon & \varepsilon \\ \delta^7 & \varepsilon & \varepsilon & \varepsilon & e & \varepsilon & \varepsilon & \varepsilon & \varepsilon & \varepsilon \\ \varepsilon & \delta & \varepsilon & \varepsilon & \varepsilon & \varepsilon & \varepsilon & \varepsilon & \varepsilon & \varepsilon \\ \varepsilon & \varepsilon & \varepsilon & \varepsilon & \varepsilon & \varepsilon & \gamma^2 & \varepsilon & \varepsilon & \varepsilon \\ \varepsilon & e & \varepsilon & \delta & \varepsilon & \varepsilon & \varepsilon & \varepsilon & \varepsilon & \varepsilon \\ \varepsilon & \varepsilon & \varepsilon & \varepsilon & \delta & \varepsilon & \varepsilon & \varepsilon & e & \varepsilon \\ \varepsilon & \varepsilon & \varepsilon & \varepsilon & \varepsilon & \delta^{12} & \varepsilon & \varepsilon & \varepsilon & \varepsilon \\ \varepsilon & \varepsilon & \varepsilon & \varepsilon & \varepsilon & \varepsilon & \varepsilon & \varepsilon & \varepsilon & \gamma \\ \varepsilon & \varepsilon & \varepsilon & \varepsilon & \varepsilon & \varepsilon & \varepsilon & \delta^5 & \varepsilon & \varepsilon \\ \varepsilon & \varepsilon & \varepsilon & \varepsilon & \varepsilon & \varepsilon & \varepsilon & \varepsilon & \delta & \varepsilon \end{bmatrix} \boldsymbol{x}(\gamma,\delta) \oplus \begin{bmatrix} e & \varepsilon & \varepsilon \\ \varepsilon & \varepsilon & \varepsilon \\ \varepsilon & \varepsilon & \varepsilon \\ \varepsilon & e & \varepsilon \\ \varepsilon & \varepsilon & \varepsilon \\ \varepsilon & \varepsilon & \varepsilon \\ \varepsilon & \varepsilon & \varepsilon \\ \varepsilon & \varepsilon & e \\ \varepsilon & \varepsilon & \varepsilon \\ \varepsilon & \varepsilon & \varepsilon \end{bmatrix} \boldsymbol{u}(\gamma,\delta) \\ y(\gamma,\delta) &= \begin{bmatrix} \varepsilon & \varepsilon & \varepsilon & \varepsilon & \varepsilon & \varepsilon & e & \varepsilon & \varepsilon & \varepsilon \end{bmatrix} \boldsymbol{x}(\gamma,\delta) \end{aligned} \tag{22.35}$$

This model leads to the following transfer matrices, where the entries are periodic series in $\mathscr{M}_{in}^{ax}[\![\gamma,\delta]\!]$:

$$
\boldsymbol{x}(\gamma,\delta) = \begin{bmatrix}
(e)(\gamma\delta^8)^* & (\gamma\delta^2)(\gamma\delta^8)^* & (\gamma^3\delta^{19})(\gamma\delta^8)^* \\
(\delta^7)(\gamma\delta^8)^* & (\delta)(\gamma\delta^8)^* & (\gamma^2\delta^{18})(\gamma\delta^8)^* \\
(\delta^8)(\gamma\delta^8)^* & (\delta^2)(\gamma\delta^8)^* & (\gamma^2\delta^{19})(\gamma\delta^8)^* \\
(\gamma^2\delta^{20})(\gamma\delta^8)^* & e\oplus(\gamma^2\delta^{14})(\gamma\delta^8)^* & \gamma^2\delta^{17}\oplus(\gamma^3\delta^{23})(\gamma\delta^8)^* \\
(\delta^7)(\gamma\delta^8)^* & (\delta)(\gamma\delta^8)^* & (\gamma^2\delta^{18})(\gamma\delta^8)^* \\
(\delta^8)(\gamma\delta^8)^* & (\delta^2)(\gamma\delta^8)^* & \delta^5\oplus(\gamma\delta^{11})(\gamma\delta^8)^* \\
(\delta^{20})(\gamma\delta^8)^* & (\delta^{14})(\gamma\delta^8)^* & \delta^{17}\oplus(\gamma\delta^{23})(\gamma\delta^8)^* \\
\varepsilon & \varepsilon & (e)(\gamma\delta^6)^* \\
\varepsilon & \varepsilon & (\delta^5)(\gamma\delta^6)^* \\
\varepsilon & \varepsilon & (\delta^6)(\gamma\delta^6)^*
\end{bmatrix} \boldsymbol{u}(\gamma,\delta)
$$

$$
y(\gamma,\delta) = \begin{bmatrix}(\delta^{20})(\gamma\delta^8)^* & (\delta^{14})(\gamma\delta^8)^* & \delta^{17}\oplus(\gamma\delta^{23})(\gamma\delta^8)^*\end{bmatrix}\boldsymbol{u}(\gamma,\delta). \tag{22.36}
$$

The following reference trajectory is assumed to be known:

$$
z(\gamma,\delta) = \delta^{37}\oplus\gamma\delta^{42}\oplus\gamma^3\delta^{55}\oplus\gamma^4\delta^{+\infty}. \tag{22.37}
$$

It models that one part is expected before or at time 37, 2 parts before (or at) time 42 and 4 parts before (or at) time 55. It must be noted that the numbering of parts starts at 0[3]. According to Equation (22.34) and the computation rules given in the previous section, the optimal firing trajectories of the three inputs are as follows:

$$
\boldsymbol{u}_{opt}(\gamma,\delta) = \begin{bmatrix}
\delta^6\oplus\gamma\delta^{14}\oplus\gamma^2\delta^{22}\oplus\gamma^3\delta^{35}\oplus\gamma^4\delta^{+\infty} \\
\delta^{12}\oplus\gamma\delta^{20}\oplus\gamma^2\delta^{28}\oplus\gamma^3\delta^{41}\oplus\gamma^4\delta^{+\infty} \\
\delta^{11}\oplus\gamma\delta^{19}\oplus\gamma^2\delta^{25}\oplus\gamma^3\delta^{38}\oplus\gamma^4\delta^{+\infty}
\end{bmatrix}, \tag{22.38}
$$

and the resulting optimal output is:

$$
y_{opt}(\gamma,\delta) = \boldsymbol{C}\boldsymbol{A}^*\boldsymbol{B}\boldsymbol{u}_{opt}(\gamma,\delta) = \delta^{28}\oplus\gamma\delta^{36}\oplus\gamma^2\delta^{42}\oplus\gamma^3\delta^{55}\oplus\gamma^4\delta^{+\infty}, \tag{22.39}
$$

which means that the first part will exit the system at time 28, the second part at time 36, the third part at time 42 and the fourth part at time 55. This output trajectory is the greatest in the image of matrix $\boldsymbol{C}\boldsymbol{A}^*\boldsymbol{B}$, that is the greatest reachable output, such that the events occur before the required dates. This example[4] was computed with library MinMaxGD (a C++ library which is interfaced as a toolbox for both Scilab and Matlab software, see [6]).

---

[3] Following the convention defined between remarks 5.22 and 5.23 in [1, Section 5.4].

[4] The sources are available at the following URL: `www.istia.univ-angers.fr/~hardouin/DISCBookChapterControlInDioid.html`.

22.3.2 Optimal Input Filtering in Dioids

In the previous section, the proposed control law leads to the optimal tracking of a trajectory, which is assumed to be *a priori* known. In this section, the aim is to track a reference model in an optimal manner. This means that the specified output is a priori unknown and the goal is to control the system so that it behaves as the reference model which was built from a specification. Hence, this problem is a model matching problem. The reference model describing the specified behavior is assumed to be a linear model in the considered dioid. The objective is to synthesize an optimal filter controlling the inputs (sometimes called a precompensator in the related literature) in order to get an output as close as possible to the one you would obtain by applying this input to the reference model.

Formally, the specified behavior is assumed to be described by a transfer matrix denoted $\boldsymbol{G} \in \mathscr{D}^{q\times m}$ leading to the outputs $\boldsymbol{z} = \boldsymbol{G}\boldsymbol{v} \in \mathscr{D}^{q}$ where $\boldsymbol{v} \in \mathscr{D}^{m}$ is the model input. The system is assumed to be described by the transfer relation $\boldsymbol{C}\boldsymbol{A}^*\boldsymbol{B} \in \mathscr{D}^{q\times p}$ and its output $\boldsymbol{y}$ is given by $\boldsymbol{y} = \boldsymbol{C}\boldsymbol{A}^*\boldsymbol{B}\boldsymbol{u} \in \mathscr{D}^{q}$. The control is assumed to be given by a filter $\boldsymbol{P} \in \mathscr{D}^{p\times m}$, *i.e.*, $\boldsymbol{u} = \boldsymbol{P}\boldsymbol{v} \in \mathscr{D}^{p}$. Therefore the aim is to synthesize this filter such that, for all $\boldsymbol{v}$:

$$\boldsymbol{C}\boldsymbol{A}^*\boldsymbol{B}\boldsymbol{P}\boldsymbol{v} \preceq \boldsymbol{G}\boldsymbol{v}, \tag{22.40}$$

which is equivalent to

$$\boldsymbol{C}\boldsymbol{A}^*\boldsymbol{B}\boldsymbol{P} \preceq \boldsymbol{G}. \tag{22.41}$$

Thanks to residuation theory, the following equivalence holds:

$$\boldsymbol{C}\boldsymbol{A}^*\boldsymbol{B}\boldsymbol{P} \preceq \boldsymbol{G} \Leftrightarrow \boldsymbol{P} \preceq (\boldsymbol{C}\boldsymbol{A}^*\boldsymbol{B}) \backslash\!\!\!\circ\, \boldsymbol{G} \tag{22.42}$$

Hence the optimal filter is given by $\boldsymbol{P}_{opt} = (\boldsymbol{C}\boldsymbol{A}^*\boldsymbol{B}) \backslash\!\!\!\circ\, \boldsymbol{G}$ and it leads to the control $\boldsymbol{u}_{opt} = \boldsymbol{P}_{opt}\boldsymbol{v}$. Furthermore, if $\boldsymbol{G}$ is in the image of the transfer matrix $\boldsymbol{C}\boldsymbol{A}^*\boldsymbol{B}$ (*i.e.*, $\exists \boldsymbol{L}$ such that $\boldsymbol{G} = \boldsymbol{C}\boldsymbol{A}^*\boldsymbol{B}\boldsymbol{L}$), then the following equality holds:

$$\boldsymbol{G} = \boldsymbol{C}\boldsymbol{A}^*\boldsymbol{B}\boldsymbol{P}_{opt}. \tag{22.43}$$

Obviously, this reference model is reachable since it is sufficient to take a filter equal to the identity matrix to satisfy the equality. The optimal filter is given by $\boldsymbol{P}_{opt} = (\boldsymbol{C}\boldsymbol{A}^*\boldsymbol{B}) \backslash\!\!\!\circ\, (\boldsymbol{C}\boldsymbol{A}^*\boldsymbol{B})$ and Equality (22.43) is satisfied. Formally $\boldsymbol{G} = \boldsymbol{C}\boldsymbol{A}^*\boldsymbol{B} = \boldsymbol{C}\boldsymbol{A}^*\boldsymbol{B}\boldsymbol{P}_{opt} = \boldsymbol{C}\boldsymbol{A}^*\boldsymbol{B}\big((\boldsymbol{C}\boldsymbol{A}^*\boldsymbol{B}) \backslash\!\!\!\circ\, (\boldsymbol{C}\boldsymbol{A}^*\boldsymbol{B})\big)$. This means that output $\boldsymbol{y}$ will be unchanged by the optimal filter, *i.e.*, $\boldsymbol{y} = \boldsymbol{C}\boldsymbol{A}^*\boldsymbol{B}\boldsymbol{v} = \boldsymbol{C}\boldsymbol{A}^*\boldsymbol{B}\boldsymbol{P}\boldsymbol{v}$ hence the output occurrences of the controlled system will be as fast as the one of the system without control. Nevertheless the control $\boldsymbol{u}_{opt} = \boldsymbol{P}_{opt}\boldsymbol{v}$ will be the greatest one leading to this unchanged output. In the framework of manufacturing systems, this means that the job will start as late as possible, while preserving the output. Hence the work-in-progress will be reduced as much as possible. This kind of controller is called neutral due to its neutral action on the output behavior. The benefit lies on a reduction of the internal stocks.

22.3.2.1 Illustration

By applying these results to the previous example, the optimal input filter is given by:

$$P_{opt}(\gamma,\delta) = (CA^*B)\backslash(CA^*B) = \begin{bmatrix} (\gamma\delta^8)^* & (\delta^{-6})(\gamma\delta^8)^* & (\delta^{-5})(\gamma\delta^8)^* \\ (\delta^6)(\gamma\delta^8)^* & (\gamma\delta^8)^* & (\delta^1)(\gamma\delta^8)^* \\ (\delta^3)(\gamma\delta^8)^* & (\delta^{-3})(\gamma\delta^8)^* & e\oplus(\gamma\delta^6)(\gamma\delta^8)^* \end{bmatrix} \tag{22.44}$$

It must be noted that some powers are negative (see the computational rule 22.27). This means that the filter is not causal, and therefore not realizable (see Theorem 22.6). In practice, it is necessary to project it in the semiring of causal series, denoted $\mathscr{M}_{in}^{ax+}[\![\gamma,\delta]\!]$. This is done thanks to the rule introduced in Theorem 22.7.

By applying this projector to the previous filter, the optimal causal filter is given by:

$$P_{opt}^{+}(\gamma,\delta) = \mathsf{Pr}_{+}(P_{opt}) = \begin{bmatrix} (\gamma\delta^8)^* & (\gamma\delta^2)(\gamma\delta^8)^* & (\gamma\delta^3)(\gamma\delta^8)^* \\ (\delta^6)(\gamma\delta^8)^* & (\gamma\delta^8)^* & (\delta^1)(\gamma\delta^8)^* \\ (\delta^3)(\gamma\delta^8)^* & (\gamma\delta^5)(\gamma\delta^8)^* & e\oplus(\gamma\delta^6)(\gamma\delta^8)^* \end{bmatrix}. \tag{22.45}$$

This filter describes a 3-input 3-output system of which a realization can be obtained either in time or in event domain. As explained in the previous chapter, in the time domain, the variables, and their associated trajectory, will represent the maximal number of events occurred at a time t. Dually, in the event domain, they will represent the earliest occurrence dates of the kth event. The adopted point of view will depend on the technological context leading to the implementation. Indeed, the control law can be implemented in a control command system or in a PLC, which can be either event driven or synchronized with a clock. In both cases, the following method may be used to obtain a realization of the control law. First, we recall the expression of the control law $u_i = \bigoplus_{j=1}^{p} (P_{opt})_{ij} v_j$ with $i \in [1..p]$, where each entry $(P_{opt})_{ij}$ is a periodic series which can be written as follows: $(P_{opt})_{ij} = p_{ij} \oplus q_{ij} r_{ij}^*$. This leads to the following control law:

$$u_i = \bigoplus_{j=1}^{p} (P_{opt})_{ij} v_j \tag{22.46}$$

with

$$(P_{opt})_{ij} v_j = p_{ij} v_j \oplus q_{ij} r_{ij}^* v_j.$$

The last line can be written in an explicit form, by introducing an internal variable ζ_{ij} (we recall that $x = a^*b$ is the least solution of $x = ax \oplus b$, see Theorem 22.3):

$$\begin{array}{ll} \zeta_{ij} & = r_{ij}\zeta_{ij} \oplus q_{ij} v_j \\ (P_{opt})_{ij} v_j & = p_{ij} v_j \oplus \zeta_{ij} \end{array}. \tag{22.47}$$

This explicit formulation may be used to obtain the control law, either in the time domain or in the event domain. Below, the control law (Equation (22.45)) is given in the time domain:

$$\begin{aligned}
\zeta_{11}(t) &= min(1+\zeta_{11}(t-8), v_1(t)) \\
\zeta_{12}(t) &= min(1+\zeta_{12}(t-8), 1+v_2(t-2)) \\
\zeta_{13}(t) &= min(1+\zeta_{13}(t-8), 1+v_3(t-3)) \\
\zeta_{21}(t) &= min(1+\zeta_{21}(t-8), v_1(t-6)) \\
\zeta_{22}(t) &= min(1+\zeta_{22}(t-8), v_2(t)) \\
\zeta_{23}(t) &= min(1+\zeta_{23}(t-8), v_3(t-1)) \\
\zeta_{31}(t) &= min(1+\zeta_{31}(t-8), v_1(t-3)) \\
\zeta_{32}(t) &= min(1+\zeta_{32}(t-8), 1+v_2(t-5)) \\
\zeta_{33}(t) &= min(1+\zeta_{33}(t-8), 1+v_3(t-6)) \\
u_1(t) &= min(\zeta_{11}(t), \zeta_{12}(t), \zeta_{13}(t)) \\
u_2(t) &= min(\zeta_{21}(t), \zeta_{22}(t), \zeta_{23}(t)) \\
u_3(t) &= min(\zeta_{21}(t), \zeta_{22}(t), \zeta_{23}(t), v_3(t)).
\end{aligned} \tag{22.48}$$

22.3.3 Closed-Loop Control in Dioids

The two previous sections have proposed open-loop control strategies. In this section, the measurement of the system outputs are taken into account in order to compute the control inputs. The closed-loop control strategy considered is given in Fig. 22.1. It aims at modifying the dynamics of system $\boldsymbol{CA}^*\boldsymbol{B} \in \mathscr{D}^{q\times p}$ by using a controller $\boldsymbol{F} \in \mathscr{D}^{p\times q}$ located between the output $\boldsymbol{y} \in \mathscr{D}^q$ and the input $\boldsymbol{u} \in \mathscr{D}^p$ of the system and a filter $\boldsymbol{P} \in \mathscr{D}^{p\times m}$ located upstream of the system, as the one considered in the previous section. The controllers are chosen in order to obtain a controlled system as close as possible to a given reference model $\boldsymbol{G} \in \mathscr{D}^{q\times m}$. The latter is assumed to be a linear model in the considered dioid.

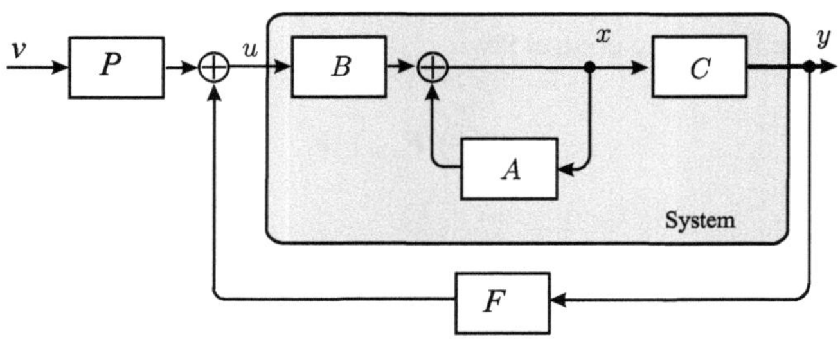

Fig. 22.1 Control architecture, the system is controlled by output feedback $\boldsymbol{F}$ and filter $\boldsymbol{P}$

Formally, the control input is equal to $\boldsymbol{u} = \boldsymbol{F}\boldsymbol{y} \oplus \boldsymbol{P}\boldsymbol{v} \in \mathscr{D}^p$ and leads to the following closed-loop behavior:

$$\begin{aligned} \boldsymbol{x} &= \boldsymbol{A}\boldsymbol{x} \oplus \boldsymbol{B}(\boldsymbol{F}\boldsymbol{y} \oplus \boldsymbol{P}\boldsymbol{v}) \\ \boldsymbol{y} &= \boldsymbol{C}\boldsymbol{x} \preceq \boldsymbol{G}\boldsymbol{v} \qquad \forall \boldsymbol{v}. \end{aligned} \tag{22.49}$$

By replacing $\boldsymbol{y}$ in the first equation, one can write

$$\begin{aligned} \boldsymbol{x} &= (\boldsymbol{A} \oplus \boldsymbol{BFC})\boldsymbol{x} \oplus \boldsymbol{BP}\boldsymbol{v} = (\boldsymbol{A} \oplus \boldsymbol{BFC})^*\boldsymbol{BP}\boldsymbol{v} \\ \boldsymbol{y} &= \boldsymbol{C}\boldsymbol{x} = \boldsymbol{C}(\boldsymbol{A} \oplus \boldsymbol{BFC})^*\boldsymbol{BP}\boldsymbol{v} \preceq \boldsymbol{G}\boldsymbol{v} \qquad \forall \boldsymbol{v}. \end{aligned} \tag{22.50}$$

Hence the problem is to find controllers $\boldsymbol{F}$ and $\boldsymbol{P}$ such that:

$$\boldsymbol{C}(\boldsymbol{A} \oplus \boldsymbol{BFC})^*\boldsymbol{BP} \preceq \boldsymbol{G}. \tag{22.51}$$

By recalling that $(a \oplus b)^* = a^*(ba^*)^* = (a^*b)^*a^*$ (see Equation (22.14)), this equation can be written as:

$$\boldsymbol{CA}^*(\boldsymbol{BFCA}^*)^*\boldsymbol{BP} = \boldsymbol{CA}^*\boldsymbol{B}(\boldsymbol{FCA}^*\boldsymbol{B})^*\boldsymbol{P} \preceq \boldsymbol{G}. \tag{22.52}$$

According to the Kleene star definition $(\boldsymbol{FCA}^*\boldsymbol{B})^* = \boldsymbol{E} \oplus \boldsymbol{FCA}^*\boldsymbol{B} \oplus (\boldsymbol{FCA}^*\boldsymbol{B})^2 \oplus \ldots$, so Equation (22.52) implies the following constraint on the filter $\boldsymbol{P}$:

$$\boldsymbol{CA}^*\boldsymbol{BP} \preceq \boldsymbol{G}, \tag{22.53}$$

which is equivalent to:

$$\boldsymbol{P} \preceq (\boldsymbol{CA}^*\boldsymbol{B}) \backslash\!\!\circ\, \boldsymbol{G} = \boldsymbol{P}_{opt}, \tag{22.54}$$

where $\boldsymbol{P}_{opt}$ is the same as the optimal filter introduced in Section 22.3.2. By considering this optimal solution in Equation (22.52), the following equivalences hold:

$$\begin{aligned} \boldsymbol{CA}^*\boldsymbol{B}(\boldsymbol{FCA}^*\boldsymbol{B})^*\boldsymbol{P}_{opt} \preceq \boldsymbol{G} &\Leftrightarrow \boldsymbol{CA}^*\boldsymbol{B}(\boldsymbol{FCA}^*\boldsymbol{B})^* \preceq \boldsymbol{G} /\!\!\circ\, \boldsymbol{P}_{opt} \\ &\Leftrightarrow (\boldsymbol{FCA}^*\boldsymbol{B})^* \preceq (\boldsymbol{CA}^*\boldsymbol{B}) \backslash\!\!\circ\, \boldsymbol{G} /\!\!\circ\, \boldsymbol{P}_{opt} = \boldsymbol{P}_{opt} /\!\!\circ\, \boldsymbol{P}_{opt}. \end{aligned} \tag{22.55}$$

By recalling that $(a /\!\!\circ\, a) = (a /\!\!\circ\, a)^*$ (see Theorem 22.4) and that $x^* \preceq a^*$ admits $x = a^*$ as its greatest solution (see Theorem 22.5), this equation can be written as:

$$(\boldsymbol{FCA}^*\boldsymbol{B}) \preceq \boldsymbol{P}_{opt} /\!\!\circ\, \boldsymbol{P}_{opt}. \tag{22.56}$$

By using the right division, the optimal feedback controller is obtained as follows:

$$\boldsymbol{F} \preceq (\boldsymbol{P}_{opt} /\!\!\circ\, \boldsymbol{P}_{opt}) /\!\!\circ\, (\boldsymbol{CA}^*\boldsymbol{B}) = \boldsymbol{P}_{opt} /\!\!\circ\, \big((\boldsymbol{CA}^*\boldsymbol{B})\boldsymbol{P}_{opt}\big) = \boldsymbol{F}_{opt}. \tag{22.57}$$

And the causal feedback is given by: $\boldsymbol{F}^+_{opt} = \mathsf{Pr}_+(\boldsymbol{F}_{opt})$. It can be noted that, as for the optimal filter, the reference model describes the required behavior. From a practical point of view, an interesting choice is $\boldsymbol{G} = \boldsymbol{CA}^*\boldsymbol{B}$. This means that the objective is to obtain the greatest control inputs, while preserving the output of the system without control, *i.e.*, to delay the input dates as much as possible while preserving the output dates.

Remark 22.5. *This kind of closed-loop control can be useful to deal with stabilization problem. This problem is not addressed here, but the reader is invited to consult [13, 8, 24]. In a few words we can recall that a timed event graph, which always admits a max-plus linear model, is stable if and only if the number of tokens in all places is bounded. A sufficient condition is fulfilled if the timed event graph is strongly connected. Hence, it is sufficient to modify the timed event graph thanks to a feedback controller in order to obtain a new graph with only one strongly connected component. Indeed, a feedback controller adds some arcs binding the output transitions and the input transitions. In [13], the author provides a static feedback ($\boldsymbol{F}_s$) allowing the graph to be strongly connected. A static feedback means that the places added between output transitions and input transitions are without delay and contain only tokens. In fact, the author gives the minimal number of tokens which are necessary to preserve the throughput of the system. He has also shown that this minimal number of tokens (which can be seen as a minimal number of resources) can be obtained by considering an integer linear programming problem. In [8, 24], the feedback is improved by considering the residuation theory and the computation of an optimal control. The idea is to use the static feedback as obtained thanks to the approach in [13] in order to compute a specification $\boldsymbol{G} = \boldsymbol{C}(\boldsymbol{A} \oplus \boldsymbol{B}\boldsymbol{F}_s\boldsymbol{C})\boldsymbol{B}$. Then this specification is considered to find a feedback controller, like the one proposed in this section, which can be seen as a dynamic feedback. It must be noted that the numbers of tokens added in the feedback arcs are the same as the ones obtained by S. Gaubert in [13]. Hence, the minimal number of resources is preserved and the dynamic feedback will only modify the dynamic behavior by adding delay in order to be optimal according to the just-in-time criterion. To summarize, this strategy ensures the stability and is optimal according to the just-in-time criterion. This kind of control leads to the greatest reduction of internal stocks in the context of manufacturing systems, as well as waiting times in a transportation network, and also avoids useless buffer saturation in a computer network.*

22.3.3.1 Illustration

The example of Section 22.3.2.1 is continued, *i.e.*, the reference model is $\boldsymbol{G} = \boldsymbol{C}\boldsymbol{A}^*\boldsymbol{B}$. Then, the optimal filter is given by Equation (22.45) and the feedback controller is obtained thanks to Equation (22.57). The practical computation yields:

$$\boldsymbol{F}^+_{opt} = \begin{bmatrix} (\gamma^3\delta^2)(\gamma\delta^8)^* \\ (\gamma^2)(\gamma\delta^8)^* \\ (\gamma^3\delta^5)(\gamma\delta^8)^* \end{bmatrix} \tag{22.58}$$

Then the control law, $\boldsymbol{u} = \boldsymbol{P}^+_{opt}\boldsymbol{v} \oplus \boldsymbol{F}^+_{opt}\boldsymbol{y}$ is obtained by adding the feedback control to law (22.48). By using the same methodology as in Section 22.3.2.1, the control law in the time domain can be obtained. An intermediate variable $\boldsymbol{\beta}$ is considered:

$$\begin{aligned}\beta_{11}(t) &= min(1+\beta_{11}(t-8), 3+y(t-2)) \\ \beta_{21}(t) &= min(1+\beta_{21}(t-8), 2+y(t)) \\ \beta_{31}(t) &= min(1+\beta_{31}(t-8), 3+y(t-5)),\end{aligned} \quad (22.59)$$

then the following control law is obtained:

$$\begin{aligned}u_1(t) &= min(\zeta_{11}(t), \zeta_{12}(t), \zeta_{13}(t), \beta_{11}(t)) \\ u_2(t) &= min(\zeta_{21}(t), \zeta_{22}(t), \zeta_{23}(t), \beta_{11}(t)) \\ u_3(t) &= min(\zeta_{21}(t), \zeta_{22}(t), \zeta_{23}(t), v_3(t), \beta_{11}(t)),\end{aligned} \quad (22.60)$$

where variable $\boldsymbol{\zeta}$ is given in Equation (22.48) and $\boldsymbol{\beta}$ in Equation (22.59). This equation can easily be implemented in a control system. It can also be realized as a timed event graph, such as the one given in Fig. 22.2.

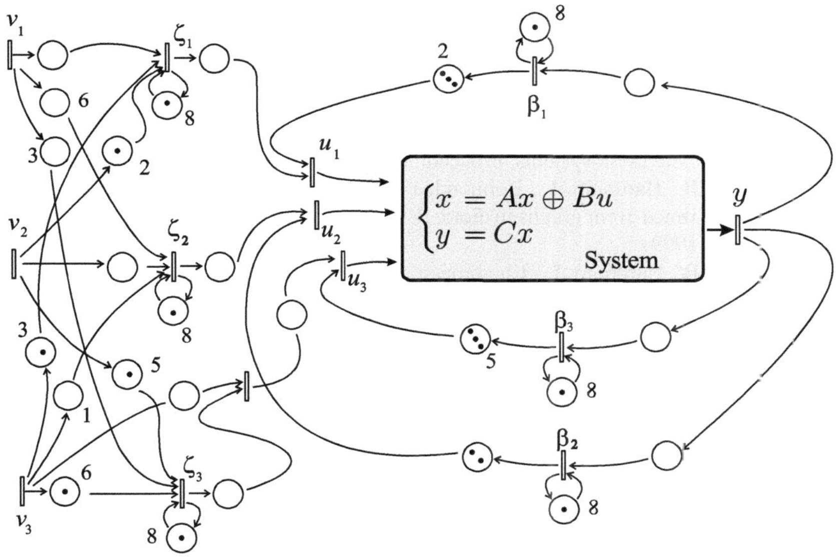

Fig. 22.2 Realization of the optimal filter $\boldsymbol{P}^+_{opt}$ and of the optimal feedback $\boldsymbol{F}^+_{opt}$

22.4 Further Reading

One should notice that some other classical standard problems can be considered from this point of view. Indeed, it is possible to synthesize a linear observer in Luenberger's sense, which allows us to estimate the unmeasured state of the system by considering its output measurement (see [17, 16]). Another related problem consists in the disturbance decoupling problem (see [18, 21]), which aims at taking into account disturbances acting on the system to compute the control law. About uncertainty, the reader is invited to consider a dioid of intervals [15, 25] allowing the synthesis of robust controllers in a bounded error context (see [22, 23]). The identification of model parameters was also considered in some previous work (see e.g., [11, 27, 33]). It can also be noted that some other points of view were considered in the related literature. In [10, 31], the authors considered the so-called model predictive control in order to minimize a quadratic criterion by using a linear programming approach. The same authors have extended their control strategy to a stochastic context (see [34]), in order to take uncertain systems into account.

To summarize, automatic control of discrete-event systems in a dioid framework is still an improving topic with many problems yet to solve.

References

1. Baccelli, F., Cohen, G., Olsder, G.J., Quadrat, J.P.: Synchronization and Linearity, An Algebra for Discrete Event Systems. John Wiley and Sons, New York (1992)
2. Blyth, T.S., Janowitz, M.F.: Residuation Theory. Pergamon Press, Oxford (1972)
3. Blyth, T.S.: Lattices and Ordered Algebraic Structures. Springer (2005)
4. Cohen, G., Moller, P., Quadrat, J.P., Viot, M.: Algebraic tools for the performance evaluation of discrete event systems. Proceedings of the IEEE 77(1), 39–58 (1989)
5. Cottenceau, B., Hardouin, L., Boimond, J.L., Ferrier, J.L.: Synthesis of greatest linear feedback for timed event graphs in dioid. IEEE Transactions on Automatic Control 44(6), 1258–1262 (1999)
6. Cottenceau, B., Hardouin, L., Lhommeau, M., Boimond, J.L.: Data processing tool for calculation in dioid. In: Proc. 5th Int. Workshop on Discrete Event Systems, Ghent, Belgium (2000)
7. Cottenceau, B., Hardouin, L., Boimond, J.L., Ferrier, J.L.: Model reference control for timed event graphs in dioids. Automatica 37(9), 1451–1458 (2001)
8. Cottenceau, B., Lhommeau, M., Hardouin, L., Boimond, J.L.: On timed event graph stabilization by output feedback in dioid. Kybernetika 39(2), 165–176 (2003)
9. Cuninghame-Green, R.A.: Minimax Algebra. Lecture Notes in Economics and Mathematical Systems, vol. 166. Springer (1979)
10. De Schutter, B., van den Boom, T.J.J.: Model predictive control for max-plus linear discrete event systems. Automatica 37(7), 1049–1056 (2001)
11. Gallot, F., Boimond, J.L., Hardouin, L.: Identification of simple elements in max-algebra: application to SISO discrete event systems modelisation. In: Proc. 4th European Control Conference, Brussels, Belgium (1997)
12. Gaubert, S.: Théorie des Systèmes Linéaires dans les Dioïdes. Thèse. École des Mines de Paris, France (1992)
13. Gaubert, S.: Resource optimization and (min,+) spectral theory. IEEE Transactions on Automatic Control 40(11), 1931–1934 (1995)
14. Hardouin, L., Menguy, E., Boimond, J.L., Ferrier, J.L.: Discrete event systems control in dioids algebra. Journal Européen des Systèmes Automatisés 31(3), 433–452 (1997)
15. Hardouin, L., Cottenceau, B., Lhommeau, M., Le Corronc, E.: Interval systems over idempotent semiring. Linear Algebra and its Applications 431(5-7), 855–862 (2009)
16. Hardouin, L., Maia, C.A., Cottenceau, B., Lhommeau, M.: Observer design for (max,plus) linear systems. IEEE Transactions on Automatic Control 55(2), 538–543 (2010)

17. Hardouin, L., Maia, C.A., Cottenceau, B., Lhommeau, M.: Max-plus linear observer: application to manufacturing systems. In: Proc. 10th Int. Workshop on Discrete Event Systems, Berlin, Germany (2010)
18. Hardouin, L., Lhommeau, M., Shang, Y.: Towards geometric control of max-plus linear systems with applications to manufacturing systems. In: Proc. 50th IEEE Conference on Decision and Control and European Control Conference, Orlando, Florida, USA (2011)
19. Heidergott, B., Olsder, G.J., van der Woude, J.: Max Plus at Work – Modeling and Analysis of Synchronized Systems: A Course on Max-Plus Algebra and Its Applications. Princeton University Press (2006)
20. Krob, D.: Complete systems of $\mathscr{B}$-rational identities. Theoretical Computer Science 89, 207–343 (1991)
21. Lhommeau, M., Hardouin, L., Cottenceau, B.: Disturbance decoupling of timed event graphs by output feedback controller. In: Proc. 6th Int. Workshop on Discrete Event Systems, Zaragoza, Spain (2002)
22. Lhommeau, M., Hardouin, L., Cottenceau, B., Jaulin, L.: Interval analysis and dioid: application to robust controller design for timed event graphs. Automatica 40(11), 1923–1930 (2004)
23. Lhommeau, M., Hardouin, L., Ferrier, J.L., Ouerghi, I.: Interval analysis in dioid: application to robust open loop control for timed event graphs. In: Proc. 44th IEEE Conference on Decision and Control and European Control Conference, Seville, Spain (2005)
24. Lhommeau, M., Hardouin, L., Santos Mendes, R., Cottenceau, B.: On the model reference control for max-plus linear systems. In: Proc. 44th IEEE Conference on Decision and Control and European Control Conference, Seville, Spain (2005)
25. Litvinov, G.L., Sobolevskiw, A.N.: Idempotent interval analysis and optimization problems. Reliable Computing 7(5), 353–377 (2001)
26. Maia, C.A., Hardouin, L., Santos Mendes, R., Cottenceau, B.: Optimal closed-loop control for timed event graphs in dioid. IEEE Transactions on Automatic Control 48(12), 2284–2287 (2003)
27. Maia, C.A., Santos Mendes, R., Hardouin, L.: Some results on identification of timed event graphs in dioid. In: Proc. 11th IEEE Mediterranean Conference on Control and Automation, Rhodes, Greece (2003)
28. Menguy, E., Boimond, J.L., Hardouin, L.: A feedback control in max-algebra. In: Proc. 4th European Control Conference, Brusselles, Belgium (1997)
29. Menguy, E., Boimond, J.L., Hardouin, L.: Optimal control of discrete event systems in case of updated reference input. In: Proc. IFAC Conference on System Structure and Control, Nantes, France (1998)
30. Menguy, E., Boimond, J.L., Hardouin, L., Ferrier, J.L.: Just in time control of timed event graphs: update of reference input, presence of uncontrollable input. IEEE Transactions on Automatic Control 45(11), 2155–2159 (2000)
31. Necoara, I., De Schutter, B., van den Boom, T.J.J., Hellendoorn, H.: Stable model predictive control for constrained max-plus-linear systems. Discrete Event Dynamic Systems: Theory and Applications 17(3), 329–354 (2007)
32. Max-Plus: Second order theory of min-linear systems and its application to discrete event systems. In: Proc. 30th IEEE Conf. on Decision and Control, Brighton, United Kingdom (1991)
33. Schullerus, G., Krebs, V., De Schutter, B., van den Boom, T.J.J.: Input signal design for identification of max-plus-linear systems. Automatica 42(6), 937–943 (2006)
34. van den Boom, T.J.J., De Schutter, B.: Model predictive control for perturbed max-plus-linear systems: A stochastic approach. Int. Journal of Control 77(3), 302–309 (2004)

Index

D

MIX
Papier aus verantwortungsvollen Quellen
Paper from responsible sources
FSC® C105338

FSC
www.fsc.org

If you have any concerns about our products,
you can contact us on
ProductSafety@springernature.com

In case Publisher is established outside the EU,
the EU authorized representative is:
Springer Nature Customer Service Center GmbH
Europaplatz 3, 69115 Heidelberg, Germany

Printed by Libri Plureos GmbH
in Hamburg, Germany